Astronomical Distance Determination in the Space Age

Richard de Grijs • Maurizio Falanga
Editors

Astronomical Distance Determination in the Space Age

Previously published in *Space Science Reviews* in the Topical Collection "Astronomical Distance Determination in the Space Age"

Springer

Editors
Richard de Grijs
Department of Physics & Astronomy
Macquarie University
Sydney, NSW, Australia

Maurizio Falanga
International Space Science Institute-Beijing
Beijing, China

ISSN 1385-7525 Space Sciences Series of ISSI
ISBN 978-94-024-1630-5

Library of Congress Control Number: 2018965609

Cover Image: Credit: NASA/JPL-Caltech

Printed on acid-free paper

This Springer imprint is published by the registered company Springer Nature B.V.
The registered company address is: Van Godewijckstraat 30, 3311 GX Dordrecht, The Netherlands

Contents

Space Sci Rev (2017) 212:1739–1741
DOI 10.1007/s11214-017-0402-4

Editorial: Topical Collection on Astronomical Distance Determination in the Space Age

Richard de Grijs[1,2] · Giuseppe Bono[3,4] · Noriyuki Matsunaga[5] · Sherry H. Suyu[6,7,8] · Maurizio Falanga[9,2]

Published online: 22 August 2017

Knowing the distance of an astrophysical object is key to understanding its formation and evolution: without an accurate distance, we do not know how bright it is, how large it is, or even when it existed. Astronomical distance measurements are challenging tasks, and indeed the typical information we have about Galactic and extragalactic sources are their

Astronomical Distance Determination in the Space Age
Edited by Richard de Grijs and Maurizio Falanga

✉ R. de Grijs
grijs@pku.edu.cn

G. Bono
Giuseppe.Bono@roma2.infn.it

N. Matsunaga
matsunaga@astron.s.u-tokyo.ac.jp

S.H. Suyu
suyu@mpa-garching.mpg.de

M. Falanga
mfalanga@issibern.ch

1 Kavli Institute for Astronomy & Astrophysics and Department of Astronomy, Peking University, Yi He Yuan Lu 5, Hai Dian District, Beijing 100871, China

2 International Space Science Institute – Beijing, 1 Nanertiao, Zhongguancun, Hai Dian District, Beijing 100190, China

3 Dipartimento di Fisica, Università di Roma Tor Vergata, via Della Ricerca Scientifica 1, 00133, Roma, Italy

4 INAF, Rome Astronomical Observatory, via Frascati 33, 00040, Monte Porzio Catone, Italy

5 Department of Astronomy, The University of Tokyo, 7-3-1 Hongo, Bunkyo-ku, Tokyo 113-0033, Japan

6 Max-Planck-Institut für Astrophysik, Karl-Schwarzschild-Str. 1, 85748 Garching, Germany

7 Institute of Astronomy and Astrophysics, Academia Sinica, P.O. Box 23-141, Taipei 10617, Taiwan

positions (perhaps as a function of time) and their brightnesses (as a function of wavelength and time).

The first modern milestone in the estimate of nearby distances dates back to the *Hipparcos* space mission (in the 1990s), which provided absolute trigonometric parallaxes at milliarcsecond-level precision across the whole sky for more than 100,000 sources. Around the same time, and to the surprise of many scientists, the Fine Guidance Sensor onboard the *Hubble Space Telescope* (which was designed to guide the telescope) was used as a scientific instrument and provided exquisite trigonometric parallaxes for a handful of primary distance indicators (including Cepheids, RR Lyrae, and δ Scuti variables). In addition, during the past 10–15 years, the use of ground-based 8–10 m-class optical and near-infrared telescopes and space observatories have provided an unprecedented wealth of accurate photometric and spectroscopic data for stars and galaxies in the Local Group (at distances $D < 1$ Mpc) and in the Local Volume ($D < 25$ Mpc). Moreover, interferometric radio observations have also achieved 10 micro-arcsecond astrometric accuracy.

This volume highlights the tremendous amount of recent and continuing research into a myriad of exciting and promising aspects of accurately pinning down the cosmic distance scale. Putting the many recent results and new developments into the broader context of the physics driving cosmic distance determination is the next logical step, which will benefit from the combined efforts of theorists, observers and modellers working on a large variety of spatial scales, and spanning a wide range of expertise. In our journey from the solar neighbourhood to the edge of the Universe, we shall encounter stars of all types, alone, in pairs and in clusters, their life cycles, and their explosive ends: binary stars, in particular, play an important role in this context; the stellar content, dynamics, and evolution of galaxies and groups of galaxies; the gravitational bending of starlight; and the expansion, geometry, and history of the Universe. As a result, this volume offers not only a multi-disciplinary, comprehensive study of distance measurements, but a tour of many recent and exciting advances on all (distance) scales in astrophysics.

The eight reviews included in this volume specifically address future efforts in this field, both theoretically and observationally. This is a critical time in the context of firming up the astronomical distance scale: Very Long Baseline Interferometry sensitivity is being expanded allowing, for example, direct measurement of distances throughout the Milky Way and to Local Group galaxies. The field will benefit tremendously in the *Gaia* era, which is now truly upon us. In addition, the next-generation ‘extremely large telescopes’ (ELTs) will play a crucial role in cosmic distance determination, since they will allow us to determine the Hubble constant only using primary distance indicators. Significant modelling efforts are currently underway to prepare the community for use of the Thirty Meter Telescope (USA, Japan, China, India), the European ELT, and the Giant Magellan Telescope (California, South Korea, Taiwan).

Space-based observatories, in particular, are allowing major advances to be made, and this will only increase in the next decade. In addition to the European *Hipparcos* and *Gaia* missions, Asian scientists, in particular our Japanese colleagues, are leading the field in relevant space missions. Nano-, micro- and (full) *JASMINE*, as well as the Japanese space interferometry missions *VSOP/VSOP-2* complement Russian efforts related to *eRosita*, in addition to US and European space missions like the *Hubble* and *Spitzer Space Telescopes*, *WFIRST*, *Herschel*, the Wilkinson Microwave Anisotropy Probe, and *Planck*. These space-

[8] Physik-Department, Technische Universität München, James-Franck-Straße 1, 85748, Garching, Germany

[9] International Space Science Institute, Hallerstrasse 6, 3012 Bern, Switzerland

and ground-based facilities will have an immediate bearing on the astronomical distance scale. Nevertheless, many uncertainties remain at the level of at least 5–10%, particularly in terms of our understanding of the physics underlying many of the methods commonly used for distance determination.

Overall, this volume provides a timely and comprehensive review of the status of ground- and space-based distance determination and the associated uncertainties. It will be particularly useful as an up-to-date reference work for postgraduate students and researchers in any area touching upon the Galactic and extragalactic distance scales. This volume is the result of an international workshop which took place at the International Space Science Institute—Beijing from 23 to 27 May 2016, where some forty leading scientists came together to discuss the current state of the art and future opportunities to make major progress.

Space Sci Rev (2017) 212:1787–1815
DOI 10.1007/s11214-017-0418-9

Modern Geometric Methods of Distance Determination

Frédéric Thévenin[1] · Maurizio Falanga[2,3] · Cheng Yu Kuo[4] · Grzegorz Pietrzyński[5] · Masaki Yamaguchi[6]

Received: 3 April 2017 / Accepted: 30 August 2017 / Published online: 18 October 2017

Abstract Building a 3D picture of the Universe at any distance is one of the major challenges in astronomy, from the nearby Solar System to distant Quasars and galaxies. This goal has forced astronomers to develop techniques to estimate or to measure the distance of point sources on the sky. While most distance estimates used since the beginning of the 20th century are based on our understanding of the physics of objects of the Universe: stars, galaxies, QSOs, the direct measures of distances are based on the geometric methods as developed in ancient Greece: the parallax, which has been applied to stars for the first time in the mid-19th century. In this review, different techniques of geometrical astrometry applied to various stellar and cosmological (Megamaser) objects are presented. They consist in parallax measurements from ground based equipment or from space missions, but also in the study of binary stars or, as we shall see, of binary systems in distant extragalactic sources us-

Astronomical Distance Determination in the Space Age
Edited by Richard de Grijs and Maurizio Falanga

✉ F. Thévenin
frederic.thevenin@oca.eu

M. Falanga
mfalanga@issibj.ac.cn

C.Y. Kuo
cykuo@mail.nsysu.edu.tw

G. Pietrzyński
pietrzynski28@gmail.com

M. Yamaguchi
masaki@ioa.s.u-tokyo.ac.jp

1 Observatoire de la Côte d'Azur, CNRS, Université Côte d'Azur, Nice, France

2 International Space Science Institute, Bern, Switzerland

3 International Space Science Institute, Beijing, China

4 National Sun Yat-Sen University, Kaohsiung, Taiwan

5 Nicolaus Copernicus Astronomical Center of the Polish Academy of Sciences, Warszawa, Poland

6 University of Tokyo, Tokyo, Japan

ing radio telescopes. The Gaia mission will be presented in the context of stellar physics and galactic structure, because this key space mission in astronomy will bring a breakthrough in our understanding of stars, galaxies and the Universe in their nature and evolution with time. Measuring the distance to a star is the starting point for an unbiased description of its physics and the estimate of its fundamental parameters like its age. Applying these studies to candles such as the Cepheids will impact our large distance studies and calibration of other candles. The text is constructed as follows: introducing the parallax concept and measurement, we shall present briefly the Gaia satellite which will be the future base catalogue of stellar astronomy in the near future. Cepheids will be discussed just after to demonstrate the state of the art in distance measurements in the Universe with these variable stars, with the objective of 1% of error in distances that could be applied to our closest galaxy the LMC, and better constrain the distances of large sub-structures around the Milky Way. Then exciting objects like X-Ray binaries will be presented in two parts corresponding to "low" or "high" mass stars with compact objects observed with X-ray satellites. We shall demonstrate the capability of these objects to have their distances measured with high accuracy with not only helps in the study of these objects but could also help to measure the distance of the structure they belong. For cosmological objects and large distances of megaparsecs, we shall present what has been developed for more than 20 years in the geometric distance measurements of MegaMasers, the ultimate goal being the estimation of the H_0 parameter.

Keywords Geometric distance · Stars · Maser · Space mission · Binary stars

1 Introduction

The geometric distance measurement has a great advantage on any other techniques: it is independent of the nature of the point source. Then it can be applied to stars in the Galaxy as well to distance point source galaxies.

To introduce the parallax (geometric distance measurement) as an help to reveal the characteristics of objects in the Universe, let's start with the an example of a benchmark star, well-known as a metal poor star belonging to the halo structure of the Galaxy, and considered as the best representative PopII star with fundamental parameters that are very well constrained. HD 122563 has been studied in photometry, spectrography for more than 50 years and had its distance measured by *Hipparcos* (Perryman et al. 1997; van Leeuwen 2007). It is a distant giant star, so the parallax is small and with a large uncertainty (4.22 ± 0.35 milli-arcsecond). Numerous spectroscopic studies of this star have attempted to measure its surface gravity which is a key point for the chemical abundance determinations of a star, abundances that are important for the study of the evolution of the Galaxy since its origin ∼13 Gyr ago. It is now well understood that metal poor stellar atmospheres suffer from departures from the Local Thermodynamic Equilibrium (LTE) which bias the spectrographic surface gravity estimates and therefore the chemical abundance estimates (see Thévenin and Idiart 1999). Recently, in a more advanced work (Ezzeddine et al. 2016) on non-LTE on the element Fe, the surface gravity of HD 122563 has been suggested to be $\log g = 1.3$ which differs from the usually accepted $\log g = 1.6$ coming from the best constrained study (Creevey et al. 2012) including effective temperature (Teff) constraint from the angular diameter measurement; note that for that star the extinction is small and could not be the source of the discrepancy between spectrography and photometric (HR diagram) studies. The only way to converge to similar $\log g$

in both techniques is to change the distance estimate of *Hipparcos* (van Leeuwen 2007; Lacroute 1974) from 4.2 to 3.2 mas as suggested by VandenBerg et al. (2014).

Once the distance of a star is accurately known, its position in the HR diagram is free of large errors and then the estimate of the surface gravity is robust, leading to accurate chemical abundance measurements: this is a key point and demonstrates that if the distance is important the *accuracy* of the distance is also very important. The tools we have in our hand to estimate the fundamental parameters of stars have now reached the point where Gaia is mandatory for future stellar studies on extensive stellar samples to reveal the large structure of the Galaxy.

Asterosesimology is another way to constrain stars and to study the galactic structure, both Gaia parallaxes and asteroseismic space mission will change our knowledge on the structure, the history and the evolution of the Milky Way.

Let's develop the discussion around the geometric distance measurements from stars to distant extra-galactique Megamaser sources.

2 Parallaxes to Measure Distances in the Universe

2.1 Distance Measurements for a Single Star Using Astrometry: The Parallax

If the observational target is a point source close enough to the Earth, we can measure the distance to the target by observing its position on the sky. The fundamental problem is the following. When we continuously observe the position of a single star for a few years, we find that the trajectory of the star shows a helical motion. This helical motion can be divided into two different kinds of motion: the uniform linear motion and the elliptical motion, which correspond to the proper motion and annual motion by parallax, respectively. The proper motion is due to a physical motion in the cosmic space, and the annual motion by parallax is due to the different direction to the star in the different seasons. Note that Gaia (see Sect. 2.3) will treat that problem for numerous stars (more than a billion) to get both accurate parallax and proper motion estimates.

The parallax p can be written using the distance to the star d as follows:

$$p = \arctan \frac{1\ \mathrm{AU}}{d} = \left(\frac{d}{1\ \mathrm{kpc}} \right)^{-1} \ \mathrm{mas}, \tag{1}$$

where mas means milli-arcsecond. The typical astrometric error for the astrometric satellite, *Hipparcos* (Perryman et al. 1997; van Leeuwen 2007), is ~0.5 to 1 mas, so that the distance within which we can measure the parallax at sufficiently high precision is ~100 pc; this of course depends on the astrophysical goals but nowadays accuracy closer to few micro-arc-seconds is mandatory for many subject matter. Whereas the trajectory by the parallax shows a different shape of ellipse depending on the ecliptic latitude of the star, the semi-major axis of the ellipse is p for any ecliptic latitude.

2.2 The Star Is in a Binary System: Effect of the Orbital Motion on the Distance Measurements

We have to take into account the orbital motion for accurate measurements of distance. About half of the stars in the Milky-Way are thought to be in a binary system. Thus, in principle, half stars show the orbital motion, although we cannot detect the orbital motion for most stars in a binary, because of too small or too large orbit. Such orbital motion is an additional motion to the helical motion of the single star, and basically, it can be detected as a deviation from the helical motion. This means that the orbital motion can affect the astrometric measurements, and therefore the distance measurements. In this section, we analytically estimate the effect of the orbital motion on the trajectory of the star on the celestial sphere and discuss this effect on the distance measurements.

When the orbital period of the binary is shorter than the observational period, the periodic deviation from the helical motion is expected. The deviation can be assumed to be the apparent semi-major axis, so that the deviation δ_s is represented as

$$\delta_s = a_*/d = 1\ \text{mas}\ M_{\text{comp},\odot}\, p_{\text{mas}} \left(\frac{P_{\text{orb,yr}}}{M_{\text{tot},\odot}}\right)^{\frac{2}{3}}, \qquad (2)$$

where $M_{\text{comp},\odot}$, $P_{\text{orb,yr}}$, and $M_{\text{tot},\odot}$ are the companion mass normalized by 1 $M_\odot$, the orbital period normalized by 1 year, and the total mass normalized by 1 $M_\odot$, respectively. The unit "mas" represents a milli-arcsecond, and p is normalized by 1 mas.

When the orbital period is longer than the observational period, we may observe the acceleration of the proper motion. In this case, the deviation from the helical motion can be estimated by the displacement of the position due to the acceleration, i.e., $\ddot{r}\Delta t^2/2$, where r and Δt represent the position along the proper motion and half of the observational period. When the acceleration is caused by the orbital motion, $\ddot{r}$ is approximately represented as $a_*\omega^2/d$, where $\omega = 2\pi/P_{\text{orb}}$. Thus, this displacement δ_l can be written as

$$\delta_l = \frac{a_*}{d}\frac{\pi\, t_{\text{obs}}}{P_{\text{orb}}} \sim 10\ \text{mas}\ p_{\text{mas}} \frac{M_{\text{comp},\odot}}{M_{\text{tot},\odot}^{2/3}} \frac{t_{\text{obs,yr}}^2}{P_{\text{orb,yr}}^{4/3}}, \qquad (3)$$

where t_{obs} means the observational period, which is normalized by 1 year in Eq. (2). We note that δ_s is proportional to a^{-2}, where a is the semi-major axis of the companion's orbit with respect to the position of the primary. This dependency reflects the fact that the gravity obeys the inverse-square law.

The orbital motion affects the distance measurements if these two kinds of deviations is larger than a parallax precision, σ_p, that is, $\delta_s > \sigma_p$ and $\delta_l > \sigma_p$. These two inequalities give a range of the orbital period in which the orbital motion can be detected. Using Eqs. (2) and (3), this range of the orbital period is represented as

$$1\ \text{yr}\left(\frac{p}{\sigma_p}\right)^{-\frac{3}{2}} \frac{M_{\text{tot},\odot}}{M_{\text{comp},\odot}^{3/2}} < P_{\text{orb}} \lesssim 5\ \text{yrs}\left(\frac{p}{\sigma_p}\right)^{\frac{3}{4}} \frac{M_{\text{comp},\odot}^{3/4}}{M_{\text{tot},\odot}^{1/2}} t_{\text{obs,yr}}^{\frac{3}{2}}. \qquad (4)$$

This means that the distance measurements are affected by the orbital motion for the normal-star binaries whose orbital period is 0.03–30 yr, where we assume that the signal-to-noise ratio of the parallax $p/\sigma_p \sim 10$ and $t_{\text{obs}} \sim 1$ yr. Therefore, we should take into account the orbital motion for the astrometric observation of binaries for which the astrometric solution departs too much from the expected accuracy estimated under the hypothesis that the target is a single star. This is done in the Gaia data treatment (Brown et al. 2016) for which all stars flagged as "bad astrometric solution" are again treated under the hypothesis of non single stars.

We should note that in the discussion above we neglect the emission from the companion. Astrometric instruments measure a photocenter of a target object. When the target is a close binary—the period is shorter than 1 year at 1 kpc—the separation between the primary and the companion star on the celestial sphere is $\lesssim$1 mas, which means that any current single telescope on Earth cannot resolve them into two sources. For Gaia, when the primary star of the binary system is not too bright to saturate the detector or to produce too strong diffraction spikes then the separation between the two components can drop to $\sim$100 mas. Thus, unless the companion is unseen, we should take into account the emission from the companion. Nevertheless, when the companion has a very faint flux compared to the primary star, we neglect its emission. This is particularly true when we focus on targets whose companion is a compact object (see 4.1) or an exoplanet.

2.3 Space Astrometry

A space mission is the only one way to accomplish a large sky survey of stars with a high accuracy in the measurement of both parallaxes and proper motions. This has been demonstrated by the *Hipparcos* spacecraft, which was launched in 1989 (Perryman et al. 1997). This success has led astronomers to foresee an ambitious follow-up mission: the Gaia Spacecraft and to realize a breakthrough in the understanding of the history of the Galaxy.

Space astrometry provides the only way to measure huge numbers of stellar positions, proper motions and parallaxes in a reference system defined over the whole sky (Perryman et al. 1997). The challenge of Gaia is to go down to few microarcsecond of accuracy for objects brighter that $V \sim 15$, with a completeness at a limiting magnitude of $V \sim 20$ to 21, e.g. more than 1 billion stars. For the faintest stars ($\sim$20) the accuracy on the parallaxes is foreseen to be similar to the best ones done by *Hipparcos*, e.g. $\sim$0.3 mas. With such numbers in our mind, it is clear that the Gaia mission is a very important one for Astronomy to answer many questions for which the distance to the source is a key element, for example: questions on the 3D structure of the Galaxy and its origin, on the age of single stars and on the stellar population structure, on the linear radius of stars and constraints on their state of evolution, on the detection of faint companion objects around stars, ... The combination of geometric distances with photometric observations, Space asteroseismic missions [like PLATO (Roxburg et al. 2007)], and interferometric diameter measurements is a key to help in a better understanding of the HR diagram. A better modelling of stars and their characterization with fundamental parameters such as their mass, helium content, or age is also very important not only for stellar populations in the Galaxy but also for exoplanetary system studies like the future space missions: TESS (Ricker et al. 2009) and PLATO (Roxburg et al. 2007). Only then will we be close to the breakthrough foreseen by astronomers in 1997 (Perryman et al. 1997).

Some astrometric space missions were studied [FAME (Horner et al. 1998), DIVA (Röser 1996), SIM (Shao et al. 1995)] but only Gaia (Brown et al. 2016; Prusti et al. 2016) and JASMINE (Gouda et al. 2002) were programmed for launch. Gaia is now working and producing results which will be published in the final catalogue at 2022, but the JASMINE project in not so in advance. In its design it is an Infra-Red Astrometric Space mission composed of 3 satellites working on the same principal as *Hipparcos* and Gaia with two field of views, the last satellite (the biggest) having the goal of 10 micro-arc second of accuracy for the parallaxes and 4 micro-arc second of accuracy for the proper motions in crowded fields with strong flux extinction like the galactic Bulge. For more detailed and description of the 3 JASMINE satellites see: www.scholarpedia.org/article/JASMINE. Beyond the 3D map of a billion stars in the Galaxy, these space missions have the challenge to collect simultaneously the photometry of the observed sources to characterize them and to produce a 3D map of extinction in the Galaxy. Extinction of sources in the Universe is an Achilles' heel that when it is unknown corrupts the source characterization and distance. With geometric distances this will no longer be a problem. The design of Gaia was even more ambitious, it requires to also collect simultaneously during the observations as many radial velocities of stars as possible. Radial velocities from Gaia has been an important challenge (Katz et al. 2004) because Gaia in its concept is a scanning machine and any spectrograph on board has to be slitless, with moving sources and overlapping spectra sensitive to the sky background.

2.4 The Gaia Mission and the Main Goals

After it became a corner stone of ESA in 2001, and a definitive acceptation of the program mission at the end of 2005, Gaia was launched on the 19th December 2013 from Kourou with a Soyuz rocket (for a detailed description of the mission and the data treatment see: Brown et al. 2016, Prusti et al. 2016). The spacecraft successfully reached its Lissajou-orbit in the L2 zone, for a scanning law of the sky starting about 7 months after the launch. The scan motion is composed of 3 rotation axis: the spin on itself, the precession axis and the rotation around the Earth, these altogether permit a complete sky survey with several passages of every source: on average 70 times.

The satellite is composed of 3 astronomical instruments: astrometric and spectrophotometric for sources brighter than $V \sim 21$, and spectrographic for stars brighter than $V{\sim}16$ mag (Brown et al. 2016). All of these data are collected on the large CCD focal plane assembly, but a key point of the success is the stability of the instrument, and its capability to control its space attitude in the L2 zone: distance to the earth, orientation and non variation of the spin velocity. Altogether, the instruments will reveal the 5–6D (dimensions) of the Galaxy and the 3D extinction map thanks to the astrophysical stellar parameters of each source: $T_{\rm eff}$, $\log g$, and metallicity (Brown et al. 2016). A large photometric accuracy accompanied by a large number of transits for each source offers the opportunity to study the variability of the sources with a detailed classification. After the launch and the stabilization of the spacecraft at L2, a period of tests was done to focus the optics and to fine tune the different on board controls, then the satellite was forced to scan the ecliptic poles during 4 weeks to validate the data acquisition and the data treatment algorithms; as a by-product, it demonstrated the powerful capability of G (astrometric) band photometry to detect more than 3000 RRLyrae and Cepheids (Clementini et al. 2016).

Regrouped in a consortium to process the data, astronomers and engineers of several European countries work together in 9 Coordination Units (CUs), a challenge of the data consortium to publish the output catalogue soon after the end of the mission.

At the end of the mission, Gaia will have observed ~2 billion sources, with a sky-completeness down to $G \sim 20.7$ mag where G is the large astrometric filter of Gaia (from UV to near IR), positions and proper motions with accuracy down to ~10 µarcsecond at magnitude 15, an accurate photometry in G band down to the milli-magnitude level at $G = 19$ and few milli-magnitude of errors at $G \sim 20$, combined with blue and red photometry with accuracy of few millimagnitude for stars at $G \sim 19$.

Using all information available from Gaia, all sources will be characterized. In particular it is foreseen to get the extinction for all sources which, combined with their distances, gives a 3D extinction map of the Galaxy. This extinction is tied together with the estimation of astrophysical parameters of the stellar sources: Teff, $\log g$, [M/H]. The radial velocities for ≈150 million of stars, the projected stellar rotational velocity $v \sin i$, the age and the mass of stars are also part of the future output catalogue. Gaia will also provide parametrized classification and redshift of 50 million galaxies and QSOs.

Finally, the detection of faint companion of stars extended to thousands of Jupiter-mass exoplanets detection will be done. The binarity detection and characterization of the system will have a strong impact on the stellar population studies.

These few examples of outputs of the Gaia catalogue demonstrate the importance of this space mission. Here we provide a non-exhaustive list of different subject matter on which the Gaia catalogue will have impacts:

What is the mass of the heaviest stars?
What is the rate of formation of stars in the milky way?
What are the cooling sequence of the white dwarfs?
Are brown-dwarfs common in binary systems?
Statistics on star binary
Are all stars ultimately variable?
How often a planetary system is found in binary systems?
Where does the halo come from?
Is there dark matter out there?
Discovery of spatially and kinematically resolved stellar distributions
History of star formation
Chemical enrichment history of the Galaxy
Fundamental parameters of stars (in particular $\log g$)
Extinction law in the Galaxy
Refined galactic rotation model
Mapping the galactic potential
Distribution of (dark) matter
Disk dynamics (bar, spirals)
Phase space (or E, Lz) structures in halo
History of galactic mergers
Stellar classification and characterization
Distance scales (geometric for <10 kpc, Cepheids and RR Lyr for >10 kpc)
Ages in the Universe (individual stars, HR diagrams)
Systems of reference (Quasars for astrometry)
Exoplanetary systems discovery (Jupiter Mass planets by astrometry anomalies)
Solar system studies with the asteroids (Taxonomy, Masses, Orbits)

However, it is important to remind the users of the Gaia catalogue for these purpose that the distance is the inverse of the parallax and that the accuracy on the parallax has to be treated with caution when used. Then, we recommend to read the following references: Bailer-Jones (2015).

2.5 Complementary Surveys from the Ground for Gaia

Surveys from the ground are developed to calibrate and/or validate the Gaia output catalogues. Our Galaxy is unique in that position and movement of numerous individual stars can be known and offer sufficient information to exhibit the fossil remains of the past formation of the Galaxy in the 6D space phase (Helmi et al. 2017). Because Gaia will not provide radial velocities for stars fainter than $V = 15$–16 mag, this survey must be complemented by follow-up ground based spectrographs. This is already on the way with APOGEE, WEAVE and MOST projects and more. If we are observing stars where they are today, they were somewhere else in the past and will evolve somewhere else in the future. The Galaxy is not a continuous distribution of stars, it is made of lot of substructures evolving and vanishing, their own histories are now part of the history of the Galaxy. Gaia will attempt to reveal some of them, and in particular in the smoothest structure of the Galaxy: the halo.

Chemical tagging of stars will also help in understanding the evolution of the Galaxy and to reconstruct the Initial Mass function and the Star Formation Rate. In particular α-element and r-process element abundances are mandatory to constrain the 13 Gyrs of the formation of the Milky Way. Note that combined with asteroseismic surveys, better age of stars and of stellar populations will come out adding a 7th dimension to the 6 phase space dimensions already foreseen. An example of the future discoveries by Gaia in this field of research is demonstrated in Simpson et al. (2017) where a young cluster ($\sim$3 Gyr) has been detected (Koposov et al. 2017) and confirmed by follow-up spectroscopy: the cluster has been named *Gaia 1*. A new area of Astronomy is starting.

2.6 Interferometry and Asteroseismology Combined Together to Test the Gaia Distances

For stars bright enough to be observed by interferometry to get their angular diameters, and with asteroseismology to get their detailed frequencies of oscillation providing the linear diameter, the combination of both measurements delivers the distance of these stars. These distances can be compared directly to those geometric ones obtained by Gaia and to validate each to other. Calibrations and validations of Gaia outputs have already started with the Gaia release 1 and TGAS (Brown et al. 2016) and (Lindegren et al. 2016), see for example the discussions in Gontcharov and Mosenkov (2017) and Huber et al. (2017). When the magnitude of stars do not permit the interferometric angular diameter measurement, one can play with the asteroseismic frequencies and the fundamental parameter of the stars to get the parallaxes (see Creevey et al. 2017).

Another possibility of calibration is to use calibrated surface brightness relations (calibrated with interferometric stellar diameters, Kervella et al. 2004a) combined with asteroseismic frequencies to provide the linear diameters and then the distances of these targets. But the use of the surface brightness is limited by the capability to estimate the extinction in the line of sight of the stars; the Gaia extinction map will then be very welcome for this kind of work.

There are other developments to test and to estimate systematics of Gaia parallaxes with comparisons with non geometric distances on cluster stars like for example by transforming the magnitude of stars in pseudo magnitude (a combination of different filters (see Madore 1982 for the definition and the first detailed application on Cepheids) to be almost independent of the extinction (see Chelli et al. 2016, Chelli and Duvert 2016). This will be used to test these parallaxes on well-known clusters (van Leeuwen et al. 2017) to validate this work and then determine the fundamental parameters and the physics of the stars at large distances (external galaxies). Similar studies are developed for standard candles like Cepheids, which will be discussed in Sect. 3.

2.7 After Gaia: Nanoastrometry Is Foreseen

After Gaia two large missions are suggested: the first idea is to construct a second Gaia mission, working in the Infra-Red, to extend the survey towards low mass stars (brown dwarfs). This future satellite (Hobbs et al. 2016) will have to be launched with enough time difference with Gaia ($\sim$2030) to improve the proper motion estimates when comparing with the Gaia output catalogue similarly to what has been done with *Hipparcos* and Gaia with the

TGAS work (Lindegren et al. 2016). This will better constrain the movement of stars and then the galactic structure studies.

A second dedicated satellite with selected fields of observation is foreseen to push down the accuracy of the parallax to the level of accuracy better than 1 μas depending on the magnitude of stars (see Fig. 3.2.1 of Boehm et al. 2017 and also Malbet et al. 2014); the idea behind this is to push the detection of secondary companion of stars as Gaia is presently doing, but reaching the smaller rocky planets (see Eq. (3) for an estimate of the displacement of the main star on the sky due to such small companion). The dark matter detection is also one of the main goal of this astrometric study. Such astrometric satellite could be developed in the next 30 years, but for the moment an extensive use of the Gaia catalog will teach us a lot on the limits of the present space astrometry and help to choose where to go in the next 20 years.

Let see now few examples of the astrometry of binary systems studied from ground based equipment. We shall first discuss the Cepheid stars which are very important candles for the distance scale in the Universe, then we shall detailed a few examples of X-ray binaries studies as astrometric distance indicators before presenting the geometric distance measurements of very distance sources in Universe: the Megamasers applied to the estimate of the parameter H_0.

3 Direct Distances of Stellar Binary Systems

3.1 Eclipsing Binaries

Eclipsing binaries are the unique way to get all the parameters of the binary system without hypothesis on the projection angle of the orbit and for which radius of both stars are well constrained by the eclipse. It is also better for the study of the system through the third Kepler's law, then the distance of the system can be constrained. To summarize, detached eclipsing doubled-lined binaries offer a unique opportunity to measure directly, and very accurately, stellar parameters like mass, luminosity, and radius (Andersen 1991) and the distance (Thompson et al. 2001; Pietrzynski et al. 2009). It has been argued in Wyithe and Wilson (2001) that with current observational facilities, and applying an appropriate surface-brightness colour relation (SBCR), eclipsing binaries have the potential to yield the most direct (one step) and the most accurate (about 1%) distance to nearby galaxies (Kruszewski and Semeniuk 1999; Wyithe and Wilson 2001). However the SBCR is well calibrated from interferometry only for late type stars (e.g. Di 2005, Kervella et al. 2004a). Unfortunately even in the closest galaxy, the LMC, the frequent main-sequence binary stars are too faint to secure high-resolution spectra for accurate radial velocity measurements even with the largest current telescopes. Therefore few attempts were made to use eclipsing binaries to determine distance to the LMC (e.g. Guinan et al. 1998). These studies have used relatively bright early-type systems, for which however an accurate empirical SBCR is currently not available. As a consequence, theoretical models had to be used in this work, which significantly limited the accuracy of these distance determinations to 5–10%. The distances to eclipsing binaries can be measured using a very simple equation:

$$d\ (\mathrm{pc}) = 1.337 * 10^{-5} r/\phi \tag{5}$$

with r in km and ϕ in mas. The 1% linear diameter (r) will come from an analysis of the light curve and spectroscopic orbit following a classical and extremely well tested approach, while the angular diameter (ϕ) can be calculated from the surface brightness—colour relation according to

$$\phi \text{ (mas)} = 10^{0.2(S-m_o)} \quad (6)$$

where S is the surface brightness in a given photometric band, and m_o is the unreddened magnitude of a given star in this band. It is a decisive advantage of this method that the sources of uncertainty including systematic errors are known and can be accurately quantified, due to the simplicity and empirical nature of this near-geometrical method. However the SBCR is well established only for late type stars. Therefore, in spite of the fact that eclipsing binaries were long recognized as the potentially best distance indicators, certainly the best to measure distances to nearby galaxies, astronomers were not able to take advantage to use the full potential of this method in the past century for an accurate determination of the Hubble constant H_0 because of the lack of adequate targets in nearby galaxies. Based on the monitoring of about 35 million stars in the Magellanic Clouds over the last 19 years in the course of the OGLE project some 50 extremely scarce eclipsing binary systems composed of helium burning giants were discovered (Soszynski et al. 2010; Graczyk et al. 2014). Given the large dimensions of such stars (some 20–40 $R_\odot$) our fully detached systems have very long orbital periods (150 to 700 days), so their discovery is a real challenge because of the long times of constant light between the eclipses. However their relatively bright ($V = 14$–17 mag) late-type components are perfectly suited for deriving accurate distances: they are bright enough to obtain very accurate orbital radial velocity curves with 4–8 m-class telescopes yielding, among other parameters, and the angular diameters of the component stars can be measured very accurately with the SBCR. The first distance results for just eight of these systems in the LMC already resulted in the most accurate distance (2.2%) determination ever to this galaxy (Pietrzynski et al. 2013). Similar studies on 5 systems in the SMC yielded a 3% distance to that galaxy (Graczyk et al. 2014). Very recent results on calibration of SBCR demonstrated that we should expect 1% and 2% distance determinations to LMC and SMC, respectively. As a result, eclipsing binaries together with the new calibration of the SBCR are a tool for 1% distance determination within the Local Group now and soon up to 10–15 Mpc. These results yield both a very precise Zero Point for Cepheid period luminosity (PL) relation and new one means to calibrate the metallicity effect.

3.2 Spectroscopic and Astrometric Binaries

Having both astrometric and spectroscopic orbits of the binary systems, we can measure this direct geometric distances with an accuracy that can be expected to be better than 1% (Torres et al. 2009; Zwahlen et al. 2004). Indeed, the interferometric observations will provide a measurement of the projection of the semi major axis of the orbit on the sky, while the spectroscopic orbit will provide its linear size. This purely geometrical technique does not require any assumptions about reddening, metallicity, etc. It was already demonstrated on δ Vel that an accuracy of 1% on the parallax (39.8 ± 0.4 mas; Kervella et al. 2004b), can be reached, confirming the *Hipparcos* value (40.5 ± 0.4 mas) with a competitive accuracy. Another example is TZ For, see Gallenne et al. (2016). However this technique is limited only to nearby systems (we need to resolve both components). Even with high resolution interferometry only distance to binaries located closer than about 100 pc can be measured with this technique. However, they provide a unique opportunity to check the absolute calibration of the SBCR (see Fig. 1).

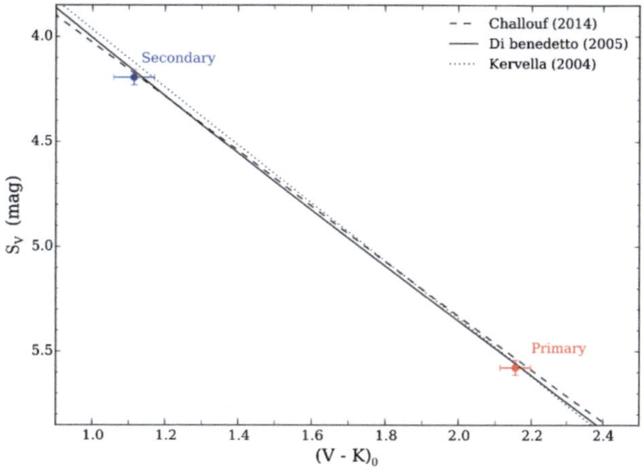

Fig. 1 SBCR relation obtained for two components of TZ For system (linear sizes come from analysis of light curve and radial velocity curves and knowing the distance we have angular diameters and thus SBCR). Now such a relation can be obtained for some 20 eclipsing binary systems which are located close enough to permit accurate determination of their interferometric orbits. This is the unique way to apply to techniques which provide 1% accuracy to the same targets and as the result check on their absolute zero point

3.3 Cepheids in Eclipsing Binaries

We already conducted an extensive program on verifying some 20 candidates for eclipsing binary systems with classical Cepheid components reported by Soszynski et al. (2008). From several high quality spectroscopic observations we confirmed 9 classical Cepheids in 8 systems [7 systems composed of a Cepheid in orbit with a helium burning red giant (Pietrzynski et al. 2010, 2013; Gieren et al., Pilecki et al. 2015, 2015), and one so far unique system composed of two classical Cepheids (Gieren et al. 2014)]. We note that these are the first-ever known Cepheids in eclipsing binary systems. All systems are very well detached without any sign of activity or additional variations in the orbital light curves except the ones caused by the regular pulsations of the Cepheids. In order to precisely model these very challenging systems it has been developed and tested a novel method for modelling eclipsing binaries containing radially pulsating components (see Pilecki et al. 2013). A first application of this method to model the high quality photometric and radial velocity data for the eclipsing system (Cep227) resulted in the determination of very precise values for all important physical parameters of this system, including the dynamical mass of the Cepheid, and its mean radius with unprecedented accuracy of 0.8% and 1% respectively. Very importantly for the Baade-Wesselink type IRSB technique, we also measured for the first time a geometrical and distant-independent projection factor for this Cepheid ($p = 1.23 \pm 0.03$ (stat.)± 0.04 (syst.)). We remind that the projection factor is in the classical Baade-Wesselink (BW) technique a multiplicative parameter used to convert radial velocities into pulsational velocities. BW distances are linearly proportional to the p-factor, then its accurate calibration for Cepheids is of critical importance for the distance scale.

These results will provide a unique opportunity to significantly improve our knowledge of the basic physics of classical Cepheids. In particular the sub-percent dynamical masses will serve to solve the long standing problem of the Cepheid mass discrepancy. They also provide an opportunity to apply three different techniques for distance determination (Cepheid P–L relation, B–W method and eclipsing binary method) to measure distances to these same objects. That should contribute significantly to check the potential systematic errors. The final output Gaia catalogue is foreseen for 2022 and will deliver distances of many Cepheids helping for the distance scale in the Universe and to better understand the details of the modelling of these pulsating stars.

4 Distance and Orbital Motions in X-Ray/Gamma-Ray Binaries

4.1 Estimation of the Compact Object Mass

The fact that the orbital motion affects the astrometric measurements means that the astrometric measurements enable us to estimate the orbital motion, thereby the orbital elements (see Sect. 2.2). Given the mass of the target star, we can estimate the companion mass. In this section, we focus on X-ray and on gamma-ray binaries and we estimate the precision of the determination of the companion mass (= compact object mass).

The precision in estimating the mass of a compact object is related to the accuracy of the semi-major axis of the orbit of a companion star. Firstly, the relation between the mass of the compact object and the semi-major axis of the star is represented as

$$M_c = \left(\frac{4\pi^2}{G}\right)^{\frac{1}{3}} \left(\frac{M_{\rm tot}}{P}\right)^{\frac{2}{3}} da, \tag{7}$$

where M_c, $M_{\rm tot}$, d and G are the mass of the compact object, the total mass of the star plus the compact object, the distance to the binary system from the Earth, and the gravitational constant, respectively. This equation means that the precision of the compact object mass σ_m is determined by the precision of the semi-major axis σ_a, that is,

$$\sigma_m = \left(\frac{4\pi^2}{G}\right)^{\frac{1}{3}} \left(\frac{M_{\rm tot}}{P}\right)^{\frac{2}{3}} d\sigma_a, \tag{8}$$

where we assume that the fraction of uncertainty of the total mass is much smaller than the ratio of σ_a to a. That is,

$$\frac{\sigma_M}{M_{\rm tot}} \ll \frac{\sigma_a}{a}, \tag{9}$$

where σ_M represents the standard deviation of the total mass. This assumption is valid for all binaries listed up in Table 1 if we adopt the mass precision given in the literature (Tetzlaff et al. 2011; Casares et al. 2012). In addition, we make a similar assumption for d in (8);

$$\frac{\sigma_d}{d} \ll \frac{\sigma_a}{a}. \tag{10}$$

where σ_d represents the standard deviation of the distance.

Next, we estimate the typical precision of the compact object mass. We normalize Eq. (8) to obtain

$$\sigma_m \sim 0.1\ \mathrm{M}_\odot \left(\frac{M_{\rm tot,20}}{P_{200d}}\right)^{\frac{2}{3}} d_{\rm kpc}\sigma_{\rm mis,10}, \tag{11}$$

where the total mass, the orbital period, the distance, and the astrometric precision of the mission $\sigma_{\rm mis}$ are normalized by 20 $\mathrm{M}_\odot$, 200 days, 1 kpc, and 10 µas, respectively. Here, we assume the equation $\sigma_a \sim \sigma_{\rm mis}$, which is valid unless the eccentricity and inclination angle are large and the argument of periapsis is close to 90°. Thus, for typical High Mass X ray Binary (HMXBs), we can measure the compact object mass by the precision of $\sim$0.1 $\mathrm{M}_\odot$ with the use of measurements by astrometric observation with the precision of $\sim$10 µas, such as Gaia and Small-JASMINE (see Sect. 2.3). We apply Eq. (11) to objects discussed in the next section.

Table 1 High-mass X-ray binary, γ Cas, and two gamma-ray binaries, whose compact objects are unknown. In M_{tot} column, we show the stellar masses M_* before the "+" symbols and M_c after the "+" symbols

Object name	M_{tot} ($M_\odot$)	Parallax (mas)	Distance (kpc)	P (days)	σ_m ($M_\odot$)
γ Cas	$19^a + 1$	5.9^b		204^c	0.03
LS I +61° 303	$12.5^d + 1.4$		1.9^e	26^e	1
HESS J0632+057	$16^d + 1.4$		1.4^e	320^e	0.5

[a] Appearing in the catalog "A catalog of young runaway *Hipparcos* stars within 3 kpc from the Sun" (Tetzlaff et al. 2011)

[b] Values appeared in *Hipparcos* catalog (van Leeuwen 2007)

[c] Appearing in the HMXB catalog (Liu et al. 2006)

[d] For M_*, we adopt the middle values of the mass ranges appearing in Casares et al. (2012)

[e] Values appearing in Casares et al. (2012)

4.2 Application: High-Mass X-Ray Binaries γ Cas

The X-ray emission mechanism has been extensively discussed for binaries such as γ Cas. The binary γ Cas is one of the peculiar Be/X-ray binaries, which generally consists of a pulsar and a Be star (B star with a circumstellar disk). Although the X-ray source is associated with the Be star, spectral features and X-ray variability of γ Cas are different from those of other typical Be/X-ray binaries. The observational features of γ Cas differ from other Be/X-ray binaries from thermal X-ray corresponding to high temperature (10–12 keV), relatively low luminosity ($L_X \sim 10^{32-33}$ erg/s), and no flare activity (Lopes de Oliveira et al. 2010 and reference therein). Such peculiar features are observed in some other Be/X-ray binaries, which are called γ Cas like objects (Lopes de Oliveira et al. 2006). Two possible scenarios can explain these X-ray features:

(1) accretion of the matter from the circumstellar disk onto a white dwarf (Lopes de Oliveira et al. 2007) or a neutron star (White et al. 1982)
(2) magnetic interaction between the stellar surface and the circumstellar disk (e.g. Smith and Robinson 1999).

The compact object has not been identified in the former scenario, so it is unknown if the accretion in the system is similar to that in a cataclysmic variable (a white dwarf) or an X-ray pulsar (a neutron star).

The precision of the compact object mass is estimated using Eq. (11), which includes the total mass, the orbital period, the distance, and the precision of the semi-major axis. We adopt masses of companion stars in the catalog "A catalog of young runaway *Hipparcos* stars within 3 kpc from the Sun" (Tetzlaff et al. 2011) and assume $M_c = 1$ $M_\odot$ for binaries whose compact objects are unknown and $M_c = 1.4$ $M_\odot$ for binaries whose compact objects are known as pulsars. Thus, each total mass of the target objects is given in Table 1. We refer to the *Hipparcos* catalog (van Leeuwen 2007) for the parallax. We calculate the precision of the semi-major axis for the four binaries in the way stated in Sect. 4.1 and show them in Table 1.

We can determine the compact object mass with an accuracy of 0.03 $M_\odot$, so we can determine whether its compact object is a white dwarf or a neutron star. More than 95% of 3000 white dwarfs have a mass less than 1.0 $M_\odot$ (Kepler et al. 2007), and 95% of neutron stars are inferred to have a mass more than 1.0 $M_\odot$ (Kiziltan et al. 2013). This means that we can set a boundary of mass which separates a white dwarf and a neutron star at 1.0 $M_{\odot}$. However, if we set the boundary, we should note that about 5% of white dwarfs have a mass more than 1 $M_{\odot}$, and their distribution extends up to $\sim$1.4 $M_{\odot}$ (Kepler et al. 2007). This extension of 0.4 $M_{\odot}$ is sufficiently larger than the precision of the compact object mass (0.03 $M_{\odot}$) that a confidence level of the identification of the compact object is determined just by the mass distribution of white dwarfs and neutron stars. Since more than 95 % of white dwarfs (neutron stars) are less (more) massive than 1.0 $M_{\odot}$, we can determine the identity of the compact object in γ Cas with about 2-σ confidence level.

4.3 Application: Gamma-Ray Binaries

The emission mechanisms in gamma-ray binary systems are not yet fully understood, which is related to the fact that compact objects in most of gamma-ray binaries have never been identified. The gamma-ray binary is a new class of HMXB, which was first discovered in 2004 (Aharonian et al. 2005), and five objects have been identified so far. Non-thermal radio to TeV gamma-ray emission has been detected from gamma-ray binaries, and some of this emission varies with the orbital period. However, the emission mechanism has not yet been explained clearly. If the compact object is a neutron star, we expect that the non-thermal emission is produced at shocks formed by the collision between the pulsar wind and the stellar wind. If it is a black hole instead, we expect that the stellar wind accretes to drive the jet, so that non-thermal emission can be produced at shocks in the jet (internal shock model) and/or shocks between the jet and the stellar wind (external shock model). Thus, the identification of the compact object allows us to constrain the emission mechanism.

In Table 1, we show the precision of the compact object mass for three gamma-ray binaries. There is a fair chance for the identification of the compact object in LS I +61° 303 and HESS J0632+057 by the astrometric measurements. About 95% of neutron stars are inferred to have a mass less than 2.0 $M_{\odot}$ (Kiziltan et al. 2013), and masses of black holes in Shaposhnikov and Titarchuk (2009) are $\gtrsim$ 3 $M_{\odot}$. These facts allow us to set a boundary at 2.5 $M_{\odot}$ which separates a neutron star and a black hole. Since there is no overlap between mass distributions of neutron stars and black holes, a confidence level of the identification of the compact object is determined by the precision of the compact object mass due to the astrometric measurements unlike the case of γ Cas. The confidence level is estimated using probabilities that the measured mass is larger than 2.5 $M_{\odot}$ for a neutron star and that it is smaller than 2.5 $M_{\odot}$ for a black hole. Because few black holes are inferred to have a mass with $\sim$3 $M_{\odot}$, we can neglect the latter. The precision of the compact object mass for LS I +61° 303 is 1 $M_{\odot}$, so that the probability that estimated mass is larger than 2.5 $M_{\odot}$ for a neutron star with a mass of 1.4 $M_{\odot}$ is roughly 30%, which corresponds to the 1-σ confidence level for the identification of the compact object. For HESS J0632+057, the corresponding probability is calculated to be roughly 5% because its precision of the compact object mass is 0.5 $M_{\odot}$, which corresponds to the 2-σ confidence level. Therefore, the astrometric observations for LS I +61° 303 and HESS J0632+057 allow us to identify the compact object in these systems.

Next, we estimate the precision of the compact object masses for nearby HMXB and gamma-ray binaries. Since the compact object mass correlates with the semi-major axis of the primary star orbit, we obtain the precision of the compact object mass by calculating the precision of the semi-major axis. We summarize the results for each binary in what follows.

- For the Be/X-ray binary γ Cas, whose mass precision is calculated to be 0.03 $M_{\odot}$, we can determine whether the compact object is a neutron star or a white dwarf with 2-σ

confidence level. Thus, we can provide important clue for revealing the X-ray emission mechanism in γ Cas.

- Regarding two gamma-ray binaries, LS I +61° 303 and HESS J0632+057, we can detect their orbits even if the compact object is a neutron star because the mass precision are respectively 1 $M_\odot$ (LS I +61° 303) and 0.5$M_\odot$ (HESS J0632+057). In addition, we can identify the compact object in these systems with the 1-σ confidence level for LS I +61° 303 and 2-σ confidence level for HESS J0632+057.

Although the emission mechanism in gamma-ray binaries have not been clarified, the astrometric data provided by high-precision (10 micro-arcsecond level) astronomical mission, such as Small-JASMINE and Gaia, will help us to reveal the emission mechanism.

5 Photospheric Radius Expansion X-Ray Bursts as Standard Candles and Distance Determination

5.1 Type-I X-Ray Bursts

As for the previous discussion HMBX objects, low-mass X-ray binary systems (LMXBs), typically host a compact object (often a neutron star, NS) that accretes matter via Roche-lobe overflow from a main-sequence donor star ($M < 1$ $M_\odot$), thus forming a disk around the compact object. The ionized accreting gas is funneled by the NS magnetic field onto the surface, reaching such hot temperatures that it emits in the X-ray energy band. Depending on the accreted mass density and temperature, thermonuclear explosions (type-I X-ray bursts) on the surface of the accreting NS can occur triggered by unstable hydrogen and/or helium burning (see for reviews Lewin et al. 1993, Strohmayer and Bildsten 2003). Type-I X-ray bursts are among the most evident signatures of the presence of a NS in low-mass X-ray binaries, and several thousand bursts, from about 110 NSs, have been observed to date (see Galloway et al. 2008). From the duration of the bursts, three main branches are distinguishable: normal bursts, intermediate long-bursts, and super-bursts (see Fig. 7 in Falanga et al. 2008). The rise time of these bursts is ≈1–100 s, whereas the decay time, τ, is approximately exponential, with a duration of a few seconds for normal bursts, tens of minutes for intermediate long-bursts, and up to several hours for super-bursts, depending on the nuclear reactions involved. These bursts can be described in terms of different fuel types and accretion rates, and therefore, also exhibit different recurrence times (Lewin et al. 1993; Strohmayer and Bildsten 2003).

The burst spectra can usually be well described as a black-body with temperature $kT_{\rm bb} \approx$ 1–3 keV. The energy-dependent decay time of these bursts is attributed to the cooling of the NS photosphere and results in a gradual softening of the burst spectrum (Lewin et al. 1993; Strohmayer and Bildsten 2003). During some type I X-ray bursts the energy release is high enough that the luminosity reaches the Eddington limit, and at that point the radiation pressure lifts the surface layers from the NS in a photospheric radius expansion (PRE) episode. Since the luminosity scales as $L_{\rm burst} \propto R^2 T^4$ for pure black-body spectrum during the PRE episode, while the bolometric luminosity remains constant at the Eddington value, the temperature, T, drops when the radius of the photosphere, R, expands (see Fig. 2). The point at which the NS atmosphere reaches the surface again, $R_{\rm bb,min} = R_{\rm NS}$, i.e., at the highest temperatures, is called *touch-down*.

When bursts undergo PREs, they act as standard candles, since the source distance can be determined based on the assumption that the bolometric peak luminosity is saturated at the Eddington limit, $L_{\rm Edd}$ (see Sect. 5.2).

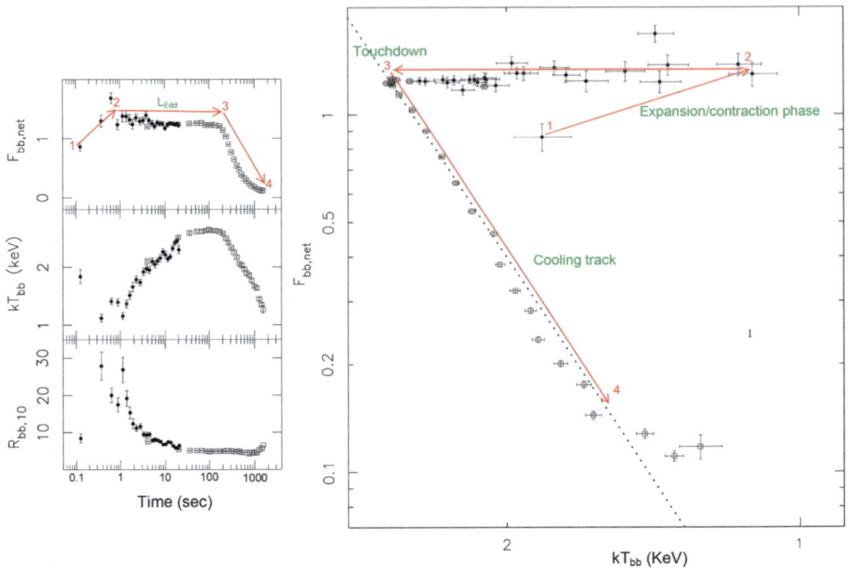

Fig. 2 During a PRE, i.e., during the expansion and contraction of the photosphere the Eddington luminosity limit is reached. During this plateau phase (constant maximal flux), the radius and temperature follows the black-body behavior as follows $L_{\text{bb,burst}} \propto R^2T^4$. This figure has been adapted from Kuulkers et al. (2002)

5.2 Bursts with Photospheric Radius Expansion: The Eddington Luminosity and Bursters Distances

During very strong bursts the NS photosphere expands as a result of radiation pressure to the Eddington limit, i.e., maximum luminosity at which the outward radiation force equals the inward force of gravity. This state of balance is called hydrostatic equilibrium, $F_{\text{grav}} + F_{\text{rad}} = 0$. During the expansion and contraction phase, with the photospheric radius, R, the Eddington luminosity remains constant as measured by a distant observer following the equation as:

$$L_{\text{Edd},\infty} = \frac{4\pi c G M_{\text{NS}}}{\kappa}(1+z)^{-1}, \tag{12}$$

where, G, is the gravitational constant, c, the light velocity, and $(1+z) = [1 - 2GM/(Rc^2)]^{-1/2}$ the gravitational redshift factor. We notice, that R is not the radius of the NS, but rather the radius of the photosphere, which at maximum expansion can be a few times the stellar radius. Here, κ, is the high temperature dependence electron scattering opacity. During the expansion phase, κ-equation is approximated as follows:

$$\kappa = 0.2(1+X)\left[1 + (\alpha T)^{0.86}\right]^{-1}. \tag{13}$$

This equation takes into account that for very high temperatures the scattering electrons become relativistic, and is valid for low-density limit as appropriate for NS atmosphere (Lewin et al. 1993), where $\alpha = 2.2 \times 10^{-9}\ \text{K}^{-1}$. Here, X, is the hydrogen mass fraction, $X = 0.73$ for solar composition or $X = 0$ for a hydrogen-poor bursts. In Fig. 3 we show in the two bands the range in the Eddington luminosity for different matter composition and for canonical NS mass 1.4 $M_\odot$ and radius 10 km. The upper and lower boundaries of each band is determined by the Eddington limit observed at infinity with very strong and very week

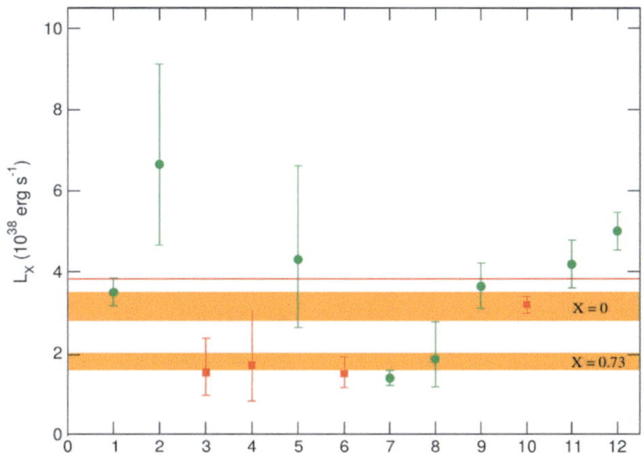

Fig. 3 Bolometric peak luminosities reached during the brightest X-ray bursts seen in the twelve globular cluster X-ray sources (see Table 2). Photospheric radius expansion X-ray bursts have been denoted with green circles; the others with red squares. With orange bands we give the expected range in the Eddington limit for matter with cosmic composition ($X = 0.73$) and hydrogen-poor matter ($X = 0$), see Sect. 5.2. This figure has been adapted from Kuulkers et al. (2003)

photospheric radius expansion episodes. Now, from the observed PRE bursts parameters (R_{bb}, T_{bb}, and F_{burst}), and the burst type, helium ($X = 0$) or hydrogen burst ($X = 0.73$), the burst PRE flux can be converted to luminosity and assumed to be at the Eddington luminosity as follows:

$$L_x = 4\pi F_x d^2 = L_{Edd}. \tag{14}$$

For PRE bursting NSs located in LMXBs, the distance to the source/system, d, can now be determined.

However, the association that the maximal observed burst flux during a PRE is equal to the theoretical Eddington luminosity (Eq. (12)) has to be observationally confirmed. Therefore, Kuulkers et al. (2003) determined the PRE luminosity from twelve burster NS sources located in globular clusters. The distance to the globular clusters are independently known, and therefore, the Eddington luminosity has been empirically determined to be $3.79 \pm 0.15 \times 10^{38}$ erg s^{-1} (see Fig. 3).

As an example, assuming a burst bolometric flux luminosity of $F_x = 8.2 \pm 2 \times 10^{-8}$ equal to the Eddington value for a helium type-I X-ray burst $L_{Edd} \approx 3.8 \times 10^{38}$ erg s^{-1}), as empirically derived by Kuulkers et al. (2003), we obtain the source distance of $d = 6.2^{+1.6}_{-0.9}$ kpc. For comparison, the theoretical value of this distance (see Eq. (12)) is found by assuming a helium atmosphere ($X = 0$), a peak flux temperature of $kT_{bb} = 2.7 \pm 0.2$ keV, and canonical NS parameters, is $5.5^{+1.3}_{-0.8}$ kpc. Alternatively, assuming the peak luminosity to be the Eddington luminosity for solar composition ($X = 0.73$) infers a distance of ≈ 4 kpc. The fiducial distance can be assumed to be $d = 6.2^{+1.6}_{-0.9}$ kpc for a helium burst.

Determining the source distances from type-I X-ray bursters with PRE is a valid method. LMXBs are old systems concentrated in the direction of the Galactic center or globular clusters. So far, 16 bursting NSs have been found to be located in 13 globular clusters, where the Eddington limit can be empirically determined, and in around 90 Galactic bursting NSs, for which the distance or an upper limit can be estimated.

In the following reference (Jonker and Nelemans 2004) systematics on distance estimates from these sources are discussed. On addition, D. Galloway and collaborators (in preparation 2017) are working on the likely contribution of systematic effects for the sources and the method, and they are writing a paper to improve the 2004 published values. The cluster distances obtained by Gaia will be an important source of calibration for this technique.

Table 2 Bursting neutron stars hosted in their globular clusters at a measured distance. The observed PRE bursts Eddington luminosity, L_{Edd}, is also reported. This Table has been adapted from Kuulkers et al. (2003)

	X-ray burster	Globular cluster	Distance kpc	L_{bol} 10^{38} erg s^{-1}
1	MX 0513−40	NGC 1851	12.1 ± 0.3	$3.5^{+0.3}_{-0.3}$
2	4U 1722−30	Terzan 2	$9.5^{+2.5}_{-2.0}$	$6.6^{+2.5}_{-2.0}$
3	MXB 1730−335	Liller 1	$8.8^{+3.3}_{-2.4}$	$1.5^{+0.8}_{-0.6}$
4	XB 1733−30	Terzan 1	$4.3^{+2.3}_{-1.5}$	$1.6^{+1.3}_{-0.8}$
5	XB 1745−25	Terzan 5	$8.7^{+3.3}_{-2.4}$	$4.3^{+2.3}_{-1.7}$
6	MX 1746−20	NGC 6440	$8.4^{+1.5}_{-1.3}$	$1.5^{+0.4}_{-0.3}$
7	4U 1746−37	NGC 6441	$11.0^{+0.9}_{-0.8}$	$1.4^{+0.2}_{-0.2}$
8	GRS 1747−312	Terzan 6	$9.5^{+3.3}_{-2.5}$	$1.8^{+0.9}_{-0.7}$
9	4U 1820−30	NGC 6624	7.6 ± 0.4	$3.6^{+0.6}_{-0.6}$
10	H1825−331	NGC 6652	9.6 ± 0.4	$3.2^{+0.2}_{-0.2}$
11	A1850−08	NGC 6712	8.2 ± 0.6	$4.2^{+0.6}_{-0.6}$
12	4U 2129+12^{b}	NGC 7078	10.3 ± 0.4	$5.0^{+0.5}_{-0.5}$

6 Direct Angular Diameter Distance Measurement with the H_2O Megamaser Method

In the previous sections, we discussed methods of distance determination for galactic objects such as stars and X-ray binaries. These methods allow one to constrain the distance scales within the Milky Way Galaxy. To establish the *extragalactic* distance scale, different distance indicators are needed because of the limitation of the astrometric measurements of stellar objects. The traditional methods of extragalactic distance measurements usually involve Cepheids, tip of the red giant branch, surface-brightness fluctuations, the Tully-Fisher relation and Type Ia supernovae. For these extragalactic distance indicators, a well-written review has been provided by Freedman and Madore (2010), and we will not elaborate on these methods here. In this section we will focus on a complete review on a relatively new method of extragalactic distance determination which has been continuously developed during the last 20 years. This technique is called "the H_2O Megamaser Method".

The H_2O Megamaser Method involves sub-milliarcsecond resolution imaging of H_2O maser emission from sub-parsec circumnuclear disks at the center of active galaxies, and this technique was first established based on the study of the nearby galaxy NGC 4258 which exhibits strong H_2O maser emissions (Herrnstein et al. 1999). Throughout this section, we will refer to such circumnuclear disks as "megamaser disks" and galaxies that contain such megamaser disks as "megamaser galaxies". Before discussing the actual application of this method to extragalactic distance measurement, let's first demonstrate how this method works.

6.1 The Concept of the H_2O Megamaser Method

The H_2O megamaser method requires sub-milliarcsec resolution imaging and single-dish monitoring of H_2O maser emission from megamaser disks at the center of active galaxies. In this technique, the key to determining the distance to a maser galaxy is measuring four orbital parameters (i.e. the radius, velocity, acceleration, and inclination of an orbit). Here,

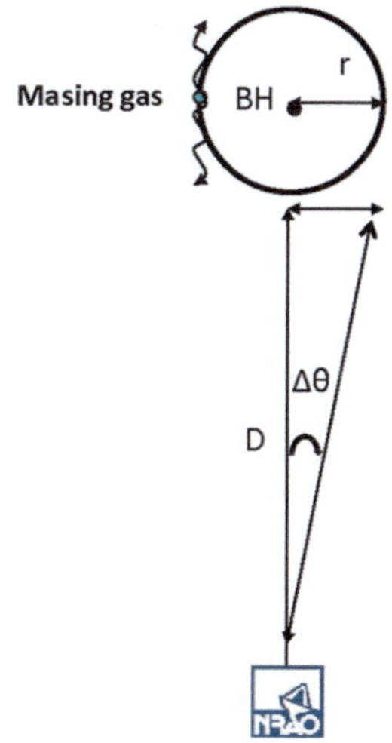

Fig. 4 A model to illustrate how we determine the distance of a galaxy with a H_2O maser disk

we use the example of NGC 4258 (Herrnstein et al. 1999), UGC 3789 (Braatz et al. 2010) and a cartoon plot to explain the principle behind this technique.

Figure 4 shows a cartoon plot of masing gas orbiting around a supermassive black hole in a circular orbit at the center of a galaxy. D is the distance from the observer to the galaxy, r is the physical radius of the orbit, and $\Delta\theta$ is apparent angular radius seen by the observer. Based on simple geometry, we can express the distance as

$$D = \frac{r}{\Delta\theta}. \tag{15}$$

According to Newton's second law, the gravitational acceleration a of the masing gas in the circular orbit is related to the orbital velocity v_0 and physical radius r with

$$a = \frac{v_0^2}{r}. \tag{16}$$

For convenience, the above equation can be re-written as

$$r = \frac{v_0^2}{a}. \tag{17}$$

By combining Eqs. (15) and (17), and correcting the maser velocity and acceleration for the inclination of the orbit, the distance D can be expressed as

$$D = \frac{v_0^2}{a\,\Delta\theta} \sin i. \tag{18}$$

Therefore, one can determine an accurate distance to a megamaser galaxy if the four parameters in Eq. (18) can be determined precisely. Now, let's use the real data from NGC 4258 and UGC 3789 to illustrate how to measure these parameters.

Figure 5 shows the maser distribution and the position-velocity (PV) diagram of the H_2O maser disk at the center of NGC 4258. The black hole is sitting at the center of the disk with a jet ejected toward the top and bottom side of the disk. The data points in the maps and rotation curves are colour-coded to indicate redshifted, blueshifted (the redshifted/blueshifted masers are also called "high velocity" masers throughout the section), and systemic masers, where the "systemic" masers refer to the maser spectral components having velocities close to the systemic velocity of the galaxy. Note that the redshifted and blueshifted masers are

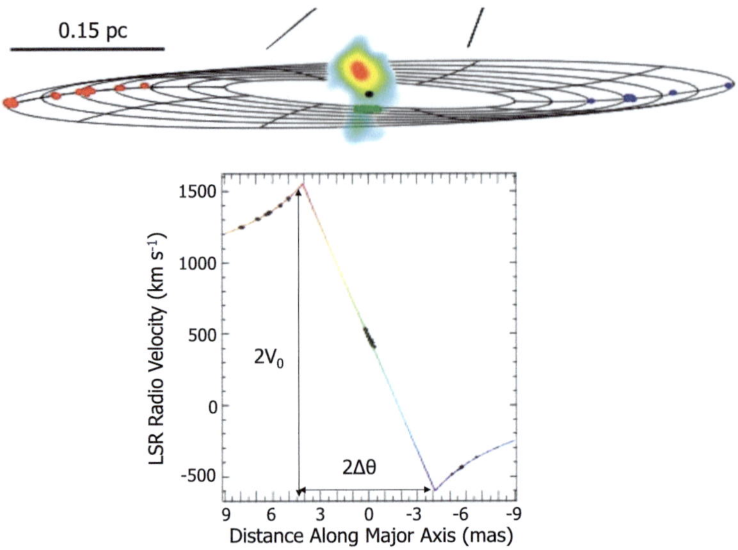

Fig. 5 The maser distribution (*the top panel*) and the PV diagram (*the bottom panel*) of the H_2O maser disk at the center of NGC 4258. This figure was made based on the work done by Herrnstein et al. (1999)

located very close to the mid-line (the intersection of the plane of the sky and the disk) of the disk and can be traced nearly perfectly with a Keplerian rotation curve in the PV diagram. The systemic masers are located in front of the black hole,[1] and in the PV diagram they can be beautifully fit with a straight line, a feature that indicates that these masers lie in a single narrow ring in the disk.

As shown in Fig. 5, the intersection between the straight line and the Keplerian rotation curve that fit the systematic and high velocity masers directly give the orbital velocity V_0 and angular size $\Delta\theta$ of the ring in which the systemic masers reside. In addition, with the angular offset y of the systemic masers from the black hole in the vertical direction, one can obtain the inclination i of the orbit for systemic masers

$$i = \cos^{-1}\left(\frac{y}{\Delta\theta}\right). \tag{19}$$

Therefore, one can obtain three parameters for the distance measurement from the images of megamaser disks. The remaining parameter, a, has to be measured from multi-epoch monitoring of maser spectra. Here, we use UGC 3789 from Braatz et al. (2010) to illustrate how to measure the centripetal acceleration of H_2O masers.

It is believed that the origin of the accelerations of H_2O masers that we observe is the centripetal acceleration due to the gravity of the black hole at the center of maser disk. That the maser is accelerating does not mean that the masing gas is spiraling toward the black hole. Rather, they stay in a circular orbit because the acceleration due to gravity is balanced by the centrifugal acceleration that keeps the object in orbit. While the orbital speed remains the same, the centripetal force causes the masing gas to keep changing directions in order to maintain the circular orbit, and therefore the line-of-sight velocity seen by the observer

[1]The mid-line of the disk and places in front of the black hole are particularly favorable for maser emission because in these locations the path lengths for maser amplification reach maximum values (Lo 2005).

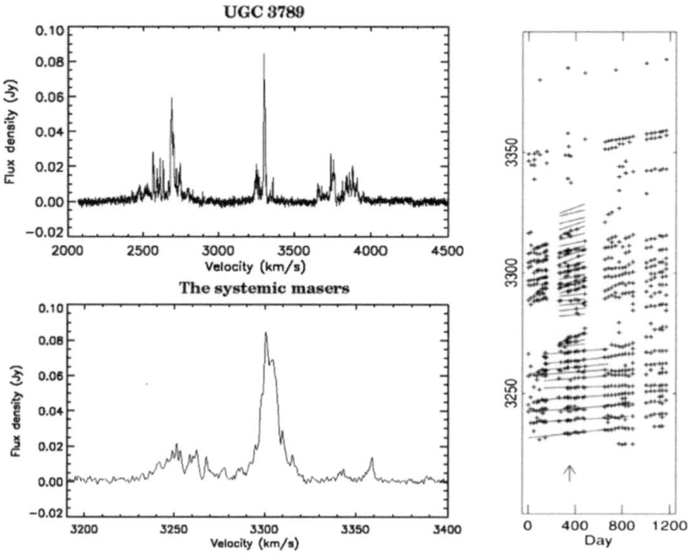

Fig. 6 *The top-left plot* shows an H_2O maser spectrum of UGC 3789, with the spectrum for the systemic masers at *the bottom*. In *the right panel* we plot the velocities of the systemic maser lines as a function of time. The slopes of the fitted straight lines (*the solid lines*) directly give the accelerations of the masers. Illustrations and figures are from Braatz et al. (2010)

will keep changing. The rate of change in the observed maser velocity is the centripetal acceleration along the line of sight that we want to measure.

In the top-left panel of Fig. 6 is an H_2O maser spectrum for UGC 3789 with the zoom-in spectrum for the systemic masers on the bottom. We can see several distinct lines in the spectrum. Because of the gravitational acceleration, we expect that the systemic maser lines will drift toward higher velocities (i.e. toward the right) with time, and this is indeed seen in the right panel of Fig. 6. Here, we plot the velocities of maser lines we can identify in the spectra from multi-epoch observations as a function of time. As we expect, we do see clear positive trends, and the slopes of the lines that fit the data directly measure the maser accelerations. For maser disks of sub-pc size with $\sim 10^7$ $M_\odot$ black holes, the acceleration is typically a few $\mathrm{km\,s^{-1}\,yr^{-1}}$. Since the maser lines are often blended and can have significant variability over timescales of a few months to years, in order to reduce systematic error caused by blending and variability, it is usually necessary to improve the acceleration measurement by detailed modelling of maser lines with a global fitting program that fits all lines in all epochs simultaneously. The detailed discussion on maser line modelling can be found in Reid et al. (2013) and we will describe some details in Sect. 6.2 of this article.

One should note that while the conceptual idea of measuring the orbital parameters as described above is simple, the actual process of measuring these parameters is more complicated. The major complexity comes from the fact that in all well-behaved megamaser disks that have been published (e.g. Braatz et al. 2010; Reid et al. 2013; Kuo et al. 2015; Gao et al. 2016), systemic masers do not usually reside in a single narrow ring as in the case of NGC 4258. The systemic masers in our sample often have different accelerations, indicating that systemic masers in a maser galaxy are located at multiple radii. Therefore, one cannot determine the galaxy distance by simply applying Eq. (18). More sophisticated 3-dimensional modelling of megamaser disks that takes the disk warp and radial distribution of systemic masers into account is needed to determine the distance accurately. We will discuss the details of the 3-dimensional modelling in our project in Sect. 6.3.

6.2 The Application of the H_2O Megamaser Method

After the establishment of the methodology based on the study of the megamaser galaxy NGC 4258, the first systematic effort of applying the maser technique to determine distances to galaxies beyond the Local Group was made in the *Megamaser Cosmology Project* (MCP; Braatz et al. 2009, Reid et al. 2009, Braatz et al. 2010, Reid et al. 2013, Kuo et al. 2015, Gao et al. 2016). The primary goal of this project is to determine the Hubble constant H_0 to $\sim$3% accuracy in order to constrain the equation of state parameter w of dark energy (Hu 2005).

In order to determine the Hubble constant to 3% accuracy, the MCP team aims to measure accurate (i.e. $\leq$10%) distances to about 10 maser galaxies in the first stage of this project (i.e. 2007–2017). To achieve this goal, it is important to first conduct sensitive large surveys to find more megamaser galaxies similar to NGC 4258, followed by imaging the megamaser disks with high resolutions. In the MPC, a four step approach is followed:

(1) Survey with the Green Bank Telescope (GBT) to identify additional high-quality disk masers in the Hubble flow for $\sim$10% distance measurements.
(2) Image the sub-pc megamaser disks to obtain their rotation curves with the High Sensitivity Array (HSA; i.e. VLBA+GBT+EB).[2]
(3) Measure accelerations of masers accurate to $< 10\%$ by monitoring their spectra over a timescale of a few years.
(4) Determine the distance to maser host galaxies by modeling the maser disk kinematics.

As the MCP continues to discover and image megamaser disks, an important result, in addition to the distance determination, is the accurate measurement of the masses of black holes at the centers of these megamaser galaxies. These accurate black hole masses provide an important basis for testing and understanding the nature of the famous correlation between black hole mass and stellar velocity dispersion of a galactic bulge, the M_{BH}-σ relation.

6.3 The MCP Survey

In the MCP survey, one searches H_2O megamaser galaxies with the GBT. By the end of 2016, the MCP has attempted to look for megamaser emissions from a sample of $\sim$4860 active galaxies, which are selected primarily from the SDSS, 6dF, and 2MRS survey. The majority of the sample galaxies in the MCP survey are Seyfert 2 galaxies because it is in Seyfert 2 galaxies that the probability of finding edge-on circumnuclear molecular disks is the highest.

In total, $\sim$140 new megamaser galaxies are detected in the MCP and the detection rate is $\sim$3%. Including the megamaser galaxies detected by other surveys, the total number of reported megamaser detection is 162 by the time this article is written. The entire sample of the MCP maser survey and the list of detected megamasers from all maser surveys can be found from the MCP website: safe.nrao.edu/wiki/bin/view/Main/MegamaserCosmologyProject.

[2]The VLBA is a facility of the National Radio Astronomy Observatory, which is operated by the Associated Universities, Inc. under a cooperative agreement with the National Science Foundation (NSF). The Effelsberg 100-m telescope is a facility of the Max-Planck-Institut für Radioastronomie.

Among the detected megamaser systems, ∼50 galaxies show spectral features consistent with masers being in the circumnuclear disks and the detection rate of these "disk megamasers" is ∼1%. Most of other systems show only one or two chunks of maser lines. It is quite likely that the megamasers in these systems are not associated with the circumnuclear disks, but are triggered by jet-cloud interactions in AGNs or are excited by intense star formation. In either cases, the gas distribution and kinematics cannot be easily modelled. Therefore, the MCP only focuses on the ∼1% detection of the disk maser systems for Hubble constant determination (Kuo et al. 2013).

6.4 The Megamaser Images

Once a high quality sample of disk megamasers is obtained, the next important step for distance measurement is to image the megamaser disks and measure their rotation curves using the technique of very long baseline interferometry (VLBI). The VLBI technique plays a crucial role in measuring precise maser positions and the rotation curves of the megamaser disks because the angular resolution provided by VLBI is two orders of magnitude higher than the best optical resolution. The typical angular resolution of the HSA used by the MCP team is ∼0.3–1 mas, which is comparable to the angular sizes of the megamaser disks observed in the MCP. The typical position accuracy that can be achieved when observing a maser spot can be as high as 0.01 mas if the signal-to-noise ratio of the corresponding maser feature is ∼25.

By the time this review is written, the MCP has published images of H_2O megamaser emissions from 13 galaxies in Kuo et al. (2011), Gao et al. (2016, 2017), respectively. Figure 7 shows the example VLBI maps of the seven megamaser disks published by Kuo et al. (2011). The data points in the VLBI maps are color-coded to indicate redshifted, blueshifted, and systemic masers, where the "systemic" masers refer to the maser spectral components having velocities close to the systemic velocity of the galaxy. Among the 13 megmaser disks published by the MCP, eleven of them show well-ordered maser disks with rotation curves consistent with being Keplerian. The Keplerian nature of the eleven systems allows one to determine black hole masses with percentage level of accuracy, and the black hole masses are typically on the order of $\sim 1 \times 10^7\ M_\odot$ (see Gao et al. 2017). The typical angular size of the megamasers disks seen in these papers are ∼1 mas and their physical sizes are in general less than a parsec.

While the MCP obtained a sample of eleven megamaser systems that show well-ordered thin disks with nearly perfect Keplerian rotation curves, not all of them have sufficient data quality for accurate distance measurement. In some galaxies, such as NGC 1194 (Fig. 7), the numbers of systemic maser features and their velocity ranges are too small to measure the rotation curves precisely, and this in turn limit the accuracy of the distance measurements. Therefore, the MCP only focuses on those maser galaxies in which the properties of the maser disks are good enough for us to determine the distance to ∼10% accuracy. These galaxies include UGC 3789 (Reid et al. 2013), NGC 6264 (Kuo et al. 2013), NGC 6323 (Kuo et al. 2015), NGC 5765b (Gao et al. 2016), and Mrk 1419 (Impellizzeri et al. in prep.). When we discuss the acceleration measurements below, it is toward these five systems that we attempt to measure accurate acceleration for distance and Hubble constant determination.

6.5 The Acceleration Measurements of Megamasers

Water masers in circumnuclear disks experience strong centripetal acceleration as they orbit the central black hole, and the acceleration can be measured by tracking the line-of-sight

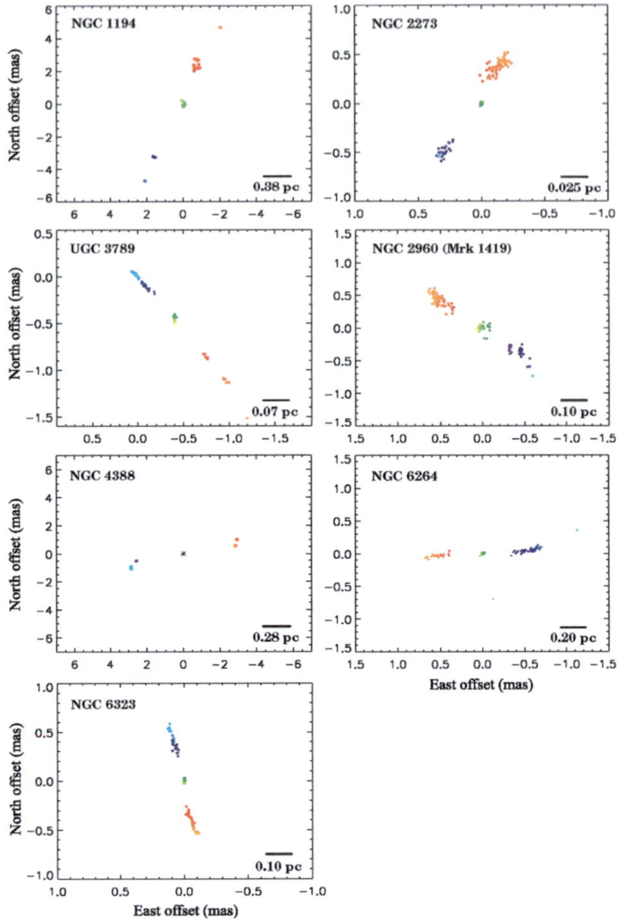

Fig. 7 VLBI maps for the seven 22 GHz H_2O masers megamasers analyzed. The maps are color-coded to indicate *redshifted*, *blueshifted*, and systemic masers, where the "systemic" masers refer to the maser components having recessional velocities close to the systemic velocity of the galaxy. This figure is adopted from Kuo et al. (2011)

velocity of each maser line as they change with time. In the MCP, two different methods are adopted to determine the accelerations of maser spots. The first method is called the *Eye-tracking* method. In this method, one first measures the peak velocities of maser spectra at different epochs and plots the spectral peak velocities as a function of time (see Fig. 8). One then identifies a group of data points within a particular time segment by eye that is believed to be from a particular maser line that persists over time and fit the trend of data points in the plot with straight lines. The slopes of the fitted lines directly give the accelerations.

The second method is the *global least-squares fitting* (Kuo et al. 2013; Reid et al. 2013) method. The global least-squares fitting method fits the amplitudes for a range of channels in all selected spectra simultaneously, with a model consisting of multiple Gaussian-shaped lines that drift linearly in time. This method is particularly important for maser lines that have high degree of blending and variability (e.g. Kuo et al. 2013). In such cases, there will be significant systematic errors caused by line-blending and variability in the acceleration measurements if one does not fit multiple line components simultaneously using methods similar to the global least-squares fitting method presented here (see Fig. 9).

Among the two methods mentioned above, the global least-squares fitting technique is applied for the systemic masers because this method can better handle the line-blending effect of the systemic masers and determine their accelerations more accurately. On the other

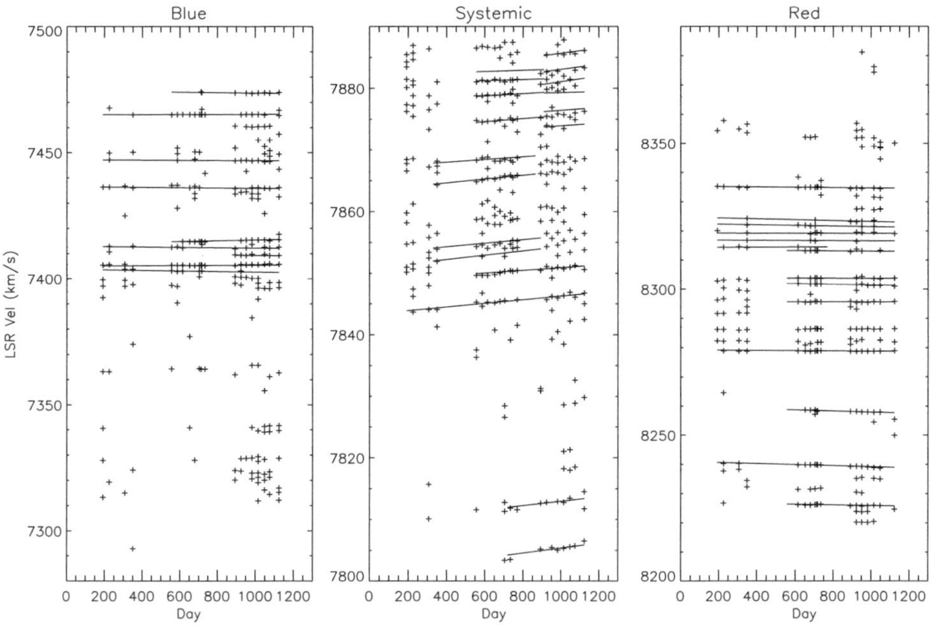

Fig. 8 The radial velocities of NGC 6323 maser peaks as a function of time (*the crosses*) for the blue-shifted masers (*the left panel*), the systemic masers (*the middle panel*), and the red-shifted masers (*the right panel*). *On top of these plots*, we overplot the fitting results from the eye-tracking method for the high-velocity masers and from the least-squares fitting for the systemic masers

hand, since many high-velocity maser lines are relatively isolated, and very precise acceleration measurements of these masers are not critical for accurate distance determination, the quicker eye-tracking method is usually used to provide estimates of accelerations for the high-velocity masers.

In general, the accelerations of the systemic masers in the MCP maser galaxies range from $\sim$1 $\mathrm{km\,s^{-1}\,Mpc^{-1}}$ to $\sim$6 $\mathrm{km\,s^{-1}\,Mpc^{-1}}$. It is found that in a single system, the accelerations usually vary as a function of velocities, indicating that the systemic masers often have radial distribution in the disk. As a result, the simple conceptual idea of the Megamaser technique mentioned in Sect. 6 cannot be directly applied to maser systems in the MCP because it requires the systemic masers to reside on a single narrow ring. To determine galaxy distance and Hubble constant, one has to allow the systemic masers to have radial distribution in the model. As a result, modeling the maser disk in three dimensions becomes necessary.

6.6 The Hubble Constant Determination: The Bayesian Approach

To determine the Hubble constant using the position, velocity, and acceleration of maser spots in a gas disk, the MCP adopts a Bayesian approach to model the maser disk in three dimensions (Reid et al. 2013). The disk model is described by 15 global parameters including the Hubble constant and black hole mass. The galaxy distance D is calculated as a local variable in the fitting program. The model also allows the masers to have eccentric orbits and the disk can be warped along the line-of-sight and along the mid-line of the disk (see Reid et al. 2013 for details). Note that the parameters associated with eccentricity are generally not used in the actual fitting because it is believed that the deviation of the masers from circular orbits are negligible (Reid et al. 2013).

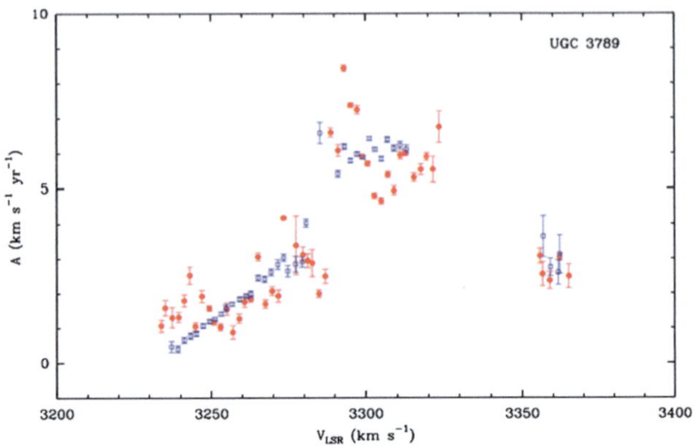

Fig. 9 The acceleration measurements for UGC 3789 from the global least-squares fitting method. *The filled red squares* are the measurements from Method 1 whereas *the open blue squares* are the accelerations measured from Method 2. The differences between Method 1 and Method 2 are used to estimate the systematic errors in the acceleration measurement

The program adopts a Markov chain Monte Carlo (McMC) approach to obtain the *posteriori* probability distribution function of the model parameters. Optimum values of the model parameters were estimated from the *posteriori* probability distributions, which were usually generated from a total of $\sim 3 \times 10^9$ McMC trials. The large number of McMC trials is needed here to ensure the convergence of the Bayesian fitting, given that there are 15 global parameters (e.g. H_0) and more than 100 internal parameters associated with the radii and azimuthal angles of maser spots.

6.7 Hubble Constant Derived from the Bayesian Fitting

Among the five galaxies for the Hubble constant determination, the MCP have published the results from four of them. The Hubble constant determined from UGC 3789, NGC 6264, NGC 6323, and NGC 5765b are $H_0 = 68.9 \pm 7.1$, $H_0 = 68 \pm 9$, H_0 is 73^{+26}_{-22}, $66 \pm 6.0\ \mathrm{km\,s^{-1}\,Mpc^{-1}}$, respectively. The weighted average of these values is $H_0 = 68.0 \pm 4.4\ \mathrm{km\,s^{-1}\,Mpc^{-1}}$. The uncertainties of the H_0 determination shown here include both the measurement errors and systematic errors, and the total uncertainties are dominated by measurement errors for all sources.

An interesting point to note for the H_0 measurements presented here is that all measurements are quite close to each other and are close to the value determined by the Planck Collaboration (Ade et al. 2015). Given the uncertainties of these measurements, one would expect that the scatter between the H_0 measurements from different galaxies would be larger than what we see here. As a result, it is not unreasonable that one may suspect whether there are any *unconscious bias* involved in the data analysis so that all of the measurements are biased toward the preferred value. We must admit that this is not impossible because our current analysis is not *blind*. Here, blind analysis means that one only discloses the value of H_0 after making sure all of the analysis before fitting H_0 are correct and does not go back to modify the data analysis after finding the value of H_0 from the fitting.

In the current phase of the MCP, when the H_2O megamaser technique was being developed, such blind analysis was impossible. The megamaser technique evolved from assuming the systemic masers residing in a single narrow ring to allowing the masers to have radial distributions. During the development process, the MCP team continuously evaluates the dependence of the H_0 uncertainty on the data sensitivities and uncertainty in acceleration measurement. As a result, all subsequent data analysis on the same galaxy will not be totally blind.

While unconscious bias due to unblind analysis cannot be ruled out in the data analysis, the apparent closeness of the H_0 measurements is more likely to be caused by *the convergence problem* of the fitting. In the earlier stage of the MCP (Reid et al. 2013; Kuo et al. 2013), the slow convergence rate of the Bayesian fitting was not well recognized. Therefore, the H_0 measurements from the MCP presented before 2015 are not results from converging fittings and it is expected that the value of the Hubble constant measured from UGC 3789 and NGC 6264 would change by at least a few percent when the fittings actually converge. The MCP team is currently re-analyzing all the data-sets for the H_0 determination by considering the convergence problem carefully and will present the new results in 2018 (Braatz et al. in prep.).

7 concluding Remarks

Recent decencies of development of instrumentation give to astronomers the capability to set at the right position in distance the objects of the Universe. This is mandatory to understand how the Universe structured and evolved with time. Progresses are still foreseen but the tools to study stars or more complex objects like active galaxies in Universe are now at a point that a new astronomy is starting. By products like ages or fundamental parameters of stars with unprecedent accuracies for example for studies of exoplanetary systems for which we need to tag in age and distance the stars hosting these systems. The few examples given in this review have summarized the state of the art in this field of research even if it is just few aspects of these developments, producing already numerous discoveries and more with for example the future releases of the Gaia catalogue.

One have to keep in mind that this important step in the distance measurements done by Gaia is also a good benchmark Space instrumentation to learn for the future missions at horizon of 30 years, not only for star studies but also for distant objects like nucleus of point source galaxies for which the astrometry could be done in the future as it is today done for distant stars in the Milky Way.

References

P.A.R. Ade, N. Aghanim et al. (Planck Collaboration), Astron. Astrophys. **580**, A22 (2015)
F. Aharonian, A.G. Akhperjanian, K.-M. Aye et al., Astron. Astrophys. **442**, 1 (2005)
J. Andersen, Astron. Astrophys. Rev. **3**, 91 (1991)
C.A.L. Bailer-Jones, Publ. Astron. Soc. Pac. **127**, 994 (2015)
C. Boehm, A. Krone-Martins et al. (The Theia Collaboration), arXiv:1707.01348 (2017)
J. Braatz, J. Condon, M. Reid, C. Henkel, K.Y. Lo, C.Y. Kuo, C. Impellizzeri, L. Hao, The Megamaser Cosmology Project Large Proposal, submitted to NRAO, 1 June 2009
J.A. Braatz, M.J. Reid, E.M.L. Humphreys et al., Astrophys. J. **718**, 657 (2010)
A.G.A. Brown, A. Vallenari et al. (Gaia Collaboration), Astron. Astrophys. **595**, A2 (2016)
J. Casares, M. Ribó, I. Ribas, J.M. Paredes, F. Vilardell, I. Negueruela, Mon. Not. R. Astron. Soc. **421**, 1103 (2012)
A. Chelli, G. Duvert, Astron. Astrophys. **593**, L18 (2016)
A. Chelli, G. Duvert, L. Bourgès et al., Astron. Astrophys. **589**, A112 (2016)
G. Clementini, V. Ripepi, S. Leccia et al., Astron. Astrophys. **595**, A133 (2016)
O.L. Creevey, F. Thévenin, T.S. Boyajian et al., Astron. Astrophys. **545**, A17 (2012)
O. Creevey, T. Metcalfe, D. Salabert et al., arXiv:1703.01084 (2017)
B. Di, Mon. Not. R. Astron. Soc. **357**, 174 (2005)
R. Ezzeddine, B. Plez, T. Merle, M. Gebran, F. Thévenin, arXiv:1612.09302 (2016)
M. Falanga, J. Chenevez, A. Cumming et al., Astron. Astrophys. **484**, 43 (2008)
W.L. Freedman, B.F. Madore, Astrophys. J. **719**, 335 (2010)

A. Gallenne, G. Pietrzyński, D. Graczyk et al., Astron. Astrophys. **586**, A35 (2016)
D.K. Galloway, M.P. Muno, J.M. Hartman, D. Psaltis, D. Chakrabarty, Astrophys. J. Suppl. **179**, 360–422 (2008)
F. Gao, J.A. Braatz, M.J. Reid et al., Astrophys. J. **817**, 128 (2016)
F. Gao, J.A. Braatz, M.J. Reid et al., Astrophys. J. **834**, 52 (2017)
W. Gieren et al., Astrophys. J. **786**, 80 (2014)
W. Gieren et al., Astrophys. J. **815**, 28 (2015)
G.A. Gontcharov, A.V. Mosenkov, Mon. Not. R. Astron. Soc. **470**, L97 (2017)
N. Gouda, T. Tsujimoto, Y. Kobayashi et al., in *8th Asian-Pacific Regional Meeting*, vol. II (2002), p. 5
D. Graczyk et al., Astrophys. J. **780**, 59 (2014)
E.F. Guinan, E.L. Fitzpatrick, L.E. DeWarf et al., Astrophys. J. Lett. **509**, L21 (1998)
A. Helmi, J. Veljanoski, M.A. Breddels, H. Tian, L.V. Sales, Astron. Astrophys. **598**, A58 (2017)
J.R. Herrnstein, J.M. Moran, L.J. Greenhill et al., Nature **400**, 539 (1999)
D. Hobbs, E. Høg, A. Mora et al., arXiv:1609.07325 (2016)
S.D. Horner, M.E. Germain, T.P. Greene et al., Astron. Astrophys. Suppl. **193**, 1206 (1998)
W. Hu, Obs. Dark Energy **339**, 215 (2005)
D. Huber, J. Zinn, M. Bojsen-Hansen et al., Astrophys. J. **844**, 102 (2017)
P.G. Jonker, G. Nelemans, Mon. Not. R. Astron. Soc. **354**, 355 (2004)
D. Katz, U. Munari, M. Cropper et al., Mon. Not. R. Astron. Soc. **354**, 1223 (2004)
S.O. Kepler, S.J. Kleinman, A. Nitta, D. Koester et al., Mon. Not. R. Astron. Soc. **375**, 1315 (2007)
P. Kervella et al., Astron. Astrophys. **426**, 297 (2004a)
P. Kervella et al., Astron. Astrophys. **428**, 587 (2004b)
B. Kiziltan, A. Kottas, M. De Yoreo, S.E. Thorsett, Astrophys. J. **778**, 66 (2013)
S.E. Koposov, V. Belokurov, G. Torrealba, Mon. Not. R. Astron. Soc. **470**, 2702 (2017)
A. Kruszewski, I. Semeniuk, Acta Astron. **49**, 561 (1999)
C.Y. Kuo, J.A. Braatz, J.J. Condon, C.M.V. Impellizzeri et al., Astrophys. J. **727**, 20 (2011)
C.Y. Kuo, J.A. Braatz, M.J. Reid, K.Y. Lo, J.J. Condon, C.M.V. Impellizzeri, C. Henkel, Astrophys. J. **767**, 155 (2013)
C.Y. Kuo, J.A. Braatz, K.Y. Lo, M.J. Reid et al., Astrophys. J. **800**, 26 (2015)
E. Kuulkers, J. Homan, M. van der Klis, W.H.G. Lewin, M. Méndez, Astron. Astrophys. **382**, 947 (2002)
E. Kuulkers, P.R. den Hartog, J.J.M. in't Zand et al., Astron. Astrophys. **399**, 663 (2003)
P. Lacroute, *Revue Scientifique et Technique CECLES CERS*, vol. 6 1974), p. 253
W.H.G. Lewin, J. van Paradijs, R.E. Taam, Space Sci. Rev. **62**, 223 (1993)
L. Lindegren, U. Lammers, U. Bastian et al., Astron. Astrophys. **595**, A4 (2016)
Q.Z. Liu, J. van Paradijs, E.P.J. van den Heuvel, Astron. Astrophys. **455**, 1165 (2006)
K.Y. Lo, Annu. Rev. Astron. Astrophys. **43**, 625 (2005)
R. Lopes de Oliveira, C. Motch, F. Haberl, I. Negueruela, E. Janot-Pacheco, Astron. Astrophys. **454**, 265 (2006)
R. Lopes de Oliveira, C. Motch, M.A. Smith, I. Negueruela, M. Torrejòn, Astron. Astrophys. **474**, 983 (2007)
R. Lopes de Oliveira, M.A. Smith, C. Motch, Astron. Astrophys. **512**, 22 (2010)
B.F. Madore, Astrophys. J. **253**, 575 (1982)
F. Malbet, A. Crouzier, A. Léger, M. Shao, R. Goullioud Proc. SPIE **9143**, 91432L (2014)
M.A.C. Perryman, L. Lindegren, J. Kovalevsky, E. Hoeg et al., Astron. Astrophys. **323**, 49 (1997)
G. Pietrzynski et al., Astrophys. J. **697**, 862 (2009)
G. Pietrzynski et al., Nature **468**, 542 (2010)
G. Pietrzynski et al., Nature **495**, 76 (2013)
B. Pilecki et al., Mon. Not. R. Astron. Soc. **436**, 953 (2013)
B. Pilecki et al., Astrophys. J. **806**, 29 (2015)
T. Prusti, J.H.J. de Bruijne et al. (Gaia Collaboration), Astron. Astrophys. **595**, A1 (2016)
M.J. Reid, J.A. Braatz, J.J. Condon, L.J. Greenhill, C. Henkel, K.Y. Lo, Astrophys. J. **695**, 287 (2009)
M.J. Reid, J.A. Braatz, J.J. Condon, K.Y. Lo, C.Y. Kuo, C.M.V. Impellizzeri, C. Henkel, Astrophys. J. **767**, 154 (2013)
G. Ricker, D.W. Latham, R.K. Vanderspeck, et al., Astron. Astrophys. Suppl. **41**, 729 (2009)
S. Röser, AGAb **12**, 71 (1996)
I. Roxburg, C. Catala, PLATO Consortium, Commun. Astroseismol. **150**, 357 (2007)
M. Shao, T.R. Livermore, D.M. Wolff, J.W. Yu, M.M. Colavita, Astron. Astrophys. Suppl. **187**, 7103 (1995)
N. Shaposhnikov, L. Titarchuk, Astrophys. J. **699**, 453 (2009)
J.D. Simpson, G.M. De Silva, S.L. Martell et al., arXiv:1703.03823 (2017)
M.A. Smith, R.D. Robinson, Astrophys. J. **517**, 866 (1999)
I. Soszynski et al., Acta Astron. **58**, 293 (2008)
I. Soszynski et al., Acta Astron. **60**, 17 (2010)

T. Strohmayer, L. Bildsten, arXiv:astro-ph/0301544 (2003)
N. Tetzlaff, R. Neuhäuser, M.M. Hohle, Mon. Not. R. Astron. Soc. **410**, 190 (2011)
F. Thévenin, T.P. Idiart, Astrophys. J. **521**, 753 (1999)
I.B. Thompson et al., Astron. J. **121**, 3089 (2001)
G. Torres et al., Astrophys. J. **700**, 1349 (2009)
F. van Leeuwen, Astron. Astrophys. **474**, 653 (2007)
F. van Leeuwen A. Vallenari et al. (Gaia Collaboration), arXiv:1703.01131 (2017)
D.A. VandenBerg, H.E. Bond, E.P. Nelan et al., Astrophys. J. **792**, 110 (2014)
N.E. White, J.H. Swank, S.S. Holt, A. Parmar, Astrophys. J. **263**, 277 (1982)
J.S.B. Wyithe, R.E. Wilson, Astrophys. J. **559**, 260 (2001)
N. Zwahlen, P. North, Y. Debernardi et al., Astron. Astrophys. **425**, L45 (2004)

Space Sci Rev (2017) 212:1817–1869
DOI 10.1007/s11214-017-0427-8

Young and Intermediate-Age Distance Indicators

Smitha Subramanian[1] · Massimo Marengo[2] · Anupam Bhardwaj[3] · Yang Huang[4] · Laura Inno[5] · Akiharu Nakagawa[6] · Jesper Storm[7]

Received: 9 June 2017 / Accepted: 23 September 2017 / Published online: 30 October 2017

Abstract Distance measurements beyond geometrical and semi-geometrical methods, rely mainly on standard candles. As the name suggests, these objects have known luminosities by virtue of their intrinsic properties and play a major role in our understanding of modern cosmology. The main caveats associated with standard candles are their absolute calibration, contamination of the sample from other sources and systematic uncertainties. The absolute calibration mainly depends on their chemical composition and age. To understand the

Astronomical Distance Determination in the Space Age
Edited by Richard de Grijs and Maurizio Falanga

✉ S. Subramanian
smitha@pku.edu.cn

M. Marengo
mmarengo@iastate.edu

A. Bhardwaj
abhardwaj@eso.org

Y. Huang
yanghuang@pku.edu.cn

L. Inno
inno@mpia.de

A. Nakagawa
nakagawa@sci.kagoshima-u.ac.jp

J. Storm
jstorm@aip.de

1 Kavli Institute for Astronomy and Astrophysics, Peking University, Beijing, China

2 Department of Physics and Astronomy, Iowa State University, Ames, IA, USA

3 European Southern Observatory, 85748, Garching, Germany

4 Department of Astronomy, Kavli Institute for Astronomy and Astrophysics, Peking University, Beijing, China

5 Max-Planck-Institut für Astronomy, 69117, Heidelberg, Germany

6 Faculty of Science, Kagoshima University, Korimoto 1-1-35, Kagoshima 890-0065, Japan

impact of these effects on the distance scale, it is essential to develop methods based on different sample of standard candles. Here we review the fundamental properties of young and intermediate-age distance indicators such as Cepheids, Mira variables and Red Clump stars and the recent developments in their application as distance indicators.

Keywords Stars: distances · Stars: variables: Cepheids · Stars: AGB and post-AGB · Stars: horizontal branch stars · Cosmology: distance scale

1 Introduction

Distance to celestial objects is one of the key astrophysical parameters. Measurements of many fundamental properties, such as luminosities, masses, sizes and ages, depend on distances. Moreover, distance measurements play an important role in our understanding of modern cosmology. We have various methods based on geometrical, semi-geometrical, photometric and kinematic methods, to estimate the distances to celestial objects. Each method has limited range of applicability, such as the radar techniques for solar system objects, parallax methods for stars within a few kpc, Cepheids standard candles to a few Mpc and Type Ia supernovae to several 1000 Mpc. So each method which makes each rung in the distance ladder, is used to calibrate the next most distant method. This approach enables us to determine the scale of the Universe.

The parallax method of determining distance is the most direct and fundamental to all others. As the Earth revolves around the Sun, there is an apparent change in position of nearby stars with respect to the distant stars. This apparent change in position is a measure of the distance to that object. These distances are purely based on geometrical calculation and no assumptions about the physical or chemical properties of the sources are required. Hence, parallax distances are considered to be highly reliable measurements. However, the parallax measurements of giant stars which have large radius are not very precise. For distant sources, the parallax becomes smaller and this limits the application of parallax method to very nearby objects.

Distance measurement using eclipsing binaries, which is based on a semi-geometrical method, can reach up to the edge of the Local Group with an accuracy of 5%. We can measure the sizes of the stars from the velocities and the eclipse durations, the luminosities from the radii and surface temperatures, and the distance to the system from the magnitudes. This requires good modelling of the binary system. For early-type stars, this is a challenge due to the problems with accurate flux calibration and due to the degeneracy between stellar temperature and reddening. For late-type stars, these problems are minimal as very accurate flux calibration is available and accurate measurements of linear and angular sizes are possible. Thus, we can derive a distance which is primarily limited by the accuracy with which we can measure the light curve and the radial velocities. Such a distance is completely independent of the usual distance ladder. The most accurate distances, to date, to the Magellanic Clouds are estimated using eclipsing binaries (Pietrzyński et al. 2013, Graczyk et al. 2014).

Beyond the distances which can be reached using the geometrical and semi-geometrical methods, objects with known luminosity are adopted. They are called standard candles and they include Cepheids, RR Lyrae stars, Mira stars and Red Clump (RC) stars. Whereas, Type Ia supernovae are standardisable candle as their luminosities are not known. What we know

[7] Leibniz-Institut für Astrophysik Potsdam (AIP), An der Sternwarte 16, 14482, Potsdam, Germany

from nearby Type Ia supernovae is that there is a specific relation between their peak brightness and the time it takes for them to decay (Phillips 1993). Intrinsically bright supernovae shine longer than faint supernovae. From the ratio of peak to width of their light curve, we can determine their absolute magnitude. Here, we restrict our review to distance indicators whose luminosities are known, which are standard candles. The main problems associated with standard candles are their calibration, selection and other systematic uncertainties. To mitigate these issues, several detailed studies are going on.

In this chapter we review the fundamental properties of young and intermediate-age stellar populations such as, Cepheids (15–300 Myr), Mira variables (1–7 Gyr) and RC stars (2–9 Gyr) and the recent developments in their application as distance indicators.

2 Cepheids

Since Henrietta Leavitt discovered their Period-Luminosity (PL) relation (*Leavitt law*, Leavitt and Pickering 1912), Classical Cepheid variables have assumed a fundamental role as distance indicators. The Leavitt law was indeed the crucial ingredient allowing Edwin Hubble to determine the first reliable distance to M31 (Hubble 1929a). In doing so he established the technique that in short order would lead to the discovery of the eponymous *Hubble law* (Hubble 1929b), and modern cosmology.

Almost a century later, Cepheids PL relations are still the cornerstone of the cosmological distance ladder. Cepheids in the Large Magellanic Cloud (LMC), the Small Magellanic Cloud (SMC) and other Local Group galaxies, provide the link between the local universe, distant standard candles (e.g. Supernovae Type Ia) and the Hubble Flow. The Hubble Space Telescope (HST) Key Project on the Extragalactic Distance (Freedman et al. 2001), in particular, tied all this together by delivering the first precise (10% accuracy) determination for the value of the Hubble constant.

Despite their importance, Cepheid distances are still challenged by several factors. On the observational side, in addition to the statistical photometric uncertainties related to the measurement of Cepheid's mean magnitudes, the main errors in Cepheid distances are due to interstellar extinction and the systematic errors in the chosen PL relation. These systematic errors are related to the uncertainty in the PL slope and zero point, as well as its intrinsic scatter σ_{IS}, which in turn is the consequence of PL relations not taking into account the width in temperature of the Cepheid Instability Strip (IS). Due to this intrinsic scatter, PL relations should be really considered as statistical relations that can be correctly applied only to a statistical ensemble of stars, and not to individual Cepheids. As such, when Cepheids are used as galactic tracers to derive the three-dimensional structure of an underlying stellar population, the PL relation σ_{IS} becomes the dominant source of error. Instead when we measure the distance of a group of Cepheid stars (such as when using Cepheids to derive the distance of a galaxy used to calibrate the next step in the cosmological distance ladder) interstellar extinction and the empirical calibration of the PL zero point remain the largest uncertainties.

The last decade has seen great progress to overcome these limitations, with observations moving towards infrared wavelengths (where extinction is reduced) and the adoption of HST parallaxes for a number of Cepheid calibrators. Great progress has also been made in bringing under control issues regarding the linearity of the PL relation, and its dependence on metallicity. This has allowed a new determination of the Hubble constant, anchored on Cepheids, with accuracy better than 3% (Freedman et al. 2012). The adoption of Period-Wesenheit (PW) relations has reduced the uncertainties related to the σ_{IS} and to extinction at optical and near-infrared wavelengths, while templates have been produced to allow greater photometric accuracy even for sparsely sampled Cepheids light-curves.

In the theoretical arena the outstanding issue directly affecting the accuracy of Cepheid distances is the so-called "Cepheid mass discrepancy" between evolutionary models and Cepheid masses determined from pulsation theory and in binary systems. If this discrepancy is related to mass loss, it could lead to the formation of dusty circumstellar clouds, that would increase local extinction in the visible, and/or excess thermal emission in the infrared. Detailed calculations of the *projection factor* and *limb darkening*, necessary to precisely measure the distance of nearby individual Cepheid calibrators with variants of the Baade–Wesselink method, such as the infrared brightness method, had also become available (Marengo et al. 2002, 2003, Storm et al. 2011a, Ngeow et al. 2012 and references therein).

In this section we discuss how these issues are dealt with modern measurements of Cepheid distances. In Sect. 2.1 we discuss the techniques enabling the measurement of accurate distances for individual Cepheids in the near-IR, with particular emphasis to PW relations, independent determinations of reddening and the use of light curve templates. In Sect. 2.2 we analyse the status of mid-infrared Cepheid distances as part of new efforts to improve the cosmological distance scale. In Sect. 2.3 we explore the issue of linearity in the PL relation and its role for the determination of the Hubble constant. In Sect. 2.4 we discuss the effect of metallicity in the slope of the PL relation of Galactic and Magellanic Clouds Cepheids.

2.1 Accurate Individual Cepheid Distances from Near-Infrared Observations

A comprehensive theoretical framework has been developed over the past 20 years to account for all the phenomenological features of pulsating stars (Bono et al. 1999a,b, 2000a; Marconi et al. 2005, 2015; Fiorentino et al. 2013). These models allow us to predict σ_{IS} and, in turn, to pinpoint the best tool to derive Cepheids' individual distances. The left panel of Fig. 1 shows the predicted slope (green dots) of PL relations in optical (*VI*) and near-infrared (NIR, JHK_s) bands for $\sim$1,300 synthetic Cepheids in the LMC presented in Inno et al. (2016). We note that non-linearity in the PL relations is not considered here. As we can see later (Sect. 2.3), the implications of non-linearity on the distance scale or the value of Hubble constant is not statistically significant. The LMC is indeed a perfect workbench to test the accuracy of PL relations, since it is the closest star-forming galaxy almost face-on (inclination $\sim$25°–35°: Jacyszyn-Dobrzeniecka et al. 2016, Inno et al. 2016, Subramanian and Subramaniam 2013), with a relatively small extension along the line-of-sight ($\sim$0.5–10 kpc, Subramanian and Subramaniam 2009), and for which more than 3,000 Cepheids are known, thanks to the extensive observational campaigns of micro-lensing surveys (e.g. MACHO: Alcock et al. 2001, EROS: Beaulieu et al. 1995, OGLE: Soszyński et al. 2008, 2010). Moreover, the metallicity content of the LMC young stellar population is well constrained Romaniello et al. (2008), and we can adopt the mean metallicity $Z = 0.008$ to produce a synthetic population of Cepheids. The green error-bars in Fig. 1 indicate the dispersion around each relation. In the case of the theoretical PL relations, the dispersion corresponds exactly to σ_{IS}. The error bars clearly show how σ_{IS} decreases when moving from optical to the NIR bands. This trend is a simple consequence of the dependence of bolometric corrections on temperature, which produces a narrowing of the IS in the Colour–Magnitude–Diagram (CMD) at longer wavelengths. In the same figure, the observed slopes and dispersions for the LMC Cepheids are also shown (Inno et al. 2016). While the slopes are

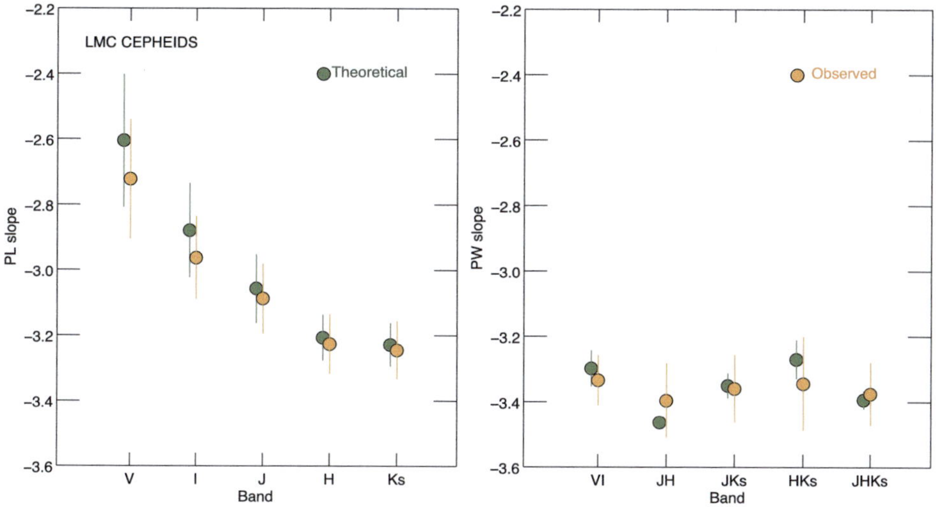

Fig. 1 Predicted (*green*) and observed (*orange*) slopes and dispersions ($\pm 1\sigma$) of Period-Luminosity (PL, *left panel*) and Period-Wesenheit (PW, *right panel*) relations of Cepheids in the LMC. The models and the data used to produce this plot are described in Inno et al. (2016). The slope and dispersion of PL relations decrease with increasing wavelength. The smallest dispersion is found around the K_S band PL relation, which is the less affected by systematics and by uncertainty on the reddening correction (i.e. $A_K \sim A_V/10$). However, (JH) and (JHK_S) PW relations show a smaller predicted dispersion than any PL, but they are reddening free by construction. The use of these relations to determine accurate Cepheids' individual distances is still limited by the lower photometric precision of NIR data, which propagates to uncertainty on the colour term in the Wesenheit definition

remarkably in agreement with theoretical predictions, especially in the NIR bands, the observed dispersions are systematically larger. Such difference is due to the additional sources of scatter that affect the observed mean magnitudes: the photometric error, the uncertainty on the reddening correction and the dispersion due to the line-of-sight depth of the LMC. The scatter contributing to the latter term is indeed the quantity we want to measure. Thus, by reducing the scatter due to the PL intrinsic dispersion to the reddening effects and to the photometric error, we are able to improve the accuracy on the Cepheid's individual distances.

2.1.1 Period-Wesenheit Relations in the NIR

Reddening corrections can be performed by adopting different approaches. If the reddening law for a given galaxy or line of sight is known, then multi-band photometry of Cepheids allows us to obtain their distances and colour excess. For instance, we can use the B and V band PL relations to obtain the $E(B-V)$, and then use the predicted total to selective extinction ratio $A_V/E(B-V)$, to determine A_V. The above approach is similar to the use of the so-called PW relations, where Wesenheit magnitudes (van den Bergh 1975; Madore 1982) are constructed to be reddening-free. In fact, if we combine the two observed magnitudes m_{λ_1} and m_{λ_2}, into the Wesenheit magnitude

$$W(\lambda_2, \lambda_1) = m_{\lambda_1} - \left[\frac{A(\lambda_1)}{E(m_{\lambda_2} - m_{\lambda_1})} \right] \times (m_{\lambda_2} - m_{\lambda_1}), \qquad (1)$$

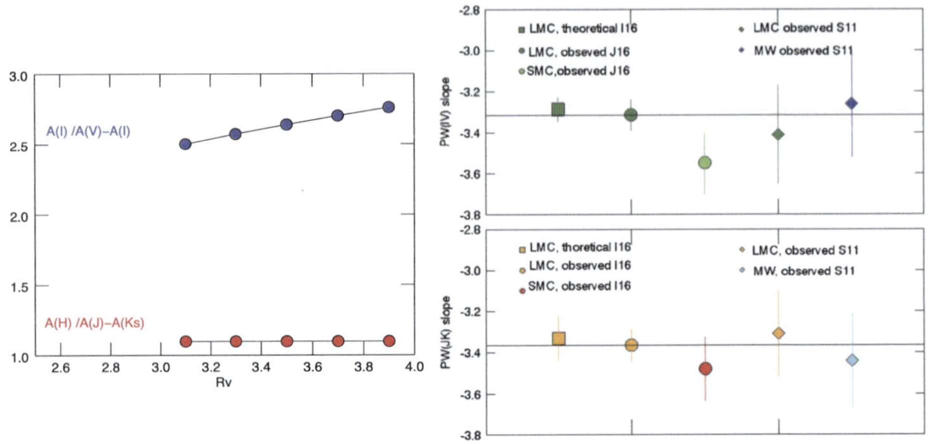

Fig. 2 *Left*—The variation of the Wesenheit coefficient in the optical I and V bands (*blue dots*) and in the NIR J, H and K_S bands as a function of the assumed total-to-selective absorption ratio (R_V). The NIR coefficient is insensitive to the changes in R_V. *Right*—Comparison between slopes of PW relations in the optical (*top*) and in the NIR (*bottom*) as measured in the LMC, SMC and in the Milky Way. The observed slopes in the optical are from Jacyszyn-Dobrzeniecka et al. (2016) for the LMC (*green dot*) and SMC (*lime dot*), and from Storm et al. (2011b) for the LMC (*green diamond*) and the Milky Way (*blue diamond*). The predicted slope from Inno et al. (2016) is also shown, while the *black line* correspond to the observed slope of LMC Cepheids and can be used as a reference. The NIR slopes are also taken from Inno et al. (2016) and Storm et al. (2011b). The comparison shows that the NIR PW relations are in very good agreement inside the dispersion, thus they seem to be minimally affected by metallicity, at least in the metallicity range covered by the Milky Way and the Magellanic Clouds

where $A(\lambda_i)/E(m_{\lambda_2} - m_{\lambda_1})$ is the total to selective extinction for the given filters, we find that the observed Wesenheit magnitude is identical to the de-reddened one. Wesenheit magnitudes can be defined with any combination of photometric bands, but recent theoretical predictions (Bono et al. 2000a; Marconi et al. 2005; Fiorentino et al. 2007) and empirical results (Bono et al. 2010; Inno et al. 2013, 2016) show that NIR PW relations are indeed the best suited tool to derive accurate Cepheids' distances. In fact, they are linear over the entire period range and, because they mimic a period-luminosity-colour relation, they have a smaller σ_{IS} when compared to optical and NIR PL relations, as demonstrated by comparing the right and left panel of Fig. 1. In the right panel the predicted slopes (green dots) of the optical (*VI*) and NIR PW relations are plotted together with their intrinsic dispersions (error bars). We find that the predicted dispersion around the *VI* ($\sigma_{IS} = 0.05$ mag) and the *JHK* ($\sigma_{IS} = 0.03$ mag) PW relations are 30–60% smaller than the one around the K_s band PL relation ($\sigma_{IS} = 0.07$ mag).

Moreover, when comparing optical to NIR PW relations, we also find that they are minimally affected by uncertainties on the adopted reddening law and marginally affected by metallicity effects (Bono et al. 2010; Freedman and Madore 2010a). For instance, in the left panel of Fig. 2 we show the change of the Wesenheit coefficients with values of the total-to-selective absorption ratio (R_V), ranging from 3.1 to 4 (McCall 2004; De Marchi et al. 2016), in the case of Cardelli's reddening law (Cardelli et al. 1989). Because the ratio $A(\lambda)/A(V)$ is independent of R_V for $\lambda \geq 0.9$ μm (Cardelli et al. 1989) and the reddening law can then be approximated by a power law, the NIR coefficient results do not change, while the optical coefficient changes by ~10%. In the right panel of Fig. 2 the comparison between slopes of PW relations for the Milky Way (Storm et al. 2011b) and the Magellanic

Cloud Cepheids (Jacyszyn-Dobrzeniecka et al. 2016; Inno et al. 2016) for the optical (top) and the NIR (bottom) is also shown. In the case of the NIR PW (JK_S) relation, the slopes are common within the error given by their scatter, thus indicating a small (if any) dependence on the metallicity, at least in the metallicity range covered by these galaxies.

In the right panel of Fig. 1 the observed slopes and dispersion of optical and NIR PW relations for LMC Cepheids (orange dots) are also shown Inno et al. (2016). Once again, we find that the observed dispersions ($\pm 1\sigma$) are larger than the predicted ones. In fact, while the uncertainty on the reddening correction vanishes for the PW relations, the photometric error becomes larger, since it propagates on the mean colour, which appears on the right side of (1). Improving the photometric accuracy on NIR measurement is then a crucial step in order to take full advantage of the use of NIR PW relations as tools to derive accurate Cepheids' individual distances.

2.1.2 Independent Reddening Estimates

An alternative approach based on the use of multi-band photometry of Cepheids is the so-called reddening-law fitting method (Freedman et al. 1985; Rich et al. 2014; Inno et al. 2016), which allows to derive individual Cepheids' distances and extinction independently at the same time. The above method is based on the following expression of the Cepheid true distance modulus as a function of wavelength:

$$\mu_0 = \mu_{obs}(x) + \left[a(x)R_V + b(x)\right] \times E(B-V) \tag{2}$$

where $x \equiv \lambda^{-1}$, $a(x)$ and $b(x)$ are the coefficients of the adopted reddening law. Thus, if we obtain μ_{obs} in different filters by using the PL relations, the fit will provide simultaneously μ_0 and the extinction $E(B-V)$. However, these quantities would still be affected by systematics due to the degeneracy in intrinsic colour and extinction.

Reddening maps can also be used to predict the amount of extinction in a given band for a specific line-of-sight. In the case of an external system, such as the Magellanic Clouds, two-dimensional reddening maps can be adopted to correct the observed magnitude for the individual reddening. This method was applied in Haschke et al. (2012), Macri et al. (2015), Ripepi et al. (2016), Bhardwaj et al. (2016b), who used the reddening maps from Haschke et al. (2011). In the Galaxy, individual reddening along the line-of-sight can be obtained from up-to-date three-dimensional reddening maps (Marshall et al. 2006; Green et al. 2015; Schlafly et al. 2016). This approach is very promising, especially when accurate three-dimensional map of the dust distribution in the Milky Way will become available thanks to Gaia (Rezaei et al. 2017). Individual reddening of Cepheids can also be determined from their spectra, by using empirically calibrated relations between the equivalent width of Diffuse Interstellar Bands (DIB) and $E(B-V)$ (Friedman et al. 2011). For instance, this method has been applied to derive de-reddened magnitudes of Cepheids in the inner disk by Martin et al. (2015), Andrievsky et al. (2016). Note that all above methods still rely on the assumption of the reddening law and thus total-to-selective absorption ratio, thus they are still prone to systematic errors.

2.1.3 NIR Templates as Tools to Optimize NIR Variability Surveys

The main drawback in using NIR PL or PW relations to determine Cepheid distances is that NIR light curves are available for a limited number of stars. For instance, in the Galaxy only $\lesssim$200 (Laney and Stobie 1992; Monson and Pierce 2011) out of $\gtrsim$600 (Dambis et al. 2015) known Cepheids have accurate NIR photometry. The reasons is that variability surveys have been traditionally performed in optical bands, where the pulsation amplitude is typically larger, (e.g. $\sim$1.2 mag in B vs. $\sim$0.2 mag in K_S), thus allowing a more robust classification of the variables. On the other hand, the lower amplitudes in the NIR bands enable us to determine accurate mean magnitudes from a smaller number of phase points ($\sim$10–12 vs. $\sim$30). This is a sensitive issue, especially for surveys aiming at discovering new Cepheids in the most extincted region of the Galaxy, where NIR photometry is necessary to cope with such high extinction (e.g. VISTA Variables in the Via Lactea (VVV), Minniti et al. 2010, InfraRed Survey Facility (IRSF), Matsunaga et al. 2011, 2013, 2016). However, NIR ground-based observations are still more time-consuming than optical observations, because of sky subtraction. A cheap solution to this problem in terms of observing time, is the use of light-curve templates to determine accurate NIR mean magnitudes from a few or even only one measurement. Such templates are available in the literature for both RR Lyrae (Jones et al. 1996; Sesar et al. 2010) and classical Cepheids (Soszyński et al. 2005; Pejcha and Kochanek 2012; Inno et al. 2015; Ripepi et al. 2016) and rely on empirical calibrations.

NIR Templates to Fit Sparsely Sampled Light-Curves Recent and ongoing surveys, such as the VVV, the VISTA Survey of Magellanic Clouds System (VMC, Cioni et al. 2011), the NIR Synoptic Survey (Macri et al. 2015) and the IRSF surveys, use NIR bands to access farther or more obscured stars. In order to optimise their scientific outcome respect to the surveyed area and the cadence, these surveys collect sparsely sampled light-curves in one or more NIR bands. However, if the period of a given Cepheid is known with sufficient accuracy, it is possible to perform a template-fitting to the multi-epoch observations in order to accurately determine their mean magnitudes. For instance, VMC provides NIR magnitudes measured at on average $\sim$5.7, $\sim$6.3 and $\sim$16.7 epochs in the Y, J and K_S band respectively, for the SMC Cepheids in the OGLE-III catalog. Ripepi et al. (2016) used the best sampled light-curves in the VMC dataset to build new templates, interpolated the data of the sparsely sampled curves with splines and then applied a template-fitting procedure in three steps. They obtained first the optical-to-NIR scaled amplitudes, then the phase-lags between the OGLE optical and the NIR light curves and finally the mean magnitudes. Inno et al. (2016) applied directly the templates from Inno et al. (2015) to similar VMC data for LMC Cepheids and solved independently for amplitudes, mean magnitudes and phase-lags at the same time. A comparison between of the two mentioned template-fitting techniques applied to the VMC light curves for Cepheids in the SMC is shown in the left and middle panels of Fig. 3. The templates allow to predict the shape of the light curve and then determine accurate mean-magnitudes from six epochs. Differences on the mean magnitude obtained by adopting the templates by Ripepi et al. (2016) (red line) and the ones by Inno et al. (2016) (green line) are smaller than 0.01 mag, or 0.5% in distance to the SMC. The template-fitting approach is also to be preferred when the photometric error on individual observations is so large ($\gtrsim$0.1 mag) to introduce spurious features in the light-curve shape. This is for instance the case of light curves from the Near-Infrared Synoptic Survey. Macri et al. (2015) uses the templates by Soszyński et al. (2005) and a template-fitting procedure similar to the one described above to determine LMC Cepheids J, H and K_S mean magnitudes. For completeness, we also applied the same approach to the VMC data in Fig. 3 (right panel). The fit obtained by using the templates by Soszyński et al. (2005) (blue line) seems to be less accurate in reproducing the light-curve shape of the SMC Cepheids, with a difference on the mean magnitude of $\sim$0.03 mag, or 1.5% in distance to the SMC.

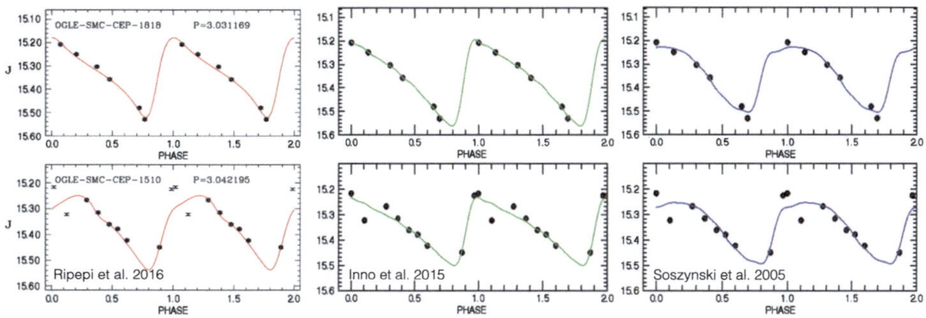

Fig. 3 A comparison between J band templates available in the literature applied to the VMC data of two SMC Cepheids with periods of about 3 days. The *left panels* are directly taken from Fig. 11 in Ripepi et al. (2016), while the *middle* and *right panels* have been obtained using the approach by Inno et al. (2016) and by Macri et al. (2015) respectively. In the case of the Cepheid OGLE-SMC-CEP-1818, shown in the *top row*, the templates by Ripepi et al. (2016) (*left panel*) and by Inno et al. (2016) (*middle panel*) are more accurate than the one from Soszyński et al. (2005), which predicts a shallow bump on the rising branch, that is not consistent with the data. The differences in the fit produces a difference in the resulting mean magnitude of 0.03 mag, or $\gtrsim$1.5% in distance. The *bottom row* shows similar results in the case of the Cepheid OGLE-SMC-CEP-1510. Note that the method implemented by Ripepi et al. (2016) rely on an outlier cut, which removes three observation (*crosses*), while this interaction is not necessary in the case of the template-fitting method used by Inno et al. (2016), which provides a mean-magnitude smaller of $\lesssim$0.01 mag, or 0.5% in distance. The use of the new accurate templates available in the literature allows us to derive accurate mean magnitude, and hence distances, even for sparsely observed Cepheids

NIR Templates to Correct for Random-Phase Effects If the light curve in the optical band is available, the NIR templates can also be used to obtain mean-magnitude from NIR single-epoch observations (Soszyński et al. 2005; Inno et al. 2015). The accuracy on these mean-magnitudes is limited by i) the photometric error on the NIR single-epoch observation; *ii*) the precision to which we can predict the phase of the NIR observation and the NIR amplitude; and *iii*) the accuracy of the template itself. Indeed, the templates by Inno et al. (2015) provide mean magnitude with an error lower than 2% for single-epoch measurements with 1% photometric precision or better. Thus, the uncertainty on Cepheid's individual distance determination is already dominated by the σ_{IS} ($\gtrsim$3%). This means that these templates allow us to already reach the precision limit of the method, even with single-epoch NIR observations, if the photometric precision is sufficient.

2.1.4 Summary

NIR PW relations allow us to derive Cepheid distances with the accuracy to meet the precision of ongoing cosmological experiments. In fact, they appear to be linear over the entire period range and their slopes appear to be common between the Magellanic Clouds, indicating that they are universal in the considered metallicity range (see Sect. 2.3 and 2.4 for an in-depth discussion of the linearity of the NIR PL relations and their dependence on metallicity). Moreover, PW relations are minimally affected by uncertainties in the adopted reddening law. For instance, by using NIR PW relations, Inno et al. (2013) obtained the most precise estimate of distances to the Magellanic Clouds based on Cepheids, with random errors at 1% level and systematics at 5% level. The systematic errors are dominated by the uncertainties in the PW zero point. This means that the new independent calibrations from Gaia and HST will play a crucial role to reach a better accuracy.

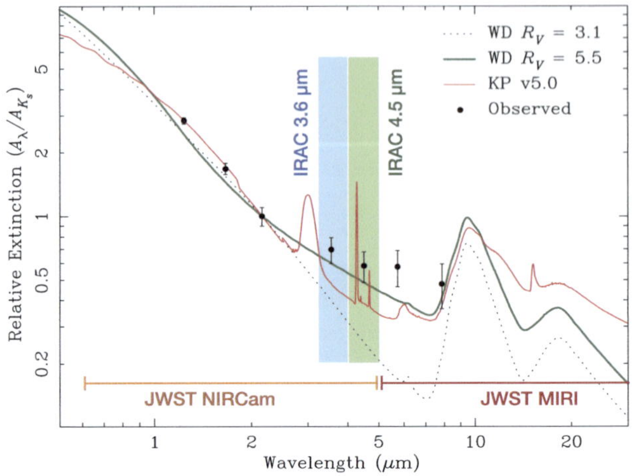

Fig. 4 Infrared average extinction law in the Perseus molecular cloud (data), compared with a variety of dust models. Warm Spitzer/IRAC and JWST bandpasses are indicated on the plot. The minimum extinction is found in the spectral region covered by warm IRAC and the red edge of JWST NIRCam. Adapted from Chapman et al. (2009)

Moreover, new NIR light-curve templates can provide accurate mean magnitudes from a few or even one observation, allowing a significant saving of telescope time and an optimal usage of data already available (e.g. from 2MASS, VVV, VMC, etc.). The use of templates to predict the amplitude and the shape of the variation of physical quantities of Cepheids along the pulsation cycle, such as luminosity and radial velocity, is a powerful tool. In fact, this approach (single-epoch photometry and spectroscopy + light and radial velocity curve templates for Cepheids) is very efficient to investigate the spatial distribution and the kinematics of the resolved stellar population in nearby galaxies, especially in the approaching E-ELT era, since we will be able to apply it to more distant spiral galaxies in the Local Group and beyond (>4 Mpc).

2.2 Mid-Infrared Cepheid Distances and the Cosmological Distance Scale

Observing Cepheids in the mid-infrared (MIR, L, M, N and Q photometric bands) leads to significant advantages with respect to optical wavelengths. Recent determinations of the interstellar and extragalactic extinction curve (see e.g. Chapman et al. 2009, Rieke and Lebofsky 1985, Indebetouw et al. 2005, Flaherty et al. 2007, Román-Zúñiga et al. 2007, Nishiyama et al. 2009, Fig. 4) show that in the thermal infrared, the dust extinction is reduced by more than an order of magnitude compared to the V band ($A_\lambda/A_V \simeq 0.065$ and $\simeq 0.052$ in the L and M spectral bands respectively Indebetouw et al. 2005). Furthermore, for stars with G and K spectral type the shape of the spectrum closely follows the temperature-independent Rayleigh–Jeans distribution. As a consequence, the brightness variations related to the pulsations are largely insensitive to the effective temperature variation, being dominated by the smaller changes in surface area consequence of the stars' radial pulsations. This leads to small amplitudes for wavelengths longer than the K band (∼0.3–0.4 mags for the average 10 days period), which reduces the number of epochs necessary to measure accurate average magnitudes. Finally, infrared Cepheid spectra are mostly free from line blanketing, in principle reducing the dependence of the Leavitt law from metallicity, even though CO absorption can still play a significant role (see discussion in Sect. 2.2.2 below). For these reasons the last decade has seen multiple efforts aimed to develop robust Cepheid distance determinations using their PL relation at MIR wavelengths.

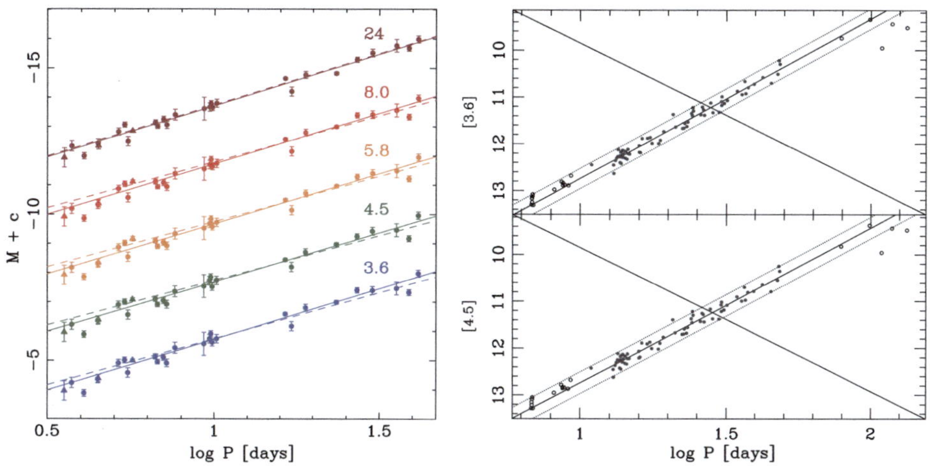

Fig. 5 Empirical Leavitt law of Classical Cepheids in the Galaxy (*left panel*, adapted from Marengo et al. 2010) and in the LMC (*right panel*, adapted from Scowcroft et al. 2011). These Galactic Classical Cepheids relations have been obtained with the Spitzer space telescope IRAC (3.6 to 8.0 μm) and MIPS (24 μm) instruments during the cryogenic mission, and are based on two epochs measurements. *Solid lines* show the best fit PL relation for all stars, while the *dashed line* was derived using only stars with astrometric distances. The period of first overtone Cepheids was "fundamentalised" using the relation from Feast and Catchpole (1997). The LMC relations have instead been obtained during the Spitzer warm mission, and are based on carefully reconstructed light curves (24 epochs each) for each Cepheid. A similar strategy was used by Monson et al. (2012) to derive accurate zero points for the Galactic 3.6 and 4.5 μm Classical Cepheids PL relation

2.2.1 *Empirical PL Relations and the CHP Program*

Mid-infrared astronomy received a boost in sensitivity and accuracy by many orders of magnitude with the launch of the Spitzer space telescope Werner et al. (2004), and in particular its InfraRed Array Camera (IRAC, Fazio et al. 2004). IRAC is currently operating in the extended "warm mission" following the exhaustion of its cryogenic liquid Helium (LHe), still acquiring images and providing better than 1% photometry in its 3.6 and 4.5 μm bands. IRAC enabled obtaining accurate magnitudes of Cepheid stars as far as the Magellanic Clouds, in addition to the Milky Way Galaxy. These capabilities have been exploited by several teams that derived an empirical calibration of MIR Cepheids PL relations (Ngeow and Kanbur 2008; Marengo et al. 2010; Scowcroft et al. 2011; Monson et al. 2012; Ngeow et al. 2015b).

Figure 5 shows the 3.6 to 24 μm Leavitt law (Marengo et al. 2010) measured for Galactic Classical Cepheids with IRAC and MIPS (Rieke et al. 2004) during the Spitzer cryogenic mission. Each star was observed in two random epochs six months apart. The PL relation zero point was then determined using parallaxes measured with the HST Fine Guidance Sensors (Benedict et al. 2007) and Hipparcos (for Polaris, van Leeuwen et al. 2007a), or distances derived with the InfraRed Surface Brightness (IRSB) technique by Fouqué et al. (2007) (see detailed description of the method in Sect. 2.4.1). The calibration of the Classical Cepheids PL relation was then refined at 3.6 and 4.5 μm during the Spitzer/IRAC warm mission with the Carnegie Hubble Program (CHP, Freedman et al. 2011). As part of this program Monson et al. (2012) measured well sampled (24 epochs) light-curves of 37 Galactic Classical Cepheids with accurate astrometric or IRSB distances, calibrating the PL zero points with a ~3% accuracy. Scowcroft et al. (2011, 2016b) then used the Galactic zero

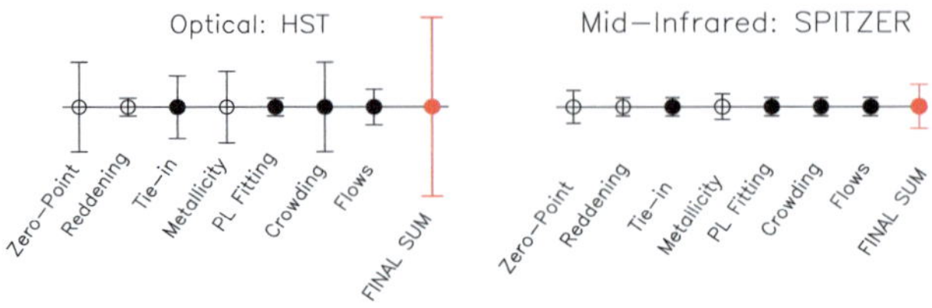

Fig. 6 Comparison of individual error sources in the Hubble constant determination between the HST Key project (*left*) and the CHP (*right*). Open symbols are the uncertainties most directly affected by the change in wavelength from optical to infrared bands. The final sum shows how the overall uncertainty decreased from ~10% ($H_0 = 72 \pm 7$ km s^{-1} Mpc^{-1}) in the HST Key Project to ~2.8% ($H_0 = 74.3 \pm 2.1$ km s^{-1} Mpc^{-1}) in the CHP. Adapted from Freedman et al. (2012)

points to measure the distance and geometry of the LMC and SMC with a ~4% accuracy. This provided a new, direct way of anchoring the extragalactic distance scale to Classical Cepheid distances, allowing a new measurement of the Hubble constant and other cosmological parameters (Freedman et al. 2012).

Figure 6 breaks down the error budget in the CHP (right) and HST Key Project determination of H_0. The open symbols indicate the sources of uncertainty more directly related to the change in wavelength from the optical to the MIR range. The largest improvement is in the PL zero point calibration. This is the consequence of being able to use Galactic calibrators with astrometric distances to anchor the Leavitt law, rather than relying to independent estimates of the LMC distance. As mentioned above, this is only possible in the infrared where the extinction of individual calibrator stars is one order of magnitude less than in the visible. The effect of reddening itself on the overall H_0 determination is however not dramatically reduced, since it contributes in the form of a statistical uncertainty, and was already limited to less than ~1% in the HST Key Project by observing a large number of distance indicators. The uncertainty due to metallicity effects is instead halved from ~3.5% to ~1.4%, again a consequence of being able to directly calibrate the Cepheids PL relation at high metallicity in the Galaxy, rather then relying to the intermediate step of calibrating the zero point using the lower metallicity LMC (see detailed discussion in Freedman et al. 2012).

The overall effect of moving from visible to infrared bands results in both the reduction of the uncertainty in the H_0 determination (from ~10% in the HST Key Project to ~2.8% in the CHP) and an increase in the value of the constant (from $H_0 = 72 \pm 3 \pm 7$ to 74.3 ± 2.1 km s^{-1} Mpc^{-1} in the HST Key Project and CHP, respectively). The increase, while still within the uncertainties, is significant because it goes in the opposite direction of recent determinations of H_0 from the analysis of Planck Cosmic Microwave Background (CMB) radiation and Baryonic Acoustic Oscillations (BAO), which are instead favouring lower values (e.g. 67.8 ± 0.9 km s^{-1} Mpc^{-1} and 68.8 ± 2.2 km s^{-1} Mpc^{-1} in Planck Collaboration 2016 and Cuesta et al. 2015 respectively). A careful assessment of the systematics present in the "direct" determination of the distance scale (anchored on stellar indicators such as Cepheids and SN Type Ia) vs. the CMB/BAO analysis, reveals that this *tension* in the value of H_0 is approaching 3σ (Riess et al. 2016). Solving this issue has important consequences for cosmology, since the CMB analysis is dependent on the specifics of the ΛCDM cosmological model used to fit the data. Any discrepancy with a direct determination of H_0 and the other cosmological parameters would be an indication of new physics, such as the number or relativistic species and/or the mass of neutrinos (see e.g. Dvorkin et al. 2014, Wyman et al. 2014). For this reason it is worth analysing in detail what are the

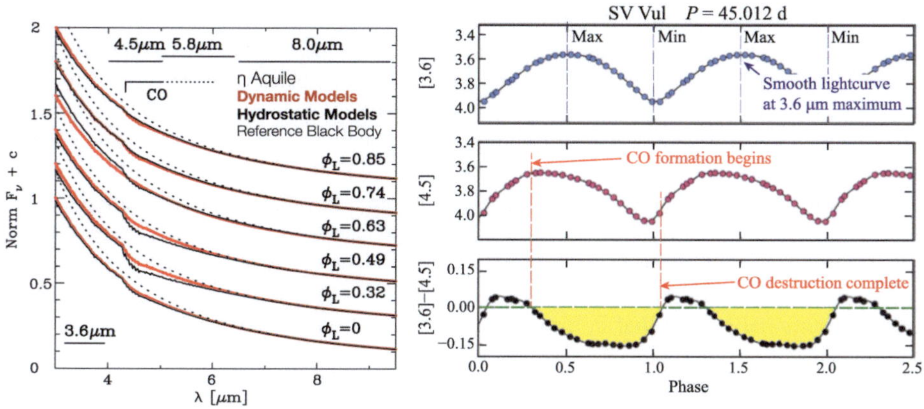

Fig. 7 *Left*—Series of time-dependent models of the Galactic Classical Cepheids η Aql showing the changes in the broad CO absorption feature straddling the IRAC 4.5 μm band (adapted from Marengo et al. 2010). *Right*—Colour curve of the long period ($P = 45.012$ days) Galactic Classical Cepheid star SV Vul, adapted from Scowcroft et al. (2016a). Note the flattening of the 4.5 μm curve at maximum light, with a pronounced [3.6]–[4.6] blue colour as the star progresses towards minimum radius

possible systematics that may still affect the role of Cepheids as infrared standard candles, as well as to develop complementary "direct" routes to the distance scale, anchored to Population II distance indicators such as RR Lyrae and the Tip of red giant branch (RGB) in Globular Clusters, or other standard candles such as RC stars and Miras.

2.2.2 *Effects of Variable CO Absorption on Cepheids Infrared Distances*

The method followed by the CHP to calibrate the Leavitt law for Classical Cepheids relies on the assumption that the PL slope (determined by fitting LMC Cepheids) and zero point (measured from Galactic Cepheids with known distances) do not depend in a significant way from metallicity. An in-depth discussion of metallicity effects on Cepheids PL relations, with special emphasis on the NIR, is presented in Sect. 2.4. In the MIR, as mentioned before, the influence of metallicity is generally believed to be lower, due to the far less importance of line blanketing at wavelengths where the stellar spectrum approaches the Rayleigh–Jeans limit. Indeed, tests performed as part of the CHP showed no significant correlation in the residuals between the adopted PL relations and the 3.6 and 4.5 μm magnitudes of individual stars over the metallicity range $-0.6 < [\mathrm{Fe/H}] < +0.2$ (see Freedman et al. 2012) of the observed stars.

Figure 7 however shows that the spectral brightness in the *M* band (straddling across the IRAC 4.5 and 5.8 μm bands) is severely affected by variable CO absorption. Time-dependent hydrodynamic models of pulsating Cepheids (left panel, calculated for η Aql by Marengo et al. 2002) reveal that CO absorption is greatly increased for pulsation phases following maximum radius, where the atmosphere is expanding and the effective temperature is at its minimum. As the star approaches minimum radius (corresponding to minimum infrared luminosity) the CO molecule is chemically destroyed, removing a significant fraction of its absorption feature in the star's spectrum. The effect of this variable CO abundance on the infrared light curve and colour is studied in great detail by Scowcroft et al. (2016a), and shown in the right panel of Fig. 7. Since at 3.6 μm the stellar spectrum is largely unaffected by CO absorption, the light curve is smooth during the entire pulsation cycle. At 4.5 μm, however, the formation of the CO molecule sets a ceiling in how bright the star could become at maximum light, flattening the light curve, until CO is destroyed near minimum radius. This behaviour leads to a well defined negative-slope period-colour-metallicity relation for Classical Cepheids with period between ∼6 and 60 days ($0.8 < \log P < 1.8$), characterised by Scowcroft et al. (2016a) for stars in the Galaxy, LMC and SMC. For Cepheids with period shorter than 6 days the relation flattens, due to the higher effective temperature ($T_{\mathrm{eff}} \simeq 6309$ K) that prevents the formation of CO spectral headbands for the entire pulsation cycle. The period-colour relation for ultra-long period Cepheids (P greater than ≈60 days) is instead dramatically inverted (large positive slope). The reasons why this happen is not well understood. Several lines of evidence points towards the effects of lower rotation, suppressing rotational mixing that in turn would decrease the abundance of CNO elements in the photosphere (Anderson et al. 2014), as well as the lack of first dredge-up for Cepheids in this period range (Bono et al. 2000b). Both phenomena would prevent the formation of significant amount of CO.

Based on this analysis, Scowcroft et al. (2016a) concluded that the IRAC band at 3.6 µm is the safest choice for measuring precision Cepheids distances in the MIR, since at this wavelength the effects of metallicity (in terms of absorption features from CO and other molecules) are minimised. The 4.5 µm band dependence on a variable CO absorption, on the other hand, suggests the enticing possibility of adopting the [3.6]–[4.5] period-colour-metallicity relation as a metallicity indicator. The expected dispersion of such indicator appears to be on the order of ≈0.20 dex (Scowcroft et al. 2016a), comparable with the current precision in [Fe/H] measurements for individual Cepheids in the Galaxy, LMC and SMC.

2.2.3 Mid-IR Excess Around Cepheids

Cepheids are intermediate mass stars (typically 4–6 $M_{\odot}$), and as such they enter the core He-burning phase at a relatively young phase, before they have the chance to wander far from the regions where they form. This means that a large fraction of them is found at small Galactic latitude (of the 10 Cepheid calibrators with HST parallax, all except β Dor are found within ≈±10 deg from the Galactic plane). As a consequence, Galactic Cepheids, as well as Cepheids detected in other galaxies seen edge-on, are projected over regions with high extinction (in the visible) or contaminated by "cirrus" of diffuse infrared emission. MIR is the ideal wavelength range to limit the impact of both effects, since at these wavelengths extinction is minimised (see discussion above) while the emission from galactic filaments starts to pick-up only at wavelengths longer than ∼25 µm. The situation is however different in case of circumstellar emission, produced by circumstellar dust heated at temperatures warm enough to emit copious amount of thermal MIR radiation.

Circumstellar dusty envelopes can occur for two different reasons: *i*) the dust is part of a cloud that formed independently from the star, and it just so happen that the star drifted into it, or *ii*) the dust condensed in an outflow originating from the star. The existence of outflows from Cepheids is controversial, but can be related to the "Cepheids mass discrepancy" mentioned before. As was first noted more than 40 years ago by Fricke et al. (1972), current evolutionary models predict Cepheid masses 10–15% larger than masses derived from pulsation theory (see e.g. Caputo et al. 2005, Keller et al. 2006) or measured from Cepheids in binary systems (Evans et al. 2008, 2013; Pietrzyński et al. 2010). This discrepancy could be at least in part explained if the mass deficit is the result of mass loss before or during the Cepheid phase (see e.g. Neilson et al. 2012 and references therein). Mass loss during earlier evolutionary phases could result in circumstellar clouds still lingering as the star enters the instability strip, while active outflows would interact with the surrounding

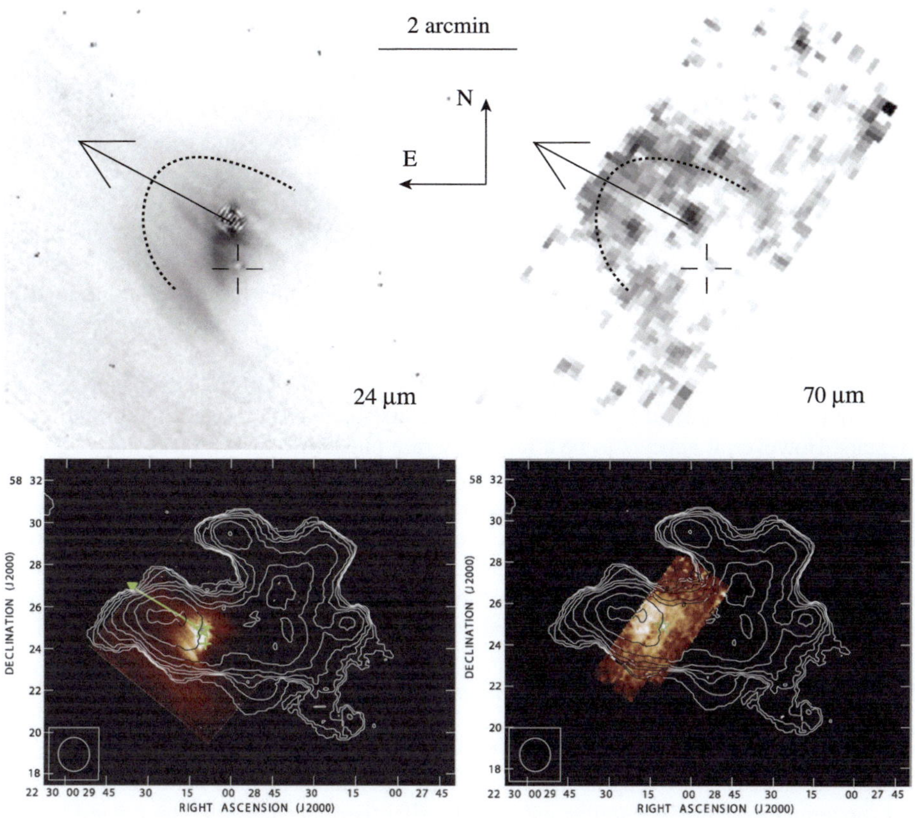

Fig. 8 δ Cep circumstellar nebula observed by Spitzer at 24 and 60 μm (*top row*, adapted from Marengo et al. 2010), and at the VLA in the H I 21-cm line (*bottom row*, adapted from Matthews et al. 2012). The shape of the far-IR emission resembles a bow shock aligned with the direction of the space motion of the star (indicated by *an arrow*). The same morphology is found in the much larger structure detected in the radio

interstellar medium (ISM), piling-up ISM dust and gas in a "bow shock" aligned with the direction of motion of the star. It is not clear what could be triggering Cepheids mass loss: perhaps a pulsation-driven wind, or the influence of a close-in binary companion (Anderson et al. 2015). This range of phenomena is however observed in several Cepheids and should be taken into consideration when assessing the accuracy of the Cepheids PL relation (and its calibration with Galactic Cepheids) at MIR wavelengths.

The extreme case of a Cepheid embedded in a thick circumstellar envelope is RS Pup, a long period Classical Cepheid ($P = 41.39$ days) located at the centre of a well known reflection nebula $2'$ across (Westerlund 1961). The nebula is also bright at infrared wavelengths as short as 5.8 μm (McAlary and Welch 1986; Kervella et al. 2009; Barmby et al. 2011). The bulk of this nebula is believed to be comprised of cold, dusty ISM material, shaped by a stellar wind that originated from the star, possibly in the earlier evolutionary phase when RS Pup was a rapidly rotating B dwarf (Kervella et al. 2009). Warm emission detected at 10 μm, however, hints to the possibility of ongoing mass loss (Kervella et al. 2009; Gallenne et al. 2011).

The most convincing case for ongoing mass loss is however offered by the Cepheid class namesake itself, δ Cep. This star has been found to posses a bright circumstellar nebula, extending as much as $\sim 1.3'$ ($\sim 2 \times 10^4$ AU at the distance of the star; top panels in Fig. 8, Marengo et al. 2010). Observed at 70 μm, the shape of the nebula resembles a bow shock aligned with the space velocity of the star. Subsequent radio observations with the VLA found that this morphology repeats at much larger spatial scales ($\sim 13'$ or 1 pc) in the 21-cm H I line (Matthews et al. 2012), consistent with a stellar wind with outflow velocity of $\sim 35 \pm 1.2$ km/s. Combined with the infrared observations, this detection suggest the presence of ongoing mass loss with a rate up to 10^{-7}–10^{-6} $M_\odot$/yr. The estimated gas to dust mass ratio is very high (~2300 Marengo et al. 2010, Matthews et al. 2012), suggesting that the dust-poor wind is originated by some dynamic process different than the dust-driven winds found in Miras. Assuming that a significant portion of this circumstellar gas is originating from the star (as opposed to have been collected from the local ISM), such mass loss rates can help explain the Cepheids mass discrepancy described earlier.

Excess emission, if not taken properly into account, could bias Cepheid distances, or the calibration of the empirical Leavitt law, by artificially increasing the infrared brightness of the stars. However, it appears to be a relatively rare phenomenon, with less than ~25% of Cepheids surveyed with Spitzer showing any evidence of infrared emission Barmby et al. (2011) (this result is supported by a similar rate of detection of H I nebulae in a follow-up VLA survey, Matthews et al. 2016). Even in extreme cases like RS Pup, the infrared excess emission is not detected at wavelengths shorter than 5.8 μm: this suggests that any bias due to circumstellar emission is not affecting observations at 3.6 and 4.5 μm, which remain safe for determining distances with PL relations with the techniques explained in Sect. 2.2.1.

2.2.4 The Future of MIR Cepheid Distances

The advent of MIR space astronomy has finally opened the possibility of measuring accurate Cepheid distances at wavelengths less affected by interstellar extinction, metallicity effects and contamination from interstellar and circumstellar thermal emission. The L band appears to be the ideal wavelength range for this measurements, lacking the presence of variable CO absorption that is instead affecting in the M band. The CHP project (Freedman et al. 2011) has demonstrated that the 3.6 μm Cepheid PL relation can be effectively used to measure the Hubble constant and other cosmological parameters with a precision comparable to the accuracy (better than ~3%) in Planck's CMB and BAO analysis (Freedman et al. 2012). A "tension" between Cepheid distances and the Planck analysis, however, remain, and needs to be investigated for its potential to reveal new physics in the accepted ΛCDM model. Accurate (μas) parallaxes for hundreds of Galactic calibrators, expected by 2022 with the final Gaia data release, will provide the next significant jump in reducing the systematic errors in the calibration of the Cepheid Leavitt law. Further improvement will come with the launch, in 2018, of the James Webb Space telescope (JWST), whose NIRCam instrument will allow L band observation of Cepheids in host galaxies of Supernovæ Type Ia beyond the Local Group. Combined together, these improvement will enable 1% measurements of H_0 as well as competitive measurements of other cosmological parameters such as the equation of state of dark energy w_0, key to validate current cosmological models (Gardner 2012).

2.3 Linearity of the Cepheids PL Relation

The linearity of the Cepheid PL relation has been under debate in the past decade with empirical evidences of fundamental-mode Cepheid PL relation displaying a break at 10 days in optical bands (Tammann et al. 2003; Kanbur and Ngeow 2005; Ngeow et al. 2005; García-Varela et al. 2013). It was suggested that the LMC Cepheid PL relation is best described by two separate slopes for short ($P < 10$ days) and long ($P > 10$ days) period Cepheids, therefore, a "Non-linear relation". The argument for a fiducial break period at 10 days was attributed to the resonance $P_2/P_0 = 0.5$, in the normal mode spectrum (Simon and Lee 1981; Ngeow et al. 2005). The non-linearity analysis of PL relations was extended to multiple wavelengths BVI_cJHK_s (Ngeow et al. 2008), advocating that PL relations at longer wavelengths are supposedly linear, being less sensitive to metallicity and extinction. Recently, a statistical framework to study the non-linearity in LMC Cepheid PL relations at multiple wavelengths was developed (Bhardwaj et al. 2016a) and it was found that optical band fundamental-mode Cepheid PL relations are indeed non-linear at 10 days while NIR PL relations provide evidence of a statistically significant non-linearity around 18 days.

In this section, we will discuss the application of a statistical framework on the multi-wavelength data for Cepheids in the LMC. We use optical V and I band mean magnitudes for Cepheids in the LMC from the Optical Gravitational Lensing Experiment (OGLE—III) survey (Soszyński et al. 2008). We include 1849 fundamental (FU) and 1238 first-overtone (FO) mode Cepheids in this analysis. The near-infrared (JHK_S) mean magnitudes for 775 FU and 474 FO mode Cepheids are adopted from LMC near-infrared synoptic survey (Macri et al. 2015). The extinction corrections were applied using reddening maps of Haschke et al. (2011) and Cardelli et al. (1989) extinction law.

2.3.1 The Statistical Framework

We discuss the test-statistics in brief. The detailed mathematical formalism is provided in Bhardwaj et al. (2016a).

F-Test The F-test works under the null hypothesis that a single regression line over the entire period range is a better model to fit the data. The alternative hypothesis consists of a full model with two different slopes for short and long period Cepheids:

$$\begin{aligned} m_0 &= a_S + b_S \log(P); \quad \text{where } P < P_b, \\ &= a_L + b_L \log(P); \quad \text{where } P \geq P_b, \end{aligned} \tag{3}$$

where m_0 is the extinction-corrected mean magnitude and P_b is the fiducial break period. The difference in the residual sum of squares of the two fits, given the number of degrees of freedom, is estimated as the observed F-statistics. This F-value is compared with theoretical F-distribution with 95% significance level. An observed F-value greater than the theoretical F-value leads to the rejection of null hypothesis.

Random-Walk The robust non-parametric random-walk test generates the distribution of the residuals from the data itself. At first, the data is sorted according to the period and a single regression line is fitted over the entire period range. The amplitude of the partial sum of residuals is taken as observed R-statistics. If the partial sums are a random-walk, R will be small, suggesting a linear relation. In order to generate a theoretical R-distribution, the residuals are permuted 10^4 times to estimate R-values. The fraction of permuted R-values that exceed the observed R-value, gives the probability of acceptance of a linear relation.

Testimator For testimator analysis, the sample is sorted and divided into several subsets of period bins. Slope of the PL relation in the first subset is adopted as initial estimate for the next subset. The null hypothesis is that the two slopes under consideration are equal. The standard t-test is applied to test the consistency of the slopes. If the slopes are equal, a smoothed slope over the two subsets is adopted as initial estimate for next subset. The process is repeated for all subsets unless the null-hypothesis is rejected.

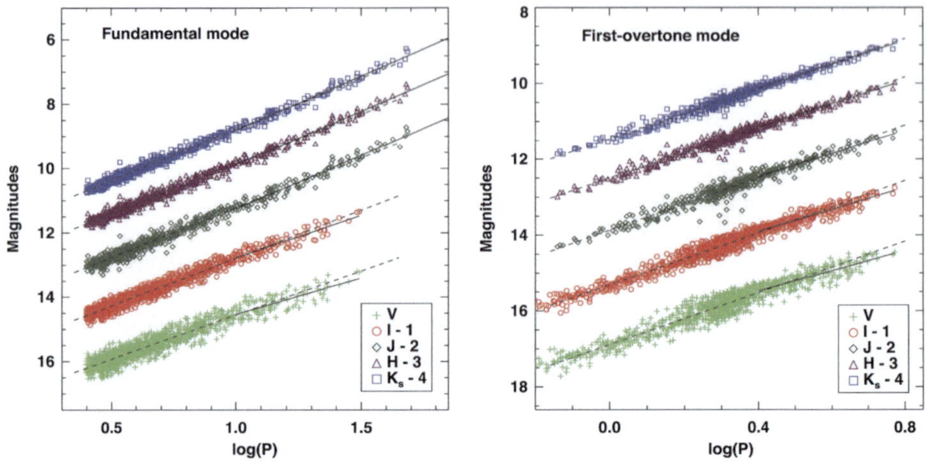

Fig. 9 Optical and near-infrared PL relations for LMC FU and FO mode Cepheids. The *dashed/solid lines* represent the best fit regression for Cepheids with periods below and above the break period (10 days for FU and 2.5 days for FO mode Cepheids)

Segmented Lines and the Davies Test An alternative approach to the problem of identifying the existence of a break point is presented in Muggeo (2003). The method first performs a linear piecewise regression considering the existence of the break. Thereafter, the Davies test is used to evaluate if the two segments are different enough to account for two separate linear behaviours. The method is implemented in the R package segmented (Muggeo 2008).

2.3.2 Analysis and Results

The optical and near-infrared band PL relations for FU and FO mode Cepheids are displayed in Fig. 9 and the results of statistical analysis are presented in Table 1. If the probability of the acceptance of null hypothesis is less than 0.05, the PL relation under consideration is non-linear. For FU mode Cepheids, optical band PL relations are found to be non-linear at 10 days, according to the F-test, testimator and the Davis test. The random walk does not provide evidence to support a non-linearity at 95% significance level. If more than two statistics suggest a break, we consider the PL relation to be non-linear. At NIR wavelengths, all four test statistics suggest a statistically significant deviation in the slope for short and long period Cepheids, except for random walk in K_S band. Interestingly, the break period is around $\log(P) \sim 1.25$–1.30 in the near-infrared PL relations. It is important to note that a distinct variation in Fourier amplitude parameters around 20 days was also observed for redder bands as compared to optical bands (Bhardwaj et al. 2015). This suggest that the changes in light curve structure can also be associated with these non-linear features in PL relations. Although the multi-wavelength LMC PL relations exhibit a non-linearity, the SMC PL relations are found to be linear at 10 days (Ngeow et al. 2015a).

We also test the FO Cepheid data in the LMC for possible statistically significant non-linearities. The optical band PL relations present evidence of a non-linearity based on all four test statistics. The Davis test suggests a break at $\log(P) = 0.45$ ($P \sim 2.8$) days in PL relations. The evidence of non-linearity decreases for longer wavelengths and HK_S bands are found to be linear with all four test-statistics. The robustness of these results was also validated under various assumptions, such as independent and identically distributed random observations, normality of residuals and homogeneity of variances.

Table 1 The results of statistical tests on LMC Cepheid PL relations at multiple wavelengths

Band	$\log(P_b)$	b_S	σ_S	b_L	σ_L	$p(F)$	$p(R)$	TM	$p(D)$
FU									
V	1.004	-2.743 ± 0.032	0.172	-2.285 ± 0.185	0.232	0.002	0.168	Reject	0.006
I	1.014	-2.973 ± 0.022	0.117	-2.628 ± 0.127	0.161	0.001	0.128	Reject	0.003
J	1.262	-3.118 ± 0.031	0.109	-3.336 ± 0.071	0.144	0.004	0.037	Reject	0.003
H	1.259	-3.106 ± 0.026	0.090	-3.320 ± 0.060	0.120	0.001	0.044	Reject	0.000
K_s	1.275	-3.192 ± 0.022	0.079	-3.361 ± 0.055	0.111	0.003	0.076	Reject	0.001
FO									
V	0.446	-3.410 ± 0.045	0.187	-2.688 ± 0.108	0.155	0.000	0.000	Reject	0.000
I	0.449	-3.430 ± 0.032	0.132	-2.894 ± 0.078	0.112	0.000	0.000	Reject	0.000
J	0.427	-3.455 ± 0.064	0.117	-3.107 ± 0.103	0.102	0.043	0.122	Accept	0.059
H	0.422	-3.322 ± 0.049	0.097	-3.117 ± 0.075	0.075	0.136	0.284	Accept	0.156
K_s	0.019	-3.303 ± 0.042	0.087	-3.179 ± 0.062	0.062	0.376	0.277	Accept	0.735

Notes: P_b, b and σ refer to the break period, slope and dispersion, respectively. The subscripts S and L refer to the short period range and long period range, respectively. $p(F)$, $p(R)$ and $p(D)$ represent the probability of acceptance of null hypothesis i.e. linear relation. "TM" displays the results of testimator analysis: "Accept/Reject" represents a linear/non-linear relation

The LMC FU mode Cepheid sample was used together with Cepheids in supernovae host galaxies to estimate the impact of these non-linearities on the distance scale (Bhardwaj et al. 2016a). The additional Cepheid and supernovae data was adopted from the SH0ES *(Supernovae and* H_0 *for the Equation of State of dark energy)* project (Riess et al. 2011). The Cepheids in the LMC were calibrated using the late-type eclipsing binary distance (Pietrzyński et al. 2013) and a global matrix formalism was adopted similar to SH0ES. The global slope of Cepheid PL relations is found to be consistent with linear version of LMC PL relations. However, the two slope model of LMC PL relations provides a stronger constrain on the global slope and the metallicity coefficients. This model was also adopted in recent SH0ES paper with additional supernovae hosts, leading to a 2.4% local determination of Hubble constant.

2.3.3 Conclusions

The statistical analysis of non-linearities in the LMC Cepheid PL relations provides evidence of a break in fundamental-mode Cepheid PL relations at $VIJHK_S$ wavelengths. The first-overtone mode Cepheids display a break at 2.8 days, only at optical wavelengths. The observed non-linearities suggest a correlation with sharp changes in the light curve structure of Cepheids. This needs a theoretical investigation to look for additional resonance feature in the light curves for Cepheids. The global slope for Cepheids in SH0ES galaxies using the linear and non-linear versions of the LMC PL relations are found to be very similar. Therefore, these non-linearities will not have any statistically significant impact on the distance scale or the value of Hubble constant. This is due to the fact that the dispersion of the *HST*-based PL relations for Cepheids in supernovae host galaxies (∼0.3–0.4 mag) is more than three times the dispersion of the calibrator LMC PL relation. Therefore, such small but statistically significant changes in slope of PL relation may only have a major impact on distance parameters with more precise PL relations, for example with the *JWST*.

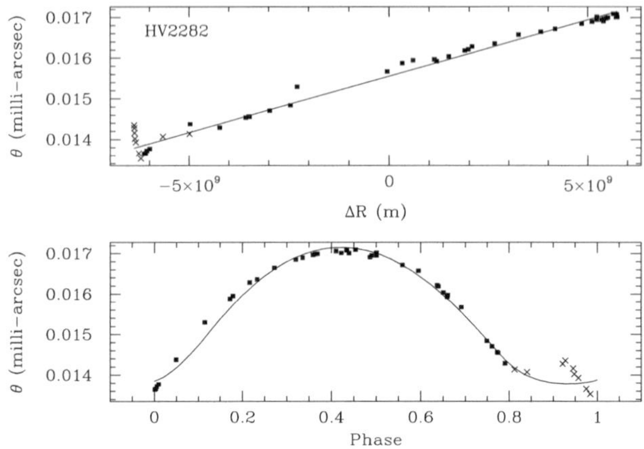

Fig. 10 Example of the fit of the angular diameters derived from photometry and spectroscopy for the LMC star HV2282 (Storm et al. 2011b)

2.4 Metallicity Effects

The metallicity of a star usually affects its luminosity to a certain degree and there is no reason to believe that Cepheids should differ from other stars in this regard. The question is if the effect is significant and, if significant, whether it can be properly corrected for. Riess et al. (2011) pointed out that to determine the Hubble constant to better than 2% one of the major obstacles is the effect of metallicity on the Cepheid luminosities.

In preparation for the HST Key project on the extragalactic distance scale (Freedman et al. 2001) the metallicity effect was determined empirically by comparing high metallicity Cepheids in the inner parts of M101 with low metallicity Cepheids in the outer part of the galaxy (Kennicutt Jr. et al. 1998). They found an effect of 0.1 ± 0.1 mag/dex in the V band, metal-poor stars being brighter. Many different empirical efforts trying different avenues followed in the next years but even the sign of the effect was disputed (e.g. Romaniello et al. 2008).

The approach used by Kennicutt Jr. et al. (1998) which relies on extragalactic Cepheids has the obvious advantage that the stars are at the same distance and span a large range of metallicities but it suffers from the problem of disentangling reddening and blending effects and the necessity to infer the metallicities from the local gas in the galaxy. The Gaia satellite will surely provide very accurate distance to galactic Cepheids but the sample will be limited to Cepheids with metallicities not too different from solar, so only a limited metallicity range can be studied and this will eventually limit the accuracy with which the slope of the period-luminosity relation can be determined.

Here we present the status of an alternative approach based on the infrared surface brightness method, which is a Baade–Wesselink type method. In this approach the stellar luminosity is determined directly for each star individually and a PL relation can be constructed on this basis. The added advantage of this method is that it can be applied to Cepheids not only in the Milky Way but also in the Magellanic Clouds (e.g. Storm et al. 2004). In this way we can compare stars spanning a significant range of metallicities in a direct way: in fact we can delineate period-luminosity relations for Cepheids of three different metallicities representing the young populations in the SMC, LMC and solar neighbourhood.

2.4.1 The Near-IR Surface Brightness Method

The IRSB method is a geometrical method as it relates the stellar angular diameter, Θ, at a given phase, ϕ, with the stellar radius, R, and the distance, d through

$$\Theta(\phi) = 2R(\phi)/d = 2\big[R_0 - \Delta R(\phi)\big]/d \quad (4)$$

If we can observe the angular diameter, $\Theta(\phi)$, and the radius variation, $\Delta R(\phi)$ at different phases we can solve the equation for the two variables, radius R_0, and distance d.

The angular diameter is related to the stellar surface brightness F_V, through the relation $F_V = 4.2207 - 0.1V_0 - 0.5\log(\Theta)$. Barnes and Evans (1976) found that a linear relation between colour and stellar surface brightness could be established. Fouque and Gieren (1997) expanded this to the NIR $(V - K)$ colour, based on interferometric measurements of static stars. This relation was refined by di Benedetto (1998) and Kervella et al. (2004) first deriving the relation:

$$F_V = -0.1336(V - K)_0 + 3.9530 \quad (5)$$

based on interferometric measurements of Cepheids.

Next we can measure the absolute radius variation ΔR, by integrating the pulsation velocity curve. This differs from the directly observed radial velocity curve as it takes out the systemic velocity v_{sys}, and correct for the projection effect as the radial velocity measures a weighted mean across the surface of the star and not just the velocity exactly on the point nearest to the observer. The projection effect is corrected by multiplying by the so-called p-factor. For a more detailed discussion of the calibration of the p-factor, see Storm et al. (2011a). This can then be written:

$$\Delta R(\phi) = \int -p\big(v_{rad}(\phi) - v_{sys}\big)d\phi \quad (6)$$

These equations shows that all that is needed is good light curves in V and K and a good radial velocity curve. These are all obtainable for stars in the Milky Way as well as in the Magellanic Clouds and potentially in M31.

In Fig. 10 an example of a fit to (4) for the LMC star HV2282 is shown. We note that the phase interval between 0.8 and 1.0 is usually disregarded as the stars often shows certain spectral lines in emission suggesting shocks passing through the atmosphere. This can lead to the surface-brightness relation being invalid and it can affect the radial velocity measurement at some level.

2.4.2 The Data Set

Storm et al. (2011a) presented an analysis based on 76 Milky Way stars, 36 LMC stars and 5 SMC stars. In Fig. 11 (left panel) a typical data set for one of these LMC stars is shown. The project has been expanded to include an additional 25 SMC stars to bring down the error bar on the lowest metallicity data point. In Fig. 11 (right panel) an example of this new data set is shown. The data acquisition is still in progress so the phase coverage will eventually be better, but the high quality of the data is evident. The optical data is taken from the OGLE-IV Udalski (2015) survey while the K band data is being acquired at the ESO-NTT and the radial velocities with HARPS at the ESO 3.6 m telescope.

In addition to the work on the SMC Cepheids we have also been acquiring new accurate radial velocity curves for more than 50 Milky Way Cepheids for which good V and K band light curves exist. This should also further constrain the most metal rich data point.

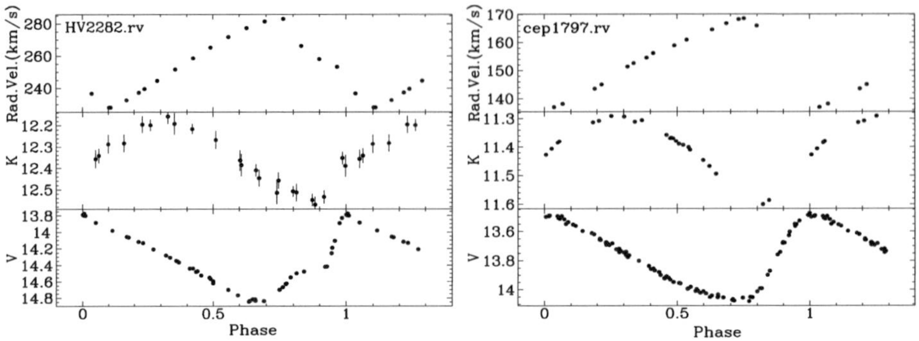

Fig. 11 *Left*—Example of data set for the LMC Cepheid HV2282. The V band data from OGLE-III Soszyński et al. (2008), the K-band data is from Persson et al. (2004), and the radial velocity data is from the HARPS spectrograph at ESO Storm et al. (2011b). *Right*—Example of a data set for the SMC Cepheid SMC-1797. The K band data is a preview from an on-going program at the ESO-NTT and the radial velocity data is a preview of an ongoing program with HARPS at the ESO 3.6 m telescope

2.4.3 The Period-Luminosity Relations at Different Metallicities

In Fig. 12 (top panels) we have plotted the PL relations in the K band for the samples of Milky Way and LMC stars from Storm et al. (2011a,b). In both plots the line over-plotted is the best fit to their complete sample of stars. It is evident that the two relations are very similar. In Storm et al. (2011b) only 5 SMC stars with a period of around 15 days were available so the low metallicity relation was poorly constrained. Work is now ongoing to remedy this limitation by extending the sample by an additional 25 stars (PI: Gieren). In the bottom left panel of Fig. 12 we plot the preliminary PL relation for the sub-sample of stars where the data is sufficient to perform an initial analysis. Again the over-plotted line represents the fit to the global sample in Storm et al. (2011b). We note that at least the preliminary agreement is very good.

We can then determine the offset for each sample at a period of 10 days with respect to the reference relation in each of the plots. We have adopted metallicities of $[\mathrm{Fe/H}] = +0.03$ for the Milky Way, -0.34 for the LMC, and -0.68 for the SMC, following Luck et al. (1998). In the last panel of Fig. 12 we have over-plotted all the stars with the reference relation. In the inset the individual offsets for the three samples is plotted against the adopted metallicity. It is evident that the effect of metallicity in the K band appears to be small, consistent with zero, and that the error estimate on the slope will be significantly less than 0.1 mag/dex. The final estimate and associated error bar will have to wait the completion of the data acquisition and final analysis, but the preliminary results are looking very promising.

3 Miras

Mira variables are M-type stars pulsating with periods of 100 to 1000 days, showing high mass loss rates (10^{-5}–10^{-7} $M_{\odot}$ yr^{-1}). They are thought to be experiencing a pulsating phase while on the asymptotic giant branch (AGB) evolutionary phase. Mira variables are also well known for their remarkable periodicity and large amplitude in V-band magnitude. Mira variables in the LMC exhibit a PL relation between their pulsation period and K-band apparent

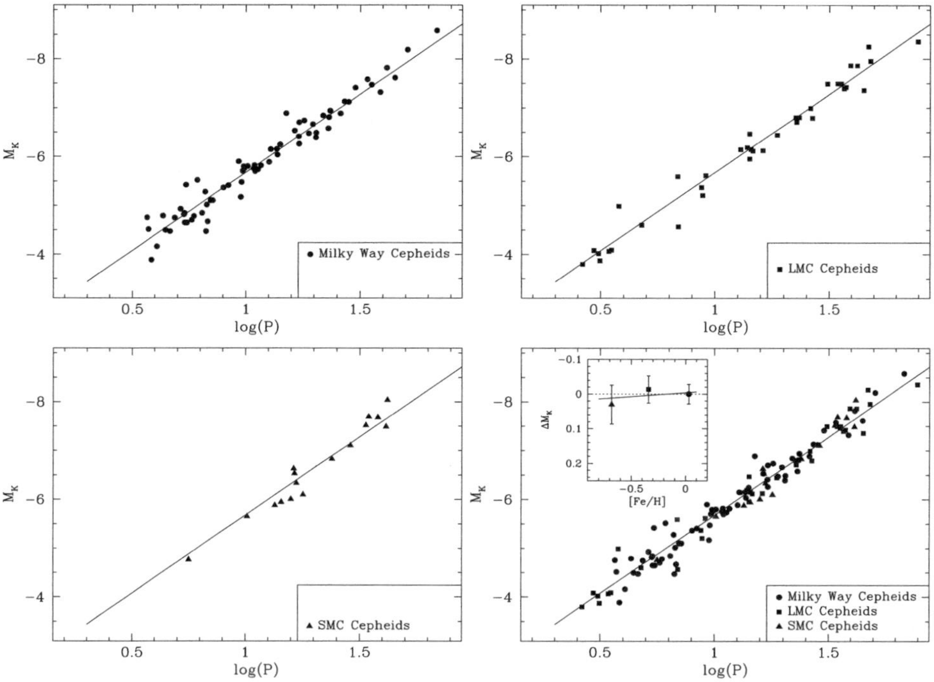

Fig. 12 *Top*—The PL relation in the *K* band for a sample of Cepheids in the Milky Way (*left*) and in the LMC (*right*), from Storm et al. (2011a,b). *Bottom left*—Preliminary PL relation in the *K* band for part of the new sample of SMC Cepheid. *Bottom right*—Combined PL relation with all the stars presented in the previous three panels. The *inset* shows the offset of the three different samples with respect to the overall PL relation at a period of 10 days

magnitude (K). This PL relation allows these stars to be used as distance estimators, and accurate distance of these sources helps us to understand the nature of their variability.

Although a narrow PL relation for Miras in the LMC was found by Feast et al. (1989), a similar relation for the Galactic Miras has not been precisely obtained because of large errors in absolute magnitudes. Such large errors arise from ambiguities of absolute magnitudes suffering directly from errors in distances estimations. van Leeuwen et al. (1997) reported the relation of Galactic Miras based on the parallax measured with Hipparcos satellite (Perryman et al. 1997). Although the Hipparcos mission has been an historical milestone in astrometry, distances of the Galactic Mira variables are too far for an accurate determination of their parallaxes. Therefore, a calibration of the PL relation based on Galactic Mira variables is not well established even today.

The very long baseline interferometer (VLBI) consists of a number of antennas separated by several thousand kilometers. Angular resolution of the order of milli-arc seconds is obtained by imaging synthesis of VLBI at radio wave bands of cm to mm wavelength. Because of its high resolution, the VLBI has a strong advantage for parallax measurements compared to other current astronomical instruments. Using absolute magnitudes derived from accurate distances measured with astrometric VLBI observations, we can investigate precise PL relation in the Galaxy. In this sub-section, we present our recent on-going efforts to calibrate the PL relation of Galactic Mira variables using the VLBI Exploration of Radio Astronomy (VERA) array.

Fig. 13 Optical and radio variations of a Mira variable 'R UMa'. *Filled squares* and *crosses* indicate H_2O maser and V-band variations of the source. Magnitudes from American Association of Variable Star Observers (AAVSO) are used for the V-band light curve

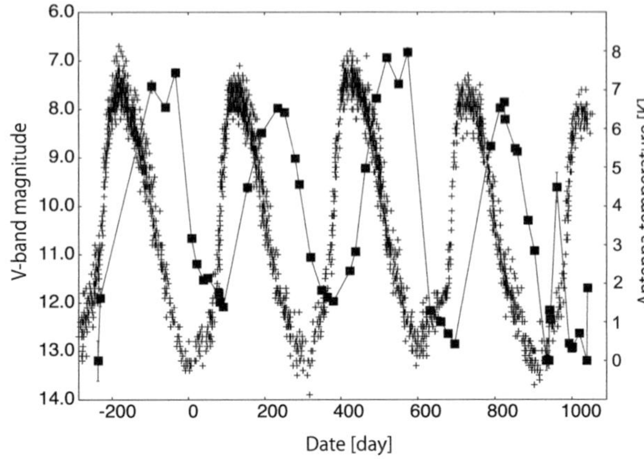

3.1 Source Selection and Single-Dish Monitoring

Mira variables show maser emission around them. Maser stands for Microwave Amplification by Stimulated Emission of Radiation and hence is a non-thermal radiation representing very compact structure and high brightness temperature. Many different masers occur in astrophysical environments. SiO, H_2O, OH, and CH_3OH masers are detected around evolved stars or star forming regions. They are frequently used as probes to study kinematics of matter around the stellar objects. The milli arc second level high resolution achieved by the VLBI technique is well suited for observation of the maser emissions. H_2O maser emission around Mira variables is so bright and compact that they can be a good target for VLBI astrometry. But the intensity of the maser emission usually shows large variability. Like the periodic pulsation at optical band, intensity of the maser emission sometimes represents quasi-periodic variability with the same pulsation periods defined in the visible. Figure 13 shows a result from monitoring the Mira variable 'R UMa'. This is a case that represents a good period consistency in optical (V band) and maser emission (22 GHz).

In order to derive the parallax from maser emission, we have to find sources with bright maser emission (all through their cycle) enough to be detected with the VLBI method. We are monitoring a sample of candidate targets with the IRIKI 20 m telescope in VERA array with one month interval. Early results of the monitoring program are reported by Shintani et al. (2008). The results from the single-dish monitoring are used to select the sample for our VLBI program.

The distribution of the pulsation periods of our targets is presented in Fig. 14. Open blocks indicates ~800 nearby Mira variables studied by Feast and Whitelock (2000). Filled blocks indicate the sources of our VLBI study which are accompanied by H_2O maser emission. We fitted these two distributions with Gaussian models and obtained peak periods of 338 days ($\log P = 2.53$) and 407 days ($\log P = 2.61$) for optically defined Miras and maser emitting Miras, respectively. Although there is a difference between two peaks, a selection bias can not be ruled out at present. More careful consideration about both populations is needed before we conclude them as two separate populations.

The diffuse and weak components of the maser emission are resolved by the small synthesised beam of the VLBI. Therefore, it is important to find target sources that have bright and compact components that can be detected with the VLBI method. As a threshold of a total power intensity, ~10 Jy is adopted for detection in our VLBI monitoring program. We

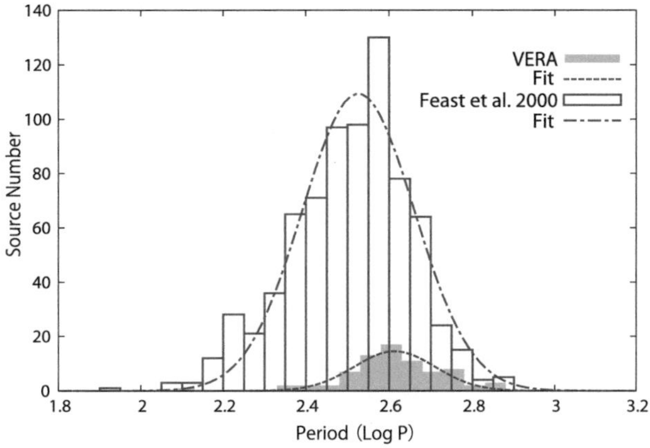

Fig. 14 Period distribution of the Galactic Mira variables. *Open blocks* indicates ~800 Mira variables studied in Feast and Whitelock (2000). The period distribution of Mira variables accompanied with H_2O maser emission seems to have slight shift to longer period

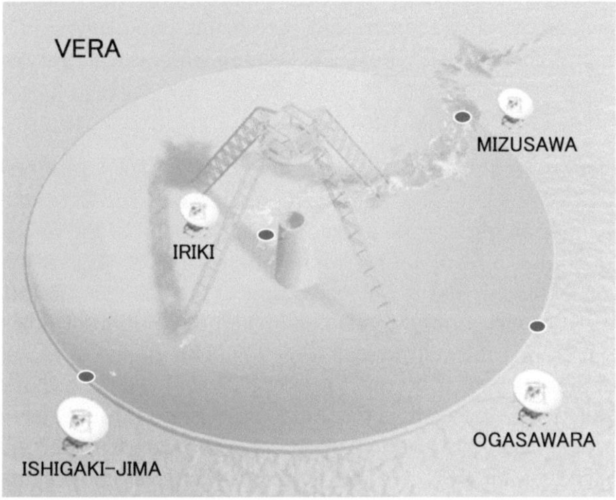

Fig. 15 Schematic figure of the VLBI Exploration of Radio Astrometry (VERA) antennas. Four antennas with 20 m aperture spanning Japan are dedicated to astrometric observation at frequencies of 22 and 43 GHz

are observing ~250 maser sources to find targets whose maser emission stably shows intensity with $\geq \sim 10$ Jy. From the point of view of pulsation periods, sampling a wide periods range is required to obtain better solutions of the PL relation.

3.2 Astrometric VLBI Observations with VERA

We started a series of VLBI observations with VERA after a careful selection of the targets. The VERA array consists of four antennas with 20 m aperture located in Japan (Kobayashi et al. 2003). The maximum baseline length is ~2300 km (Fig. 15).

We observe target maser source and reference continuum source simultaneously using the dual beam system installed in the VERA antenna (Kawaguchi et al. 2000). At first, we solve phase solutions using a reference source, which usually is a bright QSO. Data reduction of the target source can be realised by applying the phase solution from the QSO to the target. Two dimensional Fourier transformation of the data provide a map of the target maser source, from which we measure the positions of the maser spots. Since core emission of QSO can be treated as a position reference fixed to the sky plane in milli arc second scale astrometry, the motion of the maser spots detected on the phase reference map can be considered as a combination of the linear proper motion and parallactic oscillation. As the effect of the solar motion is same in the target and reference source, this motion is cancelled out in the phase referencing technique. By numerical fitting of the data using a function consisting of an oscillating term and a linear term, we derive the parallax. We note that the linear motion term is a combination of internal motion of the maser spots and systemic motion of the star. Both these motions are treated as linear motion in the data analysis. Typically, we need 1.5 to 2.0 years to measure a parallax for one source. As part of this program we have observed 33 sources, with 6 more under monitoring. Table 2 shows the target sources of our study. Coordinates, variable types, pulsation periods, and the logarithms of the periods are also presented.

3.3 Results of Our Observation with VERA

The main goal of our VLBI observations with VERA is measuring the parallaxes (and hence, the distances) to our sources. Once the distance is determined, we study the properties of the target sources and their PL relation. Here, we present some results on stellar properties, such as effective temperature, size of the photosphere, and kinematics of the maser spots. A preliminary result of the PL relation study is also presented.

3.3.1 Semi-regular Variable S Crt

Figures 16 and 17 show the results of our VERA observation of the semi-regular variable S Crt. We see a clear serpentine motion of a maser spot on the sky plane (Fig. 16). Since the motion of the maser spot is a combination of the linear proper motion and the parallactic ellipse, a resultant motion is observed as the curvature. A parallax of 2.33 ± 0.13 mas was obtained, and it was converted to the distance of 430^{+25}_{-23} pc. From these series of our VLBI observations, we also estimated the motion of dozens of maser spots in a fixed system around the central star (Fig. 17). This figure shows the distribution of maser spots on an area of $\sim$40 $\times$ 40 mas^2 ($\sim$17 $\times$ 17 au^2). The distribution of the maser spots was found to be relatively compact as much as $\sim$10 au in diameter. A cross symbol on the map centre indicates the position of the central star, estimated from the distribution and motions of the maser spots. Outwards motions and inhomogeneous distribution of maser spots are described in Nakagawa et al. (2008). Aringer et al. (1999) determined an effective temperature of the photosphere ($T_{\rm BB}$) of S Crt to be 3097 ± 100 K by fitting two blackbodies to its infrared spectrum. Another temperature of 496 K was attributed to the circumstellar gas and dust shell. A shaded circle at the center indicate the photosphere size as estimated from the effective temperature of the photosphere ($T_{\rm BB}$) and observed infrared magnitudes. The estimated photosphere radius (R_*) is $1.81 \pm 0.14 \times 10^{13}$ cm (260 ± 20 R$_\odot$). If we assume that the optical magnitude variation is attributed only to the photospheric radius variation and also assume a constant temperature, then an acceptable radius range of 213–309 R$_\odot$ can be estimated. In previous studies by Haniff et al. (1995) and van Leeuwen et al. (1997), the photospheric radii of dozen of Mira variables are reported to be larger than 300 R$_\odot$. This suggest that the photospheric radius of S Crt may be close to the lower limits to those of Mira variables reported in the literature. This result shows once more, how accurate distance measurements are important for better understanding of the difference between Mira variables and semi-regular variables.

3.3.2 Mira Variable SY Scl

The next example of our study is related to the kinematics of a Mira variable SY Scl. The parallax of this source is measured to be 0.75 ± 0.03 mas and its corresponding distance is

Table 2 Target LPVs of our study

ID.	Source	RA [h m s]	DEC [° ′ ″]	Period [day]	Maser
1	SY Scl	00 07 36.25	−25 29 39.9	413	H_2O
2	WX Psc	01 06 25.984	+12 35 53.05	660	H_2O
3	RU Ari	02 44 45.50	+12 19 03.0	354	H_2O
4	IK Tau	03 53 28.87	+11 24 21.7	470	SiO
5	V637 Per	03 54 02.28	+36 32 17.6	...	H_2O
6	R Tau	04 28 18.0004	+10 09 44.770	321	H_2O
7	BX Eri	04 40 32.754	−14 12 02.39	165	H_2O
8	T Lep	05 04 50.84569	−21 54 16.5172	368	H_2O
9	BW Cam	05 19 52.56	+63 15 55.8	...	H_2O
10	RW Lep	05 38 52.737	−14 02 26.84	145	H_2O
11	BX Cam	05 46 44.29	+69 58 24.2	454	H_2O
12	U Ori	05 55 49.16994	+20 10 30.6872	368	H_2O
13	AP Lyn	06 34 33.92	+60 56 26.2	450	H_2O
14	U Lyn	06 40 46.487	+59 52 01.64	434	H_2O
15	NSV17351	07 07 49.38	−10 44 05.9	...	H_2O
16	Z Pup	07 32 38.05674	−20 39 29.0936	509	H_2O
17	OZ Gem	07 33 57.75	+30 30 37.8	598	H_2O
18	QX Pup	07 42 16.947	−14 42 50.20	551	H_2O
19	V353 Pup	07 46 34.151	−32 18 16.26	...	H_2O
20	HU Pup	07 55 40.160	−28 38 54.84	238	H_2O
21	R Cnc	08 16 33.82789	+11 43 34.4557	362	H_2O
22	X Hya	09 35 30.26615	−14 41 28.6002	301	H_2O
23	R LMi	09 45 34.28304	+34 30 42.7839	372	H_2O
24	R Leo	09 47 33.48791	+11 25 43.6650	310	SiO
25	R UMa	10 44 38.47292	+68 46 32.7016	302	H_2O
26	VX UMa	10 55 39.88	+71 52 09.8	215	H_2O
27	HS UMa	11 35 30.70408	+34 52 04.1775	517	H_2O
28	S Crt	11 52 45.09863	−07 35 48.0810	155	H_2O
29	T UMa	12 36 23.46459	+59 29 12.9746	257	H_2O
30	U CVn	12 47 19.61	+38 22 30.5	346	H_2O
31	W Hya	13 49 01.99810	−28 22 03.4881	361	H_2O
32	RU Hya	14 11 34.39861	−28 53 07.4089	332	H_2O
33	RX Boo	14 24 11.62662	+25 42 13.4091	278	H_2O
34	RS Vir	14 27 16.38997	+04 40 41.1432	354	H_2O
35	FV Boo	15 08 25.766	+09 36 18.19	306	H_2O
36	Y Lib	15 11 41.30861	−06 00 41.3727	276	H_2O
37	S CrB	15 21 23.95608	+31 22 02.5730	360	H_2O
38	WX Ser	15 27 47.043	+19 33 51.71	425	SiO
39	SW Lib	15 55 33.42	−12 51 05.5	292	H_2O
40	FS Lib	16 00 23.76	−12 20 57.6	415	H_2O
41	IRC+10374	18 43 36.47	+13 57 22.8	514	H_2O
42	IRC−20540	19 08 54.62	−22 14 19.4	510	H_2O
43	SY Aql	20 07 05.40	+12 57 06.3	356	H_2O
44	SV Peg	22 05 42.08385	+35 20 54.5280	145	H_2O
45	R Peg	23 06 39.16689	+10 32 36.0892	378	H_2O
46	R Aqr	23 43 49.46201	−15 17 04.1385	387	SiO
47	R Cas	23 58 24.87336	+51 23 19.7011	430	SiO

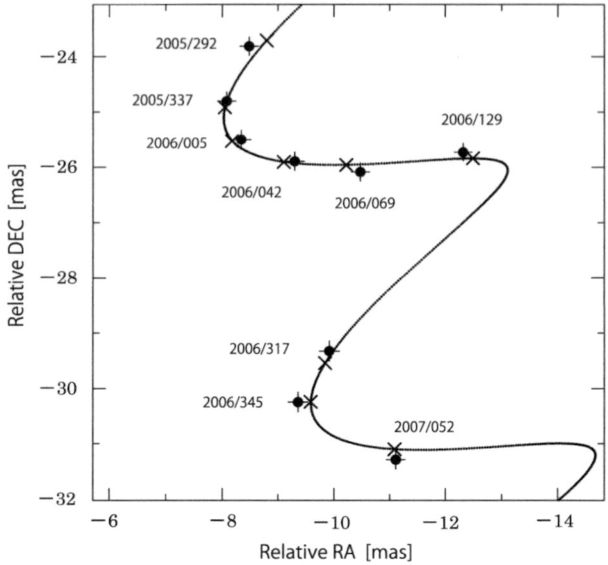

Fig. 16 Serpentine trajectory of a maser spot in S Crt. *Filled circles* indicate observed positions, *cross symbols* indicate best-fit model positions. Combination of a linear proper motion and a parallactic motion gives the serpentine motion

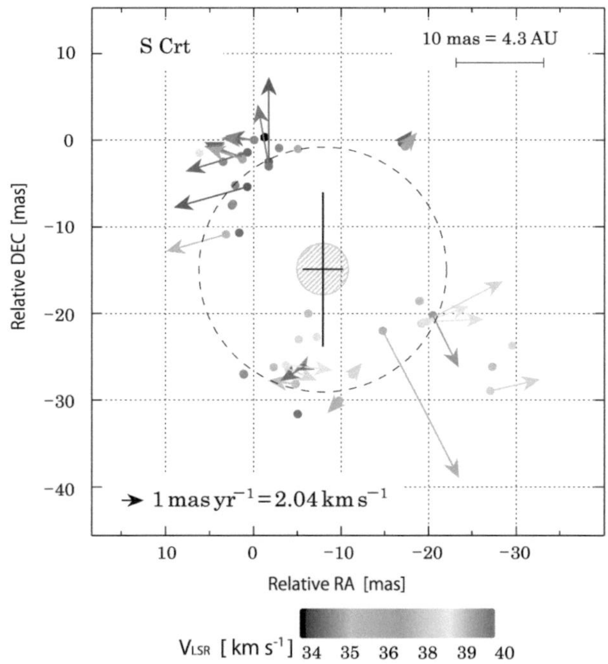

Fig. 17 Distribution and motions of the maser spots detected around a Mira variable S Crt. *Inner small circle* on the map centre indicates the estimated photospheric size of the star. *Arrows* indicate internal motion of the maser spots, and radial velocity is indicated by *gray scale*. The *arrow at the bottom-left corner* indicates a proper motion of 1 mas yr^{-1}, corresponding to 2.04 km s^{-1} at the distance of 430 pc. The *dashed line* represents a circular fit to the maser distribution, and has a radius of 14.9 mas (see Nakagawa et al. 2008 for more details)

1.33 ± 0.05 k pc Nyu et al. (2011). We also derived the source systemic motion $(\mu_x^\star, \mu_y^\star) = (5.57 \pm 0.04, -7.32 \pm 0.12)$ [mas yr^{-1}].

Using these results and radial velocity of 22 km s^{-1} for this source, we obtained a 3-dimensional motion from a numerical simulation. This spatial systemic motion is converted to that with respect to the Local Standard of Rest (LSR), by adopting a solar motion of $(U_\cdot, V_\cdot, W_\cdot) = (10.0 \pm 0.4, 5.2 \pm 0.6, 7.2 \pm 0.4)$ [km s^{-1}] (Dehnen and Binney 1998). We

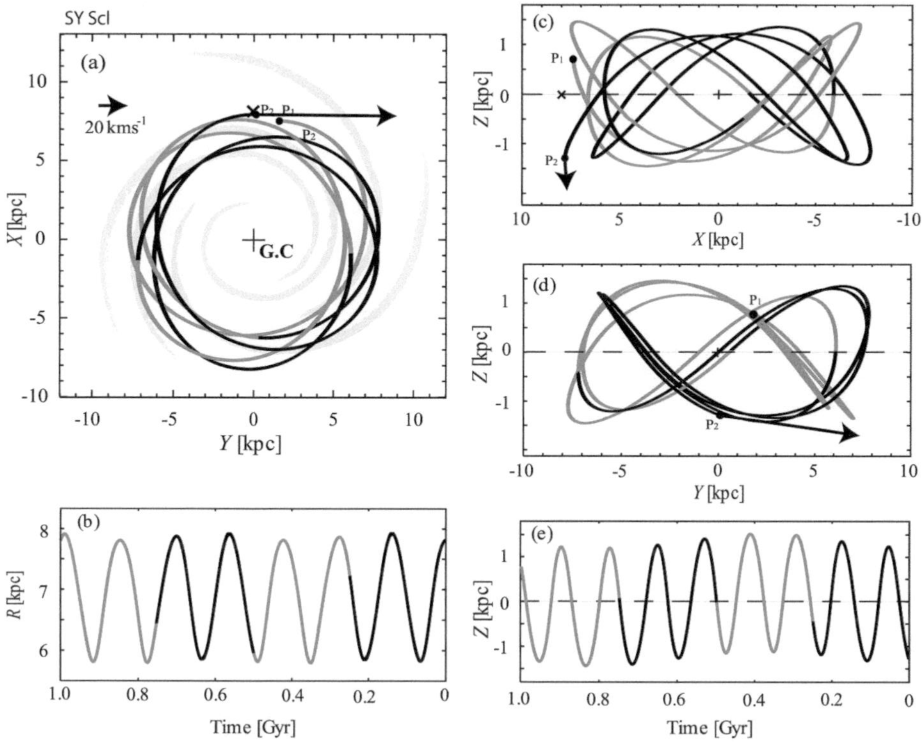

Fig. 18 Three dimensional orbit of the Mira variable SY Scl during the last 1 Gyr on the schematic view of the Galaxy. *Black* and *gray lines* indicate the orbit of SY Scl over 1 Gyr in which *black/gray line* changes every 250 Myr. (**a**) orbit in the face-on view of the Galaxy. P_1 and P_2 indicate the position of SY Scl 1 Gyr ago and present, respectively. The Sun's location is presented with a cross at $(X_\odot, Y_\odot, Z_\odot) = (8.0, 0, 0)$ kpc. Spiral arms reported by Nakanishi and Sofue (2006) are shown as *light-gray lines* in the background of the figure. (**b**) Time variation of radial distance of SY Scl from the Galactic center. (**c**) Orbit in the edge-on view (X–Z plane). (**d**) Same as (**c**) but in the Y–Z plane. (**e**) Time variation of hight from the Galactic plane. *Dotted lines* at $Z = 0$ in (**a**), (**b**), and (**c**) represent the Galactic plane

found the star has a motion with respect to the LSR $(U\text{s}, V\text{s}, W\text{s}) = (-5.2 \pm 0.3, -53.8 \pm 2.1, -33.3 \pm 0.3)$ [km s^{-1}], where U points toward the Galactic centre, V in the direction of the Galactic rotation, and W toward the northern Galactic pole. We can see that SY Scl is moving approximately southward from the Galactic plane with a peculiar velocity of 63 ± 2 km s^{-1}. Figure 18 illustrates the orbits of SY Scl in the (X, Y, Z) coordinate system in the last 1 Gyr. One can see that the orbit shows large distortions and oscillations of the order of kpc scale. The radial oscillation has a period of 142 Myr and shows an amplitude of 1.6 kpc (Fig. 18b). The orbit projected in the Z-direction has an amplitude of 1.5 kpc and the period is 120 Myr (Fig. 18e). This spatial motion suggests that SY Scl is apparently orbiting the Galaxy as a member of the Galactic thick disk rather than the thin disk.

3.3.3 Semi-regular Variable RX Boo

Long period variables sometime show multiple pulsation periods. The semi-regular variable RX Boo is an example of this case. We determined a parallax of 7.31 ± 0.50 mas and its corresponding distance of 136^{+10}_{-9} pc. An apparent K-band magnitude of -1.85 is converted

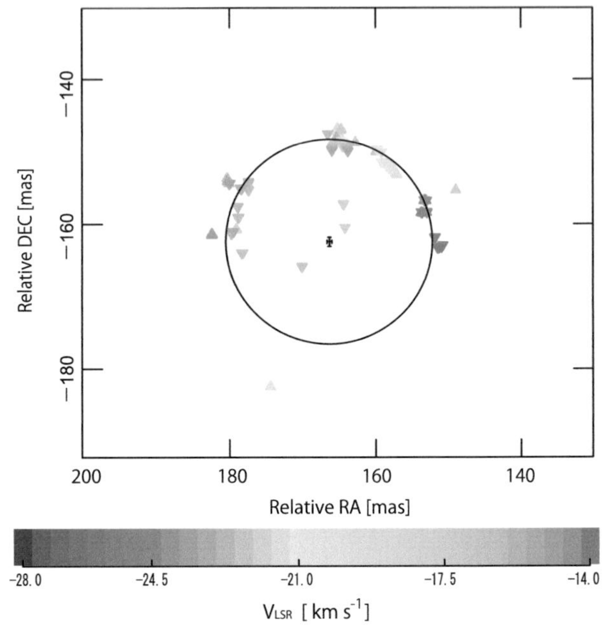

Fig. 19 Distribution of SiO masers around R Aqr (Kamohara et al. 2010). Relative positions of maser spots are given with respect to the J2000 Hipparcos coordinates. Triangles pointing upwards represent the distribution of the v = 1, J = 1–0 transition and those pointing downwards show the spots of the v = 2, J = 1–0 transition. The *circle* fitted to the maser spots is shown and the *central cross* marks its centre and the uncertainty of the central position (at 2σ-level)

to an absolute magnitude of $-7.35^{+0.15}_{-0.14}$ mag based on the parallax measurement. This source is reported to have different pulsation periods pulsation periods, which are summarised in Table 4 of Kamezaki et al. (2012). We found that they can be classified into two typical values, ~160 days and over ~320 days. From the position of the source on $\log P - M_K$ plane, one can study the stellar structure and pulsation properties. Kamezaki et al. (2012) compared the M_K and dual periods with the P-L relation in the Magellanic Clouds (Ita et al. 2004). Then it was found that RX Boo belongs to the sequence C (pulsating in fundamental mode) at longer period, and belongs to the sequence C' (pulsating in first-over tone mode) at shorter period. This fact indicates that both periods of RX Boo appear to be related to the properties of RX Boo stellar structure, with each period corresponding to a different pulsation mode. Kamezaki et al. (2012) concluded that the simultaneous enhancement of two modes of pulsation may be evidence for the transient nature of RX Boo between the two modes.

3.3.4 43 GHz Observation of a Mira Variable R Aqr

Figure 19 represents a result of VLBI observation of SiO maser in the Mira variable R Aqr observed on 24 December 2005 (Kamohara et al. 2010). Although the observing duration of R Aqr was longer than one year, it was difficult to find long-lived maser spots which can be used to solve an annual parallax. Kamohara et al. (2010) fitted the spatial distribution of the maser spots with a ring and derived the center position at each observation. Then, by tracing the ring centres, a parallax of 4.7 ± 0.8 mas was determined. The method worked well and a distance of 214^{+45}_{-32} pc was derived. Several years later, Min et al. (2014) re-analysed the same data set and successfully detected one long-lived maser spot with a lifetime of 1.1 year. Then, the parallax of R Aqr was revised to the more accurate value of 4.59 ± 0.24 mas with its corresponding distance of 218^{+12}_{-11} pc. The two measurements show consistency within their errors.

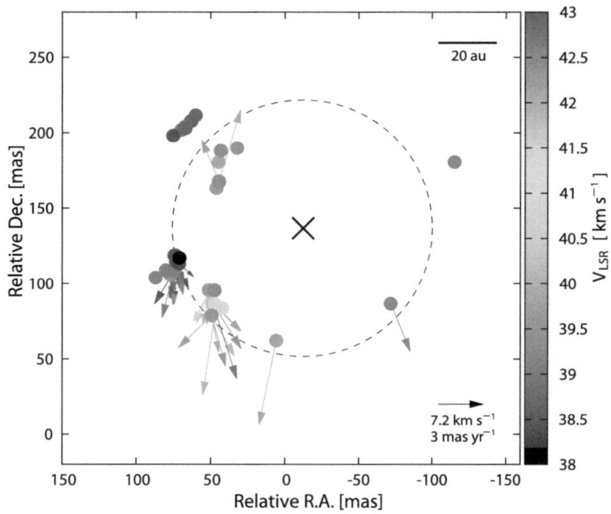

Fig. 20 H_2O maser distribution and internal motion in R UMa (Nakagawa et al. 2016). A 300 mas square area, which corresponds to 152 au square, is presented. Maser spots are indicated with *filled circles*. H_2O maser shell with a radius of 85 mas is presented with a *dotted circle*. This motion is derived by subtracting the VLBI astrometric result from the Hipparcos data. See Nakagawa et al. (2016) for more details

3.4 Combined Analysis of VERA and Hipparcos

With our VLBI method, it is very difficult to reveal maser spot motions at a rest frame fixed to the central star because we cannot detect any emission directly from the star. Therefore, we introduce some reasonable assumption in order to reveal the kinematics of the maser spots. For example, the isotropy of the circumstellar kinematics was usually assumed. In this subsection, we show the results obtained from a new method which combines astrometric measurements from the Hipparcos satellite (Perryman et al. 1997) and astrometric VLBI.

In VLBI observations, the detected proper motions of each maser spot (μ^{VLBI}) consist of three kinds of motions, such as the Galactic rotation, a systemic motion of the star, and their circumstellar motions. As for the Hipparcos data, proper motions (μ^{HIP}) also include the same kinematics as μ^{VLBI} except for circumstellar motions of the maser spots. So, if we subtract μ^{VLBI} from μ^{HIP}, we can obtain the circumstellar motions of the maser spots on the rest frame fixed to the central star.

Figure 20 shows the distribution and circumstellar motions of maser spots around the Mira variable R UMa derived with the procedure described above. In the revised Hipparcos catalog (van Leeuwen 2007b), the proper motion of R UMa is reported to be $\mu^{\mathrm{HIP}} = (-40.51 \pm 0.79, -22.66 \pm 0.78)$ mas yr^{-1}. By subtracting this value from the proper motions of each maser spot obtained from VERA observations, we obtained circumstellar motions for 38 out of all 72 maser spots in R UMa. The motions are indicated with arrows in the figure. Angular motion of 1 mas yr^{-1} corresponds to 2.41 km s^{-1} at the source distance of 508 pc. Parallax of this source was measured to be 1.97 ± 0.05 mas which corresponds to a distance of 508 ± 13 pc (Nakagawa et al. 2016). Figure 21 represents oscillating terms of the parallactic motion in RA (a) and DEC (b) of the 15 maser spots.

3.5 Period Luminosity Relation of the Galactic Mira Variables

In Table 3, we summarise the parallax measurements of the Galactic LPVs derived from astrometric VLBI observations in the order of increasing pulsation period. It includes not only Mira variables, but also semi-regular (SRa, SRb) and red supergiants (SRc). Maser molecules used in the parallax measurements are also shown. References for parallaxes and

Fig. 21 Parallactic oscillation along the axes of RA (**a**) and DEC (**b**) of the maser spots in a Mira variable R UMa (Nakagawa et al. 2016). *Filled circles* and *solid lines* represent observed data and best fit models, respectively

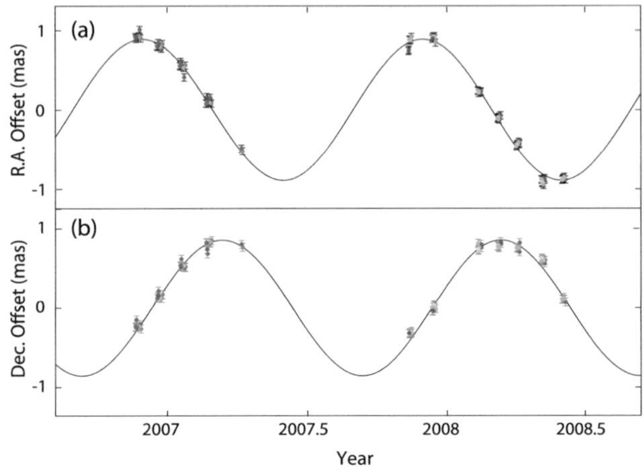

apparent magnitudes, K are given in the Table 3 footnote. Absolute magnitudes M_K are obtained from the distances and K values. Only distance errors were considered for error estimation of M_K.

Here, we define a P-L relation in the form of $M_K = -3.52 \log P + \delta$, where we assume a fixed slope of -3.52 as determined by Ita et al. (2004). Using Miras and four semi-regular variables (RW Lep, S Crt, RX Boo, and W Hya) in Table 3, we solved for the constant δ. Unweighted and weighted least squares fitting to the data gives δ of 1.09 ± 0.14 and 1.45 ± 0.07, respectively. In this fitting, the periods of semi-regular variables are being 'fundamentalised' by being multiplied by two (Kamezaki et al. 2012). Since M_K errors of a few sources are quite small compared to other source, it gives a δ discrepancy of 0.36 between two fittings.

In Fig. 22, filled squares represent sources in Table 3 on $\log P - M_K$ plane. The $\log P - M_K$ relation obtained from unweighted (upper) and weighted (lower) fitting is presented with two solid lines. Two dashed lines indicate relations derived by Ita et al. (2004) for sequence C' (first-overtone) and C (fundamental tone), respectively. The LPVs in LMC in Ita et al. (2004) are also presented with small dots in a shaded area. We used a distance modulus of 18.49 (van Leeuwen 2007b, Pietrzyński et al. 2013 and de Grijs et al. 2014) to estimate M_K values of LMC stars. Since R UMa falls on the sequence C, it is likely that the star pulsates in a fundamental mode. We find a consistency of the relations for Mira variables between our Galaxy and the LMC within the accuracy of δ.

In summary, we have presented the VLBI measured astrometric results of LPVs, mainly focusing on Mira variables in our Galaxy. Better calibration of period-luminosity relation of the Galactic Mira variables is important to estimate their distances free from an assumption that the relation of the LMC and our Galaxy is same. We are continuing our VLBI observations for this purpose.

4 Red Clump Stars

RC stars are intermediate-age (∼2–9 Gyr) core helium burning stars, which are easily identifiable in the CMD, close to the RGB, of star clusters and nearby galaxies. The location of RC in the $(Y - K_s)$ vs K_s CMD of a region in the SMC is shown in the upper panel of Fig. 23. The absolute magnitude (M_{Ks}) vs $(J - K_s)$ CMD of Galactic stars which are common in Large Sky Area Multi-Object Fibre Spectroscopic Telescope (LAMOST) survey

Table 3 VLBI parallaxes of the Galactic long period variables

Source	Type	Parallax [mas]	P [day]	Log$P$	K [mag]	M_K [mag]	Maser	Reference∗ (Parallax, K)
RW Lep	SRa	1.62 ± 0.16	150	2.176	0.639	-8.31 ± 0.22	H_2O	Kamezaki et al. (2014), a
S Crt	SRb	2.33 ± 0.13	155	2.190	0.786	-7.38 ± 0.12	H_2O	Nakagawa et al. (2008), a
RX Boo	SRb	7.31 ± 0.5	162	2.210	−1.96	-7.64 ± 0.15	H_2O	Kamezaki et al. (2012), b
R UMa	Mira	1.97 ± 0.05	302	2.480	1.19	-7.34 ± 0.06	H_2O	Nakagawa et al. (2016)
FV Boo	Mira	0.97 ± 0.06	340	2.531	3.836	-6.23 ± 0.13	H_2O	Kamezaki et al. (2016b), a
W Hya	SRa	10.18 ± 2.36	361	2.558	−3.16	-8.12 ± 0.51	OH	Vlemmings et al. (2003), c
S CrB	Mira	2.39 ± 0.17	360	2.556	0.21	-7.90 ± 0.15	OH	Vlemmings and van Langevelde (2007), c
T Lep	Mira	3.06 ± 0.04	368	2.566	0.12	-7.45 ± 0.03	H_2O	Nakagawa et al. (2014), c
R Aqr	Mira	4.7 ± 0.8	390	2.591	−1.01	-7.65 ± 0.37	SiO	Kamohara et al. (2010), c
R Aqr	Mira	4.59 ± 0.24	390	2.591	−1.01	-7.70 ± 0.11	SiO	Min et al. (2014), c
RR Aql	Mira	1.58 ± 0.40	396	2.598	0.46	-8.55 ± 0.56	OH	Vlemmings and van Langevelde (2007), c
U Her	Mira	3.76 ± 0.27	406	2.609	−0.27	-7.39 ± 0.16	OH	Vlemmings and van Langevelde (2007), c
SY Scl	Mira	0.75 ± 0.03	411	2.614	2.55	-8.07 ± 0.09	H_2O	Nyu et al. (2011), b
R Cas	Mira	5.67 ± 1.95	430	2.633	−1.80	-8.03 ± 0.78	OH	Vlemmings et al. (2003), c
U Lyn	Mira	1.27 ± 0.06	434	2.637	1.533	-7.95 ± 0.10	H_2O	Kamezaki et al. (2016a), a
UX Cyg	Mira	0.54 ± 0.06	565	2.752	1.40	-9.94 ± 0.24	H_2O	Kurayama et al. (2005), a
S Per	SRc	0.413 ± 0.017	822	2.915	1.33	-10.59 ± 0.09	H_2O	Asaki et al. (2010), b
PZ Cas	SRc	0.356 ± 0.026	925	2.966	1.00	-11.24 ± 0.16	H_2O	Kusuno et al. (2013), b
VY CMa	SRc	0.88 ± 0.08	956	2.980	−0.72	-11.00 ± 0.20	H_2O	Choi et al. (2008), b
NML Cyg	–	0.62 ± 0.047	1280	3.107	0.791	-10.25 ± 0.16	H_2O	Zhang et al. (2012), a

References of the apparent magnitudes (K) are as follows: (a) The IRSA 2MASS All-Sky Point Source Catalog (Cutri et al. 2003), (b) Catalogue of Stellar Photometry in Johnson's 11-colour system (Ducati 2002), (c) Photometry by Whitelock and Feast (2000)

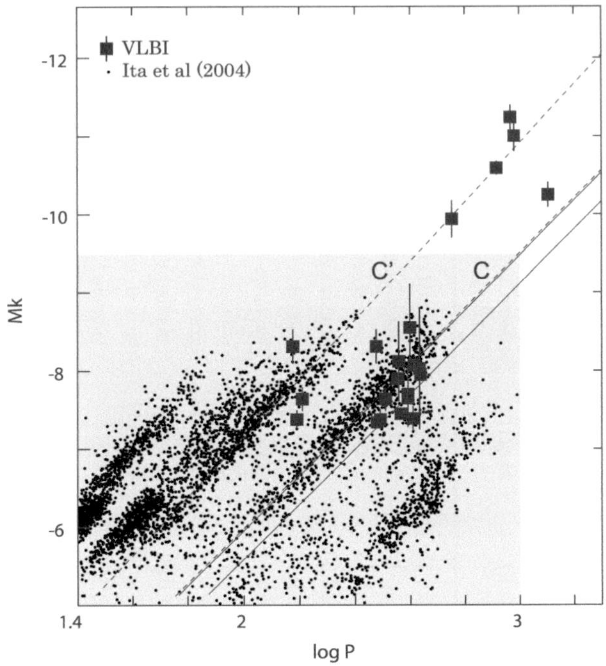

Fig. 22 Absolute magnitudes (M_K) − log P diagram of the Galactic long period variables. *Filled squares* indicate sources in Table 3, whose distances are derived from VLBI astrometry. *Solid lines* (*upper*: unweighted, *lower*: weighted) show the fits of M_K − log P relations for Galactic Mira variables. *Small dots* in the *shaded area* represent the LPVs in the LMC (Ita et al. 2004). *Dashed lines* corresponding to the labels, C and C' represent the fundamental mode and first over-tone sequences given in Ita et al. (2004)

data and Gaia DR1 Tycho-Gaia Astrometric Solution (TGAS) data, with the location of RC (selection of RC stars is described in Sect. 4.2), is shown in the bottom panel of Fig. 23. RC stars have a mass range of 1–3 solar masses and are the younger and metal-rich counter parts of horizontal branch stars.

Stellar evolutionary models indicate that RC stars should have constant luminosity. The near constancy of the clump absolute magnitude is the result of He ignition in an electron-degenerate core for low mass stars. He burning cannot start until the stellar core mass attains a critical value of about 0.45 $M_\odot$. Hence, all these stars have similar core masses at the beginning of He burning, and hence similar luminosities.

Cannon (1970) compared the observed CMDs of intermediate-age clusters with that derived from stellar evolutionary models and identified RC as a distance indicator. Benefited from the *Hipparcos* (ESA 1997), the absolute magnitudes of RC stars are well calibrated with accurate parallaxes (typically with errors smaller than 10%) of several hundred local RC stars. Since then, RC stars (which are numerous and bright) have been widely used as standard candles, to estimate distances to the Galactic Centre (Paczyński and Stanek 1998) and the Local Group of galaxies (e.g., LMC by Laney et al. 2012; SMC by Cole 1998; M31 by Stanek and Garnavich 1998). The study of RC stars in the Galactic Bulge revealed double RC feature, which indicated populations at different distances, implying the X-shaped structure of the Bulge (Wegg and Gerhard 2013). Apart from being used as absolute distance indicators, RC stars are also used to map the extinction (e.g., Haschke et al. 2011, Tatton et al. 2013) and to study the three dimensional structure (e.g., Olsen and Salyk 2002; Koerwer 2009; Subramanian and Subramaniam 2010, 2012, 2013) of Magellanic Clouds.

The general method adopted in the estimation of distance using RC stars is by modelling the RC luminosity function (LF) with a Gaussian + quadratic polynomial term. The mean and dispersion of the Gaussian function represent the mean RC magnitude and its dispersion respectively. The quadratic polynomial term is to model the RGB stars. Instead of quadratic

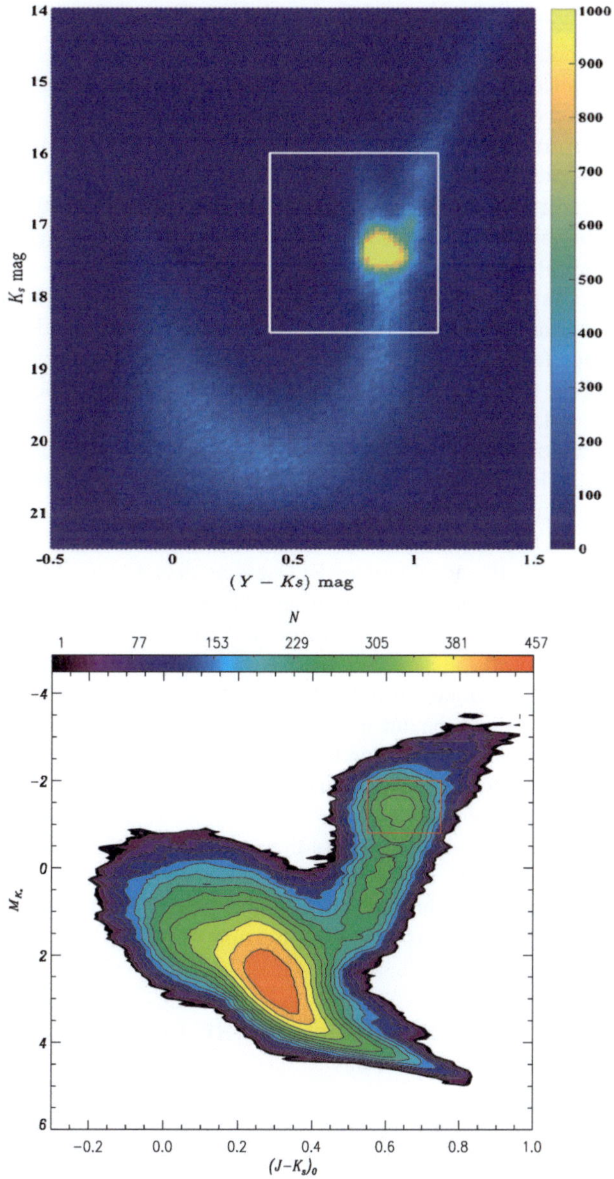

Fig. 23 *Top panel*: Hess diagram representing the stellar density in one of the sub-regions of the central VMC tile, SMC4_3. The *colour code* represents the increasing stellar density. The *white box* represents the initial selection box of the RC stars. *Bottom panel*: Hess diagram corresponding to the Galactic stars that are common in LAMOST spectroscopic survey and Gaia DR1 TGAS data (TGAS parallax err $<30\%$ and magnitudes from 2 MASS). The *red box* represents the location of spectroscopically selected RC stars

polynomial, various studies also use power law or exponential function. The mean RC magnitude is compared with the absolute magnitude of solar neighbourhood sample to estimate the distance to star clusters and nearby galaxies.

Despite the advantages of being numerous, easily identifiable in the CMD and a simple approach to measure the distances using single epoch photometric data in two bands, distance estimation using RC suffers from various caveats and limitations.

The main caveat of the RC method is the effect on the mean absolute magnitude of RC stars due to the differences in ages and metallicities of the sample with respect to the local Hipparcos sample. This effect is termed as population effect and it is also dependent on the wavelength in which RC stars are observed. One of the earlier striking results from the RC method (with similar modelling of the I band LF) was the determination of 15% shorter distance to the Large Magellanic Cloud (LMC) (Stanek et al. 1998, Udalski et al. 1998) than the values assumed to date. This created concerns on the Cepheid distance scale and determination of Hubble constant. The theoretically predicted population corrections, for the Hipparcos–LMC system, in the I band were ∼0.2–0.3 mag (Girardi et al. 1998, Cole 1998, Girardi and Salaris 2001) and hence reconciling with the LMC distance of about 50 kpc. This suggests the importance of population correction in the estimation of absolute distance using RC stars.

The other caveat is related to the accuracy of the estimation of the mean RC magnitude from the LF. The LF is contaminated by RGB stars and it is also skewed towards brighter magnitudes due to the presence of large number of young (∼400 Myr) RC stars (which appear as vertical extension above the classical RC in the CMD). The presence of 1 Gyr old secondary clump (Girardi et al. 1998, compared to classical RC they occupy a fainter and bluer location in the CMD) also makes it difficult to accurately measure the mean magnitude of the RC. The secondary RC stars are those in which the He burning is initiated in a non-degenerate core. Refer the detailed review by Girardi (2016) on the properties of RC stars and its use as distance indicator.

Again, the advantage of RC stars as distance indicator using photometric observations is mainly restricted to star clusters, Galactic bulge and local group galaxies, where they are identified as a clear clump feature in CMDs of these regions due to their large number and located nearly at same distance. However, for numerous field stars of the Milky Way, it is not easy to single out individual RC stars. Field RC stars are spread over a wide range of distances along the line of sight causing a vertical structure in the CMD rather than a clump. The differential extinction makes this feature more complicated to be identified easily from the CMD. Spectroscopic surveys are required for the reliable identification and study of Milky Way field RC stars.

In this section we discuss the recent studies in the application of RC method for distance estimation, which addresses some of the above mentioned limitations. In Sect. 4.1 we describe a potential method to remove the contamination of the RGB stars from the RC LF. Then we discuss the identification of Milky Way field RC stars using the LAMOST spectroscopic sample in Sect. 4.2.

4.1 Modelling the Near Infrared RC Luminosity Function of Crowded Regions in External Galaxies

As afore-mentioned, the main caveats of applying RC method for distance estimation to external galaxies are the population effects on their absolute magnitude calibration and how well we can model the observed RC LF by incorporating the effects of contaminants, especially the RGB stars.

The population correction is estimated as follows: the solar neighbourhood RC sample and the external galaxy RC sample are modelled using stellar population models, assuming the star formation rate and age–metallicity relation of the respective systems. The difference between the theoretically calculated absolute magnitudes of the solar neighbourhood sample and the external galaxy sample is taken as the population correction term. This correction is applied to the observed absolute magnitude of the Hipparcos RC sample and compared with the apparent magnitude of the RC in the external galaxy to estimate the distance modulus. The estimation of population correction term requires knowledge of detailed star formation history of the external galaxy. We note here that the theoretically calculated absolute RC magnitude of external galaxy sample is not directly used as reference for distance estimation due to the limitations in the models (Salaris and Girardi 2002) and for absolute calibration we still rely on highly precise observational values. These population effects strongly affect both V and I band magnitudes in a complex way (as demonstrated by Salaris and Girardi 2002, Pietrzyński et al. 2003, Girardi 2016) and hence, limits the application of RC as distance indicator in optical bands.

The population effects of RC stars in near infrared (NIR) bands, especially in K_s-band, are less or even negligible compared to optical bands (Salaris and Girardi 2002, Pietrzyński et al. 2003, Girardi 2016). The effect of extinction is also less in K_s-band. These properties make RC a better standard candle in K_s band. But in the NIR CMDs of crowded regions, the RGB and RC stars are not well separated. Subramanian and Subramaniam (2013) studied the NIR data (*JH* bands) of RC stars in the Large Magellanic Cloud (LMC) and showed that the profile fits to the observed RC LF in the crowded central regions are not satisfactory.

The best way to remove the effect of contaminants in the observed LF of the RC and to obtain their apparent magnitude, is to model the LF using stellar population models, including the star formation rate and age–metallicity relation obtained from the star formation history of the external galaxies. An attempt of such an elaborate modelling of the LF of the RC stars was preformed by Wegg and Gerhard (2013) in their study on X shaped bulge of our Galaxy. Apart from the knowledge of the star formation history of the galaxy, this detailed modelling limits the simple application of the RC stars as standard candles.

In this sub-section we discuss a simple method to remove the contamination of RGB stars from the RC stars, applicable in crowded fields, using the NIR data from the VISTA survey of the Magellanic Cloud system (VMC). Nidever et al. (2013) had used a similar technique to remove the contamination of the RGB stars from RC stars from the optical CMD in sparse outer fields of the SMC.

The VMC survey is a continuous and homogeneous ongoing survey of the Magellanic system in the YJK_s (central wavelengths, $\lambda_c = 1.02$ μm, 1.25 μm and 2.15 μm, respectively) near-infrared (NIR) bands using the 4.1 m VISTA telescope located at Paranal Observatory in Chile. The limiting magnitudes with signal-to-noise (S/N) ratio of 5, for single-epoch observations of each tile, in the Y, J and K_s bands are ~21.1 mag, 20.5 mag and 19.2 mag, respectively, in the Vega system. The stacked images can provide sources with limiting magnitudes of up to 21.5 mag in K_s with S/N = 5. A detailed description of the VMC survey is given by Cioni et al. (2011). The RC feature in the SMC (K_s ~17.3 mag) is around 2 mag brighter than the 5σ detection limit of single-epoch observations.

As the $(Y - K_s)$ colour gives the widest colour separation possible in the VMC data, which allows a better separation between RC and RGB stars, we use the Y and K_s band photometric data to do the RC selection and further analysis. Hess diagrams, with bin sizes of 0.01 mag in $(Y - K_s)$ colour and 0.04 mag in K_s, representing the stellar density in the observed $(Y - K_s)$ vs K_s CMD of one of the central VMC tiles, SMC 4_3 is shown in Fig. 23. The RC stars, at $(Y - K_s) \sim 0.7$ mag and $K_s \sim 17.3$ mag, is easily identifiable in the CMD.

First we defined a box with size, $0.5 \leq (Y - K_s) \leq 1.1$ mag in colour and $16.0 \leq K_s \leq 18.5$ mag in magnitude, for the selection of the RC stars from the CMD. The selection box is also shown Fig. 23. The range in magnitude was chosen to include the vertical extent (clearly seen in eastern regions of the SMC, refer Subramanian et al. 2017) of the RC regions and also to incorporate the shift towards fainter magnitudes due to interstellar extinction.

The colour cut which separates the RC and RGB is not very well defined in the CMD of the crowded central region. Hence, RGB stars are also included in the initial selection box. Later we will model the RGB density and subtract it from the Hess diagrams to obtain the

RC distribution. Before modelling the RGB stars, the RC region in the CMD is cleaned by removing the closest matching star corresponding to the colour and magnitude of each foreground Milky Way (MW) star from the TRILEGAL MW stellar population model (Girardi et al. 2005) which includes a model of extinction in the Galaxy. The cleaned CMD region is then corrected for foreground and internal extinction using the extinction map of Rubele et al. (2015).

We performed a careful analysis to remove RGB contamination in order to analyse the RC LF. Tile SMC 4_3 was divided into four sub-regions. The RC stars in each sub-region were identified and Hess diagrams corresponding to the RC region were constructed. The bin sizes in colour and magnitude were 0.01 mag and 0.04 mag, respectively. The separation of the RC from the RGB based on their colour may not be reliable for the fainter magnitudes and we restrict the analysis of luminosity function up to $K_{s,0} = 18$ mag. This fainter magnitude cut off of $K_{s,0} = 18$ mag is ∼0.6–0.7 mag lower than the peak of the Classical RC. In a given magnitude bin (along a row in the Hess diagram), we performed a double Gaussian profile fit to the colour distribution and obtained the peak colours corresponding to the RC and the RGB distributions. This was repeated for all magnitude bins, over the magnitude range of 16.0–18.0 mag in the $K_{s,0}$ band. After this first step, we performed a linear least-squares fit to the RGB colours with the magnitude corresponding to each bin. Again, the first step was repeated with the constraint that the colour of the RGB corresponding to each magnitude is the same as that obtained from the linear fit. From the resultant Gaussian parameters corresponding to the RGB, we obtained a model for the RGB density distribution. This model was subtracted from the Hess diagram and a clean RC distribution is obtained. These steps are demonstrated in Fig. 24.

The figure clearly demonstrates that the contamination from the RGB stars is minimal in the final RC distribution. The final RC density distribution is summed in the colour range $0.55 \leq (Y - K_s)_0 \leq 0.85$ mag to obtain the LF of the RC stars. The resultant LF is initially modelled with a Gaussian profile, representing the RC, and a quadratic polynomial term to include the presence of any remaining RGB component. The fit improves by 30% when an additional broad component is included. The LF and the profile fits are shown in the lower-right panel of Fig. 24. The narrow component represents the Classical RC. The broad component could be due to the presence of young stars and/or the intrinsic line of sight depth of the SMC central regions. If the broad component is due to the presence of young stars, further division of the sub-regions into smaller regions would suppress this component owing to the smaller fraction of young stars compared to Classical RC. On the other hand, if this component is due to the intrinsic line of sight depth of the SMC then it would still show up. Earlier studies of the SMC central regions (using the RC as well as other tracers such as RR Lyrae stars, Kapakos and Hatzidimitriou 2012) show large line of sight depth in the central region. So the observed broad component could be due to the line of sight depth effect. In any case, for distance estimation we are only interested in the peak of the Classical RC, which is dominant and represented by the narrow component.

Thus, the method described above effectively removes the contamination of RGB stars, even in the crowded fields, and provides a clean RC distribution to estimate the apparent magnitude of Classical RC. The final step in distance estimation is to apply the population correction term and compare with absolute RC magnitude of Hipparcos sample.

Subramanian et al. (2017) applied the above procedure to the NIR data of ∼20 $\deg^2$ region of the SMC and estimated a mean $K_{s,0}$ band peak magnitude corresponding to Classical RC as 17.36 ± 0.01 mag. This magnitude is used to estimate the distance modulus to the central SMC by using the absolute magnitude of the RC stars provided by Laney et al. (2012). Using high-precision observations of solar neighbourhood RC stars, they provide the absolute RC magnitudes in the J, H and K bands in 2MASS system. We converted them into

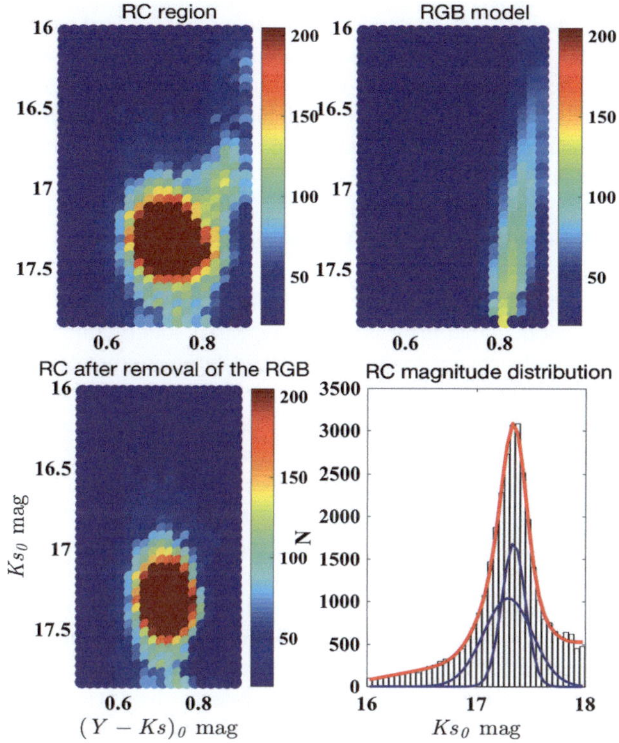

Fig. 24 Steps involved in deriving the luminosity function of the RC stars and the profile fitting to the distribution for one sub-field of tile SMC 4_3. The *top-left* and *top-right* plots show the reddening-corrected and MW-foreground-subtracted RC region and the RGB model obtained for the sub-field respectively. The *bottom-left* and *bottom-right* plots show the RGB subtracted RC region and the luminosity function of the RC stars with the profile fits respectively. The total fit (*thick-red line*) to the distribution and the separate components (*thin-blue line*) are also shown in the *bottom-right plot*

the VISTA K_S system using the transformation relations provided by Rubele et al. (2015). The absolute magnitude of RC stars in the VISTA K_S band, $M_{Ks} = -1.604 \pm 0.015$ mag. As indicated earlier, the absolute magnitudes of RC stars in the solar neighbourhood and in the SMC are expected to be different owing to the differences in metallicity, age and star formation rate between the two regions. Salaris and Girardi (2002) estimated this correction term in the K band to be -0.07 mag. We applied this correction and estimated the mean distance modulus to the SMC as 18.89 ± 0.04 mag.

Our estimate is higher than the previous estimates of SMC distance based on RC LF in optical bands, including those which incorporated population corrections (Cole 1998). But it is lower than the recent accurate estimates of distance to the SMC (18.96 ± 0.02 mag, Graczyk et al. 2014, de Grijs and Bono 2015). This difference could be mainly due to the inaccuracy of the population correction term. As suggested by Pietrzyński et al. (2003) the population corrections calculated from models are not accurate enough for high-precision distance measurements. Better models and accurate star formation history estimates of the observed galaxy are essential to determine the population correction term and hence use of RC as accurate distance indicators.

Now we have stellar population models (such as PARSEC Bressan et al. 2012) with improved knowledge of opacities, solar abundance and microphysics. Better estimates of star formation history of nearby galaxies, using ground as well as space based deep observations, are now available. Gaia parallaxes and stellar parameters will improve the absolute calibration of the RC. Better estimates of population correction, combined with more accurate absolute RC magnitude will provide better results in future to use RC as a high-precision distance indicator.

The best way to use RC as an accurate standard candle would be to perform complete CMD fitting procedure using improved and complete set of models. This fitting procedure will be naturally driven by the dominant intermediate-age population in the CMD and hence, the RC. This is not a simple procedure and for galaxies like the SMC, which has considerable amount of depth along the line of sight, requires a distance distribution instead of discrete distance grid for better results.

4.2 Selection of Milky Way Field RC Stars

For Galactic centre or the Local Group of galaxies, the advantage of RC stars as distance indicators is that there is a striking, easily identifiable feature in the CMD of stars in those distant objects due to the presence of a large number of RC stars at nearly the same distance. However, it is not easy to identify individual RC stars from the numerous field stars of the Milky Way. Field RC stars are spread over a wide range of distances and there is no apparent over-density in the CMD of field stars that can be used to identify the RC members.

Fortunately, high-resolution spectroscopic analysis of nearby RC stars show that they distribute in a relative "small box" in the $T_{\rm eff}$–$\log g$ diagram (i.e. HR diagram) of $4800 \leq T_{\rm eff} \leq 5200$ K and $2.0 \leq \log g \leq 3.0$ (e.g. Puzeras et al. 2010). This is simply because $T_{\rm eff}$ is an excellent proxy of stellar colour whereas $\log g$ is sensitive to the stellar absolute luminosity. Recently, with values of $T_{\rm eff}$ and $\log g$ available from large-scale spectroscopic surveys for large numbers of field stars, it has become feasible to select a large number of field RC candidates. RC candidates thus selected have been widely used to study the metallicity gradients and stellar kinematics of the Galactic disk(s) (e.g. Bilir et al. 2012, Siebert et al. 2012, Williams et al. 2013, Bienaymé et al. 2014).

However, stars falling inside that "small box" are not purely RC stars and have significant (∼60%; Williams et al. 2013) contamination from RGB. The differences in absolute magnitudes between RC and RGB stars can be larger than 1 mag and this can lead to large systematic errors in distances of the selected RC sample stars. More recently, a new method has been proposed by Bovy et al. (2014) (hereafter B14) to select a clean RC sample from large-scale spectroscopic data. The method first separates the RC-like and RGB stars using cuts on a metallicity-dependent $T_{\rm eff}$–$\log g$ diagram, assisted by theoretical stellar isochrones and calibrated using *Kepler* high quality asteroseismic $\log g$ (e.g. Creevey et al. 2013), as well as high precision stellar atmospheric parameters from the APOGEE survey (Pinsonneault et al. 2014). Secondary RC stars are then removed from the RC-like stars via cuts on a metallicity (Z)-colour $(J - K_{\rm s})_0$ diagram. The expected purity of the final RC sample is ≥93% and the typical distance errors are within 5–10%. The key point of this new method in selecting a clean RC sample is the surface gravity $\log g$.

Unfortunately, it is difficult to apply this method to the LAMOST spectroscopic survey of the galactic anti-centre (LSS-GAC, Liu et al. 2014, Yuan et al. 2015) data given the relatively large systematic plus random errors (∼0.2–0.4 dex) of $\log g$ estimates delivered by the LAMOST Stellar Parameter Pipeline at Peking University (LSP3; Xiang et al. 2015), as shown by a recent comparison of LSP3 and asteroseismic $\log g$ values for common objects in the LAMOST–*Kepler* fields (Ren et al. 2016). To apply the similar method developed by B14 to select pure RC stars from LSS-GAC data, we have improved $\log g$ measurements using a Kernel Principal Component Analysis (KPCA, Schölkopf et al. 1998) method trained by accurate asteroseismic data from the LAMOST–Kepler fields (Huang et al. 2015, hereafter H15). First, the training sample is constructed by cross-matching our sub-sample of about 50 000 stars in the LAMOST–Kepler fields with the currently available largest asteroseismic $\log g$ sample (∼16 000 stars) from Huber et al. (2014). In total, 3562 common

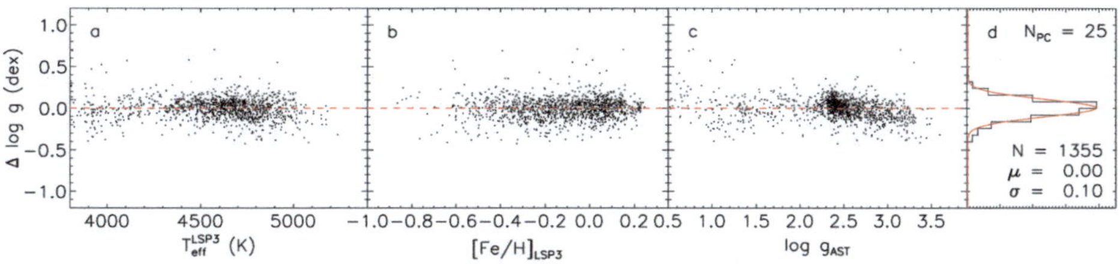

Fig. 25 Distributions of values of residuals, $\log g_{\rm KPCA} - \log g_{\rm AST}$, of the training sample as a function of LSP3 $T_{\rm eff}$ and [Fe/H] (panels **a** and **b**, respectively) and of asteroseismic $\log g$ (panel **c**). Panel (**d**) shows a histogram distribution of the residuals (*black line*). Also over plotted in *red* is a Gaussian fit to the distribution. The mean $\mu = 0.00$ dex and dispersion $\sigma = 0.10$ dex of the fit, as well as the number of stars used, $N = 1355$, are marked. For the current KPCA analysis, $N_{\rm PC} = 25$ is assumed. See the text for details. The figure is taken from H15

sources with SNR ≥ 10 are identified. To further select red giant stars and ensure high quality of the stellar parameters yielded by LSP3, we take a sub-sample of 1355 stars from the 3562 common sources. Secondly, we apply the KPCA method to the training sample to construct the relations between seismic $\log g$ and the LAMOST spectra (see Fig. 25). Finally, We apply the constructed relations to the LAMOST spectra and re-determine $\log g$ values for all LSS-GAC DR2 red giants. The accuracy of the newly estimated KPCA $\log g$ is as high as 0.1–0.15 dex and has no significant systematic error. With the high accurate, newly estimated KPCA $\log g$, we can apply the similar technique developed by B14 to select main RC from LAMOST data. To show how well the technique works for LAMOST data, we present the two steps of the selections as mentioned above in Figs. 26 and 27.

4.2.1 The LSS-GAC RC Sample and Its Applications

With $T_{\rm eff}$, [Fe/H] estimated with the LSP3, $\log g$, derived with the KPCA method and $(J - K_{\rm s})_0$ calculated from 2MASS photometry[1] after corrected for extinction as estimated with the 'star pair' method (Yuan et al. 2015), we apply the similar technique developed by B14 to LSS-GAC DR2 red giant sample and obtain a clean RC sample of over 0.11 million stars of S/N (4650 Å) ≥ 10 (see Fig. 28). The distances of these RC stars are derived using the recent calibration, $M_{K_{\rm s}} = -1.61$ mag, for nearby RC sample (Laney et al. 2012). Since the intrinsic scatter of absolute magnitudes of RC stars is estimated to be within 0.1 mag, the distances derived are expected to have uncertainties no more than 5–10%, given a typical photometric error of ~0.05 mag in $K_{\rm s}$-band and an extinction error of ~0.04 mag in $E(B - V)$ (see Yuan et al. 2015). Proper motions of the sample stars are taken from the U.S. Naval Observatory CCD Astrograph Catalog (UCAC4, Zacharias et al. 2013) and PPMXL (Roeser et al. 2010) catalogues Given the high distance precision and large spatial coverage (Fig. 27) of the current RC star sample, it is very useful to tackle a variety of problems with regard to the Galactic chemistry (e.g. stellar metallicity gradients of Galactic disk, listed below), structures and dynamics (e.g. Galactic rotation curve, listed below). The sample is publicly available at website: (http://162.105.156.249/site/RC_Sample).

Galactic rotation curve: Using the pure RC sample selected from LSS-GAC and the APOGEE surveys, combined with a sample of halo K giants selected from the SDSS/SEGUE, the Galactic rotation curve (RC) between 8 and 100 kpc in Galactocentric radius has been constructed (Huang et al. 2016). The newly constructed RC has a generally flat

[1]Only stars with *ph_qual* flagged as 'A' in both J and $K_{\rm s}$ bands are included.

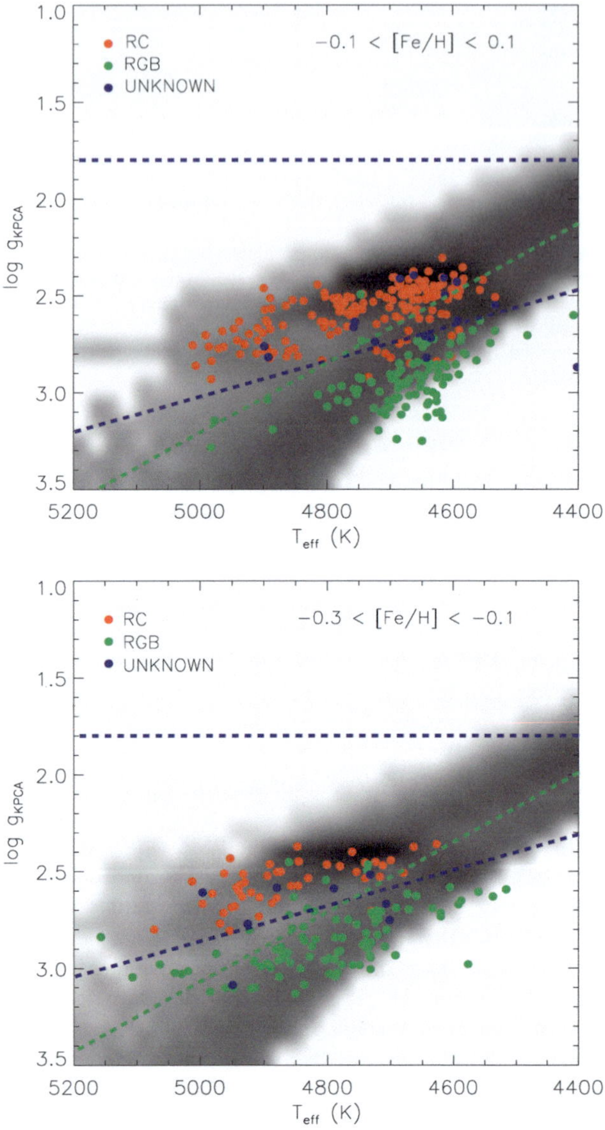

Fig. 26 Distribution of stars on the $T_{\rm eff}$–$\log g$, HR diagram predicted by the PARSEC stellar evolution model (Bressan et al. 2012), compared to that of the asteroseismic sample with evolution stages classified (*coloured points*, Stello et al. 2013, for two metallicity bins as marked. Values of $T_{\rm eff}$ and [Fe/H] of the asteroseismic sample stars are from LSP3 and those of $\log g$ from the KPCA method. *Red*, *magenta* and *blue points* represent stars of RC, RGB and unknown evolutionary stages, as classified by Stello et al. (2013), respectively. The *blue dashed lines* represent the cuts that separate RC stars and the less luminous RGB stars as given in Eqs. (1) and (2) in H15. The *green dashed lines* give the cuts adopted by B14. The figure is taken from H15

value of 240 km s^{-1} within a Galactocentric radius r of 25 kpc and then decreases steadily to 150 km s^{-1} at $r \sim 100$ kpc. On top of this overall trend, the RC exhibits two prominent localised dips, one at $r \sim 11$ kpc and another at $r \sim 19$ kpc. The dips could be explained by assuming the existence of two massive (dark) matter rings in the Galactic plane. From the newly constructed RC, combined with other data, we have built a mass model of the Galaxy, yielding a virial mass of the Milky Way's dark matter halo of $0.90^{+0.07}_{-0.08} \times 10^{12}$ $M_\odot$ and a local dark matter density, $\rho_{\odot,\rm dm} = 0.32^{+0.02}_{-0.02}$ GeV cm^{-3}.

Metallicity gradients of the Galactic disk: We also have determined the radial and vertical metallicity gradients of the Galactic disk(s), again using the LSS-GAC pure RC sample (H15). Our analysis shows that, near the solar circle ($7 \leq R \leq 11.5$ kpc), the radial metal-

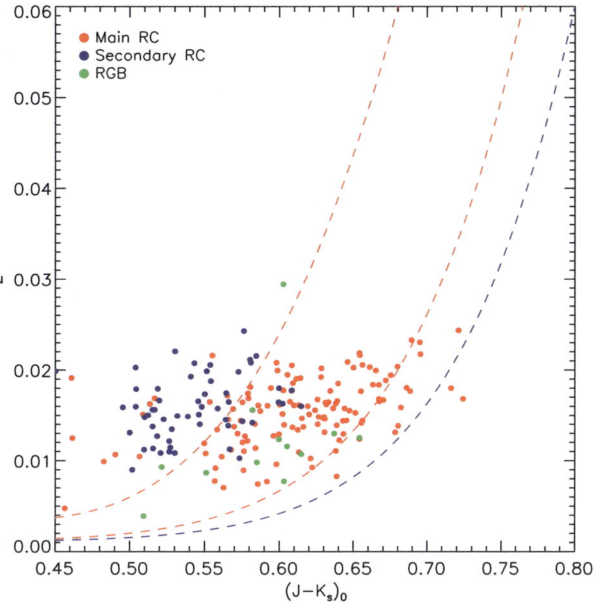

Fig. 27 Distribution of RC-like stars in the colour $(J - K_S)_0$–Z plane, using 2MASS photometric measurements and LSP3 metallicities. *Dots* in *red*, *blue* and *green* represent main RC, secondary RC and RGB stars, respectively. The *red dashed lines* are developed by B14 to eliminate the contamination of secondary RC stars. The *blue dashed line* is the revised cut developed in H15 in replacement of the lower *red dashed line*. The figure is taken from H15

licity gradient exhibits a negative slope that flattens with increasing $|Z|$. In contrast, in the outer disk ($11.5 < R \leq 14$ kpc), the radial metallicity gradient has an essentially constant, much less steep slope at all heights from the mid-plane, suggesting that the outer disk may have experienced an evolutionary path different from that of the inner disk. In addition, the vertical metallicity gradient is found to flatten roughly with increasing R. The flattening of the gradient in the lower disk ($0 \leq |Z| \leq 1$ kpc) is much quicker than in the upper disk ($1 < |Z| \leq 3$ kpc).

The parallaxes and stellar parameters of Galaxy RC sample from Gaia will improve the zero-point calibration. Combing the improved zero-point calibration with the results from on-going and future asterosesmic surveys such as PLAnetary Transits and Oscillations of stars (Rauer et al. 2014), the Transiting Exoplanet Survey Satellite (Ricker et al. 2014), and the second phase of Kepler (K2; Stello et al. 2015) will provide large sample of RC stars to perform Galactic archaeology.

5 Final Remarks

We provided an overview of the properties of young and intermediate-age stars, such as Cepheids, Mira stars and RC stars in the context of their application as distance indicators. Below given are the salient points.

- Cepheids
  - *NIR Cepheid distances*
    - NIR PW relations allow us to derive Cepheid distances with the accuracy to meet the precision of ongoing cosmological experiments. They are less affected by extinction and metallicity effects. By using NIR PW relations, Inno et al. (2013) obtained the most precise estimate of distances to the Magellanic Clouds based on Cepheids, with random errors at 1% level and systematics at 5% level. Independent calibrations from Gaia and HST will play a crucial role to reach a better accuracy.

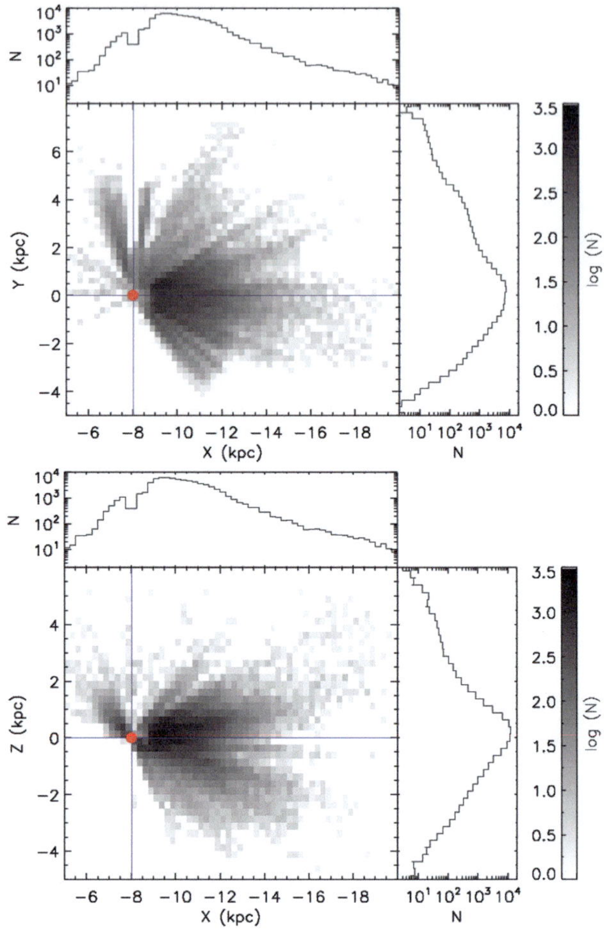

Fig. 28 Greyscale number density distribution of the 0.11 million LSS-GAC RC sample stars in the X–Y (*left panel*) and X–Z (*right panel*) planes. The Sun is located at $(X, Y, Z) = (-8.0, 0.0, 0.0)$ kpc. The stars are binned by 0.25×0.25 kpc^2 in both diagrams. The densities are shown on a logarithmic scale. Histogram distributions along the X, Y and Z axes are also plotted

- New NIR light-curve templates can provide accurate mean magnitudes from a few or even one observation, allowing a significant saving of observing time and an optimal usage of data already available.

– *MIR Cepheid distances*

- MIR studies of Cepheids lead to significant advantages with respect to optical wavelengths. The L band appears to be ideal wavelength range for these measurements, lacking the presence of variable CO absorption that is instead affecting the M band.
- The CHP project (Freedman et al. 2011) has demonstrated that the 3.6 μm Cepheid PL relation can be effectively used to measure the Hubble constant and other cosmological parameters with a precision comparable to the accuracy (better than ∼3%) in Planck's CMB and BAO analysis (Freedman et al. 2012).
- Accurate parallaxes for Galactic Cepheids expected with the final Gaia data release, combined with future JWST L band observations of Cepheids in host galaxies of Type Ia Supernovae will enable 1% measurements of H_0 as well as competitive measurements of other cosmological parameters.

  - *Linearity of PL relation*
    - The statistical analysis of non-linearities in the LMC PL relations provides evidence of a break in both fundamental mode and first-overtone mode relations.
    - The observed non-linearities suggest a correlation with sharp changes in the light curve structure of Cepheids and needs a theoretical investigation to look for additional resonance features.
    - The effect of these non-linearities in the slope of HST-based PL relations is minimal. Hence, does not have any statistically significant effect on the distance scale or the value of Hubble constant. But they will have a significant impact on the accuracies of distance scale with more precise PL relations, for example from James Webb Space Telescope.
  - *Metallicity effects in PL relation*
    - NIR Surface Brightness method which is a geometrical method, enables us to independently estimate the distances to nearby Cepheids.
    - The PL relations constructed from these independent distance estimates for Cepheids in Milky Way, LMC and SMC are ideal to investigate the effect of metallicity in Cepheid PL relations.
    - Preliminary results indicate that the effect of metallicity, especially in K band PL relation, is very small.
- Mira variables
  - For better calibration of Galactic PL relation of Mira variables and to understand their physical properties, very precise parallax measurements are essential. We present the strategy and initial results of an on-going VLBI astrometric survey of Galactic Miras.
  - The initial results show consistent PL relations for Mira variables between our Galaxy and the LMC.
- Red Clump stars
  - *Galactic Bulge and External galaxies*
    - We present a simple method to remove the contamination of RGB stars from the RC LF function in crowded fields and hence provide better measurements of the apparent magnitude of Classical RC. This technique is used to estimate the mean distance modulus of the SMC as 18.89 ± 0.04 mag (including a population correction term). We note that the population correction terms obtained from models may not be accurate enough to use RC as high precision absolute distance indicator.
    - New stellar population models, improved star formation history measurements of the galaxies and improved calibration of absolute RC magnitude from Gaia will all together provide better results to use RC as accurate distance indicator.
    - Though not simple, another way to use RC as accurate standard candle is to perform complete CMD fitting procedure using improved and complete set of stellar population models.
  - *Field RC stars in Milky Way*
    - Recent spectroscopic surveys allow us to select a large number of field RC stars in Milky Way. A technique to identify a clean sample of RC stars from field stars in Milky Way using LAMOST spectroscopic survey is presented. This sample is used to derive the metallicity gradient and rotation curve of our Galaxy.
    - The parallaxes and stellar parameters from Gaia, combined with the results from ongoing and future asterosesmic surveys will provide a large sample of RC stars to perform Galactic archaeology.

Acknowledgements SS acknowledges research funding support from Chinese Postdoctoral Science Foundation (grant number 2016M590013). YH thanks the support by the by the National Key Basic Research Program of China 2014CB845700, the National Natural Science Foundation of China 11473001 and the LAMOST FELLOWSHIP. LI acknowledge support from the Sonderforschungsbereich SFB 881 *The Milky Way System* (subproject A3) of the German Research Foundation (DFG).

References

C. Alcock et al., The MACHO project: microlensing detection efficiency. Astrophys. J. Suppl. Ser. **136**, 439 (2001)

R.I. Anderson et al., On the effect of rotation on populations of classical Cepheids. I. Predictions at solar metallicity. Astron. Astrophys. **564**, A100 (2014)

R.I. Anderson et al., Revealing δ Cephei's secret companion and intriguing past. Astrophys. J. **804**, 144 (2015)

S.M. Andrievsky, R.P. Martin, V.V. Kovtyukh, S.A. Korotin, J.R.D. Lépine, Oxygen, α-element and iron abundance distributions in the inner part of the Galactic thin disc—II. Mon. Not. R. Astron. Soc. **461**, 4256 (2016)

B. Aringer, S. Höfner, G. Wiedemann, J. Hron, U.G. Jørgensen, H.U. Käufl, W. Windsteig, SiO rotation-vibration bands in cool giants II. The behaviour of SiO bands in AGB stars. Astron. Astrophys. **342**, 799 (1999)

Y. Asaki, S. Deguchi, H. Imai et al., Distance and proper motion measurement of the Red Supergiant, S Persei, with VLBI H2O maser astrometry. Astrophys. J. **721**, 267 (2010)

P. Barmby et al., Galactic Cepheids with Spitzer. II. Search for extended infrared emission. Astron. J. **141**, 42 (2011)

T.G. Barnes, D.S. Evans, Stellar angular diameters and visual surface brightness—I. Late spectral types. Mon. Not. R. Astron. Soc. **174**, 489 (1976)

J.P. Beaulieu et al., EROS variable stars: fundamental-mode and first-overtone Cepheids in the bar of the Large Magellanic Cloud. Astron. Astrophys. **303**, 137 (1995)

G.F. Benedict et al., Hubble Space Telescope fine guidance sensor parallaxes of Galactic Cepheid variable stars: period-luminosity relations. Astron. J. **133**, 1810 (2007)

A. Bhardwaj et al., On the variation of Fourier parameters for Galactic and LMC Cepheids at optical, near-infrared and mid-infrared wavelengths. Mon. Not. R. Astron. Soc. **447**, 3342 (2015)

A. Bhardwaj et al., Large Magellanic Cloud near-infrared synoptic survey—III. A statistical study of non-linearity in the Leavitt laws. Mon. Not. R. Astron. Soc. **457**, 1644 (2016a)

A. Bhardwaj et al., Period-luminosity relations derived from the OGLE-III first-overtone mode Cepheids in the Magellanic Clouds. Mon. Not. R. Astron. Soc. **458**, 3705 (2016b)

O. Bienaymé, B. Famaey, A. Siebert et al., Weighing the local dark matter with RAVE Red Clump stars. Astron. Astrophys. **571**, 92 (2014)

S. Bilir, S. Karaali, S. Ak et al., Local stellar kinematics from RAVE data—III. Radial and vertical metallicity gradients based on Red Clump stars. Mon. Not. R. Astron. Soc. **421**, 3362 (2012)

G. Bono et al., Theoretical models for classical Cepheids. II. Period-luminosity, period-color, and period-luminosity-color relations. Astrophys. J. **512**, 711 (1999a)

G. Bono et al., Intermediate-mass star models with different helium and metal contents. Astrophys. J. **543**, 955 (2000b)

G. Bono et al., Insights into the Cepheid distance scale. Astrophys. J. **715**, 277 (2010)

G. Bono, M. Marconi, R.F. Stellingwerf, Classical Cepheid pulsation models. I. Physical structure. Astrophys. J. Suppl. Ser. **122**, 167 (1999b)

G. Bono, M. Marconi, R.F. Stellingwerf, Classical Cepheid pulsation models—VI. The Hertzsprung progression. Astron. Astrophys. **360**, 245 (2000a)

J. Bovy, D.L. Nidever, H.-W. Rix et al., The APOGEE red-clump catalog: precise distances, velocities, and high-resolution elemental abundances over a large area of the Milky Way's disk. Astrophys. J. **790**, 127 (2014)

A. Bressan, P. Marigo et al., PARSEC: stellar tracks and isochrones with the PAdova and TRieste Stellar Evolution Code. Mon. Not. R. Astron. Soc. **427**, 127 (2012)

R.D. Cannon, Red giants in old open clusters. Mon. Not. R. Astron. Soc. **150**, 111 (1970)

F. Caputo et al., Pulsation and evolutionary masses of classical Cepheids. I. Milky Way variables. Astrophys. J. **629**, 1021 (2005)

J.A. Cardelli, G.C. Clayton, J.S. Mathis, The relationship between infrared, optical, and ultraviolet extinction. Astrophys. J. **345**, 245 (1989)

N.L. Chapman et al., The mid-infrared extinction law in the Ophiuchus, Perseus, and Serpens molecular clouds. Astrophys. J. **690**, 496–511 (2009)

Y.K. Choi, T. Hirota, M. Honma et al., Astrometry of H2O masers in nearby star-forming regions with VERA. II. SVS13 in NGC1333. Publ. Astron. Soc. Jpn. **60**, 1007 (2008)

M.-R.L. Cioni et al., The VMC survey. I. Strategy and first data. Astron. Astrophys. **527**, A116 (2011)

A.A. Cole, Age, Metallicity, and the distance to the Magellanic Clouds from Red Clump stars. Astrophys. J. **500**, 137 (1998)

O.L. Creevey, F. Thévenin, S. Basu et al., A large sample of calibration stars for Gaia: log g from Kepler and CoRoT fields. Mon. Not. R. Astron. Soc. **431**, 2419 (2013)

A.J. Cuesta et al., Calibrating the cosmic distance scale ladder: the role of the sound-horizon scale and the local expansion rate as distance anchors. Mon. Not. R. Astron. Soc. **448**, 3463 (2015)

R.M. Cutri, M.F. Skrutskie, S. van Dyk et al., *VizieR Online Data Catalog 2MASS All-Sky Catalog of Point Sources (Cutri+ 2003)* (2003), p. 2246

A.K. Dambis, L.N. Berdnikov, Y.N. Efremov, A.Y. Kniazev, A.S. Rastorguev, E.V. Glushkova, V.V. Kravtsov, D.G. Turner, D.J. Majaess, R. Sefako, Classical Cepheids and the spiral structure of the Milky Way. Astron. Lett. **41**, 489 (2015)

R. de Grijs, G. Bono, Clustering of local group distances: publication bias or correlated measurements? III. The Small Magellanic Cloud. Astron. J. **149**, 179 (2015)

R. de Grijs, J.E. Wicker, G. Bono, Clustering of local group distances: publication bias or correlated measurements? I. The Large Magellanic Cloud. Astron. J. **147**, 122 (2014)

G. De Marchi et al., Hubble tarantula treasury project—IV. The extinction law. Mon. Not. R. Astron. Soc. **455**, 4373 (2016)

W. Dehnen, J.J. Binney, Local stellar kinematics from HIPPARCOS data. Mon. Not. R. Astron. Soc. **298**, 387 (1998)

G.P. di Benedetto, Towards a fundamental calibration of stellar parameters of A, F, G, K dwarfs and giants. Astron. Astrophys. **339**, 858 (1998)

J.R. Ducati, VizieR Online Data Catalog: Catalogue of Stellar Photometry in Johnson's 11-color system. VizieR Online Data Catalog (2002), p. 2237

C. Dvorkin et al., Neutrinos help reconcile Planck measurements with both the early and local universe. Phys. Rev. D **90**, 083503 (2014)

ESA, *The HIPPARCOS and TYCHO Catalogues. Astrometric and Photometric Star Catalogues Derived from the ESA HIPPARCOS Space Astrometry Mission*. ESA Special Publication, vol. 1200 (1997)

N.R. Evans et al., Direct detection of the close companion of polaris with the Hubble Space Telescope. Astron. J. **136**, 1137 (2008)

N.E. Evans et al., Binary Cepheids: separations and mass ratios in 5 $M_{\odot}$ binaries. Astron. J. **146**, 93 (2013)

G.G. Fazio et al., The Infrared Array Camera (IRAC) for the Spitzer Space Telescope. Astrophys. J. Suppl. Ser. **154**, 10 (2004)

M.W. Feast, R.M. Catchpole, The Cepheid period-luminosity zero-point from HIPPARCOS trigonometrical parallaxes. Mon. Not. R. Astron. Soc. **286**, L1 (1997)

M.W. Feast, I.S. Glass, P.A. Whitelock, R.M. Catchpole, A period-luminosity-colour relation for Mira variables. Mon. Not. R. Astron. Soc. **241**, 375 (1989)

M.W. Feast, P.A. Whitelock, Hipparcos parallaxes for Mira-like long-period variables. Mon. Not. R. Astron. Soc. **317**, 460 (2000)

G. Fiorentino et al., Classical Cepheid pulsation models. XI. Effects of convection and chemical composition on the period-luminosity and period-Wesenheit relations. Astron. Astrophys. **476**, 863 (2007)

G. Fiorentino, I. Musella, M. Marconi, Cepheid theoretical models and observations in HST/WFC3 filters: the effect on the Hubble constant H_0. Mon. Not. R. Astron. Soc. **434**, 2866 (2013)

K.M. Flaherty et al., Infrared extinction toward nearby star-forming regions. Astrophys. J. **663**, 1069 (2007)

P. Fouqué et al., A new calibration of Galactic Cepheid period-luminosity relations from B to K bands, and a comparison to LMC relations. Astron. Astrophys. **476**, 73 (2007)

P. Fouque, W.P. Gieren, An improved calibration of Cepheid visual and infrared surface brightness relations from accurate angular diameter measurements of cool giants and supergiants. Astron. Astrophys. **320**, 799 (1997)

W.L. Freedman et al., The Carnegie Hubble program. Astron. J. **142**, 192 (2011)

W.L. Freedman et al., Carnegie Hubble program: a mid-infrared calibration of the Hubble Constant. Astrophys. J. **758**, 24 (2012)

W.L. Freedman, G.R. Grieve, B.F. Madore, BVRI photometry of extragalactic Cepheids and new insights for the distance scale. Astrophys. J. Suppl. Ser. **59**, 311 (1985)

W.L. Freedman, B.F. Madore, The Hubble Constant. Annu. Rev. Astron. Astrophys. **48**, 673 (2010a)

W.L. Freedman, B.F. Madore, B.K. Gibson, L. Ferrarese, D.D. Kelson, S. Sakai, J.R. Mould, R.C. Kennicutt Jr., H.C. Ford, J.A. Graham, J.P. Huchra, S.M.G. Hughes, G.D. Illingworth, L.M. Macri, P.B. Stetson, Final results from the Hubble Space Telescope key project to measure the Hubble Constant. Astrophys. J. **553**, 47 (2001)

K. Fricke, R.S. Stobie, P.A. Strittmatter, The masses of Cepheid variables. Astrophys. J. **171**, 593 (1972)

S.D. Friedman et al., Pairwise correlations of eight strong DIBs and N(H), N(H2), and E(B-V). Bull. Am. Astron. Soc. **43**, 129.10 (2011)

A. Gallenne et al., Spatially extended emission around the Cepheid RS Puppis in near-infrared hydrogen lines. Adaptive optics imaging with VLT/NACO. Astron. Astrophys. **527**, A51 (2011)

A. García-Varela, B.E. Sabogal, M.C. Ramírez-Tannus, A study on the universality and linearity of the Leavitt law in the LMC and SMC galaxies. Mon. Not. R. Astron. Soc. **431**, 2278 (2013)

J.P. Gardner, The James Webb Space Telescope: extending the science. SPIE **8442**, 844228 (2012)

L. Girardi, Red Clump Stars. Annu. Rev. Astron. Astrophys. **54**, 95 (2016)

L. Girardi, M.A.T. Groenewegen, E. Hatziminaoglou, L. da Costa, Star counts in the Galaxy. Simulating from very deep to very shallow photometric surveys with the TRILEGAL code. Mon. Not. R. Astron. Soc. **436**, 895 (2005)

L. Girardi, M.A.T. Groenewegen, A. Weiss, M. Salaris, Fine structure of the red giant clump from HIPPARCOS data, and distance determinations based on its mean magnitude. Mon. Not. R. Astron. Soc. **301**, 149 (1998)

L. Girardi, M. Salaris, Population effects on the red giant clump absolute magnitude and distance determinations to nearby galaxies. Mon. Not. R. Astron. Soc. **323**, 109 (2001)

D. Graczyk, G. Pietrzyński, I.B. Thompson et al., The Araucaria Project. The distance to the Small Magellanic Cloud from late-type eclipsing binaries. Astrophys. J. **780**, 59 (2014)

G.M. Green et al., A three-dimensional map of Milky Way dust. Astrophys. J. **810**, 25 (2015)

C.A. Haniff, M. Scholz, P.G. Tuthill, New diameter measurements of 10 Mira variables—implications for effective temperatures atmospheric structure and pulsation modes. Mon. Not. R. Astron. Soc. **276**, 640 (1995)

R. Haschke, E.K. Grebel, S. Duffau, New optical reddening maps of the Large and Small Magellanic Clouds. Astron. J. **141**, 158 (2011)

R. Haschke, E.K. Grebel, S. Duffau, Three-dimensional maps of the Magellanic Clouds using RR Lyrae stars and Cepheids. I. The Large Magellanic Cloud. Astron. J. **144**, 106 (2012)

Y. Huang, X.-W. Liu, H.-B. Yuan et al., The Milky Way's rotation curve out to 100 kpc and its constraint on the Galactic mass distribution. Mon. Not. R. Astron. Soc. **463**, 2623 (2016)

Y. Huang, X.-W. Liu, H.-W. Zhang et al., On the metallicity gradients of the Galactic disk as revealed by LSS-GAC Red Clump stars. Res. Astron. Astrophys. **15**, 1240 (2015)

E.P. Hubble, A spiral nebula as a stellar system, Messier 31. Astrophys. J. **69**, 103 (1929a)

E. Hubble, A relation between distance and radial velocity among extra-galactic nebulae. Proc. Natl. Acad. Sci. **15**, 168 (1929b)

D. Huber, V. Silva Aguirre, J.M. Matthews et al., Revised stellar properties of Kepler targets for the quarter 1–16 transit detection run. Astrophys. J. Suppl. Ser. **211**, 2 (2014)

R. Indebetouw et al., The wavelength dependence of interstellar extinction from 1.25 to 8.0 μm using GLIMPSE data. Astrophys. J. **619**, 931 (2005)

L. Inno et al., On the distance of the Magellanic Clouds using Cepheid NIR and optical-NIR period-Wesenheit relations. Astrophys. J. **764**, 84 (2013)

L. Inno et al., New NIR light-curve templates for classical Cepheids. Astron. Astrophys. **576**, A30 (2015)

L. Inno et al., The panchromatic view of the Magellanic Clouds from classical Cepheids. I. Distance, reddening and geometry of the Large Magellanic Cloud disk. Astrophys. J. **832**, 1761 (2016)

Y. Ita, T. Tanabé, N. Matsunaga et al., Variable stars in the Magellanic Clouds: results from OGLE and SIRIUS. Mon. Not. R. Astron. Soc. **347**, 720 (2004)

A.M. Jacyszyn-Dobrzeniecka et al., OGLE-ing the Magellanic System: three-dimensional structure of the clouds and the bridge using classical Cepheids. Acta Astron. **66**, 149 (2016)

R.V. Jones, B.W. Carney, J.P. Fulbright, Template K light curves for RR Lyrae Stars. Publ. Astron. Soc. Pac. **108**, 877 (1996)

T. Kamezaki, T. Kurayama, A. Nakagawa et al., Annual parallax measurements of a semi-regular variable star, RW Leporis. Publ. Astron. Soc. Jpn. **66**, 107 (2014)

T. Kamezaki, A. Nakagawa, T. Omodaka et al., VLBI astrometry of the semiregular variable RX Bootis. Publ. Astron. Soc. Jpn. **64**, 7 (2012)

T. Kamezaki, A. Nakagawa, T. Omodaka et al., Annual parallax measurements of a Mira variable star, U Lyncis. Publ. Astron. Soc. Jpn. **68**, 71 (2016a)

T. Kamezaki, A. Nakagawa, T. Omodaka et al., Annual parallax and a dimming event of a Mira variable star, FV Bootis. Publ. Astron. Soc. Jpn. **68**, 75 (2016b)

R. Kamohara, V. Bujarrabal, M. Honma et al., VERA observations of SiO maser emission from R Aquarii. Astron. Astrophys. **510**, A69 (2010)
S. Kanbur, C. Ngeow, The physics behind the non-linearity of the Cepheid period-luminosity relation, in *The Three-Dimensional Universe with Gaia*, ed. by C. Turon, K.S. O'Flaherty, M.A.C. Perryman. ESA Special Publication, vol. 576 (2005), p. 691
E. Kapakos, D. Hatzidimitriou, RR Lyrae variables in the Small Magellanic Cloud—II. The extended area: chemical and structural analysis. Mon. Not. R. Astron. Soc. **426**, 2063 (2012)
N. Kawaguchi, T. Sasao, S. Manabe, Dual-beam VLBI techniques for precision astrometry of the VERA project. Proc. SPIE **4015**, 544 (2000)
S.C. Keller, P.R. Wood, Bump Cepheids in the Magellanic Clouds: metallicities, the distances to the LMC and SMC, and the pulsation–evolution mass discrepancy. Astrophys. J. **642**, 834 (2006)
R.C. Kennicutt Jr. et al., The Hubble Space Telescope key project on the extragalactic distance scale. XIII. The metallicity dependence of the Cepheid distance scale. Astrophys. J. **498**, 181 (1998)
P. Kervella et al., Cepheid distances from infrared long-baseline interferometry. III. Calibration of the surface brightness-color relations. Astron. Astrophys. **428**, 587 (2004)
P. Kervella, A. Mérand, A. Gallenne, The circumstellar envelopes of the Cepheids ℓ Carinae and RS Puppis. Comparative study in the infrared with Spitzer, VLT/VISIR, and VLTI/MIDI. Astron. Astrophys. **498**, 425 (2009)
D.E. Knuth, *Seminumerical Algorithms, the Art of Computer Programming*, 2nd edn., vol. 2 (Addison-Wesley, Reading, 1981)
H. Kobayashi et al., VERA project, in *ASP Conference Series*, vol. 306 (2003), 48P
J.F. Koerwer, Large Magellanic Cloud distance and structure from near-infrared red clump observations. Astron. J. **138**, 1 (2009)
T. Kurayama, T. Sasao, H. Kobayashi, Parallax measurements of the Mira-Type Star UX Cygni with phase-referencing VLBI. Astrophys. J. Lett. **627**, L49 (2005)
K. Kusuno, Y. Asaki, H. Imai, T. Oyama, Distance and proper motion measurement of the red supergiant, PZ Cas, in very long baseline interferometry H2O maser astrometry. Astrophys. J. **774**, 107 (2013)
C.D. Laney, M.D. Joner, G. Pietrzyński, A new Large Magellanic Cloud K-band distance from precision measurements of nearby Red Clump stars. Mon. Not. R. Astron. Soc. **419**, 1637 (2012)
C.D. Laney, R.S. Stobie, JHKL observations of galactic Cepheids. Astron. Astrophys. Suppl. Ser. **93**, 93 (1992)
H.S. Leavitt, E.C. Pickering, Periods of 25 Variable Stars in the Small Magellanic Cloud. Circ. - Harv. Coll. Obs. **173**, 1 (1912)
X.-W. Liu, H.-B. Yuan, Z.-Y. Huo et al., LSS-GAC—a LAMOST Spectroscopic Survey of the Galactic Anti-center, in *Proc. IAU Symp. 298, Setting the Scene for Gaia and LAMOST*, ed. by S. Feltzing, G. Zhao, N. Walton, P. Whitelock (Cambridge University Press, Cambridge, 2014), pp. 310–321
R.E. Luck et al., Magellanic Cloud Cepheids—Abundances. Astron. J. **115**, 605 (1998)
L.M. Macri et al., Large Magellanic Cloud near-infrared synoptic survey. I. Cepheid Variables and the Calibration of the Leavitt Law. Astron. J. **149**, 117 (2015)
B.F. Madore, The period-luminosity relation. IV—Intrinsic relations and reddenings for the Large Magellanic Cloud Cepheids. Astrophys. J. **253**, 575 (1982)
M. Marconi et al., On a new theoretical framework for RR Lyrae stars. I. The metallicity dependence. Astrophys. J. **808**, 50 (2015)
M. Marconi, I. Musella, G. Fiorentino, Cepheid pulsation models at varying metallicity and $\Delta y/\Delta z$. Astrophys. J. **632**, 590 (2005)
M. Marengo et al., Theoretical limb darkening for pulsating Cepheids. Astrophys. J. **567**, 1131 (2002)
M. Marengo et al., Theoretical limb darkening for classical Cepheids. II. Corrections for the geometric Baade–Wesselink method. Astrophys. J. **589**, 968 (2003)
M. Marengo et al., Galactic Cepheids with Spitzer. I. Leavitt law and colors. Astrophys. J. **709**, 120 (2010)
M. Marengo et al., An infrared nebula associated with δ Cephei: evidence of mass loss? Astrophys. J. **725**, 2392 (2010)
D.J. Marshall et al., Modelling the Galactic interstellar extinction distribution in three dimensions. Astron. Astrophys. **453**, 635 (2006)
R.P. Martin et al., Oxygen, α-element and iron abundance distributions in the inner part of the Galactic thin disc. Mon. Not. R. Astron. Soc. **449**, 4071 (2015)
N. Matsunaga et al., Three classical Cepheid variable stars in the nuclear bulge of the Milky Way. Nature **477**, 188 (2011)
N. Matsunaga et al., Cepheids and other short-period variables near the Galactic centre. Mon. Not. R. Astron. Soc. **429**, 385 (2013)
N. Matsunaga et al., A lack of classical Cepheids in the inner part of the Galactic disc. Mon. Not. R. Astron. Soc. **462**, 414 (2016)

L.D. Matthews et al., New evidence for mass loss from δ Cephei from H I 21 cm line observations. Astrophys. J. **744**, 53 (2012)
L.D. Matthews, M. Marengo, N.R. Evans, A Search for Mass Loss on the Cepheid Instability Strip using HI 21-cm Line Observations. Astrophys. J. **152**, 200 (2016). arXiv:1609.06611
C.W. McAlary, D.L. Welch, Detection of Cepheid variables by the Infrared Astronomical Satellite. Astron. J. **91**, 1209 (1986)
M.L. McCall, On determining extinction from reddening. Astron. J. **128**, 2144 (2004)
C. Min, N. Matsumoto, M.K. Kim et al., Accurate parallax measurement toward the symbiotic star R Aquarii. Publ. Astron. Soc. Jpn. **66**, 38 (2014)
D. Minniti et al., VISTA variables in the Via Lactea (VVV): the public ESO near-IR variability survey of the Milky Way. New Astron. **15**, 433 (2010)
A.J. Monson et al., The Carnegie Hubble Program: the Leavitt law at 3.6 and 4.5 μm in the Milky Way. Astrophys. J. **759**, 146 (2012)
A.J. Monson, M.J. Pierce, Near-infrared (JHK) photometry of 131 Northern Galactic classical Cepheids. Astrophys. J. Suppl. Ser. **193**, 12 (2011)
V.M.R. Muggeo, Estimating regression models with unknown break-points. Stat. Med. **22**, 3055 (2003)
V.R.M. Muggeo, An R package to fit regression models with broken-line relationships. R News **8**(1), 20 (2008)
A. Nakagawa, T. Kurayama, M. Matsui et al., Parallax of a Mira variable R Ursae Majoris studied with astrometric VLBI. Publ. Astron. Soc. Jpn. **68**, 78 (2016)
A. Nakagawa, T. Omodaka, T. Handa et al., VLBI astrometry of AGB variables with VERA: a Mira-type variable T Lepus. Publ. Astron. Soc. Jpn. **66**, 101 (2014)
A. Nakagawa, M. Tsushima, K. Ando et al., VLBI astrometry of AGB variables with VERA – A semiregular variable S Crateris –. Publ. Astron. Soc. Jpn. **60**, 1013 (2008)
H. Nakanishi, Y. Sofue, Three-dimensional distribution of the ISM in the Milky Way Galaxy: II. The molecular gas disk. Publ. Astron. Soc. Jpn. **58**, 847 (2006)
H.R. Neilson et al., Classical Cepheids require enhanced mass loss. Astrophys. J. **760**, L18 (2012)
C.-C. Ngeow et al., Further empirical evidence for the non-linearity of the period-luminosity relations as seen in the Large Magellanic Cloud Cepheids. Mon. Not. R. Astron. Soc. **363**, 831 (2005)
C.-C. Ngeow et al., Calibrating the projection factor for Galactic Cepheids. Astron. Astrophys. **543**, A55 (2012)
C.-C. Ngeow et al., Period-luminosity relations derived from the OGLE-III fundamental mode Cepheids. II. The Small Magellanic Cloud Cepheids. Astrophys. J. **808**, 67 (2015a)
C.-C. Ngeow et al., Updated 24 μm period-luminosity relation derived from Galactic Cepheids. Astrophys. J. **813**, 57 (2015b)
C. Ngeow, S.M. Kanbur, The period-luminosity relation for the Large Magellanic Cloud Cepheids derived from Spitzer archival data. Astrophys. J. **679**, 76–85 (2008)
C. Ngeow, S.M. Kanbur, A. Nanthakumar, Testing the nonlinearity of the BVIcJHKs period-luminosity relations for the Large Magellanic Cloud Cepheids. Astron. Astrophys. **477**, 621 (2008)
D.L. Nidever, A. Monachesi, E.F. Bell et al., A tidally stripped stellar component of the Magellanic Bridge. Astrophys. J. **779**, 145 (2013)
S. Nishiyama et al., Interstellar extinction law toward the Galactic Center III: J, H, K_S bands in the 2MASS and the MKO systems, and 3.6, 4.5, 5.8, 8.0 μm in the Spitzer/IRAC system. Astrophys. J. **696**, 1407 (2009)
D. Nyu, A. Nakagawa, M. Matsui et al., Astrometry of AGB variables with VERA: annual parallax and the orbit of SY Sculptoris in the galaxy. Publ. Astron. Soc. Jpn. **63**, 63 (2011)
K.A.G. Olsen, C. Salyk, A warp in the Large Magellanic Cloud disk? Astron. J. **124**, 2045 (2002)
W.V. Oz, M. Yannakakis (eds.), *All ACM Conferences* (Academic Press, Boston, 1983)
B. Paczyński, K.Z. Stanek, Galactocentric distance with the optical gravitational lensing experiment and HIPPARCOS Red Clump stars. Astrophys. J. **494L**, 219 (1998)
O. Pejcha, C.S. Kochanek, A global physical model for Cepheids. Astrophys. J. **748**, 107 (2012)
M.A.C. Perryman et al., The HIPPARCOS catalogue. Astron. Astrophys. **323**, L49 (1997)
S.E. Persson et al., New Cepheid period-luminosity relations for the Large Magellanic Cloud: 92 near-infrared light curves. Astron. J. **128**, 2239 (2004)
M.M. Phillips, The absolute magnitudes of Type IA supernovae. Astrophys. J. **413L**, 105 (1993)
G. Pietrzyński et al., The dynamical mass of a classical Cepheid variable star in an eclipsing binary system. Nature **468**, 542 (2010)
G. Pietrzyński et al., An eclipsing-binary distance to the Large Magellanic Cloud accurate to two per cent. Nature **495**, 76 (2013)
G. Pietrzyński, W. Gieren, A. Udalski, The Araucaria project: dependence of mean K, J, and I absolute magnitudes of Red Clump stars on metallicity and age. Astron. J. **125**, 2494 (2003)

M.H. Pinsonneault, Y. Elsworth, C. Epstein et al., The APOKASC catalog: an asteroseismic and spectroscopic joint survey of targets in the Kepler fields. Astrophys. J. Suppl. Ser. **215**, 19 (2014)
Planck Collaboration et al., Planck 2015 results. XIII. Cosmological parameters. Astron. Astrophys. **594**, A13 (2016)
E. Puzeras, G. Tautvaišienė, J.G. Cohen et al., High-resolution spectroscopic study of Red Clump stars in the Galaxy: iron-group elements. Mon. Not. R. Astron. Soc. **408**, 1225 (2010)
H. Rauer et al., The PLATO 2.0 mission. Exp. Astron. **38**, 249 (2014)
J.-J. Ren, X.-W. Liu, M.-S. Xiang et al., On the LSP3 estimates of surface gravity for LAMOST-Kepler stars with asteroseismic measurements. Res. Astron. Astrophys. **16**, 45 (2016)
Kh.S. Rezaei et al., Inferring the three-dimensional distribution of dust in the Galaxy with a non-parametric method: Preparing for Gaia. Astron. Astrophys. **598**, A125 (2017). doi:10.1051/0004-6361/201628885
J.A. Rich et al., A new Cepheid distance measurement and method for NGC 6822. Astrophys. J. **794**, 107 (2014)
G.R. Ricker et al., Transiting Exoplanet Survey Satellite (TESS). SPIE **9143**, 20 (2014)
G.H. Rieke et al., The Multiband Imaging Photometer for Spitzer (MIPS). Astrophys. J. Suppl. Ser. **154**, 25 (2004)
G.H. Rieke, M.J. Lebofsky, The interstellar extinction law from 1 to 13 microns. Astrophys. J. **288**, 618 (1985)
A.G. Riess et al., A 3% solution: determination of the Hubble constant with the Hubble Space Telescope and Wide Field Camera 3. Astrophys. J. **730**, 119 (2011)
A.G. Riess et al., A 2.4% determination of the local value of the Hubble constant. Astrophys. J. **826**, 56 (2016)
V. Ripepi et al., The VMC survey. XIX. Classical Cepheids in the Small Magellanic Cloud. Astrophys. J. Suppl. Ser. **224**, 21 (2016)
S. Roeser, M. Demleitner, E. Schilbach, The PPMXL catalog of positions and proper motions on the ICRS. Combining USNO-B1.0 and the Two Micron All Sky Survey (2MASS). Astron. J. **139**, 2440 (2010)
C.G. Román-Zúñiga et al., The infrared extinction law at extreme depth in a dark cloud core. Astrophys. J. **664**, 357 (2007)
M. Romaniello et al., The influence of chemical composition on the properties of Cepheid stars. II. The iron content. Astron. Astrophys. **488**, 731 (2008)
S. Rubele, L. Girardi, L. Kerber et al., The VMC survey—XIV. First results on the look-back time star formation rate tomography of the Small Magellanic Cloud. Mon. Not. R. Astron. Soc. **449**, 639 (2015)
M. Salaris, L. Girardi, Population effects on the red giant clump absolute magnitude: the K band. Mon. Not. R. Astron. Soc. **337**, 332 (2002)
E.F. Schlafly et al., The optical-infrared extinction curve and its variation in the Milky Way. Astrophys. J. **821**, 78 (2016)
B. Schölkopf, A. Smola, K.-R. Müller, Nonlinear component analysis as a kernel eigenvalue problem. Neural Comput. **10**, 1299 (1998)
V. Scowcroft et al., The Carnegie Hubble program: the Leavitt law at 3.6 μm and 4.5 μm in the Large Magellanic Cloud. Astrophys. J. **743**, 76 (2011)
V. Scowcroft et al., The Carnegie Chicago Hubble program: the mid-infrared colours of Cepheids and the effect of metallicity on the CO band-head at 4.6 μm. Mon. Not. R. Astron. Soc. **459**, 1170 (2016a)
V. Scowcroft et al., The Carnegie Hubble program: the distance and structure of the SMC as revealed by mid-infrared observations of Cepheids. Astrophys. J. **816**, 49 (2016b)
B. Sesar et al., Light curve templates and galactic distribution of RR Lyrae stars from Sloan Digital Sky Survey Stripe 82. Astrophys. J. **708**, 717 (2010)
M. Shintani, H. Imai, K. Ando et al., Statistical properties of stellar H2O masers—results of three-year single-dish observations with the VERA Iriki telescope. Publ. Astron. Soc. Jpn. **60**, 1077 (2008)
A. Siebert, B. Famaey, I. Minchev et al., Detection of a radial velocity gradient in the extended local disc with RAVE. Mon. Not. R. Astron. Soc. **412**, 2026 (2012)
N.R. Simon, A.S. Lee, The structural properties of Cepheid light curves. Astrophys. J. **248**, 291 (1981)
I. Soszyński et al., The optical gravitational lensing experiment. The OGLE-III catalog of variable stars. I. Classical Cepheids in the Large Magellanic Cloud. Acta Astron. **58**, 163 (2008)
I. Soszyński et al., The optical gravitational lensing experiment. The OGLE-III catalog of variable stars. VII. Classical Cepheids in the Small Magellanic Cloud. Acta Astron. **60**, 17 (2010)
I. Soszyński, W. Gieren, G. Pietrzyński, Mean JHK magnitudes of fundamental-mode Cepheids from single-epoch observations. Publ. Astron. Soc. Pac. **117**, 823 (2005)
K.Z. Stanek, P.M. Garnavich, Distance to M31 with the Hubble Space Telescope and HIPPARCOS Red Clump stars. Astrophys. J. **503**, 131 (1998)
K.Z. Stanek, D. Zaritsky, J. Harris, A "short" distance to the Large Magellanic Cloud with the Hipparcos calibrated Red Clump stars. Astrophys. J. **500**, 141 (1998)

D. Stello et al., Oscillating red giants observed during Campaign 1 of the Kepler K2 mission: new prospects for Galactic archaeology. Astrophys. J. **809**, 3 (2015)
D. Stello, D. Huber, T.R. Bedding et al., Asteroseismic classification of stellar populations among 13,000 red giants observed by Kepler. Astrophys. J. **765**, 41 (2013)
J. Storm et al., The effect of metallicity on the Cepheid period-luminosity relation from a Baade–Wesselink analysis of Cepheids in the Galaxy and in the Small Magellanic Cloud. Astron. Astrophys. **415**, 531 (2004)
J. Storm et al., Calibrating the Cepheid period-luminosity relation from the infrared surface brightness technique. I. The p-factor, the Milky Way relations, and a universal K-band relation. Astron. Astrophys. **534**, A94 (2011a)
J. Storm et al., Calibrating the Cepheid period-luminosity relation from the infrared surface brightness technique. II. The effect of metallicity and the distance to the LMC. Astron. Astrophys. **534**, A95 (2011b)
S. Subramanian, S. Rubele, N.-C. Sun et al., The VMC survey—XXIV. Signatures of tidally stripped stellar populations from the inner Small Magellanic Cloud. Mon. Not. R. Astron. Soc. **467**, 2980 (2017)
S. Subramanian, A. Subramaniam, Depth estimation of the Large and Small Magellanic Clouds. Astron. Astrophys. **496**, 399 (2009)
S. Subramanian, A. Subramaniam, An estimate of the structural parameters of the Large Magellanic Cloud using Red Clump stars. Astron. Astrophys. **520**, 135 (2010)
S. Subramanian, A. Subramaniam, The three-dimensional structure of the Small Magellanic Cloud. Astrophys. J. **744**, 128 (2012)
S. Subramanian, A. Subramaniam, Structure of the Large Magellanic Cloud from near infrared magnitudes of Red Clump stars. Astron. Astrophys. **552**, A144 (2013)
G.A. Tammann, A. Sandage, B. Reindl, New period-luminosity and period-color relations of classical Cepheids: I. Cepheids in the Galaxy. Astron. Astrophys. **404**, 423 (2003)
B.L. Tatton, J.Th. van Loon, M.-R. Cioni et al., The VMC survey. VII. Reddening map of the 30 Doradus field and the structure of the cold interstellar medium. Astron. Astrophys. **554**, 33 (2013)
A. Udalski, Variable stars toward the Magellanic Clouds and the galactic bulge, in *ASP Conference Series*, vol. 491 (2015), p. 278
A. Udalski, M. Szymanski, M. Kubiak, G. Pietrzynski, P. Wozniak, K. Zebrun, Optical gravitational lensing experiment. Distance to the Magellanic Clouds with the Red Clump stars: are the Magellanic Clouds 15% closer than generally accepted? Acta Astron. **48**, 1 (1998)
S. van den Bergh, *The Extragalactic Distance Scale*, vol. 509 (1975). gaun.book
F. van Leeuwen, *Hipparcos the New Reduction of the Raw Data* Astrophys. Space Sci. Library (Cambridge University Press, Cambridge, 2007b)
F. van Leeuwen et al., Cepheid parallaxes and the Hubble constant. Mon. Not. R. Astron. Soc. **379**, 723 (2007a)
F. van Leeuwen, M.W. Feast, P.A. Whitelock, B. Yudin, First results from HIPPARCOS trigonometrical parallaxes of Mira-type variables. Mon. Not. R. Astron. Soc. **287**, 955 (1997)
W.H.T. Vlemmings, H.J. van Langevelde, Improved VLBI astrometry of OH maser stars. Astron. Astrophys. **472**, 547 (2007)
W.H.T. Vlemmings, H.J. van Langevelde, P.J. Diamond, H.J. Habing, R.T. Schilizzi, VLBI astrometry of circumstellar OH masers: proper motions and parallaxes of four AGB stars. Astron. Astrophys. **407**, 213 (2003)
C. Wegg, O. Gerhard, Mapping the three-dimensional density of the galactic bulge with VVV Red Clump stars. Mon. Not. R. Astron. Soc. **435**, 1874 (2013)
M.W. Werner et al., The Spitzer Space Telescope mission. Astrophys. J. Suppl. Ser. **154**, 1 (2004)
B. Westerlund, The nebulosity around RS Puppis. Publ. Astron. Soc. Pac. **73**, 72 (1961)
P. Whitelock, M. Feast, Hipparcos parallaxes for Mira-like long-period variables. Mon. Not. R. Astron. Soc. **319**, 759 (2000)
M.E.K. Williams, M. Steinmetz, J. Binney et al., The wobbly Galaxy: kinematics north and south with RAVE red-clump giants. Mon. Not. R. Astron. Soc. **436**, 101 (2013)
P.R. Wood, C. Alcock, R.A. Allsman et al., MACHO observations of LMC red giants: Mira and semi-regular pulsators, and contact and semi-detached binaries. Symp. - Int. Astron. Union **191**, 151 (1999)
M. Wyman et al., Neutrinos help reconcile Planck measurements with the local universe. Phys. Rev. Lett. **112**, 051302 (2014)
M.S. Xiang, X.W. Liu, H.B. Yuan et al., The LAMOST stellar parameter pipeline at Peking University—LSP3. Mon. Not. R. Astron. Soc. **448**, 822 (2015)
H.-B. Yuan, X.-W. Liu, Z.-Y. Huo et al., LAMOST Spectroscopic Survey of the Galactic Anticentre (LSS-GAC): target selection and the first release of value-added catalogues. Mon. Not. R. Astron. Soc. **448**, 855 (2015)

N. Zacharias, C.T. Finch, T.M. Girard et al., The fourth US Naval Observatory CCD Astrograph Catalog (UCAC4). Astron. J. **145**, 44 (2013)
B. Zhang, M.J. Reid, K.M. Menten, X.W. Zheng, A. Brunthaler, The distance and size of the red hypergiant NML Cygni from VLBA and VLA astrometry. Astron. Astrophys. **544**, AA42 (2012)

Space Sci Rev (2018) 214:113
https://doi.org/10.1007/s11214-018-0542-1

Old-Aged Primary Distance Indicators

Rachael L. Beaton[1,2] · Giuseppe Bono[3,4] · Vittorio Francesco Braga[3,5] · Massimo Dall'Ora[6] · Giuliana Fiorentino[7] · In Sung Jang[8] · Clara E. Martínez-Vázquez[9] · Noriyuki Matsunaga[10] · Matteo Monelli[11,12] · Jillian R. Neeley[13] · Maurizio Salaris[14]

Received: 28 February 2018 / Accepted: 28 August 2018 / Published online: 18 October 2018

Abstract Old-aged stellar distance indicators are present in all Galactic structures (halo, bulge, disk) and in galaxies of all Hubble types and, thus, are immensely powerful tools for understanding our Universe. Here we present a comprehensive review for three primary standard candles from Population II: (i) RR Lyrae type variables (RRL), (ii) type II Cepheid variables (T2C), and (iii) the tip of the red giant branch (TRGB). The discovery and use of these distance indicators is placed in historical context before describing their theoretical foundations and demonstrating their observational applications across multiple wavelengths.

Astronomical Distance Determination in the Space Age
Edited by Richard de Grijs and Maurizio Falanga

✉ R.L. Beaton
rbeaton@princeton.edu

1 Department of Astrophysical Sciences, Princeton University, 4 Ivy Lane, Princeton, NJ 08544, USA

2 The Observatories of the Carnegie Institution for Science, 813 Santa Barbara Street, Pasadena, CA 91101, USA

3 Department of Physics, University of Rome Tor Vergata, Rome, Italy

4 INAF-Osservatorio Astronomico di Roma, Rome, Italy

5 ASDC, Rome, Italy

6 INAF-Osservatorio Astronomico di Capodimonte, Naples, Italy

7 INAF—OAS Osservatorio di Astrofisica & Scienza dello Spazio di Bologna, Bologna, Italy

8 Leibniz-Institut für Astrophysics Potsdam, 14482 Potsdam, Germany

9 Cerro Tololo Inter-American, National Optical Astronomy Observatory, Casilla 603, La Serena, Chile

10 Department of Astronomy, School of Science, The University of Tokyo, Tokyo, Japan

11 IAC-Instituto de Astrofísica de Canarias, Calle Vía Lactea s/n, 38205 La Laguna, Tenerife, Spain

12 Departmento de Astrofísica, Universidad de La Laguna, 38206 La Laguna, Tenerife, Spain

13 Department of Physics, Florida Atlantic University, 777 Glades Rd, Boca Raton, FL 33431, USA

14 Astrophysics Research Institute, Liverpool John Moores University, 146 Brownlow Hill, L3 5RF Liverpool, UK

The methods used to establish the absolute scale for each standard candle is described with a discussion of the observational systematics. We conclude by looking forward to the suite of new observational facilities anticipated over the next decade; these have both a broader wavelength coverage and larger apertures than current facilities. We anticipate future advancements in our theoretical understanding and observational application of these stellar populations as they apply to the Galactic and extragalactic distance scale.

Keywords Distance scale · Population II stars · RR Lyrae · Type II Cepheids · Tip of the red giant branch · Old stars · Standard candles · Variable stars

1 Introduction

Standard candles drawn from old stellar populations have a significant advantage for distance scale as, with the exception of young Galactic star clusters, old stellar populations are found in every galactic structural component (disk, halo, bulge), galaxies of all Hubble types, and galaxies of all luminosities (from ultra-faint to ultra-luminous). Most importantly, from these distance indicators it is possible to map both Galactic and extra-galactic objects using tracers pulled from the same underlying stellar population, if not the same class of star. Moreover, due to the presence of old stars in most structural components of galaxies, it is possible to study nearby galaxies in three dimensions (e.g., measuring depths, orientations, etc.) and then to evaluate if there systematics in mean distances due to these complex structures (see e.g., Kunder et al. 2018). In turn, this helps us to better understand how projection effects and line-of-sight depth could bias mean distances. Thus, by virtue of being "old" standard candles have immense potential.

The goal of this chapter is to provide a comprehensive review for the primary standard candles drawn from Population II stars. The term Population II (Pop II) is an old one that originated from Baade (1944) in which the nebulous central regions of the Andromeda, M 32, and NGC 205 were first resolved into individual stars. Baade (1944) realized that these stars more closely resembled those in Galactic globular clusters (GGCs) than the "slow moving" stars in the solar neighborhood (e.g., disk stars). Later work would frame the differences as a function of age and metallicity, as Baade (1958a) concisely summarized at the Vatican conference. Interestingly, the nature of the variable stars and their proper classification into Pop I or Pop II was intimately entwined (Baade 1958b). We focus our attention in this chapter to relatively luminous tracers that can be used for a broad range of distances and can be considered "primary", in that there exist some absolute calibrations using parallaxes for these standard candles from before the onset of progressive *Gaia* data releases (DR1 and DR2 at the time of writing). These considerations result in three distance indicators: (i) the RR Lyrae (RRL), (ii) the Type II Cepheids (T2C), and (iii) the tip of the red giant branch (TRGB) stars. The basic properties of these standard candles are given in Table 1.

Both the RRL and the T2C are pulsational variables that occur when specific Pop II sequences cross the classical instability strip and these stars adhere to specific period–luminosity (PL) relationships from which distances can be determined to individual stars. In contrast, the stars that comprise TRGB are non-variable in nature[1] and as a result, the distance measurement is performed using a population of stars, which makes it statistical in nature.

[1] These stars likely have some intrinsic variability—as most stars do, but it is on a much smaller scale than the pulsational variables that have amplitudes $\sim$0.3 to $>$1 mag.

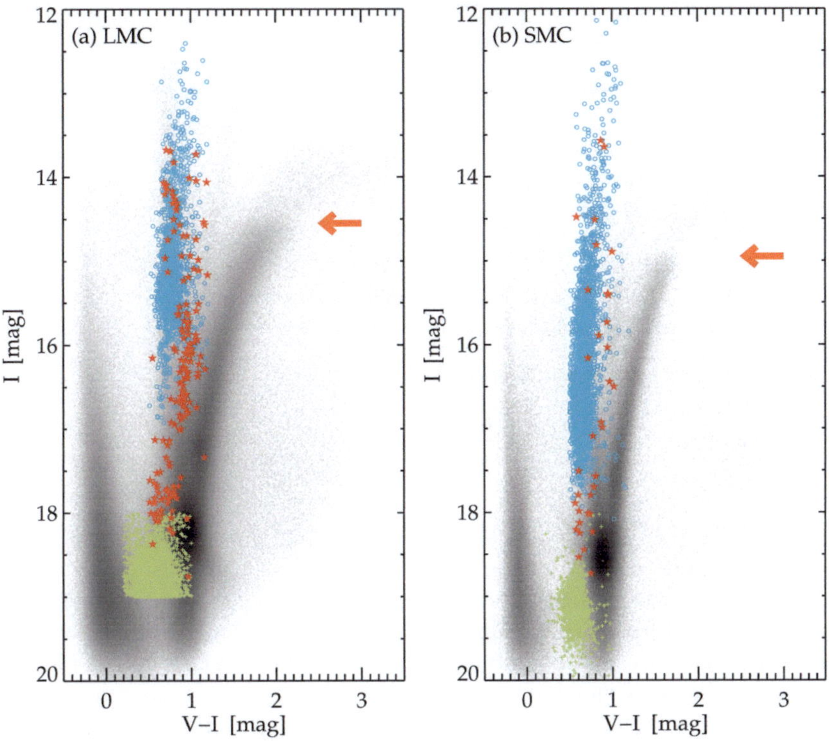

Fig. 1 Optical color magnitude diagrams for the (**a**) LMC and (**b**) SMC from the Magellanic Clouds Photometric Survey (black; MCPS Zaritsky et al. 2002, 2004) with variable star identifications from OGLE-III as follows RRLs (green pluses), T2Cs (red five-pointed stars), and the Classical Cepheids (blue open circles) over plotted (Soszyński et al. 2016b, 2008, 2015, respectively). The approximate location of the TRGB is indicated with a thick orange arrow. The axes for both panels are identical for ease of comparison. We note that no corrections for extinction (Galactic or internal to the LMC) have been applied and, thus, the color widths of the populations may be broader, to the red, than their intrinsic range

Figures 1a and 1b show the relative positions of the Classical Cepheids (blue), RRLs (green), T2Cs (red), and TRGB (orange arrow) for the Large and Small Magellanic Clouds (LMC, SMC) using variable star identifications from OGLE-III[2] (Soszyński et al. 2008, 2015, 2016b) overplotted on a color magnitude density diagram from Magellanic Clouds Photometric Survey[3] (greyscale Zaritsky et al. 2002, 2004). Figures 1a and 1b keenly demonstrate how the T2Cs could have been confused with the Classical Cepheids, as the two populations overlap in magnitude. The three sub-classifications for the T2Cs are also visible by the distinct clumps in the LMC (Fig. 1a) with these distinctions being less clear in the lower-metallicity SMC (Fig. 1b). The difference in the population size for the three variable star classes is also apparent, with the T2Cs being less abundant than either the Classical Cepheids or the RRLs. Figures 1a and 1b reinforce the mean magnitude differences given in Table 1, with the TRGB being more luminous than the bulk of the T2Cs and the T2Cs being brighter than the RRLs.

[2]The full variable star catalog can be queried here: http://ogledb.astrouw.edu.pl/~ogle/OCVS/.

[3]Data is available here: http://djuma.as.arizona.edu/~dennis/mcsurvey/.

Table 1 Basic properties of Population II distance indicators

Star	Sub-type	M_V [mag]	M_K [mag]	P [days]
RR LYRAE (RRL)				
	Fundamental mode (RRab)	$\sim +0.6$	~ -0.6	$0.3 < P < 1.0$
	First overtone mode (RRc)	$\sim +0.6$	~ -0.4	$0.2 < P < 0.5$
TYPE II CEPHEIDS (T2C)				
	BL Herulis (BL Her)	~ -0.5	~ -1.0	$1 < P < 4$
	W Virginis (W Vir)	~ -1.0	~ -4.0	$4 < P < 20$
	RV Tauri (RV Tau)	~ -2.5	~ -5.0	$20 < P < 80$
TIP OF THE RED GIANT BRANCH STARS (TRGB)				
	Metal-poor	~ -4.0	~ -5.5	
	Metal-rich	~ -3.9	~ -6.5	

As a class, the RRL have a long history in the optical, having been discovered in the 19th century in cluster diagrams, but recognition of their great potential in the infrared has come only recently (most notably, Longmore et al. 1986). The short period of RRLs make identifying them relatively simple with observations spanning only a few nights and with numerous Globular Clusters being sufficiently nearby for them to have been readily discovered by early photometric monitoring campaigns (a detailed history is given in Smith 1995). In contrast, the T2Cs were only separated from the classical Cepheids in 1956 by Baade (1956) and, despite being a solution to a difficulty in reconciling H_0 from the distance ladder and cosmological theory, have received little attention until the long-term monitoring from the OGLE project unveiled them *en masse* in the LMC (Fig. 1a, Soszyński et al. 2008). The recognition of the TRGB as a luminosity indicator came later still when the work of Da Costa and Armandroff (1990) provided high quality homogeneous photometry for a number of Galactic globular clusters (GGCs) transformed into absolute units by their RRL distances. The TRGB was first used to determine distances for galaxies by Lee et al. (1993), who developed the analysis techniques necessary to detect the tip from color–magnitude diagrams.

Each of these distance indicators, thus, has a different volume of literature accompanying them and, as a result, have different depths of both theoretical understanding and observational applications. Often these vary not only by distance indicator, but also by wavelength regimes and objects (field stars, star clusters, galaxies) in which the techniques have been employed. Thus, the depth and breath of information provided in this review varies for each distance indicator.

Generally, a single book chapter cannot fully describe any one of these standard candles. Thus, we refer the reader to more detailed discussions that are in the literature. Of particular note are the following books: Smith (1995) on RRLs and Catelan and Smith (2015) on pulsational variables of all kinds, including both RRLs and T2Cs. Additionally, McWilliam (2011) is a set of online conference articles that present reviews of many aspects of RRL beyond those that will be discussed here, as well as other discussions relating to metal-poor, old stellar populations. Salaris and Cassisi (2005) is an excellent resource on stellar evolution and stellar populations that provides insight into all three of our distance indicators, but most especially the TRGB. Lastly, Beaton et al. (2016) provides comparison of RRL and TRGB methods with Cepheids in terms of the extragalactic distance scale that may help the reader understand the recent resurgence of interest Pop II standard candles.

Our goal in this chapter is to place these Population II standard candles into the context of the distance scale by providing a sense of the current theoretical understanding and observational application of these tools. Where possible, we take a multi-wavelength approach discussing optical, near-infrared, and mid-infrared characteristics and applications. In the sections that follow we discuss each of the distance indicators in turn, with parallel discussions of theory and practice for the RRL in Sect. 2, a description and homogeneous PL relations for T2C in Sect. 3, and both a physical description and application of the TRGB is given in Sect. 4. The absolute scale for each distance indicator and inter-comparisons are described in Sect. 5. We conclude in Sect. 6 with an outlook for the future, in particular improvements to our physical understanding from *Gaia* and the observational application with future large-aperture facilities.

2 The RR Lyrae Variables

The first variable in a GGC was discovered by Williamina Fleming and reported in Pickering (1889). In 1893, Solon Bailey initiated a large scale program for imaging GGCs from the Harvard College Observatory in Arequipa, Peru with Williamina Fleming serving as his assistant. Very quickly several variables were discovered in the brightest GGCs in the Southern sky, in particular for ω Centauri (ω Cen). By the end of this project, Bailey had discovered over 500 variables in GGCs, which was equal in number to that found over the entire remainder of the sky! A more detailed story of the discovery of RRLs is given in the introduction to Smith (1995).[4] RR Lyrae, itself, was discovered by W. Flemming prior to 1899 and reported in Pickering et al. (1901).

The observations reported in Bailey (1902) define the nomenclature for RRLs that is used to this day. RRLs come in two primary types; more specifically, those that pulsate in the fundamental mode (FU) known as RRab (a combination of the original Bailey types of a and b) and those that pulsate in the first overtone (FO) known as RRc (the Bailey type c stars). The connection between the modes of pulsation and their Bailey types was first inferred by Schwarzschild (1940) using high quality photometric-plate light curves in the cluster M 3; more specifically, Schwarzschild (1940) concluded that the 'c' stars needed to pulsate in a different mode to account for the strong deviation of this class from the period-density relation for the 'a' and 'b' type variables. Schwarzschild (1940) further defined the color-edges of the instability strip concluding that no non-pulsating stars could exist in this temperature–luminosity range. Additionally, there are the RRd type stars that show two pulsation modes and an ensemble of stars discovered from the OGLE program have more complex or atypical pulsation behavior (e.g., Soszyński et al. 2016, and references therein). The mean magnitudes of the RRL place them on the horizontal branch (HB) of GGCs, whereas at any given epoch, they will scatter above or below based on their amplitudes.

Some RRLs show an amplitude-modulation phenomenon known as the Blazhko effect (first identified in Blažko 1907) that currently has no consensus for its physical origin, though its impacts on the light curve are well documented. A particularly interesting demonstration of the complexity of the Blazhko phenomenon can be found in Chadid et al. (2014,

[4]We also refer the reader to Sobel (2016) for a description of the role of Williamina Fleming in these discoveries. More specifically, Williamina Fleming was first E. C. Pickering's maid, became the first of the Harvard 'computers,' and was involved in a large number of projects at the Harvard Observatory. Her accomplishments were honored when she became the first American woman elected to the Royal Astronomical Society.

see also references therein) that presents data at very high cadence tracing a single star during Antarctic winter for 150 days. The periods of the Blazhko effect can be from tens to hundreds of pulsation cycles for a given star (see for instance, the study of Skarka 2014). The effect is well known and studied in the optical, but has recently been shown to persist into the K band (Jurcsik et al. 2018). The impact of the Blazhko effect on distances is that these stars have more uncertainty with respect to both their periods and mean magnitudes, with the impact being proportional to the level of amplitude-modulation. Thus, these stars, when identified, are not always used for PL fitting.

Bailey et al. (1919) paved the road for use of RRLs as both tracers for old stellar population (Pop. II) and as distance indicators. Subsequent investigations demonstrated that the RRL visual mean magnitude, within a given cluster, was nearly constant. The first strong evidence that RRL visual mean magnitude and the metallicity were correlated came in Baade (1958c), in which the populations of variables were compared between disk and halo star clusters and the Galactic Bulge. Additional evidence accumulated over the next few years; in particular, (i) additional probes of the HB magnitude for GGCs with metallicity estimates by Sandage and Wallerstein (1960) and (ii) comparisons between variables found in more environments, in particular the Draco dwarf Spheroidal that showed both similarities and differences to stellar sequences in the GGCs (Baade and Swope 1961). Since those realizations, RRLs have been commonly used to estimate distances by means of a calibrated V-band magnitude versus metallicity relation (e.g., Sandage 1982, and references therein), subsequently rediscussed and calibrated several times, that is still widely used. The mean magnitude of RRLs in the V-band is nearly constant ($M_V \sim 0.6$ mag), with a dependence on their chemical composition, thus making RRLs solid standard candles.

Observations of RRLs in a number of GGCs supported the evidence that the topology of the Instability Strip (IS) changes with metallicity. In particular, the first overtone blue edge (FOBE) is virtually independent of the metallicity as originally suggested by linear radiative models (Baker and Kippenhahn 1965; Cox 1963; Iben 1971). This finding and observations brought forward the opportunity to use the FOBE vs period relation as both a reddening and a distance indicator. The theoretical scenario concerning the red edge of the IS was more complex, because it is caused by the increased efficiency of the convective transport when moving toward cooler effective temperatures. Pioneering nonlinear models that included a time dependent treatment of the convective transport (Deupree 1977b; Stellingwerf 1982), suggested that the red edge becomes cooler (redder) as the metallicity increases.

The use of RRLs as distance indicators has had a quantum jump thanks to the empirical discovery by Longmore et al. (1986) that RRLs obey a linear PL relation at near–infrared wavelengths. This discovery was later soundly supported by nonlinear convective models Bono et al. (1994, 2001). More recently, the advent of mid–infrared facilities on board of space telescopes, like *WISE* and *Spitzer*, led to the derivation of empirical PL relations at longer wavelengths (e.g., $\lambda \gtrsim$ than 3.6 μm; Sect. 2.2.5). Another popular diagnostic used to derive distances is the Wesenheit function (van den Bergh 1975; Madore 1982), which is a reddening-free formulation of the less commonly used PL–color relations (or PLC), to be discussed in Sect. 2.2.7.

RRLs are readily recognizable from their light variation. Figure 2 presents an example light curve for each of the two dominant sub-classes of RRL, an RRab (Fig. 2a) and an RRc (Fig. 2b) in ten photometric broadband filters from Monson et al. (2017). Figure 2 illustrates the large amplitudes and unique shapes in the optical for both types of stars, while also demonstrating how the amplitudes decrease strongly as a function of wavelength. Indeed, the amplitude for RZ Cep is 0.1 mag in the IR compared to $\gtrsim$0.5 mag in the optical. The RRab stars, in particular, have a "sawtooth" shape in the optical, but become more sinusoidal

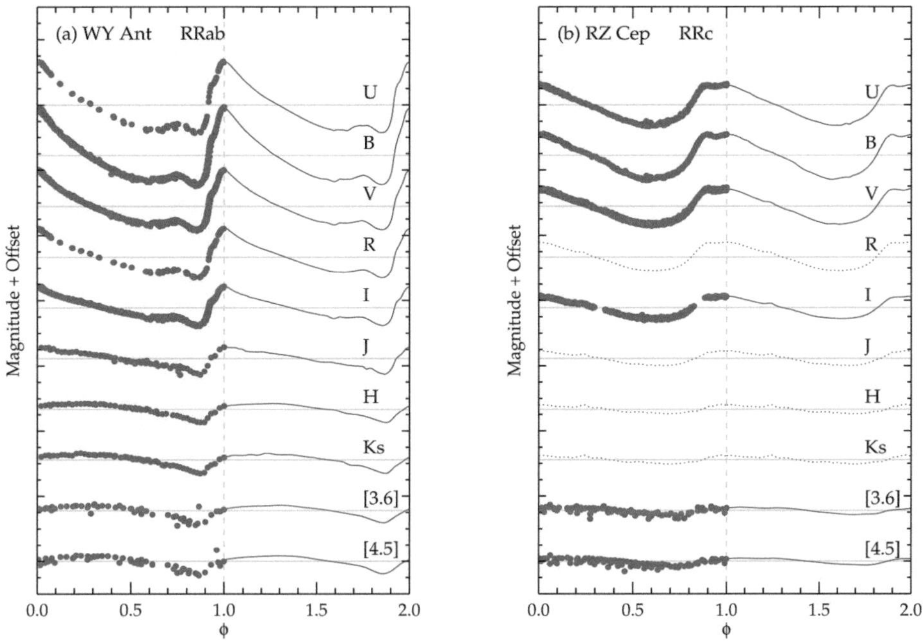

Fig. 2 (**a**) Example multi-wavelength light curves for an RRab RRL, WY Ant and (**b**) example multi-wavelength light curves for an RRc RRL, RZ Cep (data from Monson et al. 2017). In each panel, the light curves are shown in ten photometric bands, which are from bottom to top U, B, V, R, I_C, J, H, K, [3.6], and [4.5]. The optical bands are the Johnson–Cousins system, the near-infrared bands are in the 2MASS system, and the mid-infrared bands are in the *Spitzer*–IRAC system; a detailed discussion of the photometry homogenization is given in Monson et al. (2017). In addition to the raw data points, a smoothed light curve is shown in black and a template light curve is shown in gray, if no data is present. The horizontal line indicates the mean magnitude and the dashed vertical line indicates maximum light in the optical ($\phi = 1$)

at longer wavelengths as the impact of the temperature changes become less important. The RRc stars (right), in contrast, have a shape that changes comparatively little as a function of wavelength.

RRLs are nearly ubiquitously present in Local Group galaxies. Indeed, they have been identified in all the stellar systems that host an old ($\geq$10 Gyr) stellar component. This evidence makes them excellent probes to investigate the structure and the old stellar populations at the early epochs of galaxy formation. They can be used to trace the components of our Galaxy (bulge, halo, thick disk) and to determine the distance and characterize the old population in Local Group (LG, distances within $\sim$1 Mpc) galaxies. Thus, RRLs provide a crucial first step to the extragalactic distance scale for Pop II stellar systems, letting us control possible systematics affecting the commonly used distance scale based on classical Cepheids.

In the following sections, we focus our attention on the theoretical and semi-empirical background for the use of RRLs as distance indicators, with a special attention to some recent developments. Despite great progress in the determination of the RRL PL over the past decade, theoretical PL relations are often used for distance determination due to the lingering uncertainties associated with the absolute zero-point and, in particular, the effect of metallicity. Thus, the sections to follow describing theoretical efforts are quite detailed to motivate the strengths and weaknesses of the theoretical PLs.

2.1 A Physical Description of RR Lyrae

RRLs are radially pulsating low-mass (0.6–0.8 $M_\odot$) stars in their central helium-burning phase. The radial oscillations are an envelope phenomenon that takes place in a well-defined range of effective temperatures, therefore, they populate a relatively narrow region in the Hertzsprung–Russell diagram, which is the intersection between the so called *Cepheid* IS and the HB. We will shortly describe the physical mechanisms driving the radial pulsation and we will analyze with some detail the different approaches to estimate distances using RRLs, highlighting their advantages and disadvantages.

The idea of a pulsating gaseous sphere was developed for the first time by Ritter (1879),[5] who found a simple relation between pulsation period and mean density, but it was only in Shapley (1914) that the radial pulsation hypothesis was advocated for Cepheid-like stars. The dispute between binarity and radial oscillations was settled in favor of the latter by the so-called Baade–Wesselink method. Radial pulsation is a phenomenon that involves the stellar envelope for certain values of the surface effective temperature (T_{eff}) and defines in the color magnitude diagram (CMD) a region in which the stars are unstable to pulsation, the IS. It is worth mentioning that a star radially pulsates only during its crossing(s) of the IS. As a result, the most numerous pulsators are those that have long evolutionary lifetimes within the IS, such as δ Scuti (central hydrogen burning) or RR Lyrae and Cepheids (central helium burning).

The physical mechanisms underlying this phenomenon are related to the cyclic variations of the opacity and of the equation of state for regions in which H and He are partially ionized, which are referred to as the κ and γ mechanisms. The physics of radial oscillations was introduced by Eddington (1926) and in the early models by Cox (1958) and Zhevakin (1959). Contrary to what happens in the rest of the envelope, these regions trap energy during the contraction and loose energy during the expansion, thus acting similar to a mechanical valve. The initial perturbation may originate from a stochastic fluctuation in the external layers of the envelope. The mechanical work can either be driven, if the envelope mass located on top of the ionization regions is large enough (below a critical effective temperature that defines the blue edge of the IS) and it can be quenched by the efficiency of convective transport that penetrates deep into the stellar envelope (toward effective temperatures lower than the red edge of the IS).

Simple linear adiabatic models, such as the one developed by Eddington (1918), can predict the pulsation period and the pulsation mode, but they cannot predict the other observables such as the mean magnitudes, amplitudes, shapes of the light curve, and edges of the IS. This implies that non-adiabatic effects have to be considered to properly model the growth of the pulsational instability and the blue edge of the IS (Baker and Kippenhahn 1965; Cox 1963; Iben 1971). The other fundamental pulsation observables, such as the pulsation amplitudes, the morphology of the light curves, and the topology of the instability strip (modal stability), can only be predicted with accuracy with the inclusion of non-linear terms in the hydrodynamical equations (Christy 1966; Stellingwerf 1974) and by taking into account the coupling between pulsation and convection (Stellingwerf 1982, 1983; Feuchtinger et al. 1993).

Current state-of-the-art models adopt a non-linear, time dependent formalism, which also takes into account the effects of convection and its non-linear coupling with the pulsation

---

[5] Several authors also cite Ritter's work from a series of papers published between 1878 and 1883 in Wiedemann's Annalen 5–20 that the authors of the present manuscript have been unable to find. Such references begin as early as Shapley (1914) and are cited as recently as Smeyers and van Hoolst (2010).

(e.g., Bono and Stellingwerf 1994; Bono et al. 1997b; Marconi et al. 2003; Di Criscienzo et al. 2004; Szabó et al. 2004; Marconi et al. 2015). These models assume a spherical envelope (with no rotation or magnetic fields) and then solve the hydrodynamical equations; these are the conservation of mass and momentum plus a treatment for the radiative transfer that includes convection. These equations are solved as a function of time until they approach the limit cycle stability. Although this is one of the most comprehensive approaches, we need to highlight that there is still room for improvement. In particular, the treatment of the convective transfer is non-local and time dependent. Despite this limitation, these models can account for all the observables, including the red boundary of the IS. Moreover, they appear capable of explaining more complicated observables including the double–mode pulsators and the Blazhko effect (Szabó et al. 2014).

For the sake of the subsequent discussion, we note that van Albada and Baker (1971) formalized the relation between the pulsation periods and the structural parameters of mass (M), luminosity (L), effective temperature (T_{eff}), to which Marconi et al. (2015) adds composition (Z). The relations take the following form:

$$\log P = \mathfrak{F}\{\log(M/M_\odot), \log(L/L_\odot), \log T_{eff}, \log Z\}. \quad (1)$$

The most recent physical pulsation relations are given in Marconi et al. (2015) and are as follows:

$$\begin{aligned}\log P_F = {} & 11.347(\pm 0.006) + 0.860(\pm 0.003)\log\left(\frac{L}{L_\odot}\right) - 0.58(\pm 0.02)\log\left(\frac{M}{M_\odot}\right) \\ & - 3.43(\pm 0.01)\log T_{eff} + 0.024(\pm 0.002)\log Z \end{aligned} \quad (2)$$

$$\begin{aligned}\log P_{FO} = {} & 11.167(\pm 0.006) + 0.822(\pm 0.004)\log\left(\frac{L}{L_\odot}\right) - 0.56(\pm 0.02)\log\left(\frac{M}{M_\odot}\right) \\ & - 3.40(\pm 0.03)\log T_{eff} + 0.013(\pm 0.002)\log Z \end{aligned} \quad (3)$$

for FU (2) and FO pulsators (3), respectively. These physical relations are the basis for all the theoretically-determined pulsation relations (e.g., those projected into observable quantities), which are discussed in the following subsections.

We conclude here highlighting that recent theoretical investigations (e.g., Marconi et al. 2011, 2016) have shown that the pulsation properties of RRLs are also affected by the helium content. The Helium abundance impacts RRL evolutionary properties, like the total bolometric luminosity and the evolutionary timescales, which in turn impact the pulsation properties and the interpretation of observed quantities. Fig. 3 shows theoretical HB models in color–magnitude space from "a Bag of Stellar Tracks and Isochrones" (BaSTI) (most recently, Pietrinferni et al. 2004, with earlier works referenced therein)[6] for a fixed total metal content ($Z = 0.001$), but with different Helium contents of $Y = 0.24$ (solid hashing), $Y = 0.30$ (dotted hashing), and $Y = 0.40$ (dashed hashing). The First Overtone Blue Edge (FOBE) is shown for reference. From Fig. 3, the impact on the mean absolute magnitude (M_V) is demonstrated, with those stars having greater Helium enrichment being systematically brighter and having a slightly different temperature distribution. Beyond this, at fixed metal abundance, the pulsation period is expected to increase and the pulsation amplitude to decrease as the helium abundance increases. The width of the IS, however, is minimally affected.

[6]The models are publicly available: http://basti.oa-teramo.inaf.it/.

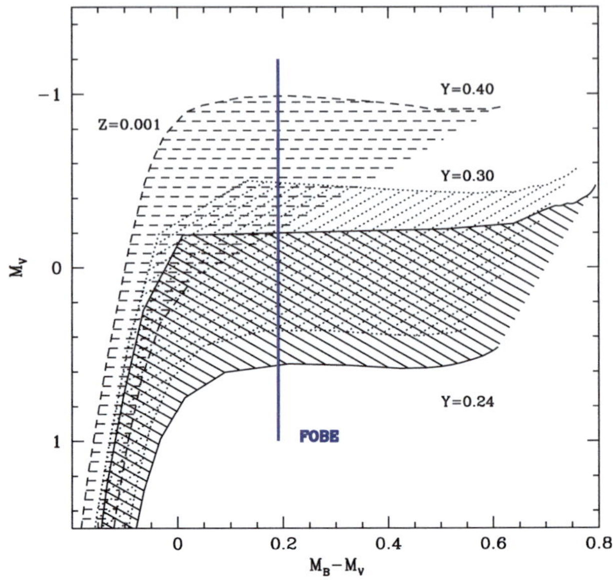

Fig. 3 Theoretical HB models from in the color–magnitude diagram. Solid, dotted and dashed regions indicate different assumptions of the helium content for a fixed overall metallicity abundance. The vertical, blue line shows the color of the First Overtone Blue Edge (FOBE) that is almost constant for the metallicity range of the RRLs. All the theoretical models used here are taken from BaSTI (Pietrinferni et al. 2004)

2.2 RR Lyrae as Standard Candles

This subsection focuses on the origin of various relationships that can be used to determine distances using RRL. First, optical relationships are explored, these being (i) the M_V-Metallicity relationship is explained in Sect. 2.2.1 and (ii) the FOBE method in Sect. 2.2.2. Then the, relatively new, NIR and MIR relationships are explained in Sect. 2.2.3. The multi-wavelength slopes are explored in Sect. 2.2.6. Lastly, reddening-free relationships constructed from multi-band observations are explored in Sect. 2.2.7.

2.2.1 The Visual Magnitude–Metallicity Relation

The original idea of a relation between the mean luminosity of RRLs and their metallicity dates back to Baade and Swope (1955) and Sandage (1958), who discovered a correlation between the mean periods and iron abundances for cluster variables (see also Arp 1955). These studies were driven by the evidence of a well defined dichotomy among the RRLs belonging to GGCs (Oosterhoff 1939). Oosterhoff (1939) found two groupings of fundamental mode RRLs, one with mean periods $P(\mathrm{RR}_{ab}) \sim 0.55$ days and [Fe/H] $\gtrsim -1.5$ and one with $P(\mathrm{RR}_{ab}) \sim 0.65$ days and [Fe/H] $\lesssim -1.5$. These two groups were later christened the Oosterhoff I (Oo I) and II (Oo II) groupings, respectively. Subsequently, this effect was also confirmed for field stars by Preston (1959). Figure 4 demonstrates the Oosterhoff dichotomy for GGCs using data from Catelan (2009). Figure 4 demonstrates two groupings of clusters in P–[Fe/H] space that separated by a region known as the "Oosterhoff gap." The dichotomy is explained as the *intrinsic luminosity* for the RRLs in two Oo groups being different, with the higher metallicity Oo I clusters being fainter (Sandage 2006). This physical understanding is largely supported by stellar evolutionary models (e.g., Cassisi and Salaris 2013), where higher metallicity stars have both a lower helium-core mass (main parameter that setting the HB luminosity) and a higher opacity in the envelope.

We demonstrate the metallicity-trend in Fig. 5 where a population of synthetic HB stars are shown for different chemical compositions (Z) from the RRL portion of the IS (drawn

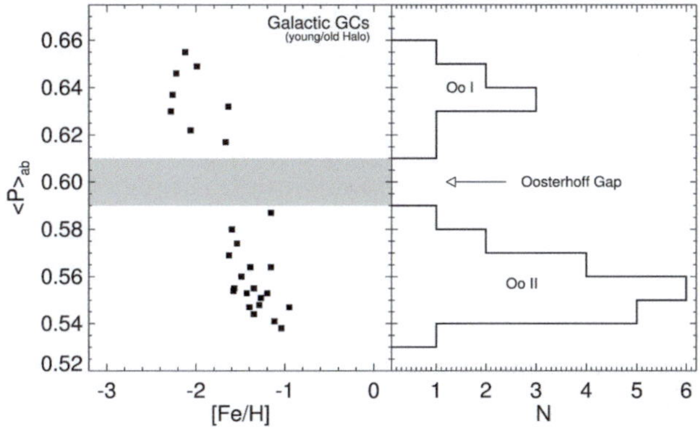

Fig. 4 Demonstration of the Oosterhoff classification for GGCs with more than ten RRab using the compilation of Catelan (2009). Both "young" and "old" halo clusters have been included in this visualization. *Left panel:* The distribution of the mean period ($\langle P \rangle_{ab}$) against the metallicity ([Fe/H]) of the cluster. *Right panel:* Histogram of the mean periods that illustrates the distinct separation between the Oo-I and Oo-II type clusters. The lack of clusters with $\langle P \rangle_{ab} \sim 0.60$ days is known as the Oosterhoff gap

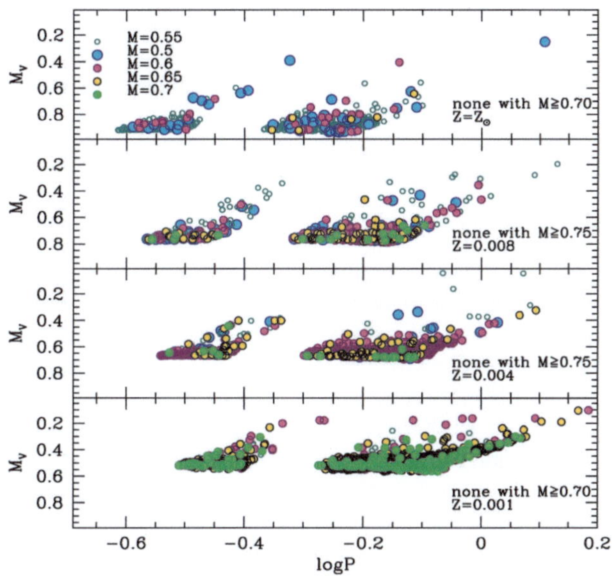

Fig. 5 Synthetic period-magnitude diagrams for a population of RRLs as a function of metallicity from $Z = 0.001$ (bottom panel) to Solar (top panel). A range of mean masses have been adopted for each simulation as indicated by color-coding and symbols shapes to highlight the population effects on both the mean magnitude and period

from the same models shown in Fig. 3). For these stars we have computed a period using the pulsation relations given in Marconi et al. (2015). We show that for metal contents ranging from $Z = 0.001$ ([Fe/H] ~ -1.6 dex) to $Z = 0.02$ ([Fe/H] ~ -0.3 dex) and for a fixed α-enhancement ([α/Fe] $\sim +0.3$ dex), synthetic periods become shorter by $\Delta \log P \sim -0.1$ and luminosities become fainter by a factor of $\Delta M_V \sim +0.3$ mag (from bottom to top in Fig. 3). However, the offset in luminosity cannot be the only cause of the difference in the mean period of fundamental RRLs between Oo I and Oo II clusters. Bono et al. (1997a, and references therein) show that the dichotomy is largely due to the different pulsation behavior in the "OR region," which is the intersection between the FU and FO IS, as a result of a hysteresis mechanism originally suggested by van Albada and Baker (1973).

Given this relation between magnitude and metallicity, it is common to assume a linear relation between the RRL V-band absolute magnitude, M_V, and the stellar metallicity, typically expressed as [Fe/H] in the following form:

$$M_V(\mathrm{RR}) = \alpha + \gamma\ [\mathrm{Fe/H}]. \tag{4}$$

Many different calibrations have been suggested in the literature for the slope (γ) and the zero point (α; e.g., Liu and Janes 1990; Carney et al. 1992; McNamara 1997; Clementini et al. 2003; Bono et al. 2003, among others).

Despite the simplicity of (4), the M_V–[Fe/H] relation as a distance diagnostic has a number of drawbacks. First of all, the reddening value and the extinction law have an impact on the absolute magnitude. The extinction coefficient is quite large in optical band, $R_V = 3.1$ (Cardelli et al. 1989), but it decreases by one order of magnitude when approaching longer wavelengths (see also Sect. 2.2.3). The reddening law, itself, may not be universal (Kudritzki and Urbaneja 2012). Moreover, several complicated astrophysical effects have a strong impact on the stability of the relationship including:

1. intrinsic deviations from the linear form of (4) (e.g., a form non-linear with iron abundance; Bono et al. 2003, 2007),
2. evolutionary effects within RRL populations,
3. measurement uncertainties for the metallicity, which include both systematic differences between metallicity scales (e.g., calibrations) and methodologies (e.g., what abundance is actually being measured),
4. measurement uncertainties for the α-enhancement, and
5. deviation from static magnitudes (see also Caputo et al. 2000).

Some additional information on these effects are given in the list below:

1. *Linearity:* The assumption of the linear M_V–[Fe/H] relation (4) has been challenged many times. There is strong empirical evidence that the M_V–[Fe/H] relation is not linear over the whole metallicity range covered by the observed GGCs (e.g., Rey et al. 2000; Caputo et al. 2000). This is also supported by both pulsational (e.g., Bono 2003; Caputo et al. 2000; Di Criscienzo et al. 2004) and evolutionary (e.g., Catelan et al. 2004) models. The slope is observed to get steeper at approximately $[\mathrm{Fe/H}] = -1.5$ and, as a result, a quadratic form of this relation has been proposed (see e.g., Caputo et al. 2000; Catelan et al. 2004; Di Criscienzo et al. 2004; Sandage and Tammann 2006; Sandage 2006; Bono et al. 2007, among others). To provide the reader with a quantitative idea of the dependence on the metallicity and of the uncertainties, we report here the calibration by Clementini et al. (2003) that is widely used:

$$V_0(\mathrm{RR}) = 19.064(\pm 0.017) + 0.214(\pm 0.047)\ [\mathrm{Fe/H}], \tag{5}$$

where the zero-point reflects the distance to LMC and the quoted uncertainties are only those on the coefficient of the linear regression. In the literature, it has been pointed out that theoretical calibrations of the M_V–[Fe/H] relation consider only the Zero Age HB (ZAHB) models, and hence they do not fully take into account evolutionary effects (e.g., Catelan et al. 2004; Sandage and Tammann 2006). However, empirical calibrations such as Caputo et al. (2000) are built on observed data, with RRL variables at different evolutionary stages, and therefore they represent a "mean" evolutionary state/status for the HB. The quadratic form of the calibration, suitable for $-2.4 < [\mathrm{Fe/H}] < 0.0$, determined in Bono et al. (2007) and applied to the Galactic Bulge and Sagittarius dwarf in Kunder and Chaboyer (2009) is as follows:

$$M_{V,\mathrm{RR}} = 1.19 + 0.5\ [\mathrm{Fe/H}] + 0.09\ [\mathrm{Fe/H}]^2. \tag{6}$$

However, it is worth noting that no deviation from linearity has been observed within the LMC over a broad metallicity range (Clementini et al. 2003). Thus, the interplay between evolutionary effects (discussed below) and metallicity is difficult to decouple in observational work. A canonical example of this difficulty is ω Centauri, which is home multiple stellar populations, and has been deemed a "red herring" in the review of Smith (1995) due to the complexity of disentangling these effects (a more recent study is Braga et al. 2016, 2018, with additional references therein).

2. *Evolutionary effects:* The evolutionary effects on the M_V–[Fe/H] relation are shown in Figs. 3 and 5. We have plotted for each selected metallicity several realizations of the HB morphology coming from different assumptions on the mean mass distribution on the HB. This is mimicking different possible values of the mass loss efficiency during the Red Giant Branch (RGB) phase, which has the result of changing the HB morphology. A comprehensive discussion of the role of mass loss on the HB morphology can be found in Salaris et al. (2002, and references therein). By an inspection of Fig. 5, it is clear that the HB width is not narrow due to the overlap in color-space of stars on the ZAHB and stars evolving off the ZAHB on their path to the Asymptotic Giant Branch (AGB) phase. The evolved-RRL have higher luminosities. The presence of both ZAHB-RRL and evolved-RRL on the HB defines the "evolutionary effect". As a result of the time spent on the HB, we observe RRLs until they have burned most of the He in their cores (up to $\sim$90%). The region where a RRL spend most of its life time is also shown in Fig. 3 for different He values and a given mean metallicity. This effect implies a broadening of the optical magnitude distribution across the HB that is also correlated with the metal content, which is supported by both theory (Bono et al. 1995) and observations (Sandage and Katem 1982).
3. *Size of the He Core:* There are other "ingredients" that affect the theoretical calibration of the M_V–[Fe/H] relation, which are related to the size of the He core reached before the He ignition. These include:
   (a) *The initial He value:* This can be summarized as the primordial He content plus a $\Delta Y/\Delta Z$, which is commonly measured from H II regions. There is evidence that $\Delta Y/\Delta Z$ may depend on the environment (e.g., Peimbert and Peimbert 2010). An increase in total He abundance of $\Delta\mathrm{He} \sim +0.2$ dex implies $\Delta M_V \sim 0.5$ mag as is demonstrated in Fig. 3.
   (b) *Details of Red Giant Evolution:* More specifically, the processes occurring along the RGB phase that delay the ignition of the 3-α reaction in the electron degenerate core. These include atomic diffusion, electron conductive opacity, loss of energy via neutrinos, and core rotation. Their impact on the HB luminosity can be up to $\Delta\log(L/L_\odot) \sim 0.1$ dex or $\Delta M_V \sim 0.2$ mag (Salaris et al. 2002; Serenelli et al. 2017).

2.2.2 The FOBE Method

The location in the period-magnitude diagram of the FOBE can be used as a distance indicator for a population of RRL independent of the M_V–[Fe/H] relationship. The FOBE method was extensively described for the first time by Caputo (1997). This is a graphical, or topological, method that produces accurate distances for those stellar systems with sizable samples of FO RRLs (Fiorentino et al. 2010). However, it can be reliably used only when the blue part of the IS is well populated (Stetson et al. 2014b; Martínez-Vázquez et al. 2017). Once the metallicity is known, preferably from spectroscopic measurements, and a reliable mass for the RRLs can be estimated (typically in the range $0.5\ M_\odot \lesssim M \lesssim 0.7\ M_\odot$; Bono et al. 2003), the following theoretical relation from Caputo et al. (2000) can be used to fit the blue edge defined by FO RRLs to predict the absolute magnitude:

$$M_V(\mathrm{FOBE}) = -0.68(\pm 0.03) - 2.25 \log P_{\mathrm{FOBE}} - 1.26 \log\left(\frac{M}{M_\odot}\right) + 0.06 \log Z, \quad (7)$$

where P_{FOBE} is the period of the bluest FO pulsators and $M_V(\mathrm{FOBE})$ is the magnitude of the bluest FO pulsators for a given composition and mass. Although, an assumption on the RRL mass has to be made (Bono et al. 2003), the possibility to use the FOBE method relies on the well understood negligible dependence of the color of the FOBE on the metallicity (Bono et al. 1997a), which is demonstrated in Fig. 3. For the bluest FO RRLs the period is most strongly dependent on luminosity and mass.

The sharp, blue edge of the IS occurs because for a given mass and luminosity, as the surface temperature increases there is less mass above the ionization zones, and in turn, their contribution to the work integral ($p\delta V$) decreases. Then, the FOBE color is essentially fixed by the minimum difference between the temperature of the stellar surface and that of the Hydrogen ionization region, within which pulsation is efficient. The minimum difference is almost constant and does not depend on metallicity, at least for the metallicity range typical of RRLs. On the contrary, the red edge of the IS strongly depends on metallicity, because the quenching mechanism for the radial pulsations is related to convection that increases with the stellar opacity, and thus by increasing the metal content (Deupree 1977a; Stellingwerf 1982). The above theoretical prediction has been observed in GGCs; more specifically, the B–V FOBE color is always ~0.2 mag. Thus, the FOBE method can also be used to estimate the mean reddening of a stellar system (see Walker 1998, for an example).

2.2.3 The NIR Period–Luminosity Relations

Unlike in the optical, the observations in the IR show a true PL relationship (first identified by Longmore et al. 1986). In Fig. 6, the mean magnitudes for RRL in the star cluster, Reticulum, are shown for eight photometric bands (B, V, I, J, H, K_s, [3.6], [4.5]) against the fundamentalized period, which is defined as follows (Bono et al. 2001):

$$\log P_{FU} = \log P_{FO} + 0.127. \quad (8)$$

Reticulum is an ideal system to visualize the behavior of the PL slope with wavelength, because it has no appreciable metallicity spread on the RGB (Grocholski et al. 2006) and, thus, scatter due to metallicity differences are minimized. In Fig. 6, preliminary PL fits are shown for each band as solid lines with the scatter about the PL shown as the shaded regions (Catelan and Smith 2015, Monson et al., in prep.). As previously discussed, the V band shows no PL slope, whereas a slope appears and becomes progressively steeper at the longest wavelengths. In the next sections, we discuss empirical and theoretical results on the RRL PL.

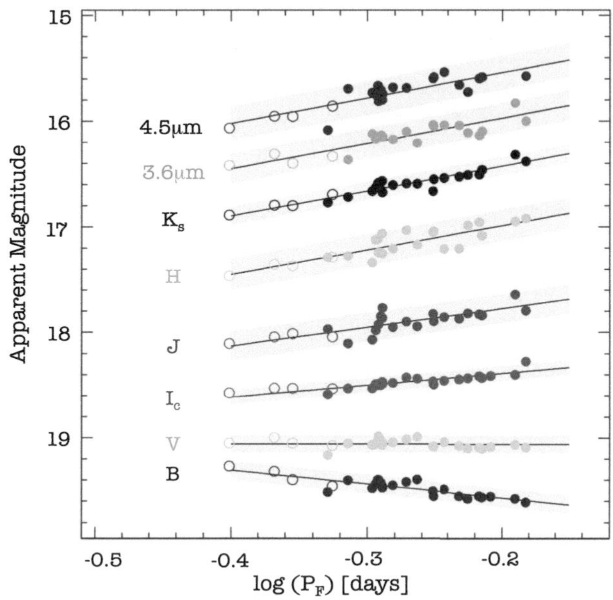

Fig. 6 Mean apparent magnitudes for RRL in the Reticulum star cluster in the optical (*B*, *V*, *I*), NIR (*J*, *H*, *K*), and MIR ([3.6], [4.5]) versus the fundamentalized period (8). PL relationships are overplotted (solid lines) for each of the bands (adapted from Catelan and Smith 2015, Monson et al., in prep.). The change in the slope from the optical into the mid-infrared is noticeable, with *V* being a nearly flat slope to the mid-infrared being quite steep

2.2.4 The K-Band Period–Luminosity Relationship

Period–luminosity–color (PLC) relations have been predicted for RRL by even the earliest theoretical models (e.g., van Albada and Baker 1971), but their use has always been hampered by several uncertainties on the structural parameters of the stars, in particular mass, effective temperature and metallicity. Evolutionary effects also play a role, because during their evolution (the He burning phase lasts at most for 100 Myr), RRLs change their luminosities and effective temperatures. Moreover, observational uncertainties, such as those associated with reddening and extinction, have to be decoupled from true astrophysical differences. Thus, empirical confirmation of the PLC has only come of late.

Longmore et al. (1986) demonstrated, on pure empirical basis, that RRLs follow a tight linear PL relation in the K-band (PL_K). They also demonstrated that the PL_K relation can be derived from the general equation of the pulsation, since the period dependence on the K-band can be derived from the $V-K$ proxy of the effective temperature (as a direct consequence of the fact that RRLs follow PLC relations). In other words, a PL_K arises because of the increasing bolometric correction as a function of the effective temperature, with the redder RRLs having K-band magnitudes brighter than the bluer RRL (see Fig. 7, for the K band). Moreover, since for a given cluster, the effective temperature distribution inside the IS is related to the period distribution, such that the stars with longer periods are also cooler, the K magnitude-effective temperature relation becomes a K magnitude-period relation. A more general empirical calibration of the PL_K was given by Janes and Liu (1992) and these authors also considered the dependence on the metallicity, which was shown to have little impact on the slopes.

A sound and complete theoretical derivation of the K-band period-luminosity–metallicity (PLZ_K) relation was subsequently explored by Bono et al. (2001, 2003), finding the following:

1. the uncertainties on the mass and luminosity (e.g., on the evolutionary effects) have only a mild impact on the PLZ_K;

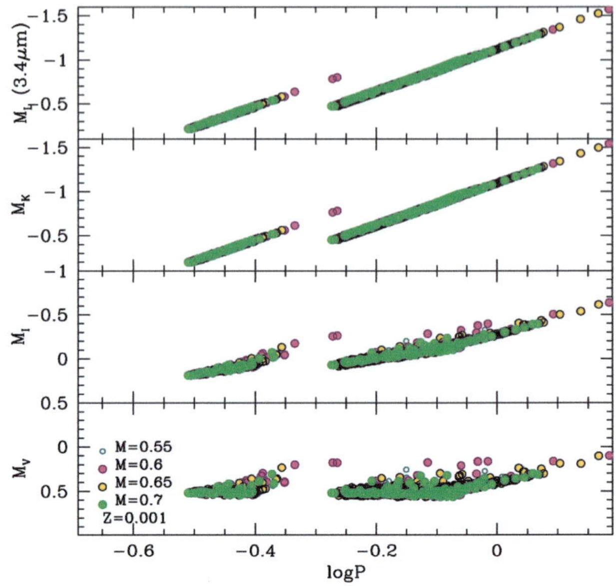

Fig. 7 PL distribution from optical (bottom panel) to mid-infrared (top panel). We have used synthetic HB models as in Figs. 3 and 5 for fixed abundance ($Z = 0.001$) and a canonical He abundance ($Y = 0.24$) with different mean mass for the population distribution

2. the overall dependence on the metallicity is lower, at $\gamma \sim 0.17$ mag dex^{-1} in [Fe/H], than that in the optical bands ($\gamma > 0.2$ mag dex^{-1}, see (5)).

On the latter aspect, we remark that Sollima et al. (2006) found an even lower metallicity term ($\gamma = 0.08$ mag dex^{-1} in [Fe/H]), on the basis of an empirical calibration. Muraveva et al. (2015) combined low-dispersion spectroscopy of 70 RRLs in the LMC, with $-2.0 \leq$ [Fe/H] ≤ -0.6 dex, with NIR photometry collected with VISTA finding a very low dependence on the metallicity ($\gamma = 0.03 \pm 0.07$ mag dex^{-1} in [Fe/H]), and a quite steep dependence on the period (-2.73 ± 0.25). All these aspects reflect on a very low dispersion of the PLZ$_K$ relation, being $\sigma = 0.035$ mag rms or <2% in distance (Bono et al. 2001).

PL$_K$ and PLZ$_K$ observational relations have been successfully tested on several stellar systems, among the others we mention the LMC cluster Reticulum (Dall'Ora et al. 2004) and the Galactic GGCs IC 4499 (Storm 2004), M 92 (Del Principe et al. 2005), Omega Centauri (Del Principe et al. 2006), M 5 (Coppola et al. 2011), and M 4 (Braga et al. 2015). The observed scatter about the PL ranges from 0.03 mag (Reticulum) to 0.09 mag (IC 4499), which suggests a single-star distance uncertainty ranging from 1.5% to 4.5% is feasible. Additional studies have occurred in dwarf galaxies such as the LMC (Ripepi et al. 2012a; Moretti et al. 2014), SMC (Szewczyk et al. 2009; Muraveva et al. 2018c), Fornax (Karczmarek et al. 2017a), Sculptor (Pietrzyński et al. 2008), Carina (Karczmarek et al. 2015), and IC 1613 (Hatt et al. 2017).

Compared to the M_V–[Fe/H] relationship, the key advantages distance determination at longer wavelengths are:

1. the impact of extinction is less by a factor of ten or more and, as a result, the impact of differential extinction is also greatly reduced;
2. the impact of the uncertainties the reddening measurement, itself, are minimized (this being what is determined to estimate extinction);
3. the pulsation amplitude is lower than in the optical bands (Fig. 2) and the light curves are more symmetrical, accurate mean magnitudes can be measured even with a few data points;

4. the availability of light curve templates (e.g., Jones et al. 1996; Monson et al. 2017; Hajdu et al. 2018), permits the estimation of the mean magnitudes even with a single data point, if the ephemerides of the variable are known from the optical photometry (see applications in Beaton et al. 2016, Rich et al. submitted).

On the other hand, finding the variables, measuring their periods, and determining their pulsation mode (FU or FO) is more difficult due to smaller amplitudes, sinusoidal shape, and the more uniform light curve shapes between the sub-classes. Moreover, templates are only available for the RRab type stars.

2.2.5 Mid-IR Period–Luminosity Relations

The advantages of using PL relations at mid-infrared wavelengths (MIR, from 3.4 to 22 μm) with Cepheids have been well-recognized in the literature since McGonegal et al. (1982), but only very recently it has been possible to explore this wavelength regime with RRLs. Indisputable advantages to the MIR PLs is that at 3.4 μm the impact of interstellar absorption is more than one order of magnitude lower than in the optical bands, which permits study of heavily obscured and differentially obscured regions like the Galactic bulge that contain a large population of RRL (see discussion in Kunder et al. 2018).

Indeed, on the basis of the Wide Field Infrared Survey Explorer (WISE) measurements of field RRL variables, Klein et al. (2011) were able to produce the first calibration of the PL relations at the W1 (3.4 μm), W2 (4.6 μm), and W3 (12 μm) bands. They subsequently refined their calibrations in Klein et al. (2012) and Klein et al. (2014), with no metallicity term. Another absolute calibration of the MIR PL relations was published by Madore et al. (2013), using four field RRLs for which individual trigonometric parallaxes were available via the *HST* Fine Guidance Sensors (*HST*–FGS; Benedict et al. 2011). Even with only four objects, the observed dispersion obtained by Madore et al. (2013) is a very promising 0.1 mag ($\sim$5% in distance). Madore et al. (2013) also found, as expected, only a marginal dependence on the metallicity, because at the MIR wavelengths stars with the temperature of RRLs have few molecular bands and metallic lines. Based on a much larger sample of 129 RRLs with *Hipparcos*-based distances, Klein et al. (2014) obtained very nice results, with observed spreads ranging from 0.01 mag to 0.05 mag, depending on the filter, on the sample (FU or FO pulsators), and assumptions on the intrinsic width of the PL. Neeley et al. (2017) expanded the *Spitzer*-analysis to a larger sample of RRL with parallaxes from the TGAS catalog in *Gaia*-DR1 (e.g., Michalik et al. 2015; Lindegren et al. 2016), finding consistent solutions within the observational uncertainties. A complementary analysis from Sesar et al. (2017) used *WISE* photometry with the TGAS parallaxes employing a Bayesian algorithm, which Muraveva et al. (2018a) has expanded upon recently using *Gaia*-DR2 (Brown et al. 2018b).

The first calibration of the MIR PL using cluster variables was published by Dambis et al. (2014), which presented an analysis of *WISE* data for 15 GGCs. Neeley et al. (2015, 2017) published data of the GGC M 4, collected with the Infrared Array Camera (IRAC) on-board the *Spitzer* Space Telescope, obtaining PL relations in the IRAC [3.6] and [4.5] bands with a scatter of $\sim$0.05 mag (2.5% in distance). Muraveva et al. (2018b) has recently published a calibration in the LMC star cluster Reticulum.

2.2.6 The RR Lyrae Period–Luminosity Slopes

Because the bolometric correction sensitivity to effective temperature is already at work starting from the R-band (see Cassisi et al. 2004; Catelan et al. 2004; Marconi et al. 2015),

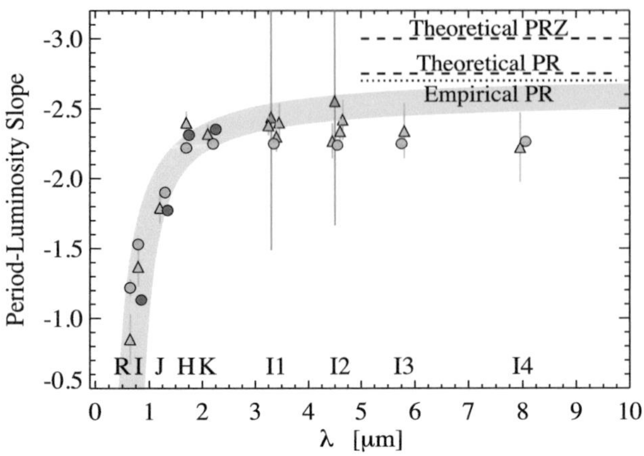

Fig. 8 Following Madore et al. (2013, their Fig. 4), the PL slopes are plotted a function of wavelength from both theoretical (circles; Catelan et al. 2004; Marconi et al. 2005a, 2015; Neeley et al. 2017) and empirical (triangles; Klein et al. 2011; Madore et al. 2013; Braga et al. 2015; Dambis et al. 2014; Neeley et al. 2015, 2017) studies. We note that studies in the same photometric band have been offset slightly in their effective wavelength for visualization purposes. The shaded region is drawn to guide the eye following Madore et al. (2013). The slopes for the long-wavelength filters asymptotically approach that predicted by period-radius relationship for both theory (dashed line Marconi et al. 2015) and empirical determinations (dotted line Burki and Meylan 1986)

PL relations have also now been derived theoretically for the photometric bands R, I, J and H, which are shown in Fig. 7. At shorter wavelengths, the effects related to the finite width of the IS and evolution of RRLs are larger and produce PL relations with larger intrinsic scatter (compare the panels of Fig. 7). The relationships in R, I, J and H are still of great interest, because these wavelengths can be observed easily from the ground.

In Fig. 8, the PL slopes determined from a subset of both empirical (triangles) and theoretical (circles) investigations are plotted versus wavelength following the example of Madore et al. (2013). These slopes are for the "global" population, e.g., the first-overtone periods have been fundamentalized following (8). The shaded region in Fig. 8 is meant to guide the eye to the behavior with wavelength and this is adapted from the visualization in Madore et al. (2013, their Fig. 4). Generally, the slopes between the different studies, both theoretical and empirical, agree well. Still, there are comparatively few studies on the PL for wavelengths longer than V and further investigations, in particular empirical studies in star clusters with differing chemical compositions and evolutionary states, are highly relevant to determining a "consensus" PL slope for a given filter.

In Fig. 8, the "predicted" slopes from the period-radius (PR) or period-radius-metallicity (PRZ) relationships are also shown following Madore et al. (2013). NIR and MIR wavelengths sample the Rayleigh–Jeans tail for RRLs and, as a result, the magnitudes are much less sensitive to temperature than to the radial variations. This behavior is readily seen in the light curves shown and, indeed, the light curves beyond H are nearly indistinguishable. Thus, predicting the PL slope from the period-radius relationship places an asymptotic limit on the PL slopes.

Starting from $L = 4\pi R^2 \sigma T_{eff}^4$ and converting into a logarithmic form yields the following,

$$\log L = \log(4\pi\sigma) + 2\log R + 4\log T_{eff}, \tag{9}$$

which can be converted into magnitudes as follows,

$$\log M = -5 \log R - 20 \log T_{eff} + \text{constant}. \quad (10)$$

From here, we can substitute either the period-radius (PR) or period-radius-metallicity (PRZ) relationship to remove the radius term. The PR has been determined in several ways, (i) empirically by Burki and Meylan (1986) using the Baade–Wesselink technique (BW), (ii) stellar evolutionary models by Marconi et al. (2005b) (evol), and (iii) pulsation models by Marconi et al. (2015) (puls). The PR/PRZ takes the form, $\log R = a + b \log P + c \log Z$, and the three studies find period slopes of $b_{BW} = 0.54$, $b_{evol} = 0.65$, and $b_{puls} = 0.55$, respectively. Substituting, these into (10), we estimate period slopes of $\beta_{BW} = -2.7$, $\beta_{evol} = -3.25$, and $\beta_{puls} = -2.75$.

If we assume that the temperature term contributes very little at long wavelengths then we can ignore that term to obtain an estimate for the slope of the PL relationship. The estimated slopes from the theoretical PR and PRZ (dashed lines; Marconi et al. 2015) and the empirical PR (dotted lines Burki and Meylan 1986) are shown in Fig. 8. The overall agreement between these asymptotic values and the slopes in Fig. 8 serves as a good cross-check on our theoretical and empirical measurements.

2.2.7 Reddening-Free Relations

Even if PLZ relations have advantages for estimating individual distances, they are still affected by the uncertainties in the reddening corrections, and the problem can be severe in regions affected by strong differential reddening (e.g., as in M4 Braga et al. 2015, or in the Galactic Bulge).

To circumvent this problem, van den Bergh (1975) and Madore (1982) developed the Wesenheit pseudo-magnitudes. They are, by construction, "reddening free" functions that redefine the observed magnitude. The fiducial form of the transformation is defined as:

$$W(A, B) = M_A + R_{A,B}(M_A - M_B) \quad (11)$$

where $R_{A,B}$ is the selective-to-total extinction ratio. In the common practice, "Cardelli's law" is adopted (Cardelli et al. 1989).

The Wesenheit period–luminosity (PW) and period–luminosity–metallicity (PWZ) relations have the following form:

$$W = \alpha + \beta \log P + \gamma \text{ [Fe/H]}, \quad (12)$$

where each combination of filters will have its own set of PWZ parameters. The advantages and disadvantages of PW and PWZ relations have been discussed several times in the literature, in particular for classical Cepheids (e.g., Inno et al. 2013; Fiorentino et al. 2007; Ngeow et al. 2005; Storm et al. 2011; Ripepi et al. 2012b) and, more recently, with RRLs (e.g., Braga et al. 2015; Marconi et al. 2015, and references therein).

Briefly, the main features can be summarized as follows:

1. the individual distances are independent of the uncertainties in the reddening or in differential reddening within an RRL population;
2. a universal reddening law is assumed, but there are indications of deviations from "Cardelli's law" in regions of high obscuration (Cardelli et al. 1989);

3. PW and PWZ relations mimic PLC relations, since they host the color term: this means that the effect of the width of the IS is reduced, and individual distances are much more precise than those obtained with simple PL relations.

Thus, in regions where the assumption of universal reddening law is reasonable, the PW/PWZ is a powerful tool.

On the practical side, however, use of the PW/PWZ are more observationally expensive than common PL and PLZ relations, because they need accurate mean magnitudes determined in two photometric bands. According to the models by Marconi et al. (2015), the dispersion around the PW and PWZ relations range from ~ 0.01 mag (in the NIR bands) to $\sim$0.07–0.08 mag (in the optical bands). Braga et al. (2015) provide complementary empirical PW/PWZ relations determined from M 4 in the optical and NIR. The $PW_{B,V}$ relation is almost independent of metallicity and is particularly useful.

We also note that, on empirical basis, Riess et al. (2011) proposed to use three-band PW relations, where the W magnitude is reconstructed from the following form,

$$W(A, B, C) = M_A + R_{A,B,C}(M_B - M_C), \tag{13}$$

where $R_{A,B,C}$ is the ratio between the selective absorption in the A band, and the color excess for the $B - C$ color $E(B - C)$. The advantages of triple band PW/PWZ relations are a smaller dispersion respect to the two band relations and lower systematics from correlated errors having used the same magnitude in the color term. The drawback is that they need accurate mean magnitudes determined for three bands. The most advanced theoretical study on this topic to date is that of Marconi et al. (2015), which provides theoretical three-band relations for RRL in the optical and NIR.

2.3 Case Studies Using RR Lyrae

In the previous section, a general physical understanding of RRLs as distance indicators was developed using both theoretical and empirical investigations. In this section, a series of case studies are presented on the practice of using RRLs for the different wavelength regimes discussed previously. In Sect. 2.3.1, the M_I–[Fe/H] relationship and the optical PW relationships are applied to the Sculptor dwarf galaxy In Sects. 2.3.2 and 2.3.4, the optical, near-infrared and mid-infrared PL relations are determined and applied to the nearby GGC M 4 that suffers from strong differential reddening in the optical.

2.3.1 An Optical Study of RR Lyrae in Sculptor

Here we present a brief summary of a comprehensive study of the RRLs in the Sculptor dwarf galaxy (Martínez-Vázquez et al. 2015, 2016a,b). Sculptor is a Milky Way dwarf spheroidal (dSph) satellite with a complex chemical enrichment history (Smith and Dopita 1983; Da Costa 1984; Majewski et al. 1999; Hurley-Keller et al. 1999; Tolstoy et al. 2004; de Boer et al. 2012; Starkenburg et al. 2013). Tolstoy et al. (2004) demonstrated that the red HB stars (RHB) are more centrally concentrated than the blue HB stars (BHB) that mirrors gradients from detailed spectroscopic studies on the RGB (Battaglia et al. 2008; Walker et al. 2007, 2009; Kirby et al. 2009; Leaman et al. 2013; Ho et al. 2015), which suggests a that range of metallicites is also present in the RRL population.

Time series photometry was produced for stars Sculptor from a set of 5149 individual images collected over 24 years. Variables were identified using the Welch–Stetson variability index (Welch and Stetson 1993) and periods obtained from a simple string-length algorithm

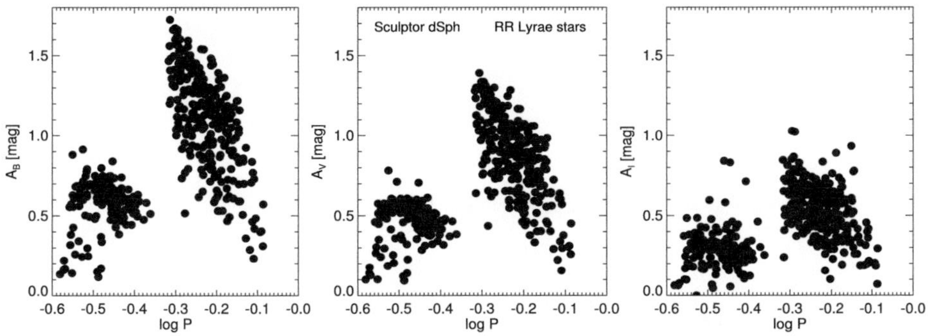

Fig. 9 Period-amplitude diagrams for the RR Lyrae stars in Sculptor in three different optical bands: *B* (left), *V* (middle) and *I* (right). RRab and RRc are shown as black circles. For the sake of clarity, RRd stars are not plotted as their periods are less certain. The RRc and RRab occupy specific regions in the Bailey diagram, with the RRc having shorter periods and smaller amplitudes (i.e., in the bottom left) than the RRab's (right). The RRab's also have a broader range of amplitudes. These primary groups in the Bailey diagram can be further refined into specific sequences for the Oo groups, with two clear sequences visible for the RRab's

(Stetson et al. 1998) from which 536 RRLs were identified using the shapes of the light curve. The final dataset contains pulsational properties and mean magnitudes for 536 RRLs in the B, V, and I bands that is complete over $\sim$70% of Sculptor's area.

Figure 9 shows the Bailey Diagrams for the B, V, and I bands. The RRab and RRc stars separate into long- and short-period groups, that have their own amplitude behavior. The overall reduction of the amplitudes from B to I is apparent. In the B and V the subdivision of the primary RRab and RRc sequences into Oo groups can also be seen, which is anticipated from the large metallicity spread.

Figure 10 shows the V magnitude distribution for the Sculptor RRLs. A zoom-in on the CMD is shown in the left panel and the corresponding luminosity function for the RRLs is shown in the right panel. The RRL stars of Sculptor show a spread in V magnitude of $\sim$0.35 mag, which is significantly larger than the typical uncertainties in the mean magnitude ($\sigma = 0.03$ mag), larger than anticipated for a mono-metallicity population showing evolutionary effects, and larger than the intrinsic spread for a mono-metallicity population.

The I-band PLZ (PLZ$_I$) can used to obtain the metallicity distribution for the RRLs following Martínez-Vázquez et al. (2016a):

1. A cross comparison of the photometric metallicities to the spectroscopic measurements from Clementini et al. (for several dozen sources; 2005).
2. Application of the technique to RRL in Reticulum, which has a well-defined spectroscopic metallicity (e.g., Mackey and Gilmore 2004) and well-sampled I light curves (Kuehn et al. 2013).

Both tests validated the photometric metallicities obtained using the PL$_I$ and the results of this analysis for the Sculptor RRLs are given in Fig. 11. In Fig. 11, the left panel shows the full distribution for Sculptor and the right panel emphasizes the difference in the population split into the Bright (Bt) and Faint (Ft) groups identified from the V-magnitudes.

The distribution of RRLs are centered at -1.90 dex, with $\gtrsim$90% of the sample between -1.2 and -2.6 dex (left panel of Fig. 11). The Bt sample is on average more metal-poor, with mean metallicity of -2.03 dex, than the Ft sample that has a mean metallicity of -1.74 dex (right panel of Fig. 11).

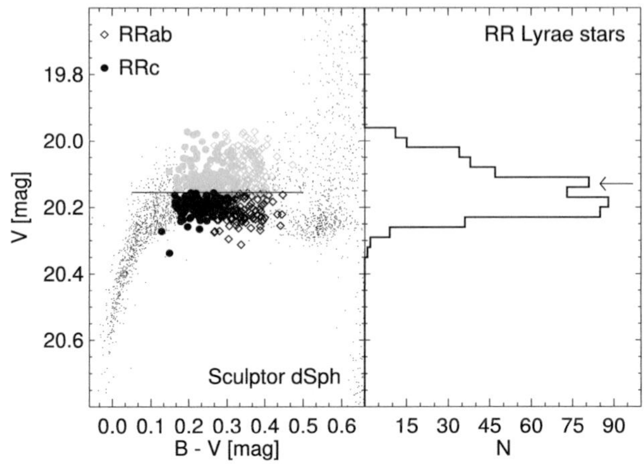

Fig. 10 The V distribution for Sculptor RRL. Left panel: Zoom-in on the HB stars in the color–magnitude diagram of Sculptor. The horizontal line (at $V = 25.155$ mag) displays the split between bright (gray) and faint (black) RRL subsamples. Open diamonds and circles represent the RRab and RRc stars, respectively. Right panel: Luminosity distribution function in V band for Sculptor RRLs. The arrow marks the magnitude adopted to split into bright and faint RRL subsamples. We note that the RRd and other more complicated RRL sub-types have been removed from this sample

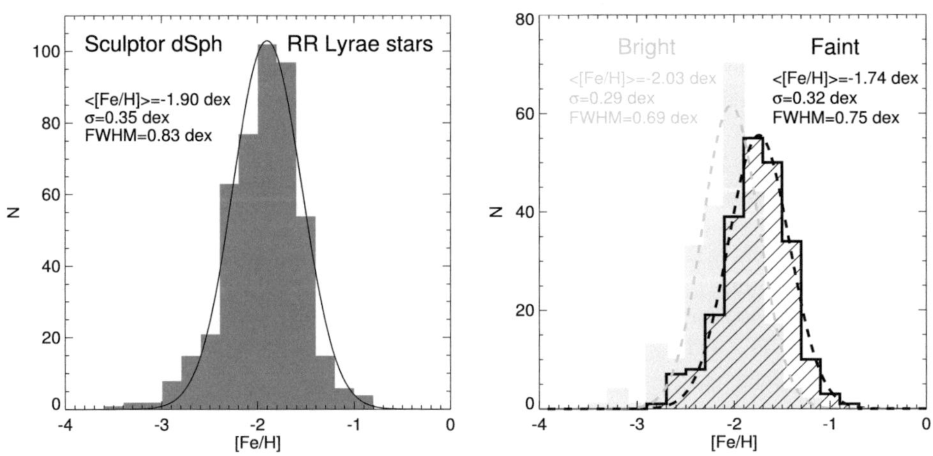

Fig. 11 *Left panel:* Metallicity distribution obtained from the PL in I band for the RRLs of Sculptor (dark grey histogram). The black solid line shows the Gaussian fit performed. The parameters of the fit are labeled in the panel. *Right panel:* Metallicity distribution for the bright (light grey histogram) and faint RRL subsample. Both distributions have been fitted to a Gaussian (dashed curves). The parameters of each fit are labeled in the panel

The distance to Sculptor was determined using the metal-independent PW relations for the V, $B{-}V$ and V, $B{-}I$ filter-color combinations (e.g., (11) and (13)) with parameters from Marconi et al. (2015, their Table 9) and Martínez-Vázquez et al. (2015, their Table 2), respectively. Marconi et al. (2015) demonstrated that these PW relations are relatively insensitive to metallicity, which is important for the large metallicity spread in Sculptor. These

PW relations were applied to each of three subsamples within the RRL population: (i) RRab, (ii) RRc, and (ii) the global sample (both RRab and fundamentalized RRc).

The zero-points of the PW relationships are not well constrained from observations as there are only good (<10%) trigonometric parallaxes for five RRL, even after the release of TGAS parallaxes from *Gaia* DR1 (Lindegren et al. 2016, and references therein) Thus, three different approaches were used to quantify the effect of both the zero-point (α) and the slope (β) of the relation on the distances. These are:

1. *Theoretical:* Using the zero-points and slopes of the predicted PW relations.
2. *Semi-empirical:* Using the same theoretical zero-points and the empirical slopes, which were obtained from the fit performed in the plane of both observational Wesenheit magnitudes.
3. *Empirical:* Adopting observed slopes used in the previous method, but using the zero-points determined from *H*ST trigonometric parallaxes (Benedict et al. 2011). More specifically, RR Lyr, itself, to calibrate both the RRab and the global samples and RZ Cep to calibrate the RRc sample.

The full details of these distance determinations can be found in Martínez-Vázquez et al. (2015, in particular, Sect. 4 and their Table 3).

The zero-points are determined for each of the four RRab, global, and RRc PW relations. For the RRab and global relations, the zero-points agree quite well with theoretical ones. However, in the case of the RRc, the new zero-points are ~30% smaller than predicted, which is in agreement with the mean RRc magnitude determined from Kollmeier et al. (2013) from statistical parallax and highlights the need to use a large sample to determine robust zero points.

The final result is $\mu_0 = 19.62$ mag with $\sigma_{mean} = 0.04$ mag, which is in good agreement with previous determinations (e.g., Rizzi 2002; Pietrzyński et al. 2008, among others). The multiple methods required to set the zero-points and slopes of the PW relationships also emphasizes the role that high-quality trigonometric parallaxes for field RRLs, and later cluster RRLs, will play in improving the accuracy of the distances determined via the RRL PL, PLZ, and PW relationships.

2.3.2 A Near-IR Study of the RRLs in M4

We present a case study using the GGC M 4. M 4 is the closest GGC to the Sun and it hosts a sizable sample of RRLs (Stetson et al. 2014b), but the cluster suffers from strong differential extinction. The mean reddening is $E(B-V) = 0.37$ mag with a range of $\Delta E(B-V) = 0.10$ mag—or a mean A_B (A_V) of 1.81 mag (1.37 mag), with variations at the 0.4 mag level across the cluster. In contrast, the IR reddening is $E(J-H) = 0.13$ mag with a variation of only $\Delta E(J-H) = 0.04$ mag—or a mean $A_H = 0.22$ mag with variation at the ~0.1 mag level.

The data for M 4 is similar in its volume and time-baseline to that for the Sculptor dwarf from Sect. 2.3.1 and is described in full in Stetson et al. (2014b). The optical data was obtained in 18 runs over 16 years for a total of over 5000 images in U, B, V, R, and I filters. The NIR data was obtained in 18 runs over 10 years in J, H, and K. The final RRL census in M 4 contains 32 RRab and 12 RRc for a total of 45 RRL.

Figure 12a shows the optical CMD for M 4 and Fig. 12b is a histogram of the mean magnitudes in V for the RRLs, both of which can be compared directly to the corresponding panels of Fig. 10 showing the same for Sculptor. M 4 has a dramatically smaller metallicity spread than Sculptor, but it suffers from strong differential extinction in the optical, which

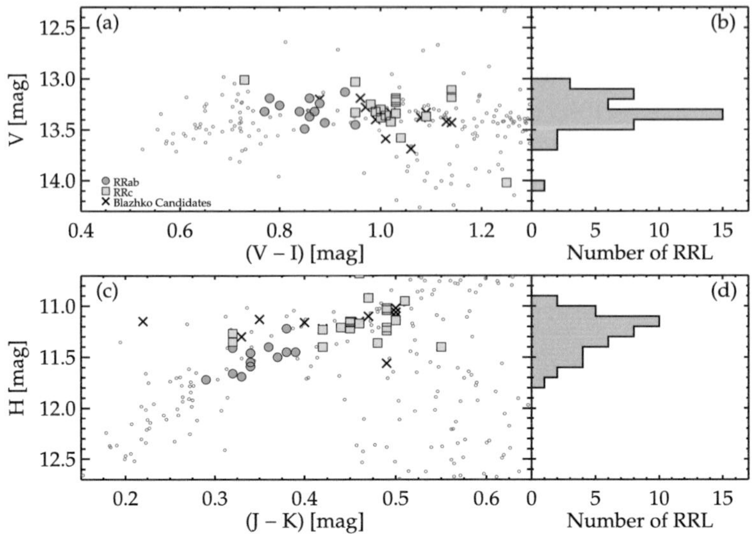

Fig. 12 Optical and NIR CMDs for M 4. (**a**): Optical $V - I$, V CMD for the HB in M 4 (data from Stetson et al. 2014a). The RRab (circles), RRc (squares), and Blazhko candidate RRLs are indicated. (**b**): V magnitude histogram. The dispersion of the distribution is ~0.18 mag, which, despite the differential reddening, is much narrower than that of Sculptor (Fig. 10) due to the lack of strong metallicity gradients. (**c**): NIR $J - K$, H CMD for the HB in M 4 (data from Stetson et al. 2014a). Unlike the optical (a), the NIR HB has a noticeable slant that gives rise to the PL. (**d**): H magnitude histogram where the width is due to the PL. The y-axis for all panels is 2.0 mag for ease of comparison

creates a broadened distribution of RRL magnitudes. Figure 12c is the NIR CMD for M 4 and Fig. 12d is a histogram of the mean magnitudes in H for the RRLs.

2.3.3 Period–Luminosity Relations

Figure 13 shows the empirical PL relations in the J (top), H (middle), and K (bottom) bands for the RRab and RRc in M 4. The RRab and RRc RRL are shown independently in Fig. 13 with their native periods. The resulting PL relations are given below. First, for the RRab stars:

$$
\begin{aligned}
J_{RRab} &= -2.030(\pm 0.204)\log P + 10.946(\pm 0.056)\sigma = 0.065 \\
H_{RRab} &= -2.215(\pm 0.176)\log P + 10.537(\pm 0.047)\sigma = 0.050 \\
K_{RRab} &= -2.372(\pm 0.142)\log P + 10.410(\pm 0.039)\sigma = 0.045
\end{aligned}
\qquad (14)
$$

For the RRc,

$$
\begin{aligned}
J_{RRc} &= -2.020(\pm 0.273)\log P + 10.634(\pm 0.148)\sigma = 0.056 \\
H_{RRc} &= -2.340(\pm 0.179)\log P + 10.232(\pm 0.097)\sigma = 0.037 \\
K_{RRc} &= -2.440(\pm 0.198)\log P + 10.058(\pm 0.108)\sigma = 0.041
\end{aligned}
\qquad (15)
$$

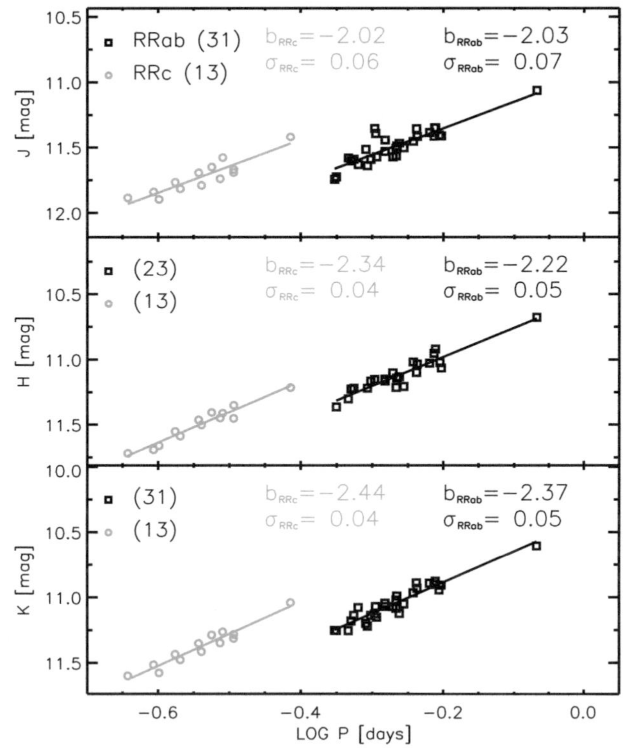

Fig. 13 NIR PL relations for RRL in M 4. Top: Empirical J-band PL relations of the RRLs of M 4. Light circles represent RRcs, while dark squares mark the position of RRabs. Slopes (b) and dispersions (σ) of the relations are labeled. Middle: same as top, but for H-band PL relations. Bottom: same as top, but for K-band PL relations

Lastly, the global PL uses fundamentalized periods for the RRc variables (e.g., (8)), which is not shown explicitly in Fig. 13:

$$\begin{aligned} J_{Glob} &= -\,1.793(\pm 0.109)\log P_F + 11.002(\pm 0.035)\sigma = 0.064 \\ H_{Glob} &= -\,2.408(\pm 0.082)\log P_F + 10.492(\pm 0.027)\sigma = 0.046 \\ K_{Glob} &= -\,2.326(\pm 0.074)\log P_F + 10.420(\pm 0.024)\sigma = 0.043 \end{aligned} \tag{16}$$

These slopes and their dispersions are also labeled on the top of each panel in Fig. 13. As expected, the slope increases with wavelength as PL approaches the PR relation (e.g., Fig. 8). The dispersion in K (0.06 mag) is half that of the R for this cluster (Braga et al. 2015), further emphasizing a strength of the NIR PL.

The distance to M 4 can be determined once the zero-point of the PL is calibrated. Similar to the methods for Sculptor, this can be done either on an empirical or on a theoretical basis, as we describe below:

1. *Empirical*: Zero points are determined by adopting the *HST* parallaxes (Benedict et al. 2011) and the mean *JHK* magnitudes (Sollima et al. 2008) for RR Lyr (an RRab). Note that the *HST* parallaxes have smaller uncertainties than those from TGAS in *Gaia* DR1 (Lindegren et al. 2016).
2. *Theoretical*: Zero points are adopted from the predictions of the theoretical PLZ relations (Marconi et al. 2015). The metallicity term is taken into account by adopting a fixed [Fe/H] $= -1.1$ dex, which is the average iron abundance of M 4 (Marino et al. 2008; Malavolta et al. 2014).

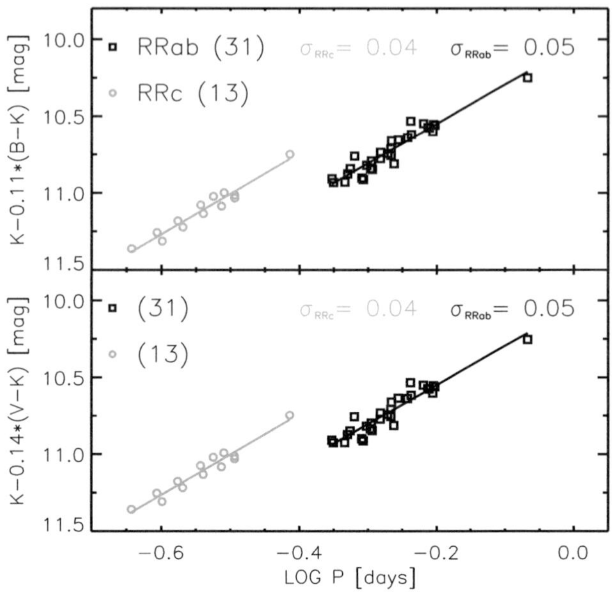

Fig. 14 Three-band PW relationships for RRL in M4. Top: Empirical PW(K,$B-K$) relations of the RRLs of M4. Light circles represent RRcs, while dark squares mark the position of RRabs. The dispersions (σ) of the relations are labeled. Bottom: same as top, but for PW(K, $V-K$) relations

The two methods provide distances of $\mu_0 = 11.35 \pm 0.03$ (standard error) ± 0.05 (standard deviation) mag and $\mu_0 = 11.283 \pm 0.010$ (standard error) ± 0.018 mag (standard deviation), respectively (see Braga et al. 2015, for additional details). Both are in agreement within 1 σ with estimates of the distance modulus from the literature, in particular that derived from eclipsing binaries of $\mu_0 = 11.30 \pm 0.05$ mag (1-σ agreement; Kaluzny et al. 2013).

Figure 14 shows two PW relationships in B, $B-K$ and K, $V-K$ for M 4 from Braga et al. (2015) that combine both optical and NIR magnitudes. Using these relationships, calibrated in a similar fashion, provide $\mu_0 = 11.272 \pm 0.004 \pm 0.013$ mag, in perfect agreement with the distance moduli determined from the PLs. For the two relationships shown in Fig. 14, the dispersions are between 0.04 and 0.05 mag. The small dispersions and the agreement between these distance moduli and those from the NIR PL provide empirical validation that the PW relationships are indeed insensitive to reddening and that the assumptions of the PW relationship (e.g., "Cardelli's Law") work well for M 4 in the NIR.

2.3.4 A MIR Study of RR Lyrae in M4

The MIR case study also uses the closest GGC, M 4. Due to the smaller amplitudes at longer wavelengths, it is not practical to identify and derive the pulsation characteristics using only MIR data (e.g., Fig. 2). Therefore we rely on ground-based optical and NIR data described previously to provide the coordinates and periods of our target RRLs, which then sets the optimal strategy to derive well sampled light curves in the MIR (see also Hendel et al. 2018; Muraveva et al. 2018b).

Intensity mean magnitudes are computed either from Fourier fitting techniques (as in Dambis et al. 2014) or the GLOESS method (as in Monson et al. 2017). Using the exquisite multi-band data available for M 4 and described in this chapter, the extinction and distance to M 4 can be measured using a semi-empirical method, as described in Neeley et al. (2017). The photometry was simultaneously fit in each band to theoretical PLZ relations, finding an extinction of $A_V = 1.45 \pm 0.12$ (assuming $R_V = 3.62$ measured directly in the foreground

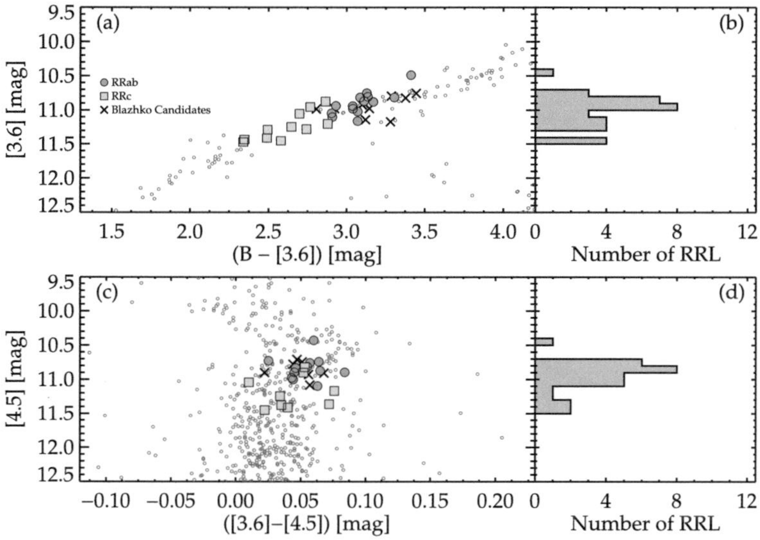

Fig. 15 MIR photometry for the HB in M 4. (**a**) Joint opt-MIR CMD for the HB in M 4 (data from Stetson et al. 2014a; Neeley et al. 2015) with the RRL common to both datasets indicated. (**b**) Distribution of [3.6]-mag for the RRLs in M 4. (**c**) MIR CMD for the HB in M 4 (data from Neeley et al. 2015) that demonstrates the HB is not readily identifiable in MIR CMDs. Identification and classification of RRLs is much more readily completed in the opt or NIR, but the RRc variables are noticeably fainter. (**d**) Histogram of [4.5] magnitudes for the RRLs in M 4

of M 4 by Hendricks et al. 2012) and a distance modulus of $\mu_0 = 11.257 \pm 0.035$ mag. Both this value and the fully empirical results above are in good agreement with other distance methods, (e.g. eclipsing binaries from Kaluzny et al. 2013) as well as from RRL in the NIR (Braga et al. 2015).

The empirical PL relations using both *Spitzer* and *WISE* data are given in Fig. 16. Periods of the RRab stars (shown as open circles) have been fundamentalized following (8). The empirical MIR PL relations for M 4 for the "global sample" are as follows:

$$
\begin{aligned}
[3.6]_{Glob} &= -2.304(\pm 0.105) \log P_F + 10.241(\pm 0.034) \\
[4.5]_{Glob} &= -2.340(\pm 0.104) \log P_F + 10.225(\pm 0.034) \\
W1_{Glob} &= -2.694(\pm 0.213) \log P_F + 10.145(\pm 0.213) \\
W2_{Glob} &= -2.540(\pm 0.248) \log P_F + 10.182(\pm 0.249).
\end{aligned}
\tag{17}
$$

The dispersion on the IRAC relations is 0.05 mag and is approaching the expected intrinsic scatter of the PL relation (~0.03 mag). The difficulty of deriving accurate photometry in clusters using WISE data (due to the larger PSF) results in a larger dispersion of ~0.09 mag. The observed slopes are in good agreement with theoretical PL relations derived from pulsation models (see discussion in Neeley et al. 2017).

To use the PL to determine a distance, the absolute zero point must be calibrated with RRL of known distances. Similar to the NIR study of M 4, there are two methods:

1. *Empirical*: Adopting observed slopes from M 4 RRLs and zero points are determined using trigonometric parallaxes for Galactic RRLs (either from *HST*-FGS or *Gaia*).

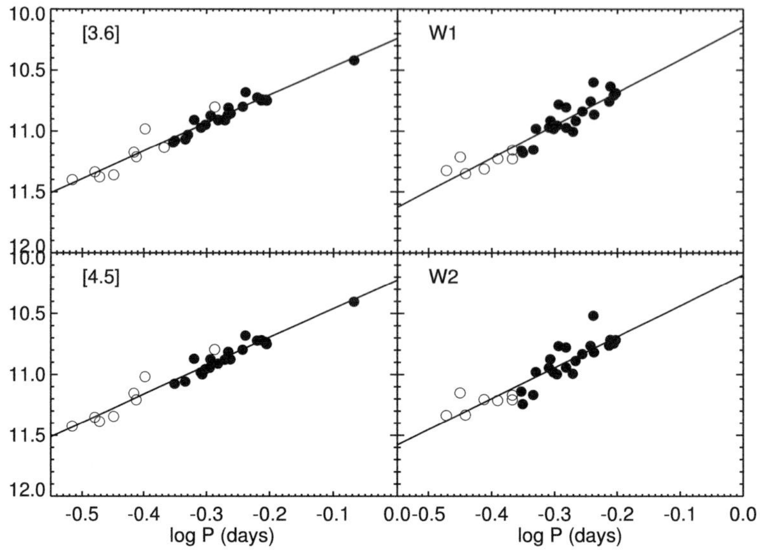

Fig. 16 IRAC (left) and WISE (right) PL relations for RRL in M 4 using data from Neeley et al. (2015, 2017) and Dambis et al. (2014). Periods of the first overtone pulsators (open circles) have been fundamentalized (e.g., (8))

2. *Semi-empirical*: The distance and extinction are measured simultaneously by fitting multi-wavelength data to the theoretical PLZ relations (Marconi et al. 2015; Neeley et al. 2017).

The distance modulus of M 4 can be derived by comparing the empirical and calibrated PL relations, as was done in Neeley et al. (2017). This results in $\mu_{[3.6]} = 11.353 \pm 0.095$ mag and $\mu_{[4.5]} = 11.363 \pm 0.095$ mag, which agrees with the Kaluzny et al. (2013) and Braga et al. (2015) measurements. Using the exquisite multi-band data available for M 4 and described in this chapter, the extinction to M 4 can be measured by fitting the distance and extinction law directly to each star as was described in Neeley et al. (2015, 2017), finding $A_V = 1.45 \pm 0.12$ mag using $R_V = 3.62$ measured directly in the foreground of M 4 by Hendricks et al. (2012, taking into account M 4 is directly behind the ρ Oph cloud).

2.4 Summary

RRLs have been so far excellent standard candles (e.g., Tammann et al. 2008). RRL can be used in individual Galactic structures, extra-galactic systems, and are likely the only relatively-luminous standard candle that can be used to trace ultra-faint galaxies and tidal streams) that populate the Galactic halo (e.g., Fiorentino et al. 2015, 2017; Andrievsky et al. 2018). Although they are typically discovered and classified in the optical, photometry of RRL can be used to determine distances in a number of passbands, from the optical to the mid-infrared using theoretical and/or empirical calibrations of the RRL PLZ or the FOBE method.

On the theoretical side, current uncertainties in the PL relations come from our lack of perfect knowledge of the evolutionary effects (mixing length, turbulent convection, mass loss, helium abundance), of the input physics (atomic diffusion, extra-mixing, neutrino losses), and of the color–temperature transformations that are adopted to transform bolometric magnitudes and effective temperatures into the observed magnitudes and colors. Given that theoretical relations are often preferred over empirical ones, these improvements are important.

As demonstrated by the case studies, the distances from the RRL are limited by the incomplete knowledge of the RRL PLZ (in all bands), in particular by the lack of a strong trigonometric parallax calibration. As such, single studies will often apply several calibrations to determine the distance. Studies using RRL are also limited by the imprecise knowledge of the metallicity for individual RRLs, although there are some forms of the RRL PLZ that have only a small dependence on metallicity.

The next generation of large telescopes will permit use of RRLs to distances of $\sim$5–6 Mpc (see e.g., Deep et al. 2011; Kalirai 2018), which have previously only been accessible via Cepheids. The proposed adaptive optics systems for 30-meter class telescopes will be optimized for the NIR wavelengths, which necessitates sound calibration of the NIR PLZ and PWZ relations.

3 The Type II Cepheids

3.1 Overview

Type II Cepheids (T2Cs) are evolved, low-mass stars (typically $<1\ M_\odot$) and have periods between 1 and 100 days, which roughly spans the range between RRLs and Miras (for the former, see the previous section and for the latter see Subramanian et al. 2017). The discovery of T2Cs played a critical role in the establishment of the modern extragalactic distance scale in the 1950s (formalized in the reviews of Baade 1958b,c, at the seminal Vatican Conference). Before its revolutionizing discovery, the stars adopted to build the PL relation of Cepheids were a mixture of both (i) young, intermediate-mass stars characterized by a radial distribution typical of thin disk stars and (ii) old, low-mass stars with a radial distribution typical of the Galactic halo or bulge. The former are the classical Cepheids (CCs) of Population-I, while the latter are known as Type-II Cepheids (T2Cs), to further emphasize the difference in age and stellar mass of their progenitors. By discovering the presence of two distinct classes of variables Baade effectively doubled the size (and the age) of the Universe (Baade 1956; Fernie 1969) by identifying and resolving a major issue in the determination of the Hubble constant at that time.

The discovery of T2Cs can be considered as one of three seminal discoveries that formed the modern distance ladder, joining (i) the measurement of stellar parallax by Bessell (1871) and (ii) the discovery of the Leavitt Law[7] or PL, first noted by Leavitt (1908) and then quantified by Leavitt and Pickering (1912). For a review of the Leavitt Law, readers are referred to Subramanian et al. (2017). After the separation of CCs and T2Cs, the latter have received far less attention than the former. Nevertheless, there are several good reviews on T2Cs and related objects which are recommended to interested readers: Harris (1985), Wallerstein and Nachr (1984), Wallerstein (2002), Sandage and Tammann (2006), Welch (2012), Feast (2010, 2013), Catelan and Smith (2015).

[7]The consensus of the scientific community present for the *Thanks to Henrietta Leavitt Symposium* held on Nov 8, 2008 was to officially adopt the term 'Leavitt Law' for the Cepheid PL relationship. The Council of the American Astronomical Society provided similar advice shortly thereafter. The reader is referred to https://www.cfa.harvard.edu/events/2008/leavitt/ and https://aas.org/archives/Newsletter/Newsletter_146_2009_05_May_June.pdf, respectively.

Dating back to almost half a century ago (Wallerstein 1970), T2Cs have been associated with GGCs characterized by a blue (hot) HB morphology. This means that the HB for these clusters is well populated on the blue tail, which is comprised of hot and extreme HB stars. Roughly one hundred of these variables have been detected in GGCs (Clement et al. 2001a). They have also been observed in stellar systems with complex star formation histories like the Magellanic Clouds (Soszyński et al. 2008, 2010; Groenewegen and Jurkovic 2017a,b) and there are a preliminary identifications in the extragalactic stellar systems with examples being: IC 1613, M 31, M 33, M 106 and NGC 4603 (Majaess et al. 2009). On the other hand, Fornax is the only nearby dSph galaxy for which even a handful of T2Cs have been detected (Bersier and Wood 2002); if these stars were a signpost for old metal-poor populations, then dSphs, which are home to many RRLs, would be expected to be hosts as well. The paucity of these objects in dSphs has been explained by their lack of blue HB morphologies. However, an observational bias cannot be excluded, since we still lack long-term photometric monitoring for the bulk of distant and diffuse stellar systems, although recent efforts have dramatically improved our variable star census (e.g., Stetson et al. 2014b, among others).

3.2 Observed Pulsational Properties

Example light curves for T2Cs for a wide span of periods are given in Fig. 17 using data from OGLE (Soszyński et al. 2008). The T2Cs show a striking degree of variation in comparison to the RRLs (Fig. 2), because T2Cs are often divided into subpopulations. The period distribution from Soszyński et al. (2008) is given in Fig. 18a and shows three groupings by period. Considering this period distribution, Soszyński et al. (2008) presented a classification scheme for the T2Cs in the LMC using the pulsation properties of the stars, including their amplitudes, light curve shapes, and periods. These pulsation properties are both distance and reddening independent and, thus, are good means for typing the stars. The three groups are divided into the classical sub-types with the following criteria:

1. BL Herculis (or BL Her) with $1 \leq P \leq 4$ days,
2. W Virginis (or W Vir) with $4 < P \leq 20$ days,
3. RV Tauri (or RV Tau) with $P > 20$ days.

The period-amplitude diagram in the I-band (A_I) for the LMC T2Cs is given in Fig. 18b with the amplitudes ranging from $\sim$0.1 to $\sim$1 mag. Figure 18b further demonstrates that there are structures in the period-amplitude diagram similar to what is seen in the Bailey diagram for RRLs (see e.g., Fig. 9), albeit the T2C groups are less well defined than the RRLs. In addition to these classical groupings, Soszyński et al. (2008) have also proposed a new group of T2Cs, the peculiar W Vir stars known as pW variables, that display peculiar light curves and are brighter than typical T2Cs at a fixed period; these stars are indicated by open circles in Fig. 18b. This classification scheme, however, has not been well settled, because the two minima of the period distribution that drive the definition of the sub-groups depends on the metal abundance and, likely, also on the environment. This means that the quoted period ranges will not universally define the three sub-groups and, thus, further investigations in a more diverse set of objects may yield further sophistication to our sub-classifications.

It is also important to note that different classification schemes have been adopted in different studies throughout the years. For example, General Catalog of Variable Stars[8] (GCVS, Samus et al. 2017) uses a boundary of 8 days to separate the T2Cs, 'CWA' for $P > 8$ days and 'CWB' for $P < 8$ days. In addition, GCVS uses the label 'RV' for RV Tau

[8] Available: http://www.sai.msu.su/gcvs/gcvs/.

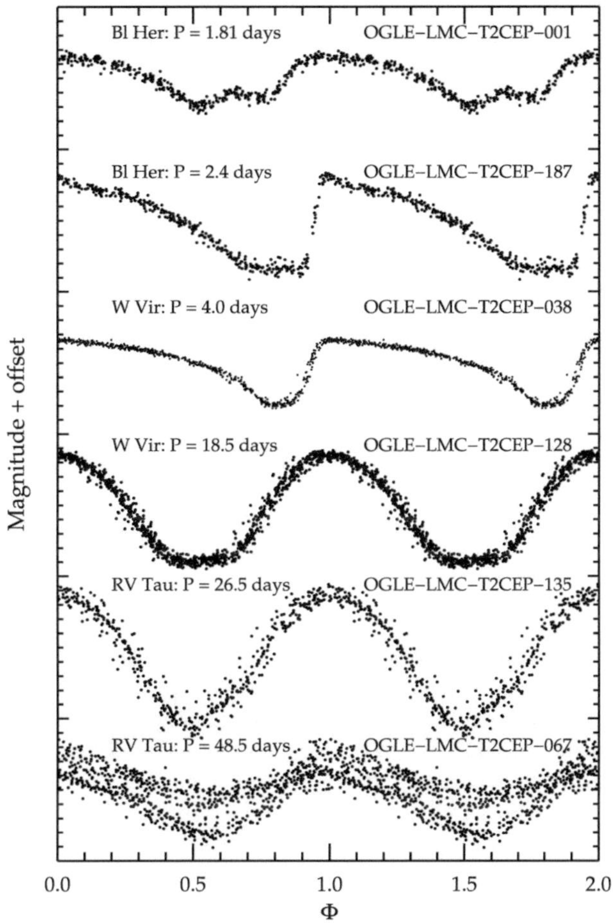

Fig. 17 Example optical, I-band, light curves for T2C in the LMC from the OGLE survey (Soszyński et al. 2008). From top to bottom there are two representative light curves for stars of each of the BL Her, W Vir, and RV Tau classes. The light curves have been normalized to amplitudes of 1 mag to allow comparison of the shapes and structures, with the minor-tick marks being 0.1 mag. The light curve shapes are easily distinguished from those of RRL with similar periods (e.g., Fig. 2)

stars for radially pulsating supergiants characterized by the presence of alternating primary and secondary minima. Different schemes were proposed by Joy (1949) and later by Diethelm (1990) that take into account factors beyond the period, such as the light curve shape and spectroscopic features (see also Sandage and Tammann 2006). Several authors have further suggested that the RV Tau stars are also heterogeneous as a class. While six objects in GGCs have been claimed to be RV Tau stars, some authors have doubted this classification from both the photometric (Zsoldos 1998) and the spectroscopic point of view (Russell 1998). The reader should be cautious, when exploring this topic, due to the complexities of these classifications.

3.3 Theoretical Pulsation Predictions

There is a general consensus that T2Cs are in a post-HB phase with a C + O degenerate core with H-shell burning as the main energy source. As is well known, Zero-Age HB stars have a wide range of effective temperatures depending on the total mass (or the envelope mass). After the He exhaustion in the core, they start to climb up in the HR diagram evolving mainly into AGB, but this climb can begin at a range of effective temperatures on the HB.

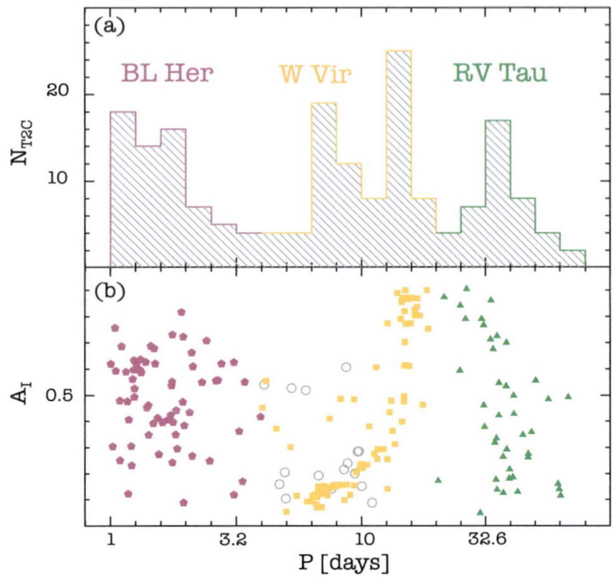

Fig. 18 Basic properties of the T2Cs as defined in the LMC from the OGLE survey (Soszyński et al. 2008). (**a**) Period distribution of the T2Cs as divided into the BL Her, W Vir, and RV Tau objects following the Soszyński et al. (2008) classification scheme. (**b**) I-band amplitude (A_I) versus period diagram (known as the Bailey diagram for the RRLs). The pentagons correspond to BL Her stars, the squares to W Vir stars, the open circles to peculiar W Vir objects, and triangles to RV Tau stars

The evolutionary paths of the post-HB phase are predicted to be highly dependent on the ZAHB position. Stars coming from the blue HB tail that go through the IS at luminosities brighter than those of RRL, are found to be pulsating stars and commonly called T2Cs. Calculated evolutionary tracks were pioneered by Gingold (1976, 1977) with Smolec (2016) and Bono et al. (2016) providing up-to-date theoretical calculations. However, we still lack a firm understanding of the production routes for T2Cs, in particular specific observational evidence for or against the postulated paths (though, we note that detailed studies such as Groenewegen and Jurkovic 2017a,b, provide a solid starting point). An interesting question is, for example, the possibility of excursions, first suggested by Schwarzschild and Härm (1970), from the AGB to the hotter side reaching the instability strip. The early investigations Gingold (1976, 1977) suggested such routes (*Gingold nose*) play an important role on explaining a fraction of T2Cs, but such excursions are not expected to take place commonly if at all according to more recent calculations (see Smolec 2016; Bono et al. 2016).

In contrast to the RRLs, theoretical investigations of the pulsation properties of T2Cs are quite limited. Early studies based on linear models were discussed by Wallerstein and Nachr (1984) together with their period distribution, light- and velocity curves, and their location in the color magnitude diagram. Subsequently, a detailed analysis of T2Cs pulsation properties was performed by Fadeev and Fokin (1985) using non linear radiative models. In particular, Fadeev and Fokin (1985) investigated the modal stability and provided theoretical constraints on the PL relation for T2Cs, but their approach neglected the convective transport and, in turn, the physical mechanism that causes the quenching of radial oscillations. This means that this approach cannot predict the location of the cool (red) boundary of the IS. Using a similar approach, a thorough analysis of the limit cycle stability of T2Cs was presented by Kovacs and Buchler (1988). They have shown that the pulsation behavior changes from single periodic, to period doubling, and eventually to chaotic when increasing the pulsation period.

A more comprehensive theoretical scenario including a time-dependent convective transport was presented for BL Her stars by Bono et al. (1997c), Marconi and Di Criscienzo

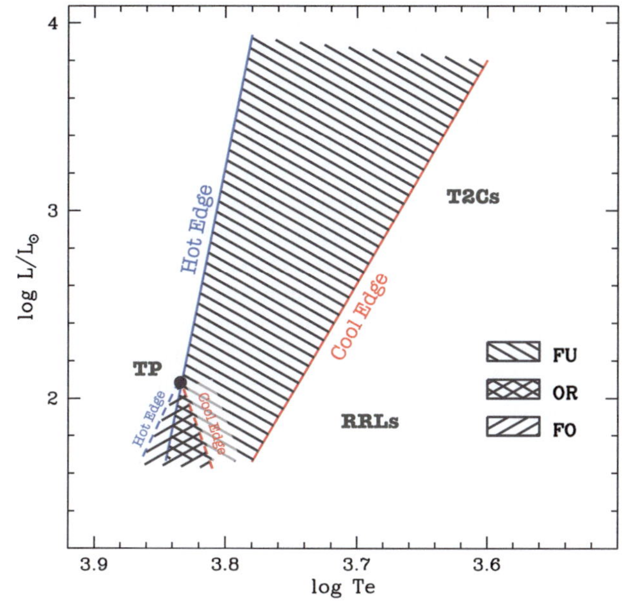

Fig. 19 Predicted topology of the instability strip in the luminosity and effective temperature range typical of RRLs and T2Cs (Marconi et al. 2015). The hot and cool edges of the strip are indicated for the FU mode (solid edges), FO mode (dashed edges), and their overlap region ("OR"; double hatched). Stars more luminous than the transition point ("TP"; see the text for definition) do not pulsate in the FO. Only in the region demarcated by the dashed edges can FO pulsators (RRc) attain a stable limit cycle. In the region labeled "OR", variables can pulsate simultaneously in the FO and in the FU mode (RRd, or mixed-mode pulsators). The region bounded by the solid edges represent the regions where FU pulsators can attain a stable limit cycle. The luminosity for the bulk of the T2Cs excludes them from the FO region

(2007), Di Criscienzo et al. (2007). In Bono et al. (1997c), a systematic theoretical investigation of T2Cs has shown that fundamental pulsators are primarily expected among this class of variable stars. Moreover, they derived a period-luminosity-amplitude relation and found that it is independent of metallicity for the metal poor regime of $-2.3 \leq [Fe/H] \lesssim -1.5$ that was explored in the study. These theoretical predictions match quite well the pulsation properties observed for GGCs.

Marconi and Di Criscienzo (2007) presented the full morphology of the theoretical IS for both fundamental and first overtone pulsators and a detailed atlas of both the light and radial velocity curves. Their models confirm that the FO IS is both narrow and limited to faint luminosities, providing new constraints on the *transition point*. The topology of the IS for the luminosity and effective temperature range of the RRLs and T2Cs is given in Fig. 19. T2Cs populate the IS at luminosities brighter than the so-called *transition point* (Stellingwerf 1979). The *transition point* corresponds to the luminosity at which the cool (red) edge of the FO IS matches the luminosity and the effective temperature of the hot (blue) edge of the FU. This means that FO pulsators do not attain a stable limit cycle for luminosities that are brighter and effective temperatures that are cooler than the *transition point*. In consequence, most T2Cs are fundamental pulsators. The first overtones, if any, are only permitted for the low luminosity or short period T2Cs. The empirical evidence, originally brought forward by McNamara (1995) and more recently by Soszyński et al. (2008, 2010) for Galactic and Magellanic T2Cs, confirms that they primarily pulsate in the fundamental mode. There is theoretical evidence that some stability islands for the FO can appear at luminosities brighter than the *transition point*, but their true nature needs to be further clarified (Marconi et al. 2015). Furthermore, the FO IS tends to vanish in the metal-rich regime (e.g., [Fe/H] ~ -1.0). Finally, using these same pulsation models and the evolutionary tracks from Pietrinferni et al. (2004), Di Criscienzo et al. (2007) derived analytical relations for the boundaries of the IS as a function of the adopted stellar parameters, as well as the PL and PW relations for these objects. They derived the T2C-based distances to GGCs and concluded that there is good agreement with the distance scale set by the RRLs.

Similarly to what happens for classical Cepheids, the PL relation for T2Cs can be derived using the fundamental pulsation relation given in (1) (van Albada and Baker 1971) valid for radial pulsators, together with the mass–luminosity relation typical of low-mass AGB evolutionary models (e.g., for BL Her stars, Di Criscienzo et al. 2007). Indeed, these AGB pulsators follow an inverse relation between stellar mass and luminosity, since lower mass pulsators are crossing the IS at a luminosity brighter than those with larger masses (Bono et al. 2016). The pulsation relation can then be averaged in temperature because the width of the IS is quite narrow ($\sim$1500 K), originating a PL relation (see also Fadeev and Fokin 1985; Matsunaga et al. 2006). Interestingly enough, the slope of the overall PL for T2Cs is less steep than that of classical Cepheids, which is also predicted by theoretical models (e.g., Bono et al. 1999; Di Criscienzo et al. 2007).

3.4 Optical and Near-IR Period-Luminosity and Period-Wesenheit Relations

It has long been recognized that T2Cs follow, at least, a loose PL relation. From previous studies on T2Cs in GGCs, it is known that the T2Cs do follow a PL in *BVI*. However, the relation has not been well established until recent work. While Harris (1985) and McNamara (1995) claimed that the optical PL slopes become steeper at around $\log P = 1$, Pritzl et al. (2003) did not find such a feature for the T2Cs in NGC 6388 and NGC 6441 based on *HST* photometric data. As Pritzl et al. (2003) claimed, many earlier studies for the T2Cs were based on magnitudes determined from photographic plates and, from these earlier works, it had been unclear if this class were useful as a distance indicator.

Breakthroughs were brought by modern photometric surveys in both the optical and the infrared. In particular, large-scale microlensing surveys that started in the late 1990s have systematically discovered T2Cs in the Magellanic Clouds. In the MACHO microlensing survey, Alcock et al. (1998) determined that W Vir and RV Tau sub-types follow a PLC relation in the form of

$$M_V = a + b \log P + c(V - R) \tag{18}$$

with a relatively small scatter of 0.15 mag, which is identical (cf. Bono and Marconi 1999) to a PW relation of the form,

$$V - c(V - R) = a + b \log P. \tag{19}$$

The photometric analysis performed by Soszyński et al. (2008) using *VI* optical data from the OGLE survey provided solid, empirical optical PL relations that support the theoretical predictions. The LMC sample is larger than the samples in GGCs and, because they are in the same system, can be studied as an ensemble without uncertainties from distances to the individual clusters. Figures 20a and 20b show the period-luminosity distribution for the T2Cs from Soszyński et al. (2008) for the *I* and *V* bands, respectively; the BL Her stars are shown as filled hexagons, the W Vir as filled squares, and the RV Tau stars as filled triangles, while the pW stars are removed (following the classification by Soszyński et al. 2008). The PL relations can be fit in three ways:

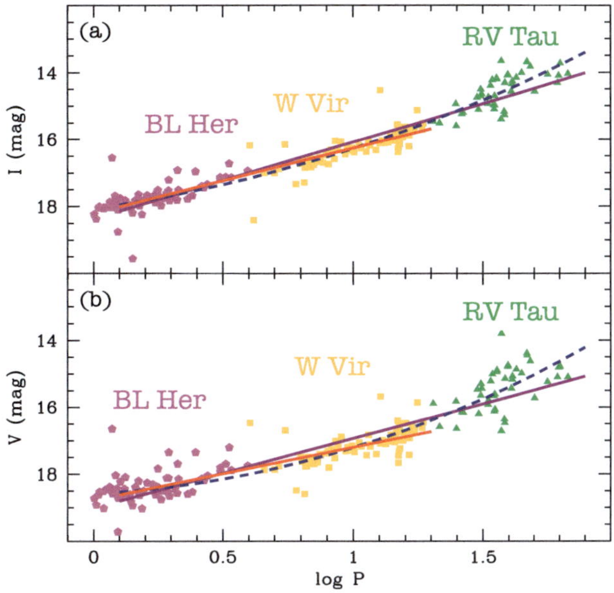

Fig. 20 The PL relations for T2Cs in the LMC: (**a**) I-band and (**b**) V-band PL relations. The Soszyński et al. (2008) T2C classification is maintained, but the peculiar W Vir stars are removed. The three PLs discussed in the text are shown for both panels: (i) a linear global fit for all three sub-types (solid line; (20)), (ii) a quadratic global fit that better accounts for the structure of the RV Tau types (dashed line; (21)), and (iii) a linear fit just to the BL Her and W Vir type stars (solid line; (22))

1. First, a linear regression can be performed for the entire sample, which results in the following relations,

$$\begin{aligned} M_V &= 19.01 - 2.07 \log P \ (\sigma_V = 0.6), \\ M_I &= 18.37 - 2.29 \log P \ (\sigma_I = 0.4). \end{aligned} \tag{20}$$

   A linear regression over the entire period range shows a decrease of the dispersion from $\sim$0.6 to $\sim$0.4 mag when moving from the V to the I band, indicating that the scatter decreases for longer wavelengths.
2. As is evident in Fig. 20, the PL relations for T2Cs in these photometric bands are far from being linear and they could be better represented by a higher order fit, more specifically a quadratic polynomial in $\log P$. Fitting this form of the PL to the Soszyński et al. (2008) data results in the following fits,

$$\begin{aligned} M_V &= 18.58 - 0.34 \log P - 1.03 (\log P)^2 \ (\sigma_V = 0.5), \\ M_I &= 18.06 - 1.04 \log P - 0.74 (\log P)^2 \ (\sigma_I = 0.3). \end{aligned} \tag{21}$$

   This reduces the scatter by 0.1 mag, corresponding to an improvement in distance precision by 5%.
3. As has been done in previous works, we also consider excluding RV Tau stars from the sample, because they show a larger spread in luminosity at fixed period for all wavelengths. Effectively, this is a cut in period for the Soszyński et al. (2008) sample to those stars with $P \leq 20$ days. Fitting a linear relation over this period range results in the following,

$$\begin{aligned} M_V &= 18.78 - 1.57 \log P \ (\sigma_V = 0.5), \\ M_I &= 18.20 - 1.94 \log P \ (\sigma_I = 0.3). \end{aligned} \tag{22}$$

   However, it is noteworthy that the scatter of these relations and those of the quadratic fits for all three sub-types are similar. Exploration of additional systems with different metallicities and in different environments may help distinguish which of the relations is more representative for a larger range of properties.

One has to keep in mind the presence of peculiar T2Cs that are scattered above the PL relation for BL Her and W Vir type T2Cs. They have some distinctive characteristics compared to W Vir stars (e.g., light curve shapes) but otherwise may be difficult to distinguish without long term monitoring. Soszyński et al. (2017) further investigated the light curves of this pW group. A significant fraction of the pW stars identified by Soszyński et al. (2008, 2010) were found to be in eclipsing binary systems. These objects are interesting in terms of evolutionary paths to produce T2Cs (and related objects), but to obtain accurate distances to stellar systems with T2Cs these atypical objects need to be identified and then excluded.

There is also evidence that T2Cs display well defined NIR PL relations that are linear over the entire period range. Matsunaga et al. (2006) reported relations with a scatter of 0.15 mag in JHK_s for BL Her to RV Tau stars in GGCs. Because they used RRL-based distances to individual GGCs to combine T2Cs onto the single relation, the well-defined relations suggest some consistency with the RRL distance scale. Ripepi et al. (2015) found smaller dispersions in both the J (0.13 mag) and K (0.09 mag) for T2Cs in the LMC, further supporting the evidence that they are good distance indicators. Recent updates on the NIR relations based on a larger photometric sample can be found in Bhardwaj et al. (2017a,b).

The slopes of these NIR PL relations do not change dramatically over the metallicity range spanned by these observations, as shown by Fig. 21 that displays the values for JHK PL relations obtained from GGCs (Matsunaga et al. 2006) and the LMC (Ripepi et al. 2015). Spectroscopic measurements are still too limited to draw firm conclusions (only a few dozen have measurements); those stars with metallicity measurements, however, span a range of metallicities similar to that of RRLs. The agreement becomes more stringent if we account for the theoretical NIR PL relations provided by Di Criscienzo et al. (2007), which are shown in Fig. 21 as the open circles. Note that a similar agreement is also found in a detailed comparison of the zero-points. Theoretical work, however, suggested a dependence of the zero-point on the metal content at the level of 0.04 to 0.06 mag dex^{-1}. Observations, on the other hand, suggest a minimal, if any, dependence (Matsunaga et al. 2006). A different result is found when considering the T2Cs in the SMC (Ciechanowska et al. 2010); more specifically, the slopes are shallower with -1.95 in the J-band ($\sigma_J = 0.24$ mag) and -2.15 in the K-band ($\sigma_K = 0.2$ mag). However, the difference with the slopes shown in Fig. 21 is within the associated 1 σ errors, and it is probably due to the high scatter of the mean NIR magnitudes derived from single epoch observations in Ciechanowska et al. (2010). In addition, the NIR sample of T2Cs in Ciechanowska et al. (2010) includes only 50% of the total optical sample for the SMC (40 objects, Soszyński et al. 2010) and thus a more complete sample with more well-sampled light curves may resolve the discrepancy.

The use of empirical relations to determine the LMC distance employed by Ripepi et al. (2015) returned a distance modulus in excellent agreement with RRLs, but with a difference of ~0.1 mag with respect to the distance modulus based on classical Cepheids (most likely due to parallax sample used, see Sect. 5.2). Even though this is not a statistical significant difference (e.g., within the formal 1 σ uncertainty), it suggests a possible population bias in the LMC distances (Inno et al. 2016). Theoretical models for the BL Her period range are in extremely good agreement with the above evidence, since they predict dispersions in these passbands equal to $\sigma_J = 0.13$ mag and $\sigma_K = 0.06$ mag (Di Criscienzo et al. 2007).

Matsunaga et al. (2006) also pointed out that the T2Cs and RRLs seem to follow the same, or continuous, PL relations in the infrared. The fact that RRLs and T2Cs follow very similar NIR PL relations (see also Feast 2011) seems to be supported by the smooth and natural transition of the evolutionary properties of evolved RRLs and BL Her type stars.

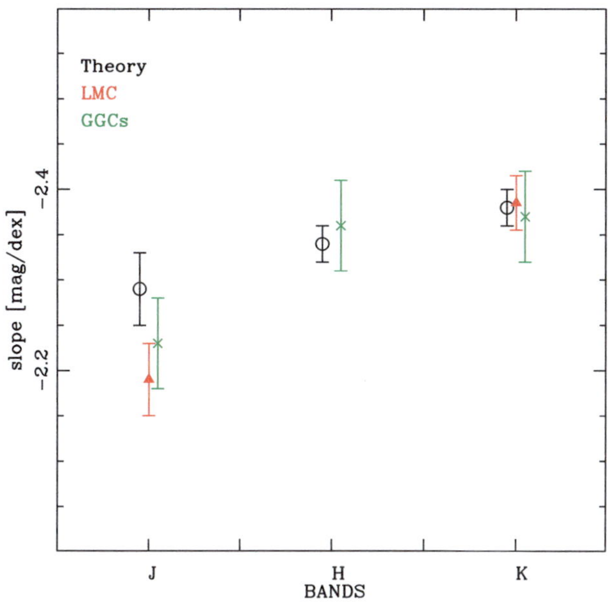

Fig. 21 Observed slopes of NIR (*JHK*) PL relations for T2Cs in GGCs (xes, Matsunaga et al. 2006) and in the LMC (triangles, Soszyński et al. 2008). The slopes predicted by Di Criscienzo et al. (2007) are plotted as open circle. The vertical bars display the uncertainties on the slopes

Although it was also discussed that BL Her and W Vir stars could have distinct NIR PL relations, recent findings based on accurate NIR mean magnitudes for LMC T2Cs suggest they are, within the errors, very similar (Ripepi et al. 2015, and references therein). This does not apply to RV Tau variables, because they typically display a larger spread compared to W Vir and BL Her stars. The increase in the spread might also mask a possible change in the slopes of the NIR PL relations (Ripepi et al. 2015). This occurrence appears to be associated with the properties of the circumstellar envelopes typical of these variables, which may also contribute to the larger scatter for this sub-type observed in the optical bands (see Fig. 20).

PW relations are also commonly used to reduce the impact of extinction in distance determination and here we derive these for T2Cs based on the most conspicuous samples available in literature (see also Bhardwaj et al. 2017a).

1. The most common Wesenheit magnitude in the optical takes the form of $W(I, V-I) = I - 1.55(V-I)$. Performing a linear regression to the BL Her and W Vir stars we find:

$$\begin{aligned} W(I, V-I)_{\text{LMC}} &= 17.31 - 2.50 \log P + \mu_0(\text{LMC}) \ (\sigma = 0.2), \\ W(I, V-I)_{\text{SMC}} &= 17.63 - 2.71 \log P + \mu_0(\text{SMC}) \ (\sigma = 0.4), \end{aligned} \tag{23}$$

   for the LMC (Soszyński et al. 2008) and SMC (Soszyński et al. 2010) data, respectively. To compare the relations on the absolute scale, we assume the consensus distance moduli, $\mu_0(\text{LMC}) = 18.49$ mag (de Grijs et al. 2014, and references therein) and $\mu_0(\text{SMC}) = 18.96$ mag (de Grijs and Bono 2015, and references therein), respectively (see also de Grijs et al. 2017, in this series). Figure 22a compares these $W(I, V-I)$ relations. It is worth noticing that when decreasing the host galaxy metallicity (from that of the LMC to that of the SMC), the slopes get steeper and the dispersion around the mean relation increases, although the difference of slopes is still within 1 σ.
2. The most common NIR Wesenheit magnitude takes the form of $W(K, J-K) = K - 0.69(J-K)$. Performing a linear regression to the BL Her and W Vir type stars

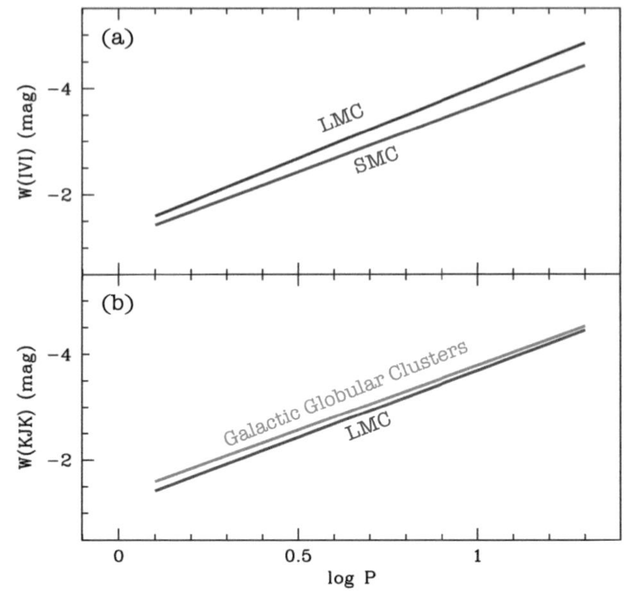

Fig. 22 (**a**) Optical $(I, V-I)$ PW relations for T2Cs in the LMC and SMC derived as described in the text using data are taken from Soszyński et al. (2008) and Soszyński et al. (2010), respectively. (**b**) NIR $(K, J-K)$ PW relations for T2Cs. The LMC data is adopted from Ripepi et al. (2015), whereas the for GGCs it has been derived using NIR magnitudes from Matsunaga et al. (2006). The precise form of the relations is given in (23) and (24) and described in the text

for the LMC (as given by Ripepi et al. 2015), SMC (IR magnitudes and periods from Ciechanowska et al. 2010) and GGCs (data from Matsunaga et al. 2006), we find the following PW relations:

$$\begin{aligned}
W(K, J-K)_{\rm LMC} &= 17.32 - 2.52\log P + \mu_0({\rm LMC})\ (\sigma = 0.08),\\
W(K, J-K)_{\rm SMC} &= 17.41 - 2.17\log P + \mu_0({\rm SMC})\ (\sigma = 0.3),\\
W(K, J-K)_{\rm GGCs} &= -1.35 - 2.44\log P\ (\sigma = 0.2).
\end{aligned} \tag{24}$$

Figure 22b compares NIR PW relations; the SMC relation is not included because the mean magnitudes have a large dispersion about the relation due to sparse sampling of the NIR light curve. As shown in Fig. 22b, the slopes for LMC and GGCs are very similar, suggestive of only a mild dependence on the metal content. However, for the SMC we note that the slope decreases down to -2.17 mag dex^{-1}, which is opposite to what happens in the optical regime. There are no doubts that a more detailed investigation of T2Cs in the SMC will be crucial to understand the universality of the NIR PW relations.

3.5 T2Cs in Context with RRLs and CCs

In this section we discuss the pulsation and evolutionary properties of T2Cs as compared to the most popular distance indicators, namely RRLs and CCs. The aim is to highlight their possible new role for these stars in the *Gaia* era that is expected to refine the extragalactic distance scale. In the following, we list their properties and assess their advantages and disadvantages compared to CCs and RRLs:

1. *NIR PL relation*: As discussed in the previous section, T2Cs follow well defined optical and NIR PL relations (Matsunaga et al. 2006, 2013). Moreover, there is evidence that their pulsation properties are either independent of, or minimally affected by, the metal abundance (Bono et al. 1997c; Di Criscienzo et al. 2007; Lemasle et al. 2015a). RRLs

also follow linear NIR PL relations, however their luminosity (and luminosity range) is much smaller, which largely limits their application to within the Local Group (e.g., distances up to ~1 Mpc). T2Cs are roughly 1.5 mag fainter than CCs at fixed period with the precise difference being period dependent. Moreover, the long period tail of CCs exceeds 100 days, while the long period tail of T2Cs approaches ~70 days.

2. *Linearity*: The current empirical evidence indicates that BL Her and W Vir stars follow NIR PL relations that are linear over the entire period range. With current data, however, it is unclear if RV Tau objects follow a long period extension of the same PL relations, or whether their PL relation is distinct. Thus, RV Tau subgroup is typically neglected when fitting PL relations and in determining distances.
3. *Brightness*: T2Cs are brighter than RRLs and cover a similar magnitude range as Classical Cepheids. The difference with RRLs ranges from at least half magnitude ($M_V \sim -0.5/-1$ mag) to roughly three magnitudes ($M_V \sim -3$ mag). This means that they can be employed to measure distances and trace old (blue HB) stellar populations beyond the Local Group.
4. *(Short) evolutionary lifetime*: T2Cs are typically approaching the AGB phase. This means that their evolutionary time scale is roughly two orders of magnitude faster than RRLs (Marconi et al. 2015; Bono et al. 2016). RRLs are low-mass stars during their core He burning phase, they have been identified in all stellar systems hosting stellar populations older than ~10 Gyr, and from a theoretical point of view stellar mass and chemical abundance ranges that can give origin to RRLs are well known (Bono et al. 1996, 1997a,b). The same is true for CCs, that are young core helium burning stars. The ratio of evolutionary times of CCs to the entire lifetime of their progenitors is longer than for T2Cs (Marconi et al. 2005a). This has a strong impact on the occurrence of T2Cs in a given system such that T2Cs are more rare than both RRLs and CCs. As shown in Fig. 1a, the LMC contains ~3,000 CCs, ~20,000 RRLs, but only ~200 T2Cs (Fiorentino and Monelli 2012).
5. *Binary or evolutionary channel*: The fraction of T2Cs in binary systems is not well constrained. Very recently Soszyński et al. (2017, 2018) presented additional monitoring of T2Cs in the LMC and in the Bulge that find new irregularities in the light curves consistent with binary companions. Maas et al. (2002) present evidence that several RV Tau stars may be in binary systems. More importantly, there is photometric evidence that the pW stars may also be binaries (most recently Pilecki et al. 2017). This working hypothesis is also supported by spectroscopic evidence of binary companions for some T2Cs (Maas et al. 2007; Soszyński et al. 2010; Jurkovic et al. 2016) and, in turn, casts some doubts on the proposed small initial mass of their progenitor stars. These kinds of objects have been dubbed *Binary Evolution Pulsators* (Karczmarek et al. 2016, 2017b). Long term monitoring of T2Cs can help identify their evolutionary path and determine if binary formation is the dominant formation path. Period changes can tell us the directions of evolution within the instability strip and this, in turn, provides additional evidence for the evolutionary paths of T2Cs (e.g., Diethelm 1996; Wallerstein 2002; Rabidoux et al. 2010).
6. *Shape of the light curve*: The coupling between shape of the light curve, e.g., the Fourier parameters and luminosity amplitudes, and the pulsation period is less distinctive for T2Cs than for RRLs (e.g., compare Fig. 17 to Fig. 2 for light curves and Fig. 18 and Fig. 9 for the amplitude–magnitude diagrams). They partially overlap with CCs (Soszyński et al. 2008).
7. *Host stellar systems*: T2Cs have been identified in a variety of stellar systems hosting an old ($t \geq 10$ Gyr) stellar population, e.g., both early and late type stellar systems.

Although this appears as a solid empirical evidence, we still lack firm identification of T2Cs in dSphs. The lack of T2Cs in dSphs can hardly be explained as an observational bias, since they are systematically brighter than RRLs and some of the classical dSphs have been surveyed for very long time scales (e.g., Stetson et al. 2014a; Coppola et al. 2015; Martínez-Vázquez et al. 2016b, among others). The working hypothesis suggested by Bono et al. (2011, 2016) to explain the desert of T2Cs in dSphs is that they typically lack hot and extreme HB stars (Salaris et al. 2013). This means that the fraction of HB stars evolving from the blue (hot) to the red (cool) side of the CMD is small. This is relevant, given that the stellar populations of dSphs do cover the same metallicity range of GGCs that do have well extended blue tails on the HB. The lack of hot/extreme HB stars, and in turn of T2Cs, may be due to an environmental effect between dSphs and GGCs, that otherwise are composed of similar stellar populations.

On the whole, T2Cs have many enticing properties and show potential as distance indicators, but also still have some more shadowy areas left to be resolved before their application en masse.

3.6 Summary and Final Remarks

In this section, the observational and theoretical properties of T2Cs have been described. Although there is no doubt that they are radially pulsating stars crossing the IS, their evolutionary origin has not yet been completely addressed (e.g., the Binary Evolution Pulsator scenario, see Karczmarek et al. 2016, 2017a). Both theoretical and observational studies demonstrate that T2Cs mainly pulsate in the fundamental mode, given that they reach luminosities brighter than the *transition point*. The possible occurrence of short-period, fainter T2Cs evolving along the so-called *blue hanger* (Bono et al. 2011) cannot be excluded. However, they are expected to add little contamination to the fundamental mode NIR PL and PW relations.

T2Cs are *largo sensu* solid distance indicators, for they follow well defined optical and NIR PL relations, and both theoretical and empirical results suggest a minimal dependence on metal-abundance. Furthermore, they also follow similar optical and NIR PW relations. The key advantages of these diagnostics is to be independent of reddening uncertainties, but rely on the assumption that the reddening law is universal. This further suggests that T2Cs can be considered *stricto sensu* ideal distance indicators, given that they can provide very accurate relative distances, independent of uncertainties in the zero-point of the PL relations. The small metallicity dependence of the slopes of the NIR PW relations further supports the use of these variables as distance indicators for stellar systems affected by differential reddening (e.g., as with M 4 in Braga et al. 2015). To obtain the tighter PL relations, only BL Her and W Vir stars (i.e., those T2Cs with periods less than 20 days and M_V fainter than -2 mag) are considered, excluding the brighter RV Tau objects. This means that one can detect and employ T2Cs as distance indicators out to $\sim$10 Mpc with *HST* (limiting magnitude $V \sim 30$ mag). However, because their periods have a wide distribution ranging from a few days to 20 days, complete photometric surveys with long temporal baseline have to be conducted in order to collect complete T2C samples.

The absolute distance scale of T2Cs, i.e., the zero point of the PL relation, is calibrated using the populations in the Magellanic Cloud and GGCs. There are only a handful of trigonometric parallax measurements for T2Cs and the uncertainties for the measurements that do exist make it difficult to establish a distance scale directly tied to parallax. The uncertainties affecting the zero-points of these relationships, and in turn, the absolute distances based on T2Cs, are going to be largely reduced during the next few years. This is thanks to the very accurate trigonometric parallaxes that will be provided by *Gaia*, not only for the nearest T2Cs, but for more than 100 field T2Cs and at least a dozen in nearby GGCs using end-of-mission predictions (Harris 1985; Matsunaga et al. 2006).

The observational outlook concerning T2Cs appears even more promising if we take account of the fact that *JWST* will be capable of producing a complete census of T2Cs for Local Group and Local Volume galaxies (Kalirai 2018). The halos of large galaxies are, indeed, marginally affected by crowding problems. Moreover, 30 meter-class telescopes will provide a unique opportunity to trace old stellar populations in the bulges and the innermost regions of galaxies in the Local Volume due to their unprecedented spatial resolution (Bono et al. 2017).

4 The Tip of the Red Giant Branch

The ultimate origin of the tip of the red giant branch (TRGB) as a standard candle is the canonical work of Baade (1944), who used red-sensitive plates to first resolve the stellar populations in early-type stellar systems, M 32, NGC 205 and the central region of M 31. From these observations developed the concept of the stellar classes of Pop I and Pop II. Baade also concluded that the brightest stars in these objects had similar magnitudes and colors. Sandage (1971) later pointed out that galaxies in the Local Group invariably contained a background sheet of red stars. Comparing the Cepheid distances to three Local Group galaxies, M 31, M 32, and IC 1613, he determined a mean absolute magnitude of the brightest red stars to be $M_V = -3.0 \pm 0.2$ mag. This work led to the early applications using RGB stars for distance determination in the early eighties (see a summary in Lee et al. 1993).

A formal method to detect the TRGB in color–magnitude-diagrams (CMDs) of galaxies, a calibration and a thorough analysis of its uncertainties were first published by Lee et al. (1993) and Madore and Freedman (1995), respectively. More specifically, Lee et al. (1993) compared empirical RGB loci for 6 GGCs (Da Costa and Armandroff 1990) with theoretical stellar evolutionary tracks (Yale Models, Green et al. 1987), and concluded that the I-band absolute magnitude of the TRGB is accurate to ± 0.1 mag for resolved stellar systems of low metallicity (e.g., [Fe/H] < -0.7). Lee et al. (1993) introduced Sobel "edge detection" kernels for detecting the discontinuity caused by the TRGB in I-band luminosity functions and this methodology was rigorously tested by Madore and Freedman (1995). Since those works, the TRGB has been used extensively in Pop II dominated systems. With current observational capabilities the TRGB can be used to determine distances out to $\sim$20 Mpc (see e.g., Jang and Lee 2015, 2017a; Jang et al. 2018, and references therein). Indeed, several authors have discussed the TRGB as an alternative to CCs for the basis of the extragalactic distance scale (e.g., Mould and Sakai 2008; Beaton et al. 2016). So far, more than 400 galaxies have been measured their distances with the TRGB (Tully et al. 2016).

The TRGB as a precise distance indicator for resolved stellar systems has held against several observational tests (Lee et al. 1993; Ferrarese et al. 2000; Bellazzini et al. 2001; Sakai et al. 2004). Several studies have shown that the TRGB distance estimates to nearby galaxies agree well with those based on other precise distance indicators, such as CCs, RRLs, and Megamasers (Freedman 1988; Lee et al. 1993; Sakai et al. 2004; Ferrarese et al. 2000; Tammann et al. 2008; Mager et al. 2008; Freedman and Madore 2010; Jang and Lee 2017b). These indicate that the precision of the TRGB is comparable to those of other primary distance indicators.

In this section, physical understanding of the TRGB is given in Sect. 4.1. Sect. 4.2 discusses empirical methods to measure the TRGB. A case study of using TRGB in the optical is described in Sect. 4.3. In Sect. 4.4, the IR properties of the TRGB are discussed. We conclude looking to future improvements of the method in Sect. 4.5.

4.1 Physical Description

The TRGB corresponds to the termination of the RGB evolution in old stellar populations, due to the onset of the He-flash in the electron degenerate cores of low-mass stars. For a given initial chemical composition, the bolometric luminosity of the TRGB is determined by the He-core mass at the He-flash. This mass is basically constant, e.g., within a few 0.001 $M_{\odot}$, for stellar masses that reach He-ignition at ages from 1.5–3 Gyr to the age of the Universe (the starting age is a function of the initial metallicity).

The use of the TRGB for distance estimates has been traditionally based on I-band photometry, in the implicit assumption that the observed RGB of a galaxy has an age comparable to the age of GGCs, or in any case larger than 4–5 Gyr and metallicity [M/H] below ~ -0.7 (see, e.g., Da Costa and Armandroff 1990; Lee et al. 1993; Salaris et al. 2002). In this age range M_{bol}^{TRGB} is roughly constant for a given [M/H], but becomes brighter with increasing [M/H]. Two effects influence M_{bol}^{TRGB} in more metal rich stars: (i) increased efficiency of the hydrogen burning shell, which makes the stars brighter, and (ii) decreasing He-core mass, which makes the stars fainter; overall, M_{bol}^{TRGB} is more luminous in metal-rich stars from these two effects. At the same time the effective temperature of the TRGB decreases with increasing [M/H] and the bolometric corrections, BC_I, decrease with decreasing T_{eff}. The resulting rate of decrease for BC_I with increasing metallicity very nearly matches the rate of increase of M_{bol}^{TRGB} with [M/H]. Given that $M_I^{TRGB} = M_{bol}^{TRGB} - BC_I$, this implies that M_I^{TRGB} stays almost constant with [M/H] for the range between ~ -2.0 and ~ -0.7 dex and for ages older than 4–5 Gyr.

From the variation of the absolute magnitude from theoretical models, it follows that applying a single-value for the absolute magnitude of the TRGB to galaxies with an extended star formation history can cause large systematic errors. Adopting a single value in a system assumes that the RGB stars in these galaxies are *all* older than 4–5 Gyr with [M/H] $<$ -0.7, which can only be assumed safely in specific conditions. This effect has been studied theoretically by Barker et al. (2004), Salaris and Girardi (2005), Cassisi and Salaris (2013), and also verified empirically by Górski et al. (2016). A recent, detailed study is given in Serenelli et al. (2017).

Figure 23 displays a theoretical $M_I^{TRGB} - (V-I)^{TRGB}$ calibration in Johnson–Cousins filters using the Pietrinferni et al. (2004, BaSTI) stellar evolution models after correcting the TRGB bolometric magnitudes for the effect of including the updated electron conduction opacities by Cassisi et al. (2007) and using the empirical bolometric corrections by Worthey and Lee (2011). In Fig. 23 [M/H] lower (larger) than -0.65, are displayed with open(filled) circles, respectively. The BaSTI theoretical models cover a range of metallicity [M/H] between -2.2 and 0.3 and ages between 1.5 and 14 Gyr. The full age and metallicity range of the models is shown in Fig. 23a, which demonstrates that decreasing age at constant [M/H] pushes M_I^{TRGB} to bluer colors, whereas increasing [M/H] at fixed age has the opposite effect. The smooth $M_I^{TRGB} - (V-I)^{TRGB}$ relation displays a maximum around $(V-I)^{TRGB} \sim 1.6$ mag, with the TRGB magnitude increasing steadily for redder colors.

Figure 23b shows TRGB magnitudes and colors for ages above 4 Gyr with [M/H] $>$ -0.65 dex and [M/H] < -0.65 dex using the same symbols as Fig. 23a compared with a 'classical' calibration (orange). This 'classical' calibration provides a median $M_I^{TRGB} =$

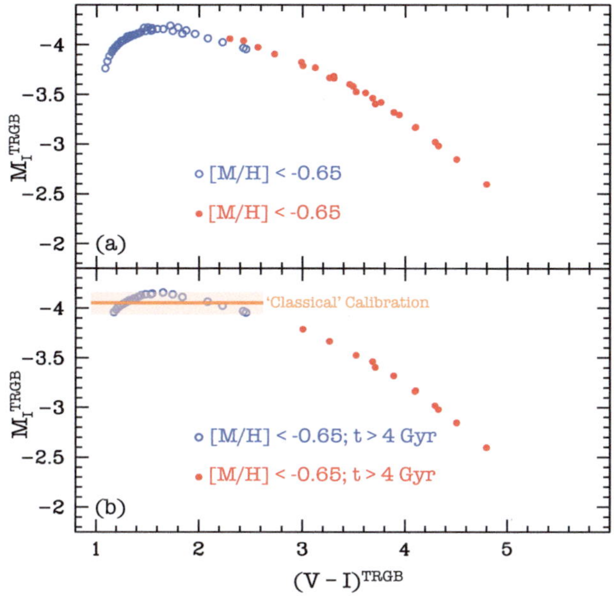

Fig. 23 $M_I^{TRGB} - (V - I)^{TRGB}$ distribution for the TRGB from theoretical models from BaSTI (Pietrinferni et al. 2004). (**a**) The full range of ages ($1.5 < t < 14$ Gyr) and metallicity ($-2.2 <$ [M/H] < 0.3 dex) for the models and (**b**) only those models with ages older than 4 Gyr. In both panels, the blue open circles are for [M/H] < 0.65 dex and the red filled circles are for [M/H] < 0.65 dex. The solid orange line represents the empirical calibration of $M_I^{TRGB} = -4.05$ mag (Lee et al. 1993; Salaris and Cassisi 1997; Tammann and Reindl 2013), with the shading giving the typical uncertainty of ± 0.10 mag. The 'classical' calibration is a reasonable approximation to the theory for stars in this age and metallicity range (corresponding to $V - I$ from 1.0 to 2.5 mag)

-4.05 mag, with typical observational uncertainties of ± 0.10 mag over the range of $V - I$ from 1.0 to 2.5 mag (shown with the shading in Fig. 23b; Lee et al. 1993; Salaris and Cassisi 1997; Tammann and Reindl 2013). The theoretical behavior of M_I^{TRGB} vs $(V - I)^{TRGB}$ in this age and metallicity range is approximately quadratic, but using a constant average value of M_I^{TRGB} is a decent approximation as is shown by the shading in Fig. 23b.

It is evident that the $M_I^{TRGB} - (V - I)^{TRGB}$ relationship is very smooth for all [M/H], and irrespective of any age sub-selection, and can be used as distance indicator for any stellar population with an RGB. Figure 24 compares the theoretical relationship of Fig. 23 (open circles) to one obtained using the same models, but applying the bolometric corrections by Gustafsson et al. (2008, open squares). For these bolometric corrections we include the TRGB only up to [M/H]$\sim$ solar, because at super-solar metallicities TRGB models have negative $\log(g)$ (surface gravity) that are not covered by these bolometric corrections (additional discussion is given in Serenelli et al. 2017). We compare the theoretical relations to the empirical relationship determined by Rizzi et al. (2007) on a sample of galaxies (solid red line) and the determination of the TRGB absolute magnitude for ω Centauri, which has a broad range of metallicities and ages (Bellazzini et al. 2001) and the 'classical' calibration from Fig. 23b.

The Rizzi et al. (2007) calibration has roughly the same slope of the theoretical ones for $(V - I)$ above $\sim$1.6–1.7 mag, but with a magnitude offset of $\sim$0.1 mag. The uncertainty for the ω Centauri datapoint is large and does not put very strong constraints on the TRGB absolute magnitude. The two sets of bolometric corrections applied to the theoretical cal-

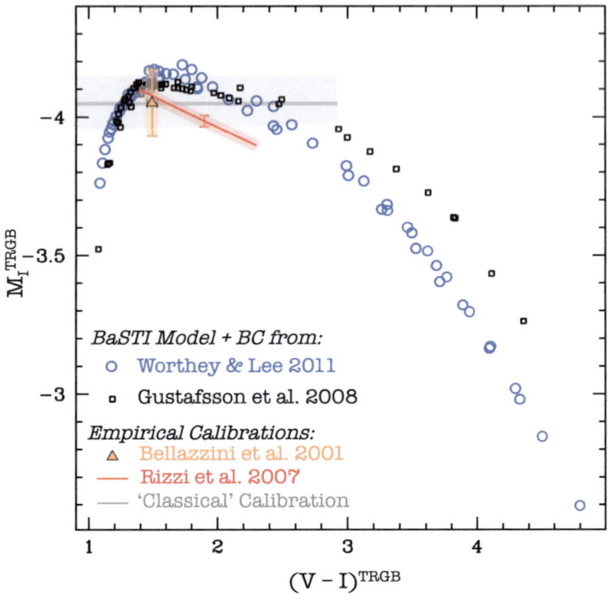

Fig. 24 Theoretical $M_I^{TRGB} - (V - I)^{TRGB}$ distribution at the TRGB for the full age range of the BaSTI models, but limited to sub-solar [M/H]. Two different sets of bolometric corrections are employed, more specifically those of Worthey and Lee (2011, blue circles) blue circles and of Gustafsson et al. (2008, black squares). We also show the empirical estimate (Bellazzini et al. 2001) for ω Centauri (orange triangle; Bellazzini et al. 2001) and the empirical relationship by Rizzi et al. (2007) based on a sample of galaxies (red line). The range of the 'classical' calibration is shown in gray. The zero point uncertainties are shown by the shading. See also Serenelli et al. (2017)

culations provide similar results as long as $(V - I)$ is below ∼2.0 mag. Increasingly larger discrepancies appear for redder colors.

The panels of Fig. 25 show theoretical TRGB absolute magnitude–color calibrations for the infrared filters K_s and J (Fig. 25a and 25b, respectively). Notice an almost linear behavior and again a smooth and tight correlation over the whole age and metallicity range covered by the models, with the added bonus of a lower sensitivity to reddening. Notice also how the dynamical range of the TRGB magnitude in J is reduced compared to the case of the K_s and I bands.

The panels of Fig. 25 also compare the results from the different bolometric corrections, as in Fig. 24. The uncertainty on the bolometric corrections is at the moment the major drawback for theoretical calibrations in these filters. Notice the different overall slope of the two sets of results, that cause systematic differences of the TRGB absolute magnitudes at fixed color up to ∼0.2 mag.

Wu et al. (2014) have also derived empirically relationships in the corresponding infrared filters of the WFC3–*HST* system, namely F110W and F160W following the same methods as Rizzi et al. (2007). Figure 26 compares their $M_{\mathrm{F110W}} - (\mathrm{F110W} - \mathrm{F160W})$ TRGB relationship (red boxes) with the theoretical one based on Gustafsson et al. (2008) bolometric corrections (filled circles). The overall shapes are different in the sense that the empirical calibration has a clear change of slope at (F110W − F160W) ∼ 0.95 mag that is not seen in the theoretical counterpart. Overall differences of the TRGB absolute magnitude at fixed color are however typically within $\sim \pm 0.05$ mag.

4.2 Detecting the TRGB

In this subsection, we discuss the different techniques employed to determine the apparent magnitude of the TRGB, which is a sharp discontinuity along the observed luminosity function (LF). In the following, we discuss the major components of the TRGB detection process: edge-detection algorithms in Sect. 4.2.1, strategies used to account for the TRGB

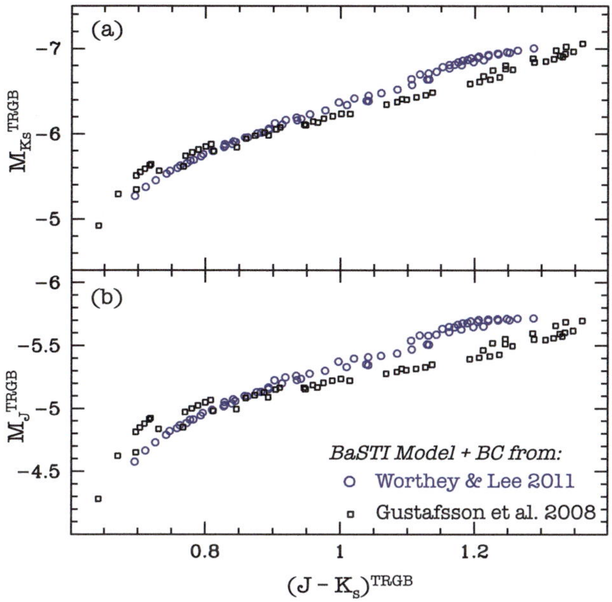

Fig. 25 Theoretical $M_{NIR}^{TRGB} - (J - K_s)^{TRGB}$ distribution at the TRGB for the full age and [M/H] range of the models, more specifically the panels are: (**a**) $M_{K_s}^{TRGB} - (J - K_s)^{TRGB}$ and (**b**) $M_J^{TRGB} - (J - K_s)^{TRGB}$. The color coding matches that of Fig. 24, more specifically open circles denote results obtained with the Worthey and Lee (2011) bolometric corrections, whilst open squares display results obtained with the Gustafsson et al. (2008) bolometric corrections. Overall, the NIR–TRGB magnitude shows an approximately linear relationship with $J - K_s$ color and, compared to the optical (Figs. 23 and 24), it more than 1 magnitude brighter, spans a range of ~1 mag over this range of [M/H] and ages, but spans a much more narrow color-range. See also Serenelli et al. (2017)

shape in Sect. 4.2.2, and brief caution regarding the overlap between the TRGB and other stellar sequences in Sect. 4.2.3.

4.2.1 Edge-Detection Techniques

Practically, the LF is constructed from binning the magnitudes of RGB stars from a selection over an appropriate color range. The TRGB is detected by finding the point of greatest change in the LF, either using tools that approximate the first derivative or tools that model the components of the LF itself (e.g., individual stellar sequences). Poisson noise occurs in the LF, especially in smaller RGB samples (like in GGCs), which in turn create spurious spikes in the edge-detection response. Thus, smoothing is generally used on the LF to mitigate this noise, which can either be (i) incorporated into the functional form of the edge-detection kernel, (ii) applied to the LF itself, or (iii) folded into the model. Edge detection techniques for the TRGB come in the following forms:

1. discrete approximations to the derivative (e.g., the Sobel kernal),
2. discrete approximations to the derivative that incorporate smoothing (e.g., a Gaussian formulation of the Sobel kernel),
3. maximum-likelihood fitting techniques (e.g., fitting a functional form to the LF with the TRGB as a parameter).

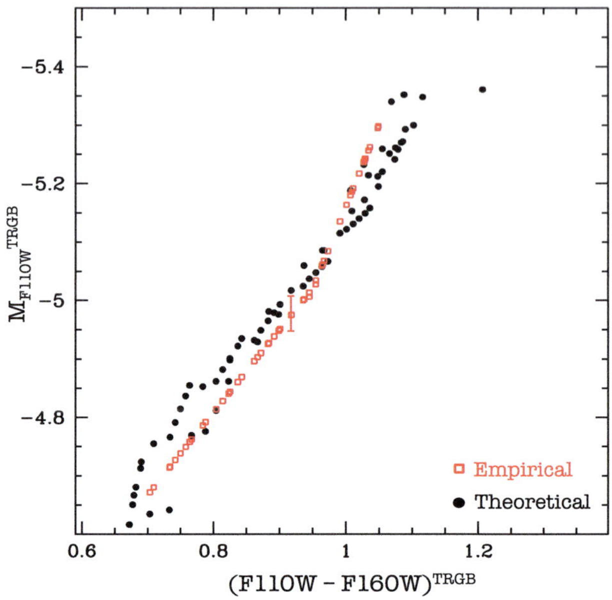

Fig. 26 Comparison of theoretical and empirical calibrations of the NIR–TRGB color–magnitude distribution for the *HST* + WFC3/IR filters F110W and F160W, which are similar to the 2MASS J and H. The filled black circles are the theoretical predictions from BaSTI using the Gustafsson et al. (2008) bolometric corrections and the open squares are the empirical measurements from Wu et al. (2014). The error bar on the zero point of the empirical calibration is also displayed

Table 2 Discrete approximations to the first derivative used for TRGB detection

Reference	Derivative approximation
Lee et al. (1993)	$[-2, 0, +2]$
Madore and Freedman (1995)	$[-1, -2, 0, +2, +1]$
Sakai et al. (1996)	$\Phi(I + \sigma_m) - \Phi(I - \sigma_m)$
Méndez et al. (2002)	$\sqrt{\Phi}[\log(\Phi(I + \sigma_m)) - \log(\Phi(I - \sigma_m))]$
Mager et al. (2008)	$\sqrt{\sum N_m}[\log(\Sigma N_m) - \log(\Sigma N_m)]$
Madore et al. (2009)	$[-1, -1, -1, 0, 0, 0, +1, +1, +1]$
Jang and Lee (2017a)	$[-1, -2, -1, 0, +1, +2, +1]$
Hatt et al. (2017)	$[-1, 0, +1]$

These methods can applied LF that are either discrete (e.g., with large bins) or a "continuous" (e.g., with bins over very small intervals). Each method has advantages and disadvantages that should be weighed for the science application in question. As an example, simulations by Madore and Freedman (1995) demonstrated that at least 100 stars within one magnitude below the TRGB to detect unambiguously the TRGB for their algorithm and similar tests should be performed for any algorithmic approach to provide guidance for its application. Popular forms of the discrete derivative approximations are given in Table 2. Alternative parametric and non-parametric methods have been also applied to determine the TRGB magnitude in a number of stellar systems; the studies of Cioni et al. (2000), Conn et al. (2011), Makarov et al. (2006) are representative examples.[9]

Figure 27 compares applications of eight different forms for TRGB detection from the literature similar to the comparison in NGC 1365 made by Jang et al. (2018, their Fig. 9). The

[9] Additional non-parametric studies are listed in Jang et al. (2018).

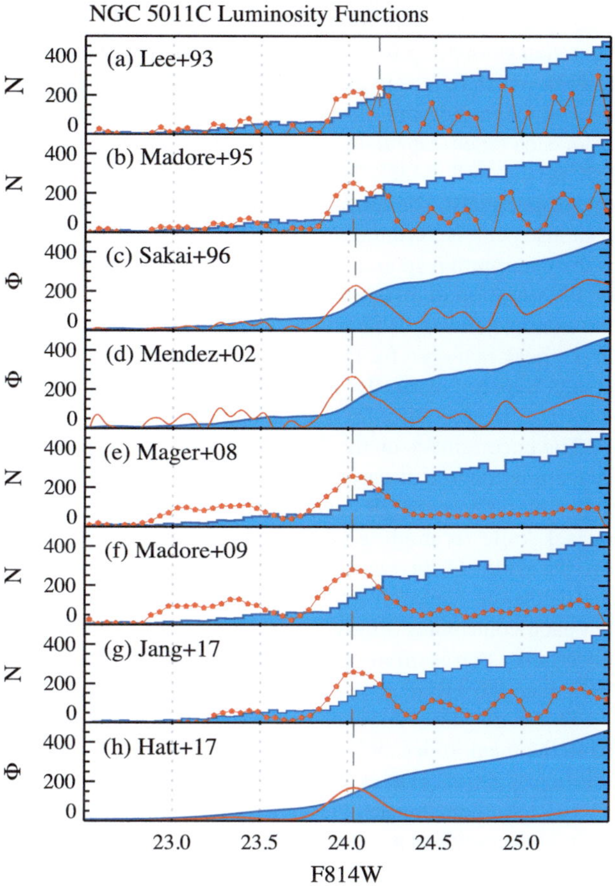

Fig. 27 Following Jang et al. (2018, their Fig. 9), F814W luminosity functions (blue histograms) and edge-detection responses (red lines) applied to NGC 5011 C. Each panel applies a specific TRGB algorithm to the same underlying data, as follows: (**a**) Lee et al. (1993), (**b**) Madore and Freedman (1995), (**c**) Sakai et al. (1996), (**d**) Méndez et al. (2002), (**e**) Mager et al. (2008), (**f**) Madore et al. (2009), (**g**) Jang and Lee (2017a), and (**h**) Hatt et al. (2017). The histograms show modifications for smoothing as employed in each of the studies with y-axis labels of N for binned starcounts and ϕ for smoothing. The TRGB is detected in each plot as the maximum of the response function, which is indicated by the dashed vertical line. Generally the algorithms agree for this dataset, but differences can occur as in Jang et al. (2018)

same photometry for NGC 5011 C,a dwarf galaxy in the Centaurus A group (Tully 2011), has been used in each panel. The histograms are slightly different owing to the prescriptions of the individual method; those that use direct binned LFs have N labels and those that smooth and/or resample the LF have are labeled ϕ. Comparison of the LF histograms, show that they look relatively similar irregardless of the smoothing, thus differences in the edge-response (red) can be largely attributed to the formulation of edge-detection algorithm.

The edge-detection response function is shown in red in each panel of Fig. 27. The form of each algorithm is given in Table 2. The Lee et al. (1993), Madore and Freedman (1995), Jang and Lee (2017a), and Madore et al. (2009) kernels are similar and are applied to binned LFs, except that additional elements are progressively added into the kernel that act to "smooth" the response; this is seen by comparing Figs. 27a, 27b, 27g, and 27f that show a progressive smoothing of the edge-detection response. The algorithms for Sakai et al. (1996), Méndez et al. (2002), and Mager et al. (2008) implement smoothing directly into the algorithm using various means of suppressing the noise; in particular, Sakai et al. (1996) uses adaptive binning and both Méndez et al. (2002) and Mager et al. (2008) use a poison-noise weight applied to a logarithmic form. The impact of these forms can be seen as Fig. 27c largely shows the same features as the previous discrete forms, but slightly broader and smoother and both Fig. 27d and 27e show a slight amplification of the smaller scale features due both their weighting and logarithmic forms. Figure 27h that was proposed in

Hatt et al. (2017) (and used subsequently by Jang et al. 2018 and Hatt et al. 2018) uses a Guassian smoothing function on the LF but applies the simple Sobel kernel from Lee et al. (1993), which results in an unambiguous single peaked edge-response.

The magnitude bin of maximum response for each panel are not significantly different from each other (especially considering the discrete forms use magnitude bins of 0.05 mag): $m^{TRGB}_{\rm F814W} = 24.18$, 24.03, 24.04, 24.02, 24.03, 24.03, 24.03, and 24.04 mag for panels (a) to (h), respectively. The total range of th detection is 0.16 mag or 8% in distance. The edge-response peak is asymmetric in many of the panels, but only in Lee et al. (1993) (Fig. 27a) is the maximum response in the secondary peak; excluding this result, the range is 0.02 mag or 1% variation in distance. This is a more consistent result than for NGC 1365, which was the example shown in Jang et al. (2018). This LF showed a much larger degree of variation in the peak response for these same algorithms, albeit the authors ultimately concluded the results were not dissimilar within the statistical uncertainties. Thus, it is important to choose the algorithm carefully for a given application.

The uncertainties on the TRGB detection are particularly difficult to determine and there is a similar degree of variation in how this measurement is made as in the TRGB detection algorithm. The strategies incorporate one or more of the following elements: boot-strap or jack-knife resampling of the LF, the bin-size, the mean photometric uncertainty at the tip, the width of the response function, the width of the kernel, and the signal-to-noise of the detection. Certainly each of these elements play a role in the underlying uncertainty, but each contributes differently in terms of random or systematic elements. One particular uncertainty that is often overlooked is the start point of the binning itself; even with bins of 0.05 mag, the TRGB result will change as the bins are shifted small amounts.

To counter this, Hatt et al. (2017) presents a detailed discussion of how the uncertainties, from the magnitudes and colors to the binning, are coupled in complex ways. As a result of this investigation, Hatt et al. (2017) developed end-to-end simulations that insert an artificial TRGB into their images, photometer it, and measure the TRGB in statistical fashion. This technique combines many of the strategies listed, but adds an ability to measure both random and systematic uncertainty as the TRGB inserted into the images was known a priori. Using these simulations, they not only produce rigorous estimates of their uncertainties, but use the statistical distributions to select the best smoothing parameters for their LF. The Hatt et al. methodology naturally incorporates the effects of incompleteness and crowding. This method has been applied thus far to five galaxies out to 20 Mpc distant with good results (Hatt et al. 2017, 2018; Jang et al. 2018).

4.2.2 Techniques to "Sharpen" the Tip Edge

The absolute magnitude of the TRGB is often approximated to a single value: $M_I^{TRGB} \sim -4.05 \pm 0.10$ mag (see, e.g., Lee et al. 1993; Salaris and Cassisi 1997; Tammann and Reindl 2013; Bellazzini et al. 2001, and references therein). However, as shown in Figs. 23 and Figs. 24, outside of the old, metal-poor regime, M_I^{TRGB} is not, in fact, constant. Fortunately, the age or metallicity effect is projected into the color of the star, e.g., more metal rich and/or younger stars are fainter and redder in I, $(V - I)$ CMDs than old their metal-poor counterparts. To both boost the signal at the TRGB and to avoid spurious detections, many authors take this effect into consideration.

The two primary ways to counteract the shape of the TRGB are as follows:

1. *Magnitude–color calibration:* Instead of determining a single-valued calibration, the calibration can be measured as a function of the mean $V - I$ color known as the $M_I^{TRGB} - (V - I)^{TRGB}$ relation. Here, the TRGB $(V - I)$ color basically accounts for

Table 3 Calibrations of the shape of the OPT-TRGB

Reference	Calibration of M_I	σ_{ZP}
Color based calibrations:		
Lee et al. (1993)	-4.0	± 0.1
Rizzi et al. (2007)	$-4.05 + 0.217[(V - I)_0 - 1.6]$	none given
Bellazzini (2008)	$-3.939 - 0.194(V - I)_0 + 0.080(V - I)_0^2$	none given
Madore et al. (2009)	$-4.05 + 0.2[(V - I)_0 - 1.5]$	none given
Jang and Lee (2017a)	$-4.015 - 0.007[(V - I)_0 - 1.5] + 0.091[(V - I)_0 - 1.5]^2$	± 0.058
Metallicity based calibrations:		
Salaris and Cassisi (1997)	$-3.732 + 0.588$ [M/H] $+ 0.193$ [M/H]^2	Theoretical
Bellazzini et al. (2001)	$-3.66 + 0.48$ [Fe/H] $+ 0.14$ [Fe/H]^2	± 0.12
Bellazzini et al. (2004)	$-3.629 + 0.679$ [M/H] $+ 0.258$ [M/H]^2	± 0.12

the star formation history of the observed population. Typically, users of this technique will use the mean color of the RGB sequence to estimate M_I^{TRGB} for their system. An alternate formulation relates M_I^{TRGB} directly to the metallicity. Given that the metallicity has to be measured in a independently, preferably from spectroscopy, but often in reference to theoretical stellar tracks (isochrone fitting), this is not necessarily an advantage.

2. *T magnitude:* Madore et al. (2009) suggested that, instead of fitting a more complex zero-point, to calibrate the slope out of the color–magnitude data. Madore et al. used theoretical tracks to produce an alternate magnitude system, known as the "T" for TRGB-magnitude system. This technique transforms the observed photometry into a system where the TRGB is flat—effectively torquing the CMD (a useful schematic can be found in Fig. 1 of Madore et al. 2009). Then, a single calibration to the metal-poor portion of the TRGB can be applied to the data over the full range. This can be done independent of priors on metallicity. The Madore et al. (2009) formulation is as follows:

$$T = I - \beta\big[(V - I)_0 - \gamma\big], \tag{25}$$

where γ is a fiducial color (preferably where the absolute zero point is determined) and β was determined relative to theoretical models to be $\beta = 0.20$ from Bellazzini et al. (2001), but values ranging from 0.15 (Mager et al. 2008) to 0.22 (Rizzi et al. 2007) have been used.

3. *QT magnitude:* More recently, Jang and Lee (2017b) expanded upon this initial formulation proposing the QT magnitude system, where the Q stands for "quadratic" because the functional form is a quadratic in color. The QT magnitude takes on the following form:

$$\mathrm{QT} = \mathrm{F814W} - \alpha(\mathrm{color} - \gamma)^2 - \beta(\mathrm{color} - \gamma) \tag{26}$$

where $\alpha = 0.159 \pm 0.010$, $\beta = -0.047 \pm 0.020$, and $\gamma = 1.1$ for the F814W, F606W–F814W magnitude–color combination (similar to the ground based I, $V - I$). Jang and Lee (2017b) provide calibrations in several common *HST* filter combinations.

Table 3 summaries several of the color–magnitude and metallicity–magnitude calibrations of the optical TRGB and their uncertainties, if available. The calibrations are compared in Fig. 28, where Fig. 28a shows the $M_I - (V - I)$ calibrations and Fig. 28b shows the M_I – [Fe/H] calibrations. The common single-value from Lee et al. (1993) and its approximate color-range is also shown for comparison and its uncertainty largely encompasses

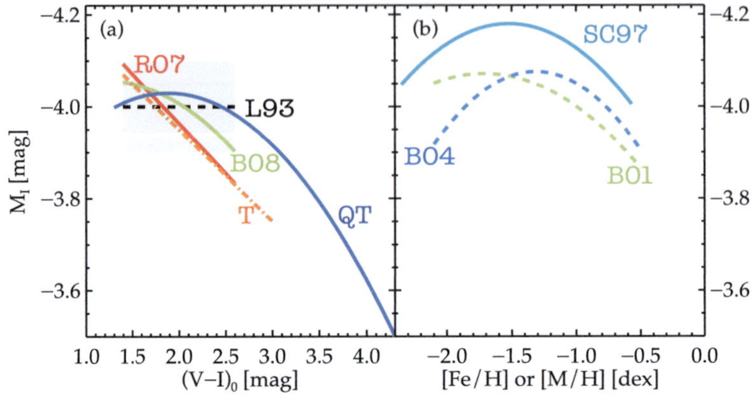

Fig. 28 Comparison of optical TRGB calibrations. (**a**) Calibrations of M_I in terms of $(V - I)_0$. The uncertainties on the zero points are $\sim\pm 0.10$ mag the span of which is shown by the gray shading, with the exception of With the exception of the QT calibration at ± 0.06 mag using multiple calibration paths (for details see Jang and Lee 2017b). (**b**) Calibrations of M_I in terms of [Fe/H] or [M/H], the uncertainty on the zeropoint of B04 and B01 is ± 0.12 mag and for SC97, which is from theoretical models, see Sect. 4.1. The labels are as follows, in panel a: L93—Lee et al. (1993); R07—Rizzi et al. (2007), B08—Bellazzini (2008), T: Madore et al. (2009), and QT: Jang and Lee (2017b); and in panel b: SC97—Salaris and Cassisi (1997); B01—Bellazzini et al. (2001), and B04 Bellazzini et al. (2004)

the M_I span of the more complex forms for the blue-side of the RGB. Moreover, Hatt et al. (2017) and Jang et al. (2018) both demonstrated that there was no formal different between the TRGB detections for in native, T or QT magnitudes for metal-poor systems. As with the edge-detectors, the application of these different techniques depends largely on the situation; in particular, only when there are well measured colors, there are significant red populations, and the uncertainties due to the transform be estimated, is it advisable to adopt more complex forms of the absolute calibration.

4.2.3 Contamination to the LF

Unfortunately, over its color–magnitude span, the RGB is not the only stellar sequence. A particularly vexing containment are the asymptotic giant branch stars (AGB), that cover a magnitude range overlapping with the brightest portion of the RGB, and eventually, depending on the object's star formation history, can extend to much brighter magnitudes than the TRGB (for more description of the AGB see Habing and Olofsson 2003; Rosenfield et al. 2014, 2016, and references therein). The AGB are particularly daunting because the RGB is their progenitor population and so, where there are strong RGB sequences, there is likely an AGB sequence, albeit at lower overall numbers.

The start of the AGB sequence for stars more luminous than the RGB can be confused for the TRGB. This can be seen in Fig. 27, especially for the Méndez et al. (2002) and Mager et al. (2008) algorithms, there are small increases in the LF and edge-response at $\sim$1 magnitude brighter than the TRGB. While in the NGC 5011 C case the effect is small, it is not so in galaxies with more recent star formation. An example being attempts to detect the TRGB in the Antennae galaxy (NGC 4038/39) by Saviane et al. (2004, 2008), who found $\mu = 30.7 \pm 0.25$ (random) ± 0.14 (systematic) mag. A later study, Schweizer et al. (2008), took into account the AGB population and used fields outside of the main star forming disk to find $\mu = 31.51 \pm 0.12$ (random) ± 0.12 (systematic) mag, which has been independently

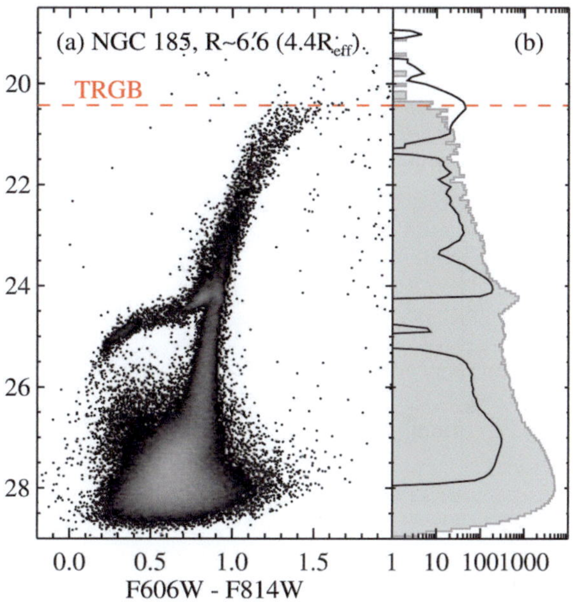

Fig. 29 Demonstration of the TRGB measurement process for the Local Group galaxy NGC 185 using *HST* + ACS data. (**a**) F814W—(F606W–F814W) CMD of resolved stars in its outer region ($R > 4.4R_{eff}$). The TRGB is marked by a dashed line. (**b**) F814W-band luminosity function of resolved stars (histogram) and corresponding edge-detection response (solid line). A strong edge-detection response is seen at the TRGB ($\text{F814W}_{TRGB} = 20.43 \pm 0.03$ mag)

confirmed by Jang and Lee (2015). The Saviane et al. physical distance was 13.3 Mpc compared to Schweizer et al. measuring 20 Mpc, which demonstrates the impact neglecting AGB stars can have.

Mitigation strategies for the AGB include: fitting the full form of the LF, including and AGB component, as in Méndez et al. (2002, among others) or working in regions of galaxies where the likelihood of significant contamination is small like in stellar halos (see discussion in Beaton et al. 2016). It is worth mentioning, that in regions that are dominated by younger stellar populations (like in disks), other populations like red super giants or highly extincted yellow super giants (Cepheids), can also cause difficulty measuring the TRGB confidently (see, e.g., in the LMC and SMC in Fig. 1). These latter populations show much more intrinsic variations as well as being embedded in dust (e.g. local changes in extinction) such that they are much more difficult to model. For the highest precision and accuracy, the TRGB is best applied to old population regimes.

4.3 Case Study for the Optical TRGB

Here we present an example of the TRGB distance measurement for the Local Group dwarf elliptical galaxy, NGC 185 (E3 pec) using the dataset described in Geha et al. (2015, Proposal ID:GO11724). Figure 29a shows the F814W–(F606W–F814W) CMD (similar to ground based I, $V - I$) for a halo field in NGC 185 and it is evident that the bulk of the bright stars are Pop II. The CMD, itself, shows multiple stellar populations, including the main sequence, blue stragglers, HB, AGB, and RGB, but few unambiguously Pop I sequences (e.g., compare to the LMC and SMC in Figs. 1a and 1b). The TRGB is the brightest part of RGB sequence and is marked by a dashed line.

The apparent magnitude of the TRGB is determined from the I (F814W) luminosity function, which is given in Fig. 29b. Because the TRGB is, effectively, the ending point of the RGB sequence, the LF should show a strong discontinuity at the TRGB. Here, we applied the edge-detection algorithm introduced in Mager et al. (2008) and plotted the response in

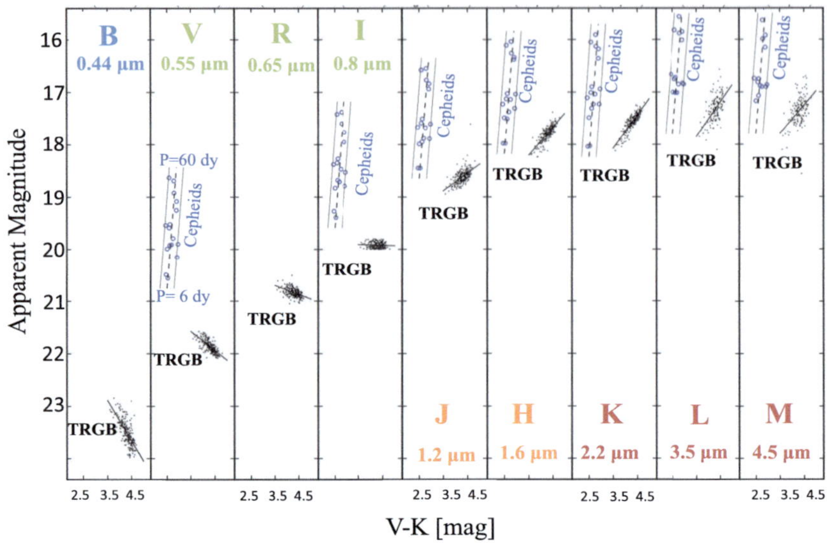

Fig. 30 The TRGB morphology as a function of wavelength for the Local Group dwarf galaxy NGC 6822. From left to right, the magnitudes in the B, V, I, J, H, K, L, and M bands are plotted against the $V - K$ color. TRGB stars are selected in the I, where the slope is flat, and the same stars are plotted in the other panels. For comparison, the Cepheid sequence in NGC 6822 is plotted when the Cepheids could be identified in the data. Figure courtesy of Erika Carlson and using data from Rich et al. (2014)

Fig. 29b as a solid line. The edge-detection response shows a strong peak at F814W = 20.43 mag. Using bootstrap resampling of the LF and subsequent TRGB detection, the uncertainty on the TRGB measurement is ±0.03 mag or a 1.5% statistical uncertainty for the TRGB measurement.

The distance modulus is then determined using a calibration of the TRGB absolute magnitude in the F814W filters and we adopt the Rizzi et al. (2007) calibration given in Table 3. In this formulation, the age-metallicity behavior of the TRGB is modeled using the color of the RGB star, effectively fitting a tilted line to the TRGB absolute magnitude (shown in Fig. 24). The median TRGB color in Fig. 29a is F606W − F814W = 1.43 ± 0.02 mag. Foreground extinction is estimated to be E(B–V) = 0.162 mag, corresponding to $A_{\mathrm{F814W}} = 0.281$ and $A_{\mathrm{F606W}} = 0.455$ mag (Schlafly and Finkbeiner 2011). Combining the absolute magnitude of the TRGB ($M_{\mathrm{F814W}} = -4.06 \pm 0.12$) with the apparent magnitude of the TRGB (F814W = 20.43 ± 0.03), we obtained a value for the distance modulus, $(m - M)_0 = 24.21 \pm 0.03$ (random) $\pm\, 0.12$ (systematic) mag.

This value is in good agreement with those based on red clump stars, $(m - M)_0 = 24.51 \sim 24.64$ mag (Pietrzyński et al. 2010), and RRLs, $(m - M)_0 = 23.93 \sim 24.15$ mag (Salaris and Cassisi 1997, 1998; Tammann et al. 2008). The distance summary for NGC 185 provided by de Grijs and Bono (2014) quotes the mean distance of 24.027 ± 0.333 mag from 26 studies for the TRGB, 23.993 ± 0.128 mag for the RR Lyrae from 8 studies, and 23.997 ± 0.119 mag for the full sample of distances.

4.4 Developments for the IR–TRGB as a Distance Indicator

Figure 30 compare the TRGB (black points) to the CC sequence (blue circles) in the Local Group dwarf NGC 6822. Each panel uses a different photometric band plotted against the

$V - K$ color, more specifically, (from left to right) B, V, I, J, H, K, L, and M. The TRGB stars are identified in the I where the tip is relatively flat and the same stars are plotted in each of the panels. The physical behavior of the TRGB described previously is seen empirically as the sloping tips in bands bluer and redder than I. While in the optical bands the CC population is significantly brighter than the TRGB, in the IR the TRGB is comparable to the shortest period stars (here $P = 10$ days). Considering that TRGB is non-variable, there are significant gains in observational efficiency using the TRGB in the IR for distances. We will focus on the NIR, but note that discussions of MIR stellar populations can be found in Boyer et al. (2015b,a).

Of course, owing to the steep slope of the TRGB in the IR, the tools developed in the optical case study are not immediately applicable. Dalcanton et al. (2012), Wu et al. (2014), and Górski et al. (2016) address this issue by using TRGB detection at a fiducial mean color for the RGB sequence (this is not dissimilar from means of calibrating the redder optical populations). Using their large sample of nearby dwarf galaxies Dalcanton et al. (2012) note, however, that there is a range of mean color across their sample and they correlate this range with the mean metallicity of the galaxies (inferred from star formation histories of the same field). Stated differently, it is not sufficient to calibrate the TRGB in the IR at one color, the slope of the TRGB needs to be determined.

Several calibrations of the sloped IR–TRGB are given in Table 4. The calibrations from both Dalcanton et al. (2012) and Wu et al. (2014) are in the *HST* flight system and are tied to optical TRGB measurements. Wu et al. (2014) rely heavily on theoretical stellar tracks to guide their calibration. Górski et al. (2016), on the other hand, transform a calibration from GGCs by Valenti et al. (2004) in 2MASS bandpasses to the color–magnitude plane. Recently, Madore et al. (2018) and Hatt et al. (2018) combine a magnitude–color slope measurement from IC 1613 with a zero-point determined in the LMC, both in 2MASS bandpasses.

These calibrations are compared in the panels of Fig. 31, with Fig. 31a showing M_H (M_{F160W}) versus $J - H$ (F110W–F160W) and Fig. 31b showing M_{K_s} versus $J - K_s$. The calibrations previously described are all shown, with the individual GGC data from Valenti et al. (2004) in both panels and a recent theoretical calibration from Serenelli et al. (2017) in Fig. 31b.

First, there is a noticeable offset between the 2MASS and *HST* filter systems shown in Fig. 31a that is likely due to bandpass differences that are described in Riess (2011) and discussed by Dalcanton et al. (2012). Both Dalcanton et al. (2012) and Riess (2011) use stellar models to transform between 2MASS and *HST* flight systems, but inconsistencies still remain for these red stars whose models are more uncertain (see Dalcanton et al. 2012, their Fig. 18 and associated discussion). Second, comparing similar filter systems, the results are not terribly dissimilar; Wu et al. (W14, 2014) and (D12, Dalcanton et al. 2012) largely agree, with the former have more objects at redder colors (they also employ very similar overall strategies). Hoyt et al. (2018) and Madore et al. (2018) (collectively called H/M18 in the figure) find results not dissimilar from the GGCS from Valenti et al. (2004), especially considering the TRGB in GGCs is not always well populated and AGB/RGB stars cannot be disentangled. Górski et al. (2016) agrees well with Valenti et al. (2004), as anticipated, as their relations are derived from those data.

Unfortunately, both Dalcanton et al. (2012) and Wu et al. (2014) predominantly use relatively distant objects for the calibration, for which precise distances are not necessarily known independent of TRGB analyses. Górski et al. (2016), Madore et al. (2018), Hoyt et al. (2018), and Valenti et al. (2004), on the other hand, focus on the Local Group dwarfs for which independent high-precision distances are known and they find reasonable agreement

Table 4 Calibrations of the Shape of the IR–TRGB

Reference	λ	Calibration of M_λ	Notes
Color based calibrations, *HST* flight system:			
Dalcanton et al. (2012)	F160W	−3.496 − 2.576 (F110W − F160W)	rms = 0.05
Wu et al. (2014)	F160W	−5.97 − 2.41 [(F110W − F160W) − 0.95]	blue
		−5.97 − 3.81 [(F110W − F160W) − 0.95]	red
	F110W	−5.02 − 1.41 [(F110W − F160W) − 0.95]	blue
		−5.02 − 2.81 [(F110W − F160W) − 0.95]	red
Serenelli et al. (2017)	F110W	−4.630 − 9.525 [(F110W − F160W) − 0.68]	blue, Theoretical
		−4.630 − 1.511 [(F110W − F160W) − 0.68]	red, Theoretical
	F160W	−5.310 − 10.525 [(F110W − F160W) − 0.68]	blue, Theoretical
		−5.310 − 2.511 [(F110W − F160W) − 0.68]	red, Theoretical
Color based calibrations, 2MASS system:			
Górski et al. (2016)	K	$-6.32 - 2.15\,[(J - K_s) - 1.00]$	ZW84 scale
	J	$-5.32 - 1.15\,[(J - K_s) - 1.00]$	ZW84 scale
Serenelli et al. (2017)	K	$-6.17 - 1.81\,[(J - K_s) - 1.00]$	theoretical
	J	$-4.96 - 0.81\,[(J - K_s) - 0.76]$	theoretical
Madore et al. (2018)	J	$-5.13(\pm 0.01_{stat} \pm 0.06_{sys}) - 1.11(\pm 0.15)\,[(J - H) - 0.80]$	
and Hoyt et al. (2018)	H	$-5.93(\pm 0.01_{stat} \pm 0.06_{sys}) - 2.11(\pm 0.26)\,[(J - H) - 0.80]$	
	K	$-6.13(\pm 0.01_{stat} \pm 0.06_{sys}) - 2.41(\pm 0.36)[(J - H) - 0.80]$	
	J	$-5.14(\pm 0.01_{stat} \pm 0.06_{sys}) - 0.85(\pm 0.12)\,[(J - K) - 1.00]$	
	H	$-5.94(\pm 0.01_{stat} \pm 0.06_{sys}) - 1.62(\pm 0.22)\,[(J - K) - 1.00]$	
	K	$-6.14(\pm 0.01_{stat} \pm 0.06_{sys}) - 1.85(\pm 0.27)\,[(J - K) - 1.00]$	
Metallicity based calibrations:			
Valenti et al. (2004)	J	−5.67 − 0.31 [Fe/H]	GGCs, CG97 scale
	H	−6.71 − 0.47 [Fe/H]	GGCs, CG97 scale
	K	−6.98 − 0.58 [Fe/H]	GGCs, CG97 scale
	M_{bol}	−3.87 − 0.18 [Fe/H]	GGCs, CG97 scale
	J	−5.64 − 0.32 [M/H]	GGCs
	H	−6.66 − 0.49 [M/H]	GGCs
	K	−6.92 − 0.62 [M/H]	GGCs
	M_{bol}	−3.85 − 0.19 [Fe/H]	GGCs

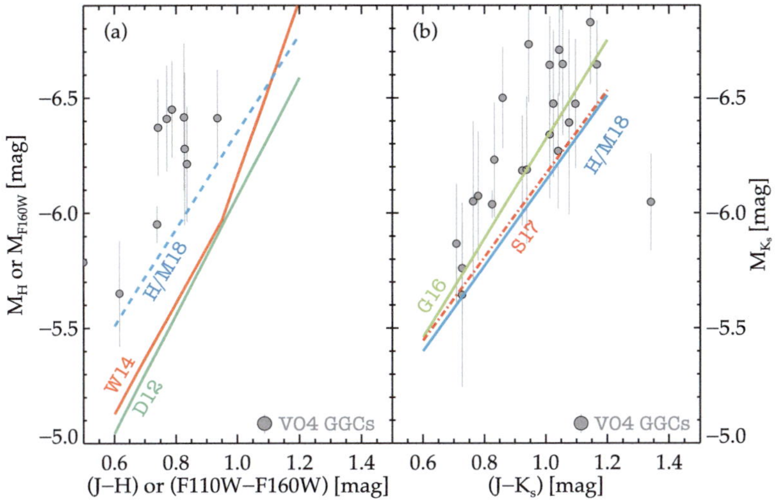

Fig. 31 Comparison of IR–TRGB slope measurements. (**a**) M_H against $J - H$ for GGCs in Valenti et al. (2004, grey circles) and the LMC zeropoint + IC 1613 slopes from Hoyt et al. (2018) and Madore et al. (2018) (dashed blue line), respectively. M_{F110W} is plotted against F110W − F160W on the same panel (without transformation) and compares Wu et al. (2014, orange) and Dalcanton et al. (2012, green). (**b**) M_{K_S} against $J - K_S$ for GGCs in Valenti et al. (2004, grey circles), theoretical calibration from Serenelli et al. (2017, dot–dash red), and LMC zeropoint + IC 1613 slopes from Hoyt et al. (2018) and Madore et al. (2018) (solid blue line). There is a clear distinction between the ground 2MASS and space WFC3/IR filter sets, but generally calibrations agree reasonably well. Functional forms are given in Table 4

between these works and their own distances. Of course the divide between filter systems still persists in these comparisons, as does the technique of application.

Dalcanton et al. (2012), Wu et al. (2014), and Górski et al. (2016) each measure the TRGB absolute magnitude at the mean-color of the RGB (similar to the procedure followed for NGC 185 in Sect. 4.3), which is slightly different from the measurements of Hoyt et al. (2018) and Madore et al. (2018) that measure the slope in the color–magnitude diagram. As in the optical, an alternate technique is to use a "T" magnitude that accounts for the color dependence. Hoyt et al. (2018) demonstrated this method to be extremely effective (precise and accurate), as they were able to see internal structure of the LMC by determining distances to individual zones across the LMC. The alternate technique is demonstrated in Figs. 32a and 32b for two galaxies in the Dalcanton et al. (2012) sample, NGC 404 and NGC 2976. Here, the sloped-TRGB is rectified into a "T" magnitude system, to which the same edge-detection tools from the optical are applied to determine the TRGB magnitude (here following the methods of Hatt et al. 2017).

4.5 Summary

In this section, we have described the TRGB as a standard candle from a theoretical and empirical perspective. The TRGB is the terminus of the RGB sequence that occurs at the transition from H-shell burning to He-core burning. In the I-band, the TRGB is observed to have a constant value of $M_I = -4$ mag with uncertainties at the 0.1 mag level for old, metal-poor stars. Many empirical and theoretical investigations have provided refinements to this value that are largely consistent, but only via secondary calibration methods; no direct calibration exists.

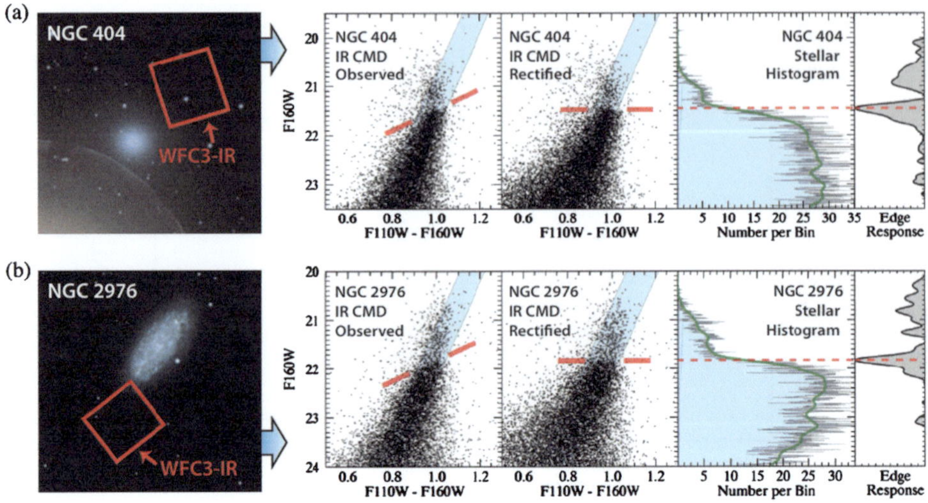

Fig. 32 Use of the NIR–TRGB as a distance indicator applied to (**a**) NGC 404 and (**b**) NGC 2976. The photometry is from Dalcanton et al. (2012). The WFC3-IR field used is shown in red for GALEX images of the galaxies in the left most panels. The CMD is shown in the left central panels and a 'rectified' CMD is shown in the next panel, from which a luminosity function is determined and an edge-detector employed. Figure courtesy of Mark Seibert and photometry courtesy of Julianne Dalcanton and Ben Williams

The TRGB method relies on detecting the star of the RGB sequence, which has been implement in a myriad of methods, some of which were compared directly in Fig. 27. Generally, the choice of the most appropriate method depends on the scientific application and the data at hand, e.g., what filters are available, the number of sources populating the TRGB, and the contamination, among other concerns. Despite the diversity of the approaches summarized here, the advantage of using a relatively-luminous non-variable stellar population comes with dramatically higher observational efficiency than that of variable stars.

5 Systematics and the Absolute Scale

The goal of this section is to compare the absolute scales amongst the old stellar population standard candles. An overall review of progress toward a self-consistent distance scale for all standard candles can be found in de Grijs et al. (2017), with significant works focusing on overall consistency for objects within ∼1 Mpc given in de Grijs et al. (2014), de Grijs and Bono (2014, 2015). A discussion of absolute scales, ultimately, rests onto two nuanced points: (i) full evaluation the systematics associated with the distance measurements, and (ii) a full understanding of the true independence of the scales utilized by each standard candles. To set the stage for this discussion, we open with a case study that compares distance measurements from CCs, RRL, and the TRGB in the Local Group galaxy IC 1613 in Sect. 5.1. In Sect. 5.2, the "primary" methods to set the absolute scale are described with summaries of progress for each of the standard candles. Section 5.3 discusses the "secondary" methods to set the absolute scale, these being the adoption of a distance to a specific object and using the standard candle in that object. With these efforts in mind, the systematic impacts of metallicity on the derived distances are discussed Sect. 5.4. Other systematic effects are discussed briefly in Sect. 5.5.

5.1 Comparison Case Study in IC1613

We use the compendium of distances to IC 1613 of Hatt et al. (2017) for CCs, RRLs, and optical TRGB standard candles; no new distances have been published since this paper (while Madore et al. 2018, uses IC 1613 to calibrate the IR–TRGB, the distance is adopted from Hatt et al.). The references for these comparisons and the shorthand used in the plot are as follows: Mac01—Macri et al. (2001), Dol01—Dolphin et al. (2001), Fre01—Freedman et al. (2001), Uda01—Udalski et al. (2001), Tik02—Tikhonov and Galazutdinova (2002), Pie06—Pietrzyński et al. (2006), Ant06—Antonello et al. (2006), Fre09—Freedman et al. (2009), Ber10—Bernard et al. (2010a), Sco13—Scowcroft et al. (2013), Dam13—Dambis et al. (2013), and Hat17—Hatt et al. (2017). Only a single T2C candidate is known in IC 1613 (Majaess et al. 2009) and, thus, we do not have a distance from T2Cs in this comparison case study; the study itself, however, still reveals the major points to be discussed in the later subsections.

Figure 33 shows independently determined distances to IC 1613 since the *HST* Key Project (Freedman et al. 2001) ordered from top to bottom as CC studies, RRL studies, and TRGB studies. The original distance moduli (μ_0) are plotted with open gray symbols with the CC distances as boxes, RRL distances as circles and TRGB distances as diamonds. As is detailed explicitly by Hatt et al. (2017, their Table 4), the studies vary considerably in their absolute calibration: for the CCs the distance modulus to the LMC varies from 18.25 to 18.54 mag (10% in distance), for RRLs the RRL zp varies from M_V^{RRL}=0.52 to 0.72 (10% in distance), and the TRGB varies from $M_I^{TRGB} = -3.91$ to -4.13 mag (10% in distance). Similar variations also occur in the Milky Way foreground extinction and the internal extinction for the CCs. Thus, in attempting to address if these Pop II distance indicators set comparable distance scales, we must homogenize for these systematics to prevent the dispersion between studies dominating the comparison.

Following Hatt et al. (2017), we normalize to $\mu_{LMC} = 18.49$ mag for the CCs, $M_V^{RRL} = 0.63$ mag and no foreground extinction ($A_V \sim 0.0$ mag); we deviate from Hatt et al. by assuming an mean extinction internal to IC 1613 of $A_I = 0.1$ mag (for the CCs) and $M_I^{TRGB} = -4.0$ (see discussion in Hoyt et al. 2018; Hatt et al. 2018). The homogenized distance moduli are shown by the filled symbols in Fig. 33 and these symbols are color coded by the wavelength regime of the study: blue for optical, orange for NIR, red for MIR, and green for reddening-free relations for the CCs and RRL. With homogenization complete, the overall dispersion within the standard candles and between the standard candles reduces. The mean distance determined from de Grijs and Bono (2014) of $24.336 \pm 0.049_{stat}$ is shown by the dashed gray line with the filled gray region showing ±0.05 mag, which is representative the typical systematic uncertainty on the zero points. The measurements shown here are all within 2-σ of this "consensus distance," from which we conclude that we see "overall" consistency between CCs, RRLs, and the TRGB in IC 1613. It is worth noting, however, that in terms of the systematic effects, IC 1613 is a simple case: it a metal-poor dwarf galaxy with low internal extinction and low source crowding even at the resolution of ground-based imaging.

Before discussing these systematic effects in detail, it is worth noting that comparisons between RRL, CCs, T2Cs, and the TRGB are currently only possible for galaxies in the Local Group; only within the Local Group are populations of for all three distance indicators identified (e.g., only the TRGB and CCs have been used for distances outside of the Local Group). These four distance indicators can only really be compared in earnest with systems of stellar masses on order of relatively massive dwarf galaxies; Fig. 1 demonstrates the difference in the numbers of CCs, T2Cs, and RRLs between the LMC and SMC.

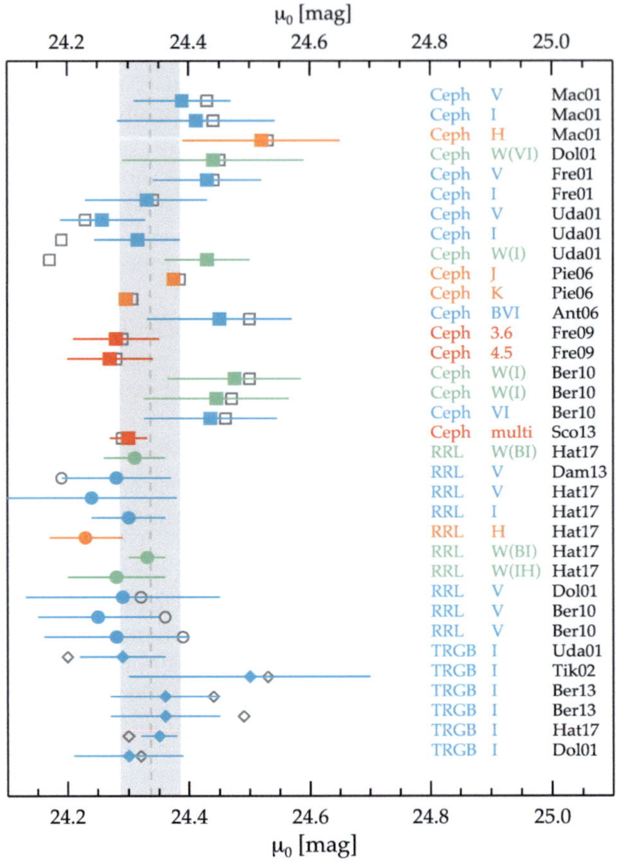

Fig. 33 Comparison of distances to IC 1613 since the *HST* Key Project using the compendium of Hatt et al. (2017, their Fig. 9, their Table 4). The original measurements are shown as open gray symbols, with Leavitt Law studies as squares, RRL as circles, and TRGB as diamonds. Each study has assumed slightly different zero points and foreground/internal extinctions. The filled symbols display homogenized distance moduli for each study, with the symbols colored by the wavelength range; optical studies in blue, NIR studies in orange, MIR studies in red, and reddening-free Wesenheit studies in green. The consensus distance modulus from de Grijs et al. (2014) of $24.336 \pm 0.049_{stat}$ mag is shown by the dashed line with a region of width ± 0.05 mag represents representative uncertainties for the absolute scales. Overall this figure demonstrates agreement for the Pop I and Pop II scales and between RRL and TRGB. The distance moduli are homogenized to an LMC distance modulus of 18.49 mag, a total $E(B-V) = 0.08$ mag for Cepheids) and $E(B-V) = 0.02$ mag for RRL and TRGB, $M_I^{TRGB} = 4.00$ mag, and $M_V^{RRL} = 0.63$ mag; original uncertainties are plotted

Farther afield, Lee and Jang (2012) provide comparisons between CCs and the TRGB in M 101 (the closest host-galaxy for a modern SN Ia). Lee and Jang find both standard candles scatter over a large range of 0.7 mag (35% in distance) and 0.4 mag (20% in distance), respectively, despite each study quoting uncertainties at the ~0.1 mag level (5% in distance)—Lee and Jang did not homogenize for zero points, but these will not fully account for the range. Recent high-precision TRGB studies such as Jang and Lee (2017a), Jang et al. (2018), and Hatt et al. (2018) compare CCs and the TRGB for SN Ia host galaxies, finding overall agreement, but the bulk of these galaxies have only a single measurement from a single team for each result. Previous work during and after the *HST* Key Project found similar agreement between scales, but the measurements at that time had larger uncertainties (e.g., Ferrarese et al. 2000; Sakai et al. 2004; Freedman and Madore 2010, among others). With these examples in mind, we now proceed to discuss the systematics affecting the absolute scales and individual distance measurements.

5.2 Absolute Scale Through Primary Calibration Techniques

Primary calibrations are produced by using individual stars to which distances are determined using geometric methods. These come in three types: (i) trigonometric parallax, (ii) statistical parallax, and, for RRLs and T2Cs, (iii) pulsational parallax. Spectroscopic parallax is another means of determining the distance to an individual star, but it requires a primary calibration from another means and will not be discussed here (e.g., Adams 1916; Sandage et al. 2016). First, we describe these methods in Sect. 5.2.1 and then discuss their application to the RRLs in Sect. 5.2.5, T2Cs in Sect. 5.2.6, and TRGB in Sect. 5.2.7.

5.2.1 The Methods

5.2.2 Trigonometric Parallax

First measured by Bessel (1838), trigonometric parallax is the apparent displacement of an object, relative to its background, due having viewed the source from two distinct positions. The size of the displacement is proportional to the separation between the two viewing positions and for astronomical applications, the typical baseline is the Earth's orbit, e.g., observations are taken $\sim$six months apart. From this setup, triangles can be constructed connecting the two viewpoints, the true location for object of interest, and its apparent displacement. Thus, once positions are robustly determined relative to a background frame, the distance is computed using simple geometry. Ground based parallaxes are feasible only for the most nearby stars, and thus, the resolution and stability of space-based observatories are required. A history of the technique and an assessment of ground-based methods is given in Vasilevskis (1966), van Altena (1983) and Upgren (1985).

Major space-based parallax-focused missions are *Hipparcos* (Lindegren et al. 1992; Perryman et al. 1997; ESA 1997; van Leeuwen 2007, and references therein) and, more recently, *Gaia* (Perryman et al. 2001; Prusti et al. 2016, and references therein). The Fine Guidance Sensor (FGS) on *HST* was designed for astrometric applications (see e.g., Duncombe et al. 1991, and references therein) and been used for parallaxes (for a comprehensive compilation and review of these measurements see Benedict et al. 2017) and, very recently using a primary *HST* instrument, WFC3 (see, e.g., Brown et al. 2018a and references therein).

5.2.3 Statistical Parallax

The method of statistical parallaxes is an old one, having been invented by Kapteyn and Kohlschütter in the early 20th century and first published in Adams and Kohlschütter (1914).[10] This technique determines a mean parallax to a group of stars whose proper motions, with the motion of the Sun removed, are analyzed. The observed proper motion, in angular units, is inversely proportional to the distance to the object, so from the analysis

[10] A detailed history of these methodological discoveries from the source material is given in Sandage (2004) whereas a briefer summary is more accessible in Sandage et al. (2016).

of an ensemble of proper motions, the parallax can be determined for the class. A discussion of the mathematics of such a determination is given in Popowski and Gould (1997, and references therein). The method was employed often in the 20th century to obtain calibrations for the distance scale, but has fallen slightly out of use as more accurate and precise trigonometric parallaxes have become available.

5.2.4 Pulsation Parallaxes

The method was proposed by Wesselink (1946) based upon earlier suggestions by Baade (1926) and, as a result, has been coined the Baade–Wesselink method. The method combines radius measurements from radial velocity curves with temperature and measurements from light curves to infer the surface brightness, and thereby, intrinsic luminosity of a pulsational variable. The method was founded upon the postulation that at points of equal color on the ascending and descending branches of the light curve, the temperature of the star should be the same, and in consequence, any difference in luminosity at these points is due to the fractional change in radius ($\Delta R/R$). The B–V color–surface brightness relationship first provided in Wesselink (1969) completed the development phase. A limitation of the method is the need for a term p coined the "projection factor" that is used to translate between the observed radial velocity and the true pulsational velocity—the radial velocity measured is the motion of the atmospheric line projected along the line-of-sight and integrated across the stellar disk. Thus, p is a multiplicative factor that can come from theoretical inferences or can be measured directly with interferometric methods (see e.g., recent work for Cepheids see Pilecki et al. 2018; Kervella et al. 2017; Breitfelder et al. 2016, among others).

5.2.5 Primary Calibration of the RRLs

All three of the direct methods can be applied to RRLs and largely find agreement, albeit most methods produce absolute calibrations at the $\sim$0.1 mag level.

Trigonometric Parallax: A selection of the major results for calibration of the RRLs via trigonometric parallax are summarized below. It is important to note that *only* in the recent *Gaia* data releases have both the period and metallicity impacts, in addition to the zeropoint, been measurable directly from parallaxes to individual stars. Thus, the future is bright for larger samples of parallaxes to produce a precise and accurate calibration of both the zero-point and the slope of the PLs.

- *Hipparcos*: Most recently, Feast et al. (2008) used 142 RRL included in the *Hipparcos* re-reduction (van Leeuwen 2007) found $M_V^{RRL} = 0.59 \pm 0.10$ mag (adjusted to [Fe/H] = -1.60 dex). For the NIR, Feast et al. found $M_{K_s}^{RRL} = -0.63 \pm 0.10$ mag at a fiducial $\log(P) = -0.252$ and assuming PL_{K_s} slope to be -2.41. Other studies using the original *Hipparcos* reduction include: Fernley et al. (1998), Feast (1998), Solano (1998), Luri et al. (1998), Koen and Laney (1998), Groenewegen and Salaris (1999), and Reid (1999), among others.
- *HST*–FGS: Benedict et al. (2011) presented *HST*–FGS parallaxes to 5 RRL (4 RRab and 1 RRc) to precisions of <10%. In Benedict et al. (2011), itself, the zero point is $M_V = 0.42$ mag at [Fe/H] = -1.58 dex, which is systematically lower than other techniques. Considering, however, the small sample size, the zero-point could be biased by evolutionary effects, e.g., even a single egregious star could bias the result. While these parallaxes were the best for the RRL, most works employed a dual theoretical–empirical procedure to determine zeropoints, which are described in the case studies in Sect. 2.3, and typically yield results more consistent with other techniques.

- *Gaia* Data Release 1 (TGAS) (Michalik et al. 2015; Lindegren et al. 2016): The first *Gaia* data release included parallaxes from the Tycho-Gaia Astrometric Solution (TGAS), which contained parallaxes for all 142 RRL in *Hipparcos* with some additional stars in the *Tycho* catalog. Neeley et al. (2017) demonstrated use of 46 well characterized RRL in TGAS provided constraints on the zero point for the MIR PL at the same precision as the smaller, but more precise, sample from *HST*–FGS. Notably, Neeley et al. (2017) argued that these data necessitated metallicity term to comply with theoretical predictions. Sesar et al. (2017), found similar results, but using larger samples and a Bayesian inference method for the $W1$ and $W2$ bands fitting for slope, metallicity, and zero-point; this paper found both consistent and inconsistent results as compared to other studies, but also issued caution on the reliability of the abundances.
- *Gaia* Data Release 2 (DR2) (Brown et al. 2018b): The second *Gaia* data release included a $<10\%$ parallaxes for several thousands of RRL. Muraveva et al. (2018a) exploited 401 of these RRL that had sufficient characterization in the literature to derive full PLs in a Bayesian framework for the V, G, K_s and $W1$ bands; here the limiting factor on the sample is the existence of non-*Gaia* measurements including photometry, extinctions, abundances, and long-baseline periods. Muraveva et al. (2018a) find: $M_V = 0.66 \pm 0.06$ mag, $M_G = 0.63 \pm 0.08$ mag, $M_{K_s} = -0.37 \pm 0.11$ mag, and $M_{W1} = -0.41 \pm 0.11$ mag for a fiducial $P = 0.5238$ days and [Fe/H] $= -1.5$ dex. Despite the larger sample and better parallaxes, the uncertainties on the zeropoints remain at the 0.1 mag level. Additional work is likely forthcoming from other teams and employing other wavebands.

Statistical parallax: Because RRLs have a small-to-negligible period dependence in the optical, the absolute magnitude can be approximated to a single value and statistical parallax can be applied. A summary of these efforts at that time is given in Smith (1995) and since that time less effort has been employed for this measurement. Popowski and Gould (1998a,b), Gould and Popowski (1998) employed a rigorous mathematical framework that incorporated observational biases directly to find $M_V^{RRab} = 0.75 \pm 0.13$ mag at [Fe/H] $= -1.61$ dex for the Layden et al. (1996) sample and $M_V^{RRab} = 0.77 \pm 0.13$ mag at [Fe/H] $= -1.60$ dex for a sample of 147 kinematically-selected halo RRLs. Using the larger ASAS all sky sample of variables, Kollmeier et al. (2013) produced the first calibration of the M_V^{RRc} using statistical parallax, finding $M_V^{RRc} = 0.59 \pm 0.10$ mag for a mean metallicity of [Fe/H] $= -1.59$ dex. Such methods do not perform as well where there is a period dependence, because the statistical samples must be restricted to period ranges, which reduces precision.

Pulsation parallax: As discussed in Smith (1995), the foundational assumption of the Baade–Wesselink method regarding equal colors implying equal temperatures fails for the RRL because, in part, of RRL shocks that perturb the color-curves in the optical. Longmore et al. (1985) noted that K observations provided better results for the RRLs and after that time the NIR results were used for such analyses. Compilations of pulsation parallaxes can be found in Liu and Janes (1990), Cacciari et al. (1992), and Skillen et al. (1993). Muraveva et al. (2015) used 23 pulsation parallaxes to derive the PL_{K_s} relationship and found the zero-point to agree well with those determined from the LMC, but disagree with that determined from Benedict et al. (2011).

There is only one star common to each of these means of measuring the parallax: SU Dra. The pulsation parallax of Liu and Janes (1990) finds a distance of 640 pc or $\pi = 1.56$ mas with an uncertainty on the absolute magnitudes of 0.12 mag, that translates to a $\sim$12% uncertainty on the parallax of $\sim$0.18 mas. The *Hipparcos* re-reduction found $\pi_{trig} = 0.20 \pm 1.13$ mas (van Leeuwen 2007) compared to the initial result of $\pi_{trig} = 1.11 \pm 1.15$ mas

(Perryman et al. 1997). The *Gaia* DR2 value (Brown et al. 2018b) is 1.4016 ± 0.0308 mas, whereas Benedict et al. (2011) finds 1.42 ± 0.16 mas. From comparison of a single star, there is overall agreement within the uncertainties as was demonstrated for the *body* of *HST*–FGS parallaxes against *Gaia* by Benedict et al. (2017, 2018). Thus, interpreting differences between zero-points determined in different contexts may be better framed as comparisons between assumptions of the PL-form, sample size, and abundances, rather than the fidelity of the parallaxes themselves.

5.2.6 Primary Calibration of the T2Cs

The body of work on the primary calibration of the T2Cs is much smaller than that of RRLs and they are discussed with respect to two key stars: VY Pyx and κ Pav that have parallax measurements from multiple techniques.

Feast et al. (2008) used *Hipparcos* trigonometric parallaxes for VY Pyx (BL Her type with $P = 1.24$ days) and κ Pav (W Vir type with $P = 9.09$ days), as well as pulsation-based parallaxes for a few T2Cs to see if their absolute magnitudes based on the parallaxes were consistent with the PL relations derived by Matsunaga et al. (2006) for T2Cs in GGCs. Their *Hipparcos* parallaxes led to the absolute infrared magnitudes more-or-less predicted by Matsunaga et al. (2006), although the pulsation parallax of κ Pav had a significant offset for which the authors gave cautions for both the pulsation and trigonometric parallax (which still had a large uncertainty). Feast (2010) later suggested that κ Pav may be of peculiar W Vir type, which are, indeed, intrinsically brighter than expected by the PL relation for the W Vir type.

Benedict et al. (2011) reported *HST*–FGS parallaxes for the same two T2Cs, VY Pyx and κ Pav, with significantly smaller parallax uncertainties than those from *Hipparcos*. Despite the lower uncertainties, these new measurements left the questions unresolved; using the *HST*–FGS parallaxes, κ Pav seems to follow the PL relation, but VY Pyx, in turn, is fainter than anticipated. Breitfelder et al. (2015) found that the *HST* parallax of κ Pav leads to p-factor consistent with other Cepheids based on their interferometric and spectroscopic data. Breitfelder et al. (2015) also found no evidence of binarity for this object, which would have been a strong clue that the star was of a peculiar type. These new results suggest that κ Pav is a normal W Vir, not a peculiar W Vir, and, thereby, should follow the PL relation of the W Vir type. The nature of VY Pyx, however, remained unclear.

Gaia DR2 offers more insight, finding for κ Pav, $\pi = 5.1992 \pm 0.3094$ mas and for VY Pyx, $\pi = 7.1173 \pm 0.0265$ mas (Brown et al. 2018b). Figure 34 compares the parallax results from *Hipparcos*, *HST*–FGS, pulsation parallaxes, and *Gaia*-DR2 for these two T2Cs. The prediction from the PL of Matsunaga et al. (2006) is plotted with the uncertainty of the zero-point in gray; the dashed line represents if the star is directly on the PL (e.g., no intrinsic dispersion). The trigonometric parallaxes for κ-Pav, Fig. 34a, are all consistent with Matsunaga et al. (2006), whereas the pulsation parallax is several sigma away. There is no pulsation parallax for VY Pyx, but the trigonometric parallaxes in Fig. 34b are all inconsistent with the Matsunaga et al. (2006)-PL. Thus, the nature and classification of VY Pyx likely requires further study. As with the RRL, this comparison highlights the need for parallaxes to a *representative* sample of stars to infer PLs instead of relying on individual stars as a prototype for a class, e.g., the *intrinsic dispersion within the class* may be large.

5.2.7 Primary Calibration of the TRGB

At the time of writing, the authors are only aware of a single study to provide a primary calibration of the TRGB. Tabur et al. (2009) combined *Hipparcos* and 2MASS

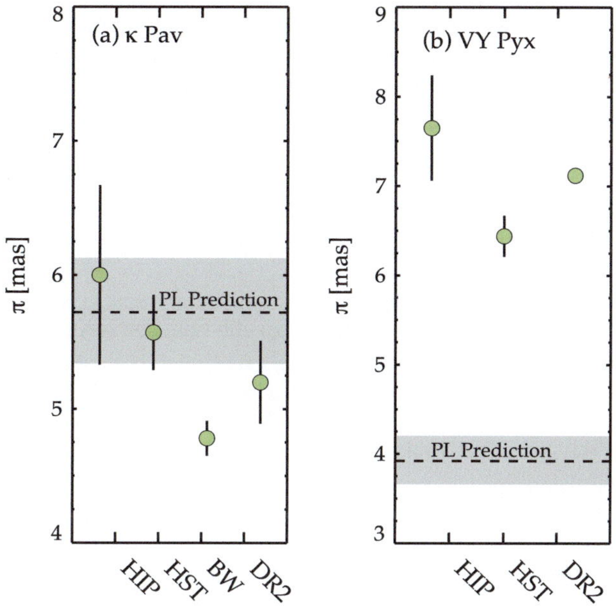

Fig. 34 Comparison of geometric distance measurements to the T2Cs (**a**) κ Pav ($P = 9.09$ dy) and (**b**) VY Pyx ($P = 1.2$ dy) to the predicted parallax from the Matsunaga et al. (2006) K_s–PL relation. The shaded region gives the scatter about the Matsunaga et al. (2006) PL (± 0.15 mag) from the GGCs. While κ Pav is largely consistent with the PL, VY Pyx is quite deviant

data to provide a preliminary calibration of the TRGB absolute magnitude in K_s, finding $M_K = -6.85 \pm 0.03$ mag. This study, however, has a number of caveats: (i) for the bright stars populating the tip, most of the magnitudes were derived from DIRBE because such stars in 2MASS had large (0.2 mag) uncertainties due to saturation, (ii) the *Hipparcos* parallaxes have extremely large errors ($\gg$20%), and (iii) the authors only fit for a single color bin. Moreover, the authors do not use a Bayesian approach to determining distances using parallaxes with large uncertainties as is recommended (see e.g., Bailer-Jones 2015). As mentioned in Tabur et al. (2009) with respect to *JHK*, the current limitation for direct measurement of the TRGB is the lack of homogeneous photometry for bright stars in the bands of interest for the TRGB—primarily, *I* and *JHK*. Such work is challenging because these stars cannot be identified a priori as tip stars and the nearby stars with the most precise parallaxes are saturated in most modern surveys. Thus, even with the wealth of parallaxes from *Gaia*-DR2, this measurement remains elusive.

5.3 The Absolute Scale via Secondary Calibration

Secondary calibration works in two steps: (i) establish the distance to an object, like a star cluster or nearby galaxy, using a geometric technique and (ii) determine the absolute magnitude of the standard candle using samples from that object. Thus, the absolute scale set has two components to the systematic uncertainty: (i) the total uncertainties from the geometric technique and (ii) the uncertainties from the standard candle. It is also possible to combine other standard candles to obtain a secondary calibration, which adds another layer of systematic uncertainty; for brevity, we will not include those results here though some discussion exist in the sections for each distance indicator.

In this volume, Thévenin et al. (2017) describe "modern" geometric techniques, which includes a number of different methods—to which methods relying on gravitational waves can be added (e.g., Chen et al. 2017), which are also described in this volume by Czerny et al. (2018). However, of those only two are applicable on the distance scales to directly calibrate the Pop II distance indicators described here, (i) eclipsing binaries and (ii) megamasers (only for the TRGB). As for the primary calibration, we give the general sense of the methods in Sect. 5.3.1 before describe application to the RRLs in Sect. 5.3.4, T2Cs in Sect. 5.3.5, and TRGB in Sect. 5.3.6.

5.3.1 *Methods for Secondary Calibration*

5.3.2 *Eclipsing Binaries*

Eclipsing binaries (EBs) are a set of binary star systems where the binary orbit is sufficiently along the line-of-sight that the one star eclipses the other. Detached eclipsing binaries (DEBs) are a subset of EBs for which both stars are entirely within their respective Roche Lobes, meaning that the stars can be modeled as a simple two-body system. Using a combination of (i) light curve analyses for the inclination, relative radii, and relative luminosity and (ii) radial velocity curves for both masses, the separation of the binary can be determined in absolute units. The best DEBs for distances have two eclipses and one of the eclipses is "complete" or nearly so (see discussion in Kaluzny et al. 2005). From the absolute separation, the radii of the two stars can be solved for, which, using a surface brightness–radius relation (e.g., di Benedetto 1998; Graczyk et al. 2017), can be converted into a luminosity, and hence, a distance. The precision of the surface brightness–radius relation is a function of the stellar type and luminosity class, with late-type binaries being those with more well-defined relations. The surface brightness–radius relation can only be calibrated for stars in the Solar neighborhood, which could impart age or metallicity biases, but so far the relationship shows only mild sensitivity to metallicity (Houdashelt et al. 2000).

A summary of EB results in GGCs can be found in Kaluzny et al. (2005). Since then EBs have been found in, monitored in, and distances derived to: NGC 6362 (14.74 ± 0.04 mag Kaluzny et al. 2015), M 4 (Kaluzny et al. (2013), see Sect. 2.3) 47 Tuc (13.35 ± 0.08 mag Thompson et al. 2010), NGC 6397 (11.65 mag Rozyczka et al. 2014), ω Centauri (14.05 ± 0.11 mag Thompson et al. 2001), and M 55 (13.94 ± 0.05 mag Kaluzny et al. 2014) within the context of the Clusters AgeS Experiment (CASE), which is also monitoring a number of other nearby GGCs.

An extragalactic census of EB results is given in Bonanos (2013) and includes sources detected in NGC 6822, IC 1613, LMC, SMC, M 31, and M 33; distances have been determined from the EBs in all but NGC 6822 and IC 1613. A 2% distance to the LMC using eight DEBs was published in Pietrzyński et al. (2013), with modeling of an additional 12 systems recently published in Graczyk et al. (2018) for a total of 20 DEBs in the LMC. A distance to M 33 with 1 DEB was given in Bonanos et al. (2006). A distance modulus of $\mu_{M33} = 24.92 \pm 0.12$ mag (964 ± 54 kpc) was found. A distance to M 31 using 2 DEBs was published in Vilardell et al. (2010). Since then, several hundred EBs have been discovered in the M 31 field (e.g., 298 from Lee et al. 2014), of which only 11 are bright enough for radial-velocity follow-up (e.g., $V < 20.5$ mag for 8–10 m class telescopes) and consistent with being in M 31. Eclipsing binaries provide an excellent means of calibrating zero points, but only in systems with well populated sequences and accompanying information on the metallicity effects.

5.3.3 *Megamasers in Accretion Disks*

The Local Volume galaxy, NGC 4258, hosts a megamaser in the accretion disk of its central black hole. The accretion disk is thought to be nearly edge-on and the emission from the megamaser traces the motion of the gas in the disk. Long-term monitoring of the megamaser emission produces a velocity field and with the assumption of Keplerian motion in the disk, a physical radius can be calculated and a geometric distance determined. Further methodological details can be found in this volume by Thévenin et al. (2017).

Since its discovery, seven distances to NGC 4258 have been published: 28.66 ± 0.47 mag (Greenhill et al. 1995), 29.03 ± 0.29 mag (Miyoshi et al. 1995), 29.29 ± 0.08 mag (Herrnstein et al. 1999), 29.29 ± 0.02 mag (Humphreys et al. 2008), 29.31 mag (no uncertainty was given) (Riess et al. 2012), 29.40 ± 0.05 mag (Humphreys et al. 2013), 29.39 ± 0.06 mag (Riess et al. 2016). While early results had large uncertainties as the methodology was being developed, the last four publications cite <2.5% uncertainties distance, but differ at the 5% level. The physical distances span from 7.2 Mpc to 7.6 Mpc (Jang and Lee 2017a).

5.3.4 Secondary Calibration of the RRLs

RRL calibration in globular clusters has been performed in M 4 in the NIR and MIR. Reasonable agreement between EBs and RRL distances from trigonometric parallaxes were found for M 4. Many of the clusters with EB distances also have samples of RRL (Clement et al. 2001a).[11] A benefit of the GGCs as calibration objects is their small age and metallicity ranges, which allows comparisons to be made for these parameters. At the same time, however, the use of mono-abundance and mono-age relations is not applicable to mixed populations; albeit ω Centauri is one exception (for detailed discussion see Braga et al. 2018). Thus, so far use of geometric EB distances for direct calibration of RRL remain poorly utilized; whereas, non-geometric distances (isochrone fitting, mean HB magnitude, and others) are used more frequently.

Performing direct calibration of the RRLs in the LMC and SMC is challenging due their line-of-sight depth and lack of clear information on how to address internal extinction; moreover, the RRL have $V \sim >18$ mag, which requires large telescopes for direct abundances. Nevertheless, Muraveva et al. (2015) performed a calibration using 70 LMC RRL, finding it to agree well with pulsation parallaxes, but disagree with the *HST*–FGS parallax sample. There are reasonably large samples of RRLs in IC 1613, M 33 and M 31, typically detected via *HST*, but these are largely still used to determine the distance to these objects rather than to calibrate the RRL themselves (see e.g., Sarajedini et al. 2006, 2009; Bernard et al. 2010b, among others). Again, the RRL at the distance of M 31 are at $V > 25$ mag and are out of range for most spectroscopic facilities as well as being difficult to obtain sufficient sampling to apply photometric techniques for metallicity.

5.3.5 Secondary Calibration of the T2C

A tabulation of T2Cs in GGCs is given by Matsunaga et al. (2006, their Table 2) and it contains 46 individual T2Cs in 26 different clusters, of which only 12 clusters contain more than a single T2C. Unfortunately, the list of clusters *with* T2Cs and *with* geometric distances is disjoint; Matsunaga et al. (2006) calibrated their relations to the magnitude of the horizontal branch. A large sample of T2Cs exist in the LMC and SMC and the geometric distances to these galaxies are the basis for relations derived in Sect. 3.4.

[11] An impressive and detailed tabulation of all known variables in clusters, which includes RRL, T2Cs, and EBs among other types, has been updated by Christine Clement regularly for many years and is available at this URL: http://www.astro.utoronto.ca/~cclement/cat/listngc.html.

5.3.6 Secondary Calibration of the TRGB

GGCs are not ideal objects to calibrate the TRGB because their stellar mass is sufficiently small that it is not guaranteed that there will be a true TRGB star. The EBs in extragalactic systems, however, make good sources of calibration; the LMC EB distance is used routinely, both Górski et al. (2016) and Hoyt et al. (2018) use the LMC to calibrate the IR–TRGB and Jang and Lee (2017a) uses the LMC in the optical.

The TRGB can be independently calibrated with NGC 4258 using *HST*. The TRGB measurements are as follows: $\mathrm{F814W}^{TRGB} = 25.20 \pm 0.06$ mag (WFPC2), $\mathrm{F814W}^{TRGB} = 25.24 \pm 0.04$ mag (ACS) (Mager et al. 2008), 25.25 (with 90% confidence interval from 25.23 to 25.38) and 25.22 ± 0.09 mag (Mouhcine et al. 2005), $\mathrm{F814W}^{TRGB} = 25.39 \pm 0.11$ mag (Madore et al. 2009), and $\mathrm{F814W}_0^{TRGB} = 25.357 \pm 0.031$ mag (Jang and Lee 2017b). Thus, the individual TRGB measurements span from 25.20 to 25.357 mag, a span of 7% in distance. For this reason, Jang and Lee (2017a) perform a through comparison using different photometry packages and different techniques for determining the point-spread-function finding differences between studies at the 0.05 mag level, which agrees with earlier comparisons in NGC 4258 in Mager et al. (2008). Taken in conjunction with the other measurement uncertainties discussed in Sect. 4.2, the difference in the TRGB magnitudes can largely be explained.

Given the variation in TRGB detections and the variation in the distances themselves, whether or not a TRGB calibration in NGC 4258 agrees with other techniques is largely dependent on which studies are compared. Rizzi et al. (2007) note that the Humphreys et al. (2005) distance to NGC 4258 calibrates the TRGB to be 2.3-σ fainter than their determination from other sources. Beaton et al. (2016) note the opposite effect via the distance from Humphreys et al. (2013), finding $M_I^{TRGB} = -4.16 \pm 0.06_{stat} \pm 0.04_{sys}$ mag, which is systematically more luminous. Jang and Lee (2017a) ultimately find $M_{\mathrm{F814W}}^{TRGB} = -4.03 \pm 0.07$ mag using Riess et al. (2016), which is consistent with other ground-based calibrations.

5.4 Metallicity

Theoretical and empirical investigations suggest that metallicity impacts the absolute magnitude for each of our Pop II standard candles. In many ways, age and metallicity are coupled. The impacts are two-fold: (i) metallicity impacts the physical structure of the star, which in turn, changes how it evolves and (ii) the metals in the atmosphere selectively absorb and thermally re-emit flux in specific bands. The impact of metallicity can be subdivided into two challenges: (i) difficulties in obtaining direct and/or independent measurements, and (ii) the use of distinct metallicity scales between calibration and application studies. The former point is distance indicator specific and will be discussed in subsections, whereas the latter is more general and will be discussed here.

The two common metallicity scales for Galactic work are those of (i) Carretta–Gratton (1997) and (ii) Zinn–West (1984). The Zinn and West scale was established using a variety of "integrated parameters" in GGCs, including the integrated line index (Q_{39}) and the infrared Calcium Triplet (CaT), to set a scale for [Fe/H]; this choice was motivated by the desire to establish a homogeneous scale that could work for both nearby and distant objects, but comes at the cost of these integrated techniques not being always correlated directly with the quantity [Fe/H], itself (e.g., see Hayden et al. 2015, for visualizations of how different elements do or do not correlate with Fe across the Galaxy). Carretta and Gratton used a smaller sample of objects (24 GGCs), but used higher resolution spectroscopy to produce metallicity measurements from Fe lines as well as other elements that contribute to integrated techniques.

A comparison between Zinn and West and Carretta and Gratton is given in Carretta and Gratton (1997, their Fig. 5) that shows a non-linear correlation between the two studies in addition to a zero-point offset due to adjustments to the Solar Fe abundance, finding:

$$[\mathrm{Fe/H}]_{CG97} = -0.618 - 0.097\ [\mathrm{Fe/H}]_{ZW} - 0.352\ [\mathrm{Fe/H}]^2_{ZW} \tag{27}$$

with a scatter of $\sigma = 0.08$ dex and uncertainties on the coefficients of ± 0.083 for the zeropoint, ± 0.0189 for the linear term, and ± 0.067 for the quadratic term. As an example, a star with $[\mathrm{Fe/H}] = -1.8$ dex on the Zinn and West scale corresponds to $[\mathrm{Fe/H}] = -1.7$ dex on the Carretta and Gratton scale, with the differences being larger on the metal-poor and metal-rich ends of the scale. Thus, the impact of using different metallicity scales may be complex.

5.4.1 Metallicity Impacts in RRL

Solid theoretical (Marconi et al. 2015) and semi-empirical (Neeley et al. 2017) evidence show that the mean magnitudes of the RRLs in all the optical-NIR–MIR photometric bands are affected by their iron abundance—the more metal-poor the variable is, the brighter. The metallicity coefficient of the PL relations ranges from 0.14 to 0.19 mag dex^{-1} (Marconi et al. 2015; Neeley et al. 2017) in the optical, NIR and MIR, showing no clear trend with wavelength. It is, however, important to note that the intrinsic dispersions for the long-wavelength PLs is smaller than the theoretical separation between $\sim$0.5 dex populations, implying the effect should be more easily distinguished for the longer-wavelength data (see discussions in Neeley et al. 2017). Thus, metallicity is fundamental for the distance determination using RRLs.

Direct spectroscopic metallicity measurements in RRL are complicated due to the short timescales over which the atmospheres are changing. Typical radial velocity amplitudes are $\sim$50 km/s (see, e.g., Sesar 2012, and references therein) such that even short observations ($\sim$10 min) can be impacted by line blurring. Thus, spectroscopic observations are either conducted at specific phases or use techniques employed at specific phases to infer the metallicity. Some common alternate techniques include:

- ΔS method: Preston (1959) developed a technique that infers metallicity using a combination hydrogen lines to infer the spectral type (e.g., temperature) and Ca II H and K lines to infer the metallicity known as the ΔS method. The method is typically only applied at minimum light. Narrow-band filters can be used for the specific lines used in this technique (see Rey et al. 2000). The recalibration by Clementini et al. (1995) using high resolution spectra, which was revisited by Carretta and Gratton (1997) to find (on their scale):

$$[\mathrm{Fe/H}] = -0.187(\pm 0.011)\Delta S - 0.088(\pm 0.04)\ (\sigma = 0.269). \tag{28}$$

- Fourier coefficients: Kovacs and Zsoldos (1995), Jurcsik and Kovacs (1996), Smolec (2005, and references therein) demonstrate a correlation between Fourier coefficients, in particular, ϕ_{31}, and metallicity. The Smolec (2005) calibration is tied to the Zinn and West (1984) metallicity scale and is based on 28 stars:

$$[\mathrm{Fe/H}]_{phot} = -3.142(\pm 0.646) - 4.902(\pm 0.375) + 0.824(\pm 0.104)\phi_{31}. \tag{29}$$

- Wallerstein (2002) demonstrated a technique based on the Calcium triplet (CaT), which is commonly used in moderate resolution spectroscopy for red giants.

Direct measurements, however, can be performed if care is taken in the data acquisition and subsequent analysis. Of particular note is a series of papers that use high cadence spectroscopic datasets obtained for the full cycle of several RRL. For et al. (2011a,b) use an iterative procedure to estimate the temperature for each spectrum and the perform direct spectral analyses for numerous chemical species. Both Govea et al. (2014) and Sneden et al. (2018) continue this work for RRc variables with continued success. These direct measurements show relative consistency for stellar parameters and abundances derived self-consistently from the spectrum at all phases. Comparison of these parameters to other, more common methods have not yet been produced, but are important.

5.4.2 Metallicity Impacts in T2Cs

The metallicity effects are less well constrained for T2Cs. A major concern, however, is the proposed shifting of P-boundaries between the sub-types apparent in the LMC and SMC. The shift in sub-population classification would be a major systematic given the differences in the PLs derived with different sub-samples.

The first chemical abundance analysis for BL Her and W Vir type T2Cs was presented in Maas et al. (2007) for 19 field stars, finding anomalous chemical compositions similar to previous studies of RV Tau stars by Giridhar et al. (2005). Subsequent analyses have focused on using spectral features and chemical abundances as means to cleanly separate the sub-populations; examples include, (i) Kovtyukh et al. (2018a) that see differences in the metal-poor populations for short period T2Cs ($P < 3$ day), (ii) Kovtyukh et al. (2018b) that compares W Vir abundances to those of the disk, suggesting the W Vir stars to be He-enhanced, and, (iii) Lemasle et al. (2015b) that compare W Vir stars to CCs. On the whole, the spectroscopic studies are more focused on the revealing the evolutionary origin of these stars rather than their use as distance indicators. Along this line-of-thinking, Welch (2012) note that additional T2C spectroscopic studies, both for radial velocity monitoring and for chemical abundance measurements, could yield significant insight into binary evolution formation channels.

5.4.3 Metallicity Impacts in TRGB

The impact of metallicity on the TRGB was discussed in Sect. 4. While theoretical studies see the effects of metallicity cleanly for bolometric quantities, there are uncertainties when translating these effects into observable quantities. Empirical studies, on the other hand, can measure the color–magnitude behavior, but largely infer the metallicity dependence from the mean metallicity of the system determined from other means (e.g., as in Valenti et al. 2004; Dalcanton et al. 2012). However, the age and metallicity effects are difficult to disentangle. A benefit, however, is that the color–magnitude relationship for the TRGB can be constrained empirically from high signal-to-noise observations irregardless of the physical origin (e.g., as in Jang and Lee 2017b; Madore et al. 2018). If the color–magnitude relationship holds in systems comprised of different stellar populations and independently determined distances such that systematics can be estimated (e.g., as in Jang and Lee 2017b), then impacts can be managed in application to distances.

5.5 Other Systematics

We list briefly other systematic terms and strategies used to compensate for them.

1. *Milky Way extinction:* The foreground extinction needs to be determined. For extragalactic objects, the Milky Way foreground is well-enough constrained by Schlegel et al. (1998) maps (noting the recalibration by Schlafly and Finkbeiner 2011). Internal to the Milky Way, the line-of-sight extinction is more difficult to constrain. Means of addressing this issue include: (i) RJCE method (Majewski et al. 2011), (ii) models (e.g., Drimmel et al. 2003), (iii) use of Wesenheit formulations (e.g., Madore 1982), and (iv) using longer wavelengths where extinction has a lower total impact.
2. *Internal extinction:* Tracers that are located inside external galaxies have an additional component from the extinction due to the external galaxy. This is difficult to constrain independently. Means of addressing this issue include: (i) use of Wesenheit magnitudes (Madore 1982), (ii) using longer wavelengths where extinction has a lower impact, (iii) using multi-wavelength observations to fit for total extinction directly, (iv) avoiding galactic structures where extinction is likely to be a significant effect (e.g., stellar halos). These techniques has been employed extensively for CCs, with the volume of literature growing for the RRLs.
3. *Non universality of the extinction law:* There is some evidence that "Cardelli's Law" is not universal, in particular in regions of high gas density (see discussions regarding M 4 in Sects. 2.3.2 and 2.3.4). Wesenheit magnitudes are not useful in this situation because they have to assume a form of the extinction law. Thus, only observations taken at longer wavelengths or in regions where dust is not anticipated (e.g., stellar halos) are robust to this systematic.
4. *Crowding or blending:* Measurements undertaken in crowded fields have to contend with flux-contamination from neighboring sources. There are algorithms that can attempt to account for this effect. Photometry packages have been designed with measurement techniques, the following packages are notable: (i) the DAOPHOT family of routines does a particularly good job in crowded regions (Stetson 1987, 1994), (ii) DoPHOT (Schechter et al. 1993), and (iii) DOLPHOT (Dolphin 2000). Riess et al. (2016, and references therein) use a modeling approach that uses higher resolution optical images to guide source isolation at lower resolutions (typically, longer wavelengths); methods such as this need to be careful because different stellar populations dominate the light in optical and IR bandpasses and corrections need to be determined independently (as in done in Riess et al. 2016). Template fitting may help to identify variable stars that have abnormal shapes due to contamination. A recent instance of crowding potentially biasing a distance measurement is discussed by van Dokkum et al. (2018), who argue how blended stars may have led Trujillo et al. (2018) to infer the wrong distance to the galaxy NGC1052–DF2, a galaxy suspected of having a low dark matter mass (van Dokkum et al. 2018b). Thus, the best means of avoiding crowding/blending is to avoid high-density regions of galaxies to which Pop II distance indicators are particularly well suited.
5. *Evolutionary effects:* These effects largely include mixing stars of the same type that are in different evolutionary states due to evolution within the class (e.g., first and second pass HB stars), evolution over time (changes to He abundance), and the possibility that some pulsators come from binary evolution formation paths. These types of effects can either add scatter to a population when their numbers are small or strongly bias a distance when their numbers are comparable to non-evolved sub-types. Sculptor (Sect. 2.3.1) and ω-Centauri (Braga et al. 2018) with their wide metallicity and age spreads are good examples of where evolutionary effects may come into play to skew the means; in populations, however, these outliers can often be identified. Strong biases are more likely when working with individual field stars, where inferring evolutionary state is more complex and, indeed, this is cited for reasons why individual parallaxes might disagree with other

forms of calibration (Sect. 5.2.1). Addressing systematics of this nature can only be accomplished by calibrating the standard candles with large statistical samples and in a variety of environments where various impacts are more or less likely to be a factor (e.g., comparison of calibrations determined using field stars, GGCs, and dwarf galaxies).

6. *Contamination:* Samples can be contaminated by non-standard candles. For the TRGB, contamination can come from AGB stars or from younger red, luminous populations. These effects are mitigated by applying the technique in Pop II dominated regions or by increasing uncertainties. Both RRL and T2C populations can be contaminated by variables with similar periods (indeed the T2Cs and RRL overlap with other variable types) or by unidentified atypical pulsators (e.g., pW Vir, multi-periodic RRLs, Blashko stars). Moreover, both period determination and variable classification are difficult with sparsely sampled data. It is not uncommon for variables to be reclassified with additional data (e.g., see notes in Clement et al. 2001b).

6 Summary and Future Prospects

We have developed complementary physical and empirical portraits of three Pop II standard candles: the RRLs, T2Cs, and TRGB. In particular, this chapter has put emphasis the history and evolution of these stars as distance indicators, with a particular focus on multi-wavelength use of these tracers. In the course of preparing this chapter, parallels between these Pop II distance indicators have become apparent. There are several "themes" that can be used to discuss our current understanding of these standard candles and these "themes", in turn, relate directly into the future potential of these objects in the coming years. Thus, we will summarize the chapter and comment on the future prospects in tandem.

First, each of these distance indicators has a deep history dating back to the early days of stellar population research; indeed, often to before the term "stellar populations" had been coined (e.g., Baade 1944, 1958a). As a consequence, empirical evidence and theoretical understanding have leap-frogged with each other over time. These Pop II standard candles have each played a significant role in our understanding of the structure of our Galaxy and of the Universe and have had their role in refining the distance scale. Moving forward, it is vital to not forget the formation of the historical nomenclature and the drivers of the historical thinking that led exploration (or a lack of exploration) for these distance indicators. Stated differently, it is important to look back to understand when and why assumptions were made or not made as we work with ever better datasets in which previously "negligible" effects may be transformed into major hurdles. In our historical discussions, we hope to have illuminated some of these previously hidden effects.

Second, none of these Pop II standard candles have a robust geometric foundation. While the RRL and T2C have some parallaxes from targeted studies, these are few in number and cannot fully encompass the intrinsic star-to-star variance nor account for systematic differences (e.g., those due to metallicity or age). Moreover, without a large geometric parallax sample, it is impossible to measure all of the desired quantities for even the simplest calibration (e.g., slopes and zero points). Instead, it is typical to combine theoretical, empirical, and semi-empirical techniques to determine distances to take into account systematics; more specifically, making combinations of slopes and zero-points either measured in different objects or that are predicted from theoretical modeling. While these are incredibly logical to the experienced practitioners, it has the impact of being confusing and inelegant to the non-expert, while also potentially imparting systematic effects between studies.

With the pending data releases from *Gaia*, however, this theme has a bright future. The end-of-mission predictions for *Gaia* are 10% parallax uncertainties at a distance of 10 kpc (de Bruijne et al. 2014), which will provide an unprecedented sample of stars from which to establish a geometric foundation for each of our Pop II indicators. In addition, *Gaia* will provide precise spectro-photometry for all sources, variability information, and, for bright ($G < 15$) sources, spectroscopic parameters (including, chemical abundances and temperatures) (Prusti et al. 2016). The enormity of the *Gaia* dataset is simply mind-boggling and has the potential to reveal not just the practical application of these tools, but also elucidate aspects of stellar theory. During the preparation of this manuscript, *Gaia* DR2 (Brown et al. 2018b) occurred and some of the early results have been included herein, which is undoubtedly only a small fraction of those to occur.

Third, the most pressing questions in both theory and in practice are related to the need to fundamentally understand the impact of metallicity, age, and evolutionary effects for using these stars as standard candles. Addressing these issues will come on two fronts: (i) observations of individual field stars and (ii) observations of populations in star clusters or nearby dwarf galaxies. Thus, addressing this issue will likely require more than just *Gaia* generated datasets.

For the field stars, determining stellar parameters (temperature, log(g), and chemical abundances) will likely involve the full pantheon of large scale spectroscopic surveys, including but not limited to APOGEE (Majewski et al. 2017), RAVE (Steinmetz et al. 2006), LAMOST (Zhao et al. 2012), GALAH (Martell et al. 2017), and their numerous successors. These surveys, however, have to properly target and process data for stars whose fundamental parameters or observed radial velocity are changing on timescales as short as 10 minutes (for the RRLs). Thus, standard processing may not be sufficient to provide high fidelity measurements. These types of data will fundamentally improve models, while also helping to push the precision of these standard candles to new ends. Moreover, fully complementary insight can be gleaned from precision high cadence photometry, in particular asteroseismic analyses enabled from *Kepler* (Borucki et al. 2010), *K2* (Howell et al. 2014), and *TESS* (Ricker et al. 2015). These photometric datasets provide an alternate means of determining some stellar parameters via asteroseismology, while also providing short and long timescale variability information to better identify contaminating populations.

For clusters and satellites, many of these large scale programs do not always have sufficient sensitivity and resolution. Here the greatest insight is likely to be gained by levering insights from the field star analyses to larger aperture facilities, from *JWST* in space (Kalirai 2018), to efficient multi-plexing on 5- to 10-meter class and high spatial and spectral resolution studies enabled on 30-meter class ground-based facilities. Moreover, continued long-term monitoring of variables, as with OGLE (e.g., Soszyński et al. 2008), Cluster AgeS Experiment (e.g., Kaluzny et al. 2005), and the homogeneous photometry series (e.g., Stetson et al. 2014b), provide additional characterization and classification diagnostics for these variables. With the ability to determine ages and star formation histories, clusters and nearby galaxies are ideal to study population-effects, while also being able to cleanly resolve individual stars.

Fourth, these standard candles are reaching to unprecedented precisions and accuracies that enable powerful scientific explorations, from the structure of our Galaxy, our galactic neighborhood, and ultimately to the measurement of fundamental cosmological parameters. By utilizing IR observations, the impact of extinction has been minimized, and, for all of the distance indicators described here, provide smaller intrinsic variations, which in turn make these distance indicators more powerful. While few percent distances are incredible (especially to individual RRL or T2C), it is also a realm where every *source of uncertainty matters*, from differences in the detailed definition of filter sets, to the unintentional inclusion of peculiar variables, to the potential for multiple formation channels, to the deviations from a standard extinction law. Stated differently, *every detail matters*. Moreover, an important consideration is the measure of the intrinsic dispersion in these relationships or for a given class of standard candle as a fundamental limit to their precision. For instance, *how stable is the TRGB magnitude?*, *how flat is the horizontal branch?*, *how standard are our standard candles?*, and *how homogeneous are our samples?* These are fundamental aspects to the measurement of distances, and quantities that can be estimated, but whose measure is nuanced and complex.

The stage for better exploiting the Pop II standard candles is set, in particular with the strong inter-relationship between the theory and observation. With the dominance of IR-focused observational facilities in the future, there is great potential for the RRL, T2Cs, and TRGB, all of which become more precise and more powerful in the IR regime. Coupled with the potential to probe nearly all galactic structures and galaxies of all Hubble types, and nearly all total luminosities, the Pop II standard candles stand to sharpen our view of the Universe.

Acknowledgements The authors warmly acknowledge the hospitality of ISSI-BJ for an engaging conference and the conference organizer Richard De Grijs for his leadership in completing these chapters. We further thank the anonymous referees for useful comments and perspectives that have improved and broadened this review. RLB acknowledges many insightful discussions with Barry Madore, Wendy Freedman, and the Carnegie–Chicago program team, as well as helpful conversations and data for figures from Mark Seibert, Erika Carlson, Andrew Monson, Victoria Scowcroft, Julianne Dalcanton, and Ben Williams. Support for this work was provided by NASA through Hubble Fellowship grant #51386.01 awarded to RLB by the Space Telescope Science Institute, which is operated by the Association of Universities for Research in Astronomy, Inc., for NASA, under contract NAS 5-26555. GB thanks A Severo Ochoa research grant at the Instituto de Astrofisica de Canarias, where part of this manuscript was written. VFB warmly thank Prof. R.P. Kudritzski for many useful discussions concerning stellar atmospheres and non-LTE effects in giant stars and Dr. P.B. Stetson for his superb-quality photometric reduction work, which has been a key component of the work presented here. VFB acknowledges PRIN–INAF 2011 "Tracing the formation and evolution of the Galactic halo with VST" (P.I.: M. Marconi), PRIN–MIUR (2010LY5N2T) "Chemical and dynamical evolution of the Milky Way and Local Group galaxies" (P.I.: F. Matteucci). VFB thank the Japan Society for the Promotion of Science for a research grant (L15518) and the support from FIRB 2013 (grant: RBFR13J716). We also thank the Education and Science Ministry of Spain (grants AYA201016717). VFB finally acknowledge the financial support from the PO FSE Abruzzo 2007–2013 through the grant "Spectrophotometric characterization of stellar populations in Local Group dwarf galaxies", prot. 89/2014/OACTe/D (PI: S. Cassisi). GF has been supported by the Futuro in Ricerca 2013 (grant RBFR13J716). CEMV and MM acknowledges support from the Spanish Ministry of Economy and Competitiveness (MINECO) under the grant (project reference AYA2014-56795-P) NM is grateful to Grant-in-Aid (KAKENHI No. 26287028) from the Japan Society for the Promotion of Science.

References

W.S. Adams, Investigations in stellar spectroscopy. III. Application of a spectroscopic method of determining stellar distances to stars of measured parallax. Proc. Natl. Acad. Sci. **2**, 152–157 (1916). https://doi.org/10.1073/pnas.2.3.152

W.S. Adams, A. Kohlschütter, Some spectral criteria for the determination of absolute stellar magnitudes. Astrophys. J. **40**, 385–398 (1914). https://doi.org/10.1086/142132

C. Alcock, R.A. Allsman, D.R. Alves, T.S. Axelrod, A. Becker, D.P. Bennett, K.H. Cook, K.C. Freeman, K. Griest, W.A. Lawson, M.J. Lehner, S.L. Marshall, D. Minniti, B.A. Peterson, K.R. Pollard, M.R. Pratt, P.J. Quinn, A.W. Rodgers, W. Sutherland, A. Tomaney, D.L. Welch, The MACHO project LMC variable star inventory. VII. The discovery of RV Tauri stars and new type II Cepheids in the Large Magellanic Cloud. Astron. J. **115**, 1921–1933 (1998). https://doi.org/10.1086/300317. arXiv:astro-ph/9708039

S. Andrievsky, G. Wallerstein, S. Korotin, D. Lyashko, V. Kovtyukh, V. Tsymbal, C.E. Davis, T. Gomez, W. Huang, E.M. Farrell, The relationship of sodium and oxygen in Galactic field RR Lyrae stars. Publ. Astron. Soc. Pac. **130**(2), 024 (2018). https://doi.org/10.1088/1538-3873/aa9783

E. Antonello, L. Fossati, D. Fugazza, L. Mantegazza, W. Gieren, Variable stars in nearby galaxies. VII. P–L relation in the BVRI bands of Cepheids in IC 1613. Astron. Astrophys. **445**, 901–906 (2006). https://doi.org/10.1051/0004-6361:20053822

H.C. Arp, Color–magnitude diagrams for seven globular clusters. Astron. J. **60**, 317 (1955). https://doi.org/10.1086/107232

W. Baade, Über eine Möglichkeit, die Pulsationstheorie der δ Cephei-Veränderlichen zu prüten. Astron. Nachr. **228**, 359 (1926). https://doi.org/10.1002/asna.19262282003

W. Baade, The resolution of Messier 32, NGC 205, and the Central Region of the Andromeda Nebula. Astrophys. J. **100**, 137 (1944). https://doi.org/10.1086/144650

W. Baade, The period-luminosity relation of the Cepheids. Publ. Astron. Soc. Pac. **68**, 5 (1956). https://doi.org/10.1086/126870

W. Baade, Galaxies and their stellar populations. (A review of the present state of affairs). Ric. Astron. **5**, 3 (1958a)

W. Baade, Physical variable stars and stellar populations. Ric. Astron. **5**, 165 (1958b)

W. Baade, The population of the galactic nucleus and the evidence for the presence of an old population pervading the whole disk of our galaxy. Ric. Astron. **5**, 303 (1958c)

W. Baade, H.H. Swope, The Palomar survey of variables in M 31 (first results). Astron. J. **60**, 151–152 (1955). https://doi.org/10.1086/107146

W. Baade, H.H. Swope, The Draco system, a dwarf galaxy. Astron. J. **66**, 300–347 (1961). https://doi.org/10.1086/108431

C.A.L. Bailer-Jones, Estimating distances from parallaxes. Publ. Astron. Soc. Pac. **127**, 994 (2015). https://doi.org/10.1086/683116. arXiv:1507.02105

S.I. Bailey, *A Discussion of Variable Stars in the Cluster ω Centauri*. Annals of Harvard College Observatory, vol. 38 (1902)

S.I. Bailey, E.F. Leland, I.E. Woods, E.C. Pickering, Variable stars in the cluster Messier 15. Ann. Harv. Coll. Obs. **78**, 195–250 (1919)

N. Baker, R. Kippenhahn, The pulsations of models of Delta Cephei stars. II. Astrophys. J. **142**, 868 (1965). https://doi.org/10.1086/148359

M.K. Barker, A. Sarajedini, J. Harris, Variations in star formation history and the red giant branch tip. Astrophys. J. **606**, 869–893 (2004). https://doi.org/10.1086/383026. arXiv:astro-ph/0401387

G. Battaglia, M. Irwin, E. Tolstoy, V. Hill, A. Helmi, B. Letarte, P. Jablonka, Analysis and calibration of CaII triplet spectroscopy of red giant branch stars from VLT/FLAMES observations. Mon. Not. R. Astron. Soc. **383**, 183–199 (2008). https://doi.org/10.1111/j.1365-2966.2007.12532.x. arXiv:0710.0798

R.L. Beaton, W.L. Freedman, B.F. Madore, G. Bono, E.K. Carlson, G. Clementini, M.J. Durbin, A. Garofalo, D. Hatt, I.S. Jang, J.A. Kollmeier, M.G. Lee, A.J. Monson, J.A. Rich, V. Scowcroft, M. Seibert, L. Sturch, S.C. Yang, The Carnegie–Chicago Hubble program. I. An independent approach to the extragalactic distance scale using only population II distance indicators. Astrophys. J. **832**, 210 (2016). https://doi.org/10.3847/0004-637X/832/2/210. arXiv:1604.01788

M. Bellazzini, The tip of the red giant branch. Mem. Soc. Astron. Ital. **79**, 440 (2008). arXiv:0711.2016

M. Bellazzini, F.R. Ferraro, E. Pancino, A step toward the calibration of the red giant branch tip as a standard candle. Astrophys. J. **556**, 635–640 (2001). https://doi.org/10.1086/321613. arXiv:astro-ph/0104114

M. Bellazzini, F.R. Ferraro, A. Sollima, E. Pancino, L. Origlia, The calibration of the RGB tip as a standard candle. Extension to near infrared colors and higher metallicity. Astron. Astrophys. **424**, 199–211 (2004). https://doi.org/10.1051/0004-6361:20035910. arXiv:astro-ph/0404572

G.F. Benedict, B.E. McArthur, M.W. Feast, T.G. Barnes, T.E. Harrison, J.L. Bean, J.W. Menzies, B. Chaboyer, L. Fossati, N. Nesvacil, H.A. Smith, K. Kolenberg, C.D. Laney, O. Kochukhov, E.P. Nelan, D.V. Shulyak, D. Taylor, W.L. Freedman, Distance scale zero points from Galactic RR Lyrae star parallaxes. Astron. J. **142**, 187 (2011). https://doi.org/10.1088/0004-6256/142/6/187. arXiv:1109.5631

G.F. Benedict, B.E. McArthur, E.P. Nelan, T.E. Harrison, Astrometry with Hubble Space Telescope Fine Guidance Sensors–a review. Publ. Astron. Soc. Pac. **129**(1), 012001 (2017). https://doi.org/10.1088/1538-3873/129/971/012001. arXiv:1610.05176

G.F. Benedict, B.E. McArthur, T.E. Harrison, An assessment of Gaia DR2 parallaxes. Res. Notes Am. Astron. Soc. **2**(2), 22 (2018). http://stacks.iop.org/2515-5172/2/i=2/a=22

E.J. Bernard, M. Monelli, C. Gallart, A. Aparicio, S. Cassisi, I. Drozdovsky, S.L. Hidalgo, E.D. Skillman, P.B. Stetson, The ACS LCID project. II. Faint variable stars in the isolated dwarf irregular galaxy IC 1613. Astrophys. J. **712**, 1259–1276 (2010a). https://doi.org/10.1088/0004-637X/712/2/1259. arXiv:1002.4178

E.J. Bernard, M. Monelli, C. Gallart, A. Aparicio, S. Cassisi, I. Drozdovsky, S.L. Hidalgo, E.D. Skillman, P.B. Stetson, The ACS LCID project. II. Faint variable stars in the isolated dwarf irregular galaxy IC

1613. Astrophys. J. **712**, 1259–1276 (2010b). https://doi.org/10.1088/0004-637X/712/2/1259. arXiv: 1002.4178
D. Bersier, P.R. Wood, Variable stars in the Fornax dwarf galaxy. Astron. J. **123**, 840–847 (2002). https://doi.org/10.1086/338315. arXiv:astro-ph/0110555
F.W. Bessel, On the parallax of 61 Cygni. Mon. Not. R. Astron. Soc. **4**, 152–161 (1838). https://doi.org/10.1093/mnras/4.17.152
F. Bessell, *Die Beweise fur die Bewegung der Erde* (1871)
A. Bhardwaj, L.M. Macri, M. Rejkuba, S.M. Kanbur, C.C. Ngeow, H.P. Singh, Large Magellanic Cloud near-infrared synoptic survey. IV. Leavitt laws for type II Cepheid variables. Astron. J. **153**, 154 (2017a). https://doi.org/10.3847/1538-3881/aa5e4f. arXiv:1702.00967
A. Bhardwaj, M. Rejkuba, D. Minniti, F. Surot, E. Valenti, M. Zoccali, O.A. Gonzalez, M. Romaniello, S.M. Kanbur, H.P. Singh, Galactic bulge population II Cepheids in the VVV survey: period-luminosity relations and a distance to the Galactic centre. Astron. Astrophys. **605**, A100 (2017b). https://doi.org/10.1051/0004-6361/201730841. arXiv:1707.03755
S. Blažko, Mitteilung über veränderliche Sterne. Astron. Nachr. **175**, 325 (1907). https://doi.org/10.1002/asna.19071752002
A.Z. Bonanos, Eclipsing binary distances to the edge of the local group, in *Advancing the Physics of Cosmic Distances, IAU Symposium*, vol. 289, ed. by R. de Grijs (2013), pp. 173–178. https://doi.org/10.1017/S1743921312021321
A.Z. Bonanos, K.Z. Stanek, R.P. Kudritzki, L.M. Macri, D.D. Sasselov, J. Kaluzny, P.B. Stetson, D. Bersier, F. Bresolin, T. Matheson, B.J. Mochejska, N. Przybilla, A.H. Szentgyorgyi, J. Tonry, G. Torres, The first DIRECT distance determination to a detached eclipsing binary in M33. Astrophys. J. **652**, 313–322 (2006). https://doi.org/10.1086/508140. arXiv:astro-ph/0606279
G. Bono, RR Lyrae distance scale: theory and observations, in *Stellar Candles for the Extragalactic Distance Scale*, ed. by D. Alloin, W. Gieren. Lecture Notes in Physics, vol. 635 (Springer, Berlin, 2003), pp. 85–104. https://doi.org/10.1007/978-3-540-39882-0_5. arXiv:astro-ph/0305102
G. Bono, M. Marconi, On the period-luminosity-color relation of Classical Cepheids, in *New Views of the Magellanic Clouds, IAU Symposium*, vol. 190, ed. by Y.H. Chu, N. Suntzeff, J. Hesser, D. Bohlender (1999), p. 527. arXiv:astro-ph/9810460
G. Bono, R.F. Stellingwerf, Pulsation and stability of RR Lyrae stars. 1: Instability strip. Astrophys. J. Suppl. Ser. **93**, 233–269 (1994). https://doi.org/10.1086/192054
G. Bono, F. Caputo, R.F. Stellingwerf, Oosterhoff dichotomy in the Galaxy and globular clusters in the Large Magellanic Cloud. Astrophys. J. **423**, 294–304 (1994). https://doi.org/10.1086/173806
G. Bono, V. Castellani, S. degl'Innocenti, L. Pulone, Advanced evolutionary phases of low-mass stars: the role of the original helium. Astron. Astrophys. **297**, 115–126 (1995)
G. Bono, R. Incerpi, M. Marconi, Metallicity effects on RR Lyrae variables. Astrophys. J. Lett. **467**, L97 (1996). https://doi.org/10.1086/310207. arXiv:astro-ph/9605143
G. Bono, F. Caputo, S. Cassisi, V. Castellani, M. Marconi, Metal-rich RR Lyrae variables. I. The evolutionary scenario. Astrophys. J. **479**, 279–289 (1997a). https://doi.org/10.1086/303872. arXiv:astro-ph/9609153
G. Bono, F. Caputo, V. Castellani, M. Marconi, Nonlinear investigation of the pulsational properties of RR Lyrae variables. Astron. Astrophys. Suppl. Ser. **121**, 327–342 (1997b). https://doi.org/10.1051/aas:1997289. arXiv:astro-ph/9606030
G. Bono, F. Caputo, P. Santolamazza, Evolutionary scenario for metal-poor pulsating stars. I. Type II Cepheids. Astron. Astrophys. **317**, 171–177 (1997c)
G. Bono, M. Marconi, R.F. Stellingwerf, Classical Cepheid pulsation models. I. Physical structure. Astrophys. J. Suppl. Ser. **122**, 167–205 (1999). https://doi.org/10.1086/313207
G. Bono, F. Caputo, V. Castellani, M. Marconi, J. Storm, Theoretical insights into the RR Lyrae K-band period-luminosity relation. Mon. Not. R. Astron. Soc. **326**, 1183–1190 (2001). https://doi.org/10.1046/j.1365-8711.2001.04655.x. arXiv:astro-ph/0105481
G. Bono, F. Caputo, V. Castellani, M. Marconi, J. Storm, S. Degl'Innocenti, A pulsational approach to near-infrared and visual magnitudes of RR Lyr stars. Mon. Not. R. Astron. Soc. **344**, 1097–1106 (2003). https://doi.org/10.1046/j.1365-8711.2003.06878.x. arXiv:astro-ph/0306142
G. Bono, F. Caputo, M. Di Criscienzo, RR Lyrae stars in Galactic globular clusters. VI. The period-amplitude relation. Astron. Astrophys. **476**, 779–790 (2007). https://doi.org/10.1051/0004-6361:20078206. arXiv:0709.3177
G. Bono, M. Dall'Ora, F. Caputo, G. Coppola, K. Genovali, M. Marconi, A.M. Piersimoni, R.F. Stellingwerf, RR Lyrae the stellar beacons of the Galactic structure, in *RR Lyrae Stars, Metal-Poor Stars, and the Galaxy*, vol. 5, ed. by A. McWilliam (2011), p. 1. arXiv:1108.5372
G. Bono, A. Pietrinferni, M. Marconi, V.F. Braga, G. Fiorentino, P.B. Stetson, R. Buonanno, M. Castellani, M. Dall'Ora, M. Fabrizio, I. Ferraro, G. Giuffrida, G. Iannicola, M. Marengo, D. Magurno, C.E. Martinez-Vazquez, N. Matsunaga, M. Monelli, J. Neeley, S. Rastello, M. Salaris, L. Short, R.F. Stellingwerf,

On the pulsation and evolutionary properties of helium burning radially pulsating variables. Commun. Konkoly Obs. Hung. Acad. Sci. **105**, 149–159 (2016)

G. Bono, V.F. Braga, I. Ferraro, G. Fiorentino, R. Gilmozzi, G. Iannicola, D. Magurno, N. Matsunaga, M. Monelli, S. Rastello, Massive stellar systems: observational challenges and perspectives in the E-ELT era, in *Formation, Evolution, and Survival of Massive Star Clusters, IAU Symposium*, vol. 316, ed. by C. Charbonnel, A. Nota (2017), pp. 36–43. https://doi.org/10.1017/S1743921315010856

W.J. Borucki, D. Koch, G. Basri, N. Batalha, T. Brown, D. Caldwell, J. Caldwell, J. Christensen-Dalsgaard, W.D. Cochran, E. DeVore, E.W. Dunham, A.K. Dupree, T.N. Gautier, J.C. Geary, R. Gilliland, A. Gould, S.B. Howell, J.M. Jenkins, Y. Kondo, D.W. Latham, G.W. Marcy, S. Meibom, H. Kjeldsen, J.J. Lissauer, D.G. Monet, D. Morrison, D. Sasselov, J. Tarter, A. Boss, D. Brownlee, T. Owen, D. Buzasi, D. Charbonneau, L. Doyle, J. Fortney, E.B. Ford, M.J. Holman, S. Seager, J.H. Steffen, W.F. Welsh, J. Rowe, H. Anderson, L. Buchhave, D. Ciardi, L. Walkowicz, W. Sherry, E. Horch, H. Isaacson, M.E. Everett, D. Fischer, G. Torres, J.A. Johnson, M. Endl, P. MacQueen, S.T. Bryson, J. Dotson, M. Haas, J. Kolodziejczak, J. Van Cleve, H. Chandrasekaran, J.D. Twicken, E.V. Quintana, B.D. Clarke, C. Allen, J. Li, H. Wu, P. Tenenbaum, E. Verner, F. Bruhweiler, J. Barnes, A. Prsa, Kepler planet-detection mission: introduction and first results. Science **327**, 977 (2010). https://doi.org/10.1126/science.1185402

M.L. Boyer, K.B.W. McQuinn, P. Barmby, A.Z. Bonanos, R.D. Gehrz, K.D. Gordon, M.A.T. Groenewegen, E. Lagadec, D. Lennon, M. Marengo, I. McDonald, M. Meixner, E. Skillman, G.C. Sloan, G. Sonneborn, J.T. van Loon, A. Zijlstra, An infrared census of DUST in Nearby Galaxies with Spitzer (DUSTiNGS). II. Discovery of metal-poor dusty AGB stars. Astrophys. J. **800**, 51 (2015a). https://doi.org/10.1088/0004-637X/800/1/51. arXiv:1412.0695

M.L. Boyer, K.B.W. McQuinn, P. Barmby, A.Z. Bonanos, R.D. Gehrz, K.D. Gordon, M.A.T. Groenewegen, E. Lagadec, D. Lennon, M. Marengo, M. Meixner, E. Skillman, G.C. Sloan, G. Sonneborn, J.T. van Loon, A. Zijlstra, An infrared census of Dust in Nearby Galaxies with Spitzer (DUSTiNGS). I. Overview. Astrophys. J. Suppl. Ser. **216**, 10 (2015b). https://doi.org/10.1088/0067-0049/216/1/10. arXiv:1411.4053

V.F. Braga, M. Dall'Ora, G. Bono, P.B. Stetson, I. Ferraro, G. Iannicola, M. Marengo, J. Neeley, S.E. Persson, R. Buonanno, G. Coppola, W. Freedman, B.F. Madore, M. Marconi, et al., On the distance of the globular cluster M4 (NGC 6121) using RR Lyrae stars. I. Optical and near-infrared period-luminosity and period-Wesenheit relations. Astrophys. J. **799**, 165 (2015). https://doi.org/10.1088/0004-637X/799/2/165. arXiv:1411.6826

V.F. Braga, P.B. Stetson, G. Bono, M. DallOra, I. Ferraro, G. Fiorentino, L.M. Freyhammer, G. Iannicola, M. Marengo, J. Neeley, E. Valenti, R. Buonanno, A. Calamida, M. Castellani, R. da Silva, S. Degl'Innocenti, A. Di Cecco, M. Fabrizio, W.L. Freedman, G. Giuffrida, J. Lub, B.F. Madore, M. Marconi, S. Marinoni, N. Matsunaga, M. Monelli, S.E. Persson, A.M. Piersimoni, A. Pietrinferni, P. Prada-Moroni, L. Pulone, R. Stellingwerf, E. Tognelli, A.R. Walker, On the RR Lyrae stars in globulars. IV. Omega Centauri optical UBVRI photometry. Astron. J. **152**, 170 (2016). https://doi.org/10.3847/0004-6256/152/6/170. arXiv:1609.04916

V.F. Braga, P.B. Stetson, G. Bono, M. DallOra, I. Ferraro, G. Fiorentino, G. Iannicola, M. Marconi, M. Marengo, A.J. Monson, J. Neeley, S.E. Persson, R.L. Beaton, R. Buonanno, A. Calamida, M. Castellani, E. Di Carlo, M. Fabrizio, W.L. Freedman, L. Inno, B.F. Madore, D. Magurno, E. Marchetti, S. Marinoni, P. Marrese, N. Matsunaga, D. Minniti, M. Monelli, M. Nonino, A.M. Piersimoni, A. Pietrinferni, P. Prada-Moroni, L. Pulone, R. Stellingwerf, E. Tognelli, A.R. Walker, E. Valenti, M. Zoccali, On the RR Lyrae stars in globulars. V. The complete near-infrared (JHKs) census of omega Centauri RR Lyrae variables. Astron. J. **155**, 137 (2018). https://doi.org/10.3847/1538-3881/aaadab. arXiv:1802.03578

J. Breitfelder, P. Kervella, A. Mérand, A. Gallenne, L. Szabados, R.I. Anderson, M. Willson, J.B. Le Bouquin, Observational calibration of the projection factor of Cepheids. I. The type II Cepheid κ Pavonis. Astron. Astrophys. **576**, A64 (2015). https://doi.org/10.1051/0004-6361/201425171. arXiv:1503.05176

J. Breitfelder, A. Mérand, P. Kervella, A. Gallenne, L. Szabados, R.I. Anderson, J.B. Le Bouquin, Observational calibration of the projection factor of Cepheids. II. Application to nine Cepheids with HST/FGS parallax measurements. Astron. Astrophys. **587**, A117 (2016). https://doi.org/10.1051/0004-6361/201527030. arXiv:1601.04727

T.M. Brown, S. Casertano, J. Strader, A. Riess, D.A. VandenBerg, D.R. Soderblom, J. Kalirai, R. Salinas, A high-precision trigonometric parallax to an ancient metal-poor globular cluster. Astrophys. J. Lett. **856**, L6 (2018a). https://doi.org/10.3847/2041-8213/aab55a. arXiv:1803.02927

A.G.A. Brown, A. Vallenari, T. Prusti, J.H.J. de Bruijne, C. Babusiaux, C.A.L. Bailer-Jones, M. Biermann, D.W. Evans et al. (Gaia Collaboration), Gaia Data Release 2. Summary of the contents and survey properties. Astron. Astrophys. **616**, A1 (2018b). https://doi.org/10.1051/0004-6361/201833051

G. Burki, G. Meylan, Rr-Lyrae Delta-Scuti Sx-Phoenicis stars and Baade–Wesselink method—part four—physical parameters and distances of Bs-Aquarii and Dy-Pegasi and the period/radius relations for pulsating stars. Astron. Astrophys. **159**, 261 (1986)

C. Cacciari, G. Clementini, J.A. Fernley, On the absolute magnitude of RR Lyrae stars—UU Ceti, RV Phoenicis, and W Tucanae. Astrophys. J. **396**, 219–237 (1992). https://doi.org/10.1086/171711

F. Caputo, The period-magnitude diagram of RR Lyrae stars—I. The controversy about the distance scale. Mon. Not. R. Astron. Soc. **284**, 994–1000 (1997). https://doi.org/10.1093/mnras/284.4.994

F. Caputo, V. Castellani, M. Marconi, V. Ripepi, Pulsational M_V versus [Fe/H] relation(s) for globular cluster RR Lyrae variables. Mon. Not. R. Astron. Soc. **316**, 819–826 (2000). https://doi.org/10.1046/j.1365-8711.2000.03591.x. arXiv:astro-ph/0003473

J.A. Cardelli, G.C. Clayton, J.S. Mathis, The relationship between infrared, optical, and ultraviolet extinction. Astrophys. J. **345**, 245–256 (1989). https://doi.org/10.1086/167900

B.W. Carney, J. Storm, R.V. Jones, The Baade–Wesselink method and the distances to RR Lyrae stars. VIII—comparisons with other techniques and implications for globular cluster distances and ages. Astrophys. J. **386**, 663–684 (1992). https://doi.org/10.1086/171047

E. Carretta, R.G. Gratton, Abundances for globular cluster giants. I. Homogeneous metallicities for 24 clusters. Astron. Astrophys. Suppl. Ser. **121**, 95–112 (1997). https://doi.org/10.1051/aas:1997116. arXiv:astro-ph/9607078

S. Cassisi, M. Salaris, *Old Stellar Populations: How to Study the Fossil Record of Galaxy Formation* (2013)

S. Cassisi, M. Castellani, F. Caputo, V. Castellani, RR Lyrae variables in Galactic globular clusters. IV. Synthetic HB and RR Lyrae predictions. Astron. Astrophys. **426**, 641–650 (2004). https://doi.org/10.1051/0004-6361:20041048. arXiv:astro-ph/0407256

S. Cassisi, A.Y. Potekhin, A. Pietrinferni, M. Catelan, M. Salaris, Updated electron-conduction opacities: the impact on low-mass stellar models. Astrophys. J. **661**, 1094–1104 (2007). https://doi.org/10.1086/516819. arXiv:astro-ph/0703011

M. Catelan, Horizontal branch stars: the interplay between observations and theory, and insights into the formation of the Galaxy. Astrophys. Space Sci. **320**, 261–309 (2009). https://doi.org/10.1007/s10509-009-9987-8. arXiv:astro-ph/0507464

M. Catelan, H.A. Smith, *Pulsating Stars* (2015)

M. Catelan, B.J. Pritzl, H.A. Smith, The RR Lyrae period-luminosity relation. I. Theoretical calibration. Astrophys. J. Suppl. Ser. **154**, 633–649 (2004). https://doi.org/10.1086/422916. arXiv:astro-ph/0406067

M. Chadid, J. Vernin, G. Preston, C. Zalian, C. Pouzenc, L. Abe, A. Agabi, E. Aristidi, L.Y. Liu, D. Mékarnia, H. Trinquet, First detection of multi-shocks in RR Lyrae stars from Antarctica: a possible explanation of the Blazhko effect. Astron. J. **148**, 88 (2014). https://doi.org/10.1088/0004-6256/148/5/88

H.Y. Chen, D.E. Holz, J. Miller, M. Evans, S. Vitale, J. Creighton, Distance measures in gravitational-wave astrophysics and cosmology (2017). arXiv:1709.08079

R.F. Christy, A study of pulsation in RR Lyrae models. Astrophys. J. **144**, 108 (1966). https://doi.org/10.1086/148593

A. Ciechanowska, G. Pietrzyński, O. Szewczyk, W. Gieren, I. Soszyński, The Araucaria project: the distance to the Small Magellanic Cloud from near infrared photometry of type II Cepheids. Acta Astron. **60**, 233–243 (2010). arXiv:1007.4217

M.R.L. Cioni, R.P. van der Marel, C. Loup, H.J. Habing, The tip of the red giant branch and distance of the Magellanic Clouds: results from the DENIS survey. Astron. Astrophys. **359**, 601–614 (2000). arXiv:astro-ph/0003223

C.M. Clement, A. Muzzin, Q. Dufton, T. Ponnampalam, J. Wang, J. Burford, A. Richardson, T. Rosebery, J. Rowe, H.S. Hogg, Variable stars in Galactic globular clusters. Astron. J. **122**, 2587–2599 (2001a). https://doi.org/10.1086/323719. arXiv:astro-ph/0108024

C.M. Clement, A. Muzzin, Q. Dufton, T. Ponnampalam, J. Wang, J. Burford, A. Richardson, T. Rosebery, J. Rowe, H.S. Hogg, Variable stars in Galactic globular clusters. Astron. J. **122**, 2587–2599 (2001b). https://doi.org/10.1086/323719. arXiv:astro-ph/0108024

G. Clementini, E. Carretta, R. Gratton, R. Merighi, J.R. Mould, J.K. McCarthy, The composition of HB stars: RR Lyrae variables. Astron. J. **110**, 2319 (1995). https://doi.org/10.1086/117692. arXiv:astro-ph/9507056

G. Clementini, R. Gratton, A. Bragaglia, E. Carretta, L. Di Fabrizio, M. Maio, Distance to the Large Magellanic Cloud: the RR Lyrae stars. Astron. J. **125**, 1309–1329 (2003). https://doi.org/10.1086/367773

G. Clementini, V. Ripepi, A. Bragaglia, A.F. Martinez Fiorenzano, E.V. Held, R.G. Gratton, The metal abundance distribution of the oldest stellar component in the Sculptor Dwarf spheroidal galaxy. Mon. Not. R. Astron. Soc. **363**, 734–748 (2005). https://doi.org/10.1111/j.1365-2966.2005.09478.x. arXiv:astro-ph/0506206

A.R. Conn, G.F. Lewis, R.A. Ibata, Q.A. Parker, D.B. Zucker, A.W. McConnachie, N.F. Martin, M.J. Irwin, N. Tanvir, M.A. Fardal, A.M.N. Ferguson, A Bayesian approach to locating the red giant branch tip magnitude. I. Astrophys. J. **740**, 69 (2011). https://doi.org/10.1088/0004-637X/740/2/69. arXiv:1107.3206

G. Coppola, M. Dall'Ora, V. Ripepi, M. Marconi, I. Musella, G. Bono, A.M. Piersimoni, P.B. Stetson, J. Storm, Distance to Galactic globulars using the near-infrared magnitudes of RR Lyrae stars—IV. The case of M5 (NGC 5904). Mon. Not. R. Astron. Soc. **416**, 1056–1066 (2011). https://doi.org/10.1111/j.1365-2966.2011.19102.x. arXiv:1105.4031

G. Coppola, M. Marconi, P.B. Stetson, G. Bono, V.F. Braga, V. Ripepi, M. Dall'Ora, I. Musella, R. Buonanno, M. Fabrizio, I. Ferraro, G. Fiorentino, G. Iannicola, M. Monelli, M. Nonino, F. Thévenin, A.R. Walker, The Carina project IX: on hydrogen and helium burning variables. Astrophys. J. **814**, 71 (2015). https://doi.org/10.1088/0004-637X/814/1/71. arXiv:1509.02687

J.P. Cox, Non-adiabatic stellar pulsation. I. Astrophys. J. **127**, 194 (1958). https://doi.org/10.1086/146451

J.P. Cox, On second helium ionization as a cause of pulsational instability in stars. Astrophys. J. **138**, 487 (1963). https://doi.org/10.1086/147661

B. Czerny, R. Beaton, M. Bejger, E. Cackett, M. Dall'Ora, R.F.L. Holanda, J.B. Jensen, S.W. Jha, E. Lusso, T. Minezaki, G. Risaliti, M. Salaris, S. Toonen, Y. Yoshii, Astronomical distance determination in the space age. Secondary distance indicators. Space Sci. Rev. **214**, 32 (2018). https://doi.org/10.1007/s11214-018-0466-9. arXiv:1801.00598

G.S. Da Costa, The age(s?) of the Sculptor Dwarf Galaxy. Astrophys. J. **285**, 483–494 (1984). https://doi.org/10.1086/162523

G.S. Da Costa, T.E. Armandroff, Standard globular cluster giant branches in the $(M_I, (V-I)_0)$ plane. Astron. J. **100**, 162–181 (1990). https://doi.org/10.1086/115500

J.J. Dalcanton, B.F. Williams, J.L. Melbourne, L. Girardi, A. Dolphin, P.A. Rosenfield, M.L. de Boyer, R.S. Jong, K. Gilbert, P. Marigo, K. Olsen, A.C. Seth, E. Skillman, Resolved near-infrared stellar populations in nearby galaxies. Astrophys. J. Suppl. Ser. **198**, 6 (2012). https://doi.org/10.1088/0067-0049/198/1/6. arXiv:1109.6893

M. Dall'Ora, J. Storm, G. Bono, V. Ripepi, M. Monelli, V. Testa, G. Andreuzzi, R. Buonanno, F. Caputo, V. Castellani, C.E. Corsi, G. Marconi, M. Marconi, L. Pulone, P.B. Stetson, The distance to the Large Magellanic Cloud cluster Reticulum from the K-band period-luminosity–metallicity relation of RR Lyrae stars. Astrophys. J. **610**, 269–274 (2004). https://doi.org/10.1086/421513. arXiv:astro-ph/0404098

A.K. Dambis, L.N. Berdnikov, A.Y. Kniazev, V.V. Kravtsov, A.S. Rastorguev, R. Sefako, O.V. Vozyakova, RR Lyrae variables: visual and infrared luminosities, intrinsic colours and kinematics. Mon. Not. R. Astron. Soc. **435**, 3206–3220 (2013). https://doi.org/10.1093/mnras/stt1514. arXiv:1308.4727

A.K. Dambis, A.S. Rastorguev, M.V. Zabolotskikh, Mid-infrared period-luminosity relations for globular cluster RR Lyrae. Mon. Not. R. Astron. Soc. **439**, 3765–3774 (2014). https://doi.org/10.1093/mnras/stu226. arXiv:1401.5523

T.J.L. de Boer, E. Tolstoy, V. Hill, A. Saha, K. Olsen, E. Starkenburg, B. Lemasle, M.J. Irwin, G. Battaglia, The star formation and chemical evolution history of the sculptor dwarf spheroidal galaxy. Astron. Astrophys. **539**, A103 (2012). https://doi.org/10.1051/0004-6361/201118378. arXiv:1201.2408

J.H.J. de Bruijne, K.L.J. Rygl, T. Antoja, Gaia astrometric science performance—post-launch predictions, in *EAS Publications Series*. EAS Publications Series, vol. 67 (2014), pp. 23–29. https://doi.org/10.1051/eas/1567004. arXiv:1502.00791

R. de Grijs, G. Bono, Clustering of local group distances: publication bias or correlated measurements? II. M31 and beyond. Astron. J. **148**, 17 (2014). https://doi.org/10.1088/0004-6256/148/1/17. arXiv:1405.2124

R. de Grijs, G. Bono, Clustering of local group distances: publication bias or correlated measurements? III. The Small Magellanic Cloud. Astron. J. **149**, 179 (2015). https://doi.org/10.1088/0004-6256/149/6/179. arXiv:1504.00417

R. de Grijs, J.E. Wicker, G. Bono, Clustering of local group distances: publication bias or correlated measurements? I. The Large Magellanic Cloud. Astron. J. **147**, 122 (2014). https://doi.org/10.1088/0004-6256/147/5/122. arXiv:1403.3141

R. de Grijs, F. Courbin, C.E. Martínez-Vázquez, M. Monelli, M. Oguri, S.H. Suyu, Toward an internally consistent astronomical distance scale. Space Sci. Rev. **212**, 1743–1785 (2017). https://doi.org/10.1007/s11214-017-0395-z. arXiv:1706.07933

A. Deep, G. Fiorentino, E. Tolstoy, E. Diolaiti, M. Bellazzini, P. Ciliegi, R.I. Davies, J.M. Conan, An E-ELT case study: colour-magnitude diagrams of an old galaxy in the Virgo cluster. Astron. Astrophys. **531**, A151 (2011). https://doi.org/10.1051/0004-6361/201116603. arXiv:1105.3455

M. Del Principe, A.M. Piersimoni, G. Bono, A. Di Paola, M. Dolci, M. Marconi, Near-infrared observations of RR Lyrae variables in Galactic globular clusters. I. The case of M92. Astron. J. **129**, 2714–2724 (2005). https://doi.org/10.1086/430148. arXiv:astro-ph/0503140

M. Del Principe, A.M. Piersimoni, J. Storm, F. Caputo, G. Bono, P.B. Stetson, M. Castellani, R. Buonanno, A. Calamida, C.E. Corsi, M. Dall'Ora, I. Ferraro, L.M. Freyhammer, G. Iannicola, M. Monelli, M. Nonino, L. Pulone, V. Ripepi, A pulsational distance to ω Centauri based on near-infrared period-luminosity relations of RR Lyrae stars. Astrophys. J. **652**, 362–369 (2006). https://doi.org/10.1086/508136. arXiv:astro-ph/0608052

R.G. Deupree, The theoretical red edge of the RR Lyrae gap. I—dependence of convection on pulsation phase. Astrophys. J. **211**, 509–526 (1977a). https://doi.org/10.1086/154958

R.G. Deupree, The theoretical red edge of the RR Lyrae gap. IV—convective steady-state models. Astrophys. J. **215**, 620–623 (1977b). https://doi.org/10.1086/155396

G.P. di Benedetto, Towards a fundamental calibration of stellar parameters of A, F, G, K dwarfs and giants. Astron. Astrophys. **339**, 858–871 (1998)

M. Di Criscienzo, M. Marconi, F. Caputo, RR Lyrae stars in galactic globular clusters. III. Pulsational predictions for metal content $Z = 0.0001$ to $Z = 0.006$. Astrophys. J. **612**, 1092–1106 (2004). https://doi.org/10.1086/422742. arXiv:astro-ph/0405415

M. Di Criscienzo, F. Caputo, M. Marconi, S. Cassisi, Synthetic properties of bright metal-poor variables. II. BL Hercules stars. Astron. Astrophys. **471**, 893–900 (2007). https://doi.org/10.1051/0004-6361:20066541. arXiv:0705.2679

R. Diethelm, Physical parameters of pulsating variables with periods between one and three days. II—fundamental parameters. Astron. Astrophys. **239**, 186–192 (1990)

R. Diethelm, Period changes of AHB1 variables. Astron. Astrophys. **307**, 803–806 (1996)

A.E. Dolphin, WFPC2 stellar photometry with HSTPHOT. Publ. Astron. Soc. Pac. **112**, 1383–1396 (2000). https://doi.org/10.1086/316630. arXiv:astro-ph/0006217

A.E. Dolphin, A. Saha, E.D. Skillman, E. Tolstoy, A.A. Cole, R.C. Dohm-Palmer, J.S. Gallagher, M. Mateo, J.G. Hoessel, Deep Hubble Space Telescope imaging of IC 1613. I. Variable stars and distance. Astrophys. J. **550**, 554–569 (2001). https://doi.org/10.1086/319809. arXiv:astro-ph/0012150

R. Drimmel, A. Cabrera-Lavers, M. López-Corredoira, A three-dimensional Galactic extinction model. Astron. Astrophys. **409**, 205–215 (2003). https://doi.org/10.1051/0004-6361:20031070. arXiv:astro-ph/0307273

R.L. Duncombe, W.H. Jefferys, P.J. Shelus, P.D. Hemenway, G.F. Benedict, Astrometry using the Hubble Space Telescope fine guidance sensors. Adv. Space Res. **11**, 87–96 (1991). https://doi.org/10.1016/0273-1177(91)90475-Y

A.S. Eddington, Stars, gaseous, on the pulsations of a gaseous star. Mon. Not. R. Astron. Soc. **79**, 2–22 (1918)

A.S. Eddington, Theories of Cepheid variation. Observatory **49**, 88 (1926)

ESA (ed.), The HIPPARCOS and TYCHO catalogues. Astrometric and photometric star catalogues derived from the ESA HIPPARCOS Space Astrometry Mission. ESA SP **1200** (1997)

I.A. Fadeev, A.B. Fokin, Hydrodynamic models for population-II Cepheids. Astrophys. Space Sci. **111**, 355–374 (1985). https://doi.org/10.1007/BF00649975

M. Feast, RR Lyraes and type II Cepheids in the Magellanic Clouds: distance scales and population gradients, in *RR Lyrae Stars, Metal-Poor Stars, and the Galaxy*, vol. 5, ed. by A. McWilliam (2011), p. 170. arXiv:1106.1449

M.W. Feast, Distance scales after HIPPARCOS. Astrophys. Space Sci. **263**, 209–214 (1998). https://doi.org/10.1023/A:1002110327632

M.W. Feast, The luminosities of type II Cepheids and RR Lyrae variables, in *Variable Stars, the Galactic Halo and Galaxy Formation*, ed. by C. Sterken, N. Samus, L. Szabados (2010). arXiv:0912.4159

M.W. Feast, *Galactic Distance Scales* (2013), p. 829. https://doi.org/10.1007/978-94-007-5612-0_16.

M.W. Feast, C.D. Laney, T.D. Kinman, F. van Leeuwen, P.A. Whitelock, The luminosities and distance scales of type II Cepheid and RR Lyrae variables. Mon. Not. R. Astron. Soc. **386**, 2115–2134 (2008). https://doi.org/10.1111/j.1365-2966.2008.13181.x. arXiv:0803.0466

J.D. Fernie, The period-luminosity relation: a historical review. Publ. Astron. Soc. Pac. **81**, 707 (1969). https://doi.org/10.1086/128847

J. Fernley, T.G. Barnes, I. Skillen, S.L. Hawley, C.J. Hanley, D.W. Evans, E. Solano, R. Garrido, The absolute magnitudes of RR Lyraes from HIPPARCOS parallaxes and proper motions. Astron. Astrophys. **330**, 515–520 (1998)

L. Ferrarese, J.R. Mould, R.C. Kennicutt Jr., J. Huchra, H.C. Ford, W.L. Freedman, P.B. Stetson, B.F. Madore, S. Sakai, B.K. Gibson, J.A. Graham, S.M. Hughes, G.D. Illingworth, D.D. Kelson, L. Macri, K. Sebo, N.A. Silbermann, The Hubble Space Telescope key project on the extragalactic distance scale. XXVI. The calibration of population II secondary distance indicators and the value of the Hubble constant. Astrophys. J. **529**, 745–767 (2000). https://doi.org/10.1086/308309. arXiv:astro-ph/9908192

M.U. Feuchtinger, E.A. Dorfi, S. Hofner, Radiation hydrodynamics in atmospheres of long-period variables. Astron. Astrophys. **273**, 513 (1993)

G. Fiorentino, M. Monelli, Anomalous Cepheids in the Large Magellanic Cloud. Insight into their origin and connection with the star formation history. Astron. Astrophys. **540**, A102 (2012). https://doi.org/10.1051/0004-6361/201118621. arXiv:1202.2752

G. Fiorentino, M. Marconi, I. Musella, F. Caputo, Classical Cepheid pulsation models. XI. Effects of convection and chemical composition on the period-luminosity and period-Wesenheit relations. Astron. Astrophys. **476**, 863–879 (2007). https://doi.org/10.1051/0004-6361:20077587. arXiv:0707.0959

G. Fiorentino, A. Monachesi, S.C. Trager, T.R. Lauer, A. Saha, K.J. Mighell, W. Freedman, A. Dressler, C. Grillmair, E. Tolstoy, RR Lyrae variables in M32 and the disk of M31. Astrophys. J. **708**, 817–833 (2010). https://doi.org/10.1088/0004-637X/708/1/817. arXiv:0911.0355

G. Fiorentino, G. Bono, M. Monelli, P.B. Stetson, E. Tolstoy, C. Gallart, M. Salaris, C.E. Martínez-Vázquez, E.J. Bernard, Weak Galactic halo-Dwarf spheroidal connection from RR Lyrae stars. Astrophys. J. Lett. **798**, L12 (2015). https://doi.org/10.1088/2041-8205/798/1/L12. arXiv:1411.7300

G. Fiorentino, M. Monelli, P.B. Stetson, G. Bono, C. Gallart, C.E. Martínez-Vázquez, E.J. Bernard, D. Massari, V.F. Braga, M. Dall'Ora, Weak Galactic halo-Fornax dSph connection from RR Lyrae stars. Astron. Astrophys. **599**, A125 (2017). https://doi.org/10.1051/0004-6361/201629501. arXiv:1612.02991

B.Q. For, G.W. Preston, C. Sneden, Radial velocities and pulsation ephemerides of 11 field RR Lyrae stars. Astrophys. J. Suppl. Ser. **194**, 38 (2011a). https://doi.org/10.1088/0067-0049/194/2/38. arXiv:1104.2668

B.Q. For, C. Sneden, G.W. Preston, The chemical compositions of variable field horizontal-branch stars: RR Lyrae stars. Astrophys. J. Suppl. Ser. **197**, 29 (2011b). https://doi.org/10.1088/0067-0049/197/2/29. arXiv:1110.0548

W.L. Freedman, Stellar content of nearby galaxies. I—BVRI CCD photometry for IC 1613. Astron. J. **96**, 1248–1306 (1988). https://doi.org/10.1086/114878

W.L. Freedman, B.F. Madore, The Hubble constant. Annu. Rev. Astron. Astrophys. **48**, 673–710 (2010). https://doi.org/10.1146/annurev-astro-082708-101829. arXiv:1004.1856

W.L. Freedman, B.F. Madore, B.K. Gibson, L. Ferrarese, D.D. Kelson, S. Sakai, J.R. Mould, R.C. Kennicutt Jr., H.C. Ford, J.A. Graham, J.P. Huchra, S.M.G. Hughes, G.D. Illingworth, L.M. Macri, P.B. Stetson, Final results from the Hubble Space Telescope key project to measure the Hubble constant. Astrophys. J. **553**, 47–72 (2001). https://doi.org/10.1086/320638. arXiv:astro-ph/0012376

W.L. Freedman, J. Rigby, B.F. Madore, S.E. Persson, L. Sturch, V. Mager, The Cepheid period-luminosity relation (the leavitt law) at mid-infrared wavelengths. IV. Cepheids in IC 1613. Astrophys. J. **695**, 996–998 (2009). https://doi.org/10.1088/0004-637X/695/2/996. arXiv:0901.2907

M. Geha, D. Weisz, A. Grocholski, A. Dolphin, R.P. van der Marel, P. Guhathakurta, HST/ACS direct ages of the Dwarf Elliptical Galaxies NGC 147 and NGC 185. Astrophys. J. **811**, 114 (2015). https://doi.org/10.1088/0004-637X/811/2/114. arXiv:1503.06526

R.A. Gingold, The evolutionary status of population II Cepheids. Astrophys. J. **204**, 116–130 (1976). https://doi.org/10.1086/154156

R.A. Gingold, Evolutionary models of metal-rich horizontal branch stars. Mon. Not. R. Astron. Soc. **178**, 533–537 (1977). https://doi.org/10.1093/mnras/178.4.533

S. Giridhar, D.L. Lambert, B.E. Reddy, G. Gonzalez, D. Yong, Abundance analyses of field RV Tauri stars. VI. An extended sample. Astrophys. J. **627**, 432–445 (2005). https://doi.org/10.1086/430265. arXiv:astro-ph/0503344

M. Górski, G. Pietrzyński, W. Gieren, M. Catelan, B. Pilecki, P. Karczmarek, K. Suchomska, D. Graczyk, P. Konorski, B. Zgirski, P. Wielgórski, The Araucaria project: on the tip of the red giant branch distance determination to the Magellanic Clouds. Astron. J. **151**, 167 (2016). https://doi.org/10.3847/0004-6256/151/6/167. arXiv:1605.06210

A. Gould, P. Popowski, Systematics of RR Lyrae statistical parallax. III. Apparent magnitudes and extinctions. Astrophys. J. **508**, 844–853 (1998). https://doi.org/10.1086/306448. arXiv:astro-ph/9805176

J. Govea, T. Gomez, G.W. Preston, C. Sneden, The chemical compositions of RR Lyrae type C variable stars. Astrophys. J. **782**, 59 (2014). https://doi.org/10.1088/0004-637X/782/2/59

D. Graczyk, P. Konorski, G. Pietrzyński, W. Gieren, J. Storm, N. Nardetto, A. Gallenne, P.F.L. Maxted, P. Kervella, Z. Kołaczkowski, The surface brightness-color relations based on eclipsing binary stars: toward precision better than 1% in angular diameter predictions. Astrophys. J. **837**, 7 (2017). https://doi.org/10.3847/1538-4357/aa5d56. arXiv:1611.09976

D. Graczyk, G. Pietrzyński, I.B. Thompson, W. Gieren, B. Pilecki, P. Konorski, S. Villanova, M. Górski, K. Suchomska, P. Karczmarek, K. Stepień, J. Storm, M. Taormina, Z. Kołaczkowski, P. Wielgórski, W. Narloch, B. Zgirski, A. Gallenne, J. Ostrowski, R. Smolec, A. Udalski, I. Soszyński, P. Kervella, N. Nardetto, M.K. Szymański, Ł. Wyrzykowski, K. Ulaczyk, R. Poleski, P. Pietrukowicz, S. Kozłowski, J. Skowron, P. Mróz, The late-type eclipsing binaries in the Large Magellanic Cloud: catalog of fundamental physical parameters. Astrophys. J. **860**, 1 (2018). https://doi.org/10.3847/1538-4357/aac2bf. arXiv:1805.04952

E.M. Green, P. Demarque, C.R. King, *The Revised Yale Isochrones and Luminosity Functions* (1987)

L.J. Greenhill, D.R. Jiang, J.M. Moran, M.J. Reid, K.Y. Lo, M.J. Claussen, Detection of a subparsec diameter disk in the nucleus of NGC 4258. Astrophys. J. **440**, 619 (1995). https://doi.org/10.1086/175301

A.J. Grocholski, A.A. Cole, A. Sarajedini, D. Geisler, V.V. Smith, Ca II triplet spectroscopy of Large Magellanic Cloud red giants. I. Abundances and velocities for a sample of populous clusters. Astron. J. **132**, 1630–1644 (2006). https://doi.org/10.1086/507303. arXiv:astro-ph/0607052

M.A.T. Groenewegen, M.I. Jurkovic, Luminosities and infrared excess in type II and anomalous Cepheids in the Large and Small Magellanic Clouds. Astron. Astrophys. **603**, A70 (2017a). https://doi.org/10.1051/0004-6361/201730687. arXiv:1705.00886

M.A.T. Groenewegen, M.I. Jurkovic, The period-luminosity and period-radius relations of type II and anomalous Cepheids in the Large and Small Magellanic Clouds. Astron. Astrophys. **604**, A29 (2017b). https://doi.org/10.1051/0004-6361/201730946. arXiv:1705.04487

M.A.T. Groenewegen, M. Salaris, The absolute magnitudes of RR Lyrae stars from BT HIPPARCOS parallaxes. Astron. Astrophys. **348**, L33–L36 (1999). arXiv:astro-ph/9907207

B. Gustafsson, B. Edvardsson, K. Eriksson, U.G. Jørgensen, B. Plez, A grid of MARCS model atmospheres for late-type stars. I. Methods and general properties. Astron. Astrophys. **486**, 951–970 (2008). https://doi.org/10.1051/0004-6361:200809724. arXiv:0805.0554

H.J. Habing, H. Olofsson (eds.), *Asymptotic Giant Branch Stars* (2003)

G. Hajdu, I. Dékány, M. Catelan, E.K. Grebel, J. Jurcsik, A data-driven study of RR Lyrae near-IR light curves: principal component analysis, robust fits, and metallicity estimates. Astrophys. J. **857**, 55 (2018). https://doi.org/10.3847/1538-4357/aab4fd. arXiv:1804.01456

H.C. Harris, Population II Cepheids, in *IAU Colloq. 82: Cepheids: Theory and Observation*, ed. by B.F. Madore (1985), pp. 232–245

D. Hatt, R.L. Beaton, W.L. Freedman, B.F. Madore, I.S. Jang, T.J. Hoyt, M.G. Lee, A.J. Monson, J.A. Rich, V. Scowcroft, M. Seibert, The Carnegie–Chicago Hubble program. II. The distance to IC 1613: the tip of the red giant branch and RR Lyrae period-luminosity relations. Astrophys. J. **845**, 146 (2017). https://doi.org/10.3847/1538-4357/aa7f73. arXiv:1703.06468

D. Hatt, W.L. Freedman, B.F. Madore, R.L. Beaton, T.J. Hoyt, I.S. Jang, M.G. Lee, A.J. Monson, J.A. Rich, V. Scowcroft, M. Seibert, The Carnegie–Chicago Hubble program. IV. The distance to NGC 4424, NGC 4526, and NGC 4356 via the tip of the red giant branch. Astrophys. J. **861**, 104 (2018). https://doi.org/10.3847/1538-4357/aac9cc. arXiv:1806.02900

M.R. Hayden, J. Bovy, J.A. Holtzman, D.L. Nidever, J.C. Bird, D.H. Weinberg, B.H. Andrews, S.R. Majewski, C. Allende Prieto, F. Anders, T.C. Beers, D. Bizyaev, C. Chiappini, K. Cunha, P. Frinchaboy, D.A. García-Herńandez, A.E. García Pérez, L. Girardi, P. Harding, F.R. Hearty, J.A. Johnson, S. Mészáros, I. Minchev, R. O'Connell, K. Pan, A.C. Robin, R.P. Schiavon, D.P. Schneider, M. Schultheis, M. Shetrone, M. Skrutskie, M. Steinmetz, V. Smith, J.C. Wilson, O. Zamora, G. Zasowski, Chemical cartography with APOGEE: metallicity distribution functions and the chemical structure of the Milky Way disk. Astrophys. J. **808**, 132 (2015). https://doi.org/10.1088/0004-637X/808/2/132. arXiv:1503.02110

D. Hendel, V. Scowcroft, K.V. Johnston, M.A. Fardal, R.P. van der Marel, S.T. Sohn, A.M. Price-Whelan, R.L. Beaton, G. Besla, G. Bono, M.R.L. Cioni, G. Clementini, J.G. Cohen, M. Fabrizio, W.L. Freedman, A. Garofalo, C.J. Grillmair, N. Kallivayalil, J.A. Kollmeier, D.R. Law, B.F. Madore, S.R. Majewski, M. Marengo, A.J. Monson, J.R. Neeley, D.L. Nidever, G. Pietrzyński, M. Seibert, B. Sesar, H.A. Smith, I. Soszyński, A. Udalski, SMHASH: anatomy of the Orphan stream using RR Lyrae stars. Mon. Not. R. Astron. Soc. **479**, 570–587 (2018). https://doi.org/10.1093/mnras/sty1455. arXiv:1711.04663

B. Hendricks, P.B. Stetson, D.A. VandenBerg, M. Dall'Ora, A new reddening law for M4. Astron. J. **144**, 25 (2012). https://doi.org/10.1088/0004-6256/144/1/25. arXiv:1204.5719

J.R. Herrnstein, J.M. Moran, L.J. Greenhill, P.J. Diamond, M. Inoue, N. Nakai, M. Miyoshi, C. Henkel, A. Riess, A geometric distance to the galaxy NGC4258 from orbital motions in a nuclear gas disk. Nature **400**, 539–541 (1999). https://doi.org/10.1038/22972. arXiv:astro-ph/9907013

N. Ho, M. Geha, E.J. Tollerud, R. Zinn, P. Guhathakurta, L.C. Vargas, Metallicity evolution of the six most luminous M31 dwarf satellites. Astrophys. J. **798**, 77 (2015). https://doi.org/10.1088/0004-637X/798/2/77. arXiv:1405.4424

M.L. Houdashelt, R.A. Bell, A.V. Sweigart, Improved color–temperature relations and bolometric corrections for Cool stars. Astron. J. **119**, 1448–1469 (2000). https://doi.org/10.1086/301243. arXiv:astro-ph/9911367

S.B. Howell, C. Sobeck, M. Haas, M. Still, T. Barclay, F. Mullally, J. Troeltzsch, S. Aigrain, S.T. Bryson, D. Caldwell, W.J. Chaplin, W.D. Cochran, D. Huber, G.W. Marcy, A. Miglio, J.R. Najita, M. Smith, J.D. Twicken, J.J. Fortney, The K2 mission: characterization and early results. Publ. Astron. Soc. Pac. **126**, 398 (2014). https://doi.org/10.1086/676406. arXiv:1402.5163

T.J. Hoyt, W.L. Freedman, B.F. Madore, M. Seibert, R.L. Beaton, D. Hatt, I.S. Jang, M.G. Lee, A.J. Monson, J.A. Rich, The near-infrared tip of the red giant branch. II. An absolute calibration in the Large Magellanic Cloud. Astrophys. J. **858**, 12 (2018). https://doi.org/10.3847/1538-4357/aab7ed. arXiv:1803.01277

E.M.L. Humphreys, A.L. Argon, L.J. Greenhill, J.M. Moran, M.J. Reid, Recent progress on a new distance to NGC 4258, in *Future Directions in High Resolution Astronomy*, ed. by J. Romney, M. Reid. Astronomical Society of the Pacific Conference Series, vol. 340 (2005), p. 466

E.M.L. Humphreys, M.J. Reid, L.J. Greenhill, J.M. Moran, A.L. Argon, Toward a new geometric distance to the active galaxy NGC 4258. II. Centripetal accelerations and investigation of spiral structure. Astrophys. J. **672**, 800–816 (2008). https://doi.org/10.1086/523637. arXiv:0709.0925

E.M.L. Humphreys, M.J. Reid, J.M. Moran, L.J. Greenhill, A.L. Argon, Toward a new geometric distance to the active galaxy NGC 4258. III. Final results and the Hubble constant. Astrophys. J. **775**, 13 (2013). https://doi.org/10.1088/0004-637X/775/1/13. arXiv:1307.6031

D. Hurley-Keller, M. Mateo, E.K. Grebel, A new culprit in the second-parameter problem in the Sculptor dwarf spheroidal galaxy? Astrophys. J. Lett. **523**, L25–L28 (1999). https://doi.org/10.1086/312248. arXiv:astro-ph/9907156

I. Iben Jr., On the specification of the blue edge of the RR Lyrae instability strip. Astrophys. J. **166**, 131 (1971). https://doi.org/10.1086/150946

L. Inno, N. Matsunaga, G. Bono, F. Caputo, R. Buonanno, K. Genovali, C.D. Laney, M. Marconi, A.M. Piersimoni, F. Primas, M. Romaniello, On the distance of the Magellanic Clouds using Cepheid NIR and optical-NIR period-Wesenheit relations. Astrophys. J. **764**, 84 (2013). https://doi.org/10.1088/0004-637X/764/1/84. arXiv:1212.4376

L. Inno, G. Bono, N. Matsunaga, G. Fiorentino, M. Marconi, B. Lemasle, R. da Silva, I. Soszyński, A. Udalski, M. Romaniello, H.W. Rix, The panchromatic view of the Magellanic Clouds from Classical Cepheids. I. Distance, reddening, and geometry of the Large Magellanic Cloud disk. Astrophys. J. **832**, 176 (2016). https://doi.org/10.3847/0004-637X/832/2/176. arXiv:1609.03554

K.A. Janes, T. Liu, K-band photometry of globular cluster RR Lyraes and other horizontal branch studies. Mem. Soc. Astron. Ital. **63**, 445–450 (1992)

I.S. Jang, M.G. Lee, The tip of the red giant branch distances to type Ia supernova host galaxies. III. NGC 4038/39 and NGC 5584. Astrophys. J. **807**, 133 (2015). https://doi.org/10.1088/0004-637X/807/2/133. arXiv:1506.03089

I.S. Jang, M.G. Lee, The tip of the red giant branch distances to typa Ia supernova host galaxies. V. NGC 3021, NGC 3370, and NGC 1309 and the value of the Hubble constant. Astrophys. J. **836**, 74 (2017a). https://doi.org/10.3847/1538-4357/836/1/74

I.S. Jang, M.G. Lee, The tip of the red giant branch distances to type Ia supernova host galaxies. IV. Color dependence and zero-point calibration. Astrophys. J. **835**, 28 (2017b). https://doi.org/10.3847/1538-4357/835/1/28. arXiv:1611.05040

I.S. Jang, D. Hatt, R.L. Beaton, M.G. Lee, W.L. Freedman, B.F. Madore, T.J. Hoyt, A.J. Monson, J.A. Rich, V. Scowcroft, M. Seibert, The Carnegie–Chicago Hubble program. III. The distance to NGC 1365 via the tip of the red giant branch. Astrophys. J. **852**, 60 (2018). https://doi.org/10.3847/1538-4357/aa9d92. arXiv:1703.10616

R.V. Jones, B.W. Carney, J.P. Fulbright, Template K light curves for RR Lyrae stars. Publ. Astron. Soc. Pac. **108**, 877 (1996). https://doi.org/10.1086/133809

A.H. Joy, Spectra of the brighter variables in globular clusters. Astrophys. J. **110**, 105 (1949). https://doi.org/10.1086/145190

J. Jurcsik, G. Kovacs, Determination of [Fe/H] from the light curves of RR Lyrae stars. Astron. Astrophys. **312**, 111–120 (1996)

J. Jurcsik, G. Hajdu, I. Dékány, J. Nuspl, M. Catelan, E.K. Grebel, Blazhko modulation in the infrared (2018). arXiv:1801.03436

M.I. Jurkovic, M. Stojanovic, S. Ninkovic, Galactic membership of BL Her type variable stars. Commun. Konkoly Obs. Hung. Acad. Sci. **105**, 175–178 (2016). arXiv:1604.06871

J. Kalirai, Scientific discovery with the James Webb Space Telescope. Contemp. Phys. **59**(3), 251–290 (2018). https://doi.org/10.1080/00107514.2018.1467648

J. Kaluzny, I.B. Thompson, W. Krzeminski, G.W. Preston, W. Pych, S.M. Rucinski, A. Schwarzenberg-Czerny, S.A. Shectman, G. Stachowski, Eclipsing binaries in globular clusters as age and distance indicators, in *Stellar Astrophysics with the World's Largest Telescopes*, ed. by J. Mikolajewska, A. Olech. American Institute of Physics Conference Series, vol. 752 (2005), pp. 70–83. https://doi.org/10.1063/1.1893337

J. Kaluzny, I.B. Thompson, M. Rozyczka, A. Dotter, W. Krzeminski, W. Pych, S.M. Rucinski, G.S. Burley, S.A. Shectman, The Cluster AgeS Experiment (CASE). V. Analysis of three eclipsing binaries in the globular cluster M4. Astron. J. **145**, 43 (2013). https://doi.org/10.1088/0004-6256/145/2/43. arXiv:1301.2946

J. Kaluzny, I.B. Thompson, A. Dotter, M. Rozyczka, W. Pych, S.M. Rucinski, G.S. Burley, The Clusters AgeS Experiment (CASE). VI. Analysis of two detached eclipsing binaries in the globular cluster M55. Acta Astron. **64**, 11–26 (2014). arXiv:1403.6325

J. Kaluzny, I.B. Thompson, A. Dotter, M. Rozyczka, A. Schwarzenberg-Czerny, G.S. Burley, B. Mazur, S.M. Rucinski, The Cluster AgeS Experiment (CASE). VII. Analysis of two eclipsing binaries in the globular cluster NGC 6362. Astron. J. **150**, 155 (2015). https://doi.org/10.1088/0004-6256/150/5/155. arXiv:1508.04894

P. Karczmarek, G. Pietrzyński, W. Gieren, K. Suchomska, P. Konorski, M. Górski, B. Pilecki, D. Graczyk, P. Wielgórski, The Araucaria project: the distance to the Carina Dwarf Galaxy from infrared photometry of RR Lyrae stars. Astron. J. **150**, 90 (2015). https://doi.org/10.1088/0004-6256/150/3/90. arXiv:1507.07713

P. Karczmarek, G. Pietrzyński, K. Belczyński, K. Stępień, G. Wiktorowicz, K. Iłkiewicz, Contamination of RR Lyrae stars from binary evolution pulsators, in *37th Meeting of the Polish Astronomical Society*, vol. 3, ed. by A. Rozanska, M. Bejger (2016), pp. 45–48

P. Karczmarek, G. Pietrzyński, M. Górski, W. Gieren, D. Bersier, The Araucaria project: the distance to the Fornax Dwarf Galaxy from near-infrared photometry of RR Lyrae stars. Astron. J. **154**, 263 (2017a). https://doi.org/10.3847/1538-3881/aa9574. arXiv:1710.10934

P. Karczmarek, G. Wiktorowicz, K. Iłkiewicz, R. Smolec, K. Stępień, G. Pietrzyński, W. Gieren, K. Belczynski, The occurrence of binary evolution pulsators in classical instability strip of RR Lyrae and Cepheid variables. Mon. Not. R. Astron. Soc. **466**, 2842–2854 (2017b). https://doi.org/10.1093/mnras/stw3286. arXiv:1612.00465

P. Kervella, B. Trahin, H.E. Bond, A. Gallenne, L. Szabados, A. Mérand, J. Breitfelder, J. Dailloux, R.I. Anderson, P. Fouqué, W. Gieren, N. Nardetto, G. Pietrzyński, Observational calibration of the projection factor of Cepheids. III. The long-period Galactic Cepheid RS Puppis. Astron. Astrophys. **600**, A127 (2017). https://doi.org/10.1051/0004-6361/201630202. arXiv:1701.05192

E.N. Kirby, P. Guhathakurta, M. Bolte, C. Sneden, M.C. Geha, Multi-element abundance measurements from medium-resolution spectra. I. The Sculptor dwarf spheroidal galaxy. Astrophys. J. **705**, 328–346 (2009). https://doi.org/10.1088/0004-637X/705/1/328. arXiv:0909.3092

C.R. Klein, J.W. Richards, N.R. Butler, J.S. Bloom, Mid-infrared period-luminosity relations of RR Lyrae stars derived from the WISE preliminary data release. Astrophys. J. **738**, 185 (2011). https://doi.org/10.1088/0004-637X/738/2/185. arXiv:1105.0055

C.R. Klein, J.W. Richards, N.R. Butler, J.S. Bloom, A Bayesian approach to calibrating period-luminosity relations of RR Lyrae stars in the mid-infrared. Astrophys. Space Sci. **341**, 83–87 (2012). https://doi.org/10.1007/s10509-012-1035-4. arXiv:1202.3990

C.R. Klein, J.W. Richards, N.R. Butler, J.S. Bloom, Mid-infrared period-luminosity relations of RR Lyrae stars derived from the AllWISE data release. Mon. Not. R. Astron. Soc. **440**, L96–L100 (2014). https://doi.org/10.1093/mnrasl/slu031. arXiv:1402.4449

C. Koen, D. Laney, On the determination of absolute magnitude zero-points from HIPPARCOS parallaxes. Mon. Not. R. Astron. Soc. **301**, 582–584 (1998). https://doi.org/10.1046/j.1365-8711.1998.02066.x

J.A. Kollmeier, D.M. Szczygieł, C.R. Burns, A. Gould, I.B. Thompson, G.W. Preston, C. Sneden, J.D. Crane, S. Dong, B.F. Madore, N. Morrell, J.L. Prieto, S. Shectman, J.D. Simon, E. Villanueva, The absolute magnitude of RRc variables from statistical parallax. Astrophys. J. **775**, 57 (2013). https://doi.org/10.1088/0004-637X/775/1/57. arXiv:1208.2689

G. Kovacs, J.R. Buchler, Regular and irregular nonlinear pulsation in population II Cepheid models. Astrophys. J. **334**, 971–994 (1988). https://doi.org/10.1086/166890

G. Kovacs, E. Zsoldos, A new method for the determination of [Fe/H] in RR Lyrae stars. Astron. Astrophys. **293**, L57–L60 (1995)

V. Kovtyukh, G. Wallerstein, I. Yegorova, S. Andrievsky, S. Korotin, I. Saviane, S. Belik, C.E. Davis, E.M. Farrell, Metal-poor type II Cepheids with periods less than three days. Publ. Astron. Soc. Pac. **054**, 201 (2018a). https://doi.org/10.1088/1538-3873/aaacf7. arXiv:1803.05041

V. Kovtyukh, I. Yegorova, S. Andrievsky, S. Korotin, I. Saviane, B. Lemasle, F. Chekhonadskikh, S. Belik, Type II Cepheids: evidence for Na–O anticorrelation for BL Her type stars? Mon. Not. R. Astron. Soc. **477**, 2276–2285 (2018b). https://doi.org/10.1093/mnras/sty671

R.P. Kudritzki, M.A. Urbaneja, Distances to galaxies from the brightest stars in the Universe. Astrophys. Space Sci. **341**, 131–142 (2012). https://doi.org/10.1007/s10509-012-1016-7. arXiv:1112.3935

C.A. Kuehn, K. Dame, H.A. Smith, M. Catelan, Y.B. Jeon, J.M. Nemec, A.R. Walker, A. Kunder, B.J. Pritzl, N. De Lee, J. Borissova, Variable stars in Large Magellanic Cloud globular clusters. III. Reticulum. Astron. J. **145**, 160 (2013). https://doi.org/10.1088/0004-6256/145/6/160. arXiv:1307.6644

A. Kunder, B. Chaboyer, Distance to the Sagittarius Dwarf Galaxy using Macho project RR Lyrae stars. Astron. J. **137**, 4478–4486 (2009). https://doi.org/10.1088/0004-6256/137/5/4478. arXiv:0903.3040

A. Kunder, E. Valenti, M. Dall'Ora, P. Pietrukowicz, C. Sneden, G. Bono, V.F. Braga, I. Ferraro, G. Fiorentino, G. Iannicola, M. Marconi, C.E. Martínez-Vázquez, M. Monelli, I. Musella, V. Ripepi, M. Salaris, P.B. Stetson, Impact of distance determinations on Galactic structure. II. Old tracers. Space Sci. Rev. **214**, 90 (2018). https://doi.org/10.1007/s11214-018-0519-0. arXiv:1806.07487

A.C. Layden, R.B. Hanson, S.L. Hawley, A.R. Klemola, C.J. Hanley, The absolute magnitude and kinematics of RR Lyrae stars via statistical parallax. Astron. J. **112**, 2110 (1996). https://doi.org/10.1086/118167. arXiv:astro-ph/9608108

R. Leaman, K.A. Venn, A.M. Brooks, G. Battaglia, A.A. Cole, R.A. Ibata, M.J. Irwin, A.W. McConnachie, J.T. Mendel, E. Starkenburg, E. Tolstoy, The comparative chemical evolution of an isolated dwarf galaxy: a VLT and Keck spectroscopic survey of WLM. Astrophys. J. **767**, 131 (2013). https://doi.org/10.1088/0004-637X/767/2/131. arXiv:1302.1879

H.S. Leavitt, 1777 variables in the Magellanic Clouds. Ann. Harv. Coll. Obs. **60**, 87–108 (1908)

H.S. Leavitt, E.C. Pickering, Periods of 25 variable stars in the Small Magellanic Cloud. Harv. Coll. Obs. Circ. **173**, 1–3 (1912)

C.H. Lee, J. Koppenhoefer, S. Seitz, R. Bender, A. Riffeser, M. Kodric, U. Hopp, J. Snigula, C. Gössl, R.P. Kudritzki, W. Burgett, K. Chambers, K. Hodapp, N. Kaiser, C. Waters, Properties of M31. V. 298 eclipsing binaries from PAndromeda. Astrophys. J. **797**, 22 (2014). https://doi.org/10.1088/0004-637X/797/1/22. arXiv:1411.1115

M.G. Lee, I.S. Jang, The distance to M101 hosting type Ia supernova 2011fe based on the tip of the red giant branch. Astrophys. J. Lett. **760**, L14 (2012). https://doi.org/10.1088/2041-8205/760/1/L14. arXiv:1210.6040

M.G. Lee, W.L. Freedman, B.F. Madore, The tip of the red giant branch as a distance indicator for resolved galaxies. Astrophys. J. **417**, 553 (1993). https://doi.org/10.1086/173334

B. Lemasle, V. Kovtyukh, G. Bono, P. François, I. Saviane, I. Yegorova, K. Genovali, L. Inno, G. Galazutdinov, R. da Silva, Type II Cepheids in the Milky Way disc. Chemical composition of two new W Virginis stars: DD Vel and HQ Car. Astron. Astrophys. **579**, A47 (2015a). https://doi.org/10.1051/0004-6361/201425541. arXiv:1505.02789

B. Lemasle, V. Kovtyukh, G. Bono, P. François, I. Saviane, I. Yegorova, K. Genovali, L. Inno, G. Galazutdinov, R. da Silva, Type II Cepheids in the Milky Way disc. chemical composition of two new W Virginis stars: DD Vel and HQ Car. Astron. Astrophys. **579**, A47 (2015b). https://doi.org/10.1051/0004-6361/201425541. arXiv:1505.02789

L. Lindegren, E. Hog, F. van Leeuwen, C.A. Murray, D.W. Evans, M.J. Penston, M.A.C. Perryman, C. Petersen, N. Ramamani, M.A.J. Snijders, The NDAC HIPPARCOS data analysis consortium—overview of the reduction methods. Astron. Astrophys. **258**, 18–30 (1992)

L. Lindegren, U. Lammers, U. Bastian, J. Hernández, S. Klioner, D. Hobbs, A. Bombrun, D. Michalik, M. Ramos-Lerate, A. Butkevich, G. Comoretto, E. Joliet, B. Holl, A. Hutton, P. Parsons, H. Steidelmüller, U. Abbas, M. Altmann, A. Andrei, S. Anton, N. Bach, C. Barache, U. Becciani, J. Berthier, L. Bianchi, M. Biermann, S. Bouquillon, G. Bourda, T. Brüsemeister, B. Bucciarelli, D. Busonero, T. Carlucci, J. Castañeda, P. Charlot, M. Clotet, M. Crosta, M. Davidson, F. de Felice, R. Drimmel, C. Fabricius, A. Fienga, F. Figueras, E. Fraile, M. Gai, N. Garralda, R. Geyer, J.J. González-Vidal, R. Guerra, N.C. Hambly, M. Hauser, S. Jordan, M.G. Lattanzi, H. Lenhardt, S. Liao, W. Löffler, P.J. McMillan, F. Mignard, A. Mora, R. Morbidelli, J. Portell, A. Riva, M. Sarasso, I. Serraller, H. Siddiqui, R. Smart, A. Spagna, U. Stampa, I. Steele, F. Taris, J. Torra, W. van Reeven, A. Vecchiato, S. Zschocke, J. de Bruijne, G. Gracia, F. Raison, T. Lister, J. Marchant, R. Messineo, M. Soffel, J. Osorio, A. de Torres, W. O'Mullane, Gaia data release 1. Astrometry: one billion positions, two million proper motions and parallaxes. Astron. Astrophys. **595**, A4 (2016). https://doi.org/10.1051/0004-6361/201628714. arXiv:1609.04303

T. Liu, K.A. Janes, The luminosity scale of RR Lyrae stars with the Baade–Wesselink method. III—the absolute magnitudes of four RR Lyrae stars in the globular cluster M4. Astrophys. J. **360**, 561–576 (1990). https://doi.org/10.1086/169145

A.J. Longmore, J.A. Fernley, R.F. Jameson, M.R. Sherrington, J. Frank, VJHK observations of the RR Lyrae star VY Serpentis. Mon. Not. R. Astron. Soc. **216**, 873–882 (1985). https://doi.org/10.1093/mnras/216.4.873

A.J. Longmore, J.A. Fernley, R.F. Jameson, RR Lyrae stars in globular clusters—better distances from infrared measurements? Mon. Not. R. Astron. Soc. **220**, 279–287 (1986). https://doi.org/10.1093/mnras/220.2.279

X. Luri, A.E. Gomez, J. Torra, F. Figueras, M.O. Mennessier, The LMC distance modulus from HIPPARCOS RR Lyrae and classical Cepheid data. Astron. Astrophys. **335**, L81–L84 (1998). arXiv:astro-ph/9805215

T. Maas, H. Van Winckel, C. Waelkens, RU Cen and SX Cen: two strongly depleted RV Tauri stars in binary systems. The RV Tauri photometric b phenomenon and binarity. Astron. Astrophys. **386**, 504–516 (2002). https://doi.org/10.1051/0004-6361:20020209

T. Maas, S. Giridhar, D.L. Lambert, The chemical compositions of the type II Cepheids—the BL Herculis and W Virginis variables. Astrophys. J. **666**, 378–392 (2007). https://doi.org/10.1086/520081. arXiv:0706.2029

A.D. Mackey, G.F. Gilmore, Photometry of Magellanic Cloud clusters with the advanced camera for surveys—I. The old Large Magellanic Cloud clusters NGC 1928, 1939 and reticulum. Mon. Not. R. Astron. Soc. **352**, 153–167 (2004). https://doi.org/10.1111/j.1365-2966.2004.07908.x. arXiv:astro-ph/0404247

L.M. Macri, D. Calzetti, W.L. Freedman, B.K. Gibson, J.A. Graham, J.P. Huchra, S.M.G. Hughes, B.F. Madore, J.R. Mould, S.E. Persson, P.B. Stetson, NICMOS observations of extragalactic Cepheids. I. Photometry database and a test of the standard extinction law. Astrophys. J. **549**, 721–744 (2001). https://doi.org/10.1086/319465. arXiv:astro-ph/0102125

B.F. Madore, The period-luminosity relation. IV—intrinsic relations and reddenings for the Large Magellanic Cloud Cepheids. Astrophys. J. **253**, 575–579 (1982). https://doi.org/10.1086/159659

B.F. Madore, W.L. Freedman, The tip of the red giant branch as a distance indicator for resolved galaxies. 2: computer simulations. Astron. J. **109**, 1645–1652 (1995). https://doi.org/10.1086/117391

B.F. Madore, V. Mager, W.L. Freedman, Sharpening the tip of the red giant branch. Astrophys. J. **690**, 389–393 (2009). https://doi.org/10.1088/0004-637X/690/1/389. arXiv:0809.2598

B.F. Madore, D. Hoffman, W.L. Freedman, J.A. Kollmeier, A. Monson, S.E. Persson, J.A. Rich Jr., V. Scowcroft, M. Seibert, A preliminary calibration of the RR Lyrae period-luminosity relation at mid-infrared wavelengths: WISE data. Astrophys. J. **776**, 135 (2013). https://doi.org/10.1088/0004-637X/776/2/135. arXiv:1308.3160

B.F. Madore, W.L. Freedman, D. Hatt, T.J. Hoyt, A.J. Monson, R.L. Beaton, J.A. Rich, I.S. Jang, M.G. Lee, V. Scowcroft, M. Seibert, The near-infrared tip of the red giant branch. I. A calibration in the isolated dwarf galaxy IC 1613. Astrophys. J. **858**, 11 (2018). https://doi.org/10.3847/1538-4357/aab7f4. arXiv:1803.01278

V.A. Mager, B.F. Madore, W.L. Freedman, Metallicity-corrected tip of the red giant branch distance to NGC 4258. Astrophys. J. **689**, 721–731 (2008). https://doi.org/10.1086/592563. arXiv:0808.2180

D. Majaess, D. Turner, D. Lane, Type II Cepheids as extragalactic distance candles. Acta Astron. **59**, 403–418 (2009). arXiv:0909.0181

S.R. Majewski, M.H. Siegel, R.J. Patterson, R.T. Rood, An internal second-parameter problem in the Sculptor dwarf spheroidal galaxy. Astrophys. J. Lett. **520**, L33–L36 (1999). https://doi.org/10.1086/312133. arXiv:astro-ph/9905238

S.R. Majewski, G. Zasowski, D.L. Nidever, Lifting the dusty veil with near- and mid-infrared photometry. I. Description and applications of the Rayleigh-jeans color excess method. Astrophys. J. **739**, 25 (2011). https://doi.org/10.1088/0004-637X/739/1/25. arXiv:1106.2542

S.R. Majewski, R.P. Schiavon, P.M. Frinchaboy, C. Allende Prieto, R. Barkhouser, D. Bizyaev, B. Blank, S. Brunner, A. Burton, R. Carrera, S.D. Chojnowski, K. Cunha, C. Epstein, G. Fitzgerald, A.E. García Pérez, F.R. Hearty, C. Henderson, J.A. Holtzman, J.A. Johnson, C.R. Lam, J.E. Lawler, P. Maseman, S. Mészáros, M. Nelson, D.C. Nguyen, D.L. Nidever, M. Pinsonneault, M. Shetrone, S. Smee, V.V. Smith, T. Stolberg, M.F. Skrutskie, E. Walker, J.C. Wilson, G. Zasowski, F. Anders, S. Basu, S. Beland, M.R. Blanton, J. Bovy, J.R. Brownstein, J. Carlberg, W. Chaplin, C. Chiappini, D.J. Eisenstein, Y. Elsworth, D. Feuillet, S.W. Fleming, J. Galbraith-Frew, R.A. García, D.A. García-Hernández, B.A. Gillespie, L. Girardi, J.E. Gunn, S. Hasselquist, M.R. Hayden, S. Hekker, I. Ivans, K. Kinemuchi, M. Klaene, S. Mahadevan, S. Mathur, B. Mosser, D. Muna, J.A. Munn, R.C. Nichol, R.W. O'Connell, J.K. Parejko, A.C. Robin, H. Rocha-Pinto, M. Schultheis, A.M. Serenelli, N. Shane, V. Silva Aguirre, J.S. Sobeck, B. Thompson, N.W. Troup, D.H. Weinberg, The Apache Point Observatory Galactic Evolution Experiment (APOGEE). Astron. J. **154**, 94 (2017). https://doi.org/10.3847/1538-3881/aa784d. arXiv:1509.05420

D. Makarov, L. Makarova, L. Rizzi, R.B. Tully, A.E. Dolphin, S. Sakai, E.J. Shaya, Tip of the red giant branch distances. I. Optimization of a maximum likelihood algorithm. Astron. J. **132**, 2729–2742 (2006). https://doi.org/10.1086/508925. arXiv:astro-ph/0603073

L. Malavolta, C. Sneden, G. Piotto, A.P. Milone, L.R. Bedin, V. Nascimbeni, Atmospheric parameters and metallicities for 2191 stars in the globular cluster M4. Astron. J. **147**, 25 (2014). https://doi.org/10.1088/0004-6256/147/2/25. arXiv:1311.2592

M. Marconi, M. Di Criscienzo, Updated pulsation models for BL Herculis stars. Astron. Astrophys. **467**, 223–227 (2007). https://doi.org/10.1051/0004-6361:20066646. arXiv:astro-ph/0701256

M. Marconi, F. Caputo, M. Di Criscienzo, M. Castellani, RR Lyrae stars in galactic globular clusters. II. A theoretical approach to variables in M3. Astrophys. J. **596**, 299–313 (2003). https://doi.org/10.1086/377641. arXiv:astro-ph/0306356

M. Marconi, I. Musella, G. Fiorentino, Cepheid pulsation models at varying metallicity and $\Delta Y/\Delta Z$. Astrophys. J. **632**, 590–610 (2005a). https://doi.org/10.1086/432790. arXiv:astro-ph/0506207

M. Marconi, T. Nordgren, G. Bono, G. Schnider, F. Caputo, Predicted and empirical radii of RR Lyrae stars. Astrophys. J. Lett. **623**, L133–L136 (2005b). https://doi.org/10.1086/430273. arXiv:astro-ph/0503382

M. Marconi, G. Bono, F. Caputo, A.M. Piersimoni, A. Pietrinferni, R.F. Stellingwerf, On the period distribution of cluster RR Lyrae stars to constrain their helium content: the case of ω Centauri. Astrophys. J. **738**, 111 (2011). https://doi.org/10.1088/0004-637X/738/1/111. arXiv:1106.4931

M. Marconi, G. Coppola, G. Bono, V. Braga, A. Pietrinferni, R. Buonanno, M. Castellani, I. Musella, V. Ripepi, R.F. Stellingwerf, On a new theoretical framework for RR Lyrae stars. I. The metallicity dependence. Astrophys. J. **808**, 50 (2015). https://doi.org/10.1088/0004-637X/808/1/50. arXiv:1505.02531

M. Marconi, G. Coppola, G. Bono, V. Braga, A. Pietrinferni, Helium abundance effects on RR Lyrae pulsation properties. Commun. Konkoly Obs. Hung. Acad. Sci. **105**, 125–a128 (2016)

A.F. Marino, S. Villanova, G. Piotto, A.P. Milone, Y. Momany, L.R. Bedin, A.M. Medling, Spectroscopic and photometric evidence of two stellar populations in the Galactic globular cluster NGC 6121 (M 4). Astron. Astrophys. **490**, 625–640 (2008). https://doi.org/10.1051/0004-6361:200810389. arXiv:0808.1414

S.L. Martell, S. Sharma, S. Buder, L. Duong, K.J. Schlesinger, J. Simpson, K. Lind, M. Ness, J.P. Marshall, M. Asplund, J. Bland-Hawthorn, A.R. Casey, G. De Silva, K.C. Freeman, J. Kos, J. Lin, D.B. Zucker, T. Zwitter, B. Anguiano, C. Bacigalupo, D. Carollo, L. Casagrande, G.S. Da Costa, J. Horner, D. Huber, E.A. Hyde, P.R. Kafle, G.F. Lewis, D. Nataf, C.A. Navin, D. Stello, C.G. Tinney, F.G. Watson, R. Wittenmyer, The GALAH survey: observational overview and Gaia DR1 companion. Mon. Not. R. Astron. Soc. **465**, 3203–3219 (2017). https://doi.org/10.1093/mnras/stw2835. arXiv:1609.02822

C.E. Martínez-Vázquez, M. Monelli, G. Bono, P.B. Stetson, I. Ferraro, E.J. Bernard, C. Gallart, G. Fiorentino, G. Iannicola, A. Udalski, Variable stars in Local Group galaxies—I. Tracing the early chemical enrichment and radial gradients in the Sculptor dSph with RR Lyrae stars. Mon. Not. R. Astron. Soc. **454**, 1509–1516 (2015). https://doi.org/10.1093/mnras/stv2014. arXiv:1508.06942

C.E. Martínez-Vázquez, M. Monelli, C. Gallart, G. Bono, E.J. Bernard, P.B. Stetson, I. Ferraro, A.R. Walker, M. Dall'Ora, G. Fiorentino, G. Iannicola, Probing the early chemical evolution of the Sculptor dSph with purely old stellar tracers. Mon. Not. R. Astron. Soc. **461**, L41–L45 (2016a). https://doi.org/10.1093/mnrasl/slw093. arXiv:1605.02768

C.E. Martínez-Vázquez, P.B. Stetson, M. Monelli, E.J. Bernard, G. Fiorentino, C. Gallart, G. Bono, S. Cassisi, M. Dall'Ora, I. Ferraro, G. Iannicola, A.R. Walker, Variable stars in Local Group galaxies—II. Sculptor dSph. Mon. Not. R. Astron. Soc. **462**, 4349–4370 (2016b). https://doi.org/10.1093/mnras/stw1895. arXiv:1607.08518

C.E. Martínez-Vázquez, M. Monelli, E.J. Bernard, C. Gallart, P.B. Stetson, E.D. Skillman, G. Bono, S. Cassisi, G. Fiorentino, K.B.W. McQuinn, A.A. Cole, A.W. McConnachie, N.F. Martin, A.E. Dolphin, M. Boylan-Kolchin, A. Aparicio, S.L. Hidalgo, D.R. Weisz, The ISLAnds project. III. Variable stars in six Andromeda dwarf spheroidal galaxies. Astrophys. J. **850**, 137 (2017). https://doi.org/10.3847/1538-4357/aa9381. arXiv:1710.09038

N. Matsunaga, H. Fukushi, Y. Nakada, T. Tanabé, M.W. Feast, J.W. Menzies, Y. Ita, S. Nishiyama, D. Baba, T. Naoi, H. Nakaya, T. Kawadu, A. Ishihara, D. Kato, The period-luminosity relation for type II Cepheids in globular clusters. Mon. Not. R. Astron. Soc. **370**, 1979–1990 (2006). https://doi.org/10.1111/j.1365-2966.2006.10620.x. arXiv:astro-ph/0606609

N. Matsunaga, M.W. Feast, T. Kawadu, S. Nishiyama, T. Nagayama, T. Nagata, M. Tamura, G. Bono, N. Kobayashi, Cepheids and other short-period variables near the Galactic centre. Mon. Not. R. Astron. Soc. **429**, 385–397 (2013). https://doi.org/10.1093/mnras/sts343. arXiv:1211.0151

R. McGonegal, C.W. McAlary, B.F. Madore, R.A. McLaren, The Cepheid distance scale—a new application for infrared photometry. Astrophys. J. Lett. **257**, L33–L36 (1982). https://doi.org/10.1086/183803

D.H. McNamara, Period-luminosity relations of population II Cepheids. Astron. J. **109**, 2134 (1995). https://doi.org/10.1086/117438

D.H. McNamara, The absolute magnitudes of the RR Lyrae stars. Publ. Astron. Soc. Pac. **109**, 857–867 (1997). https://doi.org/10.1086/133956

A. McWilliam (ed.), *RR Lyrae Stars, Metal-Poor Stars, and the Galaxy*, vol. 5 (2011). arXiv:1109.1324

B. Méndez, M. Davis, J. Moustakas, J. Newman, B.F. Madore, W.L. Freedman, Deviations from the local Hubble flow. I. The tip of the red giant branch as a distance indicator. Astron. J. **124**, 213–233 (2002). https://doi.org/10.1086/341168. arXiv:astro-ph/0204192

D. Michalik, L. Lindegren, D. Hobbs, The Tycho-Gaia astrometric solution How to get 2.5 million parallaxes with less than one year of Gaia data. Astron. Astrophys. **574**, A115 (2015). https://doi.org/10.1051/0004-6361/201425310

M. Miyoshi, J. Moran, J. Herrnstein, L. Greenhill, N. Nakai, P. Diamond, M. Inoue, Evidence for a black hole from high rotation velocities in a sub-parsec region of NGC4258. Nature **373**, 127–129 (1995). https://doi.org/10.1038/373127a0

A.J. Monson, R.L. Beaton, V. Scowcroft, W.L. Freedman, B.F. Madore, J.A. Rich, M. Seibert, J.A. Kollmeier, G. Clementini, Standard galactic field RR Lyrae. I. Optical to mid-infrared phased photometry. Astron. J. **153**, 96 (2017). https://doi.org/10.3847/1538-3881/153/3/96. arXiv:1703.01520

M.I. Moretti, G. Clementini, T. Muraveva, V. Ripepi, J.B. Marquette, M.R.L. Cioni, M. Marconi, L. Girardi, S. Rubele, P. Tisserand, R. de Grijs, M.A.T. Groenewegen, R. Guandalini, V.D. Ivanov, J.T. van Loon, The VMC survey—X. Cepheids, RR Lyrae stars and binaries as probes of the Magellanic System's structure. Mon. Not. R. Astron. Soc. **437**, 2702–2719 (2014). https://doi.org/10.1093/mnras/stt2081. arXiv:1310.6849

M. Mouhcine, H.C. Ferguson, R.M. Rich, T.M. Brown, T.E. Smith, Halos of spiral galaxies. I. The tip of the red giant branch as a distance indicator. Astrophys. J. **633**, 810–820 (2005). https://doi.org/10.1086/468177. arXiv:astro-ph/0510253

J. Mould, S. Sakai, The extragalactic distance scale without Cepheids. Astrophys. J. Lett. **686**, L75 (2008). https://doi.org/10.1086/592964

T. Muraveva, M. Palmer, G. Clementini, X. Luri, M.R.L. Cioni, M.I. Moretti, M. Marconi, V. Ripepi, S. Rubele, New near-infrared period-luminosity–metallicity relations for RR Lyrae stars and the outlook for Gaia. Astrophys. J. **807**, 127 (2015). https://doi.org/10.1088/0004-637X/807/2/127. arXiv:1505.06001

T. Muraveva, H.E. Delgado, G. Clementini, L.M. Sarro, A. Garofalo, RR Lyrae stars as standard candles in the Gaia Data Release 2 Era (2018a). arXiv:1805.08742

T. Muraveva, A. Garofalo, V. Scowcroft, G. Clementini, W.L. Freedman, B.F. Madore, A.J. Monson, The Carnegie RR Lyrae program: mid-infrared period-luminosity relations of RR Lyrae stars in reticulum. Mon. Not. R. Astron. Soc. (2018b). https://doi.org/10.1093/mnras/sty1959. arXiv:1807.07493

T. Muraveva, S. Subramanian, G. Clementini, M.R.L. Cioni, M. Palmer, J.T. van Loon, M.I. Moretti, R. de Grijs, R. Molinaro, V. Ripepi, M. Marconi, J. Emerson, V.D. Ivanov, The VMC survey—XXVI. Structure of the Small Magellanic Cloud from RR Lyrae stars. Mon. Not. R. Astron. Soc. **473**, 3131–3146 (2018c). https://doi.org/10.1093/mnras/stx2514. arXiv:1709.09064

J.R. Neeley, M. Marengo, G. Bono, V.F. Braga, M. Dall'Ora, P.B. Stetson, R. Buonanno, I. Ferraro, W.L. Freedman, G. Iannicola, B.F. Madore, N. Matsunaga, A. Monson, S.E. Persson, V. Scowcroft, M. Seibert, On the distance of the globular cluster M4 (NGC 6121) using RR Lyrae stars. II. Mid-infrared period-luminosity relations. Astrophys. J. **808**, 11 (2015). https://doi.org/10.1088/0004-637X/808/1/11. arXiv:1505.07858

J.R. Neeley, M. Marengo, G. Bono, V.F. Braga, M. Dall'Ora, D. Magurno, M. Marconi, N. Trueba, E. Tognelli, P.G. Prada Moroni, R.L. Beaton, W.L. Freedman, B.F. Madore, A.J. Monson, V. Scowcroft, M. Seibert, P.B. Stetson, On a new theoretical framework for RR Lyrae stars. II. Mid-infrared period-luminosity–metallicity relations. Astrophys. J. **841**, 84 (2017). https://doi.org/10.3847/1538-4357/aa713d. arXiv:1705.01970

C.C. Ngeow, S.M. Kanbur, S. Nikolaev, J. Buonaccorsi, K.H. Cook, D.L. Welch, Further empirical evidence for the non-linearity of the period-luminosity relations as seen in the Large Magellanic Cloud Cepheids. Mon. Not. R. Astron. Soc. **363**, 831–846 (2005). https://doi.org/10.1111/j.1365-2966.2005.09477.x. arXiv:astro-ph/0507601

P.T. Oosterhoff, Some remarks on the variable stars in globular clusters. Observatory **62**, 104–109 (1939)

A. Peimbert, M. Peimbert, On the O/H, Mg/H, Si/H, and Fe/H gas and dust abundance ratios in galactic and extragalactic H II regions. Astrophys. J. **724**, 791–798 (2010). https://doi.org/10.1088/0004-637X/724/1/791. arXiv:1006.0692

M.A.C. Perryman, L. Lindegren, J. Kovalevsky, E. Hoeg, U. Bastian, P.L. Bernacca, M. Crézé, F. Donati, M. Grenon, M. Grewing, F. van Leeuwen, H. van der Marel, F. Mignard, C.A. Murray, R.S. Le Poole, H. Schrijver, C. Turon, F. Arenou, M. Froeschlé, C.S. Petersen, The HIPPARCOS catalogue. Astron. Astrophys. **323**, L49–L52 (1997)

M.A.C. Perryman, K.S. de Boer, G. Gilmore, E. Høg, M.G. Lattanzi, L. Lindegren, X. Luri, F. Mignard, O. Pace, P.T. de Zeeuw, GAIA: composition, formation and evolution of the Galaxy. Astron. Astrophys. **369**, 339–363 (2001). https://doi.org/10.1051/0004-6361:20010085. arXiv:astro-ph/0101235

E.C. Pickering, Variable star in cluster G.C. 3636. Astron. Nachr. **123**, 207 (1889). https://doi.org/10.1002/asna.18901231306

E.C. Pickering, H.R. Colson, W.P. Fleming, L.D. Wells, Sixty-four new variable stars. Astrophys. J. **13**, 226 (1901). https://doi.org/10.1086/140808

A. Pietrinferni, S. Cassisi, M. Salaris, F. Castelli, A large stellar evolution database for population synthesis studies. I. Scaled solar models and isochrones. Astrophys. J. **612**, 168–190 (2004). https://doi.org/10.1086/422498. arXiv:astro-ph/0405193

G. Pietrzyński, W. Gieren, I. Soszyński, F. Bresolin, R.P. Kudritzki, M. Dall'Ora, J. Storm, G. Bono, The Araucaria project: the distance to the Local Group galaxy IC 1613 from near-infrared photometry of Cepheid variables. Astrophys. J. **642**, 216–224 (2006). https://doi.org/10.1086/500926. arXiv:astro-ph/0601309

G. Pietrzyński, W. Gieren, O. Szewczyk, A. Walker, L. Rizzi, F. Bresolin, R.P. Kudritzki, K. Nalewajko, J. Storm, M. Dall'Ora, V. Ivanov, The Araucaria project: the distance to the Sculptor dwarf spheroidal galaxy from infrared photometry of RR Lyrae stars. Astron. J. **135**, 1993–1997 (2008). https://doi.org/10.1088/0004-6256/135/6/1993. arXiv:0804.0347

G. Pietrzyński, M. Górski, W. Gieren, D. Laney, A. Udalski, A. Ciechanowska, The Araucaria project. Population effects on the V- and I-band magnitudes of red clump stars. Astron. J. **140**, 1038–1042 (2010). https://doi.org/10.1088/0004-6256/140/4/1038. arXiv:1008.1259

G. Pietrzyński, D. Graczyk, W. Gieren, I.B. Thompson, B. Pilecki, A. Udalski, I. Soszyński, S. Kozłowski, P. Konorski, K. Suchomska, G. Bono, P.G.P. Moroni, S. Villanova, N. Nardetto, F. Bresolin, R.P. Kudritzki, J. Storm, A. Gallenne, R. Smolec, D. Minniti, M. Kubiak, M.K. Szymański, R. Poleski, Ł. Wyrzykowski, K. Ulaczyk, P. Pietrukowicz, M. Górski, P. Karczmarek, An eclipsing-binary distance to the Large Magellanic Cloud accurate to two per cent. Nature **495**, 76–79 (2013). https://doi.org/10.1038/nature11878. arXiv:1303.2063

B. Pilecki, W. Gieren, R. Smolec, G. Pietrzyński, I.B. Thompson, R.I. Anderson, G. Bono, I. Soszyński, P. Kervella, N. Nardetto, M. Taormina, K. Stępień, P. Wielgórski, Mass and p-factor of the type II Cepheid OGLE-LMC-T2CEP-098 in a binary system. Astrophys. J. **842**, 110 (2017). https://doi.org/10.3847/1538-4357/aa6ff7. arXiv:1704.07782

B. Pilecki, W. Gieren, G. Pietrzyński, I.B. Thompson, R. Smolec, D. Graczyk, M. Taormina, A. Udalski, J. Storm, N. Nardetto, A. Gallenne, P. Kervella, I. Soszyński, M. Górski, P. Wielgórski, K. Suchomska, P. Karczmarek, B. Zgirski, The Araucaria project: high-precision Cepheid astrophysics from the analysis of variables in double-lined eclipsing binaries. Astrophys. J. **862**, 43 (2018). https://doi.org/10.3847/1538-4357/aacb32. arXiv:1806.01391

P. Popowski, A. Gould, Mathematics of Statistical Parallax and the Local Distance Scale. ArXiv Astrophysics e-prints (1997). arXiv:astro-ph/9703140

P. Popowski, A. Gould, Systematics of RR Lyrae statistical parallax. I. Mathematics. Astrophys. J. **506**, 259–270 (1998a). https://doi.org/10.1086/306216

P. Popowski, A. Gould, Systematics of RR Lyrae statistical parallax. II. Proper motions and radial velocities. Astrophys. J. **506**, 271–280 (1998b). https://doi.org/10.1086/306239

G.W. Preston, A spectroscopic study of the RR Lyrae stars. Astrophys. J. **130**, 507 (1959). https://doi.org/10.1086/146743

B.J. Pritzl, H.A. Smith, P.B. Stetson, M. Catelan, A.V. Sweigart, A.C. Layden, R.M. Rich, Hubble Space Telescope snapshot study of variable stars in globular clusters: the inner region of NGC 6441. Astron. J. **126**, 1381–1401 (2003). https://doi.org/10.1086/377024. arXiv:astro-ph/0305339

T. Prusti, J.H.J. Bruijne, A.G.A. Brown, A. Vallenari, C. Babusiaux, C.A.L. Bailer-Jones, U. Bastian, M. Biermann, D.W. Evans et al. (Gaia Collaboration), The Gaia mission. Astron. Astrophys. **595**, A1 (2016). https://doi.org/10.1051/0004-6361/201629272. arXiv:1609.04153

K. Rabidoux, H.A. Smith, B.J. Pritzl, W. Osborn, C. Kuehn, J. Randall, R. Lustig, K. Wells, L. Taylor, N. De Lee, K. Kinemuchi, A. LaCluyzé, D. Hartley, C. Greenwood, M. Ingber, M. Ireland, E. Pellegrini, M. Anderson, G. Purdum, J. Lacy, M. Curtis, J. Smolinski, S. Danford, Light curves and period changes of type II Cepheids in the globular clusters M3 and M5. Astron. J. **139**, 2300–2307 (2010). https://doi.org/10.1088/0004-6256/139/6/2300. arXiv:1003.5924

I.N. Reid, The HR diagram and the galactic distance scale after HIPPARCOS. Annu. Rev. Astron. Astrophys. **37**, 191–237 (1999). https://doi.org/10.1146/annurev.astro.37.1.191

S.C. Rey, Y.W. Lee, J.M. Joo, A. Walker, S. Baird, CCD photometry of the globular cluster ω Centauri. I. Metallicity of RR Lyrae stars from CABY photometry. Astron. J. **119**, 1824–1838 (2000). https://doi.org/10.1086/301304. arXiv:astro-ph/0001101

J.A. Rich, S.E. Persson, W.L. Freedman, B.F. Madore, A.J. Monson, V. Scowcroft, M. Seibert, A new Cepheid distance measurement and method for NGC 6822. Astrophys. J. **794**, 107 (2014). https://doi.org/10.1088/0004-637X/794/2/107. arXiv:1409.6830

G.R. Ricker, J.N. Winn, R. Vanderspek, D.W. Latham, B. GÁ, J.L. Bean, Z.K. Berta-Thompson, T.M. Brown, L. Buchhave, N.R. Butler, R.P. Butler, W.J. Chaplin, D. Charbonneau, J. Christensen-Dalsgaard, M. Clampin, D. Deming, J. Doty, N. De Lee, C. Dressing, E.W. Dunham, M. Endl, F. Fressin, J. Ge, T. Henning, M.J. Holman, A.W. Howard, S. Ida, J.M. Jenkins, G. Jernigan, J.A. Johnson, L. Kaltenegger, N. Kawai, H. Kjeldsen, G. Laughlin, A.M. Levine, D. Lin, J.J. Lissauer, P. MacQueen, G. Marcy, P.R. McCullough, T.D. Morton, N. Narita, M. Paegert, E. Palle, F. Pepe, J. Pepper, A. Quirrenbach, S.A. Rinehart, D. Sasselov, B. Sato, S. Seager, A. Sozzetti, K.G. Stassun, P. Sullivan, A. Szentgyorgyi, G. Torres, S. Udry, J. Villasenor, Transiting Exoplanet Survey Satellite (TESS). J. Astron. Telesc. Instrum. Syst. **1**(1), 014003 (2015). https://doi.org/10.1117/1.JATIS.1.1.014003

A.G. Riess, An Independent Determination of WFC3-IR Zeropoints and Count Rate Non-Linearity from 2MASS Asterisms Tech. rep. (2011)

A.G. Riess, L. Macri, S. Casertano, H. Lampeitl, H.C. Ferguson, A.V. Filippenko, S.W. Jha, W. Li, R. Chornock, A 3% solution: determination of the Hubble constant with the Hubble Space Telescope and

Wide Field Camera 3. Astrophys. J. **730**, 119 (2011). https://doi.org/10.1088/0004-637X/730/2/119. arXiv:1103.2976

A.G. Riess, L. Macri, S. Casertano, H. Lampeitl, H.C. Ferguson, A.V. Filippenko, S.W. Jha, W. Li, R. Chornock, J.M. Silverman, Erratum: "A 3% solution: determination of the Hubble constant with the Hubble Space Telescope and Wide Field Camera 3". Astrophys. J. **752**, 76 (2012). https://doi.org/10.1088/0004-637X/752/1/76

A.G. Riess, L.M. Macri, S.L. Hoffmann, D. Scolnic, S. Casertano, A.V. Filippenko, B.E. Tucker, M.J. Reid, D.O. Jones, J.M. Silverman, R. Chornock, P. Challis, W. Yuan, P.J. Brown, R.J. Foley, A 2.4% determination of the local value of the Hubble constant. Astrophys. J. **826**, 56 (2016). https://doi.org/10.3847/0004-637X/826/1/56

V. Ripepi, M.I. Moretti, G. Clementini, M. Marconi, M.R. Cioni, J.B. Marquette, P. Tisserand, Preliminary results for RR Lyrae stars and Classical Cepheids from the Vista Magellanic Cloud (VMC) survey. Astrophys. Space Sci. **341**, 51–56 (2012a). https://doi.org/10.1007/s10509-012-1021-x. arXiv:1202.5863

V. Ripepi, M.I. Moretti, M. Marconi, G. Clementini, M.R.L. Cioni, J.B. Marquette, L. Girardi, S. Rubele, M.A.T. Groenewegen, R. de Grijs, B.K. Gibson, J.M. Oliveira, J.T. van Loon, J.P. Emerson, The VMC survey—V. First results for classical Cepheids. Mon. Not. R. Astron. Soc. **424**, 1807–1816 (2012b). https://doi.org/10.1111/j.1365-2966.2012.21274.x. arXiv:1204.2273

V. Ripepi, M.I. Moretti, M. Marconi, G. Clementini, M.R.L. Cioni, R. de Grijs, J.P. Emerson, M.A.T. Groenewegen, V.D. Ivanov, T. Muraveva, A.E. Piatti, S. Subramanian, The VMC survey—XIII. Type II Cepheids in the Large Magellanic Cloud. Mon. Not. R. Astron. Soc. **446**, 3034–3061 (2015). https://doi.org/10.1093/mnras/stu2260. arXiv:1410.7817

A. Ritter, Untersuchungen über die Höhe der Atmosphäre und die Constitution gasförmiger Weltkörper. Ann. Phys. **244**, 157–183 (1879). https://doi.org/10.1002/andp.18792440910

L. Rizzi PhD thesis, University of Padova, Italy (2002)

L. Rizzi, R.B. Tully, D. Makarov, L. Makarova, A.E. Dolphin, S. Sakai, E.J. Shaya, Tip of the red giant branch distances. II. Zero-point calibration. Astrophys. J. **661**, 815–829 (2007). https://doi.org/10.1086/516566. arXiv:astro-ph/0701518

P. Rosenfield, P. Marigo, L. Girardi, J.J. Dalcanton, A. Bressan, M. Gullieuszik, D. Weisz, B.F. Williams, A. Dolphin, B. Aringer, Evolution of thermally pulsing asymptotic giant branch stars. IV. Constraining mass loss and lifetimes of low mass, low metallicity AGB stars. Astrophys. J. **790**, 22 (2014). https://doi.org/10.1088/0004-637X/790/1/22. arXiv:1406.0676

P. Rosenfield, P. Marigo, L. Girardi, J.J. Dalcanton, A. Bressan, B.F. Williams, A. Dolphin, Evolution of thermally pulsing asymptotic giant branch stars. V. Constraining the mass loss and lifetimes of intermediate-mass, low-metallicity AGB stars. Astrophys. J. **822**, 73 (2016). https://doi.org/10.3847/0004-637X/822/2/73. arXiv:1603.05283

M. Rozyczka, J. Kaluzny, I.B. Thompson, A. Dotter, W. Pych, W. Narloch, The Clusters AgeS Experiment (CASE). Analysis of the detached eclipsing binary V15 in the metal-rich open cluster NGC 6253. Acta Astron. **64**, 233–246 (2014). arXiv:1410.1324

S.C. Russell, RV Tau variables in globular clusters: clues on their nature? Publ. Astron. Soc. Aust. **15**, 189–193 (1998). https://doi.org/10.1071/AS98189

S. Sakai, B.F. Madore, W.L. Freedman, Tip of the red giant branch distances to galaxies. III. The Dwarf Galaxy Sextans A. Astrophys. J. **461**, 713 (1996). https://doi.org/10.1086/177096

S. Sakai, L. Ferrarese, R.C. Kennicutt Jr., A. Saha, The effect of metallicity on Cepheid-based distances. Astrophys. J. **608**, 42–61 (2004). https://doi.org/10.1086/386540. arXiv:astro-ph/0402499

M. Salaris, S. Cassisi, The 'tip' of the red giant branch as a distance indicator: results from evolutionary models. Mon. Not. R. Astron. Soc. **289**, 406–414 (1997). https://doi.org/10.1093/mnras/289.2.406. arXiv:astro-ph/9703186

M. Salaris, S. Cassisi, A new analysis of the red giant branch 'tip' distance scale and the value of the Hubble constant. Mon. Not. R. Astron. Soc. **298**, 166–178 (1998). https://doi.org/10.1046/j.1365-8711.1998.01598.x. arXiv:astro-ph/9803103

M. Salaris, S. Cassisi, *Evolution of Stars and Stellar Populations* (2005)

M. Salaris, L. Girardi, Tip of the red giant branch distances to galaxies with composite stellar populations. Mon. Not. R. Astron. Soc. **357**, 669–678 (2005). https://doi.org/10.1111/j.1365-2966.2005.08689.x. arXiv:astro-ph/0412156

M. Salaris, S. Cassisi, A. Weiss, Red giant branch stars: the theoretical framework. Publ. Astron. Soc. Pac. **114**, 375–402 (2002). https://doi.org/10.1086/342498. arXiv:astro-ph/0201387

M. Salaris, T. de Boer, E. Tolstoy, G. Fiorentino, S. Cassisi, The horizontal branch of the Sculptor dwarf galaxy. Astron. Astrophys. **559**, A57 (2013). https://doi.org/10.1051/0004-6361/201322501. arXiv:1309.4353

N.N. Samus, E.V. Kazarovets, O.V. Durlevich, N.N. Kireeva, E.N. Pastukhova, General catalogue of variable stars: version GCVS 5.1. Astron. Rep. **61**, 80–88 (2017). https://doi.org/10.1134/S1063772917010085

A. Sandage, The color–magnitude diagrams of galactic and globular clusters and their interpretation as age groups. Ric. Astron. **5**, 41 (1958)

A. Sandage, The distance of the local-group galaxy IC 1613 obtained from Baade's work on its stellar content. Astrophys. J. **166**, 13 (1971). https://doi.org/10.1086/150939

A. Sandage, The Oosterhoff period groups and the age of globular clusters. III—the age of the globular cluster system. Astrophys. J. **252**, 553–581 (1982). https://doi.org/10.1086/159582

A. Sandage, *The Mount Wilson Observatory* (2004)

A. Sandage, On the predicted and observed color boundaries of the RR Lyrae instability strip as a function of metallicity. Astron. J. **131**, 1750–1765 (2006). https://doi.org/10.1086/500012. arXiv:astro-ph/0508057

A. Sandage, B. Katem, On the intrinsic widths of the subgiant and horizontal branch sequences in the globular cluster M3. Astron. J. **87**, 537–554 (1982). https://doi.org/10.1086/113127

A. Sandage, G.A. Tammann, Absolute magnitude calibrations of population I and II Cepheids and other pulsating variables in the instability strip of the Hertzsprung–Russell diagram. Annu. Rev. Astron. Astrophys. **44**, 93–140 (2006). https://doi.org/10.1146/annurev.astro.43.072103.150612

A. Sandage, G. Wallerstein, Color–magnitude diagram for disk globular cluster NGC 6356 compared with Halo clusters. Astrophys. J. **131**, 598 (1960). https://doi.org/10.1086/146872

A. Sandage, R.L. Beaton, S.R. Majewski, Comparison of hipparcos trigonometric and mount Wilson spectroscopic parallaxes for 90 subgiants that defined the class in 1935. Publ. Astron. Soc. Pac. **064**, 202 (2016). https://doi.org/10.1088/1538-3873/128/964/064202. arXiv:1511.05930

A. Sarajedini, M.K. Barker, D. Geisler, P. Harding, R. Schommer, RR Lyrae variables in M33. I. Evidence for a field halo population. Astron. J. **132**, 1361–1371 (2006). https://doi.org/10.1086/506152. arXiv:astro-ph/0605580

A. Sarajedini, C.L. Mancone, T.R. Lauer, A. Dressler, W. Freedman, S.C. Trager, C. Grillmair, K.J. Mighell, RR Lyrae variables in two fields in the spheroid of M31. Astron. J. **138**, 184–195 (2009). https://doi.org/10.1088/0004-6256/138/1/184. arXiv:0904.4290

I. Saviane, J.E. Hibbard, R.M. Rich, The stellar content of the southern tail of NGC 4038/4039 and a revised distance. Astron. J. **127**, 660–678 (2004). https://doi.org/10.1086/381059. arXiv:astro-ph/0311200

I. Saviane, Y. Momany, G.S. da Costa, R.M. Rich, J.E. Hibbard, A new red giant-based distance modulus of 13.3 Mpc to the Antennae Galaxies and its consequences. Astrophys. J. **678**, 179–186 (2008). https://doi.org/10.1086/533408. arXiv:0802.1045

P.L. Schechter, M. Mateo, A. Saha, DOPHOT, a CCD photometry program: description and tests. Publ. Astron. Soc. Pac. **105**, 1342–1353 (1993). https://doi.org/10.1086/133316

E.F. Schlafly, D.P. Finkbeiner, Measuring reddening with Sloan Digital Sky Survey stellar spectra and recalibrating SFD. Astrophys. J. **737**, 103 (2011). https://doi.org/10.1088/0004-637X/737/2/103. arXiv:1012.4804

D.J. Schlegel, D.P. Finkbeiner, M. Davis, Maps of dust infrared emission for use in estimation of reddening and cosmic microwave background radiation foregrounds. Astrophys. J. **500**, 525–553 (1998). https://doi.org/10.1086/305772. arXiv:astro-ph/9710327

M. Schwarzschild, On the variables in Messier 3. Harv. Coll. Obs. Circ. **437**, 1–12 (1940)

M. Schwarzschild, R. Härm, On the evolutionary phase of Cepheids in globular clusters. Astrophys. J. **160**, 341 (1970). https://doi.org/10.1086/150433

F. Schweizer, C.R. Burns, B.F. Madore, V.A. Mager, M.M. Phillips, W.L. Freedman, L. Boldt, C. Contreras, G. Folatelli, S. González, M. Hamuy, W. Krzeminski, N.I. Morrell, S.E. Persson, M.R. Roth, M.D. Stritzinger, A new distance to the Antennae galaxies (NGC 4038/39) based on the type Ia supernova 2007sr. Astron. J. **136**, 1482–1489 (2008). https://doi.org/10.1088/0004-6256/136/4/1482. arXiv:0807.3955

V. Scowcroft, W.L. Freedman, B.F. Madore, A.J. Monson, S.E. Persson, M. Seibert, J.R. Rigby, J. Melbourne, The Carnegie Hubble program: the infrared leavitt law in IC 1613. Astrophys. J. **773**, 106 (2013). https://doi.org/10.1088/0004-637X/773/2/106. arXiv:1306.5802

A. Serenelli, A. Weiss, S. Cassisi, M. Salaris, A. Pietrinferni, The brightness of the red giant branch tip. Theoretical framework, a set of reference models, and predicted observables. Astron. Astrophys. **606**, A33 (2017). https://doi.org/10.1051/0004-6361/201731004. arXiv:1706.09910

B. Sesar, Template RR Lyrae Hα, Hβ, and Hγ velocity curves. Astron. J. **144**, 114 (2012). https://doi.org/10.1088/0004-6256/144/4/114. arXiv:1208.1997

B. Sesar, M. Fouesneau, A.M. Price-Whelan, C.A.L. Bailer-Jones, A. Gould, H.W. Rix, A probabilistic approach to fitting period-luminosity relations and validating Gaia parallaxes. Astrophys. J. **838**, 107 (2017). https://doi.org/10.3847/1538-4357/aa643b. arXiv:1611.07035

H. Shapley, On the nature and cause of Cepheid variation. Astrophys. J. **40**, 448 (1914). https://doi.org/10.1086/142137

M. Skarka, Bright Blazhko RRab Lyrae stars observed by ASAS and the SuperWASP surveys. Astron. Astrophys. **562**, A90 (2014). https://doi.org/10.1051/0004-6361/201322491. arXiv:1402.3111

I. Skillen, J.A. Fernley, R.S. Stobie, R.F. Jameson, The absolute magnitudes of Rr-Lyrae stars—part five—Wy-Antliae W-Crateris Rv-Octantis Bb-Puppis. Mon. Not. R. Astron. Soc. **265**, 301 (1993). https://doi.org/10.1093/mnras/265.2.301

P. Smeyers, T. van Hoolst (eds.), *Linear Isentropic Oscillations of Stars: Theoretical Foundations*. Astrophys. Space Sci. Library, vol. 371 (2010). https://doi.org/10.1007/978-3-642-13030-4

G.H. Smith, M.A. Dopita, The chemical inhomogeneity of the Sculptor dwarf spheroidal galaxy. Astrophys. J. **271**, 113–122 (1983). https://doi.org/10.1086/161180

H.A. Smith, *RR Lyrae Stars*. Cambridge Astrophysics Series, vol. 27 (1995)

R. Smolec, Metallicity dependence of the Blazhko effect. Acta Astron. **55**, 59–84 (2005). arXiv:astro-ph/0503614

R. Smolec, Survey of non-linear hydrodynamic models of type-II Cepheids. Mon. Not. R. Astron. Soc. **456**, 3475–3493 (2016). https://doi.org/10.1093/mnras/stv2868. arXiv:1512.01550

C. Sneden, G.W. Preston, J.A. Kollmeier, J.D. Crane, N. Morrell, J.L. Prieto, S.A. Shectman, D.M. Skowron, I.B. Thompson, Metal-rich RRc stars in the Carnegie RR Lyrae survey. Astron. J. **155**, 45 (2018). https://doi.org/10.3847/1538-3881/aa9f16. arXiv:1712.00159

D. Sobel, *The Glass Universe: How the Ladies of the Harvard Observatory Took the Measure of the Stars* (Penguin, Baltimore, 2016). https://books.google.com/books?id=wcq-CwAAQBAJ

E. Solano, The absolute magnitude of RR Lyraes: from HIPPARCOS parallaxes and proper motions. Astrophys. Space Sci. **263**, 219–222 (1998). https://doi.org/10.1023/A:1002114428541

A. Sollima, C. Cacciari, E. Valenti, The RR Lyrae period-K-luminosity relation for globular clusters: an observational approach. Mon. Not. R. Astron. Soc. **372**, 1675–1680 (2006). https://doi.org/10.1111/j.1365-2966.2006.10962.x. arXiv:astro-ph/0608397

A. Sollima, C. Cacciari, A.A.H. Arkharov, V.M. Larionov, D.L. Gorshanov, N.V. Efimova, A. Piersimoni, The infrared JHK light curves of RR Lyr. Mon. Not. R. Astron. Soc. **384**, 1583–1587 (2008). https://doi.org/10.1111/j.1365-2966.2007.12804.x. arXiv:0712.0578

I. Soszyński, A. Udalski, M.K. Szymański, M. Kubiak, G. Pietrzyński, Ł. Wyrzykowski, O. Szewczyk, K. Ulaczyk, R. Poleski, The optical gravitational lensing experiment. The OGLE-III catalog of variable stars. II. Type II Cepheids and anomalous Cepheids in the Large Magellanic Cloud. Acta Astron. **58**, 293 (2008). arXiv:0811.3636

I. Soszyński, A. Udalski, M.K. Szymański, M. Kubiak, G. Pietrzyński, Ł. Wyrzykowski, K. Ulaczyk, R. Poleski, The optical gravitational lensing experiment. The OGLE-III catalog of variable stars. VIII. Type II Cepheids in the Small Magellanic Cloud. Acta Astron. **60**, 91–107 (2010). arXiv:1005.3544

I. Soszyński, A. Udalski, M.K. Szymański, D. Skowron, G. Pietrzyński, R. Poleski, P. Pietrukowicz, J. Skowron, P. Mróz, S. Kozłowski, Ł. Wyrzykowski, K. Ulaczyk, M. Pawlak, The OGLE collection of variable stars. Classical Cepheids in the Magellanic System. Acta Astron. **65**, 297–312 (2015). arXiv:1601.01318

I. Soszyński, R. Smolec, W.A. Dziembowski, A. Udalski, M.K. Szymański, Ł. Wyrzykowski, K. Ulaczyk, R. Poleski, P. Pietrukowicz, S. Kozłowski, D. Skowron, J. Skowron, P. Mróz, M. Pawlak, Anomalous double-mode RR Lyrae stars in the Magellanic Clouds. Mon. Not. R. Astron. Soc. **463**, 1332–1341 (2016). https://doi.org/10.1093/mnras/stw1933. arXiv:1608.00576

I. Soszyński, A. Udalski, M.K. Szymański, Ł. Wyrzykowski, K. Ulaczyk, R. Poleski, P. Pietrukowicz, S. Kozłowski, D.M. Skowron, J. Skowron, P. Mróz, M. Pawlak, The OGLE collection of variable stars. Over 45 000 RR Lyrae stars in the Magellanic System. Acta Astron. **66**, 131–147 (2016b). arXiv:1606.02727

I. Soszyński, A. Udalski, M.K. Szymański, Ł. Wyrzykowski, K. Ulaczyk, R. Poleski, P. Pietrukowicz, S. Kozłowski, D.M. Skowron, J. Skowron, P. Mróz, M. Pawlak, K. Rybicki, A. Jacyszyn-Dobrzeniecka, The OGLE collection of variable stars. Classical, type II, and anomalous Cepheids toward the Galactic center. Acta Astron. **67**, 297–316 (2017). arXiv:1712.01307

I. Soszyński, A. Udalski, M.K. Szymański, Ł. Wyrzykowski, K. Ulaczyk, R. Poleski, P. Pietrukowicz, S. Kozłowski, D. Skowron, J. Skowron, P. Mróz, K. Rybicki, P. Iwanek, The OGLE collection of variable stars. Type II Cepheids in the Magellanic System. Acta Astron. **68**, 89–109 (2018). arXiv:1807.00008

E. Starkenburg, V. Hill, E. Tolstoy, P. François, M.J. Irwin, L. Boschman, K.A. de Venn, T.J.L. Boer, B. Lemasle, P. Jablonka, G. Battaglia, P. Groot, L. Kaper, The extremely low-metallicity tail of the Sculptor dwarf spheroidal galaxy. Astron. Astrophys. **549**, A88 (2013). https://doi.org/10.1051/0004-6361/201220349. arXiv:1211.4592

M. Steinmetz, T. Zwitter, A. Siebert, F.G. Watson, K.C. Freeman, U. Munari, R. Campbell, M. Williams, G.M. Seabroke, R.F.G. Wyse, Q.A. Parker, O. Bienaymé, S. Roeser, B.K. Gibson, G. Gilmore, E.K.

Grebel, A. Helmi, J.F. Navarro, D. Burton, C.J.P. Cass, J.A. Dawe, K. Fiegert, M. Hartley, K.S. Russell, W. Saunders, H. Enke, J. Bailin, J. Binney, J. Bland-Hawthorn, C. Boeche, W. Dehnen, D.J. Eisenstein, N.W. Evans, M. Fiorucci, J.P. Fulbright, O. Gerhard, U. Jauregi, A. Kelz, L. Mijović, I. Minchev, G. Parmentier, J. Peñarrubia, A.C. Quillen, M.A. Read, G. Ruchti, R.D. Scholz, A. Siviero, M.C. Smith, R. Sordo, L. Veltz, S. Vidrih, R. von Berlepsch, B.J. Boyle, E. Schilbach, The Radial Velocity Experiment (RAVE): first data release. Astron. J. **132**, 1645–1668 (2006). https://doi.org/10.1086/506564. arXiv:astro-ph/0606211

R.F. Stellingwerf, The calculation of periodic pulsations of stellar models. Astrophys. J. **192**, 139–146 (1974). https://doi.org/10.1086/153043

R.F. Stellingwerf, Pulsation in the lower Cepheid strip. I—linear survey. Astrophys. J. **227**, 935–942 (1979). https://doi.org/10.1086/156802

R.F. Stellingwerf, Convection in pulsating stars. I—nonlinear hydrodynamics. II—RR Lyrae convection and stability. Astrophys. J. **262**, 330–343 (1982). https://doi.org/10.1086/160425

R.F. Stellingwerf, Hydra—an implicit partial differential equation relaxation stability analysis package. Astrophys. J. **271**, 876 (1983). https://doi.org/10.1086/161253

P.B. Stetson, DAOPHOT—a computer program for crowded-field stellar photometry. Publ. Astron. Soc. Pac. **99**, 191–222 (1987). https://doi.org/10.1086/131977

P.B. Stetson, The center of the core-cusp globular cluster M15: CFHT and HST observations, ALLFRAME reductions. Publ. Astron. Soc. Pac. **106**, 250–280 (1994). https://doi.org/10.1086/133378

P.B. Stetson, J.E. Hesser, T.A. Smecker-Hane, Homogeneous photometry for star clusters and resolved galaxies. I. A survey of bright stars in the Fornax dwarf spheroidal galaxy. Publ. Astron. Soc. Pac. **110**, 533–552 (1998). https://doi.org/10.1086/316164

P.B. Stetson, V.F. Braga, M. Dall'Ora, G. Bono, R. Buonanno, I. Ferraro, G. Iannicola, M. Marengo, J. Neeley, Optical and near-infrared UBVRIJHK photometry for the RR Lyrae stars in the nearby globular cluster M4 (NGC 6121). Publ. Astron. Soc. Pac. **126**, 521 (2014a). https://doi.org/10.1086/677195. arXiv:1406.7531

P.B. Stetson, G. Fiorentino, G. Bono, E.J. Bernard, M. Monelli, G. Iannicola, C. Gallart, I. Ferraro, Homogeneous photometry VI: variable stars in the Leo I dwarf spheroidal galaxy. Publ. Astron. Soc. Pac. **126**, 616–641 (2014b). https://doi.org/10.1086/677352. arXiv:1406.6704

J. Storm, The distance to IC 4499 from K-band photometry of 32 RR Lyrae stars. Astron. Astrophys. **415**, 987–991 (2004). https://doi.org/10.1051/0004-6361:20034287. arXiv:astro-ph/0401237

J. Storm, W. Gieren, P. Fouqué, T.G. Barnes, I. Soszyński, G. Pietrzyński, N. Nardetto, D. Queloz, Calibrating the Cepheid period-luminosity relation from the infrared surface brightness technique. II. The effect of metallicity and the distance to the LMC. Astron. Astrophys. **534**, A95 (2011). https://doi.org/10.1051/0004-6361/201117154. arXiv:1109.2016

S. Subramanian, M. Marengo, A. Bhardwaj, Y. Huang, L. Inno, A. Nakagawa, J. Storm, Young and intermediate-age distance indicators. Space Sci. Rev. **212**, 1817–1869 (2017). https://doi.org/10.1007/s11214-017-0427-8. arXiv:1709.07265

R. Szabó, Z. Kolláth, J.R. Buchler, Automated nonlinear stellar pulsation calculations: applications to RR Lyrae stars. The slope of the fundamental blue edge and the first RRd model survey. Astron. Astrophys. **425**, 627–639 (2004). https://doi.org/10.1051/0004-6361:20035698. arXiv:astro-ph/0406373

R. Szabó, J.M. Benkő, M. Paparó, E. Chapellier, E. Poretti, A. Baglin, W.W. Weiss, K. Kolenberg, E. Guggenberger, J.F. Le Borgne, Revisiting CoRoT RR Lyrae stars: detection of period doubling and temporal variation of additional frequencies. Astron. Astrophys. **570**, A100 (2014). https://doi.org/10.1051/0004-6361/201424522. arXiv:1408.0653

O. Szewczyk, G. Pietrzyński, W. Gieren, A. Ciechanowska, F. Bresolin, R.P. Kudritzki, The Araucaria project: the distance to the Small Magellanic Cloud from near-infrared photometry of RR Lyrae variables. Astron. J. **138**, 1661–1666 (2009). https://doi.org/10.1088/0004-6256/138/6/1661. arXiv:0910.3885

V. Tabur, L.L. Kiss, T.R. Bedding, Hipparcos calibration of the tip of the red giant branch. Astrophys. J. Lett. **703**, L72–L75 (2009). https://doi.org/10.1088/0004-637X/703/1/L72. arXiv:0908.2873

G.A. Tammann, B. Reindl, The luminosity of supernovae of type Ia from tip of the red-giant branch distances and the value of H_0. Astron. Astrophys. **549**, A136 (2013). https://doi.org/10.1051/0004-6361/201219671. arXiv:1208.5054

G.A. Tammann, A. Sandage, B. Reindl, Comparison of distances from RR Lyrae stars, the tip of the red giant branch, and Classical Cepheids. Astrophys. J. **679**, 52–70 (2008). https://doi.org/10.1086/529508. arXiv:0712.2346

F. Thévenin, M. Falanga, C.Y. Kuo, G. Pietrzyński, M. Yamaguchi, Modern geometric methods of distance determination. Space Sci. Rev. **212**, 1787–1815 (2017). https://doi.org/10.1007/s11214-017-0418-9

I.B. Thompson, J. Kaluzny, W. Pych, G. Burley, W. Krzeminski, B. Paczyński, S.E. Persson, G.W. Preston, Cluster AgeS Experiment: the age and distance of the globular cluster ω Centauri deter-

mined from observations of the eclipsing binary OGLEGC 17. Astron. J. **121**, 3089–3099 (2001). https://doi.org/10.1086/321084. arXiv:astro-ph/0012493
I.B. Thompson, J. Kaluzny, S.M. Rucinski, W. Krzeminski, W. Pych, A. Dotter, G.S. Burley, The Cluster AgeS Experiment (CASE). IV. Analysis of the eclipsing binary V69 in the globular cluster 47 Tuc. Astron. J. **139**, 329–341 (2010). https://doi.org/10.1088/0004-6256/139/2/329. arXiv:0910.4262
N.A. Tikhonov, O.A. Galazutdinova, Photometry of red giants in the disk of IC 1613. Astron. Astrophys. **394**, 33–38 (2002). https://doi.org/10.1051/0004-6361:20021016
E. Tolstoy, M.J. Irwin, A. Helmi, G. Battaglia, P. Jablonka, V. Hill, K.A. Venn, M.D. Shetrone, B. Letarte, A.A. Cole, F. Primas, P. Francois, N. Arimoto, K. Sadakane, A. Kaufer, T. Szeifert, T. Abel, Two distinct ancient components in the Sculptor dwarf spheroidal galaxy: first results from the Dwarf Abundances and Radial velocities Team. Astrophys. J. Lett. **617**, L119–L122 (2004). https://doi.org/10.1086/427388. arXiv:astro-ph/0411029
I. Trujillo, M.A. Beasley, A. Borlaff, E.R. Carrasco, A. Di Cintio, M. Filho, M. Monelli, M. Montes, J. Roman, T. Ruiz-Lara, J. Sanchez Almeida, D. Valls-Gabaud, A. Vazdekis, A distance of 13 Mpc resolves the claimed anomalies of the galaxy lacking dark matter (2018). arXiv:1806.10141
R. Tully, The Geometry and Kinematics of the Local Volume (2011). HST Proposal
R.B. Tully, H.M. Courtois, J.G. Sorce, Cosmicflows-3. Astron. J. **152**, 50 (2016). https://doi.org/10.3847/0004-6256/152/2/50. arXiv:1605.01765
A. Udalski, L. Wyrzykowski, G. Pietrzynski, O. Szewczyk, M. Szymanski, M. Kubiak, I. Soszynski, K. Zebrun, The optical gravitational lensing experiment. Cepheids in the Galaxy IC1613: no dependence of the period-luminosity relation on metallicity. Acta Astron. **51**, 221–245 (2001). arXiv:astro-ph/0109446
A.R. Upgren, Trigonometric parallaxes and their calibration, in *Calibration of Fundamental Stellar Quantities, IAU Symposium*, vol. 111, ed. by D.S. Hayes, L.E. Pasinetti, A.G.D. Philip (1985), pp. 31–49
E. Valenti, F.R. Ferraro, L. Origlia, Red giant branch in near-infrared colour-magnitude diagrams—II. The luminosity of the bump and the tip. Mon. Not. R. Astron. Soc. **354**, 815–820 (2004). https://doi.org/10.1111/j.1365-2966.2004.08249.x. arXiv:astro-ph/0404403
T.S. van Albada, N. Baker, On the masses, luminosities, and compositions of horizontal-branch stars. Astrophys. J. **169**, 311 (1971). https://doi.org/10.1086/151144
T.S. van Albada, N. Baker, On the two Oosterhoff groups of globular clusters. Astrophys. J. **185**, 477–498 (1973). https://doi.org/10.1086/152434
W.F. van Altena, Astrometry. Annu. Rev. Astron. Astrophys. **21**, 131–164 (1983). https://doi.org/10.1146/annurev.aa.21.090183.001023
S. van den Bergh, *The Extragalactic Distance Scale* (University of Chicago Press, Chicago, 1975), p. 509
P. van Dokkum, S. Danieli, Y. Cohen, C. Conroy, The Distance of the Dark Matter Deficient Galaxy NGC1052–DF2 (2018). arXiv:1807.06025
P. van Dokkum, S. Danieli, Y. Cohen, A. Merritt, A.J. Romanowsky, R. Abraham, J. Brodie, C. Conroy, D. Lokhorst, L. Mowla, E. O'Sullivan, J. Zhang, A galaxy lacking dark matter. Nature **555**, 629–632 (2018b). https://doi.org/10.1038/nature25767. arXiv:1803.10237
F. van Leeuwen (ed.), *Hipparcos, the New Reduction of the Raw Data*. Astrophys. Space Sci. Library, vol. 350 (2007). https://doi.org/10.1007/978-1-4020-6342-8
S. Vasilevskis, The accuracy of trigonometric parallaxes of stars. Annu. Rev. Astron. Astrophys. **4**, 57 (1966). https://doi.org/10.1146/annurev.aa.04.090166.000421
F. Vilardell, I. Ribas, C. Jordi, E.L. Fitzpatrick, E.F. Guinan, The distance to the Andromeda galaxy from eclipsing binaries. Astron. Astrophys. **509**, A70 (2010). https://doi.org/10.1051/0004-6361/200913299. arXiv:0911.3391
A.R. Walker, CCD photometry of galactic globular clusters. IV. The NGC 1851 RR Lyrae variables. Astron. J. **116**, 220–236 (1998). https://doi.org/10.1086/300432. arXiv:astro-ph/9804339
M.G. Walker, M. Mateo, E.W. Olszewski, O.Y. Gnedin, X. Wang, B. Sen, M. Woodroofe, Velocity dispersion profiles of seven dwarf spheroidal galaxies. Astrophys. J. Lett. **667**, L53–L56 (2007). https://doi.org/10.1086/521998. arXiv:0708.0010
M.G. Walker, M. Mateo, E.W. Olszewski, B. Sen, M. Woodroofe, Clean kinematic samples in dwarf spheroidals: an algorithm for evaluating membership and estimating distribution parameters when contamination is present. Astron. J. **137**, 3109–3138 (2009). https://doi.org/10.1088/0004-6256/137/2/3109. arXiv:0811.1990
G. Wallerstein, On the incidence of Cepheids in globular clusters. Astrophys. J. **160**, 345 (1970). https://doi.org/10.1086/150434
G. Wallerstein, The Cepheids of Population II and related stars. Publ. Astron. Soc. Pac. **114**, 689–699 (2002). https://doi.org/10.1086/341698
G. Wallerstein, C.A. Nachr, The Population II Cepheids. Publ. Astron. Soc. Pac. **96**, 677–691 (1984). https://doi.org/10.1086/131406

D.L. Welch, Type 2 Cepheids in the Milky Way Galaxy and the Magellanic Clouds. J. Am. Assoc. Var. Star Obs. **40**, 492 (2012)

D.L. Welch, P.B. Stetson, Robust variable star detection techniques suitable for automated searches—new results for NGC 1866. Astron. J. **105**, 1813–1821 (1993). https://doi.org/10.1086/116556

A.J. Wesselink, The colour index of a black body with infinite temperature. Bull. Astron. Inst. Neth. **10**, 99 (1946)

A.J. Wesselink, Surface brightnesses in the U, B, V system with applications of Mv and dimensions of stars. Mon. Not. R. Astron. Soc. **144**, 297 (1969). https://doi.org/10.1093/mnras/144.3.297

G. Worthey, H.c. Lee, An empirical UBV RI JHK color–temperature calibration for stars. Astrophys. J. Suppl. Ser. **193**, 1 (2011). https://doi.org/10.1088/0067-0049/193/1/1. arXiv:astro-ph/0604590

P.F. Wu, R.B. Tully, L. Rizzi, A.E. Dolphin, B.A. Jacobs, I.D. Karachentsev, Infrared tip of the red giant branch and distances to the Maffei/IC 342 group. Astron. J. **148**, 7 (2014). https://doi.org/10.1088/0004-6256/148/1/7. arXiv:1404.2987

D. Zaritsky, J. Harris, I.B. Thompson, E.K. Grebel, P. Massey, The Magellanic clouds photometric survey: the Small Magellanic Cloud stellar catalog and extinction map. Astron. J. **123**, 855–872 (2002). https://doi.org/10.1086/338437. arXiv:astro-ph/0110665

D. Zaritsky, J. Harris, I.B. Thompson, E.K. Grebel, The Magellanic clouds photometric survey: the Large Magellanic Cloud stellar catalog and extinction map. Astron. J. **128**, 1606–1614 (2004). https://doi.org/10.1086/423910. arXiv:astro-ph/0407006

G. Zhao, Y.H. Zhao, Y.Q. Chu, Y.P. Jing, L.C. Deng, LAMOST spectral survey: an overview. Res. Astron. Astrophys. **12**, 723–734 (2012). https://doi.org/10.1088/1674-4527/12/7/002

S.A. Zhevakin, On the pulsational theory of stellar variability. VI. (A unified interpretation of various types of variability). Sov. Astron. **3**, 913 (1959)

R. Zinn, M.J. West, The globular cluster system of the galaxy. III—measurements of radial velocity and metallicity for 60 clusters and a compilation of metallicities for 121 clusters. Astrophys. J. Suppl. Ser. **55**, 45–66 (1984). https://doi.org/10.1086/190947

E. Zsoldos, No RV Tauri stars in globular clusters? Acta Astron. **48**, 775–788 (1998)

Space Sci Rev (2018) 214:74
https://doi.org/10.1007/s11214-018-0506-5

Impact of Distance Determinations on Galactic Structure. I. Young and Intermediate-Age Tracers

Noriyuki Matsunaga[1] · Giuseppe Bono[2,3] · Xiaodian Chen[4] · Richard de Grijs[5,6,7] · Laura Inno[2,8] · Shogo Nishiyama[9]

Received: 24 December 2017 / Accepted: 17 April 2018 / Published online: 1 May 2018

Abstract Here we discuss impacts of distance determinations on the Galactic disk traced by relatively young objects. The Galactic disk, $\sim$ 40 kpc in diameter, is a cross-road of studies on the methods of measuring distances, interstellar extinction, evolution of galaxies, and other subjects of interest in astronomy. A proper treatment of interstellar extinction is, for example, crucial for estimating distances to stars in the disk outside the small range of the solar neighborhood. We'll review the current status of relevant studies and discuss some new approaches to the extinction law. When the extinction law is reasonably constrained,

Astronomical Distance Determination in the Space Age
Edited by Richard de Grijs and Maurizio Falanga

✉ N. Matsunaga
matsunaga@astron.s.u-tokyo.ac.jp

G. Bono
bono@roma2.infn.it

X. Chen
chenxiaodian@nao.cas.cn

R. de Grijs
richard.de-grijs@mq.edu.au

L. Inno
inno@mpia.de

S. Nishiyama
shogo-n@staff.miyakyo-u.ac.jp

1 Department of Astronomy, The University of Tokyo, 7-3-1 Hongo, Bunkyo-ku, Tokyo 113-0033, Japan

2 Dipartimento di Fisica, Universitá di Roma Tor Vergata, Via della Ricerca Scientifica 1, 00133 Rome, Italy

3 INAF, Osservatorio Astronomico di Roma, Via Frascati 33, 00078 Monte Porzio Catone, Italy

4 Key Laboratory for Optical Astronomy, National Astronomical Observatories, Chinese Academy of Sciences, 20A Datun Road, Chaoyang District, Beijing 100012, China

5 Kavli Institute for Astronomy & Astrophysics and Department of Astronomy, Peking University, Yi He Yuan Lu 5, Hai Dian District, Beijing 100871, China

distance indicators found in today and future surveys are telling us stellar distribution and more throughout the Galactic disk. Among several useful distance indicators, the focus of this review is Cepheids and open clusters (especially contact binaries in clusters). These tracers are particularly useful for addressing the metallicity gradient of the Galactic disk, an important feature for which comparison between observations and theoretical models can reveal the evolution of the disk.

Keywords Stars: variables · Cepheids: distance scale

1 Introduction

Understanding the Galaxy is important not only because it hosts our Solar system but also because we can investigate stellar populations and interstellar components in greater detail than in any other galaxies, and the Galaxy can be a benchmark for galaxy evolution. Recent large surveys of stellar populations in various regions of the Galaxy have been revolutionizing the study of its structure and evolution (e.g. Ivezić et al. 2012; Bland-Hawthorn and Gerhard 2016). In addition to many exciting results on the halo, large-scale photometric and spectroscopic surveys also make it possible to test chemodynamical models of the disk with various physical processes like radial migration (e.g. Schönrich and Binney 2009; Casagrande et al. 2011; Haywood et al. 2013). Radial migration is the mixing process induced by the non-axisymmetric potential of the Galaxy, which brings outwards the metal-rich objects formed in the inner disk and vice-versa (Sellwood and Binney 2002; Grand et al. 2015). We need stellar tracers whose distances can be well measured over the large volume of the disk to study such a large-scale phenomenon. In this context, classical Cepheids and open clusters are useful distance indicators in order to trace the young stellar components of the disk. In particular, classical Cepheids and contact binaries found in open clusters are useful as we will see below. These objects are predominantly found in the disk.

1.1 Classical Cepheids

In this review, we focus on classical Cepheids, not type II Cepheids which are similar pulsating stars but belong to old stellar populations (Sandage and Tammann 2006; Beaton et al. 2018). Classical Cepheids (hereinafter simply called Cepheids) are pulsating stars of 4–10 $M_{\odot}$, which show well-defined relations between their intrinsic luminosity and the pulsation period. Such period–luminosity (PL) relations make them fundamental calibrators of the extragalactic distance scale (see the review by Subramanian et al. 2017). Besides this important role, Cepheids also have a number of properties that make them ideal probes of both the structure and the recent history of the Galaxy. In fact, Cepheids are luminous stars that can be seen to great distances even through the substantial interstellar extinction along the lines of sight. In addition to the distances, the ages of individual Cepheids can be determined on the basis of their periods (Bono et al. 2005). They are young stars (10–200 Myr),

[6] Department of Physics and Astronomy, Macquarie University, Balaclava Road, North Ryde, NSW 2109, Australia

[7] International Space Science Institute, Beijing, 1 Nanertiao, Zhongguancun, Hai Dian District, Beijing 100190, China

[8] Max-Planck Institute for Astronomy, Königstuhl 17, 69117 Heidelberg, Germany

[9] Miyagi University of Education, Aoba-ku, Sendai, Miyagi 980-0845, Japan

comparable with the characteristic dynamical time-scale or the orbital period (~ 240 Myr at the solar location if one assumes the IAU recommended values of the Sun–Galactic Center distance $R_0 = 8.5$ kpc and the rotation speed $V_0 = 220\ \mathrm{km\,s^{-1}}$) of the Galactic disk, thus they are still relatively close to the place where they were born as far as the distances from the Galactic Center (henceforth the GC) and to neighboring objects are concerned. These make the Cepheids ideal tracers for studying dynamical evolution of the young components like spiral arms (see e.g. Baba et al. 2018; Kawata et al. 2018). Furthermore, they are good chemical tracers. Their effective temperatures ($T_{\mathrm{eff}} \sim 5500$ K) are relatively low when compared to similarly young stars, and allow us to measure many metallic lines in their spectra giving access to precise abundances of many different elements (e.g. Lemasle et al. 2013; da Silva et al. 2016).

1.2 Open Clusters

Open clusters are aggregates of hundreds to tens of thousands of stars, all born together. Their members have similar ages, metallicities, and distances to us, which means that they usually show obvious main sequences in the Hertzsprung–Russell diagram. Based on the main sequence of a cluster, a unique theoretical isochrone can be identified to describe the evolutionary details of stars with different masses. In contrast to the Galactic globular clusters which are predominantly old, the ages of open clusters range from very young (a few Myr) to rather old (a few Gyr), and they are mostly located in the disk. Since their distances can be determined based on main-sequence or isochrone fitting, open clusters represent important tools to trace young structures such as spiral arms (see, e.g. Dias and Lépine 2005; Carraro 2014). Combining the proper motions, radial velocities, and metallicity information, open clusters can be used to study the kinematics and evolution of the young components of the Galaxy.

Furthermore, studies of variable stars in open clusters are interesting. In addition to Cepheids and other pulsating stars, eclipsing contact binaries are another type of variable often found in open clusters. One can expect PL (or PLC) relations for contact binaries (see Sect. 4.2), and recent studies actually found relatively tight relations, especially in the near-IR (Chen et al. 2016a). Such distance indicators, combined with the isochrone fitting method, can provide mutual verification of distance scales.

1.3 Other Young Tracers

There are a few other young and intermediate-age tracers which are useful to study Galactic structure.

- *Massive star-forming regions* often host methanol (CH_3OH) masers which enable us to measure the trigonometric parallaxes of the star-forming regions with VLBI (Very Long Baseline Interferometry) facilities (Reid et al. 2009; Honma et al. 2012; Reid and Honma 2014). The high accuracy of VLBI measurements, in the best cases better than 10 µas (e.g. Zhang et al. 2013; Sanna et al. 2017), allows us to map this youngest population over a large volume of the disk.
- *Long-period variables* (LPVs hereinafter) include evolved asymptotic giant branch (AGB) stars with a wide range of ages between 100 Myr and over 10 Gyr. The lower limit to their ages, or the upper limit to the initial mass with which a star evolves into an AGB star, instead of a red supergiant, is unclear. LPVs (or also called Miras) obey PL relations and thus serve as distance indicators (Whitelock et al. 2008; Subramanian et al.

2017). Their high luminosities, especially in the infrared (hereinafter IR), make them useful tracers of the Galaxy's structure. Some of them also show radio maser emission lines which are useful to study the kinematics of the Galaxy. In particular, a group of relatively massive LPVs (3–5 $M_{\odot}$, age < 1 Gyr) are often called OH/IR stars because they have OH masers and are very bright in the IR. They have provided us with good insights, for example, into the inner part of the Galactic disk (Lindqvist et al. 1992; Sevenster et al. 2000).

- *Red clump (RC) giants* are also useful tracers of stellar populations and very widely used for studying Galactic structure. They are core-helium-burning stars found spanning a wide range of ages from hundreds of Myr to 10 Gyr. Note that they are found as a clump in the color-magnitude diagram or in a luminosity function so that they can be identified only if a sizeable number of such stars in stellar population(s) are present. The specific frequency of the core-helium-burning stars, the number per unit stellar mass of the population, is maximum around 1 Gyr (Salaris and Girardi 2002). Taking the total mass of stars at different ages into account, however, the age distribution of the RC stars tends to be extended unless the star formation history is represented by a sharp peak. The brightness of the clump depends on the age and metallicity, but the dependency is mild against the rather wide range of their ages, 1.5–10 Gyr (Salaris and Girardi 2002). The error of distance estimation caused by the age distribution of stars tends to be averaged out, which is an advantage in estimating distances to systems with various stellar populations mixed, but makes it hard to separate RCs formed by stars with different ages. These complicate the use of the RC for studying the distribution of stars at a specific age in the Galaxy.

These objects are not discussed in the rest of the review. On massive star-forming regions, readers are referred to Reid et al. (2014, and references therein). On LPVs and RC giants, Subramanian et al. (2017) discuss their characteristics and applications. Some other kinds of young stars are also investigated as distance indicators, e.g. blue supergiants (Kudritzki et al. 2003, 2008) and red supergiants (Yang and Jiang 2012). In the following we focus on Cepheids and open clusters. We can measure the individual distances to these objects based on optical to IR photometric data, which is important for characterizing the interstellar extinction that shows very patchy patterns on the sky.

1.4 The Goal of This Review

The purpose of this review is to discuss the impact of distance determinations of Cepheids and open clusters (Cepheids and contact binaries in the clusters, in particular) for revealing the structure and evolution of the disk, which is a relatively unexplored frontier of Galactic archeology mainly because many previous surveys in the optical wavelengths were seriously affected by interstellar extinction.

After their birth, stars in the Galaxy change their orbits through secular evolution processes, but their chemical abundances remain almost unaltered from their compositions at birth (except for a few elements that may be affected by stellar evolution processes; for Cepheids, e.g., see Takeda et al. 2013). This motivates the science of Galactic archeology, a recently emerging field in astronomy which aims at studying the imprints of chemodynamical evolution from the chemistry and kinematics of a large number of stars. In fact, we are now experiencing a quantum jump both in the wealth and in the quality of available information for billions of stars in the Galaxy thanks to the efforts of dedicated spectroscopic surveys (e.g. APOGEE, GALAH, LAMOST, Gaia-ESO, and in the future 4MOST), capable of obtaining detailed chemical compositions, and to some extent distances and ages, of individual evolved stars (e.g. RC giants, $\sim$ 1–10 Gyr, and red giant branch stars, $\sim$ 10–13 Gyr) throughout the entire Galactic disk.

However, a similarly detailed map of the interstellar gas and young stars is still lacking. While we can obtain abundances of only a few elements (e.g. oxygen and nitrogen) for interstellar gas, we can measure the detailed abundance for the young stars, which should still reflect the composition of their birth material. A critical observable feature for investigating the large-scale disk evolution is the metallicity gradient, i.e. metallicity distribution as a function of Galactocentric distance (R_{GC}). In fact, an inside-out formation scenario for the Galactic disk will produce metal-rich stars in the inner part and metal-poor ones in the outer part of the Galactic disk (e.g. Tsujimoto et al. 1995; Chiappini et al. 2001, and references therein). Such a trend observed in interstellar gas and youngest stars reflects the present-epoch and recent chemical structure of the disk, and the time-dependent trend traced in stars with various ages can tell us the chemical structure in the past and the evolution of the disk. However, observational reconstruction of the metallicity gradient as a function of time is challenging, at least, for the following two reasons: (1) Good tracers with accurate estimates of distance, age, and metallicity need to be found and well characterized. (2) The interstellar extinction hampers accurate distance determination and detailed observations of tracers spread across the disk. Young tracers like Cepheids and open clusters embedded in the Galactic disk, and those at large distances, several kpc or more distant, are severely affected by extinction as we will see below.

In this review, we will summarize up-to-date observational results and current problems about the interstellar extinction in Sect. 2. Then, recent developments of the distance measurements and some important problems are discussed for Cepheids in the Galactic disk (Sect. 3), Cepheids in open clusters (Sect. 4.1), and contact binaries in open clusters (Sect. 4.2). Then, in Sect. 5, we discuss how the distance determination and the characterization of the interstellar extinction are important to study the metallicity gradient. Section 6 summarizes the review.

2 The Interstellar Extinction Law

A detailed understanding of the interstellar extinction law, or the wavelength dependence of the interstellar extinction, is crucial to most aspects of observational astronomy. Dust grains attenuate the light from an object seen through them, and the amount of attenuation depends on wavelength and the lines of sight. Equations that fit the observed wavelength dependence of the extinction are required to test dust grain models and to predict the amount of extinction in certain spectral regions.

A large impact of the extinction law on Galactic structure can be highlighted by comparing the *classical* law in Cardelli et al. (1989), for example, with those derived in recent studies. Nishiyama et al. (2008) reported an extinction ratio of $A_V/A_{K_s} \approx 16$, a factor of ~ 1.9 larger than the value of 8.5 in Cardelli et al. (1989) for $R_V \equiv A_V/E_{B-V} = 3.1$. Considering the column density toward the GC obtained in X-ray observations (Porquet et al. 2008), Fritz et al. (2011) derived a lower limit of $A_V/A_{K_s} \approx 15$. The difference between the new results and previous ones (Rieke and Lebofsky 1985; Cardelli et al. 1989) can be clearly seen in Fig. 8 in Fritz et al. (2011). In addition, a recent study by Hosek et al. (2018) claims that even Nishiyama et al. (2008) underestimated the optical-to-near IR extinction ratio. These results suggest that when we use the *classical* extinction law where it is not appropriate, the amount of extinction may be associated with an error of more than 50%.

In this section, we review the extinction law in wavelength order: the ultra-violet (UV) and optical range (Sect. 2.1), the near IR (Sect. 2.2), and the mid IR (Sect. 2.3). In the UV and optical wavelengths, the difference of the interstellar extinction law among various lines of sight was already known in early studies (e.g. Johnson and Morgan 1955; Johnson and

Borgman 1963; Nandy 1965). For several decades, in contrast, the extinction law in the near- and mid-IR wavelengths was believed to show a very little, if any, variation from one line of sight to another. The "universality" of the IR extinction law implies that the dust grains responsible for the near- and mid-IR extinction have almost the same size distributions along all sightlines. In addition, the IR extinction was thought to follow a power law of $A_\lambda \propto \lambda^{-\alpha}$ from 1 μm to $\sim$ 8 μm in early studies (e.g. Draine 1989). Some recent results, however, suggest that the IR extinction law is not so simple as we will see in this section, while there are still results supporting the "universality" of the IR extinction law (Sect. 6).

2.1 The Interstellar Extinction Law: UV and Optical

Johnson and Borgman (1963) found a variation of the ratio of the total-to-selective extinction in the B and V bands, R_V from 3.1 to 7.4. Bless and Savage (1972) examined the extinction law from the UV to the near-IR for bright early-type stars, and clearly demonstrated large variations of the extinction law. The parameter R_V depends on the dust properties along the lines of sight. Large grains produce gray extinction, i.e., small wavelength dependence, leading to large R_V.

In spite of such large variations of the UV/optical extinction law, Cardelli et al. (1989) suggested that the variation of the extinction law at UV–optical wavelengths can be described by a functional form with only one parameter, R_V. Their suggestion was based on the fact that there is a good linear correlation between A_λ/A_V and R_V^{-1} ($A_\lambda/A_V = a_\lambda + b_\lambda R_V^{-1}$) over an entire range of wavelengths from the UV to the near-IR. Cardelli et al. (1989) provided the λ-dependent coefficients, a_λ and b_λ, from the UV to the near-IR, which allows us to calculate A_λ/A_V if the value of R_V is known for the sight line.

The one-parameter extinction law was welcomed by the astronomical community. This makes it easier to correct the amount of extinction not only in the commonly used photometric bands actually measured by Cardelli et al. (1989) but for any band or wavelength. However, Fitzpatrick and Massa (2007) pointed out that the correlation found by Cardelli et al. (1989) is "partially illusory", and the extinction law is not a one-parameter function. The functional form of Cardelli et al. (1989) was proposed because of a good correlation between A_λ/A_V and R_V^{-1}. However, A_λ/A_V was determined based on the color-excess ratio $E_{\lambda-V}/E_{B-V}$ and R_V as

$$\frac{A_\lambda}{A_V} = \frac{E_{\lambda-V}}{E_{B-V}} R_V^{-1} + 1.$$

Since R_V is used to calculate A_λ/A_V, some degree of apparent correlation is expected between A_λ/A_V and R_V^{-1} in any case.

More than 300 curves for the extinction law from UV to optical wavelengths were obtained for different objects by Fitzpatrick and Massa (2007). The curves range from nearly gray to strongly λ-dependent. Fitzpatrick and Massa (2007) concluded that there is no functional relationship between R_V and the UV extinction law, and that the UV extinction cannot be represented by only a single parameter. It is possible to derive a typical or average extinction law; however, any mean curve is subject to a bias caused by the selection of sample used for producing the curve. Hence their conclusion is that there is *no* unique or best estimate of the mean extinction law in our Galaxy, at least, at UV–optical wavelengths.

2.2 The Interstellar Extinction Law: Near-IR

It has been considered that the near-IR extinction law can be approximated by the power-law form, $A_\lambda \propto \lambda^{-\alpha}$. Figure 1 shows the power-law index α as a function of publication year since 1984. Before $\sim$ 1995, many studies had converged to the IR extinction law that

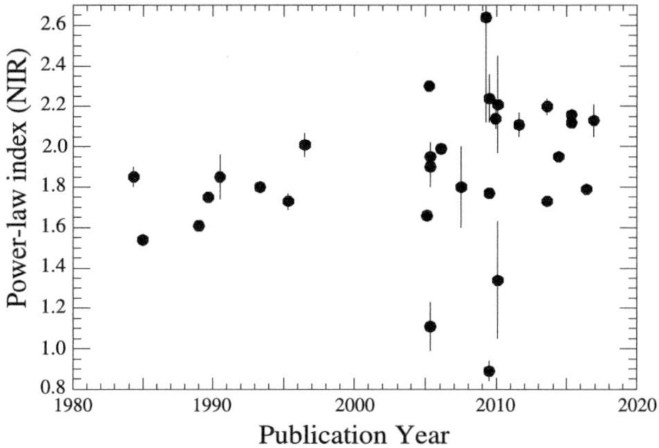

Fig. 1 The power-law index α for the near-IR extinction law is plotted against publication year since 1984. The largest and smallest α values are both from Moore et al. (2005). References: Landini et al. (1984), Rieke and Lebofsky (1985), Cardelli et al. (1989), Draine (1989), Martin and Whittet (1990), Whittet et al. (1993), He et al. (1995), Lumsden and Puxley (1996), Larson and Whittet (2005), Moore et al. (2005), Indebetouw et al. (2005), Messineo et al. (2005), Nishiyama et al. (2006), Froebrich et al. (2007), Gosling et al. (2009), Fitzpatrick and Massa (2009), Stead and Hoare (2009), Schödel et al. (2010), Fritz et al. (2011), Wang et al. (2013), Wang and Jiang (2014), Maíz Apellániz (2015), Schultheis et al. (2015), Xue et al. (2016), and Damineli et al. (2016)

is nearly constant and universal (with $\alpha \approx 1.6$–1.8). However, steeper extinction law has been derived in many recent works, e.g. $\alpha = 2.3$ for stars behind translucent clouds at high Galactic latitude (Larson and Whittet 2005), 1.9 for late-type stars with SiO maser emission (Messineo et al. 2005), 1.99 toward the GC (Nishiyama et al. 2006).

A simple comparison of the extinction law in different lines of sight can be done by using reddening vectors in color–color diagrams. To determine slopes of reddening vectors precisely, Kenyon et al. (1998) introduced a reddening probability function, in which the distribution of strongly reddened stars in a color–color diagram is compared with those of stars in a reference (off-cloud) field. The derived slopes are $E_{J-H}/E_{H-K} = 1.57 \pm 0.03$ for the ρ Ophiuchi dark cloud (Kenyon et al. 1998), 1.80 ± 0.03 for the Chameleon I cloud (Gómez and Kenyon 2001), and 2.08 ± 0.03 for Coalsack Globule 2 (Racca et al. 2002). Using the same technique but different telescopes and instruments, Naoi et al. (2006, 2007) also obtained a steeper reddening slope in E_{J-H}/E_{H-K_s} toward Coalsack Globule 2 than in other dark clouds. In addition, they found a change in the reddening slope; smaller slopes are seen toward sightlines with larger optical depth. This trend had been implied by previous studies (e.g. Kenyon et al. 1998). These results suggest spatial variation of the near-IR extinction law, at least, from a line of sight to another.

Spectroscopic studies of the near-IR extinction law for nine ultra compact H II regions and two planetary nebulae were carried out by Moore et al. (2005). The amounts of extinction for their target regions are larger than $A_V = 15$ mag. They derived the power-law index α by using hydrogen line ratios, and found that α varies from 1.11 to 1.95, with a mean of 1.6, depending on $\tau(\mathrm{Br}\gamma)$ as seen in (Fig. 2).

Fitzpatrick and Massa (2009) examined extinction law from UV to near-IR wavelengths for 14 early-type stars. They fit the extinction law in the form (their Eq. (2))

$$\frac{E_{\lambda-V}}{E_{B-V}} = k_{\mathrm{IR}} \lambda^{-\alpha} - R_V .$$

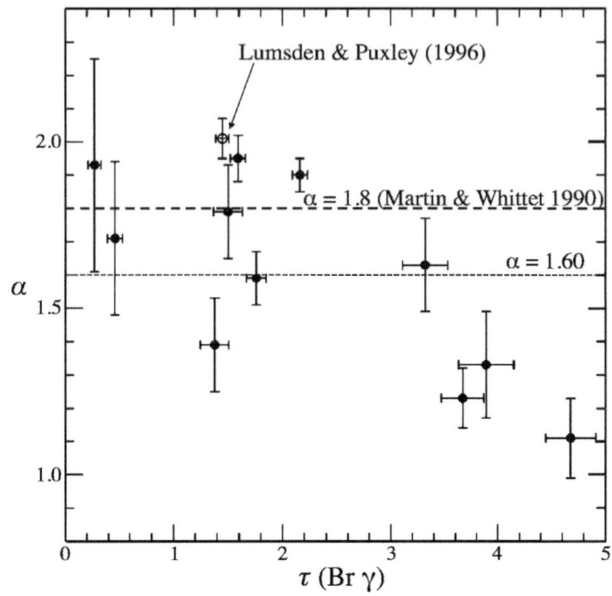

Fig. 2 The power-law index α versus $\tau(\mathrm{Br}\gamma)$. The indices are derived by using hydrogen lines for nine H II regions and two planetary nebulae. A clear variation of α is seen. Adapted from Moore et al. (2005, Fig. 4)

When the three parameters, k_{IR}, α, and R_V are left free in the fits, α varies from 0.89 to 2.24 (with a mean value of 1.77 and a typical uncertainty of 0.15). According to their conclusion, the extinction law from the I band to the K band is well represented by a simple power-law for each sightline, but α varies significantly among 14 lines of sight, and there is no universal power-law index for the near-IR extinction law. They also found that α depends on R_V, α being smaller for a larger R_V. This trend is consistent with the α dependency on $\tau(\mathrm{Br}\gamma)$ found by Moore et al. (2005) because larger R_V values are found toward lines of sight with strong extinction as mentioned above. Fitzpatrick and Massa (2009) also demonstrated that the extrapolations of the power-law fits from shorter wavelengths (0.6–0.95 μm) are inconsistent with the near-IR (JHK_{s}) data points. Including data sets in the I band or shorter wavelengths always lead to a smaller α value. On the other hand, Fitzpatrick and Massa (2009) also investigated the form (their Eq. (5))

$$\frac{E_{\lambda-V}}{E_{B-V}} = [0.349 + 2.087 R_V] \frac{1}{1 + (\lambda/0.507)^{\alpha}} - R_V$$

to represent the extinction law with two free parameters, α and R_V. This is not a power-law, but it is successful in reproducing the extinction law in the optical wavelengths longer than the V band, as well as in the near IR. They demonstrated that, unlike the UV case (Fitzpatrick and Massa 2007), the optical–near IR extinction law may be well represented by a formula with only a few parameters (based on the data towards a limited number of sightlines), but establishing such a formula needs further investigation.

2.3 The Interstellar Extinction Law: Mid-IR

The mid-IR extinction law had been scarcely studied before the launch of *Spitzer Space Telescope* in 2003. Some earlier works (Landini et al. 1984; Rieke and Lebofsky 1985; Martin and Whittet 1990) suggested that a power-law extends from the near-IR to the M

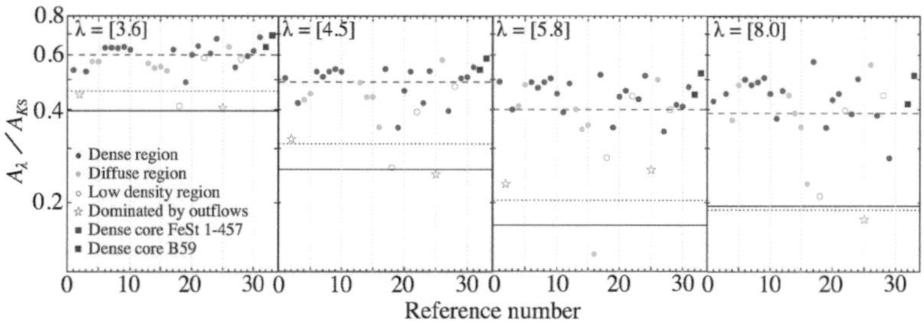

Fig. 3 Comparison of $A_\lambda/A_{K_\mathrm{s}}$ for $\lambda = [3.6]$, [4.5], [5.0], and [8.0] (from left to right), adapted from Ascenso et al. (2013, Fig. 6). The $A_\lambda/A_{K_\mathrm{s}}$ value by dust models are indicated by horizontal lines: solid line and dotted line for the model by Draine (2003a,b) for $R_V = 3.1$ and $R_V = 5.5$, respectively, and dashed line for the model by Weingartner and Draine (2001) for $R_V = 5.5$. See Ascenso et al. (2013) for references and more detail

and N bands, whereas others (Lutz et al. 1996; Lutz 1999) found shallower extinction law at $\lambda > 3$ μm.

Precise and wide-field studies of the mid-IR extinction law had become very active with *Spitzer*. The mid-IR camera, InfraRed Array Camera (IRAC), can collect images in four bands ([3.6], [4.5], [5.8], and [8,0]), and deep mid-IR photometry for strongly obscured objects has become possible thanks to its high sensitivity. Indebetouw et al. (2005) pioneered in the extinction studies with the *Spitzer*. They observed the giant H II region RCW 49 and two lines of sight, $l = 42°$ and $284°$, and derived a flat extinction law at 3–8 μm. A similar flat behavior has been supported by observations of various targets and regions: dark clouds and star forming regions (Flaherty et al. 2007; Román-Zúñiga et al. 2007; Cambrésy et al. 2011; Ascenso et al. 2013; Wang et al. 2013), the GC and Galactic bulge (Nishiyama et al. 2009; Chen et al. 2013), and different locations of Galactic plane (Gao et al. 2009; Xue et al. 2016).

The flattening of the extinction law for ~ 3–8 μm has also been confirmed by observations with other instruments. For example, Jiang et al. (2006) found $A_{7\ \mu\mathrm{m}}/A_{K_\mathrm{s}}$ values larger than suggested by the power-law extension. Schödel et al. (2010) found a clear change of the power-law index α from the H to L' bands toward the GC: $\alpha = 2.21 \pm 0.24$ inferred from data sets in H and K_s bands, and 1.34 ± 0.29 from those in K and L' bands. Fritz et al. (2011) confirmed the flattening at $\lambda \gtrsim 3$ μm toward the GC, using 18 hydrogen emission lines at $1 < \lambda < 8$ μm. The extinction law obtained by Davenport et al. (2014) shows a spatial variation but also shows a flattening in the *WISE* $W1$ (3.35 μm) and $W2$ (4.46 μm) bands.

Ascenso et al. (2013) compiled the $A_\lambda/A_{K_\mathrm{s}}$ values for the mid-IR range from the literature between 1999 and 2013 (Fig. 3). Most of the studies use *Spitzer*/IRAC data sets, but those based on other data sets at similar wavelengths are also included (Bertoldi et al. 1999; Lutz 1999; Fritz et al. 2011; Olofsson and Olofsson 2011). If the mid-IR extinction law is an extension of the near-IR power-law, Fig. 3 should show a decrease of $A_\lambda/A_{K_\mathrm{s}}$ from the left to right panels (see the solid lines, $R_V = 3.1$, Draine 2003a,b). However, the change of $A_\lambda/A_{K_\mathrm{s}}$ from [3.6] to [8.0] is rather small. This together with the above results allows us to conclude that the extinction law at $3 < \lambda < 8$ μm is not a simple extrapolation of a power-law from shorter wavelengths and the extinction is most often larger than expected from the power-law.

A color-excess ratio at mid-IR wavelengths, $E_{[3.6]-[4.5]}/E_{[4.5]-[5.8]}$, was investigated by Flaherty et al. (2007) for five star forming regions. They found that the ratio toward Ophiuchus, 1.25 ± 013, is significantly smaller than those toward the other regions, which range

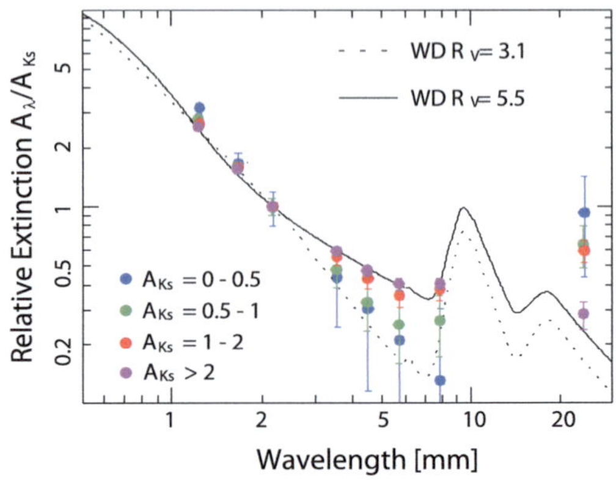

Fig. 4 Near-IR and mid-IR extinction law toward the Ophiuchus molecular cloud (Chapman et al. 2009). Four observational results for different A_{K_s} ranges are shown. The model extinction laws of Weingartner and Draine (2001) for $R_V = 3.1$ and 5.5 are also plotted as solid and dotted lines, respectively. Adapted from Chapman et al. (2009, Fig. 16)

from 2.49 to 2.61. This result strongly suggests variation of the mid-IR extinction law among dense star-forming regions.

An interesting result for the variation of the mid-IR extinction law was obtained by Chapman et al. (2009, reference #18–21 in Fig. 3). They explored changes of the near-IR and mid-IR extinction laws in three molecular clouds, Ophiuchus, Perseus, and Serpens. They found that the extinction law tends to be flatter along lines of sight with stronger extinction (Fig. 4). For $A_{K_s} < 0.5$, the data points lie almost on top of the power-law extension from shorter wavelengths. On the other hand, the extinction laws become as flat as those found in other studies for $A_{K_s} \gtrsim 1$.

Using red giants and RC giants, Gao et al. (2009) determined the color-excess ratios, $E_{K_s-\lambda}/E_{J-K_s}$ and $E_{K_s-\lambda}/E_{H-K_s}$, for the four IRAC bands ($\lambda = [3.6], [4.5], [5.8], [8.0]$). By assuming $A_J/A_{K_s} = 2.52$ and $A_H/A_{K_s} = 1.56$, they derived average A_λ/A_{K_s} values for 131 fields along the Galactic plane. The mean extinction ratio is consistent with the previous studies but shows a slightly flatter wavelength dependence. They also found a variation of A_λ/A_{K_s} along the Galactic plane. In the case of red giants, A_λ/A_{K_s} varies in the range from 0.07 to 0.16 for the four bands, with a typical uncertainty of 0.001. They also suggest systematic variations of A_λ/A_{K_s} with Galactic longitude. A_λ/A_{K_s} seems to be smaller at around the Galactic longitudes corresponding to the tangent positions of spiral arms, which can be explained by grain growth in the arms.

Zasowski et al. (2009) showed that color-excess ratios vary as a function of Galactic longitude. They used RC giants as tracers of extinction, along the Galactic plane ($|b| \lesssim 1.0°$–$1.5°$) in the first ($10° < l < 65°$), and the third to fourth quadrants ($-105° < l < -10°$). Figure 5 plots the observed color-excess ratios, $E_{H-\lambda}/E_{H-K_s}$. Here the color-excess ratios are not converted to the ratio of the amount of extinction, such as A_λ/A_{K_s}, allowing us to compare the extinction law without uncertainties from either the extinction ratio or the ratio of total-to-selective extinction. It is shown that the wavelength dependence of the extinction gets steeper with increasing Galactic longitude, while the ratio $E_{H-[3.6]}/E_{H-K_s}$ is nearly constant. The trend of the $E_{H-\lambda}/E_{H-K_s}$ ratios looks identical between positive and negative Galactic longitudes.

The mid-IR extinction law has been found to be flatter in denser molecular cloud regions (e.g. Chapman et al. 2009; McClure 2009). The flattening can be interpreted as a result of grain growth from either the coagulation of grains or the accumulation of refractory ice mantles. The steeper extinction law found by Zasowski et al. (2009) for larger angles from the GC indicates a steady decrease of the mean dust grain size in the outer Galaxy.

Fig. 5 $E_{H-\lambda}/E_{H-K_s}$ as a function of angle from the GC (Zasowski et al. 2009). The filled and open circles are data points for $l < 180°$ and $l > 180°$, respectively. Adapted from Zasowski et al. (2009, Fig. 5)

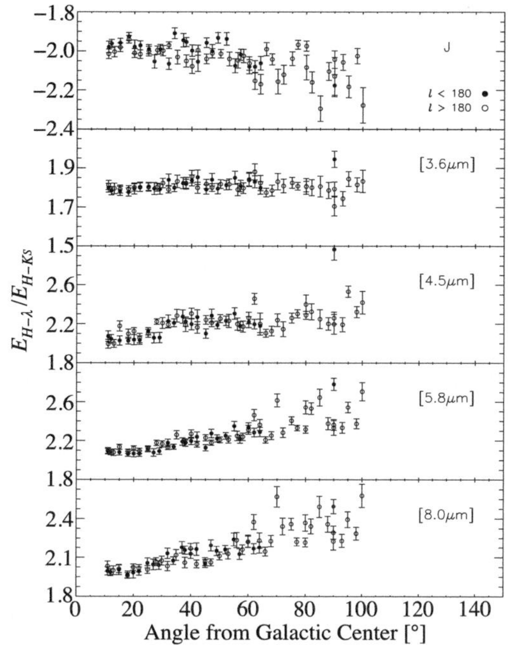

Chen et al. (2013) investigated the variation of the extinction ratio A_λ/A_{K_s} for the four *Spitzer*/IRAC bands. They combined the IRAC data with GLIMPSE-II and VVV data sets for the Galactic bulge, $-10° < l < +10°$ and $-2° < b < +2°$. Their main purpose was to construct a 3D extinction map by comparing observed photometric data (2MASS, VVV, GLIMPSE) with the Besançcon Galactic model. Since they made extinction maps using a color excess $E_{\lambda-K_s}$, they can derive the ratio of the extinction, A_λ/A_{K_s}, using the color-excess ratios, $E_{K_s-\lambda}/E_{J-K_s}$ when the near-IR ratio A_J/A_{K_s} is assumed. They found variations in A_λ/A_{K_s}, and the variations show similar behaviors in all the IRAC bands. As a function of Galactic latitude, there is a strong peak toward the Galactic plane. This suggests a stronger variation of A_λ/A_{K_s} along the Galactic latitude than along the Galactic longitude within their target region toward the bulge. The strongest variation is found at 5.8 μm, while the variation at 8.0 μm is less pronounced. The extinction at around 5.8 μm appears as continuum extinction. The extinction at around 8.0 μm, in contrast, is largely attributed to the 9.7 μm silicate absorption feature. The smaller variation of the extinction at the latter wavelength suggests that the ratio of the silicate absorption to the total extinction does not vary so much along the Galactic latitude.

Note that results on the variations in the ratio of the amount of extinctions, such as A_λ/A_{K_s} at mid-IR wavelengths, may depend not only on the mid-IR extinction law itself but also on the variation in the near-IR extinction law. In many cases, what we can determine directly from observations is a color-excess ratio, e.g. $E_{K_s-\lambda}/E_{H-K_s}$. Such a ratio can be translated into A_λ/A_{K_s} as

$$\frac{A_\lambda}{A_{K_s}} = \frac{E_{K_s-\lambda}}{E_{H-K_s}}\left(\frac{A_H}{A_{K_s}} - 1\right) + 1$$

$$= \frac{E_{K_s-\lambda}}{E_{H-K_s}}\frac{E_{H-K_s}}{A_{K_s}} + 1.$$

We thus need either a ratio of the amount of extinction (in this case, A_H/A_{K_s}) or the ratio of the total-to-selective extinction (A_{K_s}/E_{H-K_s}). These values are, however, usually difficult to determine directly from observations. As shown in Fig. 10 in Nishiyama et al. (2009), the variation of the ratio of the amount of extinction at mid-IR wavelengths, A_λ/A_{K_s}, may be attributed to differences in the near-IR law, A_H/A_{K_s}. Using color-excess ratios allows one to investigate the extinction law without obtaining or assuming the ratios with amount(s) of extinction involved (e.g. A_H/A_{K_s} and A_{K_s}/E_{H-K_s}) as demonstrated, e.g. by Zasowski et al. (2009). The color-excess ratios depend much less on the filter wavelength than the power-law index. Some observational approaches in which parameters like A_{K_s}/E_{H-K_s} are directly measured will be discussed in Sect. 6.

It should be borne in mind that the mid-IR wavelength region contains various emission and absorption features. The most prominent one is the 9.7 μm silicate absorption feature. When we observe the line of sight toward the GC, absorption features such as H_2O, CO_2, NH_3, and CH_4 are seen from ~ 3 to ~ 8 μm (Chiar et al. 2000), which are in the range of the *Spitzer*/IRAC bands. The mid-IR hosts the Polycyclic Aromatic Hydrocarbon (PAH) emission features in the range of 3–20 μm. For wavelengths longer than ~ 8 μm, the extinction may also be underestimated because of the possible presence of dust emission. The late-type giants may have thin shell(s) of silicate dust, and they may exhibit the 9.7 μm silicate emission. In addition, the observation could be affected by dust thermal emission. Although the extinction is near a minimum in the wavelength range of 3–8 μm, it is necessary to take into consideration the potential impact of the emission and absorption features described above when we try to derive the mid-IR extinction law.

In summary, the extinction law in the mid-IR is still less well established compared to that at optical and the near-IR mainly because the opportunities of obtaining precise photometric measurements are limited. However, the extinction at 3–4 μm is about half of that in the K band, and therefore the effects of the remaining uncertainties can be also small. This encourages one to further investigate the extinction law in the mid-IR and applications to various tracers.

3 Cepheids in the Galactic Disk

Cepheids have already played a prominent role in our understanding of the Galactic disk's structure and evolution: their three-dimensional distribution in the disk delineates the spiral pattern (Joy 1939; Majaess et al. 2009; Dambis et al. 2015); their kinematics provides firm estimates of the Oort constants (Kovacs et al. 1990; Metzger et al. 1991, 1998; Feast and Whitelock 1997; Pont et al. 1997), while their chemical abundances can be used for accurate measurements of the radial abundance gradient across the disk (Genovali et al. 2014, 2015; Andrievsky et al. 2016; da Silva et al. 2016).

The sample of known Cepheids in the Galaxy is not only limited (~ 600) but also strongly biased. Over 10,000 Cepheids are expected to populate the Galaxy (Windmark et al. 2011), but only $\sim 5\%$ (predominately nearby ones) have been identified so far. The currently known Cepheids are concentrated mostly within a few kpc around the Sun since they have been discovered based on optical surveys which are hampered by the high interstellar extinction toward the Galactic plane. Moreover, such surveys were performed by using different instruments and photometric techniques, producing strongly inhomogeneous data sets which can be hardly merged for a statistically coherent analysis.

Fortunately, the Cepheid sample is improving rapidly, especially thanks to the ongoing effort from large-sky time-domain surveys. Near-IR variability surveys have produced small catalogs to date, but they have discovered Cepheids in the most obscured regions of the Galaxy, challenging the current observational bias. In particular, near-IR surveys using the Infrared Survey Facility (IRSF) in South Africa toward the Galactic plane led to the discovery of four new Cepheids in the Nuclear Stellar Disk (Matsunaga et al. 2011, 2015), three at 3–5 kpc from the GC (Tanioka et al. 2017), 13 beyond the GC (Matsunaga et al. 2016), and five parallel to the Galactic central bar (Inno et al. in prep). Near-IR data collected by the ESO public survey VVV (Minniti et al. 2010) have also been used to discover 37 Cepheids located beyond the GC (Dékány et al. 2015a,b). With optical surveys, on the other hand, the Optical Gravitational Lensing Experiment (OGLE) has recently announced the discovery of $\sim 2{,}000$ Cepheids found in the Southern disk (Udalski 2017) in addition to new Cepheids in less obscured regions toward the Galactic bulge (Soszyński et al. 2017). The Pan-STARRS1 3π survey in *grizy* bands also has a great potential for identifying a large number of new Cepheids in the northern hemisphere and the exploitation of its data set is pioneering for future all-sky photometric surveys (e.g. LSST). In the near future, *Gaia* will allow us to discover a large number of Cepheids and other variables across the sky. Its limiting magnitude is deep enough to detect Cepheids to some distances, roughly a few kpc toward the inner Galaxy and more in the outer Galaxy, even in reddened regions of the Galactic disk, although it cannot reach the GC and beyond (Windmark et al. 2011).

While these surveys are widening the census of Galactic Cepheids, they (even the *Gaia*'s all-sky survey) are still characterized by unknown selection functions, given the uncertainty in the extinction throughout the Galactic disk and the different cuts in RA, Dec, magnitudes etc. For example, Matsunaga et al. (2016) discussed a gap between the optical and IR surveys toward the bulge. Moreover, the interstellar extinction not only limits the completeness of Cepheid surveys but also introduces large uncertainties on the Cepheids' distances. A relatively small error in the extinction law can lead to significant errors in distances to Cepheids in obscured regions even when IR data are used for distance estimation. A good example of such a problem is found by Matsunaga et al. (2016) who discussed conflicting estimates for distances to new Cepheids, which are more distant than the GC, in their survey and to Cepheids reported by Dékány et al. (2015b). Readers are also referred to Matsunaga (2017) about these Cepheids as well as general discussions on the effects of interstellar extinction in the context of variable stars in the disk.

Although Matsunaga et al. (2016) found that the extinction law toward the bulge should be close to the one obtained by Nishiyama et al. (2006, cf. Alonso-García et al. 2017; Hosek et al. 2018), the situation may be different for the Galactic disk due to the spatial variation as we discussed in Sect. 2. Such a spatial variation creates a very challenging problem for drawing a map across a wide range of the disk. For example, Tanioka et al. (2017) considered a large range of A_{K_s}/E_{H-K_s} values, from 1.44 (Nishiyama et al. 2006) to 1.83 (Cardelli et al. 1989), toward the inner disk, $l = +20^\circ$ and $+30^\circ$. This large uncertainty in the extinction law prevented them from obtaining an accurate distance to moderately obscured Cepheids; the error in the distance modulus is as large as 0.8 mag with $E_{H-K_s} \sim 1.8$ mag. Such large errors limit the use of Cepheids, and of any other tracers whose distances are obtained based on photometric data, to draw a map of stellar distribution in the disk. The characterization of the extinction law is an urgent task to exploit the outcomes of the ongoing large-scale surveys. Combining more multi-band magnitudes including mid-IR ones (partly done in Tanioka et al. 2017 using 3.6 μm data) would reduce such errors once the extinction coefficients in the mid-IR bands are investigated as well as in the near-IR JHK_s bands.

4 Open Clusters

A collection of 2,167 open clusters with relatively reliable parameters is found in Dias et al. (2002, see the version updated in January 2016), which encompasses an update to the previously published catalogs of Lyngå and Palous (1987) and Mermilliod (1995). Kharchenko et al. (2013) derived open cluster parameters for 3,007 clusters based on 2MASS JHK_s photometry and PPMXL proper motions. The samples contained in both of these catalogs are complete out to 1.8 kpc. With new near-IR data, such as the VVV's, hundreds of open clusters that are more distant and heavily obscured have been found (Borissova et al. 2011; Barbá et al. 2015).

Since open clusters can provide valuable information on age, distance, and extinction based on the isochrone fitting, they can serve as vital probes of the Galactic disk. Interesting results of such applications are found in various papers from long ago to very recent (Reddy et al. 2016, and references therein). In the past few years, a number of clusters containing red supergiants have been found in the interface of the Galactic (long) bar and the inner disk (e.g. Davies et al. 2009, 2012). The formation of these clusters are suggested to be triggered by the interaction between the bar and the disk. Characterizing the parameters of the clusters is, however, quite expensive. Although we already have a large number of samples which are ready to draw a comprehensive picture of the Galactic structure in the solar neighborhood (Joshi 2005; Joshi et al. 2016), samples beyond several kpc are still being discovered and a continuous effort needs to be made as to follow-up observations. An outstanding effort in the spectroscopic approach is, for example, done by the Gaia-ESO survey (Magrini et al. 2017), whose samples are still rather limited compared to the potential sample size of open clusters. In this review, we focus on Cepheids and contact binaries in open clusters, these two kinds of stellar tracers can give distances by themselves and those in the clusters are useful for establishing the distance scales (e.g. mutual calibration of different distance indicators).

4.1 Cepheids in Open Clusters

The method of estimating distances to open clusters based on the main-sequence fitting also provides us with an opportunity to calibrate (or cross-validate) the PL relations of Cepheids. Since the first Cepheids were found in the open clusters NGC 6087 and M 25 by Irwin (1955), many researchers have contributed to this field (e.g. van den Bergh 1957; Tsarevsky et al. 1966; Turner 1986, 2010; Turner et al. 1993; Baumgardt et al. 2000; Hoyle et al. 2003; Tammann et al. 2003; An et al. 2007; Majaess et al. 2008). In particular, Tammann et al. (2003) derived BVI-band Cepheid PL relations based on 25 cluster Cepheids, whereas Anderson et al. (2013) identified five new cluster Cepheids based on an eight-dimensional parameter search and obtained a V-band PL relation for 18 cluster Cepheids; the membership confidence is evaluated for each open cluster–Cepheid pair based on spatial position (two-dimensional angular distance and parallax), kinematics (radial velocity and two-dimensional proper motion), and population (age and metallicity). Recently, Chen et al. (2015, 2017) found nine new cluster Cepheids based on the near-IR 2MASS and VVV catalogs. To date, approximately 40 cluster Cepheids have been found and confirmed, of which 30 can be used to derive PL relations in various photometric filters. In the last decade, more and more attention has been paid to the near-IR or even to mid-IR PL relations, since at these longer wavelengths, the PL relations have much smaller intrinsic scatter and are less affected by interstellar extinction. Based on 31 cluster Cepheids, Chen et al. (2015, 2017) derived new JHK_s PL relations, which represented improvements of up to 40% compared with previous determinations. These relations were used by Wang et al. (2018) to derive mid-IR PL relations of hundreds of Galactic Cepheids.

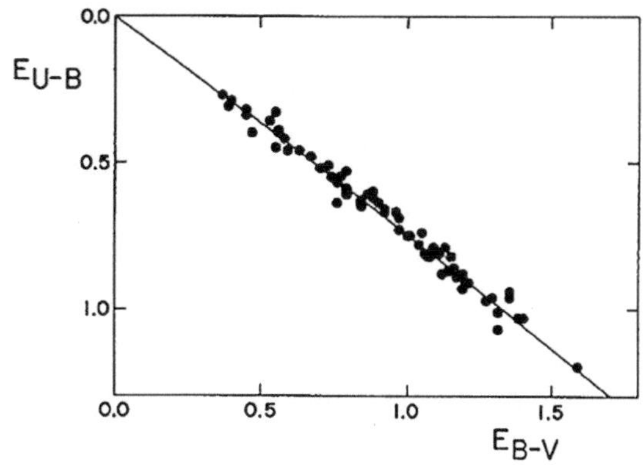

Fig. 6 69 OB-type stars in NGC 6611 and Sgr OB1 were used to derive the color-excess relation. The black line is the best-fitting line, $E_{U-B}/E_{B-V} = 0.721 + 0.025 \times E_{B-V}$. This is Fig. 1 of Turner (1989)

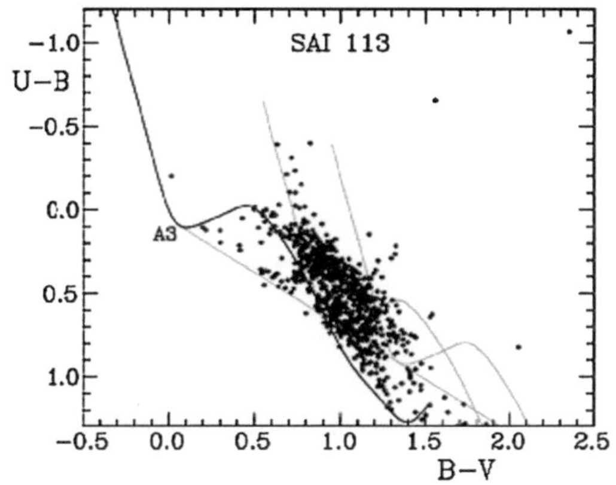

Fig. 7 *UBV* color–color diagram for stars within 6 arcmin of the center of SAI 113. The black curve is the intrinsic curve, while the two gray curves are for color excesses of $E_{B-V} = 0.88$ and 1.28 mag. The gray line shows the A3 'kink' stars for different extinction values; this provides information on extinction law, $E_{U-B}/E_{B-V} = 0.64$. This figure is the top panel of Fig. 6 from Carraro et al. (2017)

As already discussed in Sect. 3, the uncertainty in the extinction law has a large impact on distances to objects obscured by severe interstellar extinction. Here, open clusters can play an important role in disentangling this problem; they can provide an independent reddening value as well as a distance. Once the spectral types of cluster members are obtained, their color excesses can be estimated based on the difference between apparent and intrinsic colors. Based on two or more color-excess values, a more accurate extinction ratio can be estimated for stars within each particular cluster. If spectroscopic observations of most clusters are difficult to obtain, the optimal reddening value can instead be evaluated based on application of the isochrone fitting method. Turner (1989) used *UBV* photometry of six open clusters or associations to study the reddening relation, $E_{U-B}/E_{B-V} = X + Y \cdot E_{B-V}$. He found that the second-order term is almost negligible, and that X ranges from 0.62 to 0.80. This method allows one to estimate the reddening ratio for each open cluster independently based on detailed analysis of its cluster members. Figure 6 shows the reddening relation derived for early-type stars in the clusters NGC 6611 and Sgr OB1. Figure 7 presents another example which can be used to derive the reddening ratio based on the color–color diagram. The JHK_s color–color diagram also shows a 'kink' feature as the *UBV* color–color diagram (Turner 2011), so this is a potential approach for investigating variations in the near-IR extinction law. Turner (2011) provided an empirical JHK_s magnitudes and colors of ZAMS in his Table 1, and applied them to five open clusters. Chen et al. (2015) also used the JHK_s color–color diagram to study the near-IR extinction for 17 open clusters. Most of previously studied clusters have relatively small color excesses, e.g. $0.06 \leq E_{J-H} \leq 0.36$ in Chen et al.

(2015). This limits the ability to constrain the extinction law. Hopefully, with new near-IR instruments and projects such as the VVV survey, more heavily reddened open clusters can be found and investigated. More individual near-IR extinction laws for individual open clusters may be obtained in the next few years.

A particularly interesting cluster in our context is VdBH 222 in which a long-period Cepheid ($P \sim 23.325$ days) was reported by Clark et al. (2015). Prior to their work, two possible kinematic distances were suggested by Marco et al. (2014), ~ 6 kpc and ~ 10 kpc, based on the radial velocity of the cluster, but they suggested that the distance of 10 kpc is more consistent with Galactic structure. As regards the age of VdBH 222 (based on the isochrone fitting), the distance of 10 kpc would suggest 14 Myr, while the distance of 6 kpc corresponds to 20 Myr. The Cepheid provides two constraints, an independent and direct distance to the cluster and a constraint on its age. Based on the Cepheid PL relation and the best-fitting reddening value from Marco et al. (2014), a distance of $5.8^{+0.8}_{-0.7}$ kpc was estimated, while an age of $24.5^{+5.0}_{-4.1}$ Myr was determined by adoption of the Cepheid period–age relation of Bono et al. (2005). This Cepheid is thus consistent with a comparably high isochrone age and a small distance. The distance of 6 kpc places VdBH 222 in the 3-kpc inner arm, which was unexpected since star formation was considered to be suppressed in this arm (see the discussions in Clark et al. 2015). The properties of VdBH 222 also suggest that $\sim 10^4$ $M_\odot$ of stars have been formed in the last ~ 20 Myr, which provides further evidence of active star formation at the interface of the bar and the disk. As regards the problem on the extinction law, the color excess of the cluster, $E_{J-K_s} \sim 1.1$, is larger than those of other clusters investigated so far. This would allow us to give a more tight constraint on the extinction law, although such a study has not been done.

4.2 Contact Binaries in Open Clusters

For contact binaries, one can expect PL (or PLC) relations as follows. Let us start with the basic equations which govern contact binaries:

$$m_1 + m_2 \propto a^3/P^2$$
$$L \propto T_{\rm eff}{}^4 S,$$

where m_1 and m_2 are masses of the two components, a is the semi-major axis, P is the orbital period, L is the total luminosity of the system, and S is the total projected area seen by an observer. In each contact binary system, material transfer is ongoing in their common envelopes, and both components have filled their Roche lobes. The two components typically have the same temperature, $T_{\rm eff}$. For a very simple case of two stars with the identical mass, m, and radius, then a gets close to the radius of each component and S is proportional to a^2 at the maximum light. In such a case, the above two relations can be combined to give

$$L_{\rm max} \propto T_{\rm eff}{}^4 m^{2/3}(1+q)^{2/3} P^{4/3},$$

where $L_{\rm max}$ is the maximum luminosity and q is the mass ratio, m_2/m_1, but $q = 1$ for the simplest case. Main-sequence stars follow a mass-temperature relation, $\log m \sim \gamma \log T_{\rm eff}$ ($\gamma \sim 4/3$, Eker et al. 2015), which leads to a period–luminosity–temperature relation (or to a PLC relation),

$$\log L_{\rm max} = (4/3)\log P + \left[4 + (2/3)\gamma\right]\log T_{\rm eff} + \text{const.}$$

Moreover, T_{eff} is also correlated with mass and radius in case of main-sequence stars (Eker et al. 2015), and the period–luminosity–temperature relation can be reduced to a PL relation for the simple case being discussed. In realistic case of contact binaries, it has long been known since a seminal work by Eggen (1967) that they show a correlation between period and color (and thus temperature) although Eggen (1967) also suggested that the correlation is blurred by the degree of contact.

The above explanation is of course oversimplified. First, two components are no longer spheres but fill the Roche equipotentials. Furthermore, the mass ratio q and the degree of contact affect the relation between the total area S and the semi-major axis a, and the gravity-darkening effect (Lucy 1967) and other effects need to be taken into account. Nevertheless, at least a large fraction of contact binary systems show tight correlations between the physical parameters, and are governed by simple relations as demonstrated here. In his classical textbook, Kopal (1959) calculated the geometry of contact binary systems, in particular those with both components filling the Roche lobes. Table 3-3, in Sect. III.4 in the book, lists the radii of both components, r_1 and r_2, in units of the major axis ($2a$) for various mass ratios between 0 and 1. The simple projected area, S, at the maximum can be estimated as $\pi(r_1^2+r_2^2)$, and it is nearly constant for $0.2 \le q \le 1$ within $\pm 10\%$ while it significantly increases towards $q=0$ at lower q. Systems with small q tend to be missed, or can be ignored if necessary, because of their small amplitudes; see the relation between q and amplitude in Rucinski (2001). Furthermore, as discussed by Eggleton (1983), $P(\rho)^{1/2}$ shows only a little variation over a wide range of q. This constraint is analogous to the period–mean density relation of pulsating stars. Combined with the limited distribution of main-sequence stars and contact binaries themselves (Eyer and Mowlavi 2008) on the HR diagram, like the instability strip for Cepheids, one can expect a PL (or at least a PLC) relation. For more detailed calculations, readers are referred to, e.g., van Hamme and Wilson (1985), Rucinski (1994, 2004), and Chen et al. (2016a).

Contact binaries are divided into early and late types based on their spectral types; the two groups have different temperatures of the two stars in each system, and their orbital periods also tend to be different (longer for the former and shorter for the latter). Late-type contact binaries are also called W Ursa Majoris (W UMa) type, and most well-studied contact binaries belong to this type. Early-type contact binaries are relatively rare and poorly studied compared with W UMa type. Contact binaries are still ubiquitous in the Universe; their number is larger than those of any other variable stars, without solar-like oscillation or other tiny variations into account. Their density is approximately 0.2% in the solar neighborhood and the Galactic bulge but may be slightly lower in the Galactic plane, 0.1% (Rucinski 2006). In intermediate-age open clusters, contact binaries can be as abundant as 0.4%, but they are rare in young (100 Myr) and old (10 Gyr) open clusters. Because they are abundant and can be easily identified, contact binaries are useful as potential distance indicators.

To establish their PL relation, parallaxes and isochrone-fitting distances to open clusters hosting contact binaries are two independent distance estimates that can be used. Rucinski (1994) used 18 W UMa binaries in open clusters to obtain a V-band PLC relation, whereas Rucinski and Duerbeck (1997) obtained a V-band PLC relation, $M_V = -4.30 \log P + 3.28(B-V)_0 + 0.04$, $\sigma = 0.17$, based on 19 binaries with *Hipparcos* parallax accuracy better than 0.25 mag in distance modulus. The latter relation has particularly been used to anchor distances of contact binaries, although its color dependency is high. About one decade later, Rucinski (2006) tried to reduce the PLC relation to a PL relation and obtained a V-band PL relation, $M_V = -1.5\ (\pm 0.8) - 12.0\ (\pm 2.0) \log P$, $\sigma = 0.29$, based on 17 W UMa binaries. This PL relation has larger uncertainties in both coefficients, which limits its application. Since then, with some questions on the reliability and the accuracy of

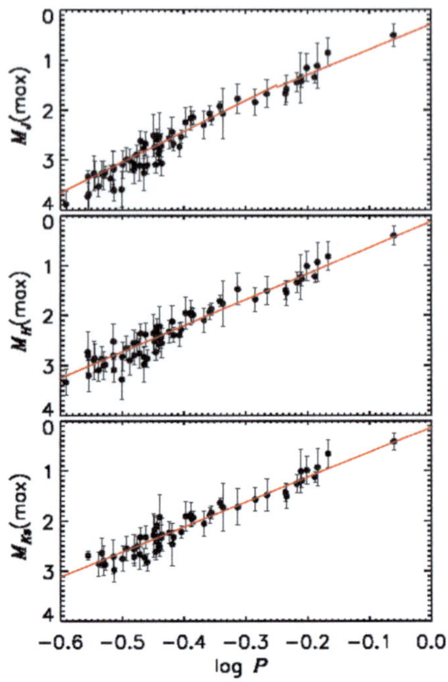

Fig. 8 46 contact binaries with open cluster or nearby moving-group distances and 20 contact binaries with high-accuracy *Hipparcos* parallaxes. The red line shows the best-fitting PL relations. Adopted from Chen et al. (2016a, Fig. 1)

W UMa binaries, their application as a distance indicator hasn't been active compared to classical distance indicators like Cepheids.

Chen et al. (2016b) discussed physical parameters of contact binaries in NGC 188 in detail, and suggested the presence of an accurate PL relation, whose slope is shallower than that derived by Rucinski (2006), considering five confirmed contact binaries in NGC 188 combined with seven in two other clusters (M 67 and Berkeley 39; see Fig. 7 in Chen et al. 2016b). Then, Chen et al. (2016a) collected data for 66 contact binaries with accurate independent distances and derived the near-IR PL relations,

$$M_{J_{\max}}^{\text{late}} = -6.15\ (\pm 0.13) \log P - 0.03\ (\pm 0.05) \qquad (\text{for } \log P < -0.25),$$

$$M_{J_{\max}}^{\text{early}} = -5.04\ (\pm 0.13) \log P + 0.29\ (\pm 0.05) \qquad (\text{for } \log P > -0.25),$$

$$M_{H_{\max}} = -5.22\ (\pm 0.12) \log P + 0.12\ (\pm 0.05),$$

$$M_{K_{\text{s,max}}} = -4.98\ (\pm 0.12) \log P + 0.13\ (\pm 0.04),$$

with 1σ residual scatter of 0.08–0.09 mag (Fig. 8). These PL relations are fairly tight and can be used to measure distances of individual objects with 5% accuracy. The metallicity effect on these relations is also discussed in Chen et al. (2016a), but it is unclear or not significantly larger than the scatters of the relations (see Fig. 6 in Chen et al. 2016a). In addition, to derive the V-band PL relation (Fig. 9), they estimated the near-IR mean distance and mean extinction of 6090 contact binaries with V-band data.

$$M_{V_{\max}} = -9.15\ (\pm 0.12) \log P - 0.23\ (\pm 0.05) \quad (\text{for } \log P < -0.25),$$

$$M_{V_{\max}} = -4.95\ (\pm 0.13) \log P + 0.85\ (\pm 0.02) \quad (\text{for } \log P > -0.25),$$

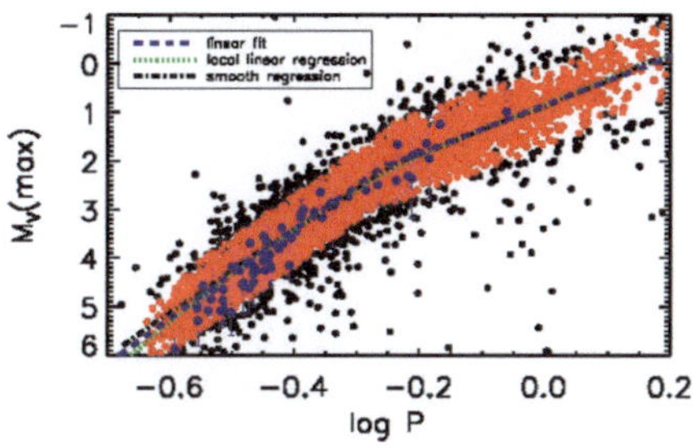

Fig. 9 V-band PL relation based on 6090 contact binaries (black dots). Red dots are stars in the 2σ distribution. The different lines denote different fitting methods; the non-parametric fits agree with the two linear line fits. The early- and late-type contact binaries have different V-band PL relations with a boundary at $\log P = -0.25$ days. Adopted from Chen et al. (2016a, Fig. 5)

with residual scatter of 0.30 and 0.35 mag. The early- and late-type contact binaries (separated at $\log P = -0.25$, Rucinski 2006) follow different PL relations in the V band, while this discrepancy disappears in the HK_s bands.

Applying the V-band PL relation to 102 early-type contact binaries in the Large Magellanic Cloud (LMC), Chen et al. (2016a) obtained an LMC distance modulus of $(m - M)_0 = 18.41 \pm 0.20$ mag. This distance modulus is in accordance with the current best distance to the LMC, $(m - M)_0 = 18.49 \pm 0.09$ mag (summarized by de Grijs et al. 2014). More than 30,000 (Chen et al. 2016b) contact binaries have been found thus far and several hundred contact binaries are found every year in individual papers. With follow-up observations such as spectroscopy, contact binaries can also provide age and kinematic information. They are thus very powerful tracers of stellar populations in the Galaxy.

5 Metallicity Gradient

When distances to many individual stellar tracers like Cepheids and open clusters become available, an important scientific step is then to derive the chemical abundance patterns of these tracers in the disk. In particular, the radial distributions of different elements (e.g. [Fe/H], [α/H], etc.) are fundamental observables to constrain the recent disk chemical enrichment, since they can be directly compared with models (Chiappini et al. 1997; Matteucci and Recchi 2001; Cescutti et al. 2007; Kudritzki et al. 2015). For example, the gradient is affected by the radial migration (see the Introduction). Constraining empirically the efficiency of the radial migration is difficult because stars in the disk quickly lose dynamical signatures of this process. On the other hand, the radial migration affects the chemical abundance patterns and increases the observed metallicity dispersion at any given radius. Observational data of the abundance gradients for stars in a wide range of ages would allow us to constrain the efficiency of orbit migration for blurring the initial abundance pattern.

Because of their young age, Cepheids should not have been mixed so much yet, thus their chemical abundances can be used to set a boundary condition to the radial migration effects. In Fig. 10, we compare the iron abundances of Cepheids in the sample of Genovali et al. (2015) with those resulting from RC stars derived on the basis of APOGEE spectra by Ness et al. (2016). On the basis of their content in α-element abundances with respect to the iron abundances, Ness et al. (2016) distinguished between RC stars in the thick and thin disk. Moreover, they were also able to determine the ages with an accuracy of better than 40% for the thin disk sample. The left panel in Fig. 10 shows the density distribution of the

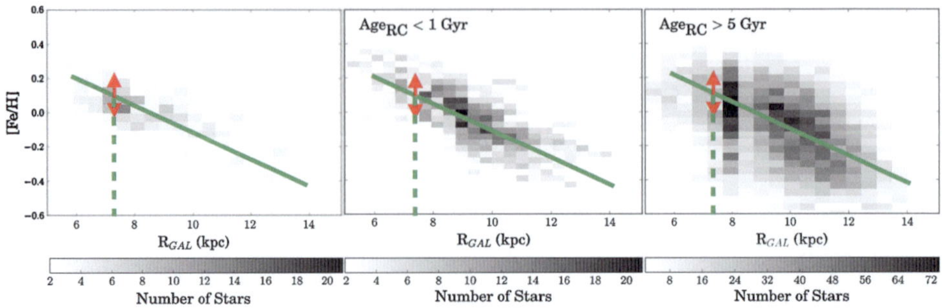

Fig. 10 Left panel: 2D histogram of the iron abundances of known Galactic Cepheids (Genovali et al. 2015) as a function of their Galactocentric radius (R_{GC}). The green solid line shows the best-fitting linear gradient, which corresponds approximately to -0.06 dex kpc^{-1}. The dashed line corresponds to a distance $R_{GC} = 7$ kpc, where the dispersion around the linear gradient: $\pm\sigma_{[Fe/H], R=7\ \mathrm{kpc}} = 0.1$ dex is indicated by the red arrows, and is consistent with the nominal error in the abundance determination of individual stars. Middle panel: Similar but for RC stars younger than 1 Gyr, with iron abundances derived from APOGEE spectra (Ness et al. 2016). The plotted lines are the same as in the left panel but they have been drawn here to guide the eye in comparing the dispersions. Left panel: Same as in the previous panels, but for older RC stars. Here the increase in $\sigma_{[Fe/H], R=7\ \mathrm{kpc}}$ is clearly detectable. Such increase in the dispersion is caused by the orbit migration in the Galactic disk

iron abundance of Cepheids as a function of R_{GC}, which follows a well established linear gradient (Genovali et al. 2015), while the middle and right panels show the distributions of young and old RC stars, respectively (Ness et al. 2016). This plot clearly illustrates the effect of radial migration in the Galactic disk. In fact, it shows how the dispersion around the radial metallicity gradient increases when we move from the younger (left) to the older stellar populations (right).

The metallicity distribution of open clusters also shows the radial gradient (Friel and Janes 1993; Twarog et al. 1997). Unlike Cepheids, there are clusters with various ages from a few Myr to a few Gyr. This makes it possible to study the time evolution of the metallicity gradient (e.g. Tsujimoto et al. 2010; Jacobson et al. 2011). Compared to Cepheids, however, the number of open clusters with their locations and metallicities well determined is still limited (see recent progress, e.g., in Magrini et al. 2017). In addition, open clusters can be disrupted because of internal (e.g. SN explosion) and external (e.g. Galactic potential) effects, thus the observed galactic population may be biased.

An interesting new feature was found in recent IR surveys revealing a rich group of massive stellar clusters which hosts several red supergiants (e.g. Davies et al. 2012, see also Sect. 4). They seem to be preferentially found around the near- and far-side ends of the Galactic bar although the global distribution of such massive clusters in the inner part of the Galaxy remains to be concluded after more complete surveys and detailed studies have been done. Surprisingly low metallicities have been reported for some of these massive clusters (Davies et al. 2009; Origlia et al. 2013, 2016), around -0.2 dex or even lower in contrast to the high metallicities, ~ 0.3 dex at a similarly inner part of the disk (Bono et al. 2013, Fig. 3 and discussions therein). No firm scenario to explain such low metallicities has been proposed yet. There is a large room of improvement for the near future both in characterization of open clusters (especially their distances and metallicities) and finding new clusters. In addition, red supergiants in the inner Galaxy and even wider areas will be useful as bright tracers of stellar populations whether or not they belong to clusters (see e.g. Messineo et al. 2016, 2017).

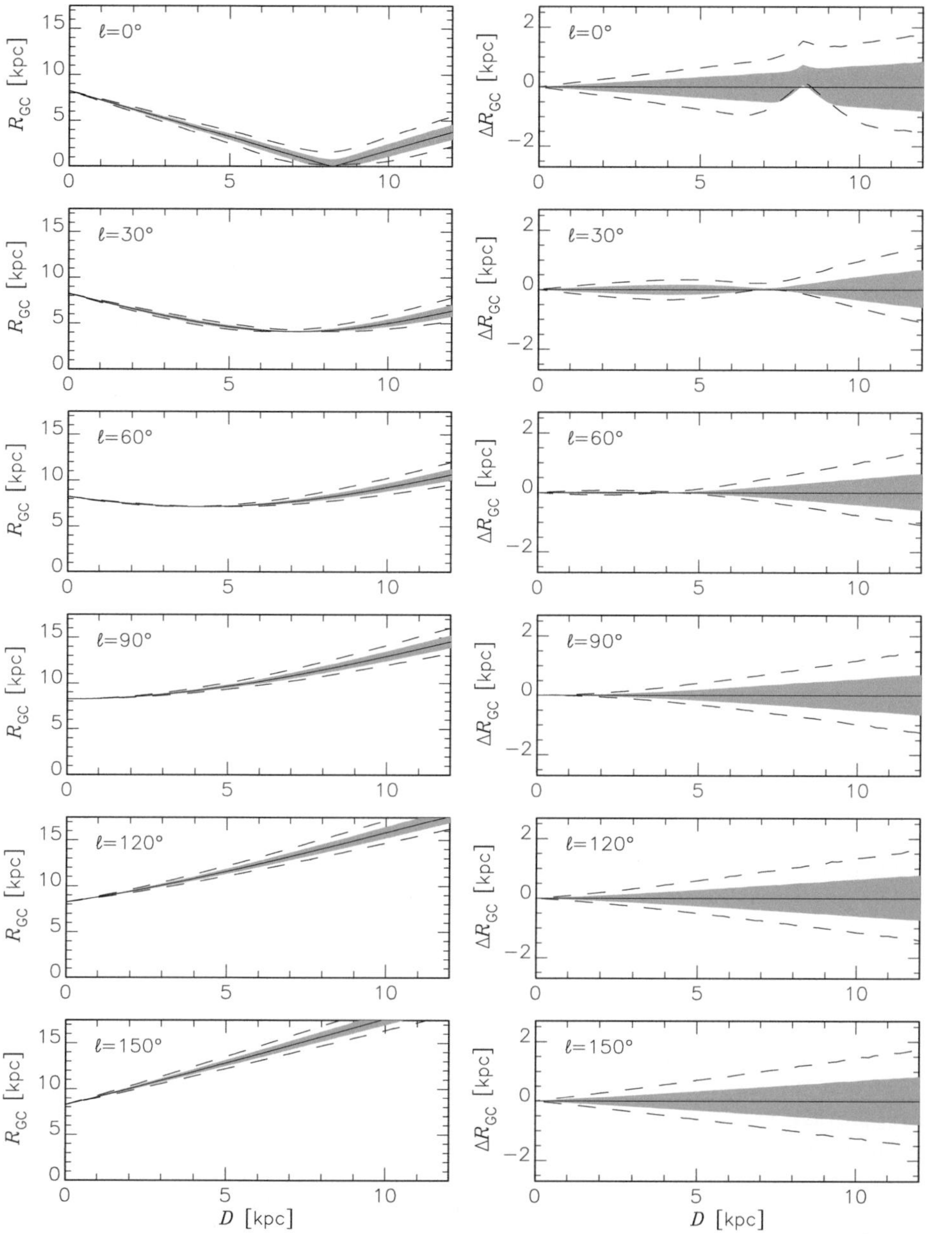

Fig. 11 The Galactocentric distance (R_{GC}) and the deviation of estimated R_{GC} are plotted against the distance, D, on the left and right side panels, respectively, for seven lines of sight with different Galactic longitudes. The solid curve in the middle indicates the true value, while gray filled area and dashed curves indicate the ranges of 1σ errors corresponding to the modulus errors in 0.15 and 0.3 mag

How is the study of the metallicity gradient affected by uncertainties in the distance determination? Fig. 11 plots how the errors in distance are translated to the errors in R_{GC}. Note that we only consider the Galactic plane, i.e. $b = 0°$, and it is assumed that the Sun is located at $R_{GC} = 8.3$ kpc (Reid and Honma 2014). The relation between the two errors depends on Galactic longitude. For example, up to ~ 5 kpc, a line of sight at $l \sim 60°$ runs at almost the same R_{GC} and thus the error in R_{GC} tends to be small. In the direction of $l = 0°$, the kink corresponds to the location of the GC where the error in R_{GC} is always positive. At around a distance of 10 kpc or beyond, the sightline dependency of ΔR_{GC} becomes small, and the errors of 0.15 and 0.30 mag in distance modulus correspond to the R_{GC} errors of approximately 0.4 and 1 kpc, respectively, while these errors are less than half at the distance of 5 kpc. The metallicity gradient is 0.06 dex kpc^{-1}, for example, in the case of Cepheids (Genovali et al. 2014) and the scatter of the metallicity of individual Cepheids around the average trend is ~ 0.1 dex. Detailed studies of substructures (see e.g. Genovali et al. 2014) would require a higher precision, ~ 0.03 dex, at each R_{GC}. Errors in R_{GC} would introduce an artificial scatter in the metallicity gradient. The Cepheids investigated by Genovali et al. (2014) and other previous investigations are not highly reddened, and the careful distance determination by Genovali et al. (2014) gives an error size of 5% or smaller, i.e. 0.1 mag in distance modulus. This error corresponds to roughly $\Delta R_{GC} \sim 0.2$ kpc at several kpc, so that the effect on the metallicity gradient is very small. For Cepheids affected by severe interstellar extinction, like those in Tanioka et al. (2017), large errors due to the extinction correction, ~ 0.5 mag or even larger, would introduce large errors in R_{GC} (1 kpc at the distance of 5 kpc and more than 2 kpc at $D = 10$ kpc). Such large errors would limit the discussions on the chemical structure of the disk traced by Cepheids, and efforts to reduce the uncertainty is required.

It is worthwhile to discuss the metallicity effect on the PL relations of Cepheids in the context of the metallicity gradient. Recent studies made use of near-IR PL relations (or period-Wesenheit relations) to estimate distances to individual Cepheids in a wide range of the disk (e.g. Genovali et al. 2014), and the same approach will be adopted even in the *Gaia* era because of the high extinction in the optical for relatively distant Cepheids. If the metallicity effect on the PL relations introduces a systematic error into distances to Cepheids depending their metallicities, such an error would introduce an additional scatter to the radial distribution of the Cepheids' metallicities and/or skew the gradient from the inner, metal-rich, to the outer, metal-poor, parts of the disk. However, the metallicity effect on the near-IR PL relations is considered to be negligible (see e.g. Bono et al. 2010; Subramanian et al. 2017). Even a very conservative upper limit, 0.1 mag dex^{-1}, for the near-IR relations would lead to the small relative error of ~ 0.1 mag, which is negligible as discussed above, when the most metal-rich ([Fe/H] $\sim +0.5$ dex) and the most metal-poor ([Fe/H] ~ -0.5 dex) Cepheids are compared. The study of the metallicity gradient is therefore not significantly affected by the possible metallicity effect on the PL relations. Instead, the metallicity gradient provides us with a sample of Cepheids with a sufficiently wide range of metallicity within a few kpc for which *Gaia* will deliver accurate parallaxes. Such a sample would be useful to provide constraints on the metallicity effect on the PL relations.

In addition to the more robust estimation of distance, the availability of new large catalogs of Cepheids in the Galaxy calls for a new approach in the determination of their chemical abundances. In fact, a comprehensive and detailed survey of element abundances for hundreds and thousands of Cepheids and young open clusters throughout the disk would offer the unique opportunity to map how chemically homogeneous the Galactic ISM is at a given time in a given place (e.g. spiral arm). This can be achieved with large (single-object or multi-object) spectroscopic surveys. However, such surveys are usually designed to be single-epoch in order to maximize the survey coverage, while Cepheids are pulsating stars affected by dynamical phenomena (pulsation, shock, etc.), and their spectra change as function of phase. Thus, we need to correctly quantify the spectral variations due to the observed pulsation phase. This can be achieved by obtaining phase-resolved spectroscopy of well-understood Cepheids of different periods, which allows for empirical calibration of their variations in effective temperature, gravity and radial velocity. Recent works (Luck and Lambert 2011) show that the metallicities of a given well-known star based on spectra at different epochs show a reasonable agreement with each other, but small scatters of about 0.1 dex may still be attributed to a systematic phase-dependent effect (see, e.g., the latest work in Vasilyev et al. 2018).

It is worthwhile to spend a few more words on the spectroscopic approach to study Cepheids. A new spin in modelling dynamical atmospheres is required to properly analyze Cepheid spectra. Even when considering simple LTE (Local Thermodynamic Equilibrium) 1D atmospheric models of giant stars, the parameter space (low gravity, cool temperature) where the Cepheids are located, is significantly under-sampled. Thus, extrapolation algorithms capable of filling the parameter-space model grid are necessary to derive Cepheid abundances. An alternative approach is to use data-driven spectral models, which can be constructed on the basis of the already available data-sets and applied blindly to new data-sets (Ness et al. 2015). Given their importance, Cepheids have already been included as a specific class of targets in one of the future largest spectroscopic surveys, the 4MOST survey (PI: C. Chiappini), which will deliver final abundances for thousands of new Cepheids (including those which will be discovered by *Gaia*).

6 Concluding Remarks

Young and intermediate-age distance indicators are crucial to study the Galactic disk. Among a few kinds of promising distance indicators, we have mainly discussed Cepheids and open clusters (and contact binaries therein). Cepheids are well established distance indicators; they have been, for example, the most successful tracers used for delineating the metallicity gradient of the disk. Open clusters and contact binaries are also very useful tracers and their advantage is that their ages are spread in a wide range (a few Myr to several Gyr) in contrast to Cepheids being 10–200 Myr old. Making use of them properly requires detailed studies on individual clusters, but they would provide us with important clues into the evolution of the Galactic disk.

When we use distance tracers discovered in large-scale surveys to investigate a large volume of the Galactic disk, the interstellar extinction poses a very challenging problem. Our review has summarized massive efforts to characterize the interstellar extinction dedicated for a long time by many astronomers, but it is also clear that many uncertainties remain. For the rest of this review, we'll discuss two promising approaches to determine the extinction law with newly available data sets today and in the future: (1) spectroscopic data, and (2) *a priori* distances of reddened distance indicators.

With the advent of the APOGEE survey, for example, much progress is going to be made in the studies of the IR extinction law. APOGEE is a high resolution ($R \approx 22{,}500$; Majewski et al. 2017), H-band ($1.51\ \mu\text{m} < \lambda < 1.70\ \mu\text{m}$) spectroscopic survey of stars in our Galaxy. Using APOGEE spectra, we can measure stellar parameters, such as the effective temperature T_{eff}, surface gravity $\log g$, and metal abundance Z. This allows us to precisely determine T_{eff} and the intrinsic colors of late-type giants. It should be noted that spectroscopic data for early-type stars has been often used for studying the extinction law (e.g. Fitzpatrick and Massa 2007). This kind of approach remains important, while the advantage of using late-type giants is they are ubiquitous and much more abundant across the disk so that we can study the spatial variation of the extinction law, if any, almost continuously. To

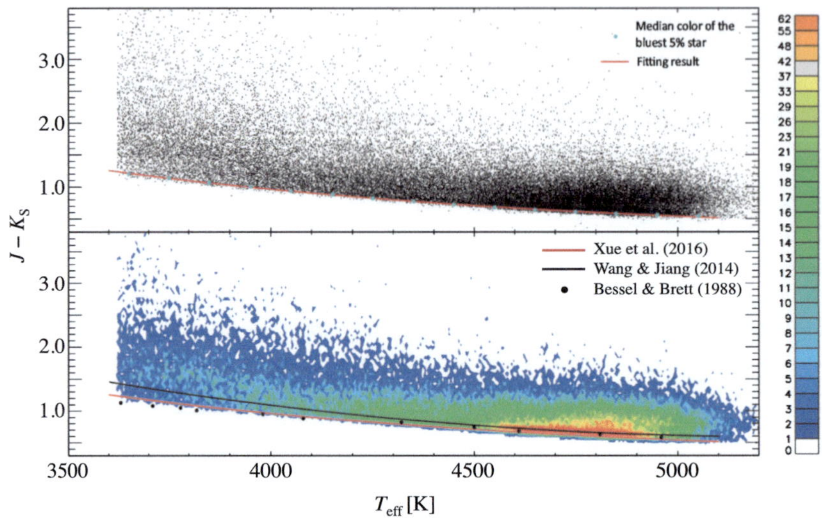

Fig. 12 $J - K_S$ versus $T_{\rm eff}$ diagram for 63,330 APOGEE stars. The intrinsic, non-reddened colors are derived by using bluest stars in this diagram. (Top) The median colors of the bluest 5% of stars are calculated in bins of $\Delta T_{\rm eff} = 100$ K (cyan dots), and they are fit with a quadratic function (red lines). (Bottom) The determined intrinsic $J - K_S$ colors by Xue et al. (2016, red line) are compared with those from Bessell and Brett (1988, black dots) and Wang and Jiang (2014, black solid line). The color scale gives the number density (number, see the scale in the legend, per 5 K in $T_{\rm eff}$ per 0.05 mag in color as indicated in the legend). Adapted from Xue et al. (2016, Fig. 1)

extend such studies to more obscured regions across the Galactic disk, new-generation near-IR high-resolution spectrographs, e.g. CRIRES$^+$ (Dorn et al. 2016) and WINERED (Ikeda et al. 2016), will be useful thanks to their high sensitivity.

By combining the stellar parameters from APOGEE and the observed colors from 2MASS, Wang and Jiang (2014) determined the intrinsic colors of K giants. The bluest edge of the color index in the $T_{\rm eff}$-color diagram is considered to represent the intrinsic colors of the red giants, i.e., the bluest stars suffer from negligible interstellar extinction (Fig. 12). The color excesses E_{J-H}, E_{H-K_s}, and E_{J-K_s} are derived for 5942 stars at $0° < l < 220°$ and $-5° \leq b \leq 5°$. Their interesting results are shown in the color-excess ratio versus color-excess plot, such as the E_{J-H}/E_{J-K_s} versus E_{J-K_s} diagram (Fig. 13). Wang and Jiang (2014) found no tendency of E_{J-H}/E_{J-K_s} with E_{J-K_s}, for $0 < E_{J-K_s} < 5$ mag (which corresponds to A_{K_s} less than 2.5 mag). A systematic change is not found in E_{J-H}/E_{J-K_s} and E_{J-H}/E_{H-K_s}, which are often used as a measure of the near-IR extinction law. The dispersion at low E_{J-K_s} is likely explained fully by observational errors. Wang and Jiang (2014) claimed that these results confirm the "universality" of the near-IR extinction law in the J, H, and K_s bands, from the diffuse to the dense interstellar medium. The variations found in the previous studies could be due partly to differences in the effective wavelength $\lambda_{\rm eff}$ in different photometric systems, or to effects of stellar metallicity, which was restricted to [Fe/H] > -1.0 in Wang and Jiang (2014). Schultheis et al. (2015) also found no evidence for any trend of the variation of the near-IR extinction law. They used stellar properties derived from the Gaia-ESO survey, mainly to probe the 3D distribution of the interstellar extinction in our Galaxy. The fields of the survey are located at high Galactic latitude, $|b| > 20°$. They show a plot of E_{J-H}/E_{H-K_s} versus angle from the GC, and claim no variation of the extinction coefficient along different lines of sight.

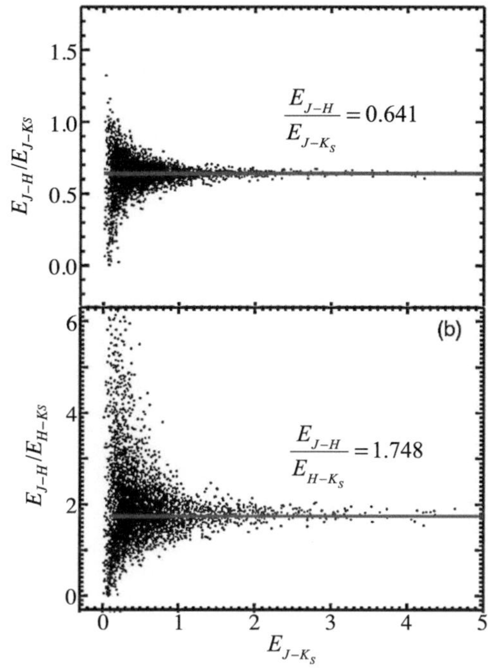

Fig. 13 Distribution of E_{J-H}/E_{J-K_s} (top) and E_{J-H}/E_{H-K_s} (bottom) as a function of E_{J-K_s} (Wang and Jiang 2014). The red lines highlight linear fitting results, $E_{J-H}/E_{J-K_s} = 0.64$ (top) and $E_{J-H}/E_{J-K_s} = 1.748$ (bottom). There is no apparent tendency in either color-excess ratios with increasing reddening. The large dispersion at smaller E_{J-K_s} is fully explained by observational uncertainties. Adapted from Wang and Jiang (2014, Fig. 3)

Such a spectroscopic approach has been extended to the mid-IR wavelengths. Xue et al. (2016) applied the method of Wang and Jiang (2014), i.e. estimating the intrinsic stellar colors in a spectroscopic way for studying the extinction law, to mid-IR data. The mid-IR data sets they used are from *Spitzer*/IRAC, *Spitzer*/MIPS, *WISE*, and *AKARI*. In their analysis, G- and K-type giants are used. To examine the variation of the extinction law, Xue et al. (2016) divided their data sets into three groups: $A_{K_s} < 0.5$, $0.5 \leq A_{K_s} < 1.0$, and $A_{K_s} \geq 1.0$. Then they compared the color-excess ratios $E_{K_s-\lambda}/E_{J-K_s}$ in the three groups, for the *WISE*, *AKARI*, and *Spitzer* nine bands. They also compared the extinction laws in the form A_λ/A_{K_s} under the assumption that $A_J/A_{K_s} = 2.72$ is universal for the three groups. Xue et al. (2016) found remarkably constant color-excess ratios among the three groups. This is also seen in A_λ/A_{K_s}.

McClure (2009), in contrast, detected variations in mid-IR extinction using *Spitzer*/MIPS spectra of 31 late-type giants behind dense molecular clouds. Their conclusion is that the median curves for specific ranges of the amount of extinction characterize the shape of the extinction law. They claimed that they found a "real" variation of the extinction law, as a function of A_K in the mid-IR waverange; the extinction law also becomes flat with increasing A_K. The difference from the results in Xue et al. (2016) may simply reflect the difference between the extinction by relatively diffuse interstellar medium and that by dense molecular clouds. The targets of McClure (2009) are molecular clouds in star forming regions, and the clouds are dense enough for grain growth to occur. As predicted by grain models (e.g. Weingartner and Draine 2001), the grain growth is likely to alter the shape of the extinction law. On the other hand, Wang and Jiang (2014) and Xue et al. (2016) do not distinguish among interstellar environments. Even though there still remain uncertainties, massive spectroscopic data available today and in the future may be critical to determine the extinction law in a precise and robust way.

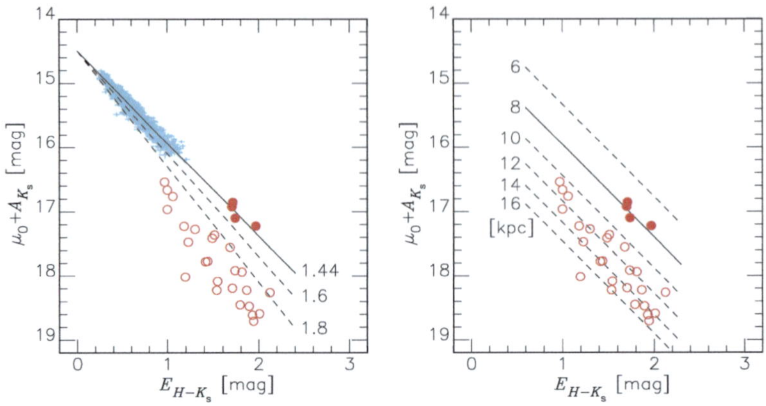

Fig. 14 (Left) Reddening vectors on the color-magnitude diagram with different coefficients of A_{K_s}/E_{H-K_s} starting from the distance of the Galactic center, 8 kpc, at the zero color excess, $E_{H-K_s}=0$ mag. The sequence of cyan points near the top-left corner, taken from Nishiyama et al. (2006), indicates the RC peaks affected by various amounts of reddening. Filled circles indicate the four classical Cepheids in the nuclear stellar disk (Matsunaga et al. 2011, 2015), and open circles indicate the other classical Cepheids reported in Matsunaga et al. (2016). (Right) Same as the left panel, but lines have the slope of $A_{K_s}/E_{H-K_s}=1.44$ and correspond to different distances from 6 to 16 kpc. Adapted from Matsunaga (2017, Fig. 6)

Besides the spectroscopic approaches, reddened objects with known distance are also useful to determine the extinction law. In order to give observational constraints on the total-to-selective extinction ratio (e.g. A_{K_s}/E_{H-K_s}), distance indicators whose distances can be addressed by more than one methods are very useful. For example, Matsunaga et al. (2016), Matsunaga (2017) adopted the extinction law in Nishiyama et al. (2006), and rejected some other laws, based on observational data for Cepheids in the Nuclear Stellar Disk, within ~ 230 pc around the GC (Launhardt et al. 2002). For these Cepheids, while distance estimation based on the PL relation of Cepheids is affected by the extinction law, their distances should agree with that of GC within the size of the Nuclear Stellar Disk because of their membership to this relatively small-sized disk. There are many independent estimates of the distance to the GC, as discussed in de Grijs and Bono (2016, 8.3 ± 0.4 kpc), including Keplerian stellar orbits around the central supermassive black hole (Boehle et al. 2016; Gillessen et al. 2017; Parsa et al. 2017) which is unaffected by the extinction law. Considering how different extinction laws (A_{K_s}/E_{H-K_s} to be more specific) lead to different estimates of distance in comparison with the assumed distance from that of GC limits an acceptable range of the extinction law. Matsunaga et al. (2016) concluded that $A_{K_s}/E_{H-K_s}=1.44$ found by Nishiyama et al. (2006) or slightly smaller values would be consistent with their observational data for the Cepheids in the Nuclear Stellar Disk. Larger A_{K_s}/E_{H-K_s} values would put the four Cepheids (with $E_{H-K_s}\sim 1.6$ mag) significantly more distant than expected; $A_{K_s}/E_{H-K_s}=1.61$ in Nishiyama et al. (2009) lead to A_{K_s} larger by $1.7\times(1.61-1.44)\sim 0.3$ mag, 15% larger in distance, or even a larger difference is introduced by $A_{K_s}/E_{H-K_s}=1.83$ in Cardelli et al. (1989).

Nishiyama et al. (2006) themselves obtained A_{K_s}/E_{H-K_s} and other parameters of the JHK_s extinction law based on a similar but slightly different principle. They compared RC peaks toward many sightlines for which the amounts of extinction are spread in a wide range, 0.3–1.2 in E_{H-K_s}. They assumed that the RC peaks along the sightlines toward the bulge trace the same distance, because of the concentration toward the center of the bulge, even with the different amounts of extinction. The difference of K_s magnitudes ~ 1.3 between the least and most reddened RC peaks observed by Nishiyama et al. (2006) leads to their extinction law, $A_{K_s}/E_{H-K_s} = 1.44 \simeq 1.3/(1.2 - 0.3)$. Note that they didn't use the distance to the GC (or the bulge) but used the relative extinction and reddening of RC peaks assumed to be at the same distance. The results of Matsunaga et al. (2016) and Nishiyama et al. (2006) are summarized in an illustrative way in Fig. 14. These works have demonstrated that distance indicators play a crucial role in determining the extinction law especially when their distances can be estimated by more than one method; *Gaia* parallaxes in the near future, for example, will provide us with great data sets also for problems of the extinction law.

Acknowledgements We thank the organizers and participants of ISSI-BJ for the inspiring discussions therein. NM is grateful to Grant-in-Aid, KAKAENHI No. 26287028, from the Japan Society for the Promotion of Science (JSPS). RdG acknowledges research support from the National Natural Science Foundation of China (grants U1631102, 11373010, and 11633005). SN was supported by JSPS KAKENHI (Nos. 25707012, 15K13463).

References

J. Alonso-García, D. Minniti, M. Catelan, R. Contreras Ramos, O.A. Gonzalez, M. Hempel, P.W. Lucas, R.K. Saito, E. Valenti, M. Zoccali, Extinction ratios in the inner galaxy as revealed by the VVV survey. Astrophys. J. **849**, L13 (2017). https://doi.org/10.3847/2041-8213/aa92c3. arXiv:1710.04854

D. An, D.M. Terndrup, M.H. Pinsonneault, The distances to open clusters from main-sequence fitting. IV. Galactic Cepheids, the LMC, and the local distance scale. Astrophys. J. **671**, 1640–1668 (2007). https://doi.org/10.1086/522304. arXiv:0707.3144

R.I. Anderson, L. Eyer, N. Mowlavi, Cepheids in open clusters: an 8D all-sky census. Mon. Not. R. Astron. Soc. **434**, 2238–2261 (2013). https://doi.org/10.1093/mnras/stt1160. arXiv:1212.5119

S.M. Andrievsky, R.P. Martin, V.V. Kovtyukh, S.A. Korotin, J.R.D. Lépine, Oxygen, α-element and iron abundance distributions in the inner part of the Galactic thin disc—II. Mon. Not. R. Astron. Soc. **461**, 4256–4259 (2016). https://doi.org/10.1093/mnras/stw1631. arXiv:1607.00129

J. Ascenso, C.J. Lada, J. Alves, C.G. Román-Zúñiga, M. Lombardi, The mid-infrared extinction law in the darkest cores of the Pipe Nebula. Astron. Astrophys. **549**, A135 (2013). https://doi.org/10.1051/0004-6361/201220658. arXiv:1211.6556

J. Baba, D. Kawata, N. Matsunaga, R.J.J. Grand, J.A.S. Hunt, Gaia DR1 evidence of disrupting the Perseus Arm. Astrophys. J. **853**, L23 (2018). https://doi.org/10.3847/2041-8213/aaa839. arXiv:1712.04714

R.H. Barbá, A. Roman-Lopes, J.L. Nilo Castellón, V. Firpo, D. Minniti, P. Lucas, J.P. Emerson, M. Hempel, M. Soto, R.K. Saito, Hundreds of new cluster candidates in the VISTA variables in the Vía Láctea survey DR1. Astron. Astrophys. **581**, A120 (2015). https://doi.org/10.1051/0004-6361/201424048. arXiv:1505.02764

H. Baumgardt, C. Dettbarn, R. Wielen, Absolute proper motions of open clusters. I. Observational data. Astron. Astrophys. Suppl. Ser. **146**, 251–258 (2000). https://doi.org/10.1051/aas:2000362. arXiv:astro-ph/0010306

R.L. Beaton, G. Bono, V.F. Braga, M. Dall'Ora, F. Fiorentino, I.S. Jang, C.E. Martínez-Vázquez, N. Matsunaga, M. Monelli, J. Neeley, M. Salariz, Old-aged primary distance indicators. Space Sci. Rev. (2018, submitted)

F. Bertoldi, R. Timmermann, D. Rosenthal, S. Drapatz, C.M. Wright, Detection of HD in the Orion molecular outflow. Astron. Astrophys. **346**, 267–277 (1999). arXiv:astro-ph/9904261

M.S. Bessell, J.M. Brett, JHKLM photometry—standard systems, passbands, and intrinsic colors. Publ. Astron. Soc. Pac. **100**, 1134–1151 (1988). https://doi.org/10.1086/132281

J. Bland-Hawthorn, O. Gerhard, The galaxy in context: structural, kinematic, and integrated properties. Annu. Rev. Astron. Astrophys. **54**, 529–596 (2016). https://doi.org/10.1146/annurev-astro-081915-023441. arXiv:1602.07702

R.C. Bless, B.D. Savage, Ultraviolet photometry from the Orbiting Astronomical Observatory. II. Interstellar extinction. Astrophys. J. **171**, 293 (1972). https://doi.org/10.1086/151282

A. Boehle, A.M. Ghez, R. Schödel, L. Meyer, S. Yelda, S. Albers, G.D. Martinez, E.E. Becklin, T. Do, J.R. Lu, K. Matthews, M.R. Morris, B. Sitarski, G. Witzel, An improved distance and mass estimate for Sgr A* from a multistar orbit analysis. Astrophys. J. **830**, 17 (2016). https://doi.org/10.3847/0004-637X/830/1/17. arXiv:1607.05726

G. Bono, M. Marconi, S. Cassisi, F. Caputo, W. Gieren, G. Pietrzynski, Classical Cepheid pulsation models. X. The period-age relation. Astrophys. J. **621**, 966–977 (2005). https://doi.org/10.1086/427744. arXiv:astro-ph/0411756
G. Bono, F. Caputo, M. Marconi, I. Musella, Insights into the Cepheid distance scale. Astrophys. J. **715**, 277–291 (2010). https://doi.org/10.1088/0004-637X/715/1/277. arXiv:1004.0363
G. Bono, N. Matsunaga, L. Inno, E.P. Lagioia, K. Genovali, Stellar populations in the galactic center, in *Cosmic Rays in Star-Forming Environments*, ed. by D.F. Torres, O. Reimer. Astrophysics and Space Science Proceedings, vol. 34 (2013), p. 115. https://doi.org/10.1007/978-3-642-35410-6-9. arXiv:1304.6211
J. Borissova, C. Bonatto, R. Kurtev, J.R.A. Clarke, F. Peñaloza, S.E. Sale, D. Minniti, J. Alonso-García, E. Artigau, R. Barbá, E. Bica, G.L. Baume, M. Catelan, A.N. Chenè, B. Dias, S.L. Folkes, D. Froebrich, D. Geisler, R. de Grijs, M.M. Hanson, M. Hempel, V.D. Ivanov, M.S.N. Kumar, P. Lucas, F. Mauro, C. Moni Bidin, M. Rejkuba, R.K. Saito, M. Tamura, I. Toledo, New Galactic star clusters discovered in the VVV survey. Astron. Astrophys. **532**, A131 (2011). https://doi.org/10.1051/0004-6361/201116662. arXiv:1106.3045
L. Cambrésy, J. Rho, D.J. Marshall, W.T. Reach, Variation of the extinction law in the Trifid nebula. Astron. Astrophys. **527**, A141 (2011). https://doi.org/10.1051/0004-6361/201015863. arXiv:1101.1089
J.A. Cardelli, G.C. Clayton, J.S. Mathis, The relationship between infrared, optical, and ultraviolet extinction. Astrophys. J. **345**, 245–256 (1989). https://doi.org/10.1086/167900
G. Carraro, The Milky Way thin disk structure as revealed by stars and young open clusters, in *Setting the Scene for Gaia and LAMOST*, ed. by S. Feltzing, G. Zhao, N.A. Walton, P. Whitelock. IAU Symposium, vol. 298 (2014), pp. 7–16. https://doi.org/10.1017/S1743921313006157. arXiv:1307.0569
G. Carraro, D.G. Turner, D.J. Majaess, G.L. Baume, R. Gamen, J.A. Molina Lera, Extinction in the star cluster SAI 113 and galactic structure in Carina. Astron. J. **153**, 156 (2017). https://doi.org/10.3847/1538-3881/aa5c3a. arXiv:1701.08319
L. Casagrande, R. Schönrich, M. Asplund, S. Cassisi, I. Ramírez, J. Meléndez, T. Bensby, S. Feltzing, New constraints on the chemical evolution of the solar neighbourhood and Galactic disc(s). Improved astrophysical parameters for the Geneva-Copenhagen Survey. Astron. Astrophys. **530**, A138 (2011). https://doi.org/10.1051/0004-6361/201016276. arXiv:1103.4651
G. Cescutti, F. Matteucci, P. François, C. Chiappini, Abundance gradients in the Milky Way for α elements, iron peak elements, barium, lanthanum, and europium. Astron. Astrophys. **462**, 943–951 (2007). https://doi.org/10.1051/0004-6361:20065403. arXiv:astro-ph/0609813
N.L. Chapman, L.G. Mundy, S.P. Lai, N.J. Evans II, The mid-infrared extinction law in the Ophiuchus, Perseus, and Serpens molecular clouds. Astrophys. J. **690**, 496–511 (2009). https://doi.org/10.1088/0004-637X/690/1/496. arXiv:0809.1106
B.Q. Chen, M. Schultheis, B.W. Jiang, O.A. Gonzalez, A.C. Robin, M. Rejkuba, D. Minniti, Three-dimensional interstellar extinction map toward the Galactic bulge. Astron. Astrophys. **550**, A42 (2013). https://doi.org/10.1051/0004-6361/201219682. arXiv:1211.3092
X. Chen, R. de Grijs, L. Deng, A search for open cluster Cepheids in the Galactic plane. Mon. Not. R. Astron. Soc. **446**, 1268–1282 (2015). https://doi.org/10.1093/mnras/stu2165. arXiv:1410.4489
X. Chen, R. de Grijs, L. Deng, Contact binaries as viable distance indicators: new, competitive $(V)JHK_S$ period-luminosity relations. Astrophys. J. **832**, 138 (2016a). https://doi.org/10.3847/0004-637X/832/2/138. arXiv:1609.02267
X. Chen, L. Deng, R. de Grijs, X. Zhang, Y. Xin, K. Wang, C. Luo, Z. Yan, J. Tian, J. Sun, Q. Liu, Q. Zhou, Z. Luo, Physical parameter study of eight W Ursae Majoris-type contact binaries in NGC 188. Astron. J. **152**, 129 (2016b). https://doi.org/10.3847/0004-6256/152/5/129. arXiv:1607.06152
X. Chen, R. de Grijs, L. Deng, New open cluster Cepheids in the VVV survey tightly constrain near-infrared period-luminosity relations. Mon. Not. R. Astron. Soc. **464**, 1119–1126 (2017). https://doi.org/10.1093/mnras/stw2390
C. Chiappini, F. Matteucci, R. Gratton, The chemical evolution of the Galaxy: the two-infall model. Astrophys. J. **477**, 765–780 (1997). https://doi.org/10.1086/303726. arXiv:astro-ph/9609199
C. Chiappini, F. Matteucci, D. Romano, Abundance gradients and the formation of the Milky Way. Astrophys. J. **554**, 1044–1058 (2001). https://doi.org/10.1086/321427. arXiv:astro-ph/0102134
J.E. Chiar, A.G.G.M. Tielens, D.C.B. Whittet, W.A. Schutte, A.C.A. Boogert, D. Lutz, E.F. van Dishoeck, M.P. Bernstein, The composition and distribution of dust along the line of sight toward the Galactic center. Astrophys. J. **537**, 749–762 (2000). https://doi.org/10.1086/309047. arXiv:astro-ph/0002421
J.S. Clark, I. Negueruela, M.E. Lohr, R. Dorda, C. González-Fernández, F. Lewis, P. Roche, A long-period Cepheid variable in the starburst cluster VdBH222. Astron. Astrophys. **584**, L12 (2015). https://doi.org/10.1051/0004-6361/201527360. arXiv:1510.06643
R. da Silva, B. Lemasle, G. Bono, K. Genovali, A. McWilliam, S. Cristallo, M. Bergemann, R. Buonanno, M. Fabrizio, I. Ferraro, P. François, G. Iannicola, L. Inno, C.D. Laney, R.P. Kudritzki, N. Matsunaga, M. Nonino, F. Primas, N. Przybilla, M. Romaniello, F. Thévenin, M.A. Urbaneja, Neutron-

capture elements across the Galactic thin disk using Cepheids. Astron. Astrophys. **586**, A125 (2016). https://doi.org/10.1051/0004-6361/201527300. arXiv:1510.06314

A.K. Dambis, L.N. Berdnikov, Y.N. Efremov, A.Y. Kniazev, A.S. Rastorguev, E.V. Glushkova, V.V. Kravtsov, D.G. Turner, D.J. Majaess, R. Sefako, Classical Cepheids and the spiral structure of the Milky Way. Astron. Lett. **41**, 489–500 (2015). https://doi.org/10.1134/S1063773715090017. arXiv:1505.01782

A. Damineli, L.A. Almeida, R.D. Blum, D.S.C. Damineli, F. Navarete, M.S. Rubinho, M. Teodoro, Extinction law in the range 0.4–4.8 μm and the 8620 Å DIB towards the stellar cluster Westerlund 1. Mon. Not. R. Astron. Soc. **463**, 2653–2666 (2016). https://doi.org/10.1093/mnras/stw2122. arXiv:1607.04639

J.R.A. Davenport, Ž. Ivezić, A.C. Becker, J.J. Ruan, N.M. Hunt-Walker, K.R. Covey, A.R. Lewis, Y. Al-Sayyad, L.M. Anderson, The SDSS-2MASS-WISE 10-dimensional stellar colour locus. Mon. Not. R. Astron. Soc. **440**, 3430–3438 (2014). https://doi.org/10.1093/mnras/stu466. arXiv:1403.1875

B. Davies, L. Origlia, R.P. Kudritzki, D.F. Figer, R.M. Rich, F. Najarro, I. Negueruela, J.S. Clark, Chemical abundance patterns in the inner Galaxy: the Scutum red supergiant clusters. Astrophys. J. **696**, 2014–2025 (2009). https://doi.org/10.1088/0004-637X/696/2/2014. arXiv:0902.2378

B. Davies, D. de La Fuente, F. Najarro, J.A. Hinton, C. Trombley, D.F. Figer, E. Puga, A newly discovered young massive star cluster at the far end of the Galactic bar. Mon. Not. R. Astron. Soc. **419**, 1860–1870 (2012). https://doi.org/10.1111/j.1365-2966.2011.19840.x. arXiv:1111.2630

R. de Grijs, G. Bono, Clustering of local group distances: publication bias or correlated measurements? IV. The Galactic center. Astrophys. J. Suppl. Ser. **227**, 5 (2016). https://doi.org/10.3847/0067-0049/227/1/5. arXiv:1610.02457

R. de Grijs, J.E. Wicker, G. Bono, Clustering of local group distances: publication bias or correlated measurements? I. The Large Magellanic Cloud. Astron. J. **147**, 122 (2014). https://doi.org/10.1088/0004-6256/147/5/122. arXiv:1403.3141

I. Dékány, D. Minniti, G. Hajdu, J. Alonso-García, M. Hempel, T. Palma, M. Catelan, W. Gieren, D. Majaess, Discovery of a pair of classical Cepheids in an invisible cluster beyond the Galactic bulge. Astrophys. J. **799**, L11 (2015a). https://doi.org/10.1088/2041-8205/799/1/L11. arXiv:1412.8658

I. Dékány, D. Minniti, D. Majaess, M. Zoccali, G. Hajdu, J. Alonso-García, M. Catelan, W. Gieren, J. Borissova, The VVV survey reveals classical Cepheids tracing a young and thin stellar disk across the Galaxy's bulge. Astrophys. J. **812**, L29 (2015b). https://doi.org/10.1088/2041-8205/812/2/L29. arXiv:1509.08402

W.S. Dias, J.R.D. Lépine, Direct determination of the spiral pattern rotation speed of the Galaxy. Astrophys. J. **629**, 825–831 (2005). https://doi.org/10.1086/431456. arXiv:astro-ph/0503083

W.S. Dias, B.S. Alessi, A. Moitinho, J.R.D. Lépine, New catalogue of optically visible open clusters and candidates. Astron. Astrophys. **389**, 871–873 (2002). https://doi.org/10.1051/0004-6361:20020668. arXiv:astro-ph/0203351

R.J. Dorn, R. Follert, P. Bristow, C. Cumani, S. Eschbaumer, J. Grunhut, A. Haimerl, A. Hatzes, U. Heiter, R. Hinterschuster, D.J. Ives, Y. Jung, F. Kerber, B. Klein, A. Lavaila, J.L. Lizon, T. Löwinger, I. Molina-Conde, B. Nicholson, T. Marquart, E. Oliva, L. Origlia, L. Pasquini, J. Paufique, N. Piskunov, A. Reiners, U. Seemann, J. Stegmeier, E. Stempels, S. Tordo, The "+" for CRIRES: enabling better science at infrared wavelength and high spectral resolution at the ESO VLT, in *Ground-Based and Airborne Instrumentation for Astronomy VI*. Proc. SPIE, vol. 9908 (2016). https://doi.org/10.1117/12.2232837

B.T. Draine, Interstellar extinction in the infrared, in *Infrared Spectroscopy in Astronomy*, ed. by E. Böhm-Vitense. ESA Special Publication, vol. 290 (1989)

B.T. Draine, Scattering by interstellar dust grains. I. Optical and ultraviolet. Astrophys. J. **598**, 1017–1025 (2003a). https://doi.org/10.1086/379118. arXiv:astro-ph/0304060

B.T. Draine, Scattering by interstellar dust grains. II. X-rays. Astrophys. J. **598**, 1026–1037 (2003b). https://doi.org/10.1086/379123. arXiv:astro-ph/0308251

O.J. Eggen, Contact binaries, II. Mon. Not. R. Astron. Soc. **70**, 111 (1967)

P.P. Eggleton, Approximations to the radii of Roche lobes. Astrophys. J. **268**, 368 (1983). https://doi.org/10.1086/160960

Z. Eker, F. Soydugan, E. Soydugan, S. Bilir, E. Yaz Gökçe, I. Steer, M. Tüysüz, T. Şenyüz, O. Demircan, Main-sequence effective temperatures from a revised mass-luminosity relation based on accurate properties. Astron. J. **149**, 131 (2015). https://doi.org/10.1088/0004-6256/149/4/131. arXiv:1501.06585

L. Eyer, N. Mowlavi, Variable stars across the observational HR diagram, in *Journal of Physics Conference Series*. Journal of Physics Conference Series, vol. 118 (2008). https://doi.org/10.1088/1742-6596/118/1/012010. arXiv:0712.3797

M. Feast, P. Whitelock, Galactic kinematics of Cepheids from HIPPARCOS proper motions. Mon. Not. R. Astron. Soc. **291**, 683 (1997). https://doi.org/10.1093/mnras/291.4.683. arXiv:astro-ph/9706293

E.L. Fitzpatrick, D. Massa, An analysis of the shapes of interstellar extinction curves. V. The IR-through-UV curve morphology. Astrophys. J. **663**, 320–341 (2007). https://doi.org/10.1086/518158. arXiv:0705.0154

E.L. Fitzpatrick, D. Massa, An analysis of the shapes of interstellar extinction curves. VI. The near-IR extinction law. Astrophys. J. **699**, 1209–1222 (2009). https://doi.org/10.1088/0004-637X/699/2/1209. arXiv:0905.0133

K.M. Flaherty, J.L. Pipher, S.T. Megeath, E.M. Winston, R.A. Gutermuth, J. Muzerolle, L.E. Allen, G.G. Fazio, Infrared extinction toward nearby star-forming regions. Astrophys. J. **663**, 1069–1082 (2007). https://doi.org/10.1086/518411. arXiv:astro-ph/0703777

E.D. Friel, K.A. Janes, Metallicities and radial velocities of old open clusters. Astron. Astrophys. **267**, 75–91 (1993)

T.K. Fritz, S. Gillessen, K. Dodds-Eden, D. Lutz, R. Genzel, W. Raab, T. Ott, O. Pfuhl, F. Eisenhauer, F. Yusef-Zadeh, Line derived infrared extinction toward the Galactic center. Astrophys. J. **737**, 73 (2011). https://doi.org/10.1088/0004-637X/737/2/73. arXiv:1105.2822

D. Froebrich, G.C. Murphy, M.D. Smith, J. Walsh, C. Del Burgo, A large-scale extinction map of the Galactic anticentre from 2MASS. Mon. Not. R. Astron. Soc. **378**, 1447–1460 (2007). https://doi.org/10.1111/j.1365-2966.2007.11886.x. arXiv:0704.2993

J. Gao, B.W. Jiang, A. Li, Mid-infrared extinction and its variation with Galactic longitude. Astrophys. J. **707**, 89–102 (2009). https://doi.org/10.1088/0004-637X/707/1/89. arXiv:0910.3037

K. Genovali, B. Lemasle, G. Bono, M. Romaniello, M. Fabrizio, I. Ferraro, G. Iannicola, C.D. Laney, M. Nonino, M. Bergemann, R. Buonanno, P. François, L. Inno, R.P. Kudritzki, N. Matsunaga, S. Pedicelli, F. Primas, F. Thévenin, On the fine structure of the Cepheid metallicity gradient in the Galactic thin disk. Astron. Astrophys. **566**, A37 (2014). https://doi.org/10.1051/0004-6361/201323198. arXiv:1403.6128

K. Genovali, B. Lemasle, R. da Silva, G. Bono, M. Fabrizio, M. Bergemann, R. Buonanno, I. Ferraro, P. François, G. Iannicola, L. Inno, C.D. Laney, R.P. Kudritzki, N. Matsunaga, M. Nonino, F. Primas, M. Romaniello, M.A. Urbaneja, F. Thévenin, On the α-element gradients of the Galactic thin disk using Cepheids. Astron. Astrophys. **580**, A17 (2015). https://doi.org/10.1051/0004-6361/201525894. arXiv:1503.03758

S. Gillessen, P.M. Plewa, F. Eisenhauer, R. Sari, I. Waisberg, M. Habibi, O. Pfuhl, E. George, J. Dexter, S. von Fellenberg, T. Ott, R. Genzel, An update on monitoring stellar orbits in the Galactic center. Astrophys. J. **837**, 30 (2017). https://doi.org/10.3847/1538-4357/aa5c41. arXiv:1611.09144

M. Gómez, S.J. Kenyon, A near-infrared imaging survey of the Chamaeleon I dark cloud. Astron. J. **121**, 974–983 (2001). https://doi.org/10.1086/318776. arXiv:astro-ph/0011112

A.J. Gosling, R.M. Bandyopadhyay, K.M. Blundell, The complex, variable near-infrared extinction towards the Nuclear Bulge. Mon. Not. R. Astron. Soc. **394**, 2247–2254 (2009). https://doi.org/10.1111/j.1365-2966.2009.14493.x. arXiv:0901.1987

R.J.J. Grand, D. Kawata, M. Cropper, Impact of radial migration on stellar and gas radial metallicity distribution. Mon. Not. R. Astron. Soc. **447**, 4018–4027 (2015). https://doi.org/10.1093/mnras/stv016. arXiv:1410.3836

M. Haywood, P. Di Matteo, M.D. Lehnert, D. Katz, A. Gómez, The age structure of stellar populations in the solar vicinity. Clues of a two-phase formation history of the Milky Way disk. Astron. Astrophys. **560**, A109 (2013). https://doi.org/10.1051/0004-6361/201321397. arXiv:1305.4663

L. He, D.C.B. Whittet, D. Kilkenny, J.H. Spencer Jones, Interstellar extinction from 0.35 to 2.2 microns: a study based on luminous southern stars. Astrophys. J. Suppl. Ser. **101**, 335 (1995). https://doi.org/10.1086/192243

M. Honma, T. Nagayama, K. Ando, T. Bushimata, Y.K. Choi, T. Handa, T. Hirota, H. Imai, T. Jike, M.K. Kim, O. Kameya, N. Kawaguchi, H. Kobayashi, T. Kurayama, S. Kuji, N. Matsumoto, S. Manabe, T. Miyaji, K. Motogi, A. Nakagawa, H. Nakanishi, K. Niinuma, C.S. Oh, T. Omodaka, T. Oyama, N. Sakai, K. Sato, M. Sato, K.M. Shibata, S. Shiozaki, K. Sunada, Y. Tamura, Y. Ueno, A. Yamauchi, Fundamental parameters of the Milky Way galaxy based on VLBI astrometry. Publ. Astron. Soc. Jpn. **64**, 136 (2012). https://doi.org/10.1093/pasj/64.6.136. arXiv:1211.3843

M.W. Hosek Jr., J.R. Lu, J. Anderson, T. Do, E.F. Schlafly, A.M. Ghez, W.I. Clarkson, M.R. Morris, S.M. Albers, The optical/near-infrared extinction law in highly reddened regions. Astrophys. J. **855**, 13 (2018). https://doi.org/10.3847/1538-4357/aaabbb. arXiv:1801.08574

F. Hoyle, T. Shanks, N.R. Tanvir, Distances to Cepheid open clusters via optical and K-band imaging. Mon. Not. R. Astron. Soc. **345**, 269–291 (2003). https://doi.org/10.1046/j.1365-8711.2003.06939.x. arXiv:astro-ph/0002521

Y. Ikeda, N. Kobayashi, S. Kondo, S. Otsubo, S. Hamano, H. Sameshima, T. Yoshikawa, K. Fukue, K. Nakanishi, T. Kawanishi, T. Nakaoka, M. Kinoshita, A. Kitano, A. Asano, K. Takenaka, A. Watase, H. Mito, C. Yasui, A. Minami, N. Izumu, R. Yamamoto, M. Mizumoto, T. Arasaki, A. Arai, N. Matsunaga, H. Kawakita, High sensitivity, wide coverage, and high-resolution NIR non-cryogenic spectrograph, WINERED, in *Ground-Based and Airborne Instrumentation for Astronomy VI*. Proc. SPIE, vol. 9908 (2016). https://doi.org/10.1117/12.2230886

R. Indebetouw, J.S. Mathis, B.L. Babler, M.R. Meade, C. Watson, B.A. Whitney, M.J. Wolff, M.G. Wolfire, M. Cohen, T.M. Bania, R.A. Benjamin, D.P. Clemens, J.M. Dickey, J.M. Jackson, H.A. Kobulnicky, A.P. Marston, E.P. Mercer, J.R. Stauffer, S.R. Stolovy, E. Churchwell, The wavelength dependence of interstellar extinction from 1.25 to 8.0 μm using GLIMPSE data. Astrophys. J. **619**, 931–938 (2005). https://doi.org/10.1086/426679. arXiv:astro-ph/0406403

J.B. Irwin, Cepheid variables and Galactic structure. Mon. Notes Astron. Soc. South. Afr. **14**, 38 (1955)

Ž. Ivezić, T.C. Beers, M. Jurić, Galactic stellar populations in the era of the Sloan Digital Sky Survey and other large surveys. Annu. Rev. Astron. Astrophys. **50**, 251–304 (2012). https://doi.org/10.1146/annurev-astro-081811-125504. arXiv:1308.6386

H.R. Jacobson, C.A. Pilachowski, E.D. Friel, A chemical abundance study of 10 open clusters based on WIYN-Hydra spectroscopy. Astron. J. **142**, 59 (2011). https://doi.org/10.1088/0004-6256/142/2/59. arXiv:1107.4139

B.W. Jiang, J. Gao, A. Omont, F. Schuller, G. Simon, Extinction at 7 μm and 15 μm from the ISOGAL survey. Astron. Astrophys. **446**, 551–560 (2006). https://doi.org/10.1051/0004-6361:20053501. arXiv:astro-ph/0509657

H.L. Johnson, J. Borgman, The law of interstellar extinction. Bull. Astron. Inst. Neth. **17**, 115 (1963)

H.L. Johnson, W.W. Morgan, Some evidence for a regional variation in the law of interstellar reddening. Astrophys. J. **122**, 142 (1955). https://doi.org/10.1086/146063

Y.C. Joshi, Interstellar extinction towards open clusters and Galactic structure. Mon. Not. R. Astron. Soc. **362**, 1259–1266 (2005). https://doi.org/10.1111/j.1365-2966.2005.09391.x. arXiv:astro-ph/0507069

Y.C. Joshi, A.K. Dambis, A.K. Pandey, S. Joshi, Study of open clusters within 1.8 kpc and understanding the Galactic structure. Astron. Astrophys. **593**, A116 (2016). https://doi.org/10.1051/0004-6361/201628944. arXiv:1606.06425

A.H. Joy, Rotation effects, interstellar absorption, and certain dynamical constants of the Galaxy determined from Cepheid variables. Astrophys. J. **89**, 356 (1939). https://doi.org/10.1086/144060

D. Kawata, J. Bovy, N. Matsunaga, J. Baba, Galactic local circular velocity from Gaia DR1 Cepheids and effects of non-axisymmetry. ArXiv e-prints (2018). arXiv:1803.05927

S.J. Kenyon, E.A. Lada, M. Barsony, The near-infrared extinction law and limits on the pre-main-sequence population of the Rho Ophiuchi dark cloud. Astron. J. **115**, 252 (1998). https://doi.org/10.1086/300188

N.V. Kharchenko, A.E. Piskunov, E. Schilbach, S. Röser, R.D. Scholz, Global survey of star clusters in the Milky Way. II. The catalogue of basic parameters. Astron. Astrophys. **558**, A53 (2013). https://doi.org/10.1051/0004-6361/201322302. arXiv:1308.5822

Z. Kopal, *Close Binary Systems* (1959)

G. Kovacs, E.G. Kisvarsanyi, J.R. Buchler, Cepheid radial velocity curves revisited. Astrophys. J. **351**, 606–616 (1990). https://doi.org/10.1086/168499

R.P. Kudritzki, F. Bresolin, N. Przybilla, A new extragalactic distance determination method using the flux-weighted gravity of late B and early A supergiants. Astrophys. J. **582**, L83–L86 (2003). https://doi.org/10.1086/367690. arXiv:astro-ph/0212042

R.P. Kudritzki, M.A. Urbaneja, F. Bresolin, N. Przybilla, W. Gieren, G. Pietrzyński, Quantitative spectroscopy of 24 A supergiants in the Sculptor galaxy NGC 300: flux-weighted gravity-luminosity relationship, metallicity, and metallicity gradient. Astrophys. J. **681**, 269–289 (2008). https://doi.org/10.1086/588647. arXiv:0803.3654

R.P. Kudritzki, I.T. Ho, A. Schruba, A. Burkert, H.J. Zahid, F. Bresolin, G.I. Dima, The chemical evolution of local star-forming galaxies: radial profiles of ISM metallicity, gas mass, and stellar mass and constraints on galactic accretion and winds. Mon. Not. R. Astron. Soc. **450**, 342–359 (2015). https://doi.org/10.1093/mnras/stv522. arXiv:1503.01503

M. Landini, A. Natta, P. Salinari, E. Oliva, A.F.M. Moorwood, A spectroscopic determination of the IR extinction curve in the direction of G333.6-0.2. Astron. Astrophys. **134**, 284–289 (1984)

K.A. Larson, D.C.B. Whittet, Reddening and the extinction law at high Galactic latitude. Astrophys. J. **623**, 897–910 (2005). https://doi.org/10.1086/428877

R. Launhardt, R. Zylka, P.G. Mezger, The nuclear bulge of the Galaxy. III. Large-scale physical characteristics of stars and interstellar matter. Astron. Astrophys. **384**, 112–139 (2002). https://doi.org/10.1051/0004-6361:20020017. arXiv:astro-ph/0201294

B. Lemasle, P. François, K. Genovali, V.V. Kovtyukh, G. Bono, L. Inno, C.D. Laney, L. Kaper, M. Bergemann, M. Fabrizio, N. Matsunaga, S. Pedicelli, F. Primas, M. Romaniello, Galactic abundance gradients from Cepheids. α and heavy elements in the outer disk. Astron. Astrophys. **558**, A31 (2013). https://doi.org/10.1051/0004-6361/201322115. arXiv:1308.3249

M. Lindqvist, H.J. Habing, A. Winnberg, OH/IR stars close to the Galactic centre. II—their spatial and kinematic properties and the mass distribution within 5–100 PC from the galactic centre. Astron. Astrophys. **259**, 118–127 (1992)

R.E. Luck, D.L. Lambert, The distribution of the elements in the Galactic disk. III. A reconsideration of Cepheids from $l = 30°$ to $250°$. Astron. J. **142**, 136 (2011). https://doi.org/10.1088/0004-6256/142/4/136. arXiv:1108.1947

L.B. Lucy, Gravity-darkening for stars with convective envelopes. Z. Astrophys. **65**, 89 (1967)

S.L. Lumsden, P.J. Puxley, Near-infrared spectroscopy of the ultracompact HII region G45.12+0.13. Mon. Not. R. Astron. Soc. **281**, 493–508 (1996). https://doi.org/10.1093/mnras/281.2.493. arXiv:astro-ph/9603115

D. Lutz, ISO observations of the Galactic centre, in *The Universe as Seen by ISO*, ed. by P. Cox, M. Kessler. ESA Special Publication, vol. 427 (1999), p. 623

D. Lutz, H. Feuchtgruber, R. Genzel, D. Kunze, D. Rigopoulou, H.W.W. Spoon, C.M. Wright, E. Egami, R. Katterloher, E. Sturm, E. Wieprecht, A. Sternberg, A.F.M. Moorwood, T. de Graauw, SWS observations of the Galactic center. Astron. Astrophys. **315**, L269–L272 (1996)

G. Lyngå, J. Palous, The local kinematics of open star clusters. Astron. Astrophys. **188**, 35–38 (1987)

L. Magrini, S. Randich, G. Kordopatis, N. Prantzos, D. Romano, A. Chieffi, M. Limongi, P. François, E. Pancino, E. Friel, A. Bragaglia, G. Tautvaišiene, L. Spina, J. Overbeek, T. Cantat-Gaudin, P. Donati, A. Vallenari, R. Sordo, F.M. Jiménez-Esteban, B. Tang, A. Drazdauskas, S. Sousa, S. Duffau, P. Jofré, G. Gilmore, S. Feltzing, E. Alfaro, T. Bensby, E. Flaccomio, S. Koposov, A. Lanzafame, R. Smiljanic, A. Bayo, G. Carraro, A.R. Casey, M.T. Costado, F. Damiani, E. Franciosini, A. Hourihane, C. Lardo, J. Lewis, L. Monaco, L. Morbidelli, G. Sacco, L. Sbordone, C.C. Worley, S. Zaggia, The Gaia-ESO survey: radial distribution of abundances in the Galactic disc from open clusters and young-field stars. Astron. Astrophys. **603**, A2 (2017). https://doi.org/10.1051/0004-6361/201630294. arXiv:1703.00762

J. Maíz Apellániz, Deriving extinction laws with O stars: from the IR to the UV, in *Highlights of Spanish Astrophysics VIII*, ed. by A.J. Cenarro, F. Figueras, C. Hernández-Monteagudo, J. Trujillo Bueno, L. Valdivielso (2015), pp. 402–410. arXiv:1410.5184

D.J. Majaess, D.G. Turner, D.J. Lane, Assessing potential cluster Cepheids from a new distance and reddening parametrization and Two Micron All Sky Survey photometry. Mon. Not. R. Astron. Soc. **390**, 1539–1548 (2008). https://doi.org/10.1111/j.1365-2966.2008.13834.x. arXiv:0808.2937

D.J. Majaess, D.G. Turner, D.J. Lane, Characteristics of the Galaxy according to Cepheids. Mon. Not. R. Astron. Soc. **398**, 263–270 (2009). https://doi.org/10.1111/j.1365-2966.2009.15096.x. arXiv:0903.4206

S.R. Majewski, R.P. Schiavon, P.M. Frinchaboy, C. Allende Prieto, R. Barkhouser, D. Bizyaev, B. Blank, S. Brunner, A. Burton, R. Carrera, S.D. Chojnowski, K. Cunha, C. Epstein, G. Fitzgerald, A.E. García Pérez, F.R. Hearty, C. Henderson, J.A. Holtzman, J.A. Johnson, C.R. Lam, J.E. Lawler, P. Maseman, S. Mészáros, M. Nelson, D.C. Nguyen, D.L. Nidever, M. Pinsonneault, M. Shetrone, S. Smee, V.V. Smith, T. Stolberg, M.F. Skrutskie, E. Walker, J.C. Wilson, G. Zasowski, F. Anders, S. Basu, S. Beland, M.R. Blanton, J. Bovy, J.R. Brownstein, J. Carlberg, W. Chaplin, C. Chiappini, D.J. Eisenstein, Y. Elsworth, D. Feuillet, S.W. Fleming, J. Galbraith-Frew, R.A. García, D.A. García-Hernández, B.A. Gillespie, L. Girardi, J.E. Gunn, S. Hasselquist, M.R. Hayden, S. Hekker, I. Ivans, K. Kinemuchi, M. Klaene, S. Mahadevan, S. Mathur, B. Mosser, D. Muna, J.A. Munn, R.C. Nichol, R.W. O'Connell, J.K. Parejko, A.C. Robin, H. Rocha-Pinto, M. Schultheis, A.M. Serenelli, N. Shane, V. Silva Aguirre, J.S. Sobeck, B. Thompson, N.W. Troup, D.H. Weinberg, O. Zamora, The Apache Point Observatory Galactic Evolution Experiment (APOGEE). Astron. J. **154**, 94 (2017). https://doi.org/10.3847/1538-3881/aa784d. arXiv:1509.05420

A. Marco, I. Negueruela, C. González-Fernández, J. Maíz Apellániz, R. Dorda, J.S. Clark, VdBH 222: a starburst cluster in the inner Milky Way. Astron. Astrophys. **567**, A73 (2014). https://doi.org/10.1051/0004-6361/201423897. arXiv:1405.7266

P.G. Martin, D.C.B. Whittet, Interstellar extinction and polarization in the infrared. Astrophys. J. **357**, 113–124 (1990). https://doi.org/10.1086/168896

N. Matsunaga, Time-series surveys and pulsating stars: the near-infrared perspective, in *European Physical Journal Web of Conferences*. European Physical Journal Web of Conferences, vol. 152 (2017). https://doi.org/10.1051/epjconf/201715201007. arXiv:1705.02547

N. Matsunaga, T. Kawadu, S. Nishiyama, T. Nagayama, N. Kobayashi, M. Tamura, G. Bono, M.W. Feast, T. Nagata, Three classical Cepheid variable stars in the nuclear bulge of the Milky Way. Nature **477**, 188–190 (2011). https://doi.org/10.1038/nature10359. arXiv:1111.6359

N. Matsunaga, K. Fukue, R. Yamamoto, N. Kobayashi, L. Inno, K. Genovali, G. Bono, J. Baba, M.S. Fujii, S. Kondo, Y. Ikeda, S. Hamano, S. Nishiyama, T. Nagata, W. Aoki, T. Tsujimoto, Kinematics of classical Cepheids in the nuclear stellar disk. Astrophys. J. **799**, 46 (2015). https://doi.org/10.1088/0004-637X/799/1/46. arXiv:1411.3789

N. Matsunaga, M.W. Feast, G. Bono, N. Kobayashi, L. Inno, T. Nagayama, S. Nishiyama, Y. Matsuoka, T. Nagata, A lack of classical Cepheids in the inner part of the Galactic disc. Mon. Not. R. Astron. Soc. **462**, 414–420 (2016). https://doi.org/10.1093/mnras/stw1548. arXiv:1606.07943

F. Matteucci, S. Recchi, On the typical timescale for the chemical enrichment from Type Ia supernovae in galaxies. Astrophys. J. **558**, 351–358 (2001). https://doi.org/10.1086/322472. arXiv:astro-ph/0105074

M. McClure, Observational 5–20 μm interstellar extinction curves toward star-forming regions derived from Spitzer IRS spectra. Astrophys. J. **693**, L81–L85 (2009). https://doi.org/10.1088/0004-637X/693/2/L81. arXiv:0810.4561

J.C. Mermilliod, The database for galactic open clusters (BDA), in *Information & On-Line Data in Astronomy*, ed. by D. Egret, M.A. Albrecht. Astrophysics and Space Science Library, vol. 203 (1995), pp. 127–138. https://doi.org/10.1007/978-94-011-0397-8_12

M. Messineo, H.J. Habing, K.M. Menten, A. Omont, L.O. Sjouwerman, F. Bertoldi, 86 GHz SiO maser survey of late-type stars in the inner Galaxy. III. Interstellar extinction and colours of the SiO targets. Astron. Astrophys. **435**, 575–585 (2005). https://doi.org/10.1051/0004-6361:20040533. arXiv:astro-ph/0503179

M. Messineo, Q. Zhu, K.M. Menten, V.D. Ivanov, D.F. Figer, R.P. Kudritzki, C.H.R. Chen, Discovery of an extraordinary number of red supergiants in the inner Galaxy. Astrophys. J. **822**, L5 (2016). https://doi.org/10.3847/2041-8205/822/1/L5

M. Messineo, Q. Zhu, K.M. Menten, V.D. Ivanov, D.F. Figer, R.P. Kudritzki, C.H.R. Chen, Red supergiants in the inner Galaxy: stellar properties. Astrophys. J. **836**, 65 (2017). https://doi.org/10.3847/1538-4357/836/1/65

M.R. Metzger, J.A.R. Caldwell, J.K. McCarthy, P.J. Schechter, Radial velocities of 26 northern Milky Way Cepheids. Astrophys. J. Suppl. Ser. **76**, 803–811 (1991). https://doi.org/10.1086/191581

M.R. Metzger, J.A.R. Caldwell, P.L. Schechter, The shape and scale of Galactic rotation from Cepheid kinematics. Astron. J. **115**, 635 (1998). https://doi.org/10.1086/300198. arXiv:astro-ph/9710055

D. Minniti, P.W. Lucas, J.P. Emerson, R.K. Saito, M. Hempel, P. Pietrukowicz, A.V. Ahumada, M.V. Alonso, J. Alonso-Garcia, J.I. Arias, R.M. Bandyopadhyay, R.H. Barbá, B. Barbuy, L.R. Bedin, E. Bica, J. Borissova, L. Bronfman, G. Carraro, M. Catelan, J.J. Clariá, N. Cross, R. de Grijs, I. Dékány, J.E. Drew, C. Fariña, C. Feinstein, E. Fernández Lajús, R.C. Gamen, D. Geisler, W. Gieren, B. Goldman, O.A. Gonzalez, G. Gunthardt, S. Gurovich, N.C. Hambly, M.J. Irwin, V.D. Ivanov, A. Jordán, E. Kerins, K. Kinemuchi, R. Kurtev, M. López-Corredoira, T. Maccarone, N. Masetti, D. Merlo, M. Messineo, I.F. Mirabel, L. Monaco, L. Morelli, N. Padilla, T. Palma, M.C. Parisi, G. Pignata, M. Rejkuba, A. Roman-Lopes, S.E. Sale, M.R. Schreiber, A.C. Schröder, M. Smith, L. Sodré Jr., M. Soto, M. Tamura, C. Tappert, M.A. Thompson, I. Toledo, M. Zoccali, G. Pietrzynski, VISTA variables in the Via Lactea (VVV): the public ESO near-IR variability survey of the Milky Way. New Astron. **15**, 433–443 (2010). https://doi.org/10.1016/j.newast.2009.12.002. arXiv:0912.1056

T.J.T. Moore, S.L. Lumsden, N.A. Ridge, P.J. Puxley, The near-infrared extinction law in regions of high A_V. Mon. Not. R. Astron. Soc. **359**, 589–596 (2005). https://doi.org/10.1111/j.1365-2966.2005.08923.x. arXiv:astro-ph/0502407

K. Nandy, Interstellar obscuration. Nature **208**, 274–275 (1965). https://doi.org/10.1038/208274a0

T. Naoi, M. Tamura, Y. Nakajima, T. Nagata, H. Suto, K. Murakawa, R. Kandori, S. Sasaki, D. Baba, D. Kato, M. Kurita, C. Nagashima, T. Nagayama, H. Nakaya, S. Nishiyama, Y. Oasa, S. Sato, K. Sugitani, Near-infrared extinction law in the ρ Ophiuchi and Chamaeleon dark clouds. Astrophys. J. **640**, 373–382 (2006). https://doi.org/10.1086/500112

T. Naoi, M. Tamura, T. Nagata, Y. Nakajima, H. Suto, K. Murakawa, R. Kandori, S. Sasaki, S. Nishiyama, Y. Oasa, K. Sugitani, Near-infrared extinction in the Coalsack Globule 2. Astrophys. J. **658**, 1114–1118 (2007). https://doi.org/10.1086/512030. arXiv:astro-ph/0612620

M. Ness, D.W. Hogg, H.W. Rix, A.Y.Q. Ho, G. Zasowski, The Cannon: a data-driven approach to stellar label determination. Astrophys. J. **808**, 16 (2015). https://doi.org/10.1088/0004-637X/808/1/16. arXiv:1501.07604

M. Ness, D.W. Hogg, H.W. Rix, M. Martig, M.H. Pinsonneault, A.Y.Q. Ho, Spectroscopic determination of masses (and implied ages) for red giants. Astrophys. J. **823**, 114 (2016). https://doi.org/10.3847/0004-637X/823/2/114. arXiv:1511.08204

S. Nishiyama, T. Nagata, N. Kusakabe, N. Matsunaga, T. Naoi, D. Kato, C. Nagashima, K. Sugitani, M. Tamura, T. Tanabé, S. Sato, Interstellar extinction law in the J, H, and K_s bands toward the Galactic center. Astrophys. J. **638**, 839–846 (2006). https://doi.org/10.1086/499038. arXiv:astro-ph/0601174

S. Nishiyama, T. Nagata, M. Tamura, R. Kandori, H. Hatano, S. Sato, K. Sugitani, The interstellar extinction law toward the Galactic center. II. V, J, H, and K_s bands. Astrophys. J. **680**, 1174–1179 (2008). https://doi.org/10.1086/587791. arXiv:0802.3559

S. Nishiyama, M. Tamura, H. Hatano, D. Kato, T. Tanabé, K. Sugitani, T. Nagata, Interstellar extinction law toward the Galactic center III: J, H, K_S bands in the 2MASS and the MKO systems, and 3.6, 4.5, 5.8, 8.0 μm in the Spitzer/IRAC system. Astrophys. J. **696**, 1407–1417 (2009). https://doi.org/10.1088/0004-637X/696/2/1407. arXiv:0902.3095

S. Olofsson, G. Olofsson, The mid-infrared extinction in molecular clouds. Case study of B 335. Astron. Astrophys. **534**, A127 (2011). https://doi.org/10.1051/0004-6361/201015564. arXiv:1112.0155

L. Origlia, E. Oliva, R. Maiolino, A. Mucciarelli, C. Baffa, V. Biliotti, P. Bruno, G. Falcini, V. Gavriousev, F. Ghinassi, E. Giani, M. Gonzalez, F. Leone, M. Lodi, F. Massi, P. Montegriffo, I. Mochi, M. Pedani, E. Rossetti, S. Scuderi, M. Sozzi, A. Tozzi, GIANO-TNG spectroscopy of red supergiants in the young star cluster RSGC2. Astron. Astrophys. **560**, A46 (2013). https://doi.org/10.1051/0004-6361/201322586. arXiv:1311.1639

L. Origlia, E. Oliva, N. Sanna, A. Mucciarelli, E. Dalessandro, S. Scuderi, C. Baffa, V. Biliotti, L. Carbonaro, G. Falcini, E. Giani, M. Iuzzolino, F. Massi, M. Sozzi, A. Tozzi, A. Ghedina, F. Ghinassi, M. Lodi, A. Harutyunyan, M. Pedani, GIANO-TNG spectroscopy of red supergiants in the young star cluster RSGC3. Astron. Astrophys. **585**, A14 (2016). https://doi.org/10.1051/0004-6361/201526649. arXiv:1510.06870

M. Parsa, A. Eckart, B. Shahzamanian, V. Karas, M. Zajaček, J.A. Zensus, C. Straubmeier, Investigating the relativistic motion of the stars near the supermassive black hole in the Galactic center. Astrophys. J. **845**, 22 (2017). https://doi.org/10.3847/1538-4357/aa7bf0. arXiv:1708.03507

F. Pont, D. Queloz, P. Bratschi, M. Mayor, Rotation of the outer disc from classical Cepheids. Astron. Astrophys. **318**, 416–428 (1997)

D. Porquet, N. Grosso, P. Predehl, G. Hasinger, F. Yusef-Zadeh, B. Aschenbach, G. Trap, F. Melia, R.S. Warwick, A. Goldwurm, G. Bélanger, Y. Tanaka, R. Genzel, K. Dodds-Eden, M. Sakano, P. Ferrando, X-ray hiccups from Sagittarius A* observed by XMM-Newton. The second brightest flare and three moderate flares caught in half a day. Astron. Astrophys. **488**, 549–557 (2008). https://doi.org/10.1051/0004-6361:200809986. arXiv:0806.4088

G. Racca, M. Gómez, S.J. Kenyon, A near-infrared imaging survey of Coalsack Globule 2. Astron. J. **124**, 2178–2184 (2002). https://doi.org/10.1086/342738. arXiv:astro-ph/0207099

A.B.S. Reddy, D.L. Lambert, S. Giridhar, The evolution of the Milky Way: new insights from open clusters. Mon. Not. R. Astron. Soc. **463**, 4366–4382 (2016). https://doi.org/10.1093/mnras/stw2287. arXiv:1609.02619

M.J. Reid, M. Honma, Microarcsecond radio astrometry. Annu. Rev. Astron. Astrophys. **52**, 339–372 (2014). https://doi.org/10.1146/annurev-astro-081913-040006. arXiv:1312.2871

M.J. Reid, K.M. Menten, X.W. Zheng, A. Brunthaler, L. Moscadelli, Y. Xu, B. Zhang, M. Sato, M. Honma, T. Hirota, K. Hachisuka, Y.K. Choi, G.A. Moellenbrock, A. Bartkiewicz, Trigonometric parallaxes of massive star-forming regions. VI. Galactic structure, fundamental parameters, and noncircular motions. Astrophys. J. **700**, 137–148 (2009). https://doi.org/10.1088/0004-637X/700/1/137. arXiv:0902.3913

M.J. Reid, K.M. Menten, A. Brunthaler, X.W. Zheng, T.M. Dame, Y. Xu, Y. Wu, B. Zhang, A. Sanna, M. Sato, K. Hachisuka, Y.K. Choi, K. Immer, L. Moscadelli, K.L.J. Rygl, A. Bartkiewicz, Trigonometric parallaxes of high mass star forming regions: the structure and kinematics of the Milky Way. Astrophys. J. **783**, 130 (2014). https://doi.org/10.1088/0004-637X/783/2/130. arXiv:1401.5377

G.H. Rieke, M.J. Lebofsky, The interstellar extinction law from 1 to 13 microns. Astrophys. J. **288**, 618–621 (1985). https://doi.org/10.1086/162827

C.G. Román-Zúñiga, C.J. Lada, A. Muench, J.F. Alves, The infrared extinction law at extreme depth in a dark cloud core. Astrophys. J. **664**, 357–362 (2007). https://doi.org/10.1086/518928. arXiv:0704.3203

S.M. Rucinski, $M_V = M_V(\log P, \log T_e)$ calibrations for W Ursae Majoris systems. Publ. Astron. Soc. Pac. **106**, 462–471 (1994). https://doi.org/10.1086/133401

S.M. Rucinski, The photometric amplitude and mass ratio distributions of contact binary stars. Astron. J. **122**, 1007–1022 (2001). https://doi.org/10.1086/321153. arXiv:astro-ph/0104483

S.M. Rucinski, Contact binary stars of the W UMa-type as distance tracers. New Astron. Rev. **48**, 703–709 (2004). https://doi.org/10.1016/j.newar.2004.03.005. arXiv:astro-ph/0311085

S.M. Rucinski, Luminosity function of contact binaries based on the All Sky Automated Survey (ASAS). Mon. Not. R. Astron. Soc. **368**, 1319–1322 (2006). https://doi.org/10.1111/j.1365-2966.2006.10207.x. arXiv:astro-ph/0602045

S.M. Rucinski, H.W. Duerbeck, Absolute magnitude calibration for the W UMa-type systems based on HIPPARCOS data. Publ. Astron. Soc. Pac. **109**, 1340–1350 (1997). https://doi.org/10.1086/134014. arXiv:astro-ph/9710214

M. Salaris, L. Girardi, Population effects on the red giant clump absolute magnitude: the K band. Mon. Not. R. Astron. Soc. **337**, 332–340 (2002). https://doi.org/10.1046/j.1365-8711.2002.05917.x. arXiv:astro-ph/0208057

A. Sandage, G.A. Tammann, Absolute magnitude calibrations of population I and II Cepheids and other pulsating variables in the instability strip of the Hertzsprung-Russell diagram. Annu. Rev. Astron. Astrophys. **44**, 93–140 (2006). https://doi.org/10.1146/annurev.astro.43.072103.150612

A. Sanna, M.J. Reid, T.M. Dame, K.M. Menten, A. Brunthaler, Mapping spiral structure on the far side of the Milky Way. Science **358**, 227–230 (2017). https://doi.org/10.1126/science.aan5452. arXiv:1710.06489

R. Schödel, F. Najarro, K. Muzic, A. Eckart, Peering through the veil: near-infrared photometry and extinction for the Galactic nuclear star cluster. Accurate near infrared H, K_S, and L' photometry and the near-infrared extinction-law toward the central parsec of the Galaxy. Astron. Astrophys. **511**, A18 (2010). https://doi.org/10.1051/0004-6361/200913183

R. Schönrich, J. Binney, Chemical evolution with radial mixing. Mon. Not. R. Astron. Soc. **396**, 203–222 (2009). https://doi.org/10.1111/j.1365-2966.2009.14750.x. arXiv:0809.3006

M. Schultheis, G. Kordopatis, A. Recio-Blanco, P. de Laverny, V. Hill, G. Gilmore, E.J. Alfaro, M.T. Costado, T. Bensby, F. Damiani, S. Feltzing, E. Flaccomio, C. Lardo, P. Jofre, L. Prisinzano, S. Zaggia, F. Jimenez-Esteban, L. Morbidelli, A.C. Lanzafame, A. Hourihane, C. Worley, The Gaia-ESO survey: tracing interstellar extinction. Astron. Astrophys. **577**, A77 (2015). https://doi.org/10.1051/0004-6361/201425333. arXiv:1502.03223

J.A. Sellwood, J.J. Binney, Radial mixing in galactic discs. Mon. Not. R. Astron. Soc. **336**, 785–796 (2002). https://doi.org/10.1046/j.1365-8711.2002.05806.x. arXiv:astro-ph/0203510

M.N. Sevenster, H. Dejonghe, K. Van Caelenberg, H.J. Habing, Distribution functions for evolved stars in the inner galactic plane. Astron. Astrophys. **355**, 537–551 (2000). arXiv:astro-ph/0001073

I. Soszyński, A. Udalski, M.K. Szymański, Ł. Wyrzykowski, K. Ulaczyk, R. Poleski, P. Pietrukowicz, S. Kozlowski, D. Skowron, J. Skowron, P. Mróz, M. Pawlak, K. Rybicki, A. Jacyszyn-Dobrzeniecka, The OGLE collection of variable stars. Classical, Type II, and anomalous Cepheids toward the Galactic center. ArXiv e-prints (2017). arXiv:1712.01307

J.J. Stead, M.G. Hoare, The slope of the near-infrared extinction law. Mon. Not. R. Astron. Soc. **400**, 731–742 (2009). https://doi.org/10.1111/j.1365-2966.2009.15530.x. arXiv:0908.1601

S. Subramanian, M. Marengo, A. Bhardwaj, Y. Huang, L. Inno, A. Nakagawa, J. Storm, Young and intermediate-age distance indicators. Space Sci. Rev. **212**, 1817–1869 (2017). https://doi.org/10.1007/s11214-017-0427-8. arXiv:1709.07265

Y. Takeda, D.I. Kang, I. Han, B.C. Lee, K.M. Kim, C, N, O and Na abundances of Cepheid variables: implications on the mixing process in the envelope. Mon. Not. R. Astron. Soc. **432**, 769–792 (2013). https://doi.org/10.1093/mnras/stt528. arXiv:1303.6593

G.A. Tammann, A. Sandage, B. Reindl, New period-luminosity and period-color relations of classical Cepheids: I. Cepheids in the Galaxy. Astron. Astrophys. **404**, 423–448 (2003). https://doi.org/10.1051/0004-6361:20030354. arXiv:astro-ph/0303378

S. Tanioka, N. Matsunaga, K. Fukue, L. Inno, G. Bono, N. Kobayashi, New classical Cepheids in the inner part of the northern Galactic disk, and their kinematics. Astrophys. J. **842**, 104 (2017). https://doi.org/10.3847/1538-4357/aa7260. arXiv:1705.02571

G.S. Tsarevsky, V. Ureche, Y.N. Efremov, A search for Cepheids—new possible members of open clusters: preliminary results. Astronomicheskij Tsirkulyar **367**, 1–3 (1966)

T. Tsujimoto, Y. Yoshii, K. Nomoto, T. Shigeyama, Abundance gradients in the star-forming viscous disk and chemical properties of the bulge. Astron. Astrophys. **302**, 704 (1995)

T. Tsujimoto, J. Bland-Hawthorn, K.C. Freeman, Evidence of early enrichment of the Galactic disk by large-scale winds. Publ. Astron. Soc. Jpn. **62**, 447–456 (2010). https://doi.org/10.1093/pasj/62.2.447. arXiv:1002.3181

D.G. Turner, Galactic clusters with associated Cepheid variables. I—NGC 6087 and S Normae. Astron. J. **92**, 111–118 (1986). https://doi.org/10.1086/114141

D.G. Turner, Comments on the cluster main-sequence fitting method. III—empirical UBV reddening lines for early-type stars. Astron. J. **98**, 2300–2305 (1989). https://doi.org/10.1086/115300

D.G. Turner, The PL calibration for Milky Way Cepheids and its implications for the distance scale. Astrophys. Space Sci. **326**, 219–231 (2010). https://doi.org/10.1007/s10509-009-0258-5. arXiv:0912.4864

D.G. Turner, The usefulness of 2MASS JHK_S photometry for open cluster studies. Rev. Mex. Astron. Astrofís. **47**, 127–137 (2011). arXiv:1102.0347

D.G. Turner, S. van den Bergh, P.F. Younger, T.A. Danks, D. Forbes, A search for OB associates near southern long-period Cepheids. IV—XY Carinae, GT Carinae, WZ Sagittarii, and SW Velorum. Astrophys. J. Suppl. Ser. **85**, 119–136 (1993). https://doi.org/10.1086/191756

B.A. Twarog, K.M. Ashman, B.J. Anthony-Twarog, Some revised observational constraints on the formation and evolution of the Galactic disk. Astron. J. **114**, 2556 (1997). https://doi.org/10.1086/118667. arXiv:astro-ph/9709122

A. Udalski, OGLE Cepheids and RR Lyrae stars in the Milky Way. ArXiv e-prints (2017). arXiv:1703.02980

S. van den Bergh, A search for Cepheids in galactic clusters. Astrophys. J. **126**, 323 (1957). https://doi.org/10.1086/146404

W. van Hamme, R.E. Wilson, Accuracy of close binary mass determinations from parallaxes. Astrophys. Space Sci. **110**, 169–175 (1985). https://doi.org/10.1007/BF00660626

V. Vasilyev, H.G. Ludwig, B. Freytag, B. Lemasle, M. Marconi, Spectroscopic properties of a two-dimensional time-dependent Cepheid model. II. Determination of stellar parameters and abundances. Astron. Astrophys. **611**, A19 (2018). https://doi.org/10.1051/0004-6361/201732201. arXiv:1711.00236

S. Wang, B.W. Jiang, Universality of the near-infrared extinction law based on the APOGEE survey. Astrophys. J. **788**, L12 (2014). https://doi.org/10.1088/2041-8205/788/1/L12

S. Wang, J. Gao, B.W. Jiang, A. Li, Y. Chen, The mid-infrared extinction law and its variation in the Coalsack nebula. Astrophys. J. **773**, 30 (2013). https://doi.org/10.1088/0004-637X/773/1/30. arXiv:1306.4441

S. Wang, X. Chen, R. de Grijs, L. Deng, The near-infrared optimal distances method applied to galactic classical Cepheids tightly constrains mid-infrared period–luminosity relations. Astrophys. J. **852**, 78 (2018). https://doi.org/10.3847/1538-4357/aa9d99. arXiv:1711.06966

J.C. Weingartner, B.T. Draine, Dust grain-size distributions and extinction in the Milky Way, large Magellanic cloud, and small Magellanic cloud. Astrophys. J. **548**, 296–309 (2001). https://doi.org/10.1086/318651. arXiv:astro-ph/0008146

P.A. Whitelock, M.W. Feast, F. Van Leeuwen, AGB variables and the Mira period-luminosity relation. Mon. Not. R. Astron. Soc. **386**, 313–323 (2008). https://doi.org/10.1111/j.1365-2966.2008.13032.x. arXiv:0801.4465

D.C.B. Whittet, P.G. Martin, E.L. Fitzpatrick, D. Massa, Interstellar extinction in the infrared—the molecular cloud toward HD 62542. Astrophys. J. **408**, 573–578 (1993). https://doi.org/10.1086/172615

F. Windmark, L. Lindegren, D. Hobbs, Using Galactic Cepheids to verify Gaia parallaxes. Astron. Astrophys. **530**, A76 (2011). https://doi.org/10.1051/0004-6361/201116929. arXiv:1104.2348

M. Xue, B.W. Jiang, J. Gao, J. Liu, S. Wang, A. Li, A precise determination of the mid-infrared interstellar extinction law based on the APOGEE spectroscopic survey. Astrophys. J. Suppl. Ser. **224**, 23 (2016). https://doi.org/10.3847/0067-0049/224/2/23. arXiv:1602.02928

M. Yang, B.W. Jiang, The period-luminosity relation of red supergiant stars in the small Magellanic cloud. Astrophys. J. **754**, 35 (2012). https://doi.org/10.1088/0004-637X/754/1/35. arXiv:1205.1275

G. Zasowski, S.R. Majewski, R. Indebetouw, M.R. Meade, D.L. Nidever, R.J. Patterson, B. Babler, M.F. Skrutskie, C. Watson, B.A. Whitney, E. Churchwell, Lifting the dusty veil with near- and mid-infrared photometry. II. A large-scale study of the Galactic infrared extinction law. Astrophys. J. **707**, 510–523 (2009). https://doi.org/10.1088/0004-637X/707/1/510. arXiv:0910.4403

B. Zhang, M.J. Reid, K.M. Menten, X.W. Zheng, A. Brunthaler, T.M. Dame, Y. Xu, Parallaxes for W49N and G048.60+0.02: distant star forming regions in the Perseus spiral arm. Astrophys. J. **775**, 79 (2013). https://doi.org/10.1088/0004-637X/775/1/79. arXiv:1312.3856

Space Sci Rev (2018) 214:90
https://doi.org/10.1007/s11214-018-0519-0

Impact of Distance Determinations on Galactic Structure. II. Old Tracers

Andrea Kunder[1] · Elena Valenti[2] · Massimo Dall'Ora[3] · Pawel Pietrukowicz[4] · Chris Sneden[5] · Giuseppe Bono[6,7] · Vittorio F. Braga[8,9] · Ivan Ferraro[7] · Giuliana Fiorentino[10] · Giacinto Iannicola[7] · Marcella Marconi[3] · Clara E. Martínez-Vázquez[11,12] · Matteo Monelli[13] · Ilaria Musella[3] · Vincenzo Ripepi[3] · Maurizio Salaris[14] · Peter B. Stetson[15]

Received: 19 February 2018 / Accepted: 14 June 2018 / Published online: 23 July 2018

Abstract Here we review the efforts of a number of recent results that use old tracers to understand the build up of the Galaxy. Details that lead directly to using these old tracers to measure distances are discussed. We concentrate on the following: (1) the structure and evolution of the Galactic bulge and inner Galaxy constrained from the dynamics of individual stars residing therein; (2) the spatial structure of the old Galactic bulge through photometric observations of RR Lyrae-type stars; (3) the three-dimensional structure, stellar density,

Astronomical Distance Determination in the Space Age
Edited by Richard de Grijs and Maurizio Falanga

✉ A. Kunder
amkunder@gmail.com

E. Valenti
evalenti@eso.org

M. Dall'Ora
dallora@na.astro.it

P. Pietrukowicz
pietruk@astrouw.edu.pl

C. Sneden
chris@verdi.as.utexas.edu

G. Bono
bono@roma2.infn.it

1 Saint Martin's University, 5000 Abbey Way SE, Lacey, WA 98503, USA

2 European Southern Observatory, Karl-Schwarzschild-Str. 2, 85748, Garching bei Munich, Germany

3 INAF, Osservatorio Astronomico di Capodimonte, Via Moiarello 16, 80131 Napoli, Italy

4 Warsaw University Observatory, Al. Ujazdowskie 4, 00-478 Warszawa, Poland

5 Department of Astronomy, University of Texas, 2515 Speedway, Stop C1400, Austin, TX 78712, USA

6 Universita' di Roma Tor Vergata, Via della Ricerca Scientifica 1, 00133 Rome, Italy

7 INAF—Osservatorio Astronomico di Roma, Via Frascati 33, 00078 Monte Porzio Catone, Italy

mass, chemical composition, and age of the Milky Way bulge as traced by its old stellar populations; (4) an overview of RR Lyrae stars known in the ultra-faint dwarfs and their relation to the Galactic halo; and (5) different approaches for estimating absolute and relative cluster ages.

Keywords First keyword · Second keyword · More

Using old stars as tracers to piece together the build-up of the Milky Way (MW) allows for a glimpse of the early stages of the Milky Way Galaxy. This is because old stars have existed for several billion years—when the Galaxy was in its infancy—and are therefore a foundation of the current ensemble of stellar populations that have since begun to also populate the MW. The signatures of old stars that are discussed here involve primarily RR Lyrae stars (RRLs) and Red Clump (RC) stars as probes, arguably the most widely used old stars as stellar tracers.

RC stars are core helium-burning stars that can be considered useful distance indicators because their magnitude changes slowly and smoothly with age and metallicity. Such a change is predicted very accurately by stellar evolution models (e.g., Salaris and Girardi 2002). They do not univocally trace old populations, but also intermediate ages. However, because being a bright and distinct feature in the color-magnitude diagram they are very numerous and easy to identify, they have been extensively used in the study of especially the Galactic bulge. It should be noted though that, according to the prescriptions of Salaris and Girardi (2002), when the age-metallicity relation and star formation history of a given composite stellar system are unknown the derived distance modulus by using RC stars can have an error up to ~ 0.3 mag. Girardi (2016) provides an overview on the advantages and caveats of using RC stars for distance determination.

RRLs, on the other hand, are unequivocally old ($\geq$ 10 Gyr). They are low-mass (~ 0.6–$0.8 M_{\odot}$) horizontal branch (HB) stars that experience radial pulsations because of periodic variations of the atmosphere opacity, in partial ionized regions (H, He). This causes cyclic variations of luminosity and effective temperature, with periods ranging typically from ~ 0.2 to ~ 1 days. Their typical mean luminosity in the V-band is in the range $M_V \sim 0.5$–1 mag, making them moderately bright objects, while effective temperatures range from ~ 7200 K to ~ 5600 K, which transform in typical mean colors between (B–V) ~ 0.2 mag and (B–V) ~ 0.4 mag. Historically, RRLs have been used as standard candles since their mean luminosity in the $V-$band is almost constant, with some dependency on their metallicity and evolutionary status (e.g. Bono et al. 2003; Clementini et al.

[8] Instituto Milenio de Astrofisica, Santiago, Chile

[9] Departamento de Fisica, Facultad de Ciencias Exactas, Universidad Andres Bello, Av. Fernandez Concha 700, Las Condes, Santiago, Chile

[10] IINAF—OAS Osservatorio di Astrofisica e Scienza dello Spazio di Bologna, Bologna, Italy

[11] Cerro Tololo Inter-American Observatory, National Optical Astronomy Observatory, Casilla 603, La Serena, Chile

[12] IAC—Instituto de Astrofísica de Canarias, Calle Vía Lactea s/n, 38205 La Laguna, Tenerife, Spain

[13] Instituto de Astrofisica de Canarias, Calle Via Lactea, 38205, La Laguna, Tenerife, Spain

[14] Astrophysics Research Institute, John Moores University, IC2, Liverpool Science Park, 146 Brownlow Hill, Liverpool L3 5RF, UK

[15] DAO/HIA, NRC, 5071 West Saanich Road, Victoria, BC V9E 2E7, Canada

2003). Moreover, they are moderately bright ($\sim 40L_{\odot}$), easy to detect from their light variations and practically ubiquitous. Also, the occurrence of the so-called Oosterhoff dichotomy allowed people for decades to disentangle the old component of the Galaxy in two distinct groups, putting strong constraints on the Galaxy formation mechanisms (e.g. Fiorentino et al. 2015, 2017; Martínez-Vázquez et al. 2016a). Their role as standard candles and stellar population tracers has been widely investigated also from the theoretical point of view, on the basis of extensive and detailed nonlinear convective pulsation models (see e.g. Marconi et al. 2015, and references therein).

RRLs are much less numerous and also more time-consuming to identify than the RC. Both the RC and the RRLs have their own observational and theoretical advantages and disadvantages, however, together, they have shown to be the invaluable available tracers of old population for the study of the Milky Way galaxy. In Sect. 1, Andrea Kunder discusses how RRLs and RC giants have shaped our view of the kinematics of the inner Galaxy. Similarly, Pawel Pietrukowicz focuses on using these stars to understand the spatial structure of the inner Galaxy in Sect. 2. The chemical composition and age of the inner Galaxy is then described in Sect. 3 by Elena Valenti, again utilizing RRLs and RC giants.

In Sects. 4 and 5, RRLs and RCs stars are also discussed within the context of the Milky Way satellites and globular clusters, to further place into context the build-up of the Milky Way. Dwarf spheroidal (dSph) satellites of the Milky Way, and the low brightness tail of the dSph, the ultra-faint dwarf (UFD) galaxies, are old, metal-poor, gas poor and dark matter-dominated systems. Comparing their RRL populations with that of the MW places allows a more complete picture of galaxy formation to emerge, as shown in Sect. 4, by Massimo Dall'Ora, Giuseppe Bono, Giuliana Fiorentino, Marcella Marconi, Clara E. Martinez-Vazquez, Matteo Monelli, Ilaria Musella and Vincenzo Ripepi.

Similarly, globular clusters harbor some of the oldest stars in our Galaxy and their distances, distribution throughout the Galaxy, and ages have long been used to as pillars to understand the early Galaxy. In Sect. 5, new observations that have improved the accuracy and precision of stellar populations within globular clusters, as well as better stellar models, have advanced our ability to use these old tracers to understand the early formation of the components of Galaxy. This section is contributed by Giuseppe Bono, Vittorio Braga, Giuliana Fiorentino, Massimo Dall'Ora, Ivan Ferraro, Giacinto Iannicola, Matteo Monelli, Maurizio Salaris and Peter B. Stetson.

1 Kinematics of the Galactic Bulge

The internal kinematics of the bulge using a statistical sample of stars was first analyzed by the Bulge Radial Velocity Assay (BRAVA) survey (Rich et al. 2007; Kunder et al. 2012). BRAVA targeted M-giants toward the Galactic bulge in a grid covering three strips of latitude, at $b = -4^{\circ}$, -6°, -8°, that span across $-10^{\circ} < l < 10^{\circ}$. From a total of $\sim$ 10,000 stars, they showed that the bulge is in cylindrical rotation. Kinematic models allow at most only $\sim 10\%$ of the original model disk mass to be in the form of a "classical" spheroid formed by dissipational collapse. Subsequent kinematic bulge surveys, probing closer to the plane and/or different stellar populations, have confirmed this result. However, a more complicated kinematic view of the bulge than was first able to be disentangled by the original BRAVA results has emerged, which is reviewed here by Andrea Kunder.

The Abundances and Radial velocity Galactic Origins Survey (ARGOS; Freeman et al. 2013) probed $\sim$ 17,400 RC giants in the bulge—fainter stars than probed by BRAVA, but having temperatures more favorable for the determination of metallicities. Probing the calcium triplet (CaT) at R $\sim$ 11,500, the ARGOS stars could be separated into metallicity sub-samples, which Ness et al. (2013a) believe to represent different populations in the bulge. Sample "A" consists of stars with [Fe/H] $\sim +0.15$ dex, which are proposed to belong to a relatively thin and centrally concentrated part of the boxy/peanut bulge. Sample "B" consists of stars with [Fe/H] ~ -0.25 dex, belonging to a thicker boxy/peanut bulge. Compared to "A", this sample is hotter and less compact. Sample "C" consists of stars with [Fe/H] ~ -0.7 dex and kinematically differs from component A and B in that it does not appear to have a latitude-independent velocity dispersion rotation. Sample "D" is the most metal-poor with [Fe/H] ~ -1.0 dex. It is the least understood, due to the paucity of ARGOS stars with such metallicities. For example, there are only two stars with [Fe/H] ~ -1.0 dex in the ARGOS field at $(l, b) = (-20°, -5°)$, so the velocity dispersion provided by Ness et al. (2013a) for this field is not well constrained. In all sub-samples cylindrical rotation was seen, although the most metal-poor RC giants (sample D) rotated slower than their metal rich counterparts (Ness et al. 2013a). But as this signature was seen at latitudes 10 degrees from the plane, they associated the slower rotation as the consequence of contamination from the halo and metal-weak thick disk populations creeping into the bulge.

The GIRAFFE Inner Bulge Survey (GIBS; Zoccali et al. 2014) is targetting RC giants closer to the plane than both BRAVA and ARGOS. Most of the fields were observed at a resolution of R = 6500, but a handful of fields were observed at R = 22,500. Metallicities for their $\sim$ 5000 surveyed stars (Gonzalez et al. 2015; Zoccali et al. 2017) were derived, as well as elemental abundances for $\sim$ 400 RC giants. They confirmed cylindrical rotation also at latitudes $b = -2°$, and found that throughout most of the bulge, a narrow metal-rich ([Fe/H] $= +0.26$) population of stars and a broad more metal-poor ([Fe/H] $= -0.31$) component appears to exist. Both components rotate cylindrically, although the metal-poor stars are kinematically hotter and less bar-like.

Lastly, the APOGEE survey has probed $\sim$ 19,000 RC giants at positive longitudes close to the plane (Zasowski et al. 2016). The high-resolution (R = 22,500), makes it feasible to obtain elemental abundances and the near-infrared wavelength regime ($\lambda = 1.51$–1.70 μm) allows the plane of the bulge to be probed, where dust and reddening is severe, but minimized by longer wavelengths. Zasowski et al. (2016) find that the transition from cylindrical to non-cylindrical rotation occurs gradually, and most notably at higher latitudes. At a longitude of $l \sim 7°$ the signature of cylindrical rotation fades, which is expected, as this longitude is near the end of the boxy bulge. Despite their large chemo-dynamical sample, they are not able to find distinct and separable bulge populations, although their measures of skewness is consistent with different evolutionary histories of metal-rich ([Fe/H] $= +0.26$) and metal-poor ([Fe/H] ~ -0.31) bulge populations.

All of these large surveys have shown that the bulge consists of a massive bar rotating as a solid body; the internal kinematics of these stars are consistent with at least 90% of the inner Galaxy being part of a pseudobulge and lacking a pressure supported, classical-like bulge. N-body barred galaxy models (boxy peanut bulge models) can explain the global kinematics, so our bulge appears to have formed from secular evolution of a massive early disk.

However, some finely detailed behavior of stars remain unexplained. For example, it is not clear how the kinematically cooler bulge stars (which are more metal-rich) fit together with the kinematically hotter (more metal-poor) bulge stars. Also, some bulge locations have shown no evidence for a metal-rich and metal-poor population, despite being at the same latitude as other fields which do clearly separate chemically (e.g., Zoccali et al. 2017).

Figure 1 shows the distributions of the targets in the survey's mentioned above. Most surveys have focussed on the Southern bulge, where the crowding is not as extreme. The

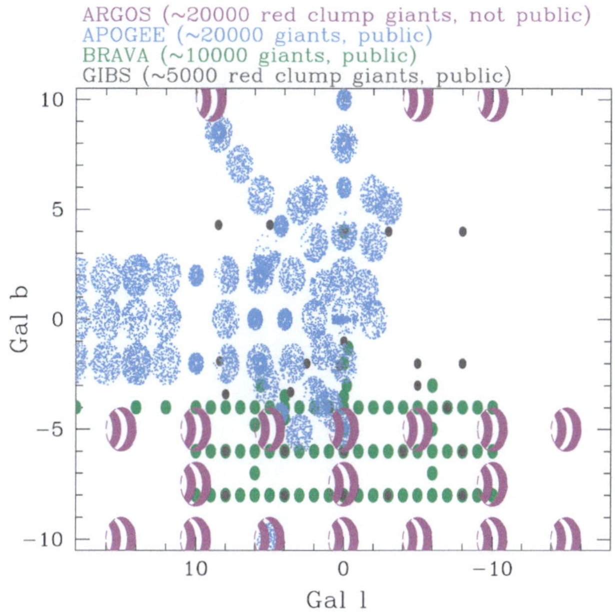

Fig. 1 Location in Galactic coordinates of the ARGOS, APOGEE, BRAVA and GIBS spectroscopic bulge surveys

APOGEE survey, in contrast, using a telescope in the North, probes more of the Northern bulge, and has not yet been able to reach the negative longitudes. With exception of the ARGOS survey, all data has been publicly released, improving the quality and value of these surveys, and providing the wider scientific community the ability to productively use data for further research with the potential to advance developments.

1.1 Targeted Kinematic Studies

Notable recent high resolution studies (R $\sim$ 20,000–30,000) of bulge field stars have been meticulously obtained by Johnson et al. (2011, 2012, 2013a, 2013b, 2014). In these papers, along with radial velocities, numerous individual elemental abundances, for some stars 27 elements ranging from oxygen to erbium, are derived for a sample of $\sim$ 500 bulge giants. Therefore, not only can the kinematics, [Fe/H] and [α/Fe] ratios of bulge stars be compared to those in the thin and thick disks, but also the light odd-Z and Fe-peak (and also neutron-capture) elements are touched on, which also provide discriminatory power between models and other stellar populations. These more detailed and targeted observations indicate that at [Fe/H] > -0.5, the bulge exhibits a different chemical composition than the local thick disk in that the bulge [α/Fe] ratios remain enhanced to a slightly higher [Fe/H] than the thick disk, and the Fe-peak elements Co, Ni, and Cu appear enhanced compared to the disk.

Further, these studies point to a bulge that formed rapidly ($<$ 1–3 Gyr), because of the enhanced [α/Fe] abundances coupled with the low [La/Eu] ratios of the bulge stars (see also McWilliam et al. 2010). This confirms a very fast chemical enrichment in the bulge put forth by the very first detailed abundance studies of red giants in the Milky Way bulge (e.g., McWilliam and Rich 1994; Zoccali et al. 2006; Fulbright et al. 2007).

Babusiaux et al. (2010, 2014) compared the velocities of metal-rich and metal-poor bulge stars and found that higher metallicity stars in the bulge show larger vertex deviations of the velocity ellipsoid than more metal-poor stars. They also found that metal-rich stars show an

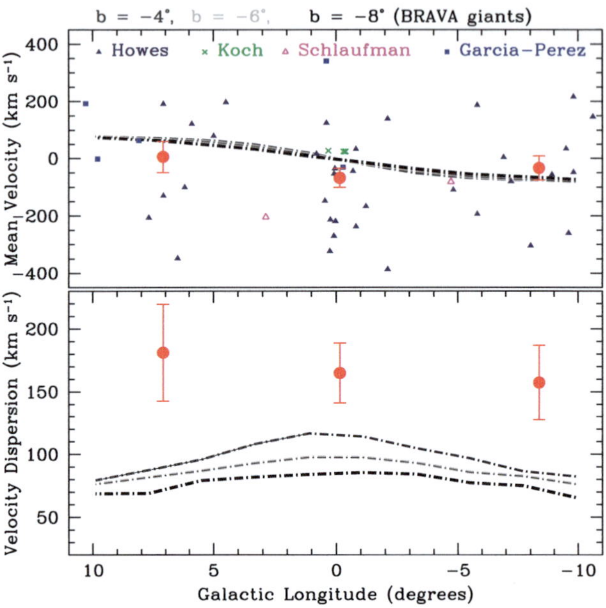

Fig. 2 The velocity dispersion profile (bottom) and rotation curve (top) for the metal-poor stars ([Fe/H] < -2.0) observed in the bulge compared to that of the BRAVA giants at $b = -4°$, $-6°$, and $-8°$ strips (Kunder et al. 2012). The individual metal-poor star measurements are given in the top panel, and the large red circles indicate the mean Galactocentric velocity (top) and velocity dispersion (bottom). The metal-poor stars have kinematics suggesting they are different from the bulge giants, although the sample size is small (~ 50). It has been put forward that these metal-poor stars are actually halo interlopers (e.g., Howes et al. 2014; Kunder et al. 2015; Koch et al. 2016)

increase in their velocity dispersion with decreasing latitude (moving closer to the Galactic plane), while metal-poor stars show no changes in the velocity dispersion profiles. They concluded that the more metal-rich stars are consistent with a barred population and the metal-poor stars with a spheroidal component. However, other high-resolution studies of bulge stars have not confirmed such trends and instead find consistent decrease in velocity dispersion with increasing [Fe/H] (e.g., Johnson et al. 2014; Uttenthaler et al. 2012; Ness et al. 2013a).

Perhaps the greatest limitation in finding possible differences between a metal-rich and a metal-poor population in the bulge is the difficulty of finding metal-poor stars in the bulge. For example, within the ARGOS survey (Ness et al. 2013a), 0.1% of the stars identified as lying in the bulge have [Fe/H] < -2.0 dex. The first metal-poor stars found close to the Galactic center was presented by Schultheis et al. (2015), who find 10 stars with [M/H] ~ -1.0 dex within ~ 200 pc from the Galactic center. García Pérez et al. (2013) used infrared spectroscopy of 2400 bulge stars to uncover five new metal-poor stars with $-2.1 <$ [Fe/H] < -1.6, and using optical photometry to first select metal-poor candidates, and Schlaufman and Casey (2014) uncovered three stars in the direction of the bulge with $-3.0 <$ [Fe/H] < -2.7. The Extremely Metal-poor BuLge stars with AAOmega (EMBLA) survey, dedicated to search for metal-poor stars in the bulge, has uncovered ~ 40 metal-poor stars (Howes et al. 2014, 2015, 2016), including a handful with [Fe/H] < -3.0. Five stars in the very metal-poor regime, at $-2.7 <$ [Fe/H] < -2.0 are presented in Koch et al. (2016), where they find that the metal-poor stars are a broad mix, and no single, homogeneous "metal-poor bulge" can yet be established.

Figure 2 shows the kinematics of these metal-poor stars compared to "normal" bulge giants from the BRAVA survey. Though the number statistics are still small, their velocity dispersion suggests either that the metal-poor stars in the bulge have different kinematics than the more metal-rich stars, or that the metal-poor stars discovered are a halo population.

Lacking a statistical sample of metal-poor stars in the bulge, understanding their kinematics and placing them in context within the Galaxy is nontrivial. Especially since the oldest and most metal poor stars (which may trace the dark matter) are thought to be found in the center of the Galaxy—in the bulge but not sharing its kinematics and abundance patterns (Tumlinson 2010), the metal-poor "bulge" stars could provide a big piece of the puzzle in understanding the formation and subsequent evolution of the Galactic bulge.

Perhaps the easiest identifiable old, metal-poor bulge population are those horizontal branch stars that pulsate as RR Lyrae stars. Their progenitors formed long ago (~ 10 Gyr), so that the RRLs we see today tell us about conditions when the halo of the Galaxy was being formed (e.g., Lee 1992). The bulge RRLs were shown to be on average ~ 1 dex more metal-poor than the majority of bulge stars residing in the bar (Walker and Terndrup 1991), although some of the bulge RRLs do appear to have metallicities that overlap in abundance with the bar population.

The ongoing Bulge Radial Velocity Assay for RR Lyrae stars, BRAVA-RR survey (Kunder et al. 2016), aims to collect spectrographic information for RRLs located toward the inner Galaxy. Their sample of RRLs are selected from the Optical Gravitational Lensing Experiment (OGLE), so the periods, amplitudes and magnitudes are already precisely known. Multi-epoch spectroscopy (typically 3 epochs per star) is used to obtain center-of-mass radial velocity with uncertainties of $\sim$ 5–10 $\mathrm{km\,s^{-1}}$. From radial velocities of about ~ 1000 RRLs surveyed by BRAVA-RR in four 2-degree fields covering approximately a Galactic latitude and longitude of $-6^\circ < b < -3^\circ$ and $-4^\circ < l < 4^\circ$, it is evident that these old and metal-poor stars are kinematically distinct from the more metal-rich red giants in the BRAVA, GIBS, ARGOS and APOGEE surveys. As seen in Fig. 3, the RRLs show null rotation and hot (high-velocity dispersion) kinematics. In the ARGOS survey one also observes a slowly-rotating metal-poor population, but these stars are believed to be contamination from disk and halo stars (as it is only seen at high Galactic latitude). In contrast, the RRLs are at low Galactic latitudes ($|b| < 7^\circ$) and have more certain distance estimates, and the larger number statistics of the RRLs makes this result quantifiable. The RR Lyrae stars trace an older, more spheroidal component in the inner Galaxy.

The mass of this 'old' bulge is estimated to be $\sim 1\%$ of the total central mass, broadly consistent with current bulge formation models, which predict that no more than $\sim 5\%$ of a merger-generated bulge (Shen et al. 2010; Ness et al. 2013a; Di Matteo et al. 2015). It may be that the RRL stars toward the bulge are actually an inner halo-bulge sample, as originally speculated in the early 1990s (e.g., Minniti 1994) and as at least one RRL orbit toward the Galactic bulge seems to indicate (Kunder et al. 2015).

Prompted by the results from the RRLs, Pérez-Villegas et al. (2017) carried out N-body simulations for the Milky Way to investigate the kinematic and structural properties of the old metal-poor stellar halo in the barred inner region of the Galaxy. They showed that the RR Lyrae population in the Galactic bulge may be the inward extension of the Galactic metal-poor stellar halo, and that especially the radial velocities of RRLs in the outer Galactic longitudes constrain a bulge/halo scenario. Unfortunately, the RRLs investigated in Kunder et al. (2016) are confined to the innermost 500 pc. This is where a slow-rotating component has the smallest velocity difference compared to the metal-rich bulge giants (~ 25 km/s), and hence where population contamination from e.g., the halo or thick disk could more easily mask the effects of rotation. Observations of RRL at further longitudes in the bulge would allow us to distinguish between a bulge or halo.

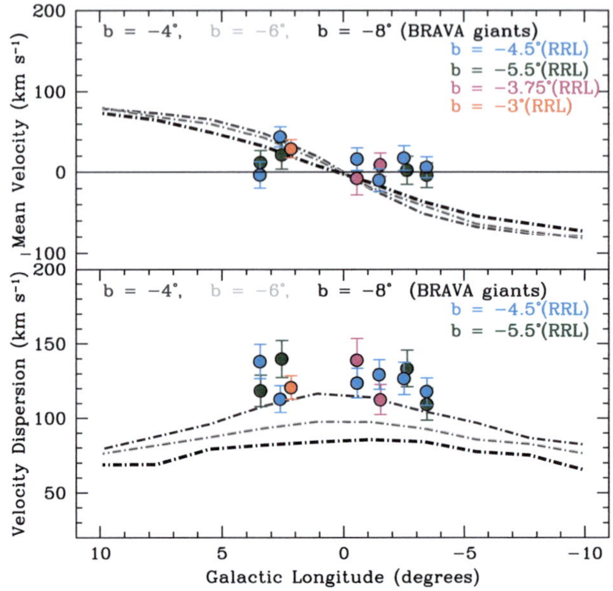

Fig. 3 The velocity dispersion profile (bottom) and rotation curve (top) for the ~ 1000 RR Lyrae stars observed in the bulge compared to that of the BRAVA giants at $b = -4°$, $-6°$, and $-8°$ strips (Kunder et al. 2012, 2016). The bulge model showing these observations are consistent with a bulge being formed from the disk is represented by the dashed lines (Shen et al. 2010). The RRLs have kinematics clearly distinct from the bulge giants, and are a non-rotating population in the inner Galaxy

1.2 Future: Gaia

Gaia has begun collecting six-dimensional space coordinates for more than 1 billion stars in the Milky Way. The bulge is a difficult target for Gaia, due to the crowding and extinction, but Gaia will still impact bulge kinematics significantly. The Radial Velocity Spectrometer (RVS), which is the spectroscopic instrument for all objects down to $G \sim 16$ mag, can cope with a crowding limit of 35,000 stars deg^{-2} (Reylé et al. 2005). In denser areas, only the brightest stars are observed and the completeness limit will be brighter than 16 mag. Therefore, we can expect the brighter giants to be surveyed throughout the bulge with RVS, but most of the red clump stars and RRLs will be lacking Gaia radial velocities.

Never-the-less, the astrometric instrument has been designed to cope with object densities up to 750,000 stars deg^{-2} and down to $G \sim 20$ mag. Therefore, for a large area of the bulge, at least some of the horizontal branch will be reachable for useful proper-motions, although no useful (5 σ) parallaxes are expected for these stars. At the end of five-year mission, Gaia will have ~ 54 Gaia transits covering the bulge (Clementini 2016).

Figure 4 shows the proper motions of the giants in the direction of the bulge surveyed in APOGEE DR13 post- and pre-Gaia DR1. Before Gaia DR1 was available, the proper motions were not of the quality that allowed one to easily distinguish between a field and bulge stellar population. Color-magnitude diagrams can help in differentiating bulge stars from field stars, since bulge stars tend to be redder, but the large and variable extinction and the sheer number of field stars along the line of sight toward the bulge, makes color cuts not always reliable.

With the precise positions of stars measured in Gaia DR1, significant improvements in the astrometric solutions were able to be obtained, leading to the release of UCAC5. For over 107 million stars, now proper motions with typical accuracies of 1 to 2 mas/yr ($R = 11$ to 15 mag), and about 5 mas/yr at 16th mag, exist. With Gaia, we are approaching the possibility of using proper motion information to separate the high degree of disk contamination from the bulge.

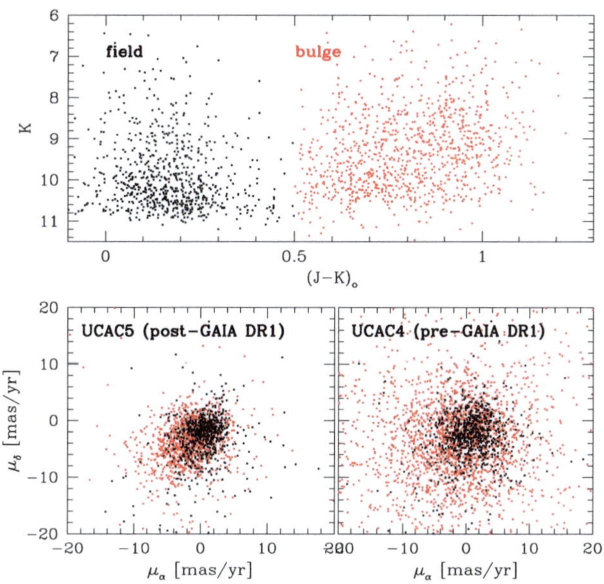

Fig. 4 *Top:* The color magnitude diagram of ∼ 2500 APOGEE giants that have UCAC5 proper motions with uncertainties smaller than 2 mas/yr. *Bottom:* The UCAC5 (left) and UCAC4 (right) proper motions of the APOGEE giants with proper motion uncertainties smaller than 2 mas/yr. Already with Gaia DR1, a sharper kinematic view of the bulge than previously feasible, is possible

2 Spatial Structure of the RR Lyrae Star Population Toward the Galactic Bulge

RR Lyrae-type variable stars can be found everywhere in the Milky Way, but they are particularly numerous in the Galactic bulge. Historically, van Gent (1932, 1933) was the first who noticed that RRLs observed close to the central regions of the Milky Way concentrate toward the Galaxy center. More than a decade later, Baade (1946) in a relatively unobscured area, today called "Baade's Window" in his honor, found a strong predominance of RRLs indicating a presence of Population II stars in the central area of the Milky Way. He assumed that the center of this population coincides with the Galactic center and assessed the distance to the center of the Galaxy using RRLs, obtaining a distance of 8.7 kpc (Baade 1951).

Until the early 1990s about one thousand RRLs inhomogeneously distributed over the Galactic bulge were known. Following the advent of massive photometric surveys, particularly focused on searches for microlensing events, the number of new RRLs toward the bulge has increased. 215 such objects were discovered during the first phase of the OGLE project (Udalski et al. 1992), conducted on the 1.0-m Swope telescope at Las Campanas Observatory, Chile, in years 1992–1995. About 1800 RR Lyrae pulsators were detected by the MACHO microlensing survey (Alcock et al. 1995) which used the 1.27-m Great Melbourne Telescope at the Mount Stromlo Observatory, Australia, in years 1992–1999. Alcock et al. (1998) examined mean magnitudes and colors of the new pulsators and found that the bulk of the population is not barred. Only stars located in the inner fields closer to the Galactic center ($l < 4°$, $b > -4°$) seem to follow the barred distribution observed for intermediate-age red clump giants (Stanek et al. 1994b). Minniti et al. (1998) used this sample to show that between about 0.3 kpc and 3 kpc from the Galactic center, the spatial density distribution of RRLs can be represented by a power law of an index of −3.0.

Analysis of the data from the second phase of the OGLE project (OGLE-II), conducted on the dedicated 1.3-m Warsaw Telescope at Las Campanas Observatory in years 1997–2000, brought a much larger set of 2713 RRLs (Mizerski 2003). Based on the sample of

1888 fundamental-mode RRLs from OGLE-II, Collinge et al. (2006) robustly detected the signature of a barred structure in the old population within the inner $\pm 3^\circ$ of Galactic longitude. Later, about 3000 fundamental-mode RRLs from the MACHO database were used to investigate the metallicity distribution of these stars (Kunder and Chaboyer 2008) and interstellar extinction toward the Galactic bulge (Kunder et al. 2008). It was determined that bulge variables have the average metallicity $[Fe/H] = -1.25$ dex, with a broad range from $[Fe/H] = -2.26$ to -0.15 dex, on the Zinn and West (1984) metallicity scale. Kunder and Chaboyer (2008) searched for the evidence of the Galactic bar and found a marginal signature of a bar at Galactic latitudes $|b| < 3.5^\circ$. The absence of a strong bar in the RR Lyrae population clearly indicated that they represent a different population than the metal-rich bulge. However, the shape of this population was far from being fully known. More data covering preferably the whole bulge area were needed.

In 2001, the OGLE project started its third phase with a new mosaic eight-CCD camera attached to the Warsaw Telescope. One of the OGLE-III results was the release of a collection of 16,836 RRLs found in an area of 69 deg^2 mostly south of the Galactic equator (Soszyński et al. 2011). The collection composed of 11,756 fundamental mode (RRab) stars, 4989 overtone pulsators (RRc), and 91 double-mode (RRd) stars. OGLE provided time-series photometry in two standard filters: V and I. This sample was promptly analyzed by Pietrukowicz et al. (2012) who demonstrated that the bulge RRLs form a metal-uniform population, slightly elongated in its inner part. The authors found that the photometrically derived metallicity distribution for RRab stars is sharply peaked at $[Fe/H] = -1.02 \pm 0.18$ dex with a dispersion of 0.25 dex, on the Jurcsik (1995) metallicity scale. This result agreed very well with the one from Kunder and Chaboyer (2008), since the Jurcsik (1995) scale is shifted roughly by $+0.24$ dex with respect to the Zinn and West (1984) scale.

Pietrukowicz et al. (2012) also estimated the distance to the Milky Way center based on the bulge RRLs to be $R_0 = 8.54 + / - 0.42$ kpc. Here, the theoretical period-luminosity-metallicity (PLZ) relations in V and I bands published by Catelan et al. (2004) were used; the zero points of these relations were calibrated to the data obtained for the well-studied representative globular cluster M3 (Catelan 2004). In their analysis, Pietrukowicz et al. (2012) made a simple assumption on a linear relation between I-band extinction A_I and reddening $E(V-I)$. At that time it was the only reasonable way to unredden mean magnitudes of the RRLs. They showed that, for RRab stars as well as for RRc stars, in the inner regions ($|l| < 3^\circ$, $|b| < 4^\circ$) the old population indeed tends to follow the barred distribution of the bulge red clump giants.

A year later Dékány et al. (2013) combined optical and near-infrared data for the OGLE-III bulge RRLs to study the RRL spatial distribution. The authors used mean I-band magnitudes from OGLE and K_s-band magnitudes from the near-infrared VISTA Variables in the Vía Láctea (VVV) survey (Minniti et al. 2010). VVV was one of the ESO (European Southern Observatory) public surveys carried out on the 4.1-m Visible and Infrared Survey Telescope for Astronomy (VISTA) in years 2010–2015. Observations were taken in $ZYJHK_s$ filters and included the Milky Way bulge and an adjacent section of the Galactic plane, covering a total area of about 562 deg^2 (Saito et al. 2012). The monitoring campaign was conducted only in the K_s band.

The approach applied by Dékány et al. (2013) is expected to bring more precise results than the ones based on optical data alone. That is because PLZ relations have decreasing metallicity dependence toward longer wavelengths and measurements in near-infrared wavebands are much less sensitive to interstellar reddening than optical ones. Dékány et al. (2013) concluded that the population of RRLs does not trace a strong bar, but have a more spheroidal, centrally concentrated distribution with only a mild elongation in its very center at an angle $i = 12.5^\circ \pm 0.5^\circ$ with respect to the line of sight from the Sun to the Galactic center.

The fourth phase of the OGLE project (OGLE-IV) (Udalski et al. 2015), which was launch in 2010, covers practically the whole bulge in V and I passbands. With the installation of a 32-CCD mosaic camera of a total field of view of 1.4 deg^2 OGLE became a truly wide-field variability survey. The OGLE-IV collection of RRLs toward the Galactic bulge was released by Soszyński et al. (2014). This collection contains data on 38,257 variables detected over 182 deg^2: 27,258 RRab, 10,825 RRc, and 174 RRd stars. The survey also includes the central part of the Sagittarius Dwarf Spheroidal Galaxy with the globular cluster M54 in its core.

Analysis of this set of data was undertaken also by the OGLE team and presented in Pietrukowicz et al. (2015). Due to some practical reasons the analysis was based only on RRab-type pulsators, and the part closest to the Galactic plane ($|b| < 3$) was avoided. RRab stars are more numerous. On average, they are intrinsically brighter in the I-band and have higher amplitudes than RRc stars. What is extremely important, RRab variables with their characteristic saw-tooth-shaped light curves, in comparison to nearly sinusoidal light curves of RRc stars, are harder to overlook. This makes the searches for these type of variables more likely to yield better completeness ratios. Another very practical property of RRab stars is that based on the pulsation period and shape of the light curve one can assess metallicity of the star (Jurcsik 1995; Jurcsik and Kovács 1996; Smolec 2005).

Pietrukowicz et al. (2015) found again that the spatial distribution of the inner bulge RRLs trace closely the barred structure formed of intermediate-age red clump giants. According to the most recent models of the Galactic bar, it is close to being a prolate ellipsoid. Based on OGLE-III data Cao et al. (2013) found the following axis ratios and inclination of the major axis: 1.0:0.43:0.40, $i = 29.4^\circ$. A similar result was obtained by Wegg and Gerhard (2013) using VVV red clump data: 1.0:0.63:0.26, $i = 27^\circ \pm 2^\circ$. This time in their analysis, Pietrukowicz et al. (2015) dereddened mean I-band magnitudes of RRLs using a new relation derived by Nataf et al. (2013). The relation was based on optical measurements from OGLE-III and near-infrared measurements from 2MASS and VVV for bulge red clump giants. After this correction the obtained distance distribution to the bulge RRLs turned out to be smoother in comparison with the previously used simple linear relation. They found the maximum of the distribution or distance to the Galactic center at $R_0 = 8.27 \pm 0.01(\text{stat}) \pm 0.40(\text{sys})$ kpc, which is in very good agreement with estimates from other measuring methods. Pietrukowicz et al. (2015) showed that the spatial distribution of the bulge RRLs has the shape of a triaxial ellipsoid with proportions 1:0.49:0.39 and the major axis located in the Galactic plane and inclined at an angle $i = 20^\circ \pm 3^\circ$ to the Sun-(Galactic center) line of sight (see Fig. 5).

Differences between the Dékány et al. (2013) results and those from OGLE-IV are the following: (1) Dékány et al. (2013) used a smaller sample of about 7700 RRab stars from the previous phase of the OGLE survey (OGLE-III), while the OGLE-IV collection (Soszyński et al. 2014) contain nearly 27,300 pulsators of this type. (2) The Dékány et al. (2013) results are based on a lower number of collected K_s-band measurements per light curve (about 40 by 2013), which could affect the average infrared brightness of the variables and could slightly smear the observed structure. However, the amplitude variation in the K_s-band is a factor of ~ 3 smaller than that in the I- or V-band. (3) Because Dékány et al. (2013) use also infrared magnitudes, the de-reddening process of the RRLs differs from that adopted in Pietrukowicz et al. (2015). Resolution of the discrepancy between the OGLE-IV RRLs presented in Pietrukowicz et al. (2015) and the VVV RRLs presented in Dékány et al. (2013) is ongoing.

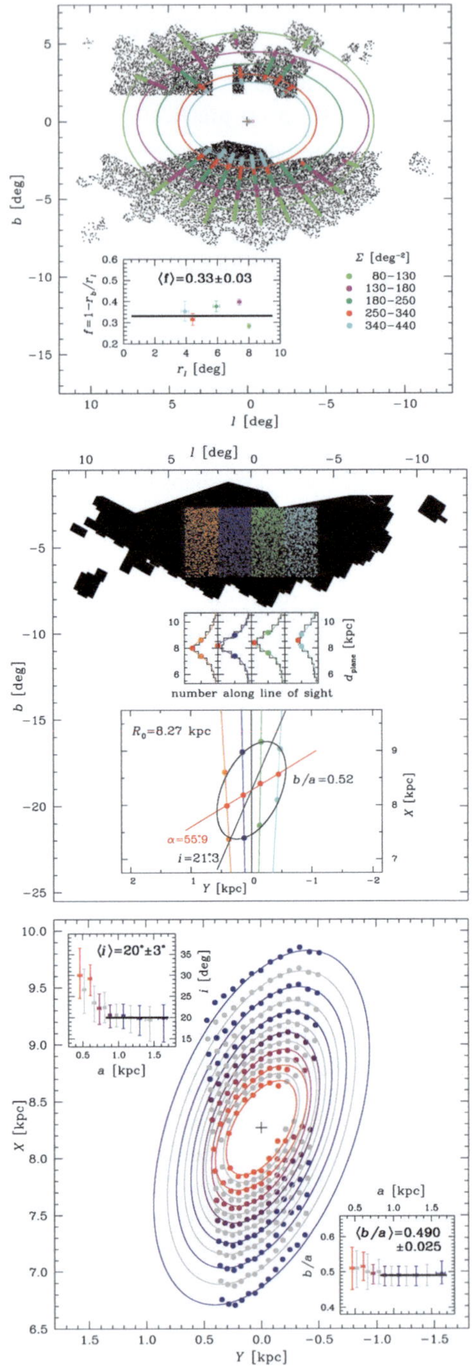

Fig. 5 Results of the analysis of density distribution of OGLE-IV bulge RRab stars (Pietrukowicz et al. 2015). Upper panel: Constant surface density lines in the sky are well represented by ellipses with a mean flattening $f = 0.33 \pm 0.03$. Middle panel: The maxima of the density distributions along four selected lines of sight clearly get closer to us with the increasing Galactic longitude. This strongly indicates the presence of a tilted axis in the plane. Lower panel: In the projection onto the Galactic plane, points of the same density level form inclined ellipses. Conclusion: the old bulge population has the shape of a triaxial ellipsoid with the major axis inclined to us at an angle $i = 20^\circ \pm 3^\circ$

The obtained sharp ellipsoidal shape does not depend on the real final distance to the studied objects. The true inclination angle as well as the axis ratios may be slightly different than reported. These values may also change with the galactocentric distance. This will be known once almost all RRLs (or at least RRab-type variables) are detected from the inner bulge to the outer Galactic halo. Unfortunately, searches for RRLs in obscured Galactic plane regions are very difficult. That is because near-infrared light curves of these pulsating stars have often symmetric, nearly sinusoidal shape. If the number of data points per light curve is small and the time coverage too short, RRLs can be easily confused with other variables, particularly with contact eclipsing binaries and spotted variables. However, the first detection of RRLs in the vicinity of the Galactic center, in the so-called nuclear bulge has been made (Minniti et al. 2016; Dong et al. 2017).

A clear result from the analysis of the OGLE-IV bulge RRab variables is that their spatial density distribution in the galactocentric distance range from about 0.2 kpc to 2.8 kpc can be described as a single power law with an index of -2.96 ± 0.03. Pietrukowicz et al. (2015) also was not able to see an X-shaped structure in the RRLs as it is observed in the case of bulge red clump giants (Nataf et al. 2010; McWilliam and Zoccali 2010a). This is expected, as it was found that only metal-rich bulge populations have this feature (Ness et al. 2012).

Another discovery by Pietrukowicz et al. (2015) is that RRab stars form two (or even more) very close sequences in the period–amplitude (or Bailey) diagram. This is interpreted as the presence of multiple old populations being likely the result of mergers in the early history of the Milky Way. So far, there are no hints that the two major old populations have different structure. They seem to be well mixed together. Lee and Jang (2016) suggest that the observed period shift between the sequences can be explained by a small difference in the helium abundance.

Recently, Pérez-Villegas et al. (2017) used N-body simulations to investigate the structural and kinematic properties of the old population in the barred inner region of the Galaxy. They showed that the RR Lyrae population in the bulge is consistent with being the inward extension of the Galactic metal-poor stellar halo, as suggested by Minniti et al. (1998) and Pietrukowicz (2016). Pérez-Villegas et al. (2017) followed the evolution of the metal-poor population through the formation and evolution of the more massive bar and boxy/peanut bulge and found the density distribution to change from oblate to triaxial. They found that at the final time of the simulations (after 5 Gyr), the axis ratios of the triaxial old stellar halo in its inner part reached $b/a \sim 0.6$ and $c/a \sim 0.5$. The ratios increase with the distance from the center to roughly 1.0 and 0.7, respectively, at a distance of 5 kpc. These results are very consistent with the observations of the bulge RRLs from OGLE-IV and also spectroscopic observations of old stars in the Milky Way halo by the SDSS survey. According to the latter studies, the Galactic halo has an oblate shape (Jurić et al. 2008; Carollo et al. 2008; Kinman et al. 2012). Ongoing photometric surveys, such as Gaia, VVV eXtended (VVVX) and OGLE-IV with extended coverage of the Galactic bulge and disk will continue in completing the picture of the old bulge population drawn by RRLs.

3 The Galactic Bulge: 3D Structure, Chemical Composition, and Age Traced by Its Old Stellar Population

The formation and evolution of galaxies is still a heavily debated question of modern astrophysics. As one of the major stellar components of the Milky Way, the bulge provides critical and unique insights on the formation and evolution of the Galaxy, as well as of the external galaxies. Indeed with the current observational facilities, it is the only bulge in which we are able to resolve stars down to the old main sequence turnoff, hence providing accurate studies of the stellar populations in almost every evolutionary stage. The physical, kinematics and chemical properties of the stellar populations in the bulge allow us to discriminate among various theoretical models for the formation and evolution of bulges at large, setting tight constraints on the role that different processes (i.e. dynamical instabilities, hierarchical mergers, gravitational collapse) may have taken place.

However, this advantage comes with the need of covering a large area of the sky (~ 500 deg^2). Therefore, to understand the global properties of the bulge one should look for reliable tracers of distance and age that can be easily observed in any region of the sky in the bulge direction. In this framework, Red Clump stars and RRLs play a crucial role, and indeed over the decades studies of these stars have been essential to build our current knowledge of the Galactic bulge. The present section put together by Elena Valenti is not intended to be a complete review of the Galactic bulge properties but an overview of the bulge structure, chemical composition, and age based on the observational results on RC stars and RRLs.

The age-metallicity relation and star formation history of the RCs in the bulge is not well-known, so an error of up to ~ 0.3 mag can be introduced when trying to use these stars to derive a distance. At the distance of the bulge this translates to ~ 1 kpc. However, considering that overall the bulge is metal rich and old ($\gtrsim 10$ Gyr), one could use the corresponding population effect correction and therefore reduce the error on the distance. RRLs represent, on the other hand, much more accurate distance candles. Unlike the RC stars, they trace univocally the oldest stellar population of any given complex system (see Sect. 2 above by P. Pietrukowicz).

In Sect. 3.1, a global view of the three-dimensional structure of the bulge addressing the observational evidences is presented, that leads to determination of the bar properties, the stellar density and the mass. The chemical composition and age of the bulge stellar populations are reviewed in Sect. 3.2, and 3.3, respectively. Finally, Sect. 3.4 summarizes the main stellar properties that combined define our current knowledge of the Milky Way bulge, and the pieces of evidence that are still lacking in order to improve and possibly complete the global picture are highlighted. Note that occasionally up to date versions of relevant figures obtained using state of the art observational data are presented.

3.1 The 3D Structure

3.1.1 The Bar Boxy/Peanut/X-Shaped Structure

Today it is well known that the Milky Way is a barred galaxy. Although the first observational evidence of the presence of a bar in the innermost region of the Galaxy was presented by Blitz and Spergel (1991) by using the stellar density profile at 2.4 μm by Matsumoto et al. (1982), its existence was hypothesized nearly 20 years earlier. Indeed, to explain the departures from circular motions seen in the HI line profile at 21 cm, de Vaucouleurs (1964) suggested for the first time that the Milky Way could host a bar in its inner regions. After then, over the decades, many different tracers—e.g. gas kinematics (Binney et al. 1991), stellar surface profile (Weiland et al. 1994; Dwek et al. 1995; Binney et al. 1997), microlensing experiments (Udalski et al. 2000; Alcock et al. 2000), and OH/IR and SiO maser kinematics (Habing et al. 2006)—have been used to confirm the presence of the bar and to constrain its properties.

Still, the strongest observational evidence for the presence of the bar comes from the use of RC stars as standard candles to deproject the stellar density distribution in the Galaxy inner region. By using the color-magnitude diagram (CMD) derived from the OGLE (Udalski et al. 1992) photometry in Baade's Window ($l = +1°, b = -3.9°$) and in two additional fields at ($\pm 5°, -3.5°$), Stanek et al. (1994a) found that the mean RC magnitude at positive

Table 1 Axis scale lengths and orientation angle of the bulge main bar based on RC stars distribution

x_0 (kpc)	y_0 (kpc)	z_0 (kpc)	Θ (deg)	Data	Area (deg^2)	Reference
1.2	0.40	0.30	24–27	OGLE-II	~ 11	Rattenbury et al. (2007)
1.0	0.41	0.38	29–32	OGLE-III	~ 90	Cao et al. (2013)
0.7	0.44	0.18	27	VVV	~ 300	Wegg and Gerhard (2013)

longitudes was brighter than the one observed at negative longitude. Under the assumption that there is no continuous metallicity and age gradient along the longitude, the observed change in mean RC magnitude across the field was interpreted in terms of distance. Stars at positive longitudes are brighter, hence closer, than those at negative longitudes. By using a triaxial model for the bulge, Stanek et al. (1994a) derived a bar pivot angle of $\Theta = 45°$.

Following this pioneering work, many studies (see Stanek et al. 1997; Bissantz and Gerhard 2002; Babusiaux and Gilmore 2005; Benjamin et al. 2005; Nishiyama et al. 2005; Rattenbury et al. 2007; López-Corredoira et al. 2007; Cabrera-Lavers et al. 2008; Cao et al. 2013; Wegg and Gerhard 2013) have used RC stars to constrain triaxial bar models. Table 1 lists the axis scale lengths and orientation angles of the bar as derived by Rattenbury et al. (2007), Cao et al. (2013), Wegg and Gerhard (2013), which among all similar studies, are those considering the largest bulge area, therefore possibly providing more accurate estimates of the bar physical properties.

Owing to the RC stars distribution across the bulge, we now know that the bar has a boxy/peanut/X-shape structure in its outer regions: a typical characteristic of all barred galaxies when seen edge-on (Laurikainen et al. 2014; Laurikainen and Salo 2017, for a recent review). McWilliam et al. (2010) and Zoccali (2010) were the first who noticed that the distribution of the RC stars in some fields in the outer regions ($|b| > 5°$) along the bulge minor axis ($l = 0°$) was bimodal, suggesting the presence of a double RC. The observed split in the RC mean magnitude was then confirmed by McWilliam and Zoccali (2010b) and Nataf et al. (2010) by using 2MASS and OGLE photometry, respectively. The authors explained the split in the RC as signature of two southern arms of an X-shape structure crossing the line of sight. An alternative explanation is that the double RC phenomenon is a manifestation of multiple populations observed in globular clusters (GCs) in the metal-rich regime (Lee et al. 2015; Joo et al. 2017).

Shortly after, the 2MASS-based 3D map of the RC distribution over a bulge area of 170 deg^2 by Saito et al. (2011) confirmed the presence of the X-shape and showed that the two RC over-densities were only visible in the outer bulge (i.e. $b < -5°$, and $b > 5°$) along the deprojected minor axis ($|l| \leq 5°$, see also Fig. 6). Thanks to the superior quality, in terms of photometric depth and spatial resolution, of the near-IR Vista Variable in the Via Lactea (VVV, Minniti et al. 2010; Saito et al. 2012) Wegg and Gerhard (2013) modelled the observed RC distribution across the whole bulge area (~ 300 deg^2) thus providing the first complete map of its X-shape structure. Although not specifically obtained through the study of RC stars, it is worth mentioning in this context the latest work by Ness and Lang (2016) based on WISE images, in which the X-shape nature of the Milky Way bulge revealed itself with unquestionable doubts (see their Fig. 2).

There is now a general consensus that the majority of the observed Milky Way bulge structure is a natural consequence of the evolution of the bar. The bar heats the disk in the vertical direction giving rise to the typical boxy/peanut shape. Dynamical instabilities cause bending and buckling of the elongating stellar orbits within the bar, resulting in an X-shape when seen edge-on (Raha et al. 1991; Merritt and Sellwood 1994; Patsis et al. 2002;

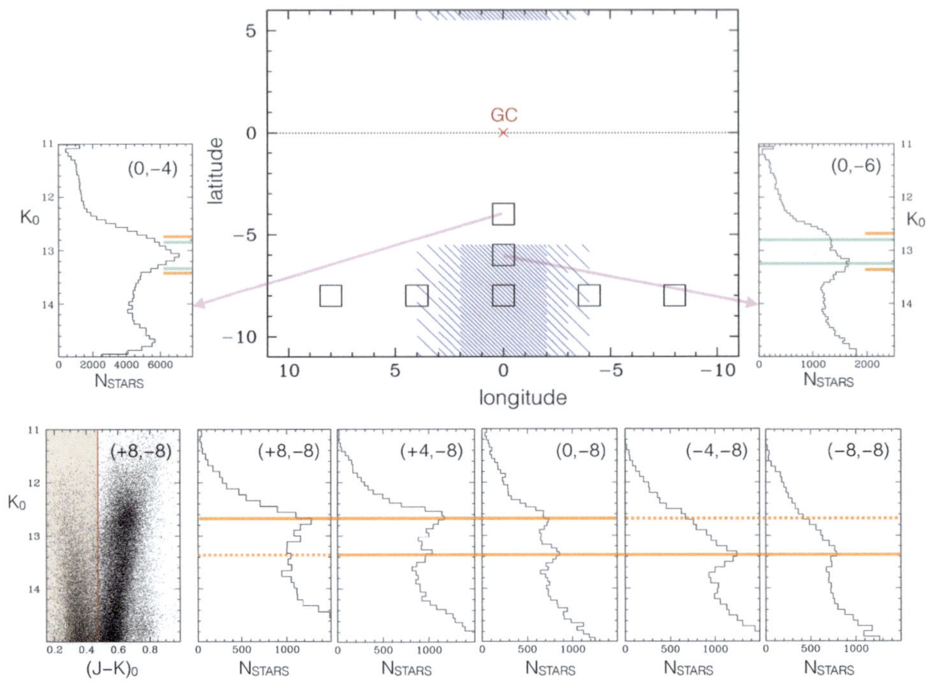

Fig. 6 Luminosity function of the bulge RGB and RC, in several fields. The large box (middle) marks the bulge area covered by the VVV survey, with the seven fields for which the RGB luminosity function is probed. The shaded blue rectangle refers to the region where the double RC is visible ($|b| > 5°$, $|l| < 4°$). The shading is finer where the double RC is clearly evident and becomes wider when one of the two RCs becomes significantly weaker than the other. The bottom left corner panel shows the CMD of the field at $(+8°, -8°)$ as an example of the colour selection applied to exclude the main sequence of the foreground disc. Orange horizontal lines mark the magnitude of the two RCs in the five fields at $b = -8°$ shown in the bottom panels. Green horizontal lines mark the magnitude of the two RCs at $b = -6°$, shown at the top right. Only one RC is visible at $b = -4°$ (Baade's Window), shown at the top left. Figure reproduced from Gonzalez et al. (2015, their Fig. 1)

Athanassoula 2005; Bureau and Athanassoula 2005; Debattista et al. 2006). However, the possible presence of a metal-poor spheroid *embedded* in the boxy/peanut bulge seems to be suggested by a number of fairly recent studies investigating the correlation between RC stars chemical and kinematics properties, as well as the spatial distribution of other stellar tracers, such as RRL, Mira and Type II Cepheids (T2C) variables. While the readers are referred to Sect. 1 above for an overview on the bulge kinematics, here it is however worth mentioning that metal-poor ($[Fe/H] \lesssim 0$) stars in the Baade's Window show negligible vertex deviation ($l_v \sim 0$) consistent with those of a spheroid. Conversely, metal-rich ($[Fe/H] \gtrsim 0$) stars exhibit significant ($l_v \sim 40$) elongated motions typical of galactic bars (Babusiaux et al. 2010). In addition, based on the spectroscopic data provided by the ARGOS survey (i.e. $\sim 14{,}000$ RC stars; Freeman et al. 2013), Ness et al. (2012) demonstrated that only the distribution of metal-rich stars shows the split in RC, a univocal signature of the bar X-shape. On the other hand, metal-poor stars show only a single RC peak. The scenario in which the metal-poor bulge stars, and therefore possibly the oldest population, do not trace the bar structure is also supported by the observed distribution of Mira (Catchpole et al. 2015), RRLs (Dékány et al. 2013), and T2C (Bhardwaj et al. 2017) variables. In particular, by using a combination of near-IR and optical data from VVV and OGLE-III, Bhardwaj et al. (2017) found that their

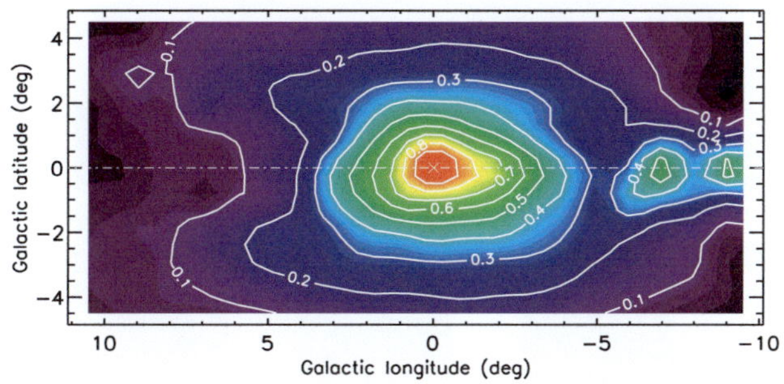

Fig. 7 Density map in Galactic coordinates based on VVV RC star counts from Valenti et al. (2016). Star counts are normalized to the maximum (*Max*). Solid contours are isodensity curves, linearly spaced by $0.1 \times Max$ deg^{-2}. Figure reproduced from Zoccali and Valenti (2016, *The 3D structure of the Galactic bulge*, their Fig. 4)

sample of T2C population in the Galactic bulge shows a centrally concentrated spatial distribution, similar to OGLE-IV counterparts of metal-poor RRLs from the VVV. Mira stars are also consistent with belonging to a boxy bulge (López-Corredoira 2017). It should be noted, however, that the spatial distribution of RRLs in the bulge is somehow still debated given that Pietrukowicz et al. (2015) by using OGLE-IV data did confirmed the presence of a bar in their spatial distribution, although the extension, pivot angle and ellipticity of the structure traced by the RRLs are significantly smaller than that traced by RC stars (see Sect. 2 above).

3.1.2 The Stellar Density Map

More than two decades ago, Weiland et al. (1994) presented the first low angular resolution map at 1.25, 2.2, 3.5 and 4.9 μm of the whole Milky Way bulge based on the COBE/DIRBE data. After correction for extinction and subtraction of an empirical model for the Galactic disk, the derived surface brightness profile of the bulge was then used to study its global morphology and structure. However, more detailed investigations of the bulge innermost region (i.e. $|b| \leq 5°$, $|l| \leq 10°$) have been possible only recently thanks to the VVV survey. Valenti et al. (2016) presented the first stellar density profile of the bulge (see Fig. 7) reaching latitude $b = 0°$. Specifically, by counting RC stars within the CMD as obtained from accurate PSF-fitting photometry of VVV data and previously corrected for extinction by using the reddening map of Gonzalez et al. (2011a), they derived a new stellar density map that allowed to investigate the morphology of the innermost regions with unprecedented accuracy. As seen from Fig. 7, the vertical extent of the isodensity contours is larger at $l > 0°$. This is an expected consequence of a bar whose closest side points towards positive longitude. The high stellar density peak in the central region (i.e. $|l| \leq 1°$ and $|b| \leq 1°$) spatially matches the σ-peak found by GIBS, the kinematics survey of RC (Zoccali et al. 2014) and by Valenti et al. (2018). The stellar density maximum is found in the region $|l| \leq 1°$ and $|b| \leq 0.5°$, and slightly asymmetric with respect to the bulge minor axis.

The observed overall elongation of the density contours towards the negative longitudes is, nevertheless, found to be progressively less pronounced when moving closer to the Galactic plane. Indeed, as shown in Fig. 8 the deprojected density maps become more and more spherically concentrated when RC stars at lower latitudes are considered, hence suggesting the presence of a quasi-axisymmetric structure in the innermost region. This evidence supports the claim by Gerhard and Martinez-Valpuesta (2012) that: the variation in the RC slope at $b = \pm 1°$ and $|l| \leq 10°$ observed by Nishiyama et al. (2005, using IRSF 1.4-m telescope) and by Gonzalez et al. (2011b, based on VVV photometry), and interpreted by these authors as evidence for the presence of a nuclear bar in the inner bulge, is instead caused by a variation of the stellar density distribution along the line of sight.

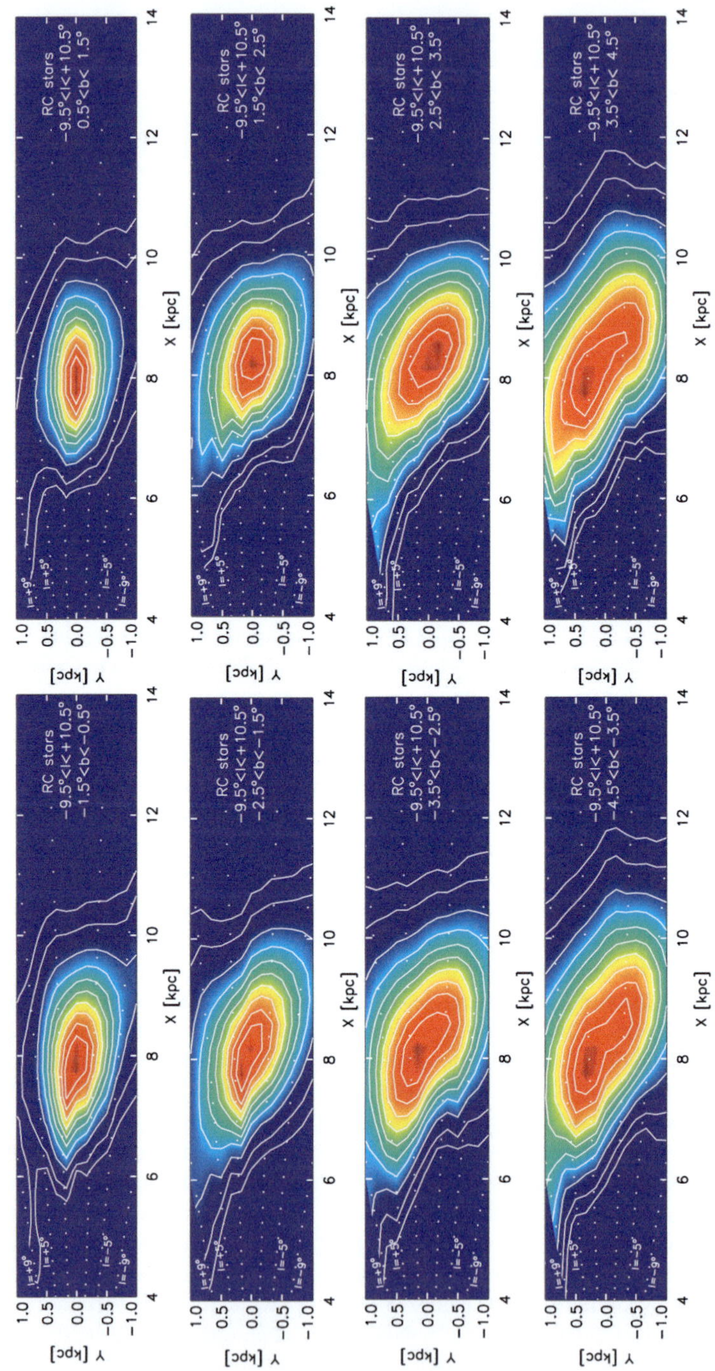

Fig. 8 Deprojection of the RC density shown in Fig. 7 at different latitudes

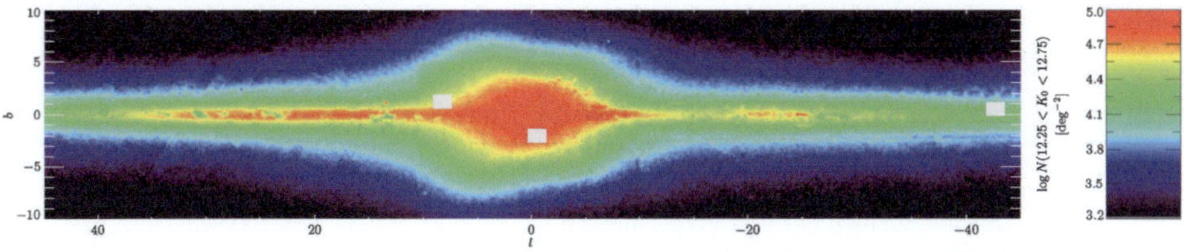

Fig. 9 Surface density map of RC stars based on VVV, UKIDSS and 2MASS K_S-band photometry. Figure reproduced from Wegg et al. (2015, *Structure of the Milky Way's bar outside the bulge*, their Fig. 1)

By using a combination of UKIDSS, VVV, 2MASS and GLIMPSE data, Wegg et al. (2015) presented the so far largest (i.e. ~ 1900 deg^2) density map of the Milky Way and long bar based on RC stars (see Fig. 9). A particularly interesting result of this study is that the orientation angle of the long bar constrained by the best-fit model to the observed density map is consistent with that of the triaxial bulge (i.e. the main bar, 28°–33°). In other words, unlike several previous studies suggesting the presence of a long bar tilted by $\sim 45°$ with respect to the Sun-Galactic centre line (i.e. in addition to the main bar; Benjamin et al. 2005; López-Corredoira et al. 2007; Cabrera-Lavers et al. 2007, 2008; Vallenari et al. 2008; Churchwell et al. 2009; Amôres et al. 2013), the long bar with a semimajor axis of ~ 4.6 kpc in length as modelled by Wegg et al. (2015) appears to be the natural extension of the bulge main bar at higher longitude. This result nicely fits the scenario proposed by Martinez-Valpuesta and Gerhard (2011) and Romero-Gómez et al. (2011) based on N-body simulations, where long and main bar are parts of the same structure. According to this, the boxy/peanut shape bulge would then *simply* be the central vertical extension of a longer and flatter single bar.

3.1.3 The Mass

One of the fundamental question of Galactic astronomy is the determination of the mass distribution in the Milky Way because in general the mass of a given system is more likely the key element driving its evolution (see Courteau et al. 2014, for a detailed review). In this context, over the past three decades many studies addressed this specific question, deriving the dynamical mass of the Galaxy bulge either by matching the galactic rotation curve inside ~ 1 kpc, or by measuring the kinematics (i.e. velocity and velocity dispersion) of a variety of different tracers (i.e. stellar and gas). Then by using an observed luminosity profile (generally in K-band) one can derive the M/L ratio, which ultimately leads to the mass of the bulge (Sellwood and Sanders 1988). Historically, the M/L ratio derived from fitting the rotation curve has been often found to be ~ 2–3, while the M/L ratio derived from stellar kinematics ~ 1. Such discrepancy has been often explained using the argument that the mass derived from the rotation curve is overestimated because of the presence of large non-circular motions that distort the rotation curve (Sofue 1990; Yoshino and Ichikawa 2008). Indeed, the accuracy on the measurement of the rotation curve strongly depends on the accuracy of the distance to the Galactic centre and the solar circular velocity. In addition, the rotation curves as derived from Hα and, in general from other gas tracers (i.e. HI, CO) are often influenced by non-circular components (i.e. inflow, outflow, streaming motions), rather than an ordered (regular) circular motion. This inevitably led to very different results for the mass of the bulge. Chemin et al. (2015) recently reviewed the uncertainties and bias affecting the determination of

Table 2 Estimates of the bulge mass over the last three decades

Mass ($\times 10^{10} M_\odot$)	Diagnostics	Reference
3.0	Rotation curve & M/L$\sim$ 3	Sellwood and Sanders (1988)
1.2	2.4 micron map IRT Spacelab 2 mission	Kent (1992)
2.0	Stellar kinematics & COBE brightness profile	Zhao et al. (1994)
1.3	COBE brightness profile	Dwek et al. (1995)
1.6	microlensing depth & COBE brightness profile	Han and Gould (1995)
2.8	Stellar kinematics & COBE brightness profile	Blum (1995)
2.4	Star counts DENIS near-IR (Besançon model)	Picaud and Robin (2004)
1.8	Gas rotation curve	Sofue et al. (2009)
0.6	Star counts, kinematics and metallicity (Besançon model)	Robin et al. (2012)
0.8	Unified rotation curve, CS & CO	Sofue (2013)
1.8	OGLE RC star counts	Cao et al. (2013)
1.8	M2M dynamical model & VVV RC density	Portail et al. (2015)
2.0	VVV RC star counts & observed IMF	Valenti et al. (2016)

the Milky Way rotation curves and the consequent effects on the derived mass distribution. Table 2 lists a number of studies, together with the adopted observables/diagnostics, that over the years tackled the problem of deriving the bulge mass. Although the reader should refrain from considering Table 2 a complete compilation, it is evident that the large spread in the listed values make the mass of the bulge still poorly constrained. Most estimates cluster to $1.5 \times 10^{10} M_\odot$, however a few authors found values as large as $3 \times 10^{10} M_\odot$ (Sellwood and Sanders 1988) or as small as $0.6 \times 10^{10} M_\odot$ (Robin et al. 2012).

In this context, a special mention is deserved for the two most recent works by Portail et al. (2015) and Valenti et al. (2016) which, although following different methodologies, they both use the distribution of the RC stars as derived by the VVV photometry. Portail et al. (2015) used made-to-measure dynamical model of the bulge, with different dark matter halo to match the stellar kinematics from BRAVA (Rich et al. 2007; Kunder et al. 2012) and the 3D surface brightness profile derived by Wegg and Gerhard (2013). Their best-fit model is consistent with the bulge having a dynamical mass of $1.8 \pm 0.07 \times 10^{10} M_\odot$, with a dark matter content that varies with the adopted initial mass function (IMF). When the observed IMF of Zoccali et al. (2000) is considered about $0.7 \times 10^{10} M_\odot$ (i.e. 40%) of dark matter is required in the bulge region. In addition, they estimated that the total stellar mass involved in the peanut shape accounts for $\sim 20\%$ of the total stellar bulge mass.

On the other hand, by scaling the observed VVV RC stellar density map (see Fig. 7) with the observed bulge luminosity function from Zoccali et al. (2000), and Zoccali et al. (2003), Valenti et al. (2016) provided the first empirical, hence no model-dependent, estimate of the bulge stellar mass. From the observed stellar mass profile shown in Fig. 10, the authors estimated that the mass in stars and remnants of the Milky Way bulge in the region $|b| < 9.5°$ and $|l| < 10°$ is $2.0 \pm 0.3 \times 10^{10} M_\odot$.

These two latest estimates are found compatible within the quoted errors, and they might be even more close when considering that the empirical estimate by Valenti et al. (2016) refers to a larger volume that is not limited along the line of sight.

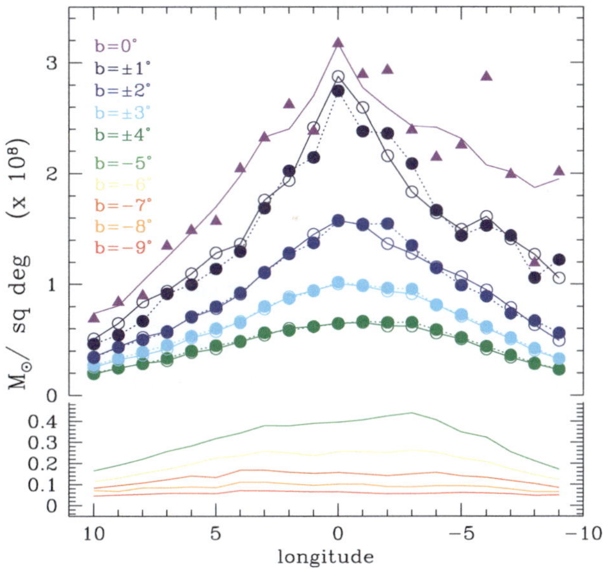

Fig. 10 Stellar mass profile across latitude in each VVV field for fixed longitudes. Filled symbols with dotted lines refer to $b < 0$, empty symbols with solid lines to $b > 0$ fields. The lower part of the plot has an expanded y-axis for a better display of the high negative longitudes, owing to the higher extinction, hence higher incompleteness in those fields. Figure reproduced from Valenti et al. (2016, *Stellar density profile and mass of the Milky Way bulge from VVV data*, their Fig. 5.)

3.2 The Chemical Composition

Because the chemical content of any given stellar system retains crucial information to unveil its origin, formation and evolution (McWilliam 2016), after the pioneering works of Frogel et al. (1984) and Rich (1988) several studies over the decades focussed on the determination of the bulge stars metallicity and abundances distribution to understand how the bulge formed. What follows is not meant to be a comprehensive compilation of all such studies for which one should dedicated an entire single review, but rather a summary of our current knowledge of the chemical composition of the bulge based on the latest results from RC and RRLs.

3.2.1 The Metallicity Distribution

As emphasized by Matteucci et al. (1999), Ferreras et al. (2003), the peak and shape of the metallicity distribution functions (MDFs) provide important constraints on the IMF, star formation efficiency, as well as to the possible gas infall timescale. However, until less than a decade ago, accurate MDF based on high-resolution spectroscopy were available only for a handful number of sparse fields mainly located along the bulge minor axis (see i.e. McWilliam and Rich 1994; Fulbright et al. 2007; Rich et al. 2007, 2012; Johnson et al. 2011; Gonzalez et al. 2011c; Hill et al. 2011, , and reference therein). The derived MDFs were consistent across various studies, which all agreed in finding the bulge population to be on average metal-rich, although spanning a fairly broad metallicity range (e.g. $-1.5 \lesssim$ [Fe/H] $\lesssim +0.5$). Our comprehension of the MDF of the bulge has improved tremendously thanks to three spectroscopic surveys, namely ARGOS (Freeman et al. 2013), GIBS (Zoccali et al. 2014), and ESO-Gaia (Rojas-Arriagada et al. 2014), that all together have provided spectra for more than 20,000 RC stars across most of the inner and outer bulge regions (see Table 3 for further details). The MDF derived by these surveys confirmed previous results although extending them on a much larger area. The mean bulge population across all fields is metal-rich with a small fraction of stars with [Fe/H] $> +0.5$ dex and

Table 3 For each spectroscopic survey, the total number of stars, the total number of targeted fields and the region within the bulge covered by the observations are given

Survey	Total RC stars	Number of fields	Bulge region
ARGOS	14, 000	27	$-10^\circ < b < -5^\circ$, $\|l\| \lesssim +30^\circ$
GIBS	5, 500	26	$-8^\circ < b < -1^\circ$, $b = +4^\circ$, $\|l\| \lesssim +30^\circ$
Gaia-ESO	1, 200	5	$-10^\circ < b < -4^\circ$, $-10^\circ < l < 7^\circ$

[Fe/H] < -1.5 dex. Only in the outermost fields ($b < -7^\circ$, $|l| > 10^\circ$) observed by ARGOS the MDF reaches metallicity as low as ~ -2.5 dex. In addition, a mild vertical gradient is found when considering the mean metallicity of each field, such as the metallicity increases moving inwards along the bulge minor axis, hence confirming what suggested previously by Minniti et al. (1995), and Zoccali et al. (2008). However, thanks to statistically robust target samples a detailed study of the MDF shape has been possible for the first time, revealing the presence of multiple components. The observed overall metallicity gradient is therefore explained as a consequence of the presence of two (see Zoccali et al. 2017; Rojas-Arriagada et al. 2014) or more (see Ness et al. 2013b) components with different mean metallicity. As evident from Fig. 11, the variations of the relative contribution of these components across the fields (i.e. metal-rich stars component becoming progressively less prominent towards the outer region) mimic the observed gradient. However, Zoccali et al. (2017) found also that at latitudes smaller than $|b| = 3^\circ$ the metal-poor component becomes important again (see first 2 top panels of Fig. 11), its relative fraction increases again close to the plane. To further investigate the spatial distribution of the two components they mapped their distribution by coupling the relative fractions derived by GIBS with the bulge stellar density from Valenti et al. (2016). The result, shown in Fig. 12, demonstrates that the metal-poor component has a spheroid-like spatial distribution, versus a boxy distribution of the metal-rich component. In addition, the metal-poor component shows a steeper radial density gradient.

Although as mentioned before the bulge kinematics is the subject of another paper in this volume (see Sect. 1 above by A. Kunder), here I will only briefly mention that the two components were found to have also different kinematics. Indeed, as already found by the BRAVA (Kunder et al. 2012) and ARGOS survey (Ness et al. 2013a), in the outer bulge ($|b| > 4^\circ$) the metal-poor component has a higher radial velocity dispersion compared to the metal-rich one, at all longitudes. However, Zoccali et al. (2017) showed that such behavior is reversed in the inner bulge. Specifically, the velocity dispersion of the metal-poor stars at $b = -3.5^\circ$, -2° becomes similar to that of the metal-rich counterpart, and progressively becomes smaller at $b = -1^\circ$.

While the chemical abundances of the RC stars in the bulge is one of the topics that received more attention in the recent years, the number of studies addressing the chemical content of the oldest bulge population, such the RRLs, is still very limited. Perhaps mostly due to the observational challenges that spectroscopic observations of RRL face, as of today there is no high-medium resolution spectroscopic measurements of a sizeable sample of RRLs in the bulge. In K-band RRLs are in general about 0.5 mag fainter than RC stars, hence their brightness makes them suitable targets at high resolution only with a 4 m-class telescope or above, depending on the bulge region. In addition, because they are much less numerous than RC stars, and so more sparsely distributed RRLs are not even suitable targets for the vast majority of the current multiplexing spectrograph facilities. An additional complication is the fact that the metallicity derived from the line equivalent width measurements strongly depends upon the pulsation phase at which the star was observed. This necessarily

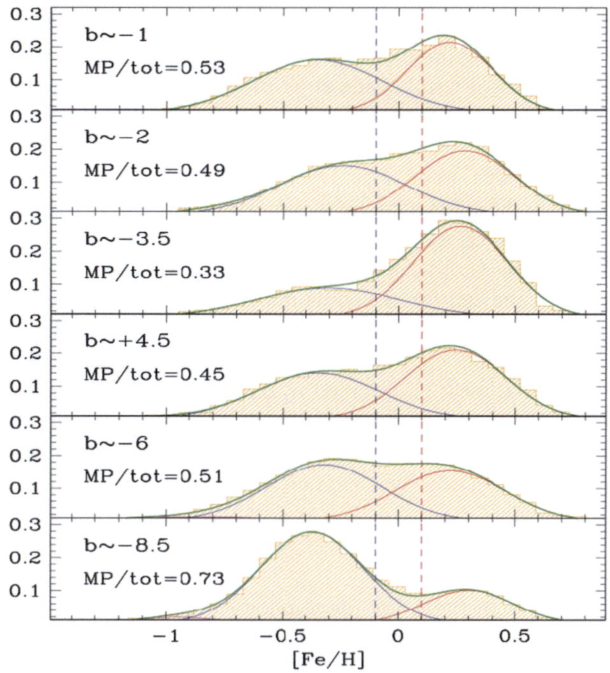

Fig. 11 MDF of RC stars at constant latitudes. The fraction of metal-poor stars, compared to the total, is given in each panel. Vertical dashed lines mark the limits of the metal-poor and metal-rich populations. Figure adapted from Zoccali et al. (2017, *GIBS-III: MDFs and kinematics for 26 fields*, their Fig. 7)

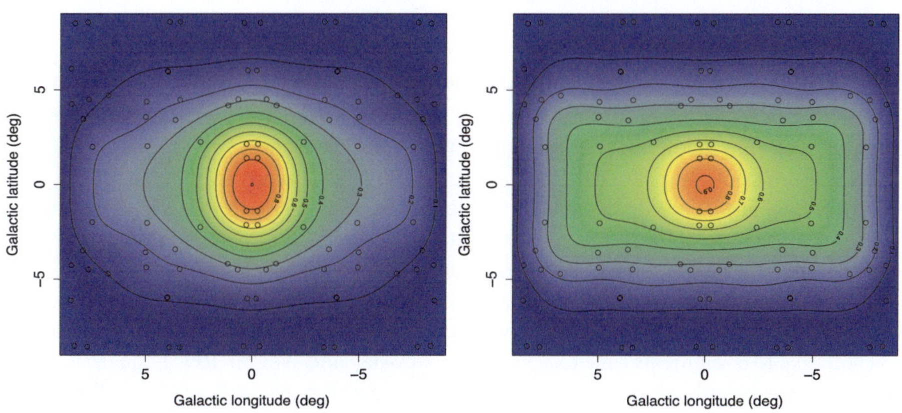

Fig. 12 Density map of metal-poor (left) and metal-rich (right) RC stars obtained using the MDF of GIBS fields and the total number of RC stars from Valenti et al. (2016). Figure adapted from Zoccali et al. (2017, *GIBS-III: MDFs and kinematics for 26 fields*, their Fig. 9)

implies a good knowledge of the variables. All of these factors make their observations very telescope time consuming.

As of today the only spectroscopic study of a sizeable sample of bulge RRLs has been presented by Walker and Terndrup (1991), who derived the MDF of 59 RRLs in the Baade's Window. The individual star metallicities were derived through the ΔS method (i.e. low-resolution, see Suntzeff et al. 1991, for a detailed description of the ΔS method) and their distribution is found to cluster around [Fe/H] $= -1$ dex. Although the MDF is relatively broad, spanning a range of about 1 dex, $-1.7 \lesssim$ [Fe/H] $\lesssim -0.5$, its very sharp peak accounts for $\approx 80\%$ of the entire sample. Based on the derived MDF, the authors concluded that the RRL are being produced by the metal-poor tail of K giants distribution (see also Fig. 11).

Recently, Pietrukowicz et al. (2015) provided a photometric MDF based on more than 27,000 RRLs from the OGLE-IV catalogs and located in the bulge region between $|l| \lesssim 10°$ and $-8° \lesssim b \lesssim -2°$, $+2° \lesssim b \lesssim +5°$. The photometric MDF is much broader than the spectroscopic one, as it spans mostly the range $-2.5 \lesssim$ [Fe/H] $\lesssim +0.5$, although the peak is found at the same metallicity, [Fe/H] $= -1$ dex. The authors showed that there is no correlation between the distance and the shape of the MDF, however they find a very mild, but statistically significant, radial metallicity gradient (i.e. the metal-rich population increases towards the centre). Based on the analysis of the Bailey diagram (i.e. period-amplitude diagram) the authors argue for the existence of two different population of RRLs with likely different metallicity, similar to the bulge RC counterparts. However, it should be mentioned that unlike what is observed in the MDF of RC stars, these 2 populations of RRLs with different metallicity do not probably change in relative fraction given that the global MDF conserves its shape throughout the total covered bulge area. Moreover, because of the lack of RRLs spectroscopic measurements in the metal-rich regime, [Fe/H] > -0.5 dex (see Walker and Terndrup 1991), one should refrain from drawing any firm conclusion from the available RRL MDF.

3.2.2 The α-Elements Abundances

The detailed study of the chemical abundances and abundance patterns in bulge stars provides a unique tool to understand the chemical evolution enrichment of the bulge, and therefore to set tight constraints on its formation scenario. The elemental abundance distributions, and the abundance ratio of certain critical elements such as Fe-peak, CNO, and α-elements (i.e. those synthesized from α particles as O, Ne, Mg, Si, Ti, Ca and S) are particularly suitable for this purpose. Indeed, these elements are synthesized in stars of different masses, hence released into the interstellar medium on different timescales.

Because most of the chemical information on RC stars comes from the analysis of the α-elements, what follows is a summary of the picture built upon those measurements. It is not meant to be a comprehensive review of the global chemical composition of the bulge, for which the readers are instead encouraged to refer to McWilliam (2016).

As mentioned above a particularly useful abundance ratio is [α/Fe]. Due to the time delay in the bulk of Fe and Fe-peak elements production (mostly due to SNe Ia, see Nomoto et al. 1984) relative to α-elements (due to SNe II, see Woosley and Weaver 1995), the [α/Fe] abundance ratio can be efficiently used as a *cosmic clock* (see e.g. McWilliam 1997; Wyse 2000, and references therein). For this reason many studies in the past addressed this question providing [α/Fe] ratios for relatively small sample of K and M giants in few bulge regions (see McWilliam and Rich 1994; Rich and Origlia 2005; Cunha and Smith 2006; Fulbright et al. 2007; Lecureur et al. 2007; Rich et al. 2007; Meléndez et al. 2008; Alves-Brito et al. 2010; Johnson et al. 2011; Gonzalez et al. 2011c; Hill et al. 2011; Rich et al. 2012; Bensby et al. 2013; Johnson et al. 2014; Bensby et al. 2017, and references therein). As it has been the case for the MDF, the advent of the recent spectroscopic surveys ARGOS (Ness et al. 2013b) and GIBS (Gonzalez et al. 2015) provided α-element abundances for thousands of RC stars over a large area, hence allowing to study the [α/Fe] trends as a function of the position in the bulge.

All previous and very recent studies agree on finding the bulge to be α-elements enhanced with respect to the solar value, thus suggesting a fast bulge formation scenario. As shown in Fig. 13, the α-element abundances of bulge stars with [Fe/H] < -0.3 are enhanced over iron

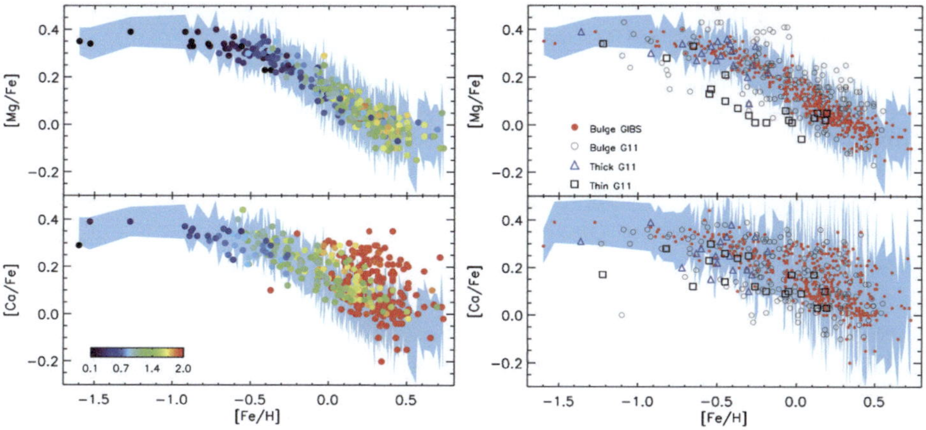

Fig. 13 Left: Distribution of [Mg/Fe] (*upper panel*) and [Ca/Fe] (*lower panel*) as a function of [Fe/H] for ~ 400 GIBS bulge giants. The abundances of [Mg/Fe] and [Ca/Fe] are color-coded according to their estimated uncertainty. The uncertainty contour of [Mg/Fe] is also shown in the background of both panels. Right: Distribution of [Mg/Fe] (*upper panel*) and [Ca/Fe] (*lower panel*) as a function of [Fe/H] used as diagnostic of the formation timescale of the bulge. Red circles refer to GIBS abundances of the bulge giants, whereas black symbols mark the abundances of bulge, thick and thin disk giants from Gonzalez et al. (2011c). Figure adapted from Gonzalez et al. (2015, *The GIRAFFE Inner Bulge Survey (GIBS). II. Metallicity distributions and alpha element abundances at fixed Galactic latitude*, their Figs. 8 and 9)

by ~ 0.3 dex, whereas metal-rich stars show a decrease in [α/Fe] reaching 0 for metallicity above the solar values. However, as discussed in Gonzalez et al. (2011c) the direct translation of this trend to absolute timescales is not easy because the SNe Ia delay time can depend on different production channels. This is the reason why a relative approach through the comparison of [α/Fe] trends observed in different Galactic components turns to be more reliable. From the comparison between bulge RC and giants in the thin and thick disk (see Fig. 13, right panels), the α-elements enhancement of the bulge with respect to the thin disk is evident across most of the entire metallicity regime. At solar metallicity, the bulge and thin disk are both α-poor. On the other hand, the thick disk giants are found to be as α-enhanced as the bulge, although they never reach the high metallicity tail of bulge stars. A possible interpretation of this relative trends is that the metal-poor bulge population experienced a fast formation scenario similar to the thick disk, whereas the metal-rich bulge population underwent a more extended (i.e. longer) star formation, on a timescale similar to that of the thin disk.

3.3 The Age

An accurate dating of the bulge stellar component allows one to gauge at which lookback (i.e., at which redshift) one should look for possible analogs of the Milky Way, when their bulge formation processes were about to start, well on their way, or even already concluded. Indeed, with an age of ~ 10 Gyr or older, it is at $z \gtrsim 2$ that such analogs can be searched, or at lower redshift if significant fraction of the stellar component is found to be several Gyr younger (see Valenti et al. 2013, for detailed discussion).

However, dating bulge stars is a very complicated task, challenged by the stellar crowding, the patchy and highly variable extinction, the uncertainties in the distance modulus, the distance spread due to the spatial depth of the bulge/bar along the line of sight, the metallicity dispersion and finally the contamination by foreground disk stars. The different

Table 4 Position of the bulge fields, in Galactic coordinates, for which an age estimate of the stellar population has been provided through MSTO determination

Longitude	Latitude	Disk decontamination	Reference
+1.13°	−3.77°	Proper motions	Kuijken and Rich (2002)
+1.25°	−2.65°	Proper motions	Kuijken and Rich (2002)
0°	−6°	Statistical	Zoccali et al. (2003)
+1.25°	−2.65°	Proper motions	Clarkson et al. (2008, 2011)
+0.25°	−2.15°	Proper motions	Brown et al. (2010)
+1.26°	−2.65°	Proper motions	Brown et al. (2010)
+1.06°	−3.85°	Proper motions	Brown et al. (2010)
−6.75°	−3.81°	Proper motions	Brown et al. (2010)
+10.3°	−4.2°	Statistical	Valenti et al. (2013)
−6.8°	−4.7°	Statistical	Valenti et al. (2013)

contribution of all these factors prevents accurate location in terms of magnitude and color of the main sequence turnoff (MSTO) of the bulge population, so far among the most reliable age diagnostics (see Renzini and Fusi Pecci 1988, and Sect. 5 by G. Bono).

Historically, the earliest age constraint by van den Bergh and Herbst (1974) in the Plaut field along the bulge minor axis at $b = -8°$ (~ 1 kpc) indicated a globular cluster (GC) like age. Terndrup (1988) fit the photometry of other bulge fields at a range of latitudes with GC isochrones of varying metallicity, but because lacking a secure distance for the bulge he derived only a weak age constraint (11–14 Gyr). Ortolani et al. (1995) solved the problem of contamination and distance uncertainties by comparing the bulge population with the NGC 6528 and NGC 6553 clusters. Forcing the bulge field and cluster luminosity function to match the HB clump luminosity level, it was possible to show for the first time that the relative ages of the bulge and metal-rich cluster population could not differ by more than 5%. Feltzing and Gilmore (2000) used HST-based photometry of Baade's Window and another low extinction field known as the Sgr-I (i.e. at $l = 1.25°$ and $b = 2.65°$) to argue that while the density of the bulge MSTO stars increases for field closer to the centre, the foreground population does not change. They concluded that the bulk of the bulge population must therefore be old. The case for an *old bulge* has been further strengthened by later and more accurate photometric studies of different bulge fields, and by tackling the problem of contamination by foreground disk stars either kinematically by using proper motions, or statistically by considering control disk fields. Table 4 lists the location of the each observed field together with the adopted decontamination approach.

As all previous studies, Valenti et al. (2013) found that the bulk stellar population of the Milky Way bar edges is over ~ 10 Gyr old (see Fig. 14), with no obvious evidence of younger population. This age is indistinguishable from the one reported for more inner bulge fields, a few degrees from the Galactic centre or lying along the bulge minor axis. From the analysis of the MSTO in the HST-based CMD kinematically decontaminated, Clarkson et al. (2011) concluded that once the blue straggler population is taken into account a significantly younger ($\lesssim 5$ Gyr) population in the bulge must be at most 3.4%.

However, there is a clear discrepancy between the ages inferred from the determination of the MSTO location in the observed CMDs and those derived for microlensed stars in a project of Bensby and collaborators, which estimates single star age from its effective temperature and gravity (i.e. from isochrones in the $T_{\rm eff}$, log g plane) as obtained from high resolution spectra. Indeed, based on a sample of 90 F and G dwarf, turnoff and subgiant stars

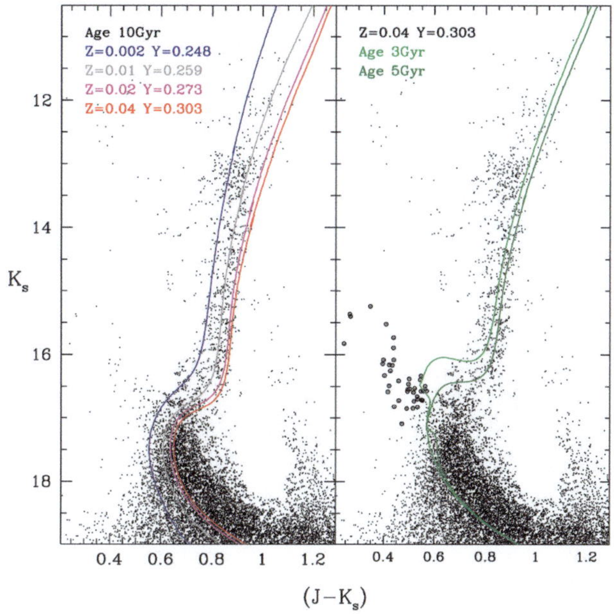

Fig. 14 Disk-decontaminated CMD of the field located at the far edge of the bar with over-plotted theoretical isochrones (Valcarce et al. 2012). Isochrone ages, metallicity and helium abundances are indicated near the top-left corner of each panel. Sub-solar isochrones are α-enhanced, and solar-scaled otherwise. The metallicity range covered by the bulk of the stars in this field is $0.002 \lesssim Z \lesssim 0.060$, as derived from the observed photometric metallicity distribution. Open circles mark stars that are most likely blue stragglers. Figure adapted from Valenti et al. (2013, *Stellar ages through the corner of the boxy bulge*, their Fig. 10)

in the bulge (i.e. $|l| \lesssim 6°$ and $-6° < b < 1°$) observed during microlensing, Bensby et al. (2017) found that about 35% of the metal-rich star ([Fe/H] > 0) are younger than 8 Gyr, whereas the vast majority of metal-poor ([Fe/H] $\lesssim -0.5$) are 10 Gyr or older. In addition, from the derived age-metallicity and age-α elements distribution the authors concluded that the bulge must have experienced several significant star formation episodes, about 3, 6, 8 and 12 Gyr ago.

As discussed by Valenti et al. (2013), each of the two approaches has its own *pros* and *cons*. The microlensing approach depends more heavily on model atmospheres that may introduce systematics especially in the metal rich regime, and it deals with small number statistics. At the same time, it has the advantage that the metallicity of each individual stars is very well constrained. Conversely, by dealing with a statistically significant number of stars, the traditional CMD method should in principle be able to reveal the presence of young populations. However, in this case the metallicities of individual stars are unknown, therefore one does not know if, for instance, some of the stars above the MSTO of the $Z = 0.060$ isochrones (see Fig. 14) are old and have lower metallicity, or whether they are metal-rich stars younger than 10 Gyr.

The effect of the age-metallicity degeneracy, specifically in terms of the color spread of the MSTO in the observed CMDs has been used by Haywood et al. (2016) to argue in favor of the scenario suggested by the microlensing results. In particular, Haywood et al. (2016) compared the MSTO color spread observed in the CMD of Clarkson et al. (2011) with that of synthetic CMDs, obtained by using two scenarios corresponding to different age-metallicity relation (AMR). In *scenario I* a simulated CMD was obtained by using the AMR presented by Bensby et al. (2013) (i.e. based on a total sample of 59 microlensed drawfs), whereas for *scenario II* an AMR that extends from [Fe/H] = −1.35 dex at 13.5 Gyr to [Fe/H] = +0.5 dex at 10 Gyr was adopted. When taking into account distance, reddening and metallicity effects, Haywood et al. (2016) showed that the MSTO color spread of a *purely* old stellar population would be wider than what observed, which in turn appears to be consistent with the simulation obtained from the *scenario I*. Unfortunately, what the Haywood et al. (2016) paper does not address is the fact that the simulation using the AMR of Bensby et al. (2013) produced a CMD that not only has a smaller MSTO color spread like the observed one but also show a remarkable number of stars just above (i.e. brighter) the MSTO, which are not matched by the observations (see their Fig. 8). In this respect, the comparison between observations and simulations presented by Zoccali et al. (2003) to infer the age of the bulge population would seem more appropriate because the synthetic CMD was obtained by using the observed luminosity function, and therefore the comparison was done such as to match not only the location and spread in color of the MSTO, but also the number of stars at the MSTO level.

3.4 Summary and Conclusions

Owing to the systematic and detailed study of RC star properties performed in the last decade by using wide area photometric surveys we have finally reached a good and complete comprehension of the 3D structure of the Milky Way bulge. The bulge, as referred as the region in the inner ~ 3 kpc is a bar with an orientation with respect to the Sun-Galactic centre line of sight of $\sim 27°$, and whose near side points in the first Galactic quadrant. The bar has a boxy/peanut/X-shape structure in its outer regions, a characteristic morphology of bulges formed out the natural evolution of disk galaxies as the consequence of disk dynamical instabilities and vertical buckling of the bar. The observed split in the RC mean magnitude distribution in the outer regions is interpreted by the dynamical models as bar growing. In the innermost region ($|l, b| < 2°$), rather than a nuclear bar, there seems to be an axisymmetric high stellar density peak, which instead may be responsible for the observed change in the bar pivot angle. In addition, RC stars trace a thinner and longer structure with a semi-major axis of ~ 4.6 kpc, known as the long bar, which according to the latest study appears to be the natural extension of the bulge main bar at higher longitude.

The bulge is the most massive stellar component of the Galaxy, with a mass ($M_B = 2 - 1.8 \times 10^{10} M_{\odot}$) close to 1/5 of the total stellar mass of the Milky Way, and about ten times larger than the mass of the halo.

The recent spectroscopic surveys (ARGOS, GIBS, Gaia-ESO) of RC stars, together with the ongoing that targets K and M giants (i.e. APOGEE-North) provided a comprehensive and detailed view of the chemical content of the stellar population over an area that corresponds to more than 80% of the entire bulge. The emerging picture is that the bulge MDF as traced by the RC is much more complex than previously thought, and it hosts two populations with different mean metallicity (i.e. metal-poor and metal-rich), spatial distribution and kinematics. The metal-poor population as traced by RC, RRLs and T2C is more spherically concentrated, whereas the RC metal-rich component traces the boxy/peanut bar. The observed properties of such metal-poor population possibly older (i.e. spatial distribution and kinematics) do not necessarily imply the presence of a *classical bulge* (i.e. a merger-driven structure dominated by gravitational collapse) embedded in the boxy bulge. Indeed, the recent N-body simulations model of Debattista et al. (2017) accounts for the presence of a metal-poor population spherically concentrated, as well as for other observed trend of densities, kinematics and chemistries, without invoking the need for a *composite bulge* scenario (i.e. the coexistence of two structures, one merger-driven and one boxy shaped formed out of disk and bar evolution). According to Debattista et al. (2017), the observed properties of the Milky Way bulge stellar populations are consistent with a bulge formed from a *continuum* of disk stellar populations kinematically separated by the bar.

Based on accurate abundances analysis of RC stars, the bulge show α-element enhancement typical of fast formation process. In particular, a possible interpretation of the observed relative trends of α-elements in the bulge, thin and thick disk is that the metal-poor bulge population experienced a fast formation scenario similar to the thick disk, whereas the metal-rich bulge population underwent a more extended (i.e. longer) star formation, on a timescale similar to that of the thin disk.

The innermost and still poorly unexplored regions (i.e. $|b| \lesssim 1°$) will be soon probed by new IR surveys planned for the near future (i.e. APOGEE-South, Multi-Object Optical and Near-infrared Spectrograph at VLT—MOONS) hence allowing us, for the first time, to complete the puzzle with a clear understanding of the chemical properties of the bulge as a whole with unprecedented accuracy.

There is no doubt that the central regions of the Milky Way host an old stellar population. The strongest evidence being the presence of a prominent population of RRLs and T2C found by OGLE and VVV (Dékány et al. 2013; Pietrukowicz et al. 2015; Gran et al. 2016, and Bhardwaj et al. 2017), which are by far the largest photometric campaigns of variable stars. Furthermore, an old age is also guaranteed by the existence of a bulge GCs system (see e.g. Valenti et al. 2010; Bica et al. 2016, and reference therein). However, what still remains to be firmly assessed is the contribution of intermediate-young (i.e. $\lesssim 5$ Gyr) stars to the global bulge stellar population. The AMR proposed by Bensby et al. (2017) should be either confirm on much statistically robust sample, or by using a methodology for the reconstruction of the star formation history more sophisticated than the approaches adopted so far. In particular, the comparison between observations and simulations should be performed by using as many as possible features of the CMDs (i.e. Gallart et al. 2005).

In the coming years, the exquisite astrometry provided by the next Gaia data releases will most probably allow us to further refine the global picture of the bulge structure. Even though a large fraction of the bulge RC population is out of Gaia range because of the crowding and high extinction, the information derived from RC stars in the low reddening regions can be used to obtain a very accurate distances map of the bulge outer regions. This can be used then as the *reference frame* upon which, through a differential analysis with the most obscured regions, we can build the entire bulge distances and structure maps.

Finally, further efforts should be put to characterize the chemical content of the RRLs and T2C, which among all tracers are those representing *univocally* and *purely* the oldest stars in the Bulge. Indeed, accurate MDF and elemental abundances from high- or medium-resolution spectroscopy for these type of stars are still missing, or largely insufficient. The future LSST project will provide the position, magnitude and colors for thousands of variable stars, spanning a variety of ages. The spectroscopic follow up of a sizeable sample of variables would literately open new frontiers of our knowledge by allowing for the first time an accurate study of the metallicity trends as a function of the stellar ages. If such analysis would be extended also outside the bulge regions we could be in position to understand the interaction among different Galaxy structures, such for instance a clear view of the transition between disk and bulge.

4 RR Lyrae Variables in the Ultra-Faint Satellites of the Milky Way

In the Λ-Cold Dark Matter (Λ-CDM) scenario, large galaxies are the result of the assembling of smaller fragments, cold-dark matter dominated (e.g. Diemand et al. 2007; Lunnan et al. 2012). The baryonic component of these fragments may eventually collapse, forming small galaxies. This idea is appealing when applied to the MW, since it echoes the early scenario envisioned by Searle and Zinn (1978), in which the outer halo of the MW may have formed by a continuous infall of protogalactic fragments onto the Galaxy, for some time after that the collapse of its central part was completed. Indeed, first attempts to link the Λ-CDM cosmology with the Galactic environment foresaw the assembling of the Galactic halo starting from a number of satellites, and producing a number of fragments and streams, which are actually observed (e.g. McConnachie 2012; Grillmair and Carlin 2016). For decades, the survivors of such a process have been identified with the dwarf spheroidal (dSph) satellites of the MW, since they are old, metal-poor, gas poor and dark matter-dominated systems. However, it was soon realized that the observed number of observed dSph was one or two order of magnitude smaller than that expected from theory. This mismatch, dubbed the "missing satellites problem" (Klypin et al. 1999; Moore et al. 1999), has been for several years a major problem in the comparison between theory and observations. A second problem, pointed out in the last few years, is that the circular velocities of the known dSph are too low, when compared to the expected values from their simulated sub-structures. In other words, the predicted densities of the massive subhaloes are too high, to host any of the bright dSphs. This mismatch, called the "too big to fail problem" (Boylan-Kolchin et al. 2012), has heavy implications, since it means either: (i) massive dark subhaloes exist as predicted, but they host faint ($L < 10^5 L_{\odot}$) satellites; (ii) massive dark subhaloes does not exist as predicted, for instance they may be less concentrated than predicted.

As a matter of fact, in the last ten years a considerable number of new and faint MW satellites has been discovered (e.g. Belokurov et al. 2007; McConnachie 2012), most of them on the basis of the SDSS data and, more recently, thanks to the ongoing large surveys conducted with OMEGACAM@VST, DECAM@CTIO and Pan-STARRS (e.g. Koposov et al. 2015; Laevens et al. 2015). These systems, called the ultra-faint dwarfs (UFDs), have integrated luminosities similar or even lower than those of the Galactic globular clusters, and are apparently dark matter dominated(see McConnachie 2012).

The large number of systems currently available (dSphs + UFDs), allowed to trace a statistically significant analysis of their spatial distribution, leading to the discovery that they actually populate a relatively thin ring, perpendicular to the MW plane, and possibly rotationally supported (Pawlowski and Kroupa 2013). Moreover, several of the recently discovered candidate MW satellites also seem to be clustered around the Magellanic Clouds, hinting that they may have fallen in as a group (e.g. Sales et al. 2015), in line with the theoretical predictions (Wetzel et al. 2016). Similar aligned structures, showing a kinematic coherence, have been discovered around the Andromeda galaxy (Ibata et al. 2013) and, outside the Local Group, around NGC 5557 (Duc et al. 2014). Similar structures, but without a clear kinematic coherence, have been reported in the literature around NGC 1097, NGC 4216, NGC 4631 (Pawlowski and Kroupa 2014, , and references therein), and possibly around the M81 and Cen A groups (Müller et al. 2016, 2018). The MW structure, dubbed the *Vast POlar Structure* (VPOS), opens a wide scenario of cosmological problems, since at the present time it is not clear if it is made of primordial (dark matter-dominated) systems, or tidal (dark matter-free) galaxies.

Interestingly, when the halo Galactic globular clusters are grouped in young halo (YH) and old halo (OH) on the basis of the variation of their HB morphology at constant [Fe/H], which is a rough approximation of the cluster age, they show up also a division by kinematics and spatial distribution (e.g. Zinn 1993; Mackey and Gilmore 2004b; Lee et al. 2007). In particular, YH clusters span a wide range in ages (~ 5 Gyr, Dotter et al. 2011) and are characterized by a hotter kinematics than the OH clusters. These occurrences suggest that YH clusters may be debris from accretion events. Finally, the discovery that YH clusters are part of the VPOS (Pawlowski and Kroupa 2013; Zinn et al. 2014), strengthens the debris hypothesis. Moreover, it also suggests that a fraction of the accreted halo may have been originated in a number of moderately massive satellites that formed GCs, similar to

Sagittarius, Fornax, or even the Magellanic Clouds (but see Fiorentino et al. 2017, for new insights on the contribution of Fornax-like systems).

The Role of the RR Lyrae Stars A fraction of the problem can be settled by carefully comparing the photometric and spectroscopic properties of the stellar populations of the halo of the MW and of its companions. Moreover, since their pulsational properties such as periods and amplitudes are a function of their structural and evolutive parameters, a detailed comparison of the pulsational properties of the RR Lyrae stars can add valuable information.

In particular, the *ensemble* pulsational properties of the RR Lyrae stars can give important hints. Indeed, it is well known that cluster and field Galactic RR Lyrae stars are affected by the so-called Oosterhoff (Oo) dichotomy, where in the Oo I group the fundamental mode variables show mean periods of $\langle P_{ab} \rangle \sim 0.55$ days, while in the Oo II group they have $\langle P_{ab} \rangle \sim 0.65$ days. In fact, the bright MW companions have $\langle P_{ab} \rangle \sim 0.6$ days and are generally classified as Oo-intermediate, which is difficult to reconcile with the dichotomy of the Galactic halo. On the other side, the RRLs hosted in the UFDs suggest an Oo II classification (Dall'Ora et al. 2012), consistent with an older population of the Galactic halo, possibly produced by an early dissipative collapse or merging (e.g. Miceli et al. 2008).

The Ultra-Faint Dwarfs UFD galaxies are, at first glance, the low brightness tail of the dSph. From this point of view, there is no structural difference between the "classical" dSphs and the low luminosity UFDs. However, a careful comparison of the central surface brightness as a function of the total luminosity, shows a "knee" around $M_V \approx -8$ mag (see McConnachie 2012, Fig. 7). The galaxies brighter than $M_V \approx -8$ mag follow a linear trend, with the brightest galaxies having a higher central brightness, while galaxies weaker than this limit follow a horizontal distribution, with a constant central brightness no matter what is the total luminosity. In this work, we will therefore consider UFDs all the galaxies that follow such a horizontal distribution.

Stated in a different way, UFDs are characterized by low luminosities and projected densities. This means that it is difficult to recognize them as stellar overdensities in the field, and the problem becomes even more severe when one wants to detect their possible tidal tails. For these reasons, RRLs become a powerful tool to study the stellar populations of the UFDs and their spatial extent. Indeed, as suggested by Baker and Willman (2015), RRLs could be the *only* method to unveil very faint satellites, with $M_V < -3.2$ mag, especially at low Galactic latitudes when both extinction and field contamination can be important. All the UFDs searched for variability so far show at least one RRL. This is not surprising, since they are composed by (at least) old, metal-poor stellar populations, which are known to produce RRLs. The small statistics must not be misleading, since if one normalizes the observed number of RRLs by the integrated luminosity (i.e. a proxy of the baryonic mass), the fraction of RRLs is even higher than that observed in the bright dSphs. Indeed, adopting the specific frequency as parametrized by Mackey and Gilmore (2004a)

$$S_{RR} = N_{RR} \times 10^{0.4(7.5+M_V)} \tag{1}$$

one finds that UFDs tend to have higher specific frequencies than dSphs, as shown in Fig. 15, which is a reproduction from Baker and Willman (2015). However, as suggested in Baker and Willman (2015), this could be due to incompleteness effects, being the census of the RRLs in the bright dSphs still not complete.

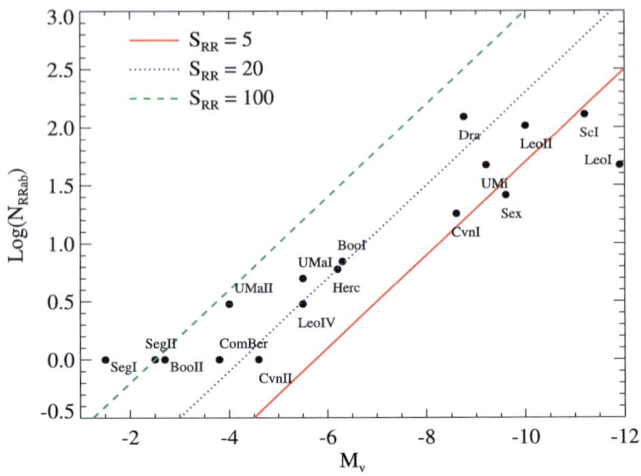

Fig. 15 Observed number of RRL stars as a function of the absolute magnitude. Overplotted are three specific frequencies (see text for details). The brightest galaxies have lower specific frequencies, possibly for incompleteness. UFDs are characterized by specific frequencies of in the range 20–100. Figure by Baker and Willman (2015), here reproduced by kind permission

4.1 Our RRL Sample

Table 1 collects all the positional and pulsational parameters of the RRLs discovered in the UFDs so far. For each variable, we list the position, the period, the mean magnitudes and the luminosity amplitude in the *BVI* bands (when available). In some cases, we merged the information on the same variable coming from different studies. We discuss these cases in the individual notes. Here, we point out that *B*-band photometry for the RRLs in Boötes I comes from Siegel (2006), and the *VI*-bands from Dall'Ora et al. (2006). We also explicitly note that, when a star was listed in both studies, we adopt the coordinates listed in Dall'Ora et al. (2006). This table does not include the RRLs hosted in CVn I, whose structural properties suggest a classification as a "classical" dSph instead as a UFD, and those hosted in Leo T, since it contains gas and a young stellar population, and in this sense is not a typical old, gas poor UFD. For these two galaxies, we refer the reader to the specific papers—CVn I, Kuehn et al. (2008); Leo T, Clementini et al. (2012). The galaxies are ordered by RA, and the sources of the data are: Segue II (Boettcher et al. 2013); UMa II (Dall'Ora et al. 2012; Vivas et al. 2016); UMa I (Garofalo et al. 2013); Leo IV (Moretti et al. 2009); Hyd II (Vivas et al. 2016); Coma (Musella et al. 2009); CVn II (Greco et al. 2008); Boo II (Sesar et al. 2014; Vivas et al. 2016); Boo III (Sesar et al. 2014; Vivas et al. 2016); Boo I (Siegel 2006; Dall'Ora et al. 2006); Hercules (Musella et al. 2012).

4.1.1 *Notes on Individual Variables*

V1 UMa II: this RRL was discovered by Dall'Ora et al. (2012), where they proposed a period of $P = 0.6593$ days. Subsequently, on the basis of a more densely sampled light curve, Vivas et al. (2016) computed a new period of $P = 0.56512$ days. This period appears to be compatible with the photometry of this variable, available in our database. Therefore, we adopt the period proposed by Vivas et al. (2016).

V1 Boo II: Sesar et al. (2014) proposed a period of $P = 0.63328$ days, while Vivas et al. (2016) computed a slightly different period of $P = 0.66349$ days. We keep the Vivas et al. (2016) estimate, together with their mean magnitude and amplitude.

V1 Boo III: Sesar et al. (2014) and Vivas et al. (2016) give practically the same period. However, here we present the mean magnitude and pulsational amplitude proposed by Vivas et al. (2016).

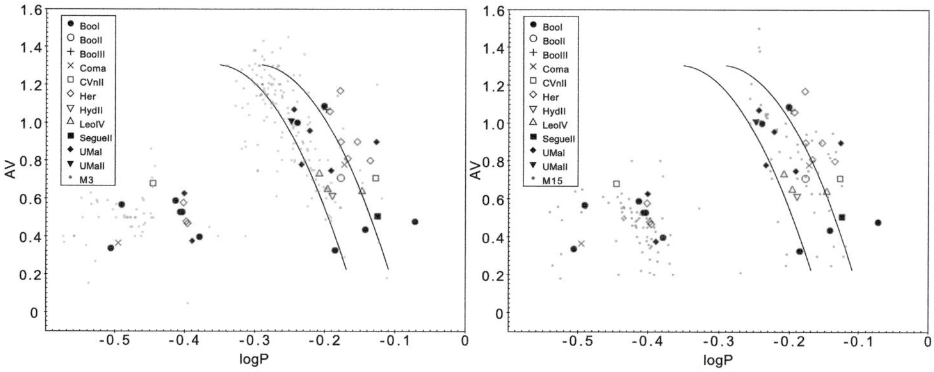

Fig. 16 Bailey's diagram of the RRL stars (black symbols of the UFD satellites of the Milky Way). Left panel shows the comparison with the Oo I Galactic globular cluster M3 (red filled circles) and with the Oo I and Oo II loci (dashed lines), according to Zorotovic et al. (2010). Right panel shows the sample comparison, but with the Oo II Galactic globular cluster M15

V5 Boo I: this star was classified as a RR_c type star by Siegel (2006), with a period of $P = 0.3863158$ days, and as a RR_{ab} star by Dall'Ora et al. (2006), with $P = 0.6506$ days. We used the Siegel (2006) period estimate to phase the data of this variable available in our database, but unfortunately we were not able to achieve a satisfactorily phased light curve. Therefore, we keep the Dall'Ora et al. (2006) estimate. It is worth noting that this variable is quite peculiar, since in the Dall'Ora et al. (2006) photometry it appears redder and brighter than the HB, and it could be blended with a companion.

V12 Boo I: this RRL was classified as RR_{ab} by Siegel (2006), with a period of $P = 0.6797488$ days, and as a double-mode pulsator RR_d by Dall'Ora et al. (2006), with periods of $P_1 = 0.3948$ days and $P_0 = 0.5296$ days. since the light curve shown by Dall'Ora et al. (2006) convincingly shows a typical double-mode behavior (see their Fig. 2), here we will keep their classification.

V1 Hyd II: the photometry of this variable was presented by Vivas et al. (2016) in the gri system. Here, for consistency we present its pulsational properties in the BVi bands, where gr magnitudes were transformed in B, V magnitudes following Jester et al. (2005).

4.2 Discussion

A glance at the data listed in Table 5 shows that, except for a few number of galaxies (namely UMa I, Boo I, Her), UFDs host a very small number of RRLs. Of course, with such small statistics, if a distance estimate can be reliable (especially in presence of a color-magnitude diagram, to check the robustness of the measured mean magnitudes and colors), some *caveats* must be recognized when using RRLs as population tracers. Indeed, since the Oo type is an *ensemble* feature, this should be declared only when a substantial number of fundamental mode RRL is available, in order to properly put them on an period-amplitude diagram (known as Bailey's diagram). Nevertheless, a comparison with the classic Oo I and Oo II lines in the Bailey's diagram can give interesting insights.

4.2.1 The Oosterhoff Classification and the Galactic Halo

In Fig. 16, we show the positions on the Bailey's diagram of the listed RRLs. For reference, we plot the loci of the Oo I and Oo II clusters, according to Zorotovic et al. (2010). At first

Table 5 Positional and pulsational parameters of the RRLs in the UFD studied so far. See text for details

ID	RA (J2000.0)	Dec (J2000.0)	Period (days)	$\langle B \rangle$ (mag)	A_B (mag)	$\langle V \rangle$ (mag)	A_V (mag)	$\langle I \rangle$ (mag)	A_I (mag)	Type
V1_SegueII	02:19:00.06	+20:06:35.2	0.748	18.62	0.62	18.25	0.51			RR_{ab}
V1_UMaII	08:50:37.43	+63:10:10.0	0.56512	18.61	1.27	18.24	1.01			RR_{ab}
V1_UMaI	10:34:59.2	+51:57:07.3	0.56924	20.72	1.29	20.47	1.07			RR_{ab}
V2_UMaI	10:35:05.5	+51:55:39.8	0.584	20.85	1.09	20.37	0.78			RR_{ab}
V3_UMaI	10:34:30.8	+51:56:28.9	0.64315	20.73	0.99	20.40	0.75			RR_{ab}
V4_UMaI	10:34:18.7	+51:58:29.2	0.74516	20.62	1.15	20.24	0.90			RR_{ab}
V5_UMaI	10:35:37.5	+52:02:35.6	0.59967	20.87	1.38	20.49	0.96			RR_{ab}
V6_UMaI	10:33:07.1	+51:50:05.1	0.39673	20.73	0.72	20.42	0.63			RR_c
V7_UMaI	10:32:37.5	+51:49:55.7	0.40749	20.54	0.37	20.44	0.38			RR_c
V1_LeoIV	11:32:59.2	−00:34:03.6	0.61895	21.82	0.99	21.47	0.73			RR_{ab}
V2_LeoIV	11:32:55.8	−00:33:29.4	0.7096	21.86	0.76	21.46	0.64			RR_{ab}
V3_LeoIV	11:33:36.6	−00:38:43.3	0.635	21.81	0.82	21.52	0.65			RR_{ab}
V1_HydII	12:21:43.51	−31:59:42.8	0.645	22.00	0.77	21.56	0.62	21.30	0.38	RR_{ab}
V1_Coma	12:27:33.50	+23:54:55.7	0.66971			18.44	0.78	17.74	0.53	RR_{ab}
V2_Coma	12:26:50.89	+23:56:00.6	0.31964			18.69	0.37	18.20	0.24	RR_c
V1_CVnII	12:57:01.6	+34:19:33.4	0.358	21.74	0.83	21.49	0.68			RR_c
V2_CVnII	12:57:11.8	+34:16:52.9	0.743	21.77	0.95	21.46	0.71			RR_{ab}
V1_BooII	13:58:07.04	+12:51:22.8	0.66349			18.23	0.71			RR_{ab}
V1_BooIII	14:00:34.52	+25:55:52.7	0.63328			18.56	1.07			RR_{ab}
V1_BooI	13:59:29.36	+14:10:43.6	0.303771	19.78	0.628					RR_{ab}
V2_BooI	13:59:51.34	+14:39:06.0	0.3119	19.73	0.67	19.63	0.34	19.32	> 0.13	RR_c
V3_BooI	14:00:26.86	+14:35:33.1	0.3232	19.79	0.64	19.58	0.57	19.25	> 0.15	RR_c
V4_BooI	14:00:08.90	+14:34:24.1	0.3860	19.82	0.58	19.57	0.59	19.19	0.2	RR_c
V5_BooI	14:00:21.56	+14:37:28.8	0.6506	19.75	0.57	19.38	0.33	18.93		RR_{ab}
V6_BooI	13:59:45.95	+14:31:40.7	0.3919	19.86	0.61	19.58	0.53	19.13	0.22	RR_c

Table 5 (*Continued*)

ID	RA (J2000.0)	Dec (J2000.0)	Period (days)	$\langle B \rangle$ (mag)	A_B (mag)	$\langle V \rangle$ (mag)	A_V (mag)	$\langle I \rangle$ (mag)	A_I (mag)	Type
V7_BooI	13:59:49.37	+14:10:05.6	0.4011623	19.76	0.74					RR_c
V8_BooI	13:59:59.69	+14:27:34.0	0.4179	19.79	0.74	19.54	0.40	19.10	0.26	RR_c
V9_BooI	13:59:47.28	+14:27:56.3	0.5755	19.84	1.28	19.55	1.00	19.07	> 0.41	RR_{ab}
V10_BooI	14:00:25.75	+14:33:08.4	0.628	19.78	1.32	19.47	1.09	18.98	> 0.24	RR_{ab}
V11_BooI	13:58:04.38	+14:13:19.3	0.6617310	19.82	1.20					RR_{ab}
V12_BooI	13:59:56.00	+14:34:55.0	0.3948	19.90	0.54	19.59	0.53	19.13	> 0.20	RR_d
V13_BooI	13:59:06.36	+14:19:00.1	0.7061108	19.77	0.82					RR_{ab}
V14_BooI	13:59:25.75	+14:23:45.3	0.7186	19.89	0.86	19.50	0.44	19.03		RR_{ab}
V15_BooI	14:00:11.09	+14:24:19.7	0.8456	19.83	0.48	19.45	0.48	18.91	0.28	RR_{ab}
V1_Her	16:31:02.17	+12:47:33.7	0.639206	21.68	1.16	21.27	1.06			RR_{ab}
V3_Her	16:30:54.93	+12:47:04.2	0.39997	21.72	0.61	21.32	0.48			RR_c
V4_Her	16:30:56.14	+12:48:29.2	0.39576	21.59	0.69	21.23	0.58			RR_c
V5_Her	16:30:52.28	+12:49:12.0	0.40183	21.64	0.61	21.30	0.47			RR_c
V6_Her	16:30:52.41	+12:49:60.0	0.69981	21.76	1.19	21.35	0.90			RR_{ab}
V7_Her	16:31:29.48	+12:47:34.9	0.67799	21.65	1.03	21.22	0.81			RR_{ab}
V8_Her	16:31:27.20	+12:44:16.7	0.66234	21.69	1.09	21.25	0.90			RR_{ab}
V9_Her	16:31:29.50	+12:40:03.1	0.72939	21.60	1.00	21.18	0.80			RR_{ab}
V10_Her	16:30:03.96	+12:52:06.3	0.6616	21.69	1.32	21.28	1.17			RR_{ab}

Table 6 RR Lyrae-based distances to the UFD satellites of the Milky Way. Distances have been estimated according to the M_V–[Fe/H] relation provided by Clementini et al. (2003). A_V absorptions are based on the calibration by Schlafly and Finkbeiner (2011)

Name	[Fe/H] (dex)	A_V (mag)	$(m-M)_0$ (mag)
SegueII	-2.00 ± 0.25	0.507	$17.31 \pm 0.03 \pm 0.06$
UMaII	-2.47 ± 0.06	0.257	$17.65 \pm 0.04 \pm 0.03$
UMaI	-2.18 ± 0.04	0.054	$19.98 \pm 0.04 \pm 0.03$
LeoIV	-2.54 ± 0.07	0.069	$21.10 \pm 0.03 \pm 0.04$
HydII	-2.02 ± 0.08	0.167	$20.96 \pm 0.04 \pm 0.03$
Coma	-2.60 ± 0.05	0.046	$18.09 \pm 0.03 \pm 0.03$
CVnII	-2.21 ± 0.05	0.027	$21.06 \pm 0.02 \pm 0.03$
BooII	-1.79 ± 0.05	0.084	$17.67 \pm 0.04 \pm 0.03$
BooIII	-2.1 ± 0.2	0.058	$18.09 \pm 0.04 \pm 0.05$
BooI	-2.55 ± 0.11	0.047	$19.18 \pm 0.06 \pm 0.04$
Her	-2.41 ± 0.04	0.171	$20.75 \pm 0.05 \pm 0.03$

glance, the positions of almost all the RRLs of the UFDs are compatible with a Oo II classification. The only apparent exception is UMa I, which was classified as Oo intermediate in Garofalo et al. (2013). However, when we compare the positions of its variables with those of the Galactic globular clusters M3 (left panel) and M15 (right panel, data made available by Clement et al. 2001), we favor a Oo II classification also for this system. Indeed, from the left panel it appears that the distribution of the UMa I RR_c variables is in good agreement with those of the other UFDs and with those of the Oo II cluster M15. Also, the mean period of the fundamental pulsators of UMa I is $\langle P_{ab} \rangle = 0.628 \pm 0.063$ days, which is in agreement with the values of other Oo II systems, such as M15 ($\langle P_{ab} \rangle = 0.643 \pm 0.063$ days), M92 ($\langle P_{ab} \rangle = 0.631 \pm 0.048$ days) and M68 ($\langle P_{ab} \rangle = 0.627 \pm 0.062$ days), where the listed values are computed on the basis of the compilation published by Clement et al. (2001). However, it should also be noted that, when discarding the variable V4, significantly brighter than the others, the mean period of the fundamental pulsators lowers to $\langle P_{ab} \rangle = 0.599 \pm 0.032$ days, as discussed in Garofalo et al. (2013). This suggest that, in general, when dealing with systems with a small number of RRLs, a correct Oo classification is a risky business, and we may consider not only to compare the positions of the RR_{ab} stars with respect to the mean Oo I and Oo II loci, but to consider the whole plane instead, with the actual distributions of the RRLs belonging to some reference Oo I and Oo II clusters.

Taken at face value, the Oo II classification could suggest a major contribution of UFD-like objects in assembling the Galactic halo. However, as pointed out in Fiorentino et al. (2015), a detailed comparison of the pulsational properties of the RRLs of the Galactic halo and of the set dSphs + UFDs, shows that the latter lack the so-called high-amplitude, short-period (HASP) variables. Fiorentino et al. (2015) argue that the HASP region is filled only when RRLs are more metal-rich than $[Fe/H] = -1.5$ dex. Thus, *present-day* dSph- and UFD-like objects seem to have played a minor role, if any, in assembling the Galactic halo.

4.2.2 A Homogeneous Distance Scale

In Table 6, we propose a homogeneous distance scale for the MW UFDs, by using the same M_V – [Fe/H] relation and the same reddening calibration from Schlafly and Finkbeiner (2011). In particular, for the RRLs luminosity we adopt the absolute magnitude of $M_V = 0.54 \pm 0.03$ mag at $[Fe/H] = -1.5$ dex (based on a LMC distance of 18.52 ± 0.09 mag from Clementini et al. 2003), with a slope of $\Delta M_V[Fe/H] = 0.214 \pm 0.047$ mag dex^{-1} (Clementini et al. 2003). We do not make any attempt to homogenize the metallicity scale, where all the values have been taken from the collection listed in McConnachie (2012), except of Boo III (Carlin et al. 2009), and Hyd II (Kirby et al. 2015). The uncertainties of the distance are split in intrinsic error (the standard deviation of the mean, when at least two RRLs are available, or the typical photometric error when only one variable is present), plus a contribution due to the uncertainty in the M_V – [Fe/H] relation. As a matter of fact, there is also another source of uncertainty, that is the internal metallicity spread observed in several UFDs, but it is difficult to manage. The spread can be of the order of ~ 0.6 dex (UMa II, Kirby et al. 2008), which means an additional uncertainty up to ~ 0.15 mag. In computing the distances of the individual galaxies, we dropped the variables V1 in Coma, V4 in UMa I, and V5 in Boo I, since they are significantly brighter (~ 0.2 mag) than the others and/or of the HB, and may be evolved variables not representative of the zero-age HB level.

The current radial distribution of both UFD and classical dwarf spheroidals (see Monelli contribution in this book) seems to be quite similar. However, the uncertainties affecting the individual distances of UFDs are still too large. Individual distances based on the use of optical and/or near-Infrared PL relations will be crucial to further constrain the similarities in radial distribution of gas poor stellar systems.

5 On the Absolute and Relative Ages of Globular Clusters

The early estimates of the ages of globular clusters date back to more than half a century ago thanks to the pioneering papers from Sandage (1958) and Arp (1962) for the observational aspects and from Hoyle (1959) and Schwarzschild (1970) for theoretical analyses. The reader interested in a more detailed discussion concerning the dawn of cluster age determination is referred to the seminal presentations and discussions of the 1957 Vatican Conference (O'Connell 1958). Particularly enlightening was the empirical evidence brought forward by Baade (1958) concerning the age difference among the different stellar populations belonging to the Galactic components (halo, disk, bulge).

5.1 Absolute Cluster Age Estimates

In the following we will focus on the most reliable classical methods used to estimate absolute ages of Galactic Globular Clusters (GGCs). The first two rely on deep photometry of individual stars of a GGC down to the MSTO and the white dwarf cooling sequence features, respectively. The second is observationally based on the detection of radioactive heavy elements in individual stellar spectra in order to use direct cosmochronometry. We will highlight strengths and weaknesses of their application.

5.1.1 The Main Sequence Turn Off

The MSTO of a cluster is identified as the bluest point along its Main Sequence. This is the most important clock to date for both open and globular stellar systems. The key advantages of this diagnostic are the following:

(i) The anti-correlation between cluster age and brightness of MSTO stars is linear over a broad range of stellar ages (see e.g., Di Cecco et al. 2015; Valle et al. 2013).

(ii) Stars in this evolutionary phase are burning hydrogen in the core. This means that they evolve on a long nuclear time scale, and therefore, the number of stars per unit magnitude tracing this evolutionary phase is quite large compared with evolved phases.

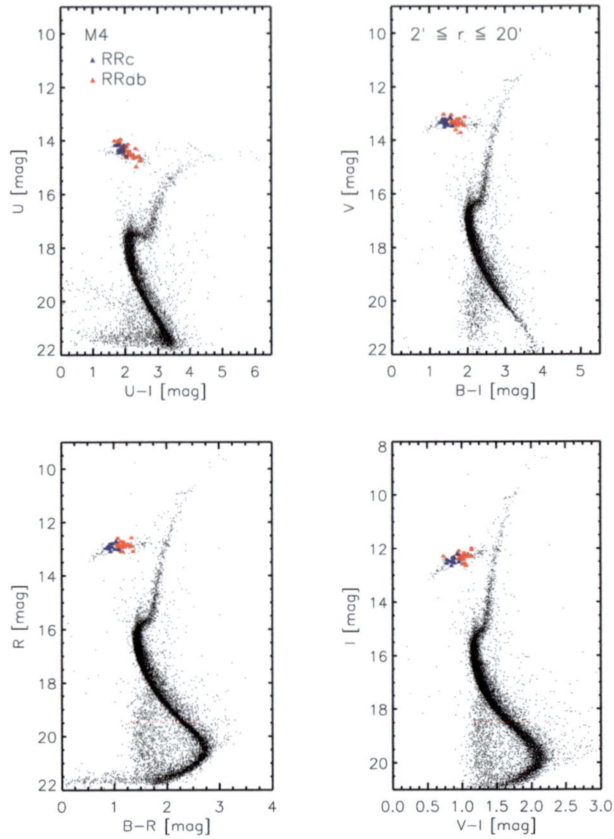

Fig. 17 Optical Color-Magnitude Diagrams (CMDs) based on UBVRI bands of the Galactic star cluster M4. Stars plotted in the CMDs were selected according to the position. Blue and red triangles mark the position of first overtone (RRc) and fundamental-mode (RRab) cluster RR Lyrae stars. Note the change in the slope of the Horizontal Branch (HB) when moving from the top left panel (U–I, U) to the bottom right panel (V–I, I). Bright red giant stars located across the tip of the red giant branch are missing due to saturation problems. The vertical plume of stars located at B–I ∼ 2.0–2.5, B–R ∼ 1.8–2.0 and V–I ∼ 1.2–2.5 mag is caused by field star contamination

(iii) Accurate apparent optical magnitudes of MSTO stars in GGCs are within the capability of 2–4 m class telescopes equipped with CCD detectors and can be easily measured.

The main cons of the MSTO are the following:

(i) The MSTO is prone to uncertainties on cluster distance and on cluster reddening. An uncertainty of 10% in the error budget (reddening plus true distance modulus) of the MSTO, implies an uncertainty of about 1 Gyr in cluster age. The problem becomes even more severe if we are dealing with stellar systems either affected by large or by differential reddening.

(ii) The identification of the MSTO is not always trivial. In a broad range of stellar ages and chemical compositions, stars across the MSTO attain in optical bands similar colors and magnitudes. In some traditional broad-band color-magnitude systems there is nearly a vertical distribution of MSTO stars (e.g., Salaris and Cassisi 2005). Figure 17 shows the optical (UBVRI) CMDs of the Galactic globular cluster M4 (Stetson et al. 2014; Braga et al. 2015). The shape of the MSTO changes from "cuspy" in the U–I, U CMD (top left panel) to "almost vertical" in the V–I, I CMD (bottom right panel).

Fortunately the variation in the color gradient of the region across the MSTO becomes more evident in the near-infrared (NIR) and in the mid-infrared (MIR) regime. Data plotted in the top panel of Fig. 18 show that cluster stars display not only a well defined bending in the region across the MSTO, but also a sharp change in the slope of the lower main sequence across the Main Sequence Knee (MSK, see Sect. 5.4). A glance at the data plotted in the bottom panels shows that the bending across the MSTO is also in NIR/MIR CMDs.

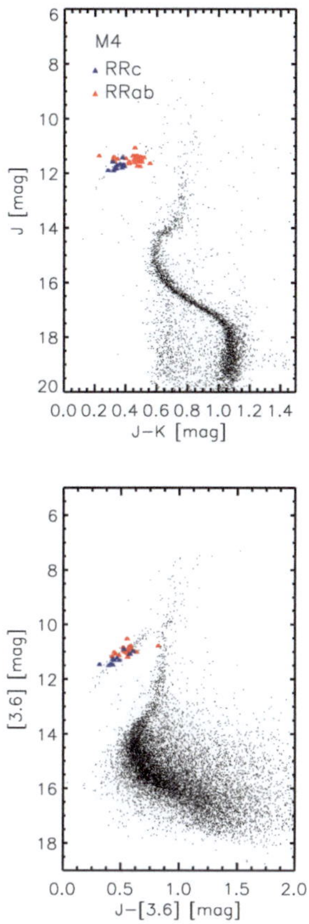

Fig. 18 Same as Fig. 17, but for NIR (JHK) and MIR ([3.6], [4.5]) bands

However, the photometric accuracy of the MIR bands do not allow us to clearly identify the MSK.

The empirical scenario concerning the shape of both the MSTO and the MSK becomes even more interesting in dealing with optical/NIR/MIR CMDs, since they bring forward several advantages: (a) They display a well defined bending across the MSTO and a sharp change in the MS slope across the MSK; (b) The broad range in central wavelengths among optical and NIR/MIR bands means also a strong sensitivity to the effective temperature. The consequence is that optical/NIR/MIR CMDs display tight stellar sequences not only along the MS, but also in advanced evolutionary phases (RGB, HB, AGB).

(iii) Optical CMDs are affected in the region across the MSTO by field star contamination. This problem is less severe in optical/NIR CMDs, since the MSTO stars attain colors that are systematically bluer than typical field stars. The bottom left panel of Fig. 19 shows that field stars are typically redder (V–K = 2.8–3.5) than MSTO stars (V–K $\simeq$2.6). However, field star contamination severely affects age and structural parameters of nearby stellar systems. Accurate and deep optical CMDs based on images collected with ACS on board of HST are less affected by the contamination of field stars. The pointing is typically located across the center of the cluster and in these regions cluster stars outnumber field stars. This rule of thumb does not apply to clusters either located or projected onto crowded Galactic regions such as the bulge and Galactic thin disk (Zoccali et al. 2001a; Ferraro et al. 2009; Lagioia et al. 2014) or nearby dwarf galaxies (Kalirai et al. 2012). A significant step forward

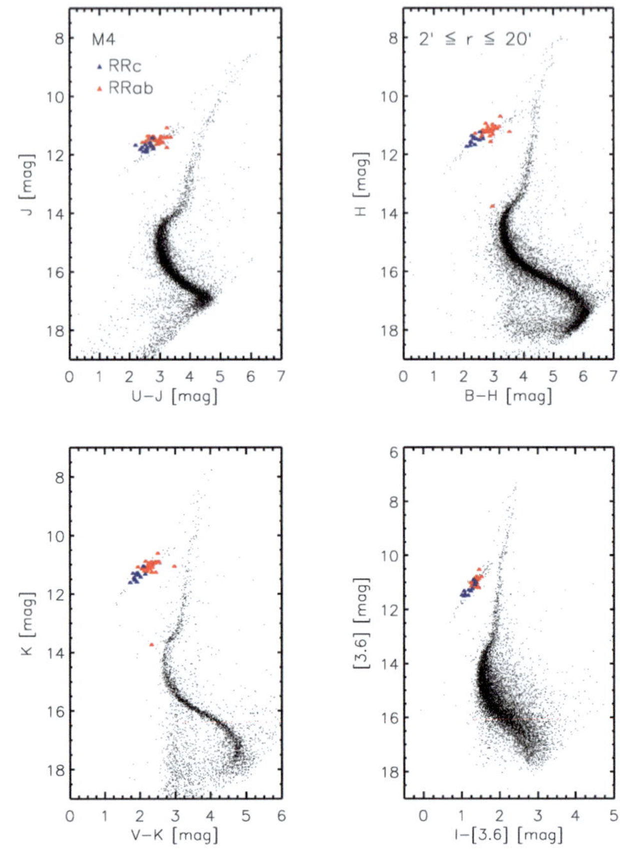

Fig. 19 Same as Fig. 17, but for optical and NIR/MIR bands

in dealing with this problem was provided by HST photometry. The superb image quality of HST optical images provided the opportunity to measure the proper motion using images collected on a time interval of the order of ten years. This approach provided the opportunity to split not only field and cluster stars, but also to clearly identify stellar populations belonging to Sagittarius (Anderson 2002; King et al. 2005; Massari et al. 2013; Milone et al. 2014). The great news in this context is that similar results can also be obtained using NIR images as a second epoch collected with AO systems available at 8–10 m class telescopes (e.g., NGC 6681, Massari et al. 2016b).

The results mentioned above are based on images that only cover a few arcminutes around the center of each cluster. The separation between cluster and field stars is much more challenging away from cluster centers, since field stars outnumber cluster stars in these regions. The reason why we are interested in tracing cluster stars in external cluster regions is twofold: (a) There is mounting empirical evidence that stellar populations change (chemical composition, age) as a function of radial distance—e.g., 47 Tuc, (Kalirai et al. 2012); Omega Cen, (Calamida et al. 2017); (b) The estimate of structural parameters depends on the star counts in the outskirts of the clusters.

These are the reasons why new photometric approaches for the separation between field and cluster stars are required. This approach does require photometric catalogs based on at least three photometric bands. To accomplish the goal we have used either multi-dimensional ridge lines as in dealing with *griz* SDSS photometry of the metal-rich globular

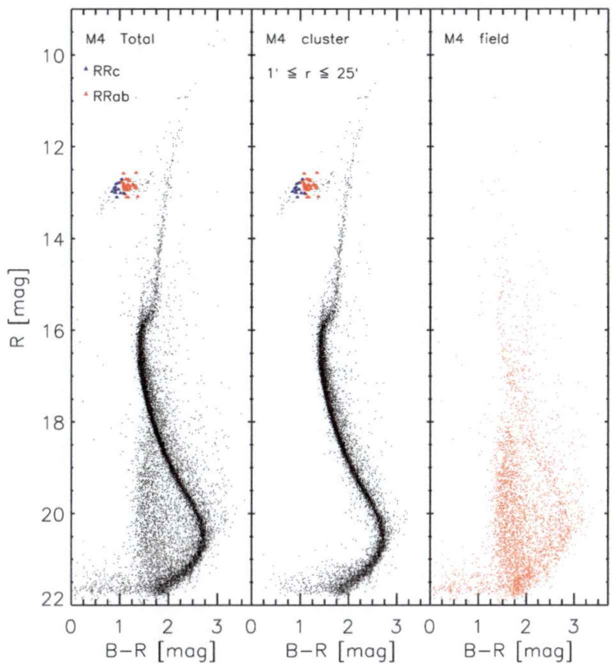

Fig. 20 Left—Optical CMD in B–R, R for stars covering a sky area of 35 × 35 arcmin squared around M4. Middle—Same as the left panel, but for candidate cluster stars. Right—Same as the left panel, but for candidate field stars

M71 (Di Cecco et al. 2015) or the difference in the spectral energy distribution in dealing with the *ugri* photometry of the complex globular ω Cen (Calamida et al. 2017). The latter approach was developed to deal with stellar systems characterized by multiple stellar populations (see e.g., Martínez-Vázquez et al. 2016a,b).

Initially, the ridge lines of the different sub-populations in ω Cen along the cluster evolutionary sequences (MS, SGB, RGB) were estimated. They are based on a 3D CMD (magnitude, color index, star counts) and the ridge lines trace the peaks of the stellar distribution. The horizontal branch (HB) stars are typically neglected, since they are either bluer (hot and extreme HB) or they can be easily distinguished (RR Lyrae stars, red HB) from the field stars. These ridge lines were estimated in an annulus neglecting stars located in the innermost and in the outermost cluster regions and using several cuts in radial distance and in photometric accuracy. Once the ridge lines were estimated we performed a linear interpolation among them and generated a continuous multi-dimensional surface. Subsequently, the separation between field and cluster stars was performed in two steps: (a) We estimated the total standard deviation among the position of individual stars and the reference surface; (b) We estimated the distance in magnitude and in color among the individual stars and the reference surface.

The approach discussed above relies on a conservative assumption, i.e. we do prefer to possibly lose some of the candidate cluster stars instead of including possible candidate field stars. Note that this assumption is fully supported by the fact that the age diagnostics we use for estimating cluster ages depend on the shape of MSTO and/or of the MSK. Stellar completeness mainly affects cluster ages based on cluster luminosity functions (Zoccali et al. 1998).

Figure 20 shows the separation between field and cluster stars for the Galactic globular cluster M4. This cluster is an acid test for the selection criteria based on photometry, since it is projected onto the Galactic bulge and it is also affected by differential reddening. To

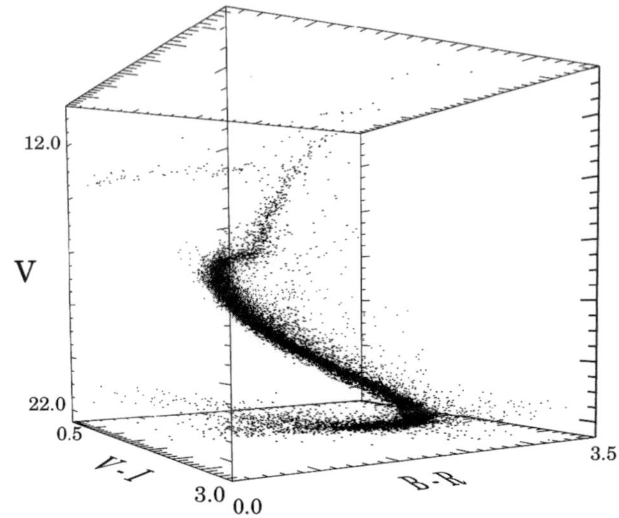

Fig. 21 Optical V, B–R, V–I color-color-magnitude diagram of candidate cluster stars for the globular M4 (see text for more details

overcome some of these problems we decided to follow the same approach we adopted for ω Cen, but the initial ridge lines were derived candidate field stars located outside the tidal radius of the cluster (Ferraro et al. 2018, in preparation). Data plotted in the left panel display stars located across the sky region covered by M4 in the B–R, R CMD. Field stars can be easily identified both along the MS and the RGB. The middle panel of the same figure shows the CMD, but for candidate cluster stars. The "cleaning" appears quite good across the cluster sequences. Note that the sequence running parallel to the cluster MS is due either to binaries or to photometric blends. The plausibility of the criteria adopted to separate field and cluster stars are further supported by the CMD of candidate field stars plotted in the right panel. Once again there is evidence that a minor fraction of candidate field stars were misidentified, but the main peak of field stars even across the MSTO was properly identified.

To further support the photometric criteria adopted to separate field and cluster stars, Fig. 21 shows a color-color-magnitude diagram of the selected candidate cluster stars. It is worth mentioning the smoothness of the cluster sequences when moving from the RGB to the MSTO and to the MSK.

5.1.2 The Error Budget

The global error budget of absolute ages of globular clusters based on the MSTO includes theoretical, empirical and intrinsic uncertainties.

Theoretical uncertainties—The precision of the clock adopted to date stellar systems depends on the precision of the input physics adopted to construct evolutionary models and, in turn, cluster isochrones. The main sources of uncertainties can be split into micro-physics (nuclear reactions, opacity, equation of state, astrophysical screening factors) and macro-physics (mixing length, mass loss, atomic diffusion, rotation, radiative levitation). The uncertainties affecting the nuclear reactions, such as the *p–p* and the *CNO* cycle, have already been discussed by degl'Innocenti et al. (2004) and Valle et al. (2013). They suggest that the uncertainties on age range from 3% for the $^{1}H(p, \nu_e, e^{+})^{2}H$ nuclear reaction to roughly 10% for the $^{14}N(p, \gamma)^{15}O$ reaction. The same authors also suggest an uncertainty of the order of 5% for the radiative opacities adopted for constructing MS and HB models. Moreover, they also suggest a similar uncertainty (5%) for the conductive opacities affecting the energy transport in electron degenerate isothermal helium cores typical of RGB structures and, in turn, on the HB luminosity level (e.g., Marta et al. 2008; Chaboyer et al. 2017).

The impact of macro-physics, and in particular the treatment of the mixing in convective regions has been addressed comprehensively in the literature (e.g., Chiosi and Maeder 1986; Castellani et al. 2013; Maeder and Meynet 2000; Salaris and Cassisi 2008; Bertelli et al. 2009; Rosenfield et al. 2017). The treatment of atomic diffusion and radiative levitation has also been investigated in many studies (Proffitt and Vandenberg 1991; Ciacio et al. 1997; Michaud et al. 2004; VandenBerg et al. 2013).

Finally, to transform theory into the observational plane, we also need predictions based on stellar atmosphere models (Gustafsson et al. 2008). We have to take account of uncertainties affecting bolometric corrections and color-temperature transformations. The impact that the quoted ingredients have on cluster isochrones has been discussed in detail in the literature (e.g., Pietrinferni et al. 2004, 2009; Cassisi et al. 2008; Sbordone et al. 2011; VandenBerg et al. 2013; Salaris 2016).

As a whole, the typical theoretical uncertainty in the adopted MSTO is of the order of 10%. In this context it is worth stressing that theoretical uncertainties mainly affect the zero point of the absolute age determinations. The relative age determinations are minimally affected by these uncertainties (see e.g., Dotter et al. 2011; VandenBerg et al. 2013; Chaboyer et al. 2017).

In this context, it is also worth mentioning that the helium to metal enrichment ratio ($\Delta Y/\Delta Z$), a fundamental ingredient in evolutionary prescriptions, is poorly known. This is used to derive the current stellar helium content through the following relation: $Y = Y_P + \Delta Y/\Delta Z \times Z$, where Y_P is the primordial helium content. However, empirical estimates range from $\Delta Y/\Delta Z = 3 \pm 2$ (Pagel and Portinari 1998) to $\Delta Y/\Delta Z = 5.3 \pm 1.4$ (Gennaro et al. 2010). Moreover and even more importantly, there are no solid reasons why this relation should be linear over the entire metallicity range and that the current local estimate is universal (Peimbert et al. 2010).

Empirical errors—The main uncertainties in the absolute age estimate of globular clusters are distances. Indeed an error of 10% in true distance modulus ($\Delta\mu_0 \sim 0.1$ mag) implies an uncertainty of ~ 1 Gyr in the absolute age. The error budget becomes even more severe once we also account for uncertainties in reddening corrections. An uncertainty of 2% in color excess implies an uncertainty of 6% in visual magnitude. The impact of this limitation becomes even more stringent if the cluster is also affected by differential reddening. In this context, it is worth mentioning that more metal-rich globular clusters are mainly distributed across the Galactic bulge, i.e., a region of the Galactic spheroid affected by large reddening and by differential reddening (Stetson et al. 2014).

The ongoing effort in trying to quantify the systematics affecting the measurements of the MSTO of a large fraction of GGCs also faces the problem of the absolute photometric calibration. In handling this thorny problem, two independent approaches have provided the opportunity to limit and/or to overcome the uncertainties associated with the zero-points of different photometric systems:

(i) Cluster photometry collected either with HST/WFPC2 or HST/ACS has provided the unique opportunity to collect accurate and deep photometry in a single well-calibrated photometric system for a sizable sample of GGCs (~ 60 out of 160, e.g., Zoccali et al. 2001b; Sarajedini et al. 2007; Marín-Franch et al. 2009; VandenBerg et al. 2013);

(ii) The major effort to provide local standard stars in GGCs also has significantly improved the accuracy of photometry and saved significant amount of telescope time (Stetson 2000).

The age diagnostic adopted to estimate the cluster age also depends on the iron content. The massive use of multi-object fiber spectrographs paved the way for the definition of a firm metallicity scale including a significant fraction of GGCs (e.g., Carretta et al. 2009). This translates to a systematic decrease in the uncertainties on the iron and on α-element abundance scale.

Intrinsic uncertainties—Dating back to more than forty years ago, spectroscopic investigations brought forward a significant star-to-star variation in C and in N among cluster stars (e.g., Osborn 1971). This evidence was complemented by discoveries of variations in Na, Al, and in O (e.g., Cohen 1978; Pilachowski et al. 1983; Leep et al. 1986) and by anti-correlations in CN–CH (e.g., Kraft 1994) as well as in O–Na and in Mg–Al (e.g., Suntzeff et al. 1991; Gratton et al. 2012).

The light element abundance variation was further strengthened by the occurrence of multiple stellar populations in more massive clusters (Bedin et al. 2004; Piotto et al. 2005, 2007). However, detailed investigations concerning the different stellar populations indicate a difference in age that is, in canonical GGCs (i.e. the most massive globulars are not included), on average shorter than 1 Gyr (Ventura et al. 2001; Cassisi et al. 2008). The intrinsic uncertainty does not seem to be the main source in the error budget of GGCs absolute ages.

5.1.3 The Cluster Distance Scale

Several approaches have been suggested in the literature to overcome some of the thorny problems affecting the estimate of the absolute age of GGCs. The time scale within which we can significantly reduce the theoretical uncertainties can be barely predicted. On the other hand, the empirical uncertainties are strongly correlated with technological developments either in the detectors or in the observing facilities or in both. The uncertainties affecting individual cluster distances would significantly benefit by the development of a *homogeneous cluster distance scale*.

Although Galactic globulars are at the cross-road of major theoretical and empirical efforts, we still lack a distance scale based on the same diagnostic and on homogeneous measurements. The problem we are facing is mainly caused by the intrinsic limitations in the diagnostics used to determine the distance. The difficulties of these methods can be summarized briefly.

(i) The tip of the Red Giant Branch (TRGB) can only be applied to two clusters (ω Cen, 47 Tuc), due to limited stellar (Poisson) statistics when approaching the tip of the RGB.

(ii) Main sequence fitting is hampered by the fact that the number of field dwarfs with accurate trigonometric parallaxes is small and covers a limited range in iron abundance (Chaboyer et al. 2017; VandenBerg et al. 2014a).

(iii) Another possible approach to constrain cluster distances is to fit the white dwarf cooling sequence. However, this method can only be applied to a few clusters (Zoccali et al. 2001b; Richer et al. 2013; Bono et al. 2013) and we are still facing some discrepancies between the distances based on this diagnostic and other distance indicators (Bono et al. 2008). Furthermore, the number of nearby White Dwarfs (WDs) for which are available accurate trigonometric parallaxes is limited.

(iv) Cluster distances based on the kinematic approach appear also to be affected by systematics. This diagnostic has the potential to be a powerful geometrical method, since it is only based on the ratio between the standard deviation of proper motions and the standard deviation of radial velocities. There is mounting evidence that the current kinematic distances are slightly larger than those based on other primary distance indicators (Bono et al. 2008).

(v) During the last few years, accurate distances have been provided for a few globulars using eclipsing binary stars. This is also a very promising approach, since it is based on a geometrical method (Thompson et al. 2001). The main limitation of this distance indicator as well as methods iii) and iv) is that they are challenging from the observational point of view. This means that they have only been applied to a very limited number of clusters.

(vi) The Horizontal Branch (HB) luminosity level is one of the most popular distance indicator for GGCs. The typical anchor for the HB luminosity is the M_V^{RR}-metallicity relation at the mean color of the RR Lyrae instability strip. This color region was selected because it is quite flat in CMDs using the V-band magnitude. This distance indicator might be affected by two possible sources of systematic errors:

(a) The HB morphology—The metallicity is the most important parameter affecting HB morphology, indeed the HB is mainly blue in the metal-poor regime and becomes mainly red in the metal-rich regime. The parameter used to trace the change in the HB morphology is $\tau_{\mathrm{HB}} = (\mathrm{B} - \mathrm{R})/(\mathrm{B} + \mathrm{V} + \mathrm{R})$, i.e. the number ratio among stars either bluer (B) or redder (R) than variable (V) RR Lyrae stars. Globulars approaching the two extrema (± 1) typically lack of RR Lyrae stars. This means that it is quite difficult from an empirical point of view to anchor the HB luminosity level. The impact of this limitation becomes even more stringent for globulars more metal-rich than 47 Tuc, since their HBs typically display a stub of red HB stars. There are two exceptions, NGC 6441 and NGC 6388, that are metal-rich but display a well populated blue and red HB, plus a good sample of RR Lyrae stars (Pritzl et al. 2003);

(b) Predicted HB luminosity level—The current evolutionary prescriptions concerning the zero age HB (ZAHB) luminosity are typically 0.10–0.15 mag brighter than the observed ones. New interior conductive opacities (Cassisi et al. 2007) contribute to alleviate the problem, but the discrepancy is still present.

(vii) Near-infrared and mid-infrared observations of cluster RR Lyrae stars appear to be very promising for providing a homogeneous cluster distance scale. The RR Lyrae stars do obey well-defined PL relations in these bands. These relations are very narrow (e.g., Braga et al. 2015; Marconi et al. 2015; Neeley et al. 2017) and are not affected by off-ZAHB evolution, thus reducing the systematics discussed above. NIR photometry has also the advantage of being less prone to uncertainties introduced by either large and/or by differential reddening corrections. This is a typical problem among the more metal-rich Bulge globulars. Indeed, the uncertainties affecting NIR/MIR bands are on average one order of magnitude smaller than the optical bands. However, the number of globulars hosting a sizable sample of RR Lyrae stars is roughly half of the entire Galactic sample. Finally, we still lack a detailed knowledge on how the reddening law changes when moving across the bulge, along the bar and into the nuclear Bulge (e.g., Indebetouw et al. 2005; Nishiyama et al. 2006, 2008, 2009; Nataf et al. 2013; Schultheis et al. 2014).

5.1.4 The White Dwarf Cooling Sequence

The white dwarf cooling sequence depends on different physics than the previously-discussed methods, and as such should be an excellent diagnostic to constrain possible systematics in absolute cluster age estimates based on the MSTO (Hansen et al. 2004; Salaris et al. 2010). This means that deep and accurate photometry of nearby globulars can provide crucial constraints on the plausibility of the physical assumptions adopted to construct either main sequence or cooling sequence models. The key advantages in using cluster WD cooling sequences is that they display a well defined blue turn off (WDBTO) in deep and accurate I, V–I CMDs, as illustrated in Fig. 22. Theory and observations indicate that this change in the slope of the WD cooling sequence is due to interplay between an opacity

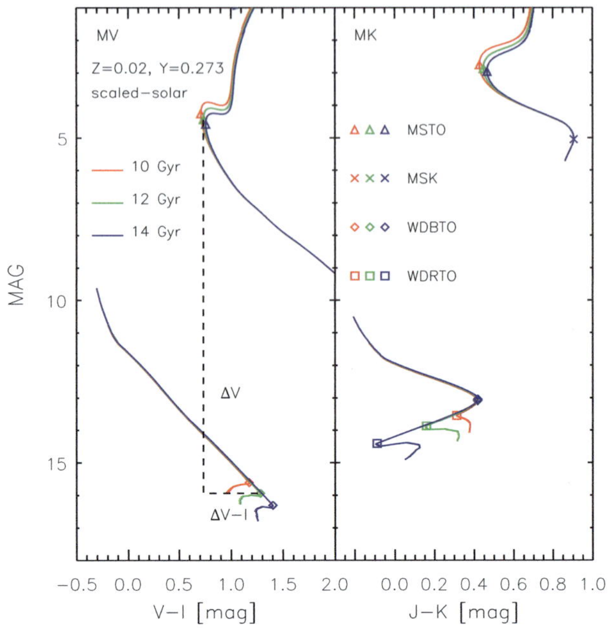

Fig. 22 Left—Predicted optical (V–I, V) CMD for cluster isochrones at solar chemical composition and at scaled solar chemical mixture. The bright solid lines show isochrones of 10 (red), 12 (green) and 14 (blue) Gyr (BaSTI data base). The triangles mark the position of the Main Sequence Turn Off (MSTO). The faint solid lines show WD isochrones of 10 (red), 12 (green) and 14 (blue) Gyr. The WD isochrones account for chemical separation (Salaris et al. 2010). The diamonds mark the position of the White Dwarf Blue Turn Off (WDBTO). The vertical and horizontal black dashed lines show the difference in magnitude and in color between the 12 Gyr MSTO and the WDBTO. Right—Same as the left, but in the NIR (K, J–K) CMD. The triangles mark the position of the MSTO, the crosses mark the MS Knee (MSK, Bono et al. 2010), the diamonds mark the WDBTO, while the squares mark the White Dwarf Red Turn Off (WDRTO)

mechanism called Collisional Induced Absorption (CIA) mainly from molecular hydrogen and/or to the cooling time of more massive WDs (Brocato et al. 1999; Hansen 2004; Richer et al. 2006; Moehler and Bono 2008; Salaris et al. 2010; O'Malley et al. 2013).

Recently, Bono et al. (2013) performed a detailed theoretical investigation of cluster WD cooling sequences and found new diagnostics along the WD cooling sequences and in NIR Luminosity Functions (LFs). The interplay between CIA and the cooling time of progressively more massive WDs causes a red turn-off along the WD cooling sequences (WDRTO). This feature is strongly correlated with the cluster age, and indeed the faint peak in the K-band increases by 0.2–0.25 mag/Gyr in the range 10–14 Gyr. Moreover, they also suggested to use the difference in magnitude between the MSTO and the WDRTO, since this age diagnostic is independent of distance and reddening. These predictions appear very promising in view of the unprecedented opportunity offered by James Webb Space Telescope (JWST) in the NIR/MIR regimes. This encouraging prospect also applies to ground-based extremely large telescopes equipped with state-of-the-art multi-conjugated adaptive optics systems (Bono et al. 2013).

5.1.5 Thorium Cosmochronometry

Elements beyond the iron-group are overwhelmingly created in neutron-capture fusion reactions by target heavy-element seed nuclei. Their syntheses mostly are either slow (β-decay timescales fast compared to neutron-capture timescales; called the s-process) or rapid (neutron-capture timescales much faster than β-decay ones; the r-process). Very heavy radioactive elements thorium (Z = 90) and uranium (Z = 92) can be created only via the r-process. The heaviest stable element is bismuth (Z = 83, with its sole natural isotope ^{209}Bi). All isotopes of elements with Z = 83–89 have very short half-lives, and therefore cannot be created in the s-process. Th and U decay on astrophysically interesting timescales: half-lives are 1.4×10^{10} yr for ^{232}Th (its only naturally-occurring isotope), 7.0×10^8 yr for ^{235}U, and 4.5×10^9 yr for ^{238}U. Therefore, derived abundances of Th and U with respect to stable r-process-dominated neutron-capture elements in low-metallicity halo stars have the potential to be translated into Galactic age estimates.

Attempts to use stellar Th abundances as chronometers began with Butcher (1987). That study included only disk stars with metallicities [Fe/H] ≥ -0.8. As is the case for most Th abundance studies, Butcher (1987) analyzed just the 4019.2 Å Th II transition. That line in high metallicity stars is at best a weak blending absorption in this crowded spectral region. Additionally, for a comparison stable neutron-capture element Nd was chosen in the Butcher (1987) study. Unfortunately, Nd in the solar system (and probably in most disk stars) has a s-process origin, accounting for $\simeq 58\%$ of its abundances; the r-process fraction is only $\simeq 42\%$ (e.g. Smolec (2005)). With these limitations, Butcher (1987) argued that the Th/Nd ratios were roughly constant in their sample of disk stars, irrespective of assumed stellar age.

Analyses of low-metallicity halo population r-process-rich stars have yielded more easily-interpreted results. The first such star, HD 115444, initially identified by Griffin et al. (1982), has [Fe/H] $\simeq -2.9$, and [Eu/Fe] $\simeq +0.9$ (e.g. Westin et al. 2000; Sneden et al. 2009). Then CS 22892-052, a red giant from the Beers et al. (1992) low resolution Galactic halo survey, was serendipitously discovered (Sneden et al. 1994) to be very r-process rich: [Fe/H] $\simeq -3.0$, [Eu/Fe] $\simeq +1.7$, and an unambiguously strong 4019 Å line, yielding [Th/Fe] $\simeq +1.4$ or [Th/Eu] $\simeq -0.3$ (e.g., Sneden et al. 2003). The relatively depressed Th compared to Eu was taken to be a sign of radioactive decay from an *assumed* initial production ratio of [Th/Eu] $\equiv 0.0$. Initial application of theoretical models suggested an ancient age for the neutron-capture elements but with a large uncertainty: $11.5 \leq t \leq 18.8$ Gyr (Cowan et al. 1997).

Detailed study of another r-process-rich star CS 31082-001 (Cayrel et al. 2001; Hill et al. 2002) revealed the first detection of U in a low metallicity star (Hill et al. 2002). But the individual abundance ratios [Th/Eu] and [U/Eu] in this star turned out to be too large to be sensibly interpreted as a straightforward radioactive decay. Instead, it was necessary to postulate an "actinide boost" with initial production ratios [Th/Eu] > 0 and [U/Eu] > 0. Fortunately, the ratio between neighboring elements Th and U should have well-understood production ratios and, for example, Schatz et al. (2002) derived an age from the U-Th abundance ratio of 15.5 ± 3.2 Gyr, consistent with but not constraining the age of the Galaxy determined from other indicators.

The problems and prospects of U and Th abundances are considered well in Hill et al. (2002) and Schatz et al. (2002) and will not be repeated here. The actinide boost problem effectively forces attention on detection of both Th and U for meaningful radioactive cosmochronometry. But the problem is the rarity of U detections even in low metallicity stars with extreme r-process enhancements (e.g. [Eu/Fe] > +1). Only a single U II transition at 3539.5 Åhas been detected to date, and it is at most a very weak bump among a clump of lines dominated by two strong Fe I lines as well as weaker Nd II and CN lines. This is shown, for example, in Fig. 10 of Hill et al. (2002), Fig. 9 of Cowan et al. (2002), Fig. 2 of Frebel et al. (2007).

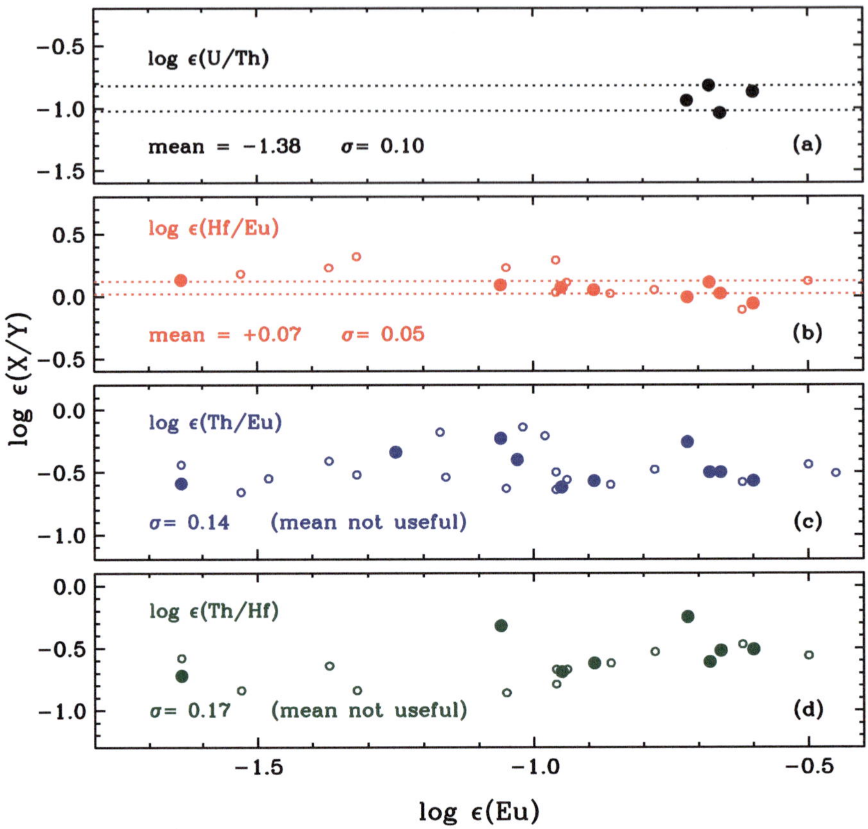

Fig. 23 Abundance ratios of elements of interest to cosmochronometry plotted as functions of the Eu abundance. The large filled circles are from individual high-resolution spectroscopic studies Westin et al. (2000), Hill et al. (2002), Cowan et al. (2002), Sneden et al. (2003), Hayek et al. (2009), Siqueira Mello et al. (2013), Mashonkina et al. (2014), Hill et al. (2016), and the small open circles are from a single survey of many r-process-rich stars Roederer et al. (2009). Means and standard deviations, computed only for the individual high-resolution studies, are written in each panel. Only in panels (**a**) and (**b**) are dotted lines drawn to indicate the σ widths around the means; the scatters in panels (**c**) and (**d**) are too large to be useful for cosmochronometry

If r-process production ratios [Th/Eu] cannot be predicted with certainty given the large Periodic Table stretch between elements 63 and 90, can another element closer to Th serve as the stable comparison element? Kratz et al. (2007) suggested that Hf ($Z = 72$) might be a good candidate, as their computations showed that Hf is made in the r-process at similar neutron densities to Th (see their Fig. 3). We tested this idea by considering Eu, Hf, Th, and U abundances reported for these kinds of stars in the literature. In Fig. 23 panel (a) we show that for radioactive elements U and Th, their "absolute" number density ratios $\log\epsilon(\mathrm{U/Th})$[1] are essentially constant in the high-resolution studies published to date, and Hill et al. (2016) report discovery of a fifth r-rich star that has a nearly identical value of this ratio. In panel (b) we show a similarly tight correlation between the stable rare-earth elements Hf and Eu. Star-to-star scatter increases markedly in ratios $\log\epsilon(\mathrm{Th/Eu})$ and

[1] $\log\epsilon \mathrm{A} \equiv \log_{10}(\mathrm{N_A}/\mathrm{N_H}) + 12.0$.

$\log\epsilon$(Th/Hf). Observations are clearly telling us that any actinide boost of the heaviest r-process elements sets in beyond Z = 72. Abundance data on 3rd r-process peak elements Os, Ir, Pt (Z = 76–78) and Pb (Z = 82); see individual papers cited above, (e.g., Cowan et al. 2005; Plez et al. 2004). But these elements: (i) have detectable transitions only from their neutral species, whereas all other neutron-capture elements with $Z \geq 56$ are represented only by ionized transitions, greatly increasing the derived abundance uncertainties in the comparison; (ii) all of their detectable transitions are in the near-UV or vacuum-UV. Their abundances do correlate with those of other very heavy neutron-capture elements, but do not add effective cosmochronometry information at present.

Moreover, predicted production ratios in the r-process can have significant uncertainties, as very little experimental data exist on nuclei far from the valley of β stability. The influence of these uncertainties have been discussed in several papers published after Th and U detections were announced (e.g., Goriely and Clerbaux 1999). A good summary of the nuclear issues is in Niu et al. (2009), who considered uncertainties in astrophysical r-process fusion conditions, nuclear mass models, and β-decay rates. They suggest that "the influence from nuclear mass uncertainties on Th/U chronometer can approach 2 Gyr."; see their Fig. 5, which shows derived ages of HE 1523-091 (Frebel et al. 2007) and CS 31082-001 (Cayrel et al. 2001; Hill et al. 2002) with variations in all of the quantities that can influence the conversion of derived observational U/Th abundance ratios into final age estimates. Their computed mean ages are 11.8 ± 3.7 Gyr for HD 1523-0901 and 13.5 ± 2.9 Gyr for CS 31082-001, consistent with current age estimates of the Galaxy, albeit with substantial error bars.

Cosmochronometry from radioactive elements U and Th is promising, but not precise enough yet to set serious age constraints. Since uncertainties abound in all phases of this exercise, one suspects that a prerequisite for further progress is simply more detections of U in very low metallicity r-process-rich stars, which should naturally have strong Th transitions. One looks forward to more discoveries of more such stars in field star surveys in the near future.

This appears as a good viaticum for future developments, since age–dating methods based on MSTO and/or on the MSK can only be applied to ensemble of stars. The key advantage of the cosmochronometric method is that it can be applied to individual cluster and field stars and it is independent of both distance and reddening. The reader interested in a more detailed discussion concerning age-dating of field stars is referred to Salaris (2016).

5.2 Relative Cluster Age Estimates

Relative cluster ages are used to address different astrophysical problems, e.g. they play a crucial role in investigating the early formation of the Galactic components dominated by old stellar populations (halo, bulge). In particular, the spread in relative ages of halo and bulge stars is tightly correlated with the timescale of the collapse of the protogalactic cloud. Classical results based on accurate and homogeneous photometry of GGCs suggested that a significant fraction of Galactic globulars are, indeed, coeval (e.g., Buonanno et al. 1998; Rosenberg et al. 1999). There are reasons, mainly kinematic, to believe that the ones that do not follow this trend are clusters that have been accreted.

Cluster ages based on a relative diagnostic (vertical, horizontal) are less prone to systematic uncertainties. The key idea in these methods is to estimate the difference in magnitude between the MSTO and a brighter point that is either independent or mildly dependent on the absolute age of the cluster. This means that they are independent of uncertainties on cluster distance and reddening. In dealing with large homogeneous photometric data sets,

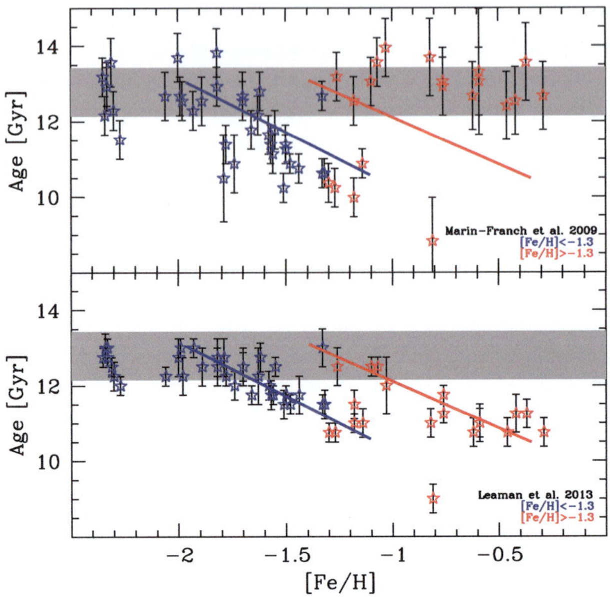

Fig. 24 Top: Relative age estimates of Galactic globular clusters according to Marín-Franch et al. (2009), these have been scaled to 12.8 Gyr, i.e. a mean age coming assuming isochrones from Dotter et al. (2007). Clusters more metal-poor than [Fe/H] $= -1.3$ were plotted as blue stars, while those that are more metal-rich as red stars. The blue and the red lines display the age-metallicity relations estimated by VandenBerg et al. (2013) and by Leaman et al. (2013). The metallicities of the GCs are in the Carretta et al. (2009) metallicity scale. The grey area shows the $\pm 1\sigma$ uncertainty in the age estimate. Bottom: Same as the top, but for the absolute age estimates by VandenBerg et al. (2013) and used in Leaman et al. (2013). This sample includes the GCs observed by Marín-Franch et al. (2009) and plotted on the top panel. However, we have plotted only GGCs in common between the two studies

the relative ages are also less affected by uncertainties on the photometric zero-point. The pros are also on the theoretical side, since the clock is used, over the entire metallicity range, in relative and not in absolute sense.

The relative cluster ages of GGCs are lively debated in the recent literature due to their impact on the early formation and evolution of the Galactic halo. The new results are based on deep and accurate photometry collected with ACS at HST. On one hand, it has been found by Marin-Franch and collaborators (Marín-Franch et al. 2009) that a sample of 64 GCs, covering a broad range in metallicity, in Galactocentric distance and in kinematic properties, show a bi-modal distribution. The former group is coeval within the errors, i.e. its mean age is ~ 12.8 Gyr with a small dispersion (5%). The latter group is more metal-poor and follows an age-metallicity relation. This group might be associated to dwarf galaxies that have been accreted into the Halo as supported by their kinematic properties. The top panel of Fig. 24 summarizes the results obtained by Marín-Franch et al. (2009).

On the other hand, based on independent ages estimated by VandenBerg et al. (2013) and Leaman et al. (2013), it was found that GGCs display a clear dichotomy, since at fixed cluster age there are two groups of GCs with roughly 0.4 dex of difference in metallicity (see bottom panel of Fig. 24). These two groups seem to follow two different, and well defined, age-metallicity relations. Leaman et al. (2013) also suggested that the metal-poor clusters were accreted, while the metal-rich ones are the truly globulars formed *in situ*. In this context it is worth mentioning that the globulars and the optical photometry adopted by

Leaman et al. (2013) significantly overlap with the sample adopted by Marín-Franch et al. (2009). We note here that data plotted in Fig. 24 are only those in common between the two studies and a glance to the data indicates a significant difference in their age determination. In particular, Fig. 24 shows several interesting features worthy of discussion.

(i) Absolute age estimates provided by VandenBerg et al. (2013) (VB) are, at fixed metal content, more precise than the relative age estimates provided by Marín-Franch et al. (2009) (MF). The dispersion in age in the former sample is $\sim 17\%$ smaller than the latter one over the entire metallicity range.

(ii) The metal–poor globulars seem to show an age-metallicity relation independently of the approach adopted to estimate the absolute cluster ages. The relation drawn through the VB's ages (blue solid line) is also a reasonable 'eye' fit to the MF's age estimates. The quoted samples also show evidence of a flatting of the quoted age-metallicity relation when moving from metal-poor ([Fe/H] ~ -2) to more metal–poor GCs.

(iii) The more metal-rich GCs display a different trend in the two different absolute age estimates. The more metal-rich GCs in the MF sample are within the errors coeval, while in the VB estimates they show a slope that is quite similar to the slope of the more metal-poor GCs, but shifted by 0.4 dex in metallicity.

The above findings clearly indicate that the main difference between the MF and the VB analysis is in the age estimates of metal-rich GCs. Note that to overcome possible deceptive uncertainties on the cluster iron abundance we adopted the homogeneous metallicity scale provided by Carretta et al. (2009). The current uncertainties on cluster iron abundances are on average smaller than 0.1 dex (Gratton et al. 2004). To unveil possible deceptive systematics either in age-dating globulars or in their metallicity scale the use of NIR diagnostics appears very promising. This applies not only to NIR photometry to use the MSK (see Sect. 5.4), but also to high-resolution NIR spectroscopy to use different sets of iron and α-element lines.

5.2.1 The Empirical Routes for Relative Age Estimates

The two most popular approaches adopted to estimate relative ages are the vertical and the horizontal photometric methods. The key advantages of these approaches are that they are independent of uncertainties affecting the distance modulus and the cluster reddening.

(i) *Vertical Method*—The vertical method relies on the difference in magnitude between the HB luminosity level and the MSTO. The HB luminosity level is typically chosen across the RR Lyrae instability strip. The former anchor is assumed to be independent of age, while the latter is age dependent. In applying this method there are a number of caveats. The main one is that it is mainly based on visual magnitudes, since at shorter (U, B) and at longer (R, I) wavelengths the HB shows a well defined negative/positive slope when moving from red to blue HB stars. This means that the anchor to the middle of the RR Lyrae instability strip might be affected by systematic errors, due to the quoted problems with the HB morphology. Moreover, this approach is also hampered by the lack either of RR Lyrae stars or of accurate and homogeneous multi-band photometry for cluster RR Lyrae stars. It is worth mentioning that the cluster age is considered to be one of the main culprits in the variation of the HB morphology (second parameter problem) when moving from more metal-poor to more metal-rich GCs. In case the cluster age is confirmed to be the second parameter, then the ages based on the vertical method would be affected by systematic errors. In particular, a decrease in the age of the progenitor causes a decrease in the HB luminosity level, therefore smaller vertical parameters, and in turn systematically younger ages. This effect becomes more relevant for stellar structures at the transition between forming or not forming an electron degenerate helium core. This means a young tail in the age distribution of GCs (see Fig. 24). Note that a systematic decrease in age typically means a redder HB morphology, thus suggesting once again that more metal-rich clusters are more prone to possible systematic age uncertainties.

(ii) *Horizontal Method*—The horizontal method relies on the difference in color between the MSTO and the base of the red giant branch. The empirical estimate of the latter reference point is not trivial in the V, I bands. Therefore, it was suggested to use the difference in color between the MSTO and the color of the RGB at a luminosity level that is 2.5 magnitudes brighter than the MSTO (Buonanno et al. 1998; Rosenberg et al. 1999). This approach has the same advantages of the vertical method, since it is independent of uncertainties on the cluster distance and reddening. Moreover, the reference point along the RGB is less affected by age effects and the RGB morphology is well defined in GGCs. However, the use of the color (horizontal method) instead of the magnitude (vertical method) together with the dependence of predicted effective temperatures at the MSTO, and in particular along the RGB, by the adopted mixing length parameter, are a source of further concern. The color-temperature relations predicted by atmosphere models are also less accurate when compared with the bolometric corrections. Note that this limitation becomes more severe when moving from the metal-poor to the metal–rich regime (Tognelli et al. 2015). Moreover, it is worth mentioning that the ratio between the duration of the sub-giant phase (hydrogen thick shell burning) and the MSTO age is not constant when moving from metal-poor to metal-rich stellar structures (Salaris and Weiss 1997). Furthermore, the slope of the RGB depends on the metal content, therefore, the difference in color with the MSTO steadily decreases when moving from more metal-poor to more metal-rich globulars.

5.3 A New Approach for Absolute Age Estimates

During the last few years it has been suggested a new vertical method to estimate the absolute age of stellar systems (Bono et al. 2010). It relies on the difference either in magnitude or in color between the MSTO and a well defined knee in the low-mass regime of the main sequence (MSK). The MSK has already been detected in several GCs—ω Cen (Pulone et al. 1998); M4 (Pulone et al. 1999; Milone et al. 2014; Braga et al. 2015; Correnti et al. 2016); NGC 3201 (Bono et al. 2010); 47 Tuc (Lagioia et al. 2014; Correnti et al. 2016); NGC 2808 (Milone et al. 2012; Massari et al. 2016a); M71 (Di Cecco et al. 2015); NGC 6752 (Correnti et al. 2016), in intermediate-age open clusters (Sarajedini et al. 2009b; An et al. 2008, 2009) and in the field of the Galactic bulge (Zoccali et al. 2000) by using either near-infrared (NIR) and/or optical–NIR photometry.

Empirical and theoretical evidence indicates that this feature is mainly caused by the collisionally induced absorption of H_2 at NIR wavelengths (Saumon et al. 1994) in the atmosphere of cool dwarfs. The shape of the bending in NIR and optical–NIR CMDs depends on the metal content, but the absolute magnitude of the knee seems to be independent of cluster age and of metallicity ($M_K \sim +4.9$ mag). This means that the difference in magnitude between the MSTO and the MSK should be a robust diagnostic to constrain the absolute cluster age. The new diagnostic shares the same positive features than the classical vertical and horizontal methods. It is independent of uncertainties in cluster distance and reddening and minimally affected by uncertainties in photometric zero-point.

The key advantages of the MSK when compared with the classical vertical and horizontal methods are the following:

(i) *The MSK is a faint but well defined feature*—The reference magnitude for the MSK is fainter than the MSTO, this implies larger photometric uncertainties for individual stars than when belonging to brighter features (RGB, HB). However, the number of faint, low mass stars ($\leq 0.5 M_{\odot}$) is intrinsically larger than that of more massive and brighter stars. Furthermore, these stars are in their slow central hydrogen burning phase, thus they are more populous than in subsequent evolved phases. As a result, old, intermediate and young stellar systems always display a well sampled faint MS, whereas: (a) the HB is less populous because it is populated by stars in their central helium burning phase and, even more, this feature shows up only in old stellar systems and depends on a number of parameters, including the metallicity; (b) the RGB is typical of stellar systems that are either old or of intermediate age, and again the difference in color with the MSTO depends on the metallicity.

The main consequence is that the MSK can be easily identified in all the stellar systems, provided that photometric data sets are pushed to reach very faint magnitudes, while the HB and the RGB detection/morphology strongly depend on the evolutionary properties of the underlying stellar population.

(ii) *The MSK is less affected by theoretical uncertainties*—The MSK only relies on the physics of hydrogen burning phase, while the vertical and the horizontal methods are affected by uncertainties in extra-mixing (RGB, HB), in microphysics (electron degeneracy, conductive opacities, 3α and $^{12}C(\alpha, \gamma)^{16}C$ nuclear reaction rates) and in dealing with the transition from the core helium flash to the core helium burning (ZAHB, Sweigart et al. 2004).

Furthermore, the MSK together with the vertical method are independent of the theoretical uncertainties plaguing the color-temperature transformations required in the horizontal method. The same outcome applies to the dependence on the adopted mixing length parameter. According to theory, MS stellar structures with a stellar mass of $M \sim 0.5 - 0.4 M_{\odot}$ are minimally affected by uncertainties in the treatment of convection, since they are almost entirely convective and the convective transport is nearly adiabatic (Saumon and Marley 2008).

(iii) *The MSK has a lower global error budget*—Recent empirical evidence indicates that the MSK provides absolute cluster ages that are a factor of two more precise when compared with the classical method of the MSTO (Bono et al. 2010; Sarajedini et al. 2009a; Di Cecco et al. 2015; Monelli et al. 2015; Massari et al. 2016a; Correnti et al. 2016). Moreover, preliminary findings suggest that the correlation between the MSK and the cluster age is linear over the age range typical of old open cluster (a few Gyrs) to GGCs. We still lack a detailed theoretical investigation to constrain the dependence of the MSK on the chemical composition (metals, helium) when moving from optical to NIR and optical/NIR CMDs.

The main cons in dealing with the MSK is that the identification of the knee requires accurate and deep photometry in crowded stellar fields. However, the advent of HST and of modern Adaptive Optics (AO) assisted NIR instrumentation overcame this problem. The MSK is a feature that can be detected and used in both optical and NIR bands, whereas the classical vertical and horizontal methods are robust age diagnostics only in optical CMDs. Accurate and deep NIR CMDs show that HB stars are far from being horizontal. They become systematically fainter when moving from cool to hot and extreme HB stars (positive slope Del Principe et al. 2006; Coppola et al. 2011; Milone et al. 2013; Stetson et al. 2014). The same problem shows up in CMDs based on near UV and far UV bands (negative slope, Ferraro et al. 2012). This means that the identification of the HB luminosity level needs a further anchor in color along the HB, difficult to be uniquely identified. On the other hand, the difference in color between the MSTO and the RGB, required by the horizontal method, is hampered by the fact that MSTO and RGB have almost the same color in NIR bands. This means that the difference in color is steadily decreasing (Coppola et al. 2011; Stetson et al. 2014) when moving from optical to NIR.

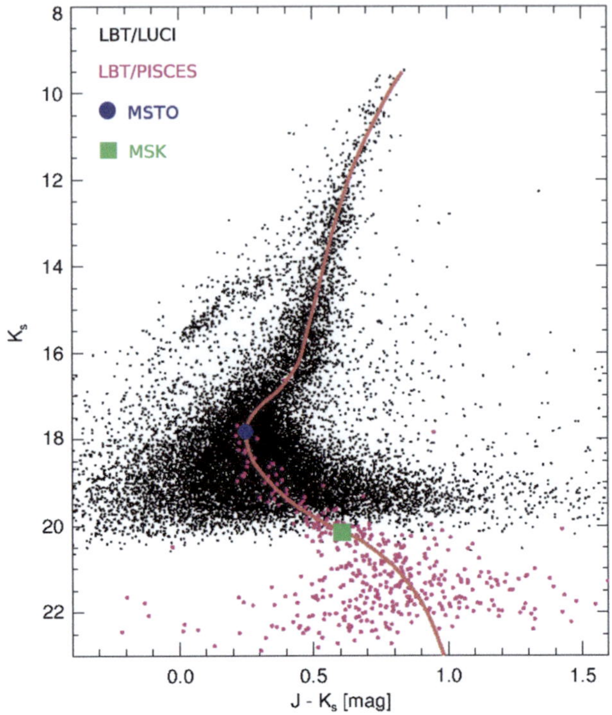

Fig. 25 Near-Infrared (K, J–K) CMD of the Galactic globular cluster M15, based on images collected with LBT (black dots: LUCI1; red dots: PISCES). The orange solid line shows the cluster ridge line adopted to determine the position of both MSTO (blue circle) and of the MSK (green square)

NIR photometry is going to be exploited even more in the near future when sophisticated AO will allow us to reach the diffraction limit of ground based extremely large telescopes (Diolaiti et al. 2016) and from the space with JWST. The use of new observables also means the opportunity to constrain possible systematics in evolutionary diagnostics currently adopted.

5.3.1 The Absolute Age of M15

To provide a new and independent absolute age estimate of the GC M15 (NGC 7078), Monelli et al. (2015) have collected collected AO images of this cluster using the First Light AO (FLAO, Esposito et al. 2012) operating at the Large Binocular Telescope (LBT). We remember here that M15 is located at ~ 10 kpc and it is affected by moderate extinction—E(B–V) = 0.08, (Harris 1996, updated 2010 version)—thus with current AO assisted 10 m telescopes we do expect to detect the MSK. Interestingly, M15 is supposed to be one of the oldest and most metal poor ([Fe/H ~ -2.4]) GCs of the Milky Way halo.

FLAO data have been taken with the NIR high resolution imager PISCES (pixel scale = $0.0193''$/pix) in J and K_s bands. Due to the highly structured and asymmetric PSF shape, the data reduction was successfully performed with ROMAFOT suite of programs (see details in Fiorentino et al. 2014; Monelli et al. 2015). A natural guide star (NGS) of R = 12.9 mag has been used to close the AO loop on a field located at $\sim 3'$ from the cluster center. This field was sufficiently uncrowded to allow us to reach a very deep K_s band magnitude of ~ 22 mag, see Fig. 25. This limiting magnitude allowed us to measure the location of the MSK; see Table 3 in Monelli et al. (2015). Note that the detection of faint stars in crowded stellar fields as the center of GCs is severely hampered by the large number of bright stars.

However, as it is shown in Fig. 25, given the small PISCES field of view (FoV) $\sim 20''$ and the radial distance from the cluster center, we do not have sufficient sampling of the MSTO magnitude. We have used LUCI1 data (FoV $= 4' \times 4'$) to properly determine the location of the MSTO. These LUCI1 data have also been used to perform a proper calibration of PISCES data to the 2MASS photometric system.

After measuring the difference MSTO-MSK, we are ready to compute the absolute age of M15. We compare this number with theoretical relations derived using a set of evolutionary isochrones provided by VandenBerg et al. (2014b) that relate the variation of MSTO-MSK with the absolute ages. Using only NIR, we derived an absolute age for M15 of 13.7 ± 1.4 Gyr, which is compatible, but with a smaller uncertainty, to that obtained using the classical MSTO method in NIR (14.0 ± 3.1 Gyr) or in purely optical HST bands (12.8 ± 2.0 Gyr). This old age provides an upper limit to the age of the Universe and a lower limit to the Hubble constant H_0, since the former parameter is roughly the inverse of the latter one (Gratton et al. 2003; Monelli et al. 2015).

5.4 Conclusions and Final Remarks

There is mounting evidence of a difference between estimates of a Hubble constant based on direct measurements (Cepheids plus supernovae: Riess et al. 2016; Freedman and Madore 2010, Beaton this conference) and indirect methods (CMD, BAO, gravitational lensing, Suyu et al. 2013; Bennett et al. 2014; Calabrese et al. 2015, Planck collaboration 2015). This critical issue has been addressed in several recent papers suggesting a difference that ranges from almost 2σ (Efstathiou 2014) to more than 3σ (Riess et al. 2011, 2016). The quoted uncertainties on the Hubble constant, once confirmed, can open the path to new physics concerning the number of relativistic species and/or the mass of neutrinos (Dvorkin et al. 2014; Wyman et al. 2014; Luković et al. 2016). Moreover, the quoted range in H_0 implies an uncertainty on the age of the Universe t_0 of the order of 2 Gyr. This uncertainty has a substantial impact not only on galaxy formation and evolution, but also on the age of the most ancient stellar systems, i.e. the globular clusters can play a crucial cosmological role.

Based on the tantalizing evidence that stellar age is not an observable, the different stellar "clocks" that can be applied to date globular clusters provide the unique opportunity to constrain the micro- and the macro-physics adopted to construct evolutionary models. This comparison becomes even more rewarding when comparing main sequence stellar structures with white dwarf models. It might be possible that in the era of "precision cosmology" we could use cosmological parameters to constrain the physics of stellar interiors. In the mean time, it is clear that calibrating clusters to which we can apply different age diagnostics (MSTO, MSK, white dwarf cooling sequence, cosmochronometry) become fundamental astrophysical and cosmological laboratories.

In this investigation we reviewed the most popular methods to estimate both relative and absolute cluster ages. We focussed our attention on the error budget and critically discussed pros and cons of the different age diagnostics. In particular, we outlined the key advantages in using a new NIR age diagnostic—the main sequence knee—and its application to M15. The main limitation being the faintness of this anchor. However, there are two new observing facilities that are going to play a fundamental role in the use of the MSK.

(i) *Multi-conjugated adaptive optics*—The development of multi-conjugated adaptive optics systems at the 8–10 m class telescopes is a real quantum jump. They provide NIR images that approach the diffraction limit of a field of view of the order of one arcmin. This means that they can provide accurate and deep NIR CMDs of the innermost crowded regions of GCs. Recent findings based on NIR images collected with GEMS at GEMINI indicate that the mix between NGS and lasers allow us to reach the MSK in a sizable sample of Galactic globulars (Massari et al. 2016a; Turri et al. 2016). The empirical scenario becomes even more compelling if we take account for the next generation of NIR detectors AO assisted at VLT (ERIS). This instrument will simultaneously cover NIR bands and the L-band, and in turn, the unique opportunity to identify the MSK in the most crowded and most reddened regions of the Galactic bulge.

(ii) *JWST*—JWST is going to revolutionize the view of resolved stellar populations in the nearby Universe. The coupling between field of view and NIR/MIR bands would provide the unique opportunity to identify the MSK in a significant fraction of nearby dwarf galaxies. This means the opportunity to determine homogeneous ages for old stars in old stellar systems (dwarfs, globulars) to investigate whether they formed, as suggested by cosmological models, at the same epoch.

The cosmic distance scale and the age-dating of nearby stellar systems have been for more than half a century the two fundamental pillars on which quantitative astrophysics build up. At the beginning of the new millennium they are waiting for massive solidification. The near future appears quite bright not only for the next Gaia data releases, but also for the near future ground-based (LSST, ELTs) and space observing facilities (JWST, WFIRST, EUCLID). The same outcome applies for the wide spectroscopic surveys in optical and NIR regimes (DESI, PFS, 4MOST, MOONS-GAL, APOGEE, WEAVE).

References

C. Alcock, R.A. Allsman, T.S. Axelrod, D.P. Bennett, K.H. Cook et al., Astron. J. **109**, 1653 (1995)

C. Alcock, R.A. Allsman, D.R. Alves, T.S. Axelrod, A.C. Becker et al., Astrophys. J. **492**, 190 (1998)

C. Alcock, R.A. Allsman, D.R. Alves, T.S. Axelrod, A.C. Becker, D.P. Bennett, K.H. Cook, A.J. Drake, K.C. Freeman, M. Geha, K. Griest, M.J. Lehner, S.L. Marshall, D. Minniti, C.A. Nelson, B.A. Peterson, P. Popowski, M.R. Pratt, P.J. Quinn, C.W. Stubbs, W. Sutherland, A.B. Tomaney, T. Vandehei, D.L. Welch (MACHO Collaboration), Astrophys. J. **541**, 734 (2000)

A. Alves-Brito, J. Meléndez, M. Asplund, I. Ramírez, D. Yong, Astron. Astrophys. **513**, 35 (2010)

E.B. Amôres, M. López-Corredoira, C. González-Fernández, Astron. Astrophys. **559**, 11 (2013)

D. An, J.A. Johnson, J.L. Clem et al., Astrophys. J. Suppl. Ser. **179**, 326 (2008)

D. An, M.H. Pinsonneault, T. Masseron et al., Astrophys. J. Suppl. Ser. **700**, 523 (2009)

J. Anderson, in *ASPCS*, vol. 35, ed. by F. van Leeuwen, J.D. Hughes, G. Piotto (2002), p. 87

H. Arp, Astrophys. J. **135**, 971 (1962)

E. Athanassoula, Mon. Not. R. Astron. Soc. **358**, 1477 (2005)

W. Baade, Publ. Astron. Soc. Pac. **58**, 249 (1946)

W. Baade, *Specola Vaticana*, Proceedings of the Conference Sponsored by the Pontifical Academy of Science and the Vatican Observatory. Ricerche Astronomiche, vol. 5 (Springer, Berlin, 1958)

W. Baade, Publ. Obs. Univ. Mich. **10**, 7 (1951)

C. Babusiaux, G. Gilmore, Mon. Not. R. Astron. Soc. **358**, 1309 (2005)

C. Babusiaux, A. Gómez, V. Hill et al., Astron. Astrophys. **519**, A77 (2010)

C. Babusiaux, D. Katz, V. Hill et al., Astron. Astrophys. **563**, A15 (2014)

M. Baker, B. Willman, Astron. J. **150**, 160 (2015)

L.R. Bedin, G. Piotto, J. Anderson, S. Cassisi, I.R. King, Y. Momany, G. Carraro, Astrophys. J. Lett. **605**, 125 (2004)

T.C. Beers, G.W. Preston, S.A. Shectman, Astron. J. **103**, 1987 (1992)

V. Belokurov, D.B. Zucker, N.W. Evans et al., Astrophys. J. **654**, 897 (2007)

R.A. Benjamin, E. Churchwell, B.L. Babler, R. Indebetouw, M.R. Meade, B.A. Whitney, C. Watson, M.G. Wolfire, M.J. Wolff, R. Ignace, T.M. Bania, S. Bracker, D.P. Clemens, L. Chomiuk, M. Cohen, J.M. Dickey, J.M. Jackson, H.A. Kobulnicky, E.P. Mercer, J.S. Mathis, S.R. Stolovy, B. Uzpen, Astrophys. J. Lett. **630**, 149 (2005)

C.L. Bennett, D. Larson, J.L. Weiland, G. Hinshaw, Astrophys. J. **794**, 135 (2014)

T. Bensby, S. Feltzing, I. Lundström, Astron. Astrophys. **410**, 527 (2003)

T. Bensby, A.R. Zenn, M.S. Oey, S. Feltzing, Astrophys. J. Lett. **663**, 13 (2007)

T. Bensby, J.C. Yee, S. Feltzing et al., Astron. Astrophys. **549**, 147 (2013)
T. Bensby, S. Feltzing, A. Gould et al., Astron. Astrophys. **605A**, 89 (2017)
G. Bertelli, E. Nasi, L. Girardi, P. Marigo, Astron. Astrophys. **508**, 355 (2009)
A. Bhardwaj, M. Rejkuba, D. Minniti, F. Surot, E. Valenti, M. Zoccali, O.A. Gonzalez, M. Romaniello, S.M. Kanbur, H.P. Singh, Astron. Astrophys. **605**, 100 (2017)
E. Bica, S. Ortolani, B. Barbuy, Publ. Astron. Soc. Aust. **33**, e028 (2016)
J. Binney, O.E. Gerhard, A.A. Stark, J. Bally, K.I. Uchida, Mon. Not. R. Astron. Soc. **252**, 210 (1991)
J. Binney, O. Gerhard, D. Spergel, Mon. Not. R. Astron. Soc. **288**, 365 (1997)
N. Bissantz, O. Gerhard, Mon. Not. R. Astron. Soc. **330**, 591 (2002)
L. Blitz, D.N. Spergel, Astrophys. J. **379**, 631 (1991)
R.D. Blum, Astrophys. J. Lett. **444**, L89 (1995)
E. Boettcher, B. Willman, R. Fadely et al., Astron. J. **146**, 94 (2013)
G. Bono, F. Caputo, V. Castellani et al., Mon. Not. R. Astron. Soc. **344**, 1097 (2003)
G. Bono, P.B. Stetson, N. Sanna et al., Astrophys. J. Lett. **686**, 87 (2008)
G. Bono, P.B. Stetson, D. VandenBerg et al., Astrophys. J. Lett. **708**, 74 (2010)
G. Bono, M. Salaris, R. Gilmozzi, Astron. Astrophys. **549**, 102 (2013)
M. Boylan-Kolchin, J.S. Bullock, M. Kaplinghat, Mon. Not. R. Astron. Soc. **422**, 1203 (2012)
V.F. Braga et al., Astrophys. J. **799**, 165 (2015)
E. Brocato, V. Castellani, M. Romaniello, Astron. Astrophys. **345**, 499 (1999)
T.M. Brown, H.C. Ferguson, E. Smith, R.A. Kimble, A.V. Sweigart, A. Renzini, R.M. Rich, D.A. VandenBerg, Astrophys. J. Lett. **592**, L17 (2003)
T.M. Brown, E. Smith, H.C. Ferguson, R.M. Rich, P. Guhathakurta, A. Renzini, A.V. Sweigart, R.A. Kimble, Astrophys. J. **652**, 323 (2006)
T.M. Brown, K. Sahu, J. Anderson, J. Tumlinson et al., Astrophys. J. Lett. **725**, L19 (2010)
G. Bruzual, S. Charlot, Mon. Not. R. Astron. Soc. **344**, 1000 (2003)
R. Buonanno, C.E. Corsi, L. Pulone, F. Fusi Pecci, M. Bellazzini, Astron. Astrophys. **333**, 505 (1998)
M. Bureau, E. Athanassoula, Astrophys. J. **626**, 159 (2005)
H.R. Butcher, Nature **328**, 127 (1987)
A. Cabrera-Lavers, P.L. Hammersley, C. González-Fernández, M. López-Corredoira, F. Garzón, T.J. Mahoney, Astron. Astrophys. **465**, 825 (2007)
A. Cabrera-Lavers, C. González-Fernández, F. Garzón, P.L. Hammersley, M. López-Corredoira, Astron. Astrophys. **491**, 781 (2008)
E. Calabrese, C. Carbone, G. Fabbian, M. Baldi, C. Baccigalupi, J. Cosmol. Astropart. Phys. **3**, 49 (2015)
A. Calamida, G. Strampelli, A. Rest et al., Astron. J. **153**, 175 (2017)
L. Cao, S. Mao, D. Nataf, N.J. Rattenbury, A. Gould, Mon. Not. R. Astron. Soc. **434**, 595 (2013)
J.L. Carlin, C.J. Grillmair, R.R. Muñoz, D.L. Nidever, S.R. Majewski, Astrophys. J. Lett. **702**, L9 (2009)
D. Carollo, T.C. Beers, Y.S. Lee et al., Nature **451**, 216 (2008)
E. Carretta, A. Bragaglia, R. Gratton, V. D'Orazi, S. Lucatello, Astron. Astrophys. **508**, 695 (2009)
S. Cassisi, A.Y. Potekhin, A. Pietrinferni, M. Catelan, M. Salaris, Astrophys. J. **661**, 1094 (2007)
S. Cassisi, M. Salaris, A. Pietrinferni, G. Piotto, A.P. Milone, L.R. Bedin, J. Anderson, Astrophys. J. **672**, L115 (2008)
V. Castellani, S. degl'Innocenti, M. Marconi, Mon. Not. R. Astron. Soc. **303**, 265 (2013)
R.M. Catchpole, P.A. Whitelock, M.W. Feast, S.M.G. Hughes, M. Irwin, C. Alard, Mon. Not. R. Astron. Soc. **455**, 2216 (2015)
M. Catelan, Astrophys. J. **600**, 409 (2004)
M. Catelan, B.J. Pritzl, H.A. Smith, Astrophys. J. Suppl. Ser. **154**, 633 (2004)
R. Cayrel, V. Hill, T.C. Beers et al., Nature **409**, 691 (2001)
B. Chaboyer et al., Astrophys. J. **835**, 152 (2017)
L. Chemin, F. Renaud, C. Soubiran, Astron. Astrophys. **578**, A14 (2015)
C. Chiosi, A. Maeder, Annu. Rev. Astron. Astrophys. **24**, 329 (1986)
E. Churchwell, B.L. Babler, M.R. Meade, B.A. Whitney, R. Benjamin, R. Indebetouw, C. Cyganowski, T.P. Robitaille, M. Povich, C. Watson, S. Bracker, Publ. Astron. Soc. Pac. **121**, 213 (2009)
F. Ciacio, S. degl'Innocenti, B. Ricci, Astron. Astrophys. Suppl. **123**, 449 (1997)
W. Clarkson, K. Sahu, J. Anderson et al., Astrophys. J. **684**, 1110 (2008)
W.I. Clarkson, K.C. Sahu, J. erson, R.M. Rich, T.E. Smith, T.M. Brown, H.E. Bond, M. Livio, D. Minniti, A. Renzini, M. Zoccali, Astrophys. J. **735**, 37 (2011)
C.M. Clement, A. Muzzin, Q. Dufton et al., Astron. J. **122**, 2587 (2001)
G. Clementini, Commun. Konkoly **105**, 3 (2016)
G. Clementini, R. Gratton, A. Bragaglia et al., Astron. J. **125**, 1309 (2003)
G. Clementini, M. Cignoni, R. Contreras Ramos et al., Astrophys. J. **756**, 108 (2012)
J.G. Cohen, Astrophys. J. **223**, 487 (1978)

M.J. Collinge, T. Sumi, D. Fabrycky, Astrophys. J. **651**, 197 (2006)
G. Coppola, M. Dall'Ora, V. Ripepi et al., Mon. Not. R. Astron. Soc. **416**, 1056 (2011)
M. Correnti, M. Gennaro, J.S. Kalirai, T.M. Brown, A. Calamida, Astrophys. J. **823**, 18 (2016)
S. Courteau, M. Cappellari, R.S. de Jong, A.A. Dutton, E. Emsellem, H. Hoekstra, L.V.E. Koopmans, G.A. Mamon, C. Maraston, T. Treu, L.M. Widrow, Rev. Mod. Phys. **86**, 47 (2014)
J.J. Cowan, A. McWilliam, C. Sneden, D.L. Burris, Astrophys. J. **480**, 246 (1997)
J.J. Cowan, C. Sneden, S. Burles et al., Astrophys. J. **572**, 861 (2002)
J.J. Cowan, C. Sneden, T. Beers et al., Astrophys. J. **627**, 238 (2005)
K. Cunha, V.V. Smith, Astrophys. J. **651**, 491 (2006)
M. Dall'Ora, G. Clementini, K. Kinemuchi et al., Astrophys. J. Lett. **653**, L109 (2006)
M. Dall'Ora, K. Kinemuchi, V. Ripepi et al., Astrophys. J. **752**, 42 (2012)
G. de Vaucouleurs, in *IAU Symposium*, vol. 20, ed. by F.J. Kerr (1964), p. 195
V.P. Debattista, L. Mayer, C.M. Carollo, B. Moore, J. Wadsley, T. Quinn, Astrophys. J. **645**, 209 (2006)
V.P. Debattista, M. Ness, O.A. Gonzalez, K. Freeman, M. Zoccali, D. Minniti, Mon. Not. R. Astron. Soc. **469**, 1587 (2017)
S. degl'Innocenti, G. Fiorentini, B. Ricci, F.L. Villante, Phys. Lett. B **590**, 13 (2004)
I. Dékány, D. Minniti, M. Catelan, M. Zoccali, R.K. Saito, M. Hempel, O.A. Gonzalez, Astrophys. J. Lett. **776**, L19 (2013)
M. Del Principe, A.M. Piersimoni, J. Storm et al., Astrophys. J. **652**, 362 (2006)
A. Di Cecco et al., Astron. J. **150**, 51 (2015)
P. Di Matteo et al., Astron. Astrophys. **577**, 1 (2015)
J. Diemand, M. Kuhlen, P. Madau, Astrophys. J. **667**, 859 (2007)
E. Diolaiti, P. Ciliegi, R. Abicca et al., SPIE **9909**, 99092D (2016)
S. Djorgovski, M. Davis, Astrophys. J. **313**, 59 (1987)
H. Dong, R. Schödel, B.F. Williams et al., Mon. Not. R. Astron. Soc. **471**, 3617 (2017)
A. Dotter, B. Chaboyer, D. Jevremović et al., Astron. J. **134**, 376 (2007)
A. Dotter, A. Sarajedini, J. Anderson, Astrophys. J. **738**, 74 (2011)
A. Dressler, D. Lynden-Bell, D. Burstein et al., Astrophys. J. **313**, 42 (1987)
P.-A. Duc, S. Paudel, R.M. McDermid et al., Mon. Not. R. Astron. Soc. **440**, 1458 (2014)
C. Dvorkin, M. Wyman, D.H. Rudd, W. Hu, Phys. Rev. D **90**, 083503 (2014)
E. Dwek, R.G. Arendt, M.G. Hauser, T. Kelsall, C.M. Lisse, S.H. Moseley, R.F. Silverberg, T.J. Sodroski, J.L. Weiland, Astrophys. J. **445**, 716 (1995)
G. Efstathiou, Mon. Not. R. Astron. Soc. **440**, 1138 (2014)
S. Esposito, A. Riccardi, E. Pinna et al., SPIE **8447**, 84470U (2012)
S. Feltzing, G. Gilmore, Astron. Astrophys. **355**, 949 (2000)
F.R. Ferraro et al., Nature **462**, 483 (2009)
F.R. Ferraro, B. Lanzoni, E. Dalessandro et al., Nature **492**, 393 (2012)
I. Ferreras, R.F.G. Wyse, J. Silk, Mon. Not. R. Astron. Soc. **345**, 1381 (2003)
G. Fiorentino, I. Ferraro, G. Bono et al., SPIE **9148**, 3U (2014)
G. Fiorentino, G. Bono, M. Monelli et al., Astrophys. J. Lett. **798**, L12 (2015)
G. Fiorentino, M. Monelli, P.B. Stetson et al., Astron. Astrophys. **599**, 125 (2017). https://doi.org/10.1051/0004-6361/201629501
A. Frebel, N. Christlieb, J.E. Norris et al., Astrophys. J. **660**, 117 (2007)
W.L. Freedman, B.F. Madore, Annu. Rev. Astron. Astrophys. **48**, 673 (2010)
K. Freeman, M. Ness, E. Wylie-de-Boer, E. Athanassoula, J. Bland-Hawthorn, M. Asplund, G. Lewis, D. Yong, R. Lane, L. Kiss, R. Ibata, Mon. Not. R. Astron. Soc. **428**, 3660 (2013)
J.A. Frogel, A.E. Whitford, R.M. Rich, Astron. J. **89**, 1536 (1984)
J.P. Fulbright, A. McWilliam, R.M. Rich, Astrophys. J. **661**, 1152 (2007)
C. Gallart, M. Zoccali, A. Aparicio, Annu. Rev. Astron. Astrophys. **43**, 387 (2005)
A.E. García Pérez, K. Cunha, M. Shetrone et al., Astrophys. J. **767**, L9 (2013)
A. Garofalo, F. Cusano, G. Clementini et al., Astrophys. J. **767**, 62 (2013)
M. Gennaro, P.G. Prada Moroni, S. Degl'Innocenti, Astron. Astrophys. **518**, 13 (2010)
O. Gerhard, I. Martinez-Valpuesta, Astrophys. J. Lett. **744**, 8 (2012)
L. Girardi, Annu. Rev. Astron. Astrophys. **54**, 95 (2016)
O.A. Gonzalez, M. Rejkuba, M. Zoccali, E. Valenti, D. Minniti, Astron. Astrophys. **534**, 3 (2011a)
O.A. Gonzalez, M. Rejkuba, D. Minniti, M. Zoccali, E. Valenti, R.K. Saito, Astron. Astrophys. **534**, 14 (2011b)
O.A. Gonzalez, M. Rejkuba, M. Zoccali, V. Hill, G. Battaglia, C. Babusiaux, D. Minniti, B. Barbuy, A. Alves-Brito, A. Renzini, A. Gomez, S. Ortolani, Astron. Astrophys. **530**, 54 (2011c)
O.A. Gonzalez, M. Zoccali, S. Vasquez et al., Astron. Astrophys. **584**, 46 (2015)
S. Goriely, B. Clerbaux, Astron. Astrophys. **346**, 798 (1999)

F. Gran, D. Minniti, R.K. Saito et al., Astron. Astrophys. **591**, 145 (2016)
R.G. Gratton et al., Astron. Astrophys. **408**, 529 (2003)
R.G. Gratton, C. Sneden, E. Carretta, Annu. Rev. Astron. Astrophys. **42**, 385 (2004)
R.G. Gratton, E. Carretta, A. Bragaglia, Astron. Astrophys. Rev. **20**, 50 (2012)
C. Greco, M. Dall'Ora, G. Clementini et al., Astrophys. J. Lett. **675**, L73 (2008)
R. Griffin, B. Gustafsson, T. Vieira, R. Griffin, Mon. Not. R. Astron. Soc. **198**, 637 (1982)
C.J. Grillmair, J.L. Carlin, Astrophys. Space Sci. Libr. **420**, 87 (2016)
B. Gustafsson, B. Edvardsson, K. Eriksson, U.G. Jørgensen, Å. Nordlund, B. Plez, Astron. Astrophys. **486**, 951 (2008)
H.J. Habing, M.N. Sevenster, M. Messineo, G. van de Ven, K. Kuijken, Astron. Astrophys. **458**, 151 (2006)
C. Han, A. Gould, Astrophys. J. **449**, 521 (1995)
B.M.S. Hansen, Phys. Rev. **399**, 1 (2004)
B.M.S. Hansen, H.B. Richer, G.G. Fahlman et al., Astrophys. J. Suppl. Ser. **155**, 551 (2004)
W.E. Harris, Astron. J. **112**, 1487 (1996)
W. Hayek, U. Wiesendahl, N. Christlieb et al., Astron. Astrophys. **504**, 511 (2009)
M. Haywood, P. Di Matteo, O. Snaith, A. Calamida, Astron. Astrophys. **593**, 82 (2016)
V. Hill, B. Plez, R. Cayrel et al., Astron. Astrophys. **387**, 560 (2002). arXiv:1608.07463
V. Hill, A. Lecureur, A. Gomez et al., Astron. Astrophys. **534**, 80 (2011)
V. Hill, N. Christlieb, T.C. Beers et al., Astron. Astrophys. **534**, 80 (2016)
L.M. Howes, M. Asplund, A.R. Casey et al., Mon. Not. R. Astron. Soc. **445**, 4241 (2014)
L.M. Howes, A.R. Casey, M. Asplund et al., Nature **527**, 484 (2015)
L.M. Howes, M. Asplund, A.C. Keller et al., Mon. Not. R. Astron. Soc. **460**, 884 (2016)
F. Hoyle, Mon. Not. R. Astron. Soc. **119**, 124 (1959)
R.A. Ibata, G.F. Lewis, A.R. Conn et al., Nature **493**, 62 (2013)
R. Indebetouw, J.S. Mathis, B.L. Babler et al., Astrophys. J. **619**, 931 (2005)
S. Jester, D.P. Schneider, G.T. Richards et al., Astron. J. **130**, 873 (2005)
C.I. Johnson, R.M. Rich, J.P. Fulbright, E. Valenti, A. McWilliam, Astrophys. J. **732**, 108 (2011)
C.I. Johnson, R.M. Rich, C. Kobayashi, J.P. Fulbright, Astrophys. J. **749**, 175 (2012)
C.I. Johnson, R.M. Rich, C. Kobayashi, A. Kunder, C. Pilachowski, A. Koch, R. De Propris, Astrophys. J. **765**, 157 (2013a)
C.I. Johnson, A. McWilliam, R.M. Rich, Astrophys. J. **775**, 27 (2013b)
C.I. Johnson, R.M. Rich, C. Kobayashi, A. Kunder, A. Koch, Astron. J. **148**, 67 (2014)
S.-J. Joo, Y.-W. Lee, C. Chung, Astrophys. J. **840**, 98 (2017)
J. Jurcsik, Revision of the [Fe/H] scales used for globular clusters and RR lyrae variables. Acta Astron. **45**, 653 (1995)
J. Jurcsik, G. Kovács, Determination of [Fe/H] from the light curves of RR Lyrae stars. Astron. Astrophys. **312**, 111 (1996)
M. Jurić, Ž. Ivezić, A. Brooks, R.H. Lupton, D. Schlegel et al., Astrophys. J. **673**, 864 (2008)
J.S. Kalirai et al., Astrophys. J. **143**, 11 (2012)
S.M. Kent, Astrophys. J. **387**, 181 (1992)
I.R. King, L.R. Bedin, G. Piotto, S. Cassisi, J. Anderson, Astron. J. **130**, 626 (2005)
T.D. Kinman, C. Cacciari, A. Bragaglia, R. Smart, A. Spagna, Mon. Not. R. Astron. Soc. **422**, 2116 (2012)
E.N. Kirby, J.D. Simon, M. Geha, P. Guhathakurta, A. Frebel, Astrophys. J. Lett. **685**, L43 (2008)
E.N. Kirby, J.D. Simon, J.G. Cohen, Astrophys. J. **810**, 56 (2015)
A. Klypin, A.V. Kravtsov, O. Valenzuela, F. Prada, Astrophys. J. **522**, 82 (1999)
A. Koch, A. McWilliam, G.W. Preston, I.B. Thompson, Astron. Astrophys. **587**, 124 (2016)
S.E. Koposov, V. Belokurov, G. Torrealba, N.W. Evans, Astrophys. J. **805**, 130 (2015)
J. Kormendy, G. Illingworth, Astrophys. J. **256**, 460 (1982)
R.P. Kraft, Publ. Astron. Soc. Pac. **106**, 553 (1994)
K.-L. Kratz, K. Farouqi, B. Pfeiffer et al., Astrophys. J. **662**, 39 (2007)
C. Kuehn, K. Kinemuchi, V. Ripepi et al., Astrophys. J. Lett. **674**, L81 (2008)
K. Kuijken, R.M. Rich, Astron. J. **124**, 2054 (2002)
A. Kunder, B. Chaboyer, Astron. J. **136**, 2441 (2008)
A. Kunder, P. Popowski, K.H. Cook, B. Chaboyer, Astron. J. **135**, 631 (2008)
A. Kunder, A. Koch, R.M. Rich et al., Astron. J. **143**, 57 (2012)
A. Kunder, R.M. Rich, K. Hawkins et al., Astrophys. J. **808**, 12 (2015)
A. Kunder, R.M. Rich, A. Koch et al., Astrophys. J. **821**, 25 (2016)
B.P.M. Laevens, N.F. Martin, E.J. Bernard et al., Astrophys. J. **813**, 44 (2015)
E.P. Lagioia et al., Astrophys. J. **782**, 50 (2014)
E. Laurikainen, H. Salo, Astron. Astrophys. **598**, 10 (2017)

E. Laurikainen, H. Salo, E. Athanassoula, A. Bosma, M. Herrera-Endoqui, Mon. Not. R. Astron. Soc. **444**, 80 (2014)
R. Leaman, D.A. VandenBerg, J.T. Mendel, Mon. Not. R. Astron. Soc. **436**, 122 (2013)
A. Lecureur, V. Hill, M. Zoccali, B. Barbuy, A. Gómez, D. Minniti, S. Ortolani, A. Renzini, Astron. Astrophys. **465**, 799 (2007)
Y.-W. Lee, Astron. J. **104**, 1780 (1992)
Y.-W. Lee, S. Jang, Astrophys. J. **833**, 236 (2016)
Y.-W. Lee, H.B. Gim, D.I. Casetti-Dinescu, Astrophys. J. Lett. **661**, L49 (2007)
Y.-W. Lee, S.-J. Joo, C. Chung, Mon. Not. R. Astron. Soc. **453**, 3906 (2015)
E.M. Leep, G. Wallerstein, J.B. Oke, Astron. J. **91**, 1117 (1986)
M. López-Corredoira, Astrophys. J. **836**, 218 (2017)
M. López-Corredoira, A. Cabrera-Lavers, T.J. Mahoney, P.L. Hammersley, F. Garzón, C. González-Fernández, Astron. J. **133**, 154 (2007)
V.V. Luković, R. D'Agostino, N. Vittorio, Astron. Astrophys. **595**, 109 (2016)
R. Lunnan, M. Vogelsberger, A. Frebel et al., Astrophys. J. **746**, 109 (2012)
A.D. Mackey, G.F. Gilmore, Mon. Not. R. Astron. Soc. **345**, 747 (2004a)
A.D. Mackey, G.F. Gilmore, Mon. Not. R. Astron. Soc. **355**, 504 (2004b)
A. Maeder, G. Meynet, Annu. Rev. Astron. Astrophys. **38**, 143 (2000)
M. Marconi, G. Coppola, G. Bono et al., Astrophys. J. **808**, 50 (2015)
A. Marín-Franch, A. Aparicio, G. Piotto et al., Astrophys. J. **694**, 1498 (2009)
M. Marta, A. Formicola, G. Gyürky et al., Phys. Rev. C **78**, 2 (2008)
I. Martinez-Valpuesta, O. Gerhard, Astrophys. J. Lett. **734**, L20 (2011)
C.E. Martínez-Vázquez, M. Monelli, C. Gallart et al., Mon. Not. R. Astron. Soc. **461**, L41 (2016a)
C.E. Martínez-Vázquez, P.B. Stetson, M. Monelli et al., Mon. Not. R. Astron. Soc. **462**, 4349 (2016b)
L. Mashonkina, N. Christlieb, K. Eriksson, Astron. Astrophys. **569**, 43 (2014)
D. Massari, A. Bellini, F.R. Ferraro et al., Astrophys. J. **779**, 81 (2013)
D. Massari, G. Fiorentino, A. McConnachie et al., Astron. Astrophys. **586**, 51 (2016a)
D. Massari, G. Fiorentino, A. McConnachie et al., Astron. Astrophys. **595**, L2 (2016b)
T. Matsumoto, S. Hayakawa, H. Koizumi, H. Murakami, K. Uyama, T. Yamagami, J.A. Thomas, in *AIP Conf. Proc*, vol. 83, ed. by G.R. Riegler, R.D. Blandford (1982), p. 48
N. Matsunaga, M.W. Feast, T. Kawadu, S. Nishiyama, T. Nagayama, T. Nagata, M. Tamura, G. Bono, N. Kobayashi, Mon. Not. R. Astron. Soc. **429**, 385 (2013)
F. Matteucci, D. Romano, P. Molaro, Astron. Astrophys. **341**, 458 (1999)
A.W. McConnachie, Astron. J. **144**, 4 (2012)
A. McWilliam, Annu. Rev. Astron. Astrophys. **35**, 503 (1997)
A. McWilliam, Publ. Astron. Soc. Aust. **33**, 40 (2016)
A. McWilliam, R.M. Rich, Astrophys. J. Suppl. Ser. **91**, 749 (1994)
A. McWilliam, M. Zoccali, Astrophys. J. **724**, 1491 (2010a)
A. McWilliam, M. Zoccali, Astrophys. J. **724**, 1491 (2010b)
A. McWilliam, J. Fulbright, R.M. Rich, in *Chemical Abundances in the Universe: Connecting First Stars to Planets*, ed. by K. Cunha, M. Spite, B. Barbuy. IAU Sym., vol. 265 (Cambridge Univ. Press, Cambridge, 2010), p. 279
J. Meléndez, M. Asplund, A. Alves-Brito, K. Cunha, B. Barbuy, M.S. Bessell, C. Chiappini, K.C. Freeman, I. Ramírez, V.V. Smith, D. Yong, Astron. Astrophys. **484**, 21 (2008)
D. Merritt, J.A. Sellwood, Astrophys. J. **425**, 551 (1994)
A. Miceli, A. Rest, C.W. Stubbs et al., Astrophys. J. **678**, 865 (2008)
G. Michaud, O. Richard, J. Richer, D.A. VandenBerg, Astrophys. J. **606**, 452 (2004)
A. Milone, A.F. Marino, S. Cassisi et al., Astrophys. J. Lett. **754**, 34 (2012)
A. Milone, A.F. Marino, G. Piotto et al., Astrophys. J. **767**, 120 (2013)
A. Milone, A.F. Marino, L.R. Bedin et al., Mon. Not. R. Astron. Soc. **439**, 1588 (2014)
D. Minniti, Publ. Astron. Soc. Pac. **106**, 813 (1994)
D. Minniti, E.W. Olszewski, J. Liebert, S.D.M. White, J.M. Hill, M.J. Irwin, Mon. Not. R. Astron. Soc. **277**, 1293 (1995)
D. Minniti, C. Alcock, D. Alves, K. Cook, S. Marshall et al., MACHO RR Lyrae stars in the Galactic bulge: the spatial distribution, in *The Central Regions of the Galaxy and Galaxies*, ed. by Y. Sofue. IAU Symp., vol. 184 (1998), p. 123
D. Minniti, P.W. Lucas, J.P. Emerson et al., New Astron. **15**, 433 (2010)
D. Minniti, R. Contreras Ramos, M. Zoccali, M. Rejkuba, O.A. Gonzalez, E. Valenti, F. Gran, Astrophys. J. **830**, L14 (2016)
T. Mizerski, Acta Astron. **53**, 307 (2003)
S. Moehler, G. Bono (2008). arXiv:0806.4456

M. Monelli, V. Testa, G. Bono et al., Astrophys. J. **812**, 25 (2015)
B. Moore, S. Ghigna, F. Governato et al., Astrophys. J. Lett. **524**, L19 (1999)
M.I. Moretti, M. Dall'Ora, V. Ripepi et al., Astrophys. J. Lett. **699**, L125 (2009)
O. Müller, H. Jerjen, M.S. Pawlowski, B. Binggeli, Astron. Astrophys. **595**, A119 (2016)
O. Müller, M.S. Pawlowski, H. Jerjen, F. Lelli et al., Science **359**, 534 (2018)
I. Musella, V. Ripepi, G. Clementini et al., Astrophys. J. Lett. **695**, L83 (2009)
I. Musella, V. Ripepi, M. Marconi et al., Astrophys. J. **756**, 121 (2012)
D.M. Nataf, A. Udalski, A. Gould, P. Fouqué, K.Z. Stanek, Astrophys. J. Lett. **721**, 28 (2010)
D.M. Nataf, A. Udalski, A. Gould, M.H. Pinsonneault, Astrophys. J. **730**, 228 (2011)
D.M. Nataf, A. Gould, P. Fouqué, O.A. Gonzalez, J.A. Johnson et al., Astrophys. J. **769**, 88 (2013)
J.R. Neeley, M. Marengo, G. Bono et al., Astrophys. J. **841**, 84 (2017)
M. Ness, D. Lang, Astron. J. **152**, 14 (2016)
M. Ness, K. Freeman, E. Athanassoula, E. Wylie-De-Boer, J. Bland-Hawthorn, G.F. Lewis, D. Yong, M. Asplund, R.R. Lane, L.L. Kiss, R. Ibata, Astrophys. J. **756**, 22 (2012)
M. Ness, K. Freeman, E. Athanassoula et al., Mon. Not. R. Astron. Soc. **432**, 2092 (2013a)
M. Ness, K. Freeman, E. Athanassoula et al., Mon. Not. R. Astron. Soc. **430**, 836 (2013b)
S. Nishiyama, T. Nagata, D. Baba, Y. Haba, R. Kadowaki, D. Kato, M. Kurita, C. Nagashima, T. Nagayama, Y. Murai, Y. Nakajima, M. Tamura, H. Nakaya, K. Sugitani, T. Naoi, N. Matsunaga, T. Tanabé, N. Kusakabe, S. Sato, Astrophys. J. Lett. **621**, 105 (2005)
S. Nishiyama, T. Nagata, N. Kusakabe et al., Astrophys. J. **638**, 839 (2006)
S. Nishiyama, T. Nagata, M. Tamura et al., Astrophys. J. **680**, 1174 (2008)
S. Nishiyama, M. Tamura, H. Hatano et al., Astrophys. J. **696**, 1407 (2009)
Z. Niu, B. Sun, J. Meng et al., Phys. Rev. C **80**, 6 (2009)
K. Nomoto, F.-K. Thielemann, J.C. Wheeler, Astrophys. J. Lett. **279**, 23 (1984)
D.J.K. O'Connell Ric. Astron. **5** (1958)
E.M. O'Malley, T. von Hippel, D.A. van Dyk, Astrophys. J. **775**, 1 (2013)
S. Ortolani, A. Renzini, R. Gilmozzi, G. Marconi, B. Barbuy, E. Bica, R.M. Rich, Nature **377**, 701 (1995)
W.H. Osborn, Yale University PhDT, Positions of Globular Cluster Stars in the Physical H-R Diagram (1971)
B.E.J. Pagel, L. Portinari, Mon. Not. R. Astron. Soc. **298**, 747 (1998)
P.A. Patsis, C. Skokos, E. Athanassoula, Mon. Not. R. Astron. Soc. **337**, 578 (2002)
M.S. Pawlowski, P. Kroupa, Mon. Not. R. Astron. Soc. **435**, 2116 (2013)
M.S. Pawlowski, P. Kroupa, Astrophys. J. **790**, 74 (2014)
M. Peimbert, A. Peimbert, L. Carigi, V. Luridiana, in *IAUS*, vol. 268 ed. by C. Charbonnel, M. Tosi, F. Primas, C. Chiappini (2010), p. 91
A. Pérez-Villegas, M. Portail, O. Gerhard, Mon. Not. R. Astron. Soc. **464**, 80 (2017)
S. Picaud, A.C. Robin, Astron. Astrophys. **428**, 891 (2004)
A. Pietrinferni, S. Cassisi, M. Salaris, F. Castelli, Astrophys. J. **612**, 168 (2004)
A. Pietrinferni, S. Cassisi, M. Salaris, S. Percival, J.W. Ferguson, Astrophys. J. **697**, 275 (2009)
P. Pietrukowicz, in *The General Assembly of Galaxy Halos: Structure, Origin and Evolution*. IAU Symp., vol. 317 (2016), p. 116
P. Pietrukowicz, A. Udalski, I. Soszyński, D.M. Nataf, Ł. Wyrzykowski et al., Astrophys. J. **750**, 169 (2012)
P. Pietrukowicz, S. Kozlowski, J. Skowron, I. Soszyński, A. Udalski, R. Poleski, Ł. Wyrzykowski, M.K. Szymański, G. Pietrzyński, K. Ulaczyk, P. Mróz, D.M. Skowron, M. Kubiak, Astrophys. J. **811**, 113 (2015)
C.A. Pilachowski, C. Sneden, G. Wallerstein, Astrophys. J. Suppl. Ser. **52**, 241 (1983)
G. Piotto, S. Villanova, L.R. Bedin et al., Astrophys. J. **621**, 777 (2005)
G. Piotto, L.R. Bedin, J. Anderson et al., Astrophys. J. Lett. **661**, 53 (2007)
Planck Collaboration, Astron. Astrophys. **580**, 22 (2015)
B. Plez, V. Hill, R. Cayrel et al., Astron. Astrophys. **428**, 9 (2004)
M. Portail, C. Wegg, O. Gerhard, I. Martinez-Valpuesta, Mon. Not. R. Astron. Soc. **448**, 713 (2015)
B.J. Pritzl, H.A. Smith, P.B. Stetson et al., Astron. J. **126**, 1381 (2003)
C.R. Proffitt, D.A. Vandenberg, Astrophys. J. Suppl. Ser. **77**, 473 (1991)
L. Pulone, G. De Marchi, F. Paresce, F. Allard, Astrophys. J. Lett. **492**, 41 (1998)
L. Pulone, G. De Marchi, F. Paresce, Astron. Astrophys. **342**, 440 (1999)
N. Raha, J.A. Sellwood, R.A. James, F.D. Kahn, Nature **352**, 411 (1991)
N.J. Rattenbury, S. Mao, T. Sumi, M.C. Smith, Mon. Not. R. Astron. Soc. **378**, 1064 (2007)
A. Renzini, F. Fusi Pecci, Annu. Rev. Astron. Astrophys. **26**, 199 (1988)
C. Reylé, A.C. Robin, M. Schultheis, S. Picaud, in *The Three-Dimensional Universe with Gaia*, ed. by E.C. Turon, K.S. O'Flaherty, M.A.C. Perryman. ESASP, vol. 576 (2005), p. 143
R.M. Rich, Astron. J. **95**, 828 (1988)
R.M. Rich, L. Origlia, Astrophys. J. **634**, 1293 (2005)

R.M. Rich, D.B. Reitzel, C.D. Howard, H. Zhao, Astrophys. J. **658**, 29 (2007)
R.M. Rich, L. Origlia, E. Valenti, Astrophys. J. **746**, 59 (2012)
H.B. Richer, J. Anderson, J. Brewer et al., Science **313**, 936 (2006)
H.B. Richer, R. Goldsbury, J. Heyl et al., Astrophys. J. **778**, 104 (2013)
A.G. Riess, L. Macri, S. Casertano et al., Astrophys. J. **730**, 119 (2011)
A.G. Riess et al., Astrophys. J. **826**, 56 (2016)
A.C. Robin, D.J. Marshall, M. Schultheis, C. Reylé, Astron. Astrophys. **538**, 106 (2012)
I.U. Roederer, K.-L. Kratz, A. Frebel et al., Astrophys. J. **698**, 1963 (2009)
A. Rojas-Arriagada, A. Recio-Blanco, V. Hill et al., Astron. Astrophys. **569**, 103 (2014)
M. Romero-Gómez, E. Athanassoula, T. Antoja, F. Figueras, Mon. Not. R. Astron. Soc. **418**, 1176 (2011)
A. Rosenberg, I. Saviane, G. Piotto, A. Aparicio, Astron. J. **118**, 2306 (1999)
P. Rosenfield, L. Girardi, B.F. Williams, Astrophys. J. **841**, 69 (2017)
R.K. Saito, M. Zoccali, A. McWilliam, D. Minniti, O.A. Gonzalez, V. Hill, Astron. J. **142**, 76 (2011)
R.K. Saito, M. Hempel, D. Minniti, P.W. Lucas, M. Rejkuba et al., Astron. Astrophys. **537**, A107 (2012)
M. Salaris, Astron. Nachr. **337**, 805 (2016)
M. Salaris, S. Cassisi, *Evolution of Stars and Stellar Populations* (Wiley, Chichester, 2005).
M. Salaris, S. Cassisi, Astron. Astrophys. **487**, 1075 (2008)
M. Salaris, L. Girardi, Mon. Not. R. Astron. Soc. **337**, 332 (2002)
M. Salaris, A. Weiss, Astron. Astrophys. **327**, 107 (1997)
M. Salaris, S. Cassisi, A. Pietrinferni, P.M. Kowalski, J. Isern, Astrophys. J. **716**, 1241 (2010)
L.V. Sales, M. Vogelsberger, S. Genel et al., Mon. Not. R. Astron. Soc. **447**, L6 (2015)
A. Sandage, Ric. Astron. **5**, 41 (1958)
A. Sarajedini, L.R. Bedin, B. Chaboyer et al., Astron. J. **133**, 1658 (2007)
A. Sarajedini, A. Dotter, A. Kirkpatrick, Astrophys. J. **698**, 1872 (2009a)
A. Sarajedini, C.L. Mancone, T.R. Lauer et al., Astron. J. **138**, 184 (2009b)
D. Saumon, M.S. Marley, Astrophys. J. **689**, 1327 (2008)
D. Saumon, P. Bergeron, J.I. Lunine, W.B. Hubbard, A. Burrows, Astrophys. J. **424**, 333 (1994)
L. Sbordone, M. Salaris, A. Weiss, S. Cassisi, Astron. Astrophys. **534**, 9 (2011)
H. Schatz, R. Toenjes, B. Pfeiffer et al., Astrophys. J. **579**, 626 (2002)
P.L. Schechter, M. Mateo, A. Saha, Publ. Astron. Soc. Pac. **105**, 1342 (1993)
E.F. Schlafly, D.P. Finkbeiner, Astrophys. J. **737**, 103 (2011)
K.C. Schlaufman, A.R. Casey, Astrophys. J. **797**, 13 (2014)
M. Schultheis, G. Zasowski, C. Allende Prieto et al., Astron. J. **148**, 24 (2014)
M. Schultheis, K. Cunha, G. Zasowski et al., Astron. Astrophys. **584**, 45 (2015)
M. Schwarzschild, Q. J. R. Astron. Soc. **11**, 12 (1970)
L. Searle, R. Zinn, Astrophys. J. **225**, 357 (1978)
J.A. Sellwood, R.H. Sanders, Mon. Not. R. Astron. Soc. **233**, 611 (1988)
B. Sesar, S.R. Banholzer, J.G. Cohen et al., Astrophys. J. **793**, 135 (2014)
J. Shen, R.M. Rich, J. Kormendy et al., Astrophys. J. **720**, L72 (2010)
M.H. Siegel, Astrophys. J. Lett. **649**, L83 (2006)
C. Siqueira Mello, M. Spite, B. Barbuy et al., Astron. Astrophys. **550**, 122 (2013)
R. Smolec, Acta Astron. **55**, 59 (2005)
C. Sneden, G.W. Preston, A. McWilliam, L. Searle, Astrophys. J. Lett. **431**, 27 (1994)
C. Sneden, J.J. Cowan, J.E. Lawler et al., Astrophys. J. **591**, 936 (2003)
C. Sneden, J.J. Cowan, R. Gallino, Annu. Rev. Astron. Astrophys. **46**, 241 (2008)
C. Sneden, J.E. Lawler, J.J. Cowan, I. Ivans, E.A. Den Harton, Astrophys. J. Suppl. Ser. **182**, 80 (2009)
Y. Sofue, Astrophys. J. Lett. **459**, 69 (1990)
Y. Sofue, Publ. Astron. Soc. Jpn. **65**, 118 (2013)
Y. Sofue, M. Honma, T. Omodaka, Publ. Astron. Soc. Jpn. **61**, 227 (2009)
I. Soszyński, W.A. Dziembowski, A. Udalski, R. Poleski, M.K. Szymański et al., Acta Astron. **61**, 1 (2011)
I. Soszyński, A. Udalski, M.K. Szymański, P. Pietrukowicz, P. Mróz et al., Acta Astron. **64**, 177 (2014)
K.Z. Stanek, M. Mateo, A. Udalski, M. Szymanski, J. Kaluzny, M. Kubiak, Astrophys. J. Lett. **429**, 73 (1994a)
K.Z. Stanek, M. Mateo, A. Udalski, M. Szymański, J. Kaluzny, M. Kubiak, Astrophys. J. **429**, L73 (1994b)
K.Z. Stanek, A. Udalski, M. Szymanski, J. Kaluzny, Z.M. Kubiak, M. Mateo, W. Krzeminski, Astrophys. J. **477**, 163 (1997)
P.B. Stetson, Publ. Astron. Soc. Pac. **112**, 925 (2000)
P.B. Stetson, V.F. Braga, M. Dall'Ora et al., Publ. Astron. Soc. Pac. **126**, 521 (2014)
N.B. Suntzeff, T.D. Kinman, R.P. Kraft, Astrophys. J. **367**, 528 (1991)
S.H. Suyu, M.W. Auger, S. Hilbert et al., Astrophys. J. **766**, 70 (2013)
A.V. Sweigart, L. Greggio, A. Renzini, Astrophys. J. **364**, 527 (1990)

A.V. Sweigart, T. Lanz, T.M. Brown, Astrophys. Space Sci. **291**, 367 (2004)
D.M. Terndrup, Astron. J. **96**, 884 (1988)
I.B. Thompson, J. Kaluzny, W. Pych et al., Astron. J. **121**, 3089 (2001)
E. Tognelli, P.G. Prada Moroni, S. Degl'Innocenti, Mon. Not. R. Astron. Soc. **454**, 4037 (2015)
J. Tumlinson, Astrophys. J. **708**, 1398 (2010)
P. Turri, A.W. McConnachie, P.B. Stetson, SPIE **9909**, 990907 (2016)
A. Udalski, M. Szymanski, J. Kaluzny, M. Kubiak, M. Mateo, Acta Astron. **42**, 253 (1992)
A. Udalski, K. Zebrun, M. Szymanski, M. Kubiak, G. Pietrzynski, I. Soszynski, P. Wozniak, Acta Astron. **50**, 1 (2000)
A. Udalski, M. Szymański, G. Szymański, Acta Astron. **65**, 1 (2015)
S. Uttenthaler, M. Schultheis, D.M. Nataf et al., Astron. Astrophys. **546**, 57 (2012)
A.A.R. Valcarce, M. Catelan, A.V. Sweigart, Astron. Astrophys. **547**, 5 (2012)
E. Valenti, F.R. Ferraro, L. Origlia, Mon. Not. R. Astron. Soc. **402**, 1729 (2010)
E. Valenti, M. Zoccali, A. Renzini, T.M. Brown, O.A. Gonzalez, D. Minniti, V.P. Debattista, L. Mayer, Astron. Astrophys. **559**, 98 (2013)
E. Valenti, M. Zoccali, O.A. Gonzalez, D. Minniti, J. Alonso-García, E. Marchetti, M. Hempel, A. Renzini, M. Rejkuba, Astron. Astrophys. **587**, L6 (2016)
E. Valenti et al., Astron. Astrophys. (2018). arXiv:1805.00275
G. Valle, M. Dell'Omodarme, P.G. Prada Moroni, S. Degl'Innocenti, Astron. Astrophys. **554**, 68 (2013)
A. Vallenari, S. Ragaini, G. Bertelli, in *Formation and Evolution of Galaxy Bulges*, ed. by M. Bureau, E. Athanassoula, B. Barbuy. IAU Sym., vol. 245 (2008), p. 371
S. van den Bergh, E. Herbst, Astron. J. **79**, 603 (1974)
H. van Gent, Bull. Astron. Inst. Neth. **6**, 163 (1932). http://adsabs.harvard.edu/abs/1932BAN.....6..163V
H. van Gent, Bull. Astron. Inst. Neth. **7**, 21 (1933). http://adsabs.harvard.edu/abs/1933BAN.....7...21V
D.A. VandenBerg, K. Brogaard, R. Leaman, L. Casagrande, Astrophys. J. **775**, 134 (2013)
D.A. VandenBerg, H.E. Bond, E.P. Nelan, P.E. Nissen, G.H. Schaefer, D. Harmer, Astrophys. J. **792**, 110 (2014a)
D.A. VandenBerg, P.A. Bergbusch, J.W. Ferguson, B. Edvardsson, Astrophys. J. **794**, 72 (2014b)
P. Ventura, F. D'Antona, I. Mazzitelli, R. Gratton, Astrophys. J. **550**, 65 (2001). https://doi.org/10.1086/319496
A.K. Vivas, K. Olsen, R. Blum et al., Astron. J. **151**, 118 (2016)
A.R. Walker, D.M. Terndrup, Astrophys. J. **378**, 119 (1991)
C. Wegg, O. Gerhard, Mon. Not. R. Astron. Soc. **450**, 4050 (2013)
C. Wegg, O. Gerhard, M. Portail, Mon. Not. R. Astron. Soc. **435**, 1874 (2015)
J.L. Weiland, R.G. Arendt, G.B. Berriman, E. Dwek, H.T. Freudenreich, M.G. Hauser, T. Kelsall, C.M. Lisse, M. Mitra, S.H. Moseley, N.P. Odegard, R.F. Silverberg, T.J. Sodroski, W.J. Spiesman, S.W. Stemwedel, Astrophys. J. Lett. **425**, 81 (1994)
J. Westin, C. Sneden, B. Gustafsson, J.J. Cowan, Astrophys. J. **530**, 783 (2000)
A.R. Wetzel, P.F. Hopkins, J-h. Kim et al., Astrophys. J. Lett. **827**, L23 (2016)
S.E. Woosley, T.A. Weaver, NASA STI/Recon Technical Report N, Black Holes (Astronomy), Stellar Evolution, Stellar Mass, Implosions, Thermonuclear Explosions, Nuclear Fusion, Stellar Activity, Supernovae, Supernova Remnants, 96 (1995)
M. Wyman, D.H. Rudd, R.A. Vanderveld, W. Hu, Phys. Rev. Lett. **112**, 051302 (2014)
R.F.G. Wyse, in *Liege International Astrophysical Colloquia*, vol. 35, ed. by A. Noels, P. Magain, D. Caro, E. Jehin, G. Parmentier, A.A. Thoul (2000), p. 305
A. Yoshino, T. Ichikawa, Publ. Astron. Soc. Jpn. **60**, 493 (2008)
N. Zacharias, C. Finch, J. Frouard (2017). arXiv:1702.05621
G. Zasowski, M.K. Ness, A.E. García Pérez, I. Martinez-Valpuesta, J.A. Johnson, S.R. Majewski, Astrophys. J. **832**, 132 (2016)
H. Zhao, D.N. Spergel, R.M. Rich, Astron. J. **108**, 2154 (1994)
R. Zinn, *The Globular Cluster-Galaxy Connection*, vol. 48 (1993), p. 38
R. Zinn, M.J. West, Astrophys. J. Suppl. Ser. **55**, 45 (1984)
R. Zinn, B. Horowitz, A.K. Vivas et al., Astrophys. J. **781**, 22 (2014)
M. Zoccali, in *Chemical Abundances in the Universe: Connecting First Stars to Planets*, ed. by K. Cunha, M. Spite, B. Barbuy. IAU Sym., vol. 265 (Cambridge Univ. Press, Cambridge, 2010), p. 265
M. Zoccali, E. Valenti, Publ. Astron. Soc. Aust. **33**, 25 (2016)
M. Zoccali, G. Piotto, S.R. Zaggia, M. Capaccioli, Astron. Astrophys. **331**, 541 (1998)
M. Zoccali, S. Cassisi, J.A. Frogel, A. Gould, S. Ortolani, A. Renzini, R.M. Rich, A.W. Stephens, Astrophys. J. **530**, 418 (2000)
M. Zoccali, A. Renzini, S. Ortolani, E. Bica, B. Barbuy, Astron. J. **121**, 2638 (2001a)
M. Zoccali, A. Renzini, S. Ortolani et al., Astrophys. J. **553**, 733 (2001b)

M. Zoccali, A. Renzini, S. Ortolani, L. Greggio, I. Saviane, S. Cassisi, M. Rejkuba, B. Barbuy, R.M. Rich, E. Bica, Astron. Astrophys. **399**, 931 (2003)
M. Zoccali, A. Lecureur, B. Barbuy et al., Astron. Astrophys. **457**, 1 (2006)
M. Zoccali, V. Hill, A. Lecureur, B. Barbuy, A. Renzini, D. Minniti, A. Gómez, S. Ortolani, Astron. Astrophys. **486**, 177 (2008)
M. Zoccali, O.A. Gonzalez, S. Vasquez et al., Astron. Astrophys. **562**, 66 (2014)
M. Zoccali, S. Vasquez, Gonzalez et al., Astron. Astrophys. **599**, 12 (2017)
M. Zorotovic, M. Catelan, H.A. Smith et al., Astron. J. **139**, 357 (2010)

Space Sci Rev (2018) 214:32
https://doi.org/10.1007/s11214-018-0466-9

Astronomical Distance Determination in the Space Age

Secondary Distance Indicators

Bożena Czerny[1] · Rachael Beaton[2] · Michał Bejger[3] · Edward Cackett[4] · Massimo Dall'Ora[5] · R.F.L. Holanda[6] · Joseph B. Jensen[7] · Saurabh W. Jha[8] · Elisabeta Lusso[9,10] · Takeo Minezaki[11] · Guido Risaliti[9] · Maurizio Salaris[12] · Silvia Toonen[13] · Yuzuru Yoshii[11,14,15]

Received: 3 July 2017 / Accepted: 3 January 2018 / Published online: 23 January 2018

Abstract The formal division of the distance indicators into primary and secondary leads to difficulties in description of methods which can actually be used in two ways: with, and without the support of the other methods for scaling. Thus instead of concentrating on the scaling requirement we concentrate on all methods of distance determination to extragalactic sources which are designated, at least formally, to use for individual sources. Among those, the Supernovae Ia is clearly the leader due to its enormous success in determina-

Astronomical Distance Determination in the Space Age
Edited by Richard de Grijs and Maurizio Falanga

✉ B. Czerny
bcz@cft.edu.pl

R.F.L. Holanda
holanda@uepb.edu.br

J.B. Jensen
jjensen@uvu.edu

T. Minezaki
minezaki@ioa.s.u-tokyo.ac.jp

M. Salaris
M.Salaris@ljmu.ac.uk

Y. Yoshii
yoshii@ioa.s.u-tokyo.ac.jp

1 Center for Theoretical Physics, Polish Academy of Sciences, Al. Lotnikow 32/46, 02-668 Warsaw, Poland

2 The Observatories of the Carnegie Institution of Washington, 813 Santa Barbara St., Pasadena, CA 91101, USA

3 Copernicus Astronomical Center, Polish Academy of Sciences, Bartycka 18, 00-716 Warsaw, Poland

4 Department of Physics & Astronomy, Wayne State University, 666 W. Hancock St., Detroit, MI 48201, USA

5 INAF, Osservatorio Astronomico di Capodimonte, via Moiariello 16, 80131 Napoli, Italy

6 Departamento de Física, Universidade Federal de Sergipe, 49100-000, Aracaju, SE, Brazil

tion of the expansion rate of the Universe. However, new methods are rapidly developing, and there is also a progress in more traditional methods. We give a general overview of the methods but we mostly concentrate on the most recent developments in each field, and future expectations.

Keywords Distance scale · Galaxies: distances and redshifts · Stars: supernovae

1 Introduction

Measurement of the distance to extragalactic objects is an important task. For objects at large distances the measurement of the redshift was frequently considered as the measurement of the distance but precision cosmology requires the knowledge of both the distance and the redshift.

The measurement of the distances to extragalactic objects was for years based on a carefully set subsequent steps of the cosmic ladder: direct distance measurement methods allowed to formulate scaling relations which then could be applied to more distant objects, and with a few such steps cosmological distances were achieved. The fundamental direct method of measuring the distance in astronomy is to measure the trigonometric parallax of a given source, reflecting the motion of the Earth around the Sun. Hipparcos mission measured parallaxes up to a few hundred parsecs, Hubble Space Telescope reached up to 5 kpc in dedicated observations. Now *Gaia* mission is collecting the data. It will cover with precisions of 10% a volume of radius 10 kpc, which encompasses $< 1\%$ of the stars in our Galaxy by number, but by volume is very large and will cover a large fraction of the Milky Way with reasonable accuracy. Parallaxes of the Cepheid stars, combined with measurements in nearby galaxies, calibrate the Leavitt Law for Cepheids, and Cepheids are accessible to 30 Mpc with the Hubble Space Telescope. Further out, SNe Ia calibrated with the help of the Cepheids take over, covering cosmological distances up to the redshift ~ 2. This method replaced older methods of covering most distant part of the ladder, like Tully-Fisher relation.

Modern astronomy still strongly relies on indirect measurements which in turn rely in some intermediate steps but in the aim to meet expectations of the precision cosmology the

[7] Utah Valley University, Orem, UT 84058, USA

[8] Department of Physics and Astronomy, The State University of New Jersey, Piscataway, NJ 08854, USA

[9] INAF, Arcetri Astrophysical Observatory, Largo E. Fermi 5, 50125 Firenze, Italy

[10] Centre for Extragalactic Astronomy, Department of Physics, Durham University, South Road, Durham, DH1 3LE, UK

[11] Institute of Astronomy, Graduate School of Science, The University of Tokyo, 2-21-1 Osawa, Mitaka, Tokyo 181-0015, Japan

[12] Astrophysics Research Institute, Liverpool John Moores University, IC2, Liverpool Science Park, 146 Brownlow Hill, Liverpool L3 5RF, UK

[13] Anton Pannekoek Institute for Astronomy, University of Amsterdam, 1090 GE Amsterdam, The Netherlands

[14] Steward Observatory, University of Arizona, 933 North Cherry Avenue, Room N204, Tucson, AZ 85721-0065, USA

[15] Institute for Cosmic Ray Research, The University of Tokyo, 5-1-5 Kashiwa-no-Ha, Kashiwa City, Chiba 277-8582, Japan

intermediate steps are refined, new shortcut developed and some of the basically indirect methods might be potentially used as direct methods under some physically well motivated assumptions. In the Local Group the use of the eclipsing binary stars allows to skip the step of the parallax measurement. A very important direct distance measurement is possible for the spatially resolved water maser in the galaxy NGC 4258. The impressive jump directly to very large distances opened with observations of gravitational lensing of quasars and the measurement of the time delays between the quasar images which allows their use as a ruler. Another direct method can be applied to galaxy clusters. This method uses the Sunyeav-Zeldovich effect. Extremely important direct methods are the Barion Acoustic Oscillation method and the observations of the fluctuations of the Cosmic Microwave Background. In both cases the (statistical) analysis of the sizes of the fluctuations allows to get directly the cosmological constraints. Another statistical method to obtain the cosmological constraints is gravitational lensing. Now, with firm detection of gravitational waves, the possibility of the direct determination of the distance to the gravitational wave source from the shape of its signal also gains importance. All these methods are described below.

We do not follow the strict definition of indirect method but in this section we concentrate on *predominantly* indirect methods which can be applied to *single specific extragalactic objects* like supernovae, galaxies, galaxy clusters, active galaxies or gamma-ray bursts. Global statistical methods, including the results based on the Cosmic Microwave Background, will be discussed in a separate Chapter. We also do not include in this Chapter strong gravitational lensing since it will be more natural to address this issue in the Chapter where weak lensing is discussed.

We outline the methods shortly, with references to the basic books or reviews, and we concentrate on most recent developments in the field. Whenever possible, we discuss the issue of the method cross-calibrations.

The order of the methods' presentation is roughly adjusted to their past and current popularity and role in astronomy, but not necessarily reflects their potential for development in the near future.

2 Supernovae Ia

Type Ia supernovae (SNe Ia) are defined by the lack of hydrogen and helium lines around the maximum light and the presence of strong SiII absorption features. Their peak absolute magnitudes are bright (M_B typically between -18.5 and -19.5 mag) and moreover for the majority of them (about 64% of all SN Ia events, denoted as 'normal', see Li et al. 2001) the light curves are remarkably similar, in the sense that they display a well defined relation between the peak absolute brightness and the width of their lightcurve (Phillips 1993). This characteristic makes SN Ia events very good cosmological distance indicators, and led to the discovery of the accelerated expansion of the universe (Riess et al. 1998a; Perlmutter et al. 1999a).

SNe Ia are intrinsically luminous (Riess et al. 2016), thus, even with modest aperture telescopes SNe Ia can be discovered over large volumes; indeed, the All Sky Survey for Supernova (ASAS-SN; Shappee et al. 2014a) uses telescopes with only a 14 cm aperture to discover a total of 242 confirmed SNe, of which 183 were SN Ia (76%) over its first 18 months of dual-hemisphere operations (01 May 2014 to 31 December 2015) alone within $m_V < 17$ (Holoien et al. 2017a).[1] Their intrinsic luminosity also permits their discovery

[1] See also Holoien et al. (2017d) and Holoien et al. (2017b) for the 2013-2014 and 2016 statistics, respectively.

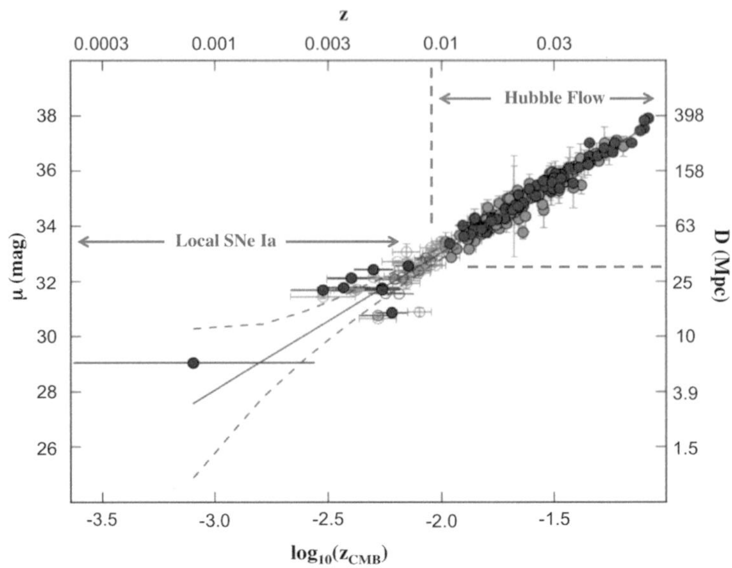

Fig. 1 Hubble diagram for SN Ia in the *Carnegie Supernova Project* (*dark grey*) and the CfA4 survey (*grey*). The scatter for the SN Ia in the 'Hubble Flow' sample ($z > 0.01$) is 0.12 mag or 6% in distance, which when averaged over the 200 SN Ia in the sample, results in a total uncertainty of 0.7% (Beaton et al. 2016). The effect of peculiar motions on the 'local' sample are evident by their large scatter. The 'local' sample is far less complete, with only eight galaxies in this visualization; Riess et al. (2016) brought another 11 SN Ia into the 'local' sample. (Adapted from Beaton et al. 2016)

at significant redshift; the previous most distant SN Ia SN UDS10Wil at $z = 1.914$, which was discovered in the Cosmic Assembly Near-infrared Deep Extragalactic Legacy Survey (CANDELS, PI: Faber & Ferguson; see Jones et al. 2013, for details), was recently replaced with another 15 in the range from $1.9 < z < 2.3$ presented in Riess et al. (2017). SNe Ia in the redshift range from $0.2 < z < 0.8$ are now routinely discovered and used for cosmological purposes; in the first two years of the ESSENCE project, 52 SNe Ia were discovered in this redshift range (Matheson et al. 2005).

SNe Ia, however, are not trivial objects to study and understand. Paramount among the difficulties of SNe Ia is that their progenitors and the exact physics of their ignition mechanisms are actively debated (see Sect. 2.7). Despite this, empirical measurements of the SNe Ia, as a class, show remarkable homogeneity and, due to their frequency and ubiquity, SNe Ia are likely to remain our tools for an en masse probe of the distant Universe. Figure 1 is a Hubble Diagram, a plot of distance versus redshift, for SNe Ia within $z < 0.1$ that demonstrates the remarkable precision ($\sigma_{\text{SNe Ia}} \sim 0.12$ mag or 6% in distance) of SNe Ia as a class. With a sample size of > 200 (and growing), this stage of the extra-galactic distance scale is, by far, the most well constrained and provides a random uncertainty of $\sim 0.7\%$ to the determination of the local expansion rate of the Universe (the Hubble constant or H_0) (for details see discussion in Beaton et al. 2016). In this Section, the key observational components and considerations required to construct the dataset in Fig. 1 are described.

2.1 Early History of SNe

Wood-Vasey et al. (2007) present a short contextual history of the role of SNe for cosmology that we will expand upon herein. Within the context of the 'Great Debate', Shapley (1919) argued that intrinsic brightness of SN 1885A found within the Andromeda galaxy of $M = -16$ mag was "out of the question" if Andromeda was, indeed, an 'extragalactic nebulae.' Thus, Shapley argued, the island universe hypothesis for the nature of the nebulae was likely incorrect owing to this extreme luminosity. In his landmark paper describing the Cepheid variables discovered in Andromeda between 1923–1928, (Hubble 1929) commented on the existence of "a mysterious class of exceptional novae which attain luminosities that are respectable fractions of the total luminosities of the systems in which they appear." Since Hubble's work provided strong evidence that the nebulae were, indeed, extra-galactic in nature (Code 1999), this new class of variables was later classified as 'super-novae' by Baade and Zwicky (1934) because the approximate amount of energy released by these objects over 25 days "is equivalent to 10^7 years of solar radiation of the present strength."

Minkowski (1941) was the first to apply a classification scheme to the supernovae, which was based on the presence or absence of hydrogen emission lines in the spectra (the type II and type I, designations respectively). As Minkowski notes, the spectra for SNe of type I seemed remarkably homogeneous in support of the postulation by Wilson (1939) that such objects could be employed for cosmological exploration. Indeed, the discovery and utilization of SNe for distances relies on their photometric properties, but the classification of SNe into phenomenological subtypes is largely reliant on their spectroscopy. The dual-need of spectroscopic *and* photometric classification is a challenge for the accumulations of the anticipated large samples to be discovered in the 2020's by large scale transient monitoring programs like the Large Synoptic Survey Telescope (LSST; see discussions in Matheson et al. 2013; LSST Science Collaboration 2009).

Moreover, with the advent of long-term spectroscopic and photometric monitoring of a large number of SNe Ia, there are now additional sub-classifications based on both spectroscopic and photometric diagnostics (see discussion in Parrent et al. 2014).

While both our theoretical understanding and experimental census of SNe Ia has increased dramatically, many of the original mysteries regarding the SNe Ia posed by Baade and Zwicky (1934), when their class was initially defined remain, as vibrant areas of research nearly a century later.

2.2 SN Ia as Standardizeable Candles

While their early homogeneity was suggestive of their utility as standard candles as early as Wilson (1939), the ability to be used as such occurred only in the last 30 years. Phillips (1993), expanding on earlier work by Pskovskii (1977), reported on an apparent empirical correlation between the absolute luminosity of a SN Ia and its the rate-of-decline in optical lightcurves (*BVI*). The decline rate is the change in brightness of a given band over the 15 days post maximum and is designated $\Delta m_{15}(\lambda)$; a typical decline rate is $\Delta m_{15}(\lambda) = 1$, but rates can vary between $0.9 < \Delta m_{15}(\lambda) < 1.8$ (Phillips 2012). The Δm_{15} parameter is relatively straight forward to measure with appropriate light curves and acts as a scale factor through which light curves of SN Ia are standardized and used as distance indicators. In works by Hamuy et al. (1995) and Riess et al. (1996), it was clear that SNe Ia corrected for their decline rate provided precise distances.

The physics underlying this relationship is that intrinsically brighter SNe Ia take longer to fade than intrinsically fainter SNe Ia and this is largely thought to be due to different amounts of ^{56}Ni contributing to the early time light curve, though the reason for this could be due to differences in either the total mass or its distribution in the explosion (see Kasen and Woosley 2007; Parrent et al. 2014, for detailed discussions). Recent energy has been focused on the near-infrared properties of SNe Ia, which show far more homogeneity than the optical and are less sensitive to dust (a detailed review is given by Phillips 2012). Indeed, in Fig. 1, the dark grey points from the Carnegie Supernova Project (CSP) show a tighter dispersion than those in grey from Harvard-Smithsonian Center for Astrophysics (CfA4), precisely because of the CSP use of near-infrared photometry.

Since the large scale demonstration of SNe Ia as standardizable candles, they have been studied at a number of regimes; these are: (i) local ($z < 0.03$), (ii) intermediate ($0.03 < z < 0.3$), (iii) distant ($z > 0.3$), and (iv) high-z ($z > 1.2$). The SN Ia discovered within each regime are used for different goals. Local SNe Ia can simultaneously constrain progenitor properties and serve as calibration sources. Intermediate SNe Ia are safely in the Hubble Flow and used to measure H_0. Distant and high-z samples are used to trace the acceleration of the Universe (w). Thus, large collaborations have been formed to find, study, and use SNe Ia and these generally separate into redshift (and brightness) regimes.

Amongst numerous efforts in the community, several of note are as follows: the CfA survey (Riess et al. 1999; Jha et al. 2006; Hicken et al. 2009), the Carnegie Supernova Project (CSP Hamuy et al. 2006; Freedman et al. 2009; Folatelli et al. 2010), ESSENCE (Miknaitis et al. 2007; Wood-Vasey et al. 2007; Foley et al. 2008, 2009), Higher-z Team and SHOES (Riess et al. 2004b,a, 2007, 2009), the Supernova Legacy Survey (SNLS Sullivan et al. 2011, and references therein), and the Foundation Survey within Pan-STARRS (e.g, Foley et al. 2017).

In addition to these largely ground based efforts, there are devoted space-based programs, one of note being the SWIFT Supernova Program that is capable of producing early- to late-time UV measurements useful for understanding the local SN Ia environs (Brown et al. 2015a).

A database of SNe discoveries (of all types), largely driven by transient monitoring programs, is maintained on the *Latest Supernovae* website (Gal-Yam et al. 2013),[2] which includes discovery imaging and classifications on a day-to-day basis. Individual teams maintain their own databases (both publicly available and proprietary), but the *Open Supernova Catalog*[3] provides a uniform access point for both original observations and derived parameters with extensive citations (Guillochon et al. 2016).

2.3 SN Ia as Cosmological Probes

The role of SNe Ia in the construction of our concordance cosmology is hard to understate. They provide the largest sample of tracers for measuring the Hubble constant (as demonstrated in Fig. 1) and their intrinsic luminosity gives us access to the evolution of the Hubble constant over time. The latter led to the discovery of the accelerating Universe (Perlmutter et al. 1997, 1998, 1999c; Riess et al. 1998b; Schmidt et al. 1998a). Of particular interest are the proceedings of the Centennial Symposium for the Carnegie Institution of Washington, entitled "Measuring and Modeling the Universe," which carefully reflects on the establishment of the concordance cosmology (Freedman 2004). More recent reviews for which SNe Ia play a central role are: the Hubble constant by Freedman and Madore (2010), for Dark Energy and the accelerating Universe by Frieman et al. (2008), and a prospectus for the long term impact on physics by Goobar and Leibundgut (2011). The hundreds of citations to these review articles indicate the speed at which these fields are evolving.

Indeed, in a span of less than 20 years the 'factor of two controversy' in the value of the Hubble constant, ultimately resolved with the *HST* Key Project (Freedman et al. 2001a),

[2] Available: http://www.rochesterastronomy.org/snimages/.

[3] Available: https://sne.space/.

has been recast as a 3σ discrepancy between the values measured from SNe Ia (e.g., by calibration of Fig. 1; Riess et al. 2011, 2016; Freedman et al. 2012) and that inferred from modeling anisotropies in the cosmic microwave background (most recently, Planck Collaboration 2016).

As is demonstrated by Beaton et al. (2016, their Fig. 1), as the uncertainties in either measurement technique have progressively decreased, the discrepancy between the two techniques has only grown. As described in Freedman (2017), the SN Ia derived cosmology 'is at a crossroads.' Thus, it is meaningful to understand the observational metrics through which we understand and characterize the SN Ia, which are driven by the wealth of data that can be collected in the near-field.

For the purposes of the distance scale, SNe Ia in the local and intermediate redshift regime are of the greatest interest, with measurements made on those samples ultimately informing their use at ever more distant regimes. Those samples used for cosmology ($z > 0.3$) ultimately drive a set of criteria that determine the "suitability" of a local or intermediate SN Ia for direct calibration. These restrictions are motivated by the desire to use the same intrinsic distance indicator in all regimes and, due to observability constraints, many of the fundamental parameters (such as the dispersion within a class of SN Ia sample) are only well-characterized in local or intermediate samples (to be discussed in a later section). The requirements from Riess et al. (2016) are: (i) observations taken on modern CCD detectors, (ii) not highly reddenned ($A_V < 0.5$), (iii) discovered before peak brightness, (iv) have 'typical' spectroscopic designations, and (v) relatively homogeneous photometric data (same filters, etc.).

2.4 SN 2017cbv: a Recent 'Local' SN Ia

On March 10, 2017, a transient source, DLT17u, was discovered in the nearby ($D =$ 16 Mpc), face-on barred spiral galaxy NGC 5643 within the context of the $D < 40$ Mpc (DLT40) one day cadence supernova search (Tartaglia et al. 2017). The top panels of Fig. 2 compare a pre- and post-discovery image of NGC 5643.

Within 20 hours, a spectrum was obtained for the source that classified it as a SN Ia from the characteristic broad emission lines (Hosseinzadeh et al. 2017). Comparison of the spectrum to a series of templates indicated this object was found at least 2 weeks before maximum light and detailed comparisons to template spectra within the Superfit code (Howell et al. 2005) confirmed this was young SN Ia. Based on this classification, the transient DLT17u was renamed to SN 2017cbv.

Additional careful photometric monitoring of the object began nearly immediately after its discovery (e.g., Coulter et al. 2017, among others) and, due to its brightness (it reached ~ 11.5 mag in V), both professional and amateur astronomers have participated in data collection. It is noteworthy that NGC 5643 was home to a SN Ia in 2013, SN 2013aa (Parrent et al. 2013), which was extensively monitored by the Carnegie Supernova Project (CSP Freedman et al. 2009; Hamuy et al. 2006). In contrast to SN 2017cbv, SN 2013aa was only classified 'a few days before maximum (Parrent et al. 2013), which limited follow-up observations by other teams.

In Fig. 3, the -11 day spectrum of SN 2017cbv is compared to the -1 day spectrum of SN 2013aa. There are marked differences between the spectral evolution over just this 10 day time span, but the broad, often asymmetric, spectral features are characteristic of SNe Ia. The ability to find these events early and begin the follow-up process not only aids in the use of a SN Ia for distance determination, but also in revealing their physical evolution. A detailed guide to optical spectral classification for all SNe is given in Filippenko (1997a)

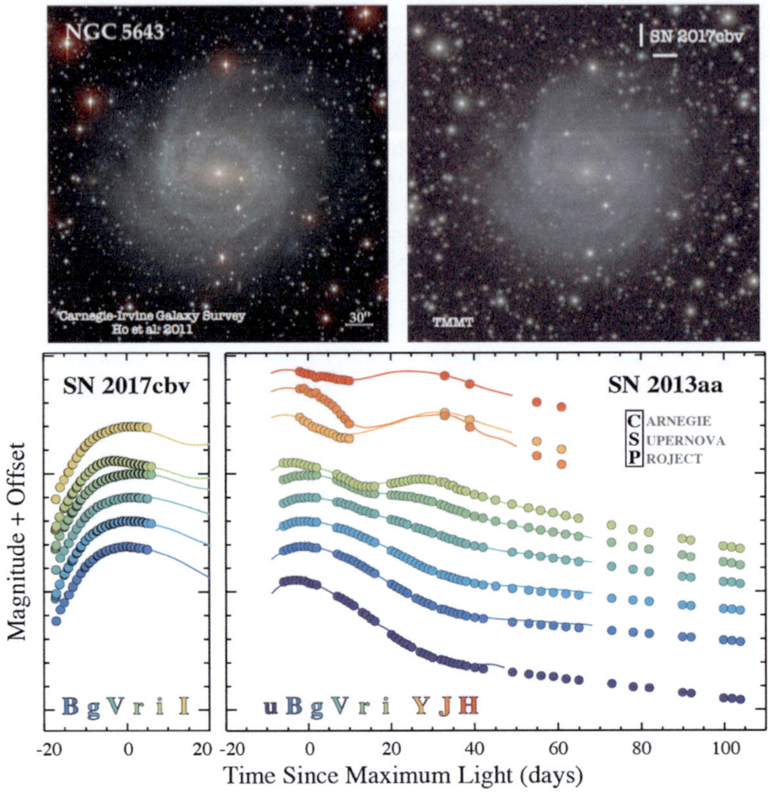

Fig. 2 Imaging data required to characterize SN Ia for use in a Hubble diagram. (*top panels*) Pre-imaging of NGC 5643 from the Carnegie-Irvine Galaxy Survey (Ho et al. 2011) as compared to a photometric monitoring image of NGC 5643 with SN 2017cbv indicated (imaging using the TMMT; Monson et al. 2017, A. Monson, priv. communication). (*bottom panels*) Daily photometric monitoring of SN 2017cbv (*left*) in six photometric bands from a pair of robotic telescopes at Las Campanas Observatory through maximum light. This exquisite pre-maximum photometry is enabled by dual-hemisphere monitoring programs and can be compared light curves from the previous SN Ia in NGC 5643, SN 2013aa (*right*). The nine-band photometric monitoring from the CSP permits detailed study of the the SN Ia, including direct fits for the extinction law local to the SN Ia. SN 2017cbv will be followed for a similar length of time as SN 2013aa from the ground and space. *Small ticks* correspond to a luminosity change by 1 mag. Light curves courtesy of A. Monson, M. Seibert, and C. Burns

and for SNe Ia by Parrent et al. (2014); in particular, Parrent et al. (2014) demonstrates the spectral evolution with time, for which the spectrum of SN 2017cbv represents one of the earliest ever observed.

The photometric monitoring data for SN 2017cbv and SN 2013aa are compared in the bottom panels of Fig. 2. Multi-band simultaneous light-curve fits from SNooPy (Burns et al. 2011, 2014) are also shown for both SNe Ia underneath the individual observations. The key difference is the (i) dense sampling permitted with robotic facilities and (ii) the >10 day difference in the starting point of the light curve. Utilizing a SN Ia for a distance measurement requires careful fitting of the time of maximum light, which is less ambiguously determined for SN 2017cbv than for SN 2013aa. Moreover, to simultaneously and self-consistently solve for both the total extinction (the combined foreground Galactic extinction and the internal host-galaxy extinction) and the reddening law (e.g., R_V) local to the SN Ia, simultaneous

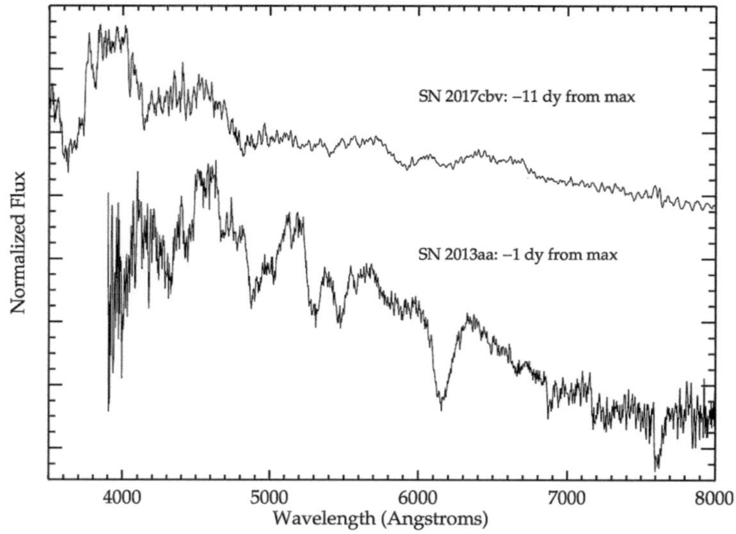

Fig. 3 Classification spectra for the SN Ia in NGC 5643. SN 2017cbv was classified at least −10 days from maximum and the same day of discovery (*top*). SN 2013aa was classified −1 day from maximum and three days after its discovery (*bottom*). These two spectra are demonstrative of the differences in SN Ia spectra on ∼ 10 day timescales. The spectrum for SN 2017cbv was retrieved from the Transient Name Server (https://wis-tns.weizmann.ac.il/object/2017cbv) and obtained by Hosseinzadeh et al. (2017). The spectrum for SN 2013aa was retrieved via the Latest Supernovae (Gal-Yam et al. 2013) compilation and obtained by Terry Bohlsen (http://users.northnet.com.au/~bohlsen/Nova/sn_2013aa.htm)

multi-band fitting is necessary. Preliminary fitting with SNooPy for both SNe Ia suggest that the total extinction for both objects is consistent with the foreground component; this is consistent with the location of either SN Ia in the outer disk of NGC 5643,

The light curves for SN 2013aa which are typical of the CSP program, show the differences between the light curves observed in blue and red wavelengths. At longer wavelengths (longward of i', there is a secondary maximum that occurs around 35 days. The strength and timing of the secondary maximum is correlated with the decline rate (with the fastest decliners showing no bump), which is suggestive that it has some physical connection with the primary maximum (see discussion in Phillips 2012, and their Fig. 1). The near-infrared properties of SNe Ia have revealed additional peculiar classes not apparently in the optical light-curves. For this reason, Phillips (2012) argues that the Y is well positioned to meet the complementary needs of signal-to-noise, insensitivity to dust, well-behaved decline-rate corrections, and simplicity of technology to provide the greatest leverage on the future cosmological measurements using SNe Ia.

The light curves for SN 2013aa also demonstrate the length of time for typical SN Ia follow-up. More specifically, to determine the stretch-factor for the Phillips-relation, Δm_{15}, observations must be sufficient to (i) determine maximum light and (ii) to follow the light curve beyond 15 days. Both SNe Ia display 'ideal' stretch factors of $\Delta m_{15} \sim 1$ and, thus, both SNe Ia are 'normal' and ideal for use as standard candles. As is demonstrated for SN 2013aa, the 'local' sample is typically followed with a dense cadence for ∼ 40 days (to detect the secondary maximum) and then a less dense cadence there after for as long as the SN Ia can be observed from the ground. When the SN Ia is no longer detectable from the ground, space-based photometry can be used to put further constraints on progenitor models; SN 2011fe in M 101 has been followed 1840 days post its B maximum (though on months long timescales) and observations are planned for at least another 600 days (Shappee et al. 2017).

The discussion of SN 2017cbv and SN 2013aa demonstrates the progress in the level of SN Ia observations in only a few years. Dual-hemisphere transient monitoring enabled by robotic, and largely autonomous, facilities permits early discovery of transient objects. A large-scale networked international community, enabled with state-of-the-art data processing tools, can classify a transient event within a day of its discovery. Classification, in turn, triggers numerous communities poised for appropriate follow-up, which for the use of SN Ia for distances requires photometric monitoring for up to 40 days post-maximum.

2.5 The 'Local' Sample in Context

The current limitation of SNe Ia as *absolute* distance indicators is the calibration of the absolute luminosity of SN Ia. We do not have theoretical predictions of their peak luminosity because there is still continuing uncertainty about their exact nature. Thus they have to be calibrated using other distance indicators, and this is one of the primary purposes for the modern distance ladder. SNe Ia are calibrated by Cepheid stars, using the Leavitt Law, which in turn have to be calibrated to an absolute scale. Currently, the Leavitt law is determined by using both Cepheids in the Milky Way and in the Small and Large Magellanic Clouds. Milky Way Cepheids have distances measured geometrically from their parallaxies, either by Hubble Space Telescope or Hipparcos (van Leeuwen et al. 2007), and new *Gaia* masurements, for a much larger number of stars (Gea 2017) are now being incorporated into the cosmic ladder. The Leavitt Law in the Large Magellanic Cloud is anchored by eclipsing binaries (e.g. Pietrzyński et al. 2013). The statistical error in the distance measurement of a single Cepheid was estimated to be around 0.3 mag (Riess et al. 2016).

However, as of Riess et al. (2016), there are in total of 19 SNe Ia host galaxies that can be calibrated via Cepheids and the Leavitt Law. Prior to 2016, there were only eight SN Ia calibrators, two of them the same SN Ia used in the *HST* Key Project (Freedman et al. 2001a). This implies that roughly ~ 1 SN Ia suitable for the calibration sample is discovered per annum, despite SNe Ia comprising 66% of the sample of all bright SNe (see Holoien et al. 2017a, and references therein). It is of interest to investigate what is limiting the build-up of the SN Ia calibration sample.

In Fig. 4a, the SNe Ia known from the 'Recent Supernova' database (Gal-Yam et al. 2013)[4] as of March 2016 are binned by the homogenized distance to their host galaxy from NASA Extragalactic Database (NED; Steer et al. 2017). This sample has been cleaned of SN Ia sub-types, but not by extinction (host galaxy or Galactic foreground) or decline rate. The average number of SNe Ia within a single 5 Mpc bin is ~ 20 SNe Ia, which suggests there are $5\times$ more SNe Ia within the volume accessible to *HST* Cepheid measurements than are currently in the calibration sample (using $\sim$ 35–40 Mpc as the limiting volume).

With the dramatic increase in efficiency for transient follow-up and classification described above, is it not unreasonable to attribute this large number entirely to recent discoveries. In Fig. 4b the 95 SNe Ia within 40 Mpc are binned by their discovery year from 1995 to 2016. For reference, the ASAS-SN all-sky project has been in operation since 2013 (Holoien et al. 2017d), but only reached peak efficiency in the latter half of 2015 (Holoien et al. 2017a), shortly before this data were compiled. From inspection of Figure 4b, the number of SNe Ia discovered per annum fluctuates widely, but, in general, the numbers do not trend strongly with the onset of all-sky monitoring programs (albeit these programs

[4] Available http://www.rochesterastronomy.org/supernova.html.

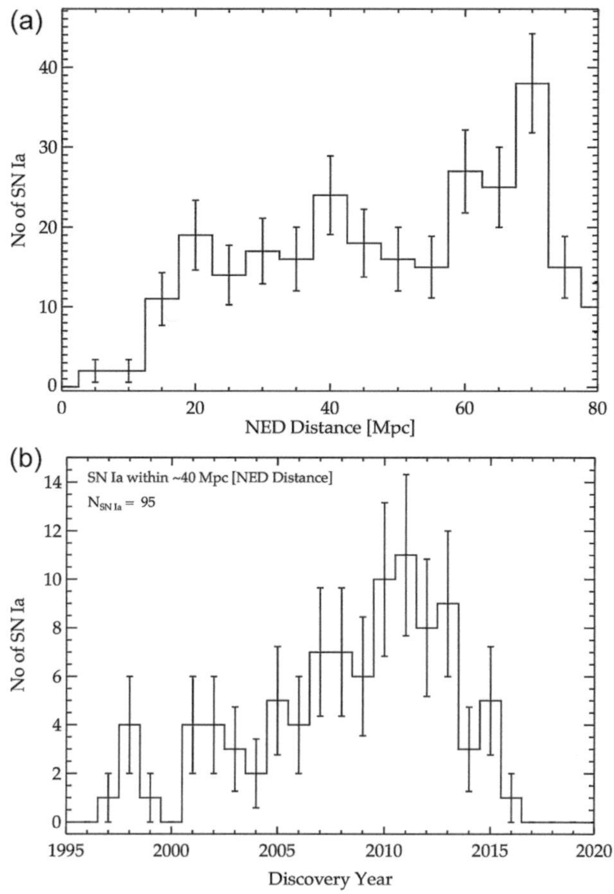

Fig. 4 'Local' SNe Ia demographics as of as of March 01, 2016. (**a**) Number of normal SNe Ia discovered since 1995 as a function of the NED Homogenized distance for their host. *Error bars* show the Poisson uncertainty for each bin. While there are few SNe Ia within 20 Mpc, the distribution beyond 20 Mpc is relatively flat. (**b**) Number of normal SNe Ia discovered per year since 1995 within 40 Mpc. There are a total of 95 SN Ia using these distances. This suggests that if primary distance techniques can push into this volume, a vastly larger number of SNe Ia can be independently calibrated to study the known biases in the SNe Ia population

permit unprecedented early detection; e.g., Fig. 2). Analyses of yearly rates over the next few years, however, will be more informative with programs like ASAS-SN having become dramatically more efficient (e.g., Holoien et al. 2017b).

Of the 95 SN Ia within 40 Mpc, nearly half (~ 40) have sufficient photometric data sufficient to be included in the calibration sample (e.g., similar in spirit to Fig. 2; C. Burns and B. Shappee, priv. communication). Despite having sufficient data to determine a SN Ia-based distance, these host galaxies lack robust independently determined distances from the Cepheid-based distance ladder. The reason for the lack of independent distances is the unsuitability of the host galaxy for Cepheid based distances, either due to its star formation history, an edge-on inclination, or its morphological type. Thus, utilization of a distance ladder constructed using a standard candle that can be applied to edge-on and non-star forming galaxies has the potential to dramatically improve the size of the calibration sample. Such a distance ladder is explored using old stellar populations in Beaton et al. (2016), and these distance methods are described elsewhere in this volume. As discussed in the following section, expanding the calibration sample to better reflect the demographics of SNe Ia in the larger, Hubble flow sample is an important step for further refinement of SNe Ia derived cosmological parameters.

In principle, the use of multiple calibrators for a single galaxy should help to control the systematic errors. For example, the use of galaxies where both SN I and II have been detected would be valuable, but unfortunately, there are only a few of such galaxies, and they lack distance determination based on the Tip of the Red Giant Branch (TRGB) and/or Cepheids. Probably, with the new facilities the number of galaxies with both SN types and a good distance determination will increase.

Apart from Cepheids, NGC 4258, an active galaxy with spatially resolved water maser emission is used as additional distance indicator since this galaxy is also a host for numerous Cepheids. A combination of four geometrical distance calibrators (Riess et al. 2016) allowed a determination of H_0 with 2.4% uncertainty, and this is the uncertainty currently underlying all cosmological measurements done with the use of SNe Ia as a systematic error.

2.6 In the Hubble Flow

Our discussion of the 'local' SNe Ia in NGC 5643 is characteristic of the attention received by these rare 'local' calibrators. The dramatically larger sample of SNe Ia in the Hubble Flow receive less attention in their data collection, but represent a much more complete sample from which to draw inferences on SN Ia as a population. More specifically, these are usually only monitored for ~ 20 days to obtain decline rates and may not have extensive multi-wavelength data or time-resolved spectroscopy.

There are three main sources of systematic uncertainty that hinder the use of SNe Ia and our understanding of the physical phenomena inferred from them (adapted from Hicken et al. 2009):

1. *Photometric accuracy*: large samples of SN Ia light curves are a largely heterogeneous ensemble of photometry coming from multiple programs operating with different sky-footprints and detectors.
2. *Host-galaxy reddening*: SN Ia are discovered in galaxies with a range of host morphologies and are found across these galaxies and thus span a large range of local conditions. Many SN Ia show anomalous R_λ extinction parameter, where $R_\lambda = A_\lambda / E_{B-V}$ is defined by the ratio of the total extinction to the color excess, but this could be attributed to many factors. Constraining the host-galaxy reddening is of paramount importance.
3. *SN Ia population differences and/or evolution*: the intrinsic variation of the SN Ia population as a function of the stellar populations of the host galaxy, etc.

The large scale programs described earlier are all aimed at addressing item (1). Item (2) will be discussed in detail later in this volume (see Sect. 2.8). The relevant results for item (3) are summarized in the points to follow.

Multi-SN Ia Systems There are only two current 'local' galaxies that are host to multiple SN Ia in the modern observation era. These are NGC 1316, a massive elliptical galaxy in the Fornax cluster, and NGC 5643, a spiral galaxy (discussed above). Stritzinger et al. (2010) used the four events in NGC 1316 and were able to determine that the three normal SNe Ia provided consistent distances at the 5% level and demonstrated that the fourth SN Ia, a fast-decliner, was discrepant at the 25–30% level. The two SNe Ia in NGC 5643 (Figs. 2 and 3) will be the first time a test of this nature can be performed on a late-type host. Moreover, both of these galaxies are sufficiently local that distances can be derived independent of SNe Ia to extend such tests in absolute terms (in lieu of differential analyses).

Intrinsic Scatter Using early data from the CSP, Folatelli et al. (2010) studied the scatter of the best studied SNe Ia in their sample; this set included 23 SNe Ia with multi-wavelength data similar to that presented for SN 2013aa in Fig. 2. A scatter of 0.11 mag (5%) in distance was reported. Folatelli et al. (2010) noted that the dispersion increased with decreasing distance, which implies that much of the observed dispersion could be accounted for with peculiar velocities. They discussed additional reasons for the observed trends, concluding that many observational and (potential) intrinsic effects could contribute to the scatter. A deeper analysis using a larger set of homogeneous data would be beneficial to understand the intrinsic scatter in SNe Ia magnitudes.

Host Mass Bias Sullivan et al. (2010) find a strong dependence on the mass of the host. Massive galaxies are systematically brighter by 0.06 to 0.09 mag (3% to 5% in distance; $> 3\sigma$ difference). Such effects will average out for cosmology if the distribution of host galaxy masses remains relatively constant in time, but it is unlikely that this is the case for distant SNe Ia samples.

Local Star Formation Bias Rigault (2013) found a dependence of derived SNe Ia magnitudes as a function of the local (1 kpc) environment around the SN Ia as probed by Hα surface brightness. SN Ia in passive (i.e., non-star forming) environments were found to be brighter by 0.094 ± 0.031 mag (5% in distance), which is termed the local star formation bias. Rigault et al. (2015) revisited this issue using both an independent sample of SN Ia (that of Hicken et al. 2009) and using *FUV* maps to trace the local region. Rigault et al. (2015) find that SN Ia in locally star-forming environments are 0.094 ± 0.025 mag (5% in distance) fainter; a 3.8σ difference. If the Hicken et al. (2009) sample is isolated to just the most massive hosts, nearly 50% of the SN Ia arise from locally passive environments. The local SN Ia calibration sample is almost entirely star-forming (Rigault et al. 2015; Riess et al. 2016). On the contrary, Jones et al. (2015) investigated this issue with a larger sample and found no evidence of a significant local star formation bias.

These effects are complicated and may bias attempts to better understand which galaxies produce SN Ia. More complex modeling as in Holoien et al. (2017c) could be of use to disentangle these effects in the general SN Ia population as well as the continued large-scale study of detailed SN Ia host demographics (e.g., Holoien et al. 2017b, and references therein).

Additional systematic error in SNe Ia measurements may come from weak lensing effect, particularly for the high redshift sources, above $z = 1.5$. Weak lensing causes additional dispersion in the measured luminosity distances, and the distances are on average effectively larger. Hilbert et al. (2011) estimated that this effect introduces systematic error of about 2% at redshift 1.5, but the appropriate simulations may allow to reduce this effect. This applies also to other measurement methods based on standard candle approach.

2.7 Progenitors

Despite their use as cosmological standard candles and their relevance also for galactic chemical evolution, consensus about the nature of SN Ia progenitor systems is still lacking, and this may affect their use as the tracers of the Universe expansion.

Any model for SN Ia progenitors must satisfy a number of observational constraints. The first obvious ones are the lack of hydrogen ($\Delta M_H < 0.01 M_\odot$) and helium in the spectra, the range of values of the peak luminosity and their correlation with the light curve width, and a total kinetic energy of the ejecta of $\sim 10^{51}$ erg. Our understanding of SN Ia lightcurves dictates that the peak luminosity is determined by the mass of radioactive ^{56}Ni, whilst the lightcurve width depend on the opacity, determined by iron-group elements like ^{56}Ni, ^{58}Ni, ^{54}Fe. These latter two elements are non-radioactive and do not contribute to the peak luminosity. The temporal evolution of the spectra also allows to reconstruct the radial composition of the ejecta. All these observations taken together imply the presence of ~ 0.1–$1.0 M_\odot$ of ^{56}Ni in the ejecta (but very little in the outer layers), ~ 0.2–$0.4 M_\odot$ of Si, S, Ca, Ar, and less than $0.1 M_\odot$ of n-rich isotopes (^{54}Fe, ^{58}Ni).

It was recognized very early on Hoyle and Fowler (1960) that SN Ia must be thermonuclear explosions in electron degenerate matter, pointing to exploding white dwarfs (WDs) as the cause of SN Ia events. Given that single WDs form with masses below the Chandrasekhar mass and are intrinsically stable objects, their evolution must be perturbed by a companion in order to cause the required explosion. Irrespective of the companion and the trigger mechanism, one can firstly investigate which WD mass+explosion mechanism can be invoked to explain 'normal' SN Ia observations (see, e.g., the review by Röpke et al. 2011). In terms of WD masses, one can divide the potential progenitors into two broad classes, namely Chandrasekhar and sub-Chandrasekhar mass progenitors.

In the Chandrasekhar mass scenario a C–O WD in a binary system approaches the Chandrasekhar mass due to accretion from the companion, and C-burning reactions set in at the centre when the density is of the order of 10^9 g cm^3. After a non-explosive initial 'simmering' phase (see, i.e., Lesaffre et al. 2006) lasting 1000–10000 yr, that burns only a small amount of C, a thermonuclear runaway starts. The number of ignition sparks of the burning front and their distribution are unknown and have a significant impact on the characteristics of the explosion. The best agreement with observations is achieved considering a combustion front propagating at first subsonically (deflagration) then supersonically (detonation—Blinnikov and Khokhlov 1986). This so-called 'delayed detonation' allows the production of a layer of intermediate-mass elements that encompasses almost all of the outer ejecta, as required by observations. The transition from deflagration to detonation is typically parametrized and not derived from first principles. Variations of the flame ignition configuration and transition to detonation can reproduce at least qualitatively the observed luminosity-decline rate relation (Kasen et al. 2009).

In the sub-Chandrasekhar mass scenario (see, i.e., Woosley and Weaver 1994) the helium layer that surrounds the CO core increases its mass due to accretion from the binary companion. When the mass of the He-layer reaches a critical value, a detonation is triggered by compression. This detonation burns the He-layer and drives a shock wave into the core, triggering a detonation of carbon and the explosion of the WD. In this scenario, a variation of the exploding WD mass could potentially explain the luminosity-decline rate relation. There are however problems related to the explosive nucleosynthesis, for not enough stable Fe and Ni isotopes to match late time spectra are produced (because of too low densities at explosion), and too much ^{56}Ni is made in the He-shell ashes (Röpke et al. 2011). If the mass of the He-layer that detonates is $\sim 0.01 M_\odot$, as possible for WDs around $1 M_\odot$ (Bildsten et al. 2007), negligible ^{56}Ni seems to be produced, although other calculations do not confirm this result (Röpke et al. 2011).

Both these WD mass+explosion mechanisms put forward to explain SN Ia events need the presence of a companion in a binary system. We can traditionally divide the progenitor systems into two broad classes, namely single-degenerate (SD) and double-degenerate (DD) systems.

In a typical SD system (Whelan and Iben 1973) the WD accretes H from a non electron-degenerate companion. Hydrogen is burned to He first, then to C, and the WD mass increases to the Chandrasekhar limit. As shown by Nomoto and Kondo (1991), stable H-burning is possible only for a finely tuned accretion rate of around $3 \times 10^{-7} M_{\odot}$ yr^{-1}. Lower rates ($< 10^{-8} M_{\odot}$ yr^{-1}) produce a Nova event, which expels the ashes of the accreted material and possibly even some of the original WD material (e.g. Yaron et al. 2005). This may not be the case at intermediate rates, where recurrent Novae systems retain some of the accreted mass (e.g. Hachisu and Kato 2000). If the accretion rate is too large, the WD is expected to expand to giant-like dimensions, and engulf the donor star in a common envelope which inhibits further accretion onto the WD. Hachisu et al. (1996) have proposed that an optically thick wind emerges from the WD that stabilizes the mass transfer. In this way the mass accretion may continue, but at a reduced rate. Generally, there is only a narrow range of accretion rates in which the WD can retain the matter and grow in mass.

If the non-degenerate companion is a He-star (produced by a previous common envelope episode) the WD can accrete He leading to either a Chandrasekhar or a sub-Chandrasekhar explosion (see, i.e., Woosley and Kasen 2011; Piersanti et al. 2014), depending on the accretion rate. Accretion rates of the order of $\approx 10^{-8}$–$10^{-9} M_{\odot}$ yr^{-1} lead to a He-detonation and a sub-Chandrasekhar explosion; in case of rates of the order of $\approx 10^{-6} M_{\odot}$ yr^{-1} steady He-burning on the WD surface can produce a Chandrasekhar mass WD that ignites C in the core.

In a DD system (Tutukov and Yungelson 1981) two WDs orbit each other and the system loses energy through gravitational waves emission. As the two components come closer and closer, eventually the least massive object fills its Roche-lobe and mass transfer sets in, with the more massive WD accreting C–O from an accretion disk/torus. If this WD can reach the Chandrasekhar mass a delayed-detonation will ensue. It is however uncertain whether this happens; even in case of a system where the combined mass of the two WDs exceeds the Chandrasekhar limit, an off-centre carbon deflagration followed by conversion to an O–Ne–Mg WD seems to be a likely outcome. Nearing the Chandrasekhar mass the O–Ne–Mg WD undergoes an electron induced collapse that leads to the formation of a neutron star. An alternative path that could lead to a SN Ia explosion is a so-called violent merger (see, i.e., Pakmor et al. 2010). At the merger, an accretion stream can dynamically produce a high temperature when hitting the WD surface. If high enough, this leads to the WD explosion according to the sub-Chandrasekhar double-detonation scenario. This scenario seems however efficient only for systems whose primary WDs are massive (> 0.8–$0.9 M_{\odot}$) and the mass ratio is high > 0.8. Pakmor et al. (2013) find that the presence of helium facilitates the ignition.

Despite the enormous efforts of the astronomical community, there has not yet been a convincing identification of a SN Ia progenitor (Li et al. 2011a; Nielsen et al. 2012). Furthermore, none of the proposed formation channels are clear matches to the observational constraints. Due to the uniformity and continuity of the observed SN Ia properties, there has been a long-time focus to find a single formation channel to explain all properties. However, from the diversity and correlations among the spectral properties of SNe Ia and their host galaxies, it is also possible that some combination of formation channels is operating.

An advantage to the SD channel is that it naturally explains the uniformity in luminosities as all WDs explode at the Chandrasekhar mass in this channel. On the other hand, the masses of the merged remnants in the DD channel display a broader range (~ 1.4–$2.0 M_{\odot}$). It is usually assumed that the DD systems with super-Chandrasekhar masses are responsible for the super-luminous SN Ia. Another expected signature of the SD channel, is that the companion is expected to survive the explosion and with an anomalous velocity, rotation, spectrum or composition. No such object has been identified unambiguously, for example in Tycho's SN remnant (Ruiz-Lapuente et al. 2004; Kerzendorf et al. 2009; Schaefer and Pagnotta 2012). Also, any mixing of the SN ejecta with that of the companion star during the explosion, has not been observed conclusively (Leonard 2007; García-Senz et al. 2012). In a minority of cases, variable NaD absorption has been detected in the SN spectra, which has been interpreted as circumstellar material from the companion in the SD channel (Patat et al. 2007; Sternberg et al. 2011). This signature, however, can also arise in the DD channel from a post-merger, pre-explosion wind (Shen et al. 2012). Furthermore, one expects soft X-ray emission from SD systems produced by the steady H-burning on the WD surface, but not enough emission has been observed from resolved nor unresolved sources (Gilfanov and Á Bogdán 2010; Di Stefano 2010). The emission might be shielded within optically thick outflows and reprocessed to UV-emission (Hachisu et al. 2010; Nielsen et al. 2013; Wheeler and Pooley 2013), but so far a population of these sources has not yet been detected (Woods and Gilfanov 2013; Lepo and van Kerkwijk 2013). Lastly, SN2011fe is a recent and close SN Ia that is well observed in optical, radio and X-ray wavelengths. These observations have ruled out most types of donor stars in the SD channel, and therefore favor a DD progenitor (e.g. Chomiuk 2013); see also Maoz et al. (2014) for a review on the observational clues to the SN Ia progenitor problem.

Additional constraints on the SN Ia progenitor channels are provided by the observed delay-time-distribution (DTD) of SN Ia events. The DTD is defined as the time interval between a star formation episode and the explosion of the related supernovae. It illustrates the distribution in evolutionary timescales of the SN Ia progenitor, which vary for the different progenitor scenarios. The DTD has been studied observationally in a range of surveys, environments and redshifts, and the emerging picture is remarkably coherent (for a review, see Maoz and Mannucci 2012). Type Ia SNe occur in young and in old stellar populations, reaching delays as long as a Hubble time. The SN Ia rate peaks at short delays of < 1 Gyr and declines at longer delay times. The DTD is best described by a power-law with an index of about -1 for delays of $1 < t < 10$ Gyr. At shorter delays, there is still some uncertainty regarding the precise shape of the DTD. Despite the consistency in the shape of the DTD from a variety of methods, the normalization or time-integrated DTD shows variations between rates in different galaxy types. SN Ia rates in galaxy clusters are found to be a factor ~ 5 higher than rates based on volumetric galaxy samples. Further research is needed to test if this is due to differences in the methods and samples, or if there is indeed an enhancement of the SN Ia rate in cluster environments (e.g. Maoz et al. 2012; Maoz and Graur 2017).

The DTDs for different progenitor channels have been estimated by analytical approaches and by modeling the evolution of binary populations (see Wang and Han 2012, for a review). For the double degenerate channel, there is a good agreement on the DTDs (e.g. Nelemans et al. 2013). The DD model gives rise to delay times ranging from a few Myr up to a Hubble time. Within these delay times, the DTD shape is a continuous power-law with a slope of roughly -1, which is remarkably similar to the observed DTD. The time-integrated DTD from the DD channel is compatible with the lower limit of the observed rate. The DTD from double WDs that undergo a violent merger is similar to that of the classical DD channel (Ruiter et al. 2013).

For the single degenerate channel, there is a large diversity in the predicted SN Ia rates spanning over several orders of magnitude (e.g. Nelemans et al. 2013). Differences arise due to ill-constrained aspects of binary formation and evolution, most importantly the accretion efficiency of WDs (Bours et al. 2013) and common-envelope evolution (e.g. Claeys et al. 2014, see Ivanova et al. 2013 for a review). The predicted rates of the SD channel tend to be (far) below that of the DD channel and the observed rate. Typically, the delay times from the SD channel range from a few-hundred Myr to a few Gyr. The models show that SD DTDs have a sharp drop after a few Gyr, in contradiction with the observed DTD. The drop can be understood by the limited range of donor masses (~ 2–$3 M_\odot$) that transfer matter to the WD at the necessary rates to ensue WD mass growth.

If the donor star is a He-star, the delay times are very short i.e. $\sim$ 50–200 Myr (Wang et al. 2009), due to the fact that He-stars evolve from more massive main-sequence stars with shorter evolutionary timescales than in the standard SD channel.

An adaptation to the SD channel has been proposed that resolved some of the observationally discrepant signatures. In the so-called 'spin-up/spin-down'-model (Justham 2011; Di Stefano et al. 2011), WD rotation due to accretion onto the WD is invoked to support against the collapse and ignition of the WD. The time that it takes for the WD to spin down and explode leads to a prolonged delay time. This may solve the issue that the classical SD progenitors do not have long delay times that are observed. Furthermore, if the timescale for spin down is sufficiently long, any traces of the mass transfer and even the companion could disappear. Lastly, the 'spin-up/spin-down'-model also allows for the possibility of super-Chandrasekhar SN Ia.

2.8 Extinction/Color Corrections for Type-Ia Supernovae

A common perception exists that the dominant parameter driving the standardization of SNe Ia for measuring distances is related to the decline-rate or light-curve shape (Phillips 1993; Riess et al. 1995; Hamuy et al. 1996; Perlmutter et al. 1997). However, corrections based on the color of the SN Ia are at least as important (Riess et al. 1996; Phillips et al. 1999). A variety of approaches has been used for this correction, and these differences ultimately relate to the cause of color variations in SN Ia, something that has been difficult to pin down.

All SN Ia luminosity distance measurements must correct for extinction by dust in the Milky Way, typically using the dust maps of Schlegel et al. (1998), or more recently, Schlafly and Finkbeiner (2011), to provide an estimate of the reddening $E(B-V)_{\mathrm{MW}}$ along the line of sight. This reddening is converted into an extinction in the observed passbands using the dust law of Cardelli et al. (1989), with updates from O'Donnell (1994) and Fitzpatrick and Massa (2007), typically assuming an extinction law parameter $R_V \approx 3.1$ as found in the diffuse ISM. Care must be taken to account for the time-evolving spectral energy distribution (SED) of the supernova. Uncertainties in the Milky Way extinction correction are correlated across all observations of an individual supernova, and systematic uncertainties in the Milky Way extinction correction can have important effects on parameter inferences from a supernova sample. Nonetheless, hereafter we discuss SN Ia extinction or color corrections under the assumption that the Milky Way component has been properly removed.

In analogy to Milky Way extinction, the light from distant SN Ia is extinguished by dust in the supernova host galaxy (which acts on the rest-frame light in contrast to the Milky Way extinction which acts after the light has been redshifted to the observer frame). One approach to this correction is to assume that the intrinsic colors (or SED) of a SN Ia are strictly determined by its light curve shape, i.e. that SN Ia are intrinsically a one-parameter family. Color variation among SN Ia is then ascribed to intrinsic stochasticity (random scatter around the nominal color) and extrinsic reddening by host-galaxy dust. This is the approach used by the Multicolor Light Curve Shape method (Riess et al. 1996, 1998b) and as extended to MLCS2k2 (Jha et al. 2007), as well as adaptations of the Δm_{15} method (Phillips et al. 1999; Burns et al. 2011). Even if the intrinsic SN Ia light curves are described with a more complex parameterization, extrinsic host-galaxy extinction can be treated separately, as in the BayeSN methodology (Mandel et al. 2011).

Applying this method gives a surprise: Hubble diagram residuals are minimized when the ratio between the inferred extinction A_V and color excess $E(B-V)$ has a value $R_V \approx$ 1–2, significantly lower than the canonical $R_V \approx 3.1$ (e.g., Conley et al. 2007). This could be interpreted as saying the typical dust in SN Ia host galaxies has smaller grains on average than Milky Way dust, with significantly more reddening for a given amount of extinction.

Indeed, we find strong evidence for "weird" low-R_V dust in heavily extinguished SN Ia (e.g., host reddening $E(B-V) \geq 1$ mag). This has been inferred using near-UV through near-infrared observations of reddened objects like SN 1999cl, 2002cv, 2003cg, 2006X, and SN 2014J (Elias-Rosa et al. 2006, 2008; Krisciunas et al. 2006, 2007; Wang et al. 2008b; Burns et al. 2014; Amanullah et al. 2014, 2015; Foley et al. 2014; Brown et al. 2015b; Gao et al. 2015). Independent, corroborating evidence comes from linear polarization measurements; dust scattering imprints a wavelength dependence to the continuum polarization fraction that reflects the grain size distribution and R_V. For several of these heavily-reddened objects, the wavelength of the continuum polarization peak implies $R_V \lesssim 2$ (Kawabata et al. 2014; Patat et al. 2015).

An intriguing possibility is that this strange dust is located near the supernova, in the circumstellar environment of the progenitor system, perhaps with multiple scattering playing a role (Wang 2005; Goobar 2008; Foley et al. 2014). This could help explain why such dust is not evidently seen in the Milky Way ISM, for example. Indeed, some of these heavily-reddened SN Ia show time-variable absorption lines (Na I D, K I) in high-resolution spectra, interpreted as arising from circumstellar gas interacting with the SN radiation field (Patat et al. 2007; Blondin et al. 2009; Graham et al. 2015).

However, the preponderance of the evidence suggests this low-R_V dust is *interstellar*. Many of these SN Ia show light echoes, with dust sheets >10 pc away from the supernova (i.e., not circumstellar Wang et al. 2008a; Crotts and Yourdon 2008; Crotts 2015; Maeda et al. 2015; Yang et al. 2017). Moreover, the dust seems to correlate best with diffuse interstellar bands (DIBs) rather than circumstellar gas absorption (Phillips et al. 2013) and moreover, for SN 2014J most of the gas absorption originates on interstellar, not circumstellar scales (Ritchey et al. 2015; Jack et al. 2015; Maeda et al. 2016).

Because of the difficulties in arriving at a physical understanding of the relationship between the observed luminosity and color of SN Ia, a leading approach is to make this an entirely empirical correction, analogous to the light curve shape correction. Tripp and Branch (1999) suggested a simple two-parameter empirical correction, with the SN Ia B-band luminosity linearly regressed against a light curve shape parameter and a color parameter. The SALT/SALT2 (Guy et al. 2005, 2007) and SiFTO (Conley et al. 2008) light curve fitters adopt this model. The single color coefficient (called β in these models) is again found to be significantly lower ($\beta \approx 2.5$) than what would be expected by standard dust ($R_V = 3.1 \leftrightarrow \beta = 4.1$).

These fitters are aimed primarily at cosmological SN Ia samples, i.e. objects typically with low extinction, not the heavily reddened SN Ia for which there is more direct evidence of unusual dust. Thus, it is not clear whether the low values of β in cosmological samples arise from the same kind of unusual dust (just less of it). For example, measurement uncertainties would lead to color "noise" that was uncorrelated to SN luminosity; if this noise level were comparable to the dust reddening (typically $E(B-V) \leq 0.3$ mag for these "cosmological" objects), the effective (observed) value of β would be lowered (della Valle and Panagia 1992). Similarly, if the SN Ia had intrinsic color variations, perhaps also correlated to luminosity with some $\beta_{\rm intrinsic}$, the effective β would be intermediate compared to $\beta_{\rm dust}$ (Mandel et al. 2016).

There is positive evidence that intrinsic color variations and the error model for SN Ia colors may be playing a role. By using spectroscopic indicators that are independent of color (line equivalent widths or velocities), it is possible to ascribe an intrinsic color to an individual SN Ia, and thus determine the color excess (observed minus intrinsic). Correlating the SN Ia luminosity against this color excess (rather than the observed color) yields an extinction law consistent with normal dust ($R_V \simeq 3$; $\beta \simeq 4$ Chotard et al. 2011; Foley and Kasen 2011; Foley et al. 2011; Sasdelli et al. 2016) for low-extinction objects. Similarly, an intrinsic variation "color-smearing" model, combined with a normal dust law, can explain the luminosity-color relation and its scatter better than a single linear model can (Scolnic et al. 2014; Mandel et al. 2016). Nonetheless, strange (strongly-reddening, low-R_V) dust is still necessary for highly-reddened SNe Ia (Mandel et al. 2011).

The emerging picture to explain the relationship between SN Ia luminosity and color has grown quite complex, including intrinsic color variations that may depend on light-curve shape, spectral features, or even host galaxy environment, normal Milky Way-like dust (in the SN host galaxy and in the Milky Way), and strange, likely interstellar but possibly also circumstellar, low-R_V host-galaxy dust. Achieving the most precise and accurate distances from SN Ia will require modeling and disentangling all of these effects, and perhaps more. These effects are likely to be more severe for high-redshift SN Ia science, for which there might be significant evolution relative to the Hubble-flow sample, and less severe in measuring H_0 from SNe Ia, where the calibrator sample is better-matched. For example, using SNe Ia as near-infrared standard candles, less sensitive to extinction corrections than in the optical, does not significantly changed the derived value of H_0 (Dhawan et al. 2017).

2.9 Summary

Future discoveries with SN Ia cosmology are now built on a framework of large-scale, homogeneous follow-up that unite teams that (i) find, (ii) classify, and then (iii) characterize the SN Ia. There are differences in the follow-up for the 'local' sample of SN Ia, which are ultimately used to set the SN Ia absolute luminosity, those SN Ia discovered in the Hubble Flow (to $z \sim 0.1$), which are used to determine the Hubble constant, and those found at higher redshifts, which are used for understanding the acceleration of the Universe. The low-redshift samples, however, remain the lens for understanding the physics of SN Ia and projecting that back in time; of key interest being how SN Ia properties may evolve with the decreasing progenitor metallicity over cosmic time and the apparent star-formation biases. Thus, further development of the 'local' sample and bolstering of the low-redshift sample remains as critical as the original questions posed by Shapley (1919) as to the size of the Universe.

3 Supernovae II

Type IIP SNe can be used as "standardized" candles to estimate distances well within the Hubble flow, with a rms precision of the order of 10–15%. This occurence makes them interesting cosmological distance indicators, since they are produced by different stellar populations than type Ia SNe, making them useful sanity check of the type Ia SNe-based results. More generally, it has demonstrated that all type II (IIP + IIL) SNe can be used as standardized candles (e.g., de Jaeger et al. 2015, 2017). Here the current status of the type IIP SNe standardized candle method is reviewed, pointing out the need for a calibration based on primary distance indicators.

3.1 Introduction

In the classical classification scheme (Filippenko 1997b), type IIP SNe are characterized spectroscopically by strong hydrogen emission features, with strong P-Cygni profiles; and

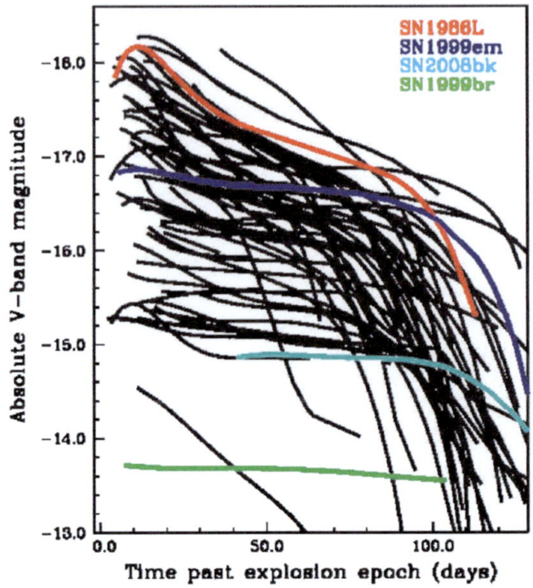

Fig. 5 Absolute V-band magnitudes of a sample of 60 type IIP SNe, as published in Anderson et al. (2014). Reproduced by kind permission of the authors

photometrically by a long plateau, lasting on average ~ 80 days, followed by a sudden drop in luminosity and a subsequent settlement on the radioactive tail.

From the physical point of view, both theoretical (e.g., Grassberg et al. 1971; Litvinova and Nadezhin 1983; Utrobin and Chugai 2008; Pumo and Zampieri 2011) and empirical investigations (e.g., Smartt et al. 2009) show that type IIP SNe are the product of a core collapse of a small to moderate massive progenitors, typically red supergiants (RSG). Interestingly, while on the basis of the empirical models (Heger et al. 2003; Walmswell and Eldridge 2012) we expect that type IIP SNe are the final fate of progenitors of masses between $8M_\odot$ and $30M_\odot$, empirical evidence found progenitors only in the range 8–$17M_\odot$. This discrepancy has been dubbed the "RSG problem" (Smartt 2009). However, some claims in the recent literature suggest higher mass limits (e.g., Dall'Ora et al. 2014), while independent studies of the massive star populations in the Local Group found RSGs with masses up to $\sim 25M_\odot$ (Massey et al. 2000, 2001).

The characteristic photometric plateau is supported by the hydrogen recombination front, which recedes in mass as the photosphere expands and cools down, producing a constant luminosity. As the density lowers down to values in correspondence of which the atmosphere becomes transparent, a sudden drop (~ 30 days) in luminosity of several magnitudes is observed, and the light curve is subsequently powered only by the radioactive decay of ^{56}Co to ^{56}Fe. In this phase, the SN luminosity depends on the amount of ^{56}Ni synthesized in the explosion (e.g., Weaver and Woosley 1980).

Since type IIP SNe are produced by a variety of progenitors, with different masses and chemical compositions, the observed features (luminosity at the plateau, length of the plateau, kinetic energy of the ejecta, amount of synthesized ^{56}Ni) can be very different from SN to SN (see Fig. 5, from Anderson et al. 2014, here reproduced by kind permission). Indeed, absolute plateau magnitudes range typically from $M_V = -15.5$ mag to $M_V = -18.5$ mag, initial velocity of the ejecta are of the order of $1\text{–}2 \times 10^4\ \mathrm{km\,s^{-1}}$, and initial photospheric temperatures are of usually around $1\text{–}2 \times 10^4$ K.

Nevertheless, interestingly and physically motivated relationships between observable quantities, such as between the linear radius (estimated from the velocity curve) and the angular radius (estimated by fitting a black body to the observed fluxes at different epochs); or between the luminosity at the middle plateau and expansion velocity, allow us to use them as "standardized" distance indicators. The first method is known as the expanding photosphere method (EPM, Kirshner and Kwan 1974), while the second has been dubbed the "standardized candle method" (SCM, Hamuy and Pinto 2002). Quite interestingly, the two methods can give consistent results, up to cosmological distances (e.g. Gall et al. 2016).

3.2 Physical Basis of the EPM and SCM Method

EPM is actually a variant of the Baade-Wesselink method, able to produce very accurate results (see Bose and Kumar 2014). The method requires the measurement of the temperature of the expanding stellar envelope, and of the envelope radius, which in turn comes from measurement of time since explosion and the expansion speed from the Doppler shift of the lines. Thus, it allows to measure the absolute luminosity, and finally to obtain the luminosity distance in a direct way. Type II SNe are intrinsicaly bright, so this method is able to provide distance estimates up to cosmological distances, independently from the adopted distance ladder, thus providing an independent check of the results obtained, for example, with the type Ia SNe, and it can be applied at any phase. However, it is observationally demanding, since it requires multi-band photometric data and good quality spectra. Moreover, some modeling is needed, especially in correctly estimating the dilution factor of SNe atmospheric models respect to a pure black body. A further improvement of the EPM is the spectral fitting expanding atmosphere method (SEAM, Mitchell et al. 2002), based on full NLTE atmospheric code. It can give quite accurate and precise results (e.g. Baron et al. 2004; Bose and Kumar 2014), but it is computationally intensive and requires high S/N spectra at early phases.

SCM was introduced by Hamuy and Pinto in 2002 (Hamuy and Pinto 2002, hereafter HP02) as an empirical correlation between the SN luminosity and the expansion velocity of the ejecta. They calibrated their luminosity-velocity relation at day 50 in the V and I photometric bands, which corresponds to a middle-plateau phase for most of the type IIP SNe. Subsequently, Kasen and Woosley (2009) provided theoretical basis on the SCM. As a matter of fact, SCM is a simple recasting of the Baade-Wesselink method. Indeed, since the expansion is homologous (and therefore the velocity is proportional to the radius), the luminosity L can be written as $L = 4\pi v_{ph}^2 t^2 \zeta^2 T_{ph}^4$, where v_{ph} and T_{ph} are the photospheric velocity and temperature, respectively; ζ is a dilution factor, which accounts for the departure from a perfect blackbody, while t is the reference epoch. Now, t can be arbitrarily chosen (50 day by construction), and T_{ph} is a good proxy of the temperature T_H of the hydrogen recombination, $T_h \approx 6000$ K, nearly a constant along the plateau. Finally, the dilution factor ζ can be estimated from NLTE models, but it is a strong function of the luminosity, and it can be absorbed in the exponent. It should be explicitly noted that SCM does not need to be applied necessarily at day 50, and that similar relations are valid all along the plateau. However, at epochs earlier than day 30 the ejecta temperature is too high, and probably the approximation $T_{ph} \sim T_h$ is not valid. Moreover, the atmospheric velocity curve is rapidly changing during the first 40–50 days, and a 10–15 days uncertainty in the explosion epoch can reflect in a substantial bias in the velocity curve.

3.3 Current Calibrations: an Overview

The first calibration was given by HP02 that, on the basis of 17 literature type IIP SNe, 8 of which well embedded in the Hubble flow, derived a relation in the V and the I band as a function of the photospheric velocity and of the redshift. Their calibration showed a scatter of the order of 9%, comparable with the precision of 7%, typical of type Ia SNe. However, an estimate of the absorption is needed, and this is usually a thorny problem when dealing with SNe. The problem was faced by Hamuy (2003), where the absorption was estimated on the basis of observed $(V - I)$ colors. Indeed, when it is assumed that the intrinsic end-of-the-plateau $(V - I)$ color is the same for all the type IIP SNe and it is a function of the photospheric temperature only, a possible $(V - I)$ color excess is due to the host galaxy extinction. The underlying physical assumption is that in type IIP SNe the opacity is mainly caused by the e^- scattering, so that they reach the same hydrogen recombination temperature as they evolve. However, some discrepancies are obtained, probably due to metallicity variations from one SN to the other. Moreover, as pointed out by Nugent et al. (2006), a color-based extinction correction is impractical for faint (i.e. basically distant) SNe, since it would require a continuous monitoring to catch the end of the plateau, before the luminosity drop.

A subsequent calibration was then proposed by Nugent et al. (2006) (N06), where an extinction correction was determined from the rest-frame $(V - I)$ color at day 50, adopting a color-stretch relationship, as done in Ia SNe studies. A standard $R_V = 3.1$ dust law was used. Moreover, since at moderate redshifts the weak Fe II $\lambda 5169$ could be hardly measured (because they can be redshifted into the OH forest), they explored the practicality of stronger lines, such as the Hβ. Also, they derived an useful empirical relation to scale the observed Fe II $\lambda 5169$ velocity at a given epoch, to the reference +50 day. They obtained the first Hubble diagram at cosmologically relevant redshifts ($z \sim 0.3$) with a rms scatter in distance of 13%, which is comparable with the rms scatter obtained with Ia SNe.

Poznanski and coworkers (Poznanski et al. 2009, hereafter P09) used a fitting method similar to those adopted by N06, but taking the extinction law as a free paramenter. This procedure yielded a mild total-to-selective absorption ratio $R_V = 1.5$ and, after discarding a few outliers with faster decline rates, they finally obtained a scatter of 10% in distance.

Subsequently, Olivares et al. (2010) adopted as a reference epoch a "custom" −30 day from the half of the luminosity drop, to take into account the different length of the plateau phase from SN to SN. By allowing R_V to vary (and confirming a low $R_V = 1.4 \pm 0.1$), they found again that SCM can deliver Hubble diagrams with rms down to 6%–9%. Interestingly, after calibrating their Hubble diagrams with the Cepheid distances to SN 1999em (Leonard et al. 2003) and SN 2004dj (Freedman et al. 2001b), they obtained a Hubble constant in the range 62–105 $\mathrm{km\,s^{-1}\,Mpc^{-1}}$, but with an average value of $69 \pm 16\ \mathrm{km\,s^{-1}\,Mpc^{-1}}$ in the V-band, and similar values in the B and I-bands. The large scatter reflects the fact that only two calibrating SNe were employed, but the average value is very close to our most precise estimate of the Hubble constant, $H_0 = 73.24 \pm 1.74\ \mathrm{km\,s^{-1}\,Mpc^{-1}}$ (Riess et al. 2016).

D'Andrea et al. (2010) adopted K-corrections to determine rest-frame magnitudes at day 50 for 15 SDSS II SNe, spanning a redshift range between $z = 0.015$ and $z = 0.12$. They also took into account the rest-frame epoch $50(1 + z)$. Their best-fit parameters differed significantly from those obtained by P09, and they attributed the discrepancy to the fact that their SNe sample could be intrinsically brighter than those of P09. Moreover, they concluded that a major source of systematic uncertainty in their analysis was probably due to the difficulty of accurately measuring the velocity of the Fe II $\lambda 5169$ feature, and to the extrapolation of the velocity measured at early epochs to later phases. Finally, they warned that the template database should be extended, in order to perform a reliable K-correction.

Maguire et al. (2010) extended the SCM to the near-infrared bands, since at those wavelengths both the extinction and the number of spectral lines are lower. The latter aspect implies that NIR magnitudes are less affected by differences in strength and width of the lines (i.e. less sensitive to metallicity effects), from SN to SN. Even though their adopted sample contained only 12 SNe, they demonstrated that using *JHK* magnitudes it is possible to reduce the scatter in the Hubble diagram down to 0.1–0.15 mag, with the error in the expansion velocity being the major source of uncertainty.

3.4 Discussion and Final Remarks

The extragalactic distance scale up to cosmological distances is intimately connected with type Ia SNe, and through type Ia SNe the acceleration of the Universe was discovered (Perlmutter et al. 1999b; Riess et al. 1998b; Schmidt et al. 1998b). At the present time, current facilities allow us to detect and study type Ia SNe up to $z \sim 1.9$ (Rubin et al. 2013; Jones et al. 2013), and recently up to 2.3 (Riess et al. 2017), while the next generation of extremely large telescopes will allow us to study type Ia SNe up to $z \sim 4$ (Hook 2013). At high z, however, the number of type Ia SNe may significantly decrease, due to the long lifetimes of their progenitors. Alternatively, the ubiquitous type II (core-collapse) SNe could be an appealing choice to probe further cosmological distances. Moreover, since type II SNe are produced essentially by young stellar populations, they may constitute a more homogeneous sample, than type Ia SNe, with respect to the age of the stellar population. However, it should be noted that they are sgnificantly fainter, and that their study could be more difficult, since they may explode in younger and dustier regions, and this especially holds at cosmological distances, in a general younger environment. On the other side, they are expected to be more abundant per unit volume (Cappellaro et al. 2005; Hopkins and Beacom 2006).

All the current calibrations of SCM basically rely on a sample of type IIP SNe spanned in a range in z, for which magnitudes and expansion velocities were available. The major uncertainties are:

- explosion epoch: unless a very early detection and very good sampling of the phenomenon is available, since the method requires a common reference epoch (day +50, but other choices are allowed, see Olivares et al. 2010), an uncertainty in the explosion epoch reflects in a scatter in the derived calibration, especially because the expansion velocity rapidly changes during the first ≈ 30 days;
- reddening correction: only for bright SNe explosions in nearby galaxies sound estimates of the local reddening are available, for example *via* the NaI D. In all the other cases, we must rely on color corrections whose precision is of the order of 0.4 mag in $(V - I)$ (Olivares et al. 2010). However, this problem has a much lower impact in the NIR bands;
- velocity of the ejecta: the Fe II $\lambda 5169$ can be difficult to be measured, especially for faint and/or distant SNe, and for the bright SNe the typical uncertainties are of the order of $150\ \mathrm{km\,s^{-1}}$, that is uncertainties of the order of 2–3% in the measure (at day 50). Given the typical slopes of the SCM, of order of 5–6 in the I-band, the final contribution to the error budget is of the order of 10–20%;
- K-corrections: when calibrating the SCM in a given photometric band using SNe at different z, it should be noted that the photons collected in the observed band, for a given distant SN, actually come from a different wavelength. This means that we are using photons coming from *intrinsically different* forms of the SCM, with different slopes. Therefore, if K-corrections are not applied, the observed relationships are likely to be affected by a larger scatter. Moreover, since the published calibrations are based on SNe spanning different ranges in z, this could also affect the coefficients. However, D'Andrea et al. (2010) correctly adopted K-corrections.

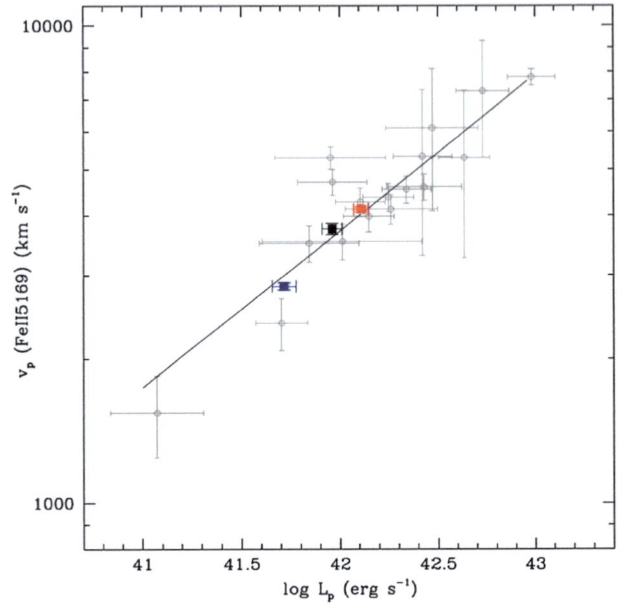

Fig. 6 Original HP02 plane, with the positions of the homogeneously studied SNe 2012A (*blue*), 2012aw (*red*) and 2012ec (*black*). Plot from Barbarino et al. (2015), reproduced by kind permission

It should be noted that, when nearby SNe are used, with a tight sampling of their evolution and a homogeneous technique of analysis is employed, the observed scatter of the SCM would be greatly reduced, as suggested in Barbarino et al. (2015). In Fig. 6 we reproduce their Fig. 20, where the position of the SNe 2012A, 2012aw and 2012ec is shown in the original Hamuy and Pinto plane.

At the present time, we still lack a homogeneous and sound calibration of the SCM based only on the primary distance indicators and, on the other side, we have only a few numbers of host galaxies where type IIP SNe have been exploded and also we detected Cepheids and/or the TRGB. A first step toward this direction was provided by Hamuy (2004) and by Jang and Lee (2014), where they adopted the calibrations of the SCM published by HP02 and Olivares et al. (2010), respectively, with the distances provided by Cepheids and TRGB, to estimate the Hubble constant. More recently, Polshaw et al. (2015) applied the SCM to SN 2014bc, exploded in the anchor galaxy NGC 4258, for which a geometric maser distance (with an uncertainty of only 3%, Humphreys et al. 2013) and a Cepheids distance (Fiorentino et al. 2013) is available. They applied almost all the currently available calibrations of the SCM to SN 2014bc, and compared the estimated distances with both the maser and the Cepheids distances. They found some discrepancies between the SCM-based distance moduli and the maser distance modulus, ranging from −0.38 mag to 0.31 mag. To further investigate the scatter among the available calibrations, they applied the SCM to a set of 6 type IIP SNe occurred in galaxies for which Cepheid distances were available. They obtained a Hubble diagram in the I-band with a small scatter ($\sigma_I \sim 0.16$ mag), and the following SCM calibration:

$$\begin{aligned} I_{50} - A_I + 5.665(\pm 0.487)\log(v_{50}/5000) \\ = 5\log(H_0 D) - 2.045(\pm 0.137) \end{aligned} \quad (1)$$

where $H_0 = 73.8\ \mathrm{km\,s^{-1}\,Mpc^{-1}}$ (Riess et al. 2011), and D is the distance.

This calibration relies on the cosmic distance ladder, even if based on only a few objects. However, the derived SCM is based on Cepheid distances based on *different* calibrations of the Cepheid period-luminosity relations. The differences among the various calibrations are typically of the order of 0.1 mag (e.g. Fiorentino et al. 2013, and references therein). Moreover, also the adopted reddenings came from different sources, even from different calibrations of the same NaI D feature. To this aspect, we point out that current calibrations of the NaI D feature in the same galaxy may provide differences up to $\Delta E(B-V)\sim 0.15$ mag (see the discussion in Dall'Ora et al. 2014).

A further development of a calibration of the SCM based on primary distance indicators (Cepheids, TRGB) is highly desirable, with a larger number of calibrators and with a *homogeneous* analysis (i.e. same estimate of the reddening and same Cepheids period-luminosity relations and TRGB calibrations). Moreover, since the SNe calibrated on primary distance indicators occur in the local Universe, they are likely to be deeply investigated. Moreover, the progenitor could be detectable on archive images. This would allow us to fully explore the space of the structural parameters that could affect the SCM. Indeed, very recently SN LSQ13fn (Polshaw et al. 2016) was found to break the standardized candle relation. A possible explanation for that could be the low metallicity of the progenitor ($\sim 0.1 Z_{\odot}$, Polshaw et al. 2016). As a matter of fact, theoretical models (Kasen and Woosley 2009) predict a metallicity dependence, but at the 0.1 mag level, not as large as observed in the case of SN LSQ13fn (almost 2 mag). However, a possible explanation could be a combined effect of low-metallicity of the ejecta and a strong circumstellar interactions. Whatever the case, detailed studies of nearby type IIP SNe, spanning a range of masses, metallicities and environments are of extreme importance to fully characterize the SCM.

As a final point we note that future facilities, such as E-ELT and NGST, will allow us to extend the range of the cosmological type IIP SNe, on which SCM could be applied, but also the range of *local* SNe, to calibrate the SCM. However, these forthcoming facilities will operate at the NIR wavelengths, therefore making it essential a sound NIR calibration of the SCM.

4 Tully-Fisher and Faber-Jackson Methods

Tully-Fisher method is a historically important method based on the empirical relation between the intrinsic luminosity of a spiral galaxy and the emission line width (Tully and Fisher 1977). This method opened a way to measure distances and proper motions for spiral galaxies and clusters of galaxies. The method was calibrated using Cepheid stars in nearby spiral galaxies, and then could be easily extended up to $\sim$ 100 Mpc. For over 20 years this was the most popular method for probing this distance range, but with invent of new methods, particularly SN Ia, it's role in cosmology strongly diminished due to considerable intrinsic scatter. This scatter can be only partially reduced when replacing the optical luminosity with IR emission, molecular hydrogen or CO. Thus, in recent years the Tully-Fisher relation was rather used to probe the mass to luminosity (M/L) variation in galaxies with known distances, rather than as a distance estimator in itself (e.g. Davis et al. 2016). In principle, once a reference sample for each morphological type is determined then one could use it to estimate distances in the usual way, but at this stage it is not yet expected to give reliable results.

Like a Tully-Fisher, Faber-Jackson relation (Faber and Jackson 1976) is an empirical relation between the intrinsic luminosity and the stellar velocity dispersion in the elliptical galaxies. However, this method has vary large intrinsic scatter although it was based on some pre-selection of morphological types, and the actual relation between the galaxy kinematics and morphology is rather complex (see e.g. Cortese et al. 2016). Nevertheless, more general version of this relation known as Fundamental plane (Dressler et al. 1987) can be applied to determine the distances in large sample of galaxies (e.g. da Cunha et al. 2017).

5 Surface Brightness Fluctuations

In 1988, Tonry and Schneider developed a new technique for measuring extragalactic distances based on the spatial luminosity variations in early-type galaxies (Tonry and Schneider 1988). The method, known as Surface Brightness Fluctuations (SBF) works because the variation in brightness from pixel-to-pixel varies as the square root of the number of stars per pixel, and thus galaxies at larger distances will have smaller variations, appearing smoother, than nearby galaxies (see Blakeslee 2013, for a recent review of the SBF technique and distance measurements).

Making accurate SBF measurements requires a solid anchor to the distance ladder for calibration. In principle, one could determine the SBF distance calibration based on theoretical modeling of stellar populations; in practice, it is more common to adopt an empirical calibration, using Cepheid distances to set the zero point, and the observed variation of SBF magnitude with color to calibrate stellar population effects. In addition to measuring distances, SBF has also been used to explore the properties of unresolved stellar populations in galaxies with known distances, which is valuable in constraining stellar population models (see Jensen et al. 2015, for details).

The ground-based SBF technique was initially calibrated for use at optical wavelengths (I and z) and in the near-IR (J, H, and K-bands). At I, the effects of age and metallicity on SBF brightness are largely degenerate, making it possible to calibrate distances for a wide variety of early-type galaxies using a single broadband color. The near-IR bands exhibit more scatter in SBF as a function of galaxy color, but the fluctuations themselves are much brighter because the stellar light in old populations is dominated by red giant branch stars, which are brighter in the near-IR. Even though the IR background is much higher, the brighter fluctuations and better seeing (from the ground) typically make it possible for IR SBF to reach much greater distances.

The first ground-based SBF surveys were largely limited to distances of about 20 Mpc (e.g., Tonry et al. 2001). During the last decade, new instruments on the *Hubble Space Telescope* (*HST*) have enabled us to achieve unprecedented precision with the SBF method and push to much larger distances, thanks to their high spatial resolution, point spread function (PSF) stability, low background levels, and relatively wide fields of view. The Advanced Camera for Surveys (ACS) was used to conduct extensive surveys of the Virgo and Fornax clusters, from which a calibration of the z-band SBF distance method with statistical scatter of 0.08 mag was achieved, corresponding to 4% in distance (Mei et al. 2007; Blakeslee et al. 2009). This puts SBF on par with the most accurate extragalactic distance indicators, including Type Ia supernovae (SNe) and Cepheids.

Jensen et al. (2015) have recently established a new SBF calibration for the F110W and F160W filters (J and H bands) of the WFC3/IR camera, which can routinely measure SBF distances to 80 Mpc in a single *HST* orbit. The new IR SBF calibration is based on the ACS SBF distances to Virgo and Fornax galaxies, and is tied to the Cepheid distance scale. They find a statistical scatter of 0.1 mag (5% in distance) per galaxy for redder ellipticals, with greater variation in bluer and lower-luminosity galaxies. Comparison with stellar population models implies that redder ellipticals contain old, metal-rich populations, as expected, and that bluer dwarf ellipticals contain a wider range of stellar population ages and lower metallicities, with the youngest populations near their centers. IR color gradients appear to be closely related to age, so IR SBF distance measurements are best limited to the reddest and oldest high mass elliptical galaxies.

A team of astronomers is now using WFC3/IR (PI J. Blakeslee) to measure IR SBF distances to a sample of 34 high-mass early-type galaxies in the MASSIVE survey (Ma et al. 2014). The goal of the MASSIVE survey is to better understand the structure and dynamics of the 100 most massive galaxies within $\sim$ 100 Mpc using a wide array of imaging and spectroscopic techniques. Of particular interest is measuring the masses of the central supermassive black holes in these galaxies, for which accurate distances are necessary. The IR SBF distances will also remove peculiar velocity errors and better constrain cosmic flows within 100 Mpc.

The power of SBF as a tool for cosmology is now being established with a new *HST* program to measure IR SBF distances to a collection of early-type supernova host galaxies (PI P. Milne). The goal of this project is to reduce systematic uncertainties in Type Ia SNe luminosities and explore possible environmental effects on the brightnesses of Ia SNe that are typically calibrated locally using Cepheids in late-type spirals, but are more often observed in early-type galaxies at high redshift. IR SBF is the only method that reaches large enough distances to observe the host galaxies and measure their distances with the requisite precision.

Accomplishing the goals of these projects relies on efficient, high-accuracy distance measurement that is currently only possible with *HST* resolution, and can be accomplished most efficiently with WFC3/IR. The future of IR SBF is not limited to *HST*, however. New AO systems on large telescopes such as the multi-conjugate AO system "GeMS" on the Gemini-South telescope provide a highly stable and uniform PSF over a wide ($\sim$ 2 arcmin^2) field of view. Initial GeMS observations of three galaxies have demonstrated that the SBF signal can be measured with high fidelity in modest exposure times out to 100 Mpc using $\sim$ 0.08 arcsec FWHM K-band images.

There are plans to continue to develop the AO SBF techniques with the expectation that the next generation of large telescope (e.g., TMT, GMT, and E-ELT) with wide-field AO systems will make reliable IR SBF distance measurements out to several hundred Mpc. The James Webb Space Telescope also has great potential to push the IR SBF technique to distances of perhaps 500 Mpc.

6 Active Galactic Nuclei

Active Galactic Nuclei (AGN) are not the objects frequently discussed in the context of distance measurements. However, an excellent example of the water maser shows their current importance, and the continuously increasing AGN samples and our understanding of those objects open new possibilities.

6.1 Water Masers

The discovery of a water maser at 22.23508 GHz in the Seyfert II galaxy NGC 4258 (Miyoshi et al. 1995; Herrnstein et al. 2005) offers an unprecedented possibility to measure the distance to this galaxy directly through geometrical methods, without any need for intermediate steps, and with high accuracy.

This coherent emission forms due to collisional excitation (for emission mechanism, see e.g. Elitzur 1992; Lo 2005), the emitted frequency corresponds to the transition between rotational energy levels in the water molecule. The emission may form in the medium with the temperature about 300–600 K, and particle number densities 10^8–10^{11} cm^{-3}. Its narrow band emission traces precisely the dynamics of the material through the line shifts.

In general, water maser emission can form in molecular clouds, comets, planetary atmospheres, stellar atmospheres, and in the case of distant galaxies can be mostly found in starburst regions. The maser emission in NGC 4258 is different, it originates in a Keplerian slightly warped disk surrounding the central black hole and it is strong enough that it can be mapped at sub-milli-arc-second resolution by Very Long Baseline Interferometry (VLBI), providing a powerful tool to probe spatial and kinematic distribution of the molecular gas at distances below 1 pc from the central black hole. The resolved image of the emission allows for the measurement of the spacial distribution of the emission, and the wavelength shifts give the velocity, and it was shown already by Miyoshi et al. (1995) that the measured radial profile corresponds to the Keplerian motion. Further measurement of the proper motion, or acceleration, thus opens a way to the measurement of the geometrical distance. Subsequent long-term monitoring of this source (Argon et al. 2007; Humphreys et al. 2013) allowed to obtain the distance to NGC 4258 with the accurracy of 3%. New analysis of the same data by Riess et al. (2016) gave the distance measurement to this source:

$$D(\text{NGC 4258}) = 7.54 \pm 0.17(\text{random}) \pm 0.10(\text{systematic})\ \text{Mpc}, \tag{2}$$

or, equivalently, magnitude distance of 29.39 ± 0.06 mag. Here the statistical error is the value coming directly from the Monte Carlo Markov Chain code, used for fitting the parameters, and the statistical error comes from the same code run for various initial values. This result, combined with a large number of Cepheid stars discovered in NGC 4258 (Fausnaugh et al. 2015) and in several hosts of SN Ia, opened a way to determine the local value of the Hubble constant $H_0 = 72.25 \pm 2.38$ km s^{-1} Mpc^{-1} with 2.4% accuracy (Riess et al. 2016).

The water maser in NGC 4258 is not the only one detected in an active galaxy, and not even the first one. The first water maser was detected in 1979 in NGC 4945 (Dos Santos and Lepine 1979), and a number of water masers are known till now, although none of those sources have such unprecedented data quality from the point of view of a distance measurement. However, statistical use of numerous sources (Megamaser Cosmology Project; Braatz et al. 2015) is expected to provide a measurement precision of $\sim 5\%$ on the Hubble constant.

6.2 Continuum Reverberation

It is widely thought that accretion onto the supermassive black holes in bright AGN takes place via an optically thick, geometry thin accretion disk described by Shakura and Sunyaev (1973). Each annulus within such a disk will emit like a blackbody with temperature T at radius R. An accretion disk dominated by heating from viscous dissipation which is around a black hole of mass M accreting at a rate $\dot{M}$, has a temperature profile given by

$$T(R) = \left(\frac{3GM\dot{M}}{8\pi R^3 \sigma} \right)^{1/4}, \tag{3}$$

when the radius is much greater than the innermost stable circular orbit. The disk's absolute flux is then given by summing up blackbodies, B_ν, over all disk annuli:

$$f_\nu(\lambda) = \int B_\nu\big(T(R), \lambda\big)\frac{2\pi R dR \cos i}{D^2}$$
$$= \left(\frac{1200G^2h}{\pi^9c^2}\right)^{1/3}\frac{\cos i\,(M\dot{M})^{2/3}\nu^{1/3}}{D^2}. \quad (4)$$

Therefore, if the temperature profile of the disk can be measured, the observed and expected fluxes can be compared in order to determine the distance. The key question, then, is how to determine the temperature profile of the disk. The accretion disk, expected to be a few light days across, is too small to be spatially resolved, and thus indirect techniques must be used. Reverberation mapping (Blandford and McKee 1982; Peterson 2014a) uses light travel time to measure spatial separations within a distant accretion flow. Much in the same way the continuum variability studies of AGN provide a means to probe the accretion disk around the central supermassive black hole and allow measurement of the temperature profile of the disk (for a detailed description see Collier et al. 1999; Cackett et al. 2007). High energy X-ray/EUV photons produced close to the compact object irradiate the surrounding gas which reprocesses this into UV/optical continuum, with the hotter inner regions emitting mainly UV photons and seeing variations in the irradiating flux before the cooler outer regions, which emit mainly optical photons. As the ionizing radiation varies erratically, so do the reprocessed components but with time delays due to light travel time within the system—the light travel time from source to reprocessing site to observer is longer than that on the direct path from source to observer. These observable delays provide indirect information on the size and structure of the surrounding accretion flows. This technique is a powerful probe of accretion flows in AGN.

Thermal radiation from a disk annulus at temperature $T(R)$ emerges with a range of wavelengths, $\lambda \sim hc/kT(R)$. Roughly speaking, each wavelength picks out a different temperature zone and the time delay $\tau = R/c$ measures the corresponding radius. Thus, in this reprocessing scenario the accretion disk reverberates and we expect to observe correlated variability between all continuum bands, with the shortest wavelength lightcurves varying first.

More specifically, the observed delays between different continuum wavelengths depend on the disk's radial temperature distribution $T(R)$, its accretion rate, and the mass of the central black hole. A disk surface with $T \propto R^{-b}$ will reverberate with a delay spectrum $\tau \propto \lambda^{-1/b}$. For the temperature distribution of a steady-state externally irradiated disk, $T(R) \propto R^{-3/4}$ (as given above), thus the wavelength-dependent continuum time delays should follow

$$\tau = \frac{R}{c} \propto (M\dot{M})^{1/3}T^{-4/3} \propto (M\dot{M})^{1/3}\lambda^{4/3}. \quad (5)$$

Accretion disk reverberation therefore allows the measurement of the disk temperature profile, and so can be used to determine the AGN distance (Collier et al. 1999) via:

$$D = 3.3\left(\frac{\tau}{\text{days}}\right)\left(\frac{\lambda}{10^4\,\text{Å}}\right)^{-3/2}\left(\frac{f_\nu/\cos i}{\text{Jy}}\right)^{-1/2}\,\text{Mpc}. \quad (6)$$

Here, f_ν must be the accretion disk flux, which can be obtained by taking difference spectra to isolate the *variable* component of AGN light.

Over the past 20 years many studies have searched for the expected signatures and found good correlations between different UV and optical wavelength-bands, but the expected interband lags were never of high significance (e.g., Edelson et al. 1996; Wanders et al. 1997; Collier et al. 1998, 2001; Sergeev et al. 2005). Importantly, though, Sergeev et al.

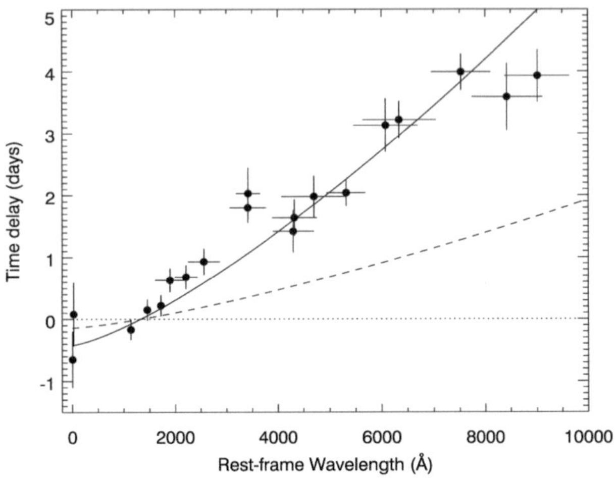

Fig. 7 Time delay versus rest-wavelength for NGC 5548 (adapted from Fausnaugh et al. 2016). The *solid line* shows the best-fitting $\lambda^{4/3}$ relation, while the *dotted line* shows the expected lags from a thin disk accreting at 10% of the Eddington rate

(2005) found that in all the 14 nearby Seyferts in their sample the time-delays are present and show an increase in time delay with increasing wavelength, and the delays increase with greater absolute luminosity of the AGN as predicted by reprocessed disk models. The first attempt to apply this disk reverberation method to determine an AGN distance yielded an H_0 value a factor of 1.66 smaller than any of the currently considered values (Collier et al. 1999). Cackett et al. (2007) applied this model to fit the lags and fluxes from all 14 AGN in the Sergeev et al. sample, but again found a value of H_0 a factor of 1.6 too small. One interpretation of this result is that the accretion disks are, on average, a factor of 1.6 larger than predicted by the standard model.

Over the last couple of years, significant progress has been made in better understanding the wavelength-dependent lags. Much improved sampling cadence and long baseline campaigns have allowed for significant lags from X-rays through to the near-IR to be measured in two objects, NGC 2617 and NGC 5548 (Shappee et al. 2014b; McHardy et al. 2014; Edelson et al. 2015; Fausnaugh et al. 2016). Both are consistent with the $\tau \propto \lambda^{4/3}$ relation, however, the lags are a factor of $\sim$3 larger than expected based on the standard disk model (McHardy et al. 2014; Edelson et al. 2015; Fausnaugh et al. 2016). Figure 7 shows the wavelength-dependent lags from NGC 5548.

The challenge now lies in understanding the discrepancy with the disk model. Strong emission lines present in broadband filters can act to lengthen the measured continuum lag, since the broad emission line lag is longer than the continuum lag (Chelouche and Zucker 2013). However, in NGC 5548 this does not appear to be a large effect (Fausnaugh et al. 2016). Furthermore, continuum emission from diffuse gas in the broad line region will also act to lengthen the lag (Korista and Goad 2001), and hints of this effect are present in the NGC 5548 lags, where the u-band (where the Balmer diffuse continuum will peak) lag is an outlier in the otherwise smooth wavelength-dependent lags (Edelson et al. 2015; Fausnaugh et al. 2016). Alternatively, our understanding of disk accretion is incomplete, and other, more complex geometries occur (e.g. Gardner and Done 2016).

Future intense monitoring campaigns will help better understand these differences with the standard disk model. If these discrepancies can be understood, then this method of measuring distances has the potential to be very powerful given the AGN monitoring that will take place in the LSST-era.

6.3 BLR Reverberation

Intense emission lines, most notably Balmer lines, with kinematic width of order of thousands km/s are the most characteristic features of AGN. The variability of the emission line intensity has been noticed already by Andrillat and Souffrin (1968), and the response of the lines to the variable intrinsic continuum opened a way to reverberation mapping of the Broad Line Region (e.g., Cherepashchuk and Lyutyi 1973; Gaskell and Sparke 1986; Kaspi et al. 2000; for a review, see Peterson 2014b; Bentz 2015). In the simplest approach, the time delay between the lines and the continuum measures the size of the BLR. Subsequent studies showed that the BLR clouds are predominantly in the Keplerian motion so their orbital velocity, measured through the line width, combined with the orbital radius allows to measure the black hole mass through the virial theorem. This application firmly established the importance of the BLR reverberation for cosmology.

The next important step has been made by Watson et al. (2011) who suggested that BLR reverberation measurement can be also used to determine the distance to the source. AGN are clearly not standard candles, as their intrinsic luminosities span orders of magnitutes, but the reverberation studies showed a tight relation between the delay of the emission line (mosty Hβ) and the intrinsic luminosity flux (e.g., Peterson et al. 1999; Kaspi et al. 2000), most frequently measured at 5100 Å since the majority of the reverberation studies were done for low redshift AGN in the optical band. Line delay can be interpreted as a mean/effective radius of the BLR region. Thus this relation, after careful subtraction of the host galaxy reads

$$\log R_{BLR} = K + \alpha \log L_{44}, \quad \text{[light days]} \tag{7}$$

where L_{44} is the monochromatic λL_λ flux expressed in units of 10^{44} erg s^{-1}, and the coefficient values from the sample version *Clean* given in Tab. 14 of Bentz et al. (2013) are $K = 1.555 \pm 0.024$, $\alpha = 0.542 \pm 0.027$. The directly measured quantities are the line delay, τ, and the observed flux, F_λ so the value $\tau/F_\lambda^{1/2}$ provides the distance indicator (Watson et al. 2011).

The dispersion in the delay–luminosity relation is only 0.13 dex, as measured by Bentz et al. (2013) in *Clean* variant. Thus, with a large number of objects and broader coverage of the redshift range the method offers a very interesting alternative to SNe Ia.

The unique aspect of the AGN Hubble diagram is that while SN distances are observationally restricted now to less than $z \sim 1.9$ (Riess et al. 2001; Jones et al. 2013), AGN diagram can be extended to much larger redshifts thus covering the full range of distances with a single method. The additional advantage of the use of quasars is that those objects, on average, are not strongly affected by extinction (they clean efficiently their environment), and they do not show significant evolution of the metallicity with redshift so in principle no hidden evolutionary bias should be present in their BLR properties across the redshift space.

The number of AGN studied in the context of reverberation mapping is not yet large, about 60 objects were monitored in Hβ line (see Du et al. 2015, 2016a, for a recent compilation), and only a handful of sources have other line delays measured. The most distant quasar with tentative time delay, measured for CIV line, is still S5 0836+71 at $z = 2.172$ studied through 7-year monitoring by Kaspi et al. (2007). However, many monitoring programs are under way (e.g., Czerny et al. 2013; King et al. 2015; Lira et al. 2016). The use of other lines than Hβ has the advantage that the measured time delay is then considerably shorter which is important for high z quasars, where the expected time delay is long due to large black hole mass as well as to the $(1 + z)$ scaling of the intrinsic timescale to the observed one.

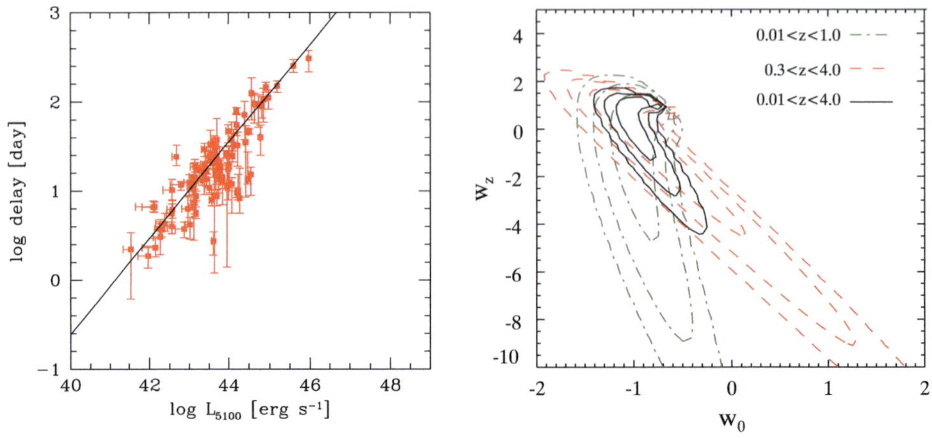

Fig. 8 *Left panel*: the time delay of Hβ line as a function of monochromatic luminosity from current reverberation measurements as compiled by Du et al. (2015, 2016a) (*red points*), and the best fit from Bentz et al. (2013) (see text). *Right panel*: the expected constraints on dark energy parameters from reverberation-measured 2000 AGN strongly depend on their redshift coverage, and the sources at very low as well as at high redshifts are needed (Fig. 6 from King et al. 2014)

Covering the broad range of redshifts with the same probe is very important for distance determination, and for subsequent cosmological constraints. This is best illustrated by King et al. (2014) where they analyze the expected results from the sample of 2000 AGN. If the dark energy is parametrized as

$$w(z) = w_0 + w_z z/(1+z) \tag{8}$$

after Chevallier and Polarski (2001), then to get the strong constraints AGN must populate the broad redshift range $0.01 < z < 4$, i.e. starting at very low redshifts (see Fig. 8). AGN can cover such a broad range while this is a problem for the methods based on gamma-ray bursts (see Sect. 8).

Nearby AGN are currently monitored by several groups (e.g., Ilić et al. 2015; Barth et al. 2015; Fausnaugh et al. 2016; Bentz et al. 2016a,b; Du et al. 2016a). SDSS quasar sample has been recently monitored for six months within the frame of the SDSS-RM project (Shen et al. 2016) which allowed to provide preliminary estimate of nine Hβ langs and six Mg II lags for objects with redshifts above 0.3, and this program is being extended now, covering 849 sources in total, with cadence of 12 epochs per year. High redshift quasars do not need such a dense sampling but then the project requires at least 5 years of monitoring (Czerny et al. 2013) to measure lags in higher z sources, and such project is under way within the Oz-DES program (King et al. 2015). Selected sources (771 in the final sample) will be monitored for six years, with 25 measurements for each quasar, on average. Individual delay measurements are expected for half of them, after carefull subtraction of the Fe II and Fe III emission, host galaxy contamination and correction for the reddening. For fainter sources stack analysis will be performed in a number of luminosity bins. Monitoring of a few $z \sim 1$ quasars with as large telescope as SALT will allow for better disentangling of the Mg II line from the underlying Fe II pseudo-continuum (Hryniewicz et al. 2014; Modzelewska et al. 2014).

Narrow-band and broad band reverberation measurements offer an interesting option since they can be done with smaller telescopes and in a more automatic way. In particular, the future Large Synoptic Sky Survey (LSST) will bring 10-year dense coverage of thousands of quasars in six photometric bands. The challenge is in disentangling the line and continuum variability but preliminary studies indicate this is possible (e.g., Chelouche et al. 2014).

However, before AGN monitoring can reach maturity in cosmological applications, several issues have to be addressed. First, the AGN Hubble diagram, like SNe Ia, currently requires calibration to obtain absolute distances. This is basically done by adopting the current value of the Hubble constant for nearby AGN. Direct comparison of AGN distances with another distance indicator is rare. One such example is through the detection of 11 Cepheid stars in a Seyfert 1 galaxy NGC 4395 (Thim et al. 2004). However, in principle the calibration issue can be avoided if the understanding of the BLR formation progresses. For example, the idea that the BLR forms as a dust-driven failed wind (Czerny and Hryniewicz 2011) combined with the theory of accretion disk sets the BLR onset at some specific value of T_{eff} provided by the dust sublimation temperature (see Sect. 6.2). Thus if the hottest dust temperature can be independently measured for a number of AGN this could allow the reverberation method to move to the class of direct methods. Recent comparison of the standard calibration with dust temperature based calibration for a whole AGN sample implies dust temperature 900 K, lower than the hottest dust temperature generally measured in AGN although much higher temperature value was found using the same model from NGC 5548 (Galianni and Horne 2013).

Second aspect is the objective removal of outliers. Some systematic departure of very high Eddington ratio sources from the overall trend has been recently noted (Du et al. 2016b). If this conclusion is supported in further studies, such super-Eddington sources have to be removed from the general sample or corrected for the departure trend.

The third problem is related to possible systematic errors when we move towards more distant, brightest quasars. The extension of the power law dependence between R_{BLR} and the monochromatic flux in a form of a simple power law has not been observationally tested. If the BLR formation is well explained by the failed dust model, expected departures from the linear trend are not strong but if the BLR radius mostly depend on the total ionizing flux then the decrease in ionizing photons with the black hole mass (larger at larger distances) may cause strong nonlinear behavior. The measured monochromatic flux is also affected by the viewing angle, usually unknown. This is not a large problem for nearby AGN since the viewing angles of type 1 AGN are limited to the range between 0 and ~ 45 deg due to the presence of the dusty torus, and the lag independence on the viewing angle was demonstrated by Starkey et al. (2016). However, if the torus opening decreases with redshift and/or with luminosity it may lead to some systematic errors. Recent studies imply some dependence of dust coverage on luminosity (e.g., Ichikawa et al. 2017) but the opposite conclusion has been reached by Mateos et al. (2017), so more future studies along this line are still needed.

Therefore, the BLR reverberation method has not yet reached maturity. On the other hand, it has a large future potential. Large samples of reverberation-studied sources are coming. Extension of the method to photometric reverberation (Haas et al. 2011) opens a way for future use of the Large Synoptic Sky Survey (LSST) data for this purpose, which will bring 10 years of quasar monitoring, with the cadence of 100 observations per year in 6 photometric channels. In the meantime, systematic problems can be studied with increasing sample of individual, well spectroscopically monitored objects.

6.4 Extragalactic Distances Based on Dust Reverberation of AGNs

The inner radius of the dust torus in an active galactic nuclei (AGN) is considered to be determined by sublimation of dust and is proportional to the square-root of the accretion-disk luminosity. If its physical size can be measured by reverberation of dust emission, then the luminosity distance of an AGN can be obtained. The Hubble constant was estimated from the distances based on the dust reverberation for local AGNs as 73 $\mathrm{km\,s^{-1}Mpc^{-1}}$, which shows good agreement with its current standard estimates.

Near-infrared interferometry recently begins to be able to measure the angular scale of the innermost dust torus for brightest AGNs, and the angular diameter distance can be obtained by comparing the reverberation radius with it.

6.4.1 Luminosity Distance Based on the Dust Reverberation

Many observations have indicated that an obscuring dust torus surrounds an accretion disk and broad emission-line region (BLR) in the center of an active galactic nucleus (Antonucci 1993). Since dust is sublimated in the vicinity of the accretion disk by absorbing its strong UV-optical continuum emission, the inner radius of the dust torus is considered to be determined by sublimation of dust. The dust sublimation radius r_{d} can be derived from the radiation-equilibrium equation for the dust grain,

$$\pi a^2 \int Q_\nu \frac{L_\nu^{\mathrm{AD}}}{4\pi r_{\mathrm{d}}^2} d\nu = 4\pi a^2 \int Q_\nu \pi B_\nu(T_{\mathrm{sub}}) d\nu \tag{9}$$

where T_{sub} is the dust sublimation temperature, a is the dust grain size, Q_ν is the absorption efficiency of dust, and B_ν is the Planck function (Barvainis 1987). The parameters T_{sub} and Q_ν at the innermost region of the dust torus would be common in AGNs, because only the dust grains with highest sublimation temperature can survive there and those parameters are determined by the properties of such dust grains. Therefore, the inner radius of the dust torus is expected to be proportional to the square-root of the accretion-disk luminosity. Inversely, the absolute luminosity can be estimated once the inner radius of the dust torus is obtained observationally, which could be used as a distance indicator for AGNs.

The dust reverberation enables us to obtain the inner radius of the dust torus by measuring the lag between the flux variation of the UV-optical continuum emission from the accretion disk and that of the near-infrared thermal emission from the dust torus. A possible application of the dust reverberation of AGNs to the cosmological distance measurement was proposed at the end of the 20th century (Kobayashi et al. 1998; Oknyanskij 1999; Yoshii 2002), and the group of the University of Tokyo started the Multicolor Active Galactic NUclei Monitoring (MAGNUM) project (PI: Yuzuru Yoshii) in 1995 to establish the distance indicator based on the dust reverberation of AGNs and to constrain the cosmological parameters (Yoshii 2002).

Figure 9 shows the radius-luminosity relation for the dust torus obtained by the largest systematic dust reverberation survey for 17 local Seyfert galaxies performed by the MAGNUM project (Koshida et al. 2014). The inner radius of the dust torus clearly correlates with the optical luminosity of AGNs as theoretically expected, which demonstrates the feasibility of the luminosity distance indicator based on the dust reverberation. Then, the dust sublimation model at the innermost dust torus was built to obtain the distances for the 17 AGNs without any distance ladder, and the distances were compared with the recession velocities to estimate the Hubble constant (Yoshii et al. 2014). Figure 10 shows the Hubble diagram. The Hubble constant was estimated approximately as $H_0 = 73\ \mathrm{km\,s^{-1}\,Mpc^{-1}}$, which shows good agreement with its current standard estimates. In addition, the distance calibration based on the dust sublimation model in Yoshii et al. (2014) was consistent with that obtained from the distances of the SNe Ia occurred in the AGN host galaxies (Koshida et al. 2017). These results indicate that the distance indicator based on the dust reverberation is a promising new tool for investigating the expanding Universe.

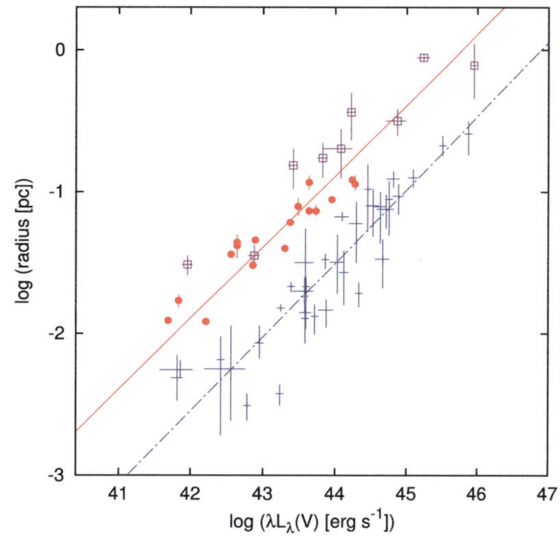

Fig. 9 Radii of the innermost dust torus and the BLR plotted against the V-band luminosity (Koshida et al. 2014). The *red filled circles* represent the reverberation radii for the dust torus obtained by the MAGNUM project, and the *red solid line* represents the best-fit regression line for them. The *blue crosses* represent the reverberation radii for the BLR and the *blue dot-dashed line* represents the best-fit regression line for them (Bentz et al. 2009). The *purple open squares* represent the interferometric radii in near-infrared for the dust torus (Kishimoto et al. 2011; Weigelt et al. 2012)

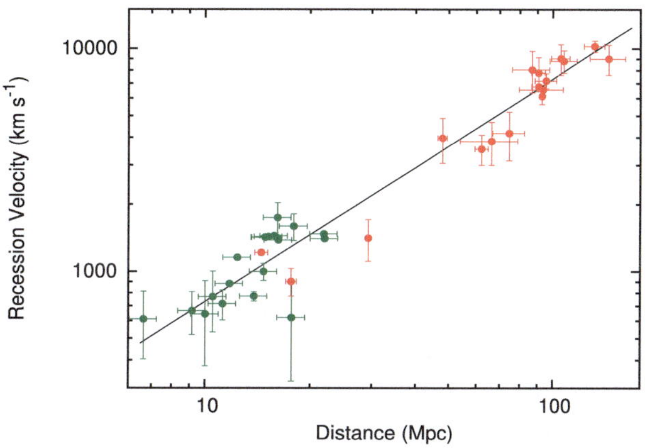

Fig. 10 Hubble diagram for galaxies with the dust reverberation distance (*red filled circles*) and the Cepheid-variable distance (*green filled circles*) from Yoshii et al. (2014). The *solid line* represents the best-fit regression line for the data based on the dust reverberation distance

6.4.2 *Trigonometric Parallax for the Dust Torus*

When both the physical size and the angular size of a certain structure in an AGN can be obtained, the angular diameter distance can be estimated by trigonometry. Although usual imaging observations cannot resolve the inner structures of AGNs whose physical sizes are obtained by reverberation mapping, their angular size might be able to be obtained by future long-baseline interferometers in optical and near-infrared wavelengths. This trigonometric parallax for AGNs was first proposed for the BLR (Elvis and Karovska 2002), but actually, it has not been applied to any AGNs yet.

Since the dust torus is more extended than the BLR, the trigonometric parallax for the dust torus would be more feasible. Indeed, recent advancement of near-infrared interferometry makes it possible to measure the angular scale of the innermost dust torus for some brightest AGNs (Kishimoto et al. 2011; Weigelt et al. 2012), and the angular diameter distance of NGC 4151 was estimated by comparing the dust reverberation radius with the interferometric angular scale in near-infrared (Hönig et al. 2014). However, as shown in Fig. 9, the interferometric radius was found to be systematically larger than the reverberation radius by about a factor of two (Koshida et al. 2014), whereas it is not the case for NGC 4151. Further studies on the structure of the dust torus are necessary to understand this systematic difference and to extend the use of the dust parallax distance of AGNs.

6.5 Single Epoch Methods

Reverberation methods are time-consuming, and much more efficient alternative is to design the method based just on a single spectrum. There are some options in consideration, but they are still in their infancy.

If Eddington ratio and black hole mass can be derived from some distance-independent measure it would be possible to derive distance-independent quasar luminosities. Sulentic et al. (2014) suggest that quasars radiating close to the Eddington limit show distinct optical and UV spectral properties that can be recognized in spectra. They propose specific criteria based on the line ratios (Al IIIλ1860/Si III]λ1892 ≥ 0.5, and (ii) Si III]λ1892/C III]λ1909 $\geq$ 1.0) which lead to a source sample with dispersion of ~ 0.13 dex in the Eddington ratio. Larger sample of such sources may give interesting constraints on cosmological parameters.

Wang et al. (2014) proposed instead to use the basic prediction of the slim accretion disk theory (Abramowicz et al. 1988): the saturation of the luminosity for high Eddington ratio sources. In slim disks the radiative efficiency drops with the Eddington ratio, so effectively the rise of the total source luminosity is only logarithmic in accretion rate. Their preliminary tests of cosmological applications are attractive but the method still requires better analysis of the saturation level as well as efficient method of selecting such AGN in large AGN catalogs.

6.6 Continuum Shape Method

Quasars can be considered a reliable cosmological tools in a similar way as for Type Ia supernovae, provided that the quasar sample is large enough and there is a relation able to standardize the quasar emission. A non-linear relation between the optical and X-ray luminosity in quasars was discovered with the first X-ray surveys in the early '80s (Tananbaum et al. 1979; Zamorani et al. 1981; Avni and Tananbaum 1986), and it has been confirmed since with various samples of a few hundred quasars observed with the main X-ray observatories over a redshift range from 0 to ~ 6.5 and about four decades in optical luminosity.

The largest samples published so far include (i) a compilation of 367 quasars from different optical surveys and observed by *ROSAT*, *XMM–Newton* and *Chandra* (333 from Steffen et al. 2006 and 34 from Just et al. 2007); (ii) 350 sources obtained from the *Sloan Digital Sky Survey* 5th quasar catalog with available *XMM–Newton* observations (Young et al. 2010); (iii) 545 objects from the XMM–COSMOS survey (Lusso et al. 2010); and (iv) 200 quasars with UV and X-ray observations from the *Swift* observatory archive (Grupe et al. 2010; Vagnetti et al. 2010). In all these works the L_X–L_{UV} relation is parametrized as a linear dependence between the logarithm of the monochromatic luminosity at 2500 Å (L_{UV}) and the α_{OX} parameter, defined as the slope of a power law connecting the monochromatic luminosity at 2 keV (L_X), and L_{UV}: $\alpha_{OX} = 0.384 \times \log(L_X/L_{UV})$. This relation implies that quasars more luminous in the optical are relatively less luminous in the X-rays. For instance, the increase by a factor of 10 in L_{UV} implies an increase by only a factor of ~ 4 in the X-ray luminosity. When expressed as a relation between X-ray and UV luminosities, the α_{OX}–L_{UV} relation becomes $\log L_X = \beta + \alpha \log L_{UV}$. All the works cited above provide consistent values (within the uncertainties) of both normalization and slope, $\beta \sim 8$–9 and $\alpha \sim 0.6$, with an observed dispersion that ranges between $\delta \sim 0.35$–0.40 dex. Luminosities are derived from fluxes through a luminosity distance (D_L) calculated assuming a standard ΛCDM model with the best estimates of the cosmological parameters Ω_M and Ω_Λ at the time of the publications.

The potential use of this relation as a cosmological probe is obvious. The observed X-ray flux is a function of the observed optical flux, the redshift, and the parameters of the adopted cosmological model as

$$\log(F_X) = \Phi(F_{UV}, D_L) = \beta' + \alpha \log(F_{UV}) + 2(\alpha - 1)\log(D_L), \tag{10}$$

where $\beta' = \beta + (\alpha - 1)\log(4\pi)$. This relation can be then fitted to a set of optical and X-ray observations of quasars in order to estimate the cosmological parameters. However, none of the samples published so far had, on their own, the size and/or homogeneity to provide useful constraints for cosmological applications.

Recently, the relation above has been employed to build a Hubble diagram for quasars to obtain an independent measure of cosmological parameters. Risaliti and Lusso (2015) have shown that it is possible to build a sample large enough to estimate the cosmological parameters and to test the ΛCDM model over the whole redshift range $z = 0$–6.5 (back to an age of the Universe of only ~ 0.8 Gyr). By using a dataset of ~ 800 quasars from the literature (basically all the ones listed at the beginning of this Section) with optical and X-ray data, Risaliti and Lusso (2015) have demonstrated that the F_X–F_{UV} relation does not exhibit any dependence with redshift, which is a necessary condition to use it as a distance estimator. It has been therefore possible to build, for the first time, a *quasar Hubble diagram* in excellent agreement with that of Type Ia supernovae over the common redshift range ($z \le 1.4$), but extending it to much higher distances. Figure 11 (left panel) shows the Hubble diagram for quasars (small black points, while red points are averages in narrow redshift bins for visual purpose only), compared to the one of Type Ia supernovae samples (cyan points, from Suzuki et al. 2012). Interestingly, since the same physical quantity (i.e. the distance modulus) is involved, this analysis based on the distance modulus–redshift relation allows us to directly merge quasar and supernova data on the same diagram, where both are used as *standardized candles*. This method offers unique access to a region of the distance modulus–redshift plane where it is extremely unlikely (if not impossible) to observe any Type Ia supernovae.

The results for the determination of the cosmological parameters Ω_M and Ω_Λ are shown in the right panel of Fig. 11. The 68% and 95% contours for Ω_M and Ω_Λ, assuming a standard ΛCDM model, are derived from the analysis of the Hubble diagram of quasars only (grey contours), from the supernovae of the Union 2.1 sample (empty black contours; Suzuki et al. 2012), and from a quasars–supernovae joint fit (orange/red contours). The present data return $\Omega_M = 0.22^{+0.08}_{-0.10}$ and $\Omega_\Lambda = 0.92^{+0.18}_{-0.30}$ ($\Omega_M = 0.28 \pm 0.04$ and $\Omega_\Lambda = 0.73 \pm 0.08$ from a joint fit). The contour plot relative to the joint sample is not a statistical intersection of the two individual contours, but the result of a simultaneous fit of the distance modulus–redshift relation for the combined sample quasar/supernovae. This also provides a significantly improved measurement of Ω_M and Ω_Λ with pure distance indicators. When all the available cosmological indicators (CMB, BAO, weak lensing, supernovae) are considered, the sample of Risaliti and Lusso (2015) does not significantly improve the constraints on the cosmological parameters. The main limitation of this work is its small predictive power, mainly due to the still large dispersion of the $\log F_X = \beta + \alpha \log F_{UV}$ relation (~ 0.30–0.35 dex). For comparison, the uncertainty in the distance of a supernova at low redshift is about 0.05 in the same units. This means that, in order to obtain the same constraints on a cosmological fit, roughly $(0.35/0.05)^2 \sim 50$ quasars per single supernova are needed.

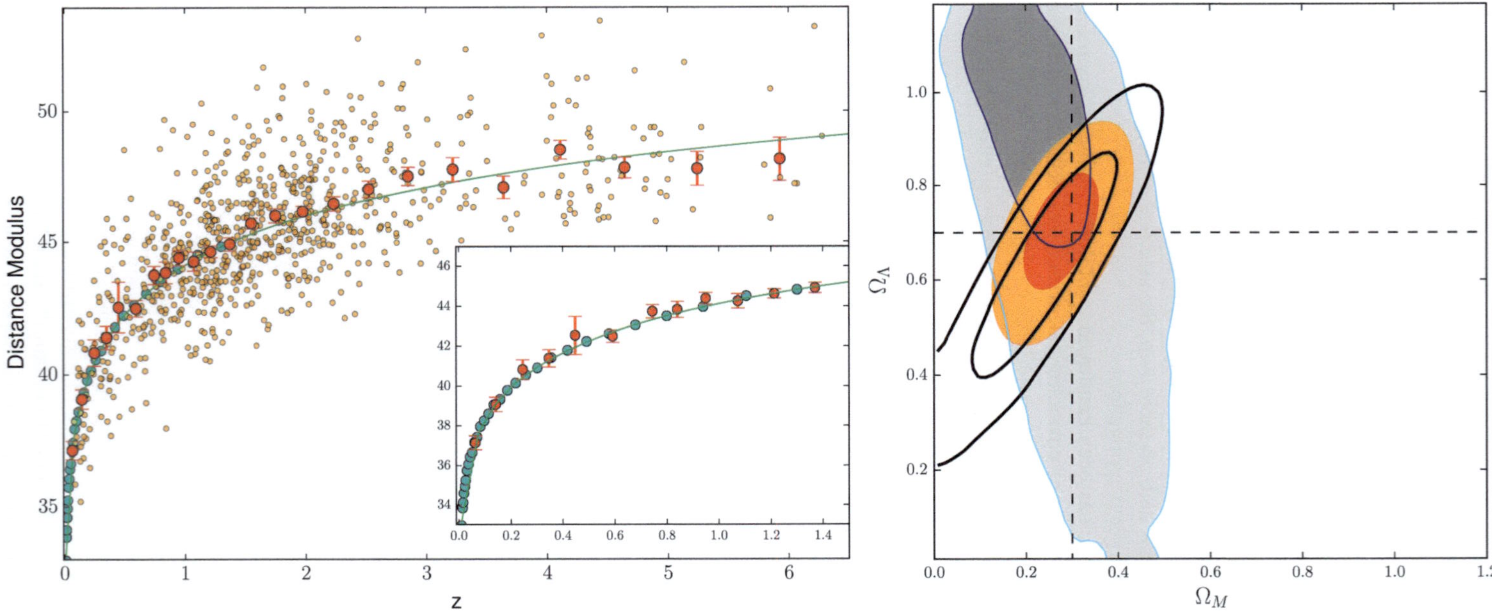

Fig. 11 *Left*: Hubble diagram for the quasar sample ($\sim$ 800 sources, *orange points*), and supernovae (*cyan points*) from Betoule et al. (2014). The *large red points* are quasar averages in narrow redshift bins. The *inner box* shows a zoom of the $z = 0$–1.5 range, in order to better visualize the match between the supernovae and quasar samples. The *continuous line* is obtained from a joint fit of the two samples assuming a standard ΛCDM cosmological model. *Right*: 68% and 95% contours for $\Omega_{\rm M}$ and Ω_Λ as derived from the Hubble diagram of quasars (*grey*, Risaliti and Lusso 2015), from the Type Ia supernovae sample (*empty black*, Betoule et al. 2014), and from a quasar–supernovae joint fit (*orange*, *red*). It is the first time that this measurement has been done using quasars as cosmological distance probes. Adapted from Risaliti and Lusso (2015)

The L_X–L_{UV} relation was recently analyzed on a sample of $> 2,600$ SDSS AGN, with X-ray data from XMM–Newton with the aim to understand the origin of the observed dispersion, and to evaluate the true intrinsic dispersion in this relation (Lusso and Risaliti 2016). This work shows that, even though the observed dispersion in the main quasar sample is ~ 0.45 dex, most of this scatter is caused by the combination of X-ray variability, poor optical/UV and X-ray data quality, and the contamination from the host galaxy and/or AGN with red continua. Once all the contaminations above are taken into account, the real physical dispersion is found to be less than 0.21 dex over roughly 4 decades in luminosity, indicating a tight empirical coupling between disc (UV) and corona (X-rays), which puts the determination of distances based on this relation on a sounder physical grounds.

Although exploratory, the analysis presented in Risaliti and Lusso (2015) already provides important results: the validation of the UV to X-ray relation and the extension of the Hubble Diagram up to $z \sim 6$. The constraints on the cosmological parameters are still loose compared with other methods, and the improvement on the measurement errors obtained by combining the quasar findings with those from supernovae, CMB, BAOs, and clusters is not significant. This work is mainly intended as a demonstration of the method, while most of its potential is still to be exploited in future work, with both available and forthcoming data (e.g. *eROSITA*, *Euclid*, *Athena*) that will significantly increase the number of quasars at $z > 2$ where the intrinsic power of this novel technique lies.

6.7 X-Ray Excess Variance Method

Another method based on a combination of X-ray observations has been proposed by La Franca et al. (2014). The method requires a single optical spectrum (determination of FWHM of one of the strong AGN lines, preferentially Hβ) and determination of the X-ray excess variance. The basic idea is simple, it relies on a combination of two methods of the black hole mass determination and subsequent elimination of the mass. The first method, derived from the reverberation campaigns (e.g., Kaspi et al. 2000; Peterson et al. 2004) shows that the black hole mass can be determined from a single epoch spectrum, from the monochromatic absolute luminosity (a proxy for the BLR size, see Sect. 6.3) and the emission line width. The second method is based on the simple relation between the black hole mass and the high frequency tail of the X-ray power spectrum, or, more conveniently, the X-ray excess variance (e.g., Nikolajuk et al. 2004; Gierliński et al. 2008). Combining the two and eliminating black hole mass one can determine the absolute luminosity from the X-ray excess variance and the FWHM. The difficulty lies in the high quality measurement of the X-ray excess variance: La Franca et al. argue that the launch of Athena X-ray Observatory (expected in 2028) will provide interesting number of measurements, and they expect more from the future generation X-ray telescopes with larger detector area, but this is not coming soon.

7 Galaxy Cluster Distance via X-Ray and Sunyaev-Zel'dovich Effect

Galaxy clusters are the largest virialized astronomical structures in the Universe and observations of their physical properties can provide important cosmological information. An important phenomenon occurring in galaxy clusters is the Sunyaev-Zel'dovich effect (Sunyaev and Zeldovich 1972; Birkinshaw 1999; Carlstrom et al. 2002), a small distortion of the cosmic microwave background radiation (CMBR) spectrum provoked by the inverse Compton scattering of the CMBR photons passing through a population of hot electrons. The

Sunyaev-Zel'dovich effect (SZE) is proportional to the electron pressure integrated along the line of sight, i.e., to the first power of the plasma density. The measured temperature decrement ΔT_{SZ} of the CMBR is given by (De Filippis et al. 2005)

$$\frac{\Delta T_{\mathrm{SZ}}}{T_{\mathrm{CMBR}}} = f(\nu, T_{\mathrm{e}})\frac{\sigma_{\mathrm{T}} k_{\mathrm{B}}}{m_{\mathrm{e}} c^2}\int_{\mathrm{l.o.s.}} n_e T_{\mathrm{e}} dl, \tag{11}$$

where T_{e} is the temperature of the intra-cluster medium, k_{B} the Boltzmann constant, $T_{\mathrm{CMBR}} = 2.728^\circ$ K is the temperature of the CMBR, σ_{T} the Thompson cross section, m_{e} the electron mass and $f(\nu, T_{\mathrm{e}})$ accounts for frequency shift and relativistic corrections (Itoh et al. 1998).

Other important physical phenomena occurring in the intra-galaxy cluster medium are the X-ray emission caused by thermal bremsstrahlung and line radiation resulting from electron-ion collisions. The X-ray surface brightness S_X is proportional to the integral along the line of sight of the square of the electron density. This quantity may be written as follows

$$S_X = \frac{D_{\mathrm{A}}^2}{4\pi D_L^2}\int_{\mathrm{l.o.s.}} n_e^2 \Lambda_{eH} dl, \tag{12}$$

where Λ_{eH} is the X-ray cooling function of the intra-cluster medium (measured in the cluster rest frame), D_A and D_L are the angular diameter and luminosity distances of the galaxy cluster, respectively, and n_e is the electron number density. It thus follows that the SZE and X-ray emission both depend on the properties (n_e, T_e) of the intra cluster medium.

As is well known, it is possible to obtain the angular diameter distance (ADD) of galaxy clusters by their SZE and X-ray surface brightness observations (SZE/X-ray technique). The calculation begins by constructing a model for the cluster gas distribution. Assuming, for instance, the spherical isothermal β-model such that n_e is given by (Cavaliere and Fusco-Femiano 1978)

$$n_e(r) = \left(1 + \frac{r^2}{r_c^2}\right)^{-3\beta/2}, \tag{13}$$

equations (11) and (12) can be integrated. Here r_c is the core radius of the galaxy cluster. This β model is based on the hydrostatic equilibrium equation and constant temperature (Sarazin 1988). In this way, we may write for the SZE

$$\Delta T_{\mathrm{SZ}} = \Delta T_0\left(1 + \frac{\theta^2}{\theta_c^2}\right)^{1/2-3\beta/2}, \tag{14}$$

where $\theta_c = r_c/D_A$ is the angular core radius and ΔT_0 is the central temperature decrement that includes all physical constants and terms resulting from the line-of-sight integration. More precisely:

$$\Delta T_0 \equiv T_{\mathrm{CMBR}} f(\nu, T_{\mathrm{e}})\frac{\sigma_{\mathrm{T}} k_{\mathrm{B}} T_{\mathrm{e}}}{m_{\mathrm{e}} c^2} n_{e0}\sqrt{\pi}\theta_c D_A g(\beta/2), \tag{15}$$

with

$$g(\alpha) \equiv \frac{\Gamma[3\alpha - 1/2]}{\Gamma[3\alpha]}, \tag{16}$$

where $\Gamma(\alpha)$ is the gamma function and the others constants are the usual physical quantities. For X-ray surface brightness, we have

$$S_X = S_{X0}\left(1+\frac{\theta}{\theta_c^2}\right)^{1/2-3\beta}, \tag{17}$$

where the central surface brightness S_{X0} reads

$$S_{X0} \equiv \frac{D_A^2 \Lambda_{eH} \mu_e/\mu_H}{D_L^2 4\sqrt{\pi}} n_{e0}^2 \theta_c D_A\, g(\beta). \tag{18}$$

Here μ is the molecular weight given by $\mu_i \equiv \rho/n_i m_p$.

One can solve (15) and (18) for the ADD by eliminating n_{e0} and taking for granted the validity of cosmic distance duality relation (CDDR), $D_L(1+z)^{-2}D_A^{-1} = \eta = 1$. In this case one obtains

$$D_A = \left[\frac{\Delta T_0{}^2}{S_{X0}}\left(\frac{m_e c^2}{k_B T_{e0}}\right)^2 \frac{g(\beta)}{g(\beta/2)^2\,\theta_c}\right] \times \left[\frac{\Lambda_{eH0}\,\mu_e/\mu_H}{4\pi^{3/2} f(\nu, T_e)^2 (T_{CMBR})^2 \sigma_T{}^2\,(1+z_c)^4}\right] \tag{19}$$

where z_c is the galaxy cluster redshift. Recently, such a technique has been applied for a fairly large number of clusters (Reese et al. 2002; De Filippis et al. 2005; Bonamente et al. 2006) with systematic and statistical errors around 20% and 13%, respectively (see Table 3 in Bonamente et al. 2006). The crucial point in the SZE/X-ray technique is the choice of morphology used to describe the galaxy cluster. The standard spherical geometry has been severely questioned, since Chandra and XMM-Newton observations have shown that clusters usually exhibit an elliptical surface brightness (Sereno et al. 2006). The assumed cluster shape can affect considerably the SZE/X-ray distances, and, consequently, the H_0 estimates and other astrophysical quantities.

7.1 SZE/X-Ray Technique Applications: the Hubble Constant

The Hubble constant, H_0, sets the scale of the size and age of the Universe and its determination from independent methods is still worthwhile to be investigated. Severeal authors used the SZE/X-ray technique to estimate the Hubble constant (see table in Holanda et al. 2012a), however, since the samples used has galaxy clusters in high redshifts (up to $z \approx 0.90$), a cosmological model had to be assumed in analysis, usually a flat ΛCDM model. Moreover, due to degeneracy on the cosmological parameters, the matter density parameter, Ω_M, was taken as being $\Omega_M = 0.3$. An interesting method was adopted by Holanda et al. (2012a). In this case, the authors used a sample of 25 ADD of galaxy clusters in redshift range $0.023 < z < 0.784$ described by an elliptical β model (De Filippis et al. 2005) to constrain H_0 in dark energy models. In order to avoid the use of priors on the cosmological parameters, a joint analysis involving the ADD, the baryon acoustic oscillations (BAO) and the CMBR Shift Parameter signature was proposed. By taking into account the statistical and systematic errors of the SZE/X-ray technique it was obtained for nonflat ΛCDM model $H_0 = 74^{+5.0}_{-4.0}\ \mathrm{km\,s^{-1}\,Mpc^{-1}}$ (1σ) whereas for a flat universe with constant equation of state parameter it was obtained $H_0 = 72^{+5.5}_{-4.0}\ \mathrm{km\,s^{-1}\,Mpc^{-1}}$ (1σ). These values are in full agreement with the latest local estimate performed by Riess et al. (2016): $H_0 = 73.24 \pm 1.74\ \mathrm{km\,s^{-1}\,Mpc^{-1}}$ (1σ).

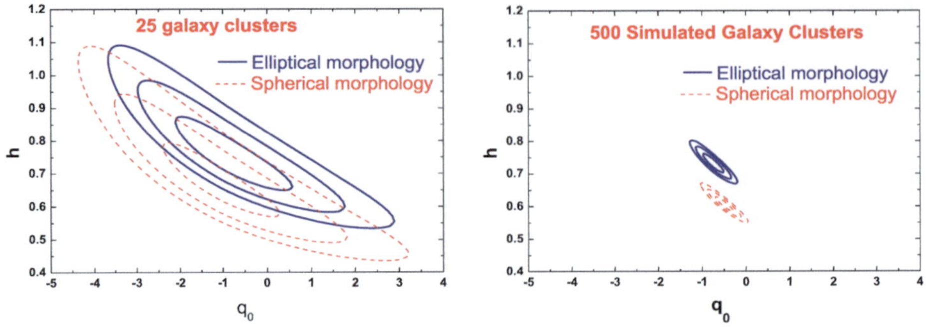

Fig. 12 Confidence contours (1, 2 and 3σ) on the plane (h–q_0). The *panel left* corresponds to the real data

7.2 The Search for Two Numbers: H_0 and q_0

A kinematic method to access cosmic acceleration and the H_0 value based exclusively on the SZE and X-ray surface brightness data from galaxy clusters also was investigated recently. This is a very important task since until very recently the type Ia supernovae observations was the unique direct access to the late time accelerating stage of the Universe. By assuming the current observational error distribution of the samples of 25 ADD from De Filippis et al. (2005), Holanda et al. (2013) performed Monte Carlo simulations based on a well-behaved parametrization for the deceleration parameter, q_0, to generate samples with different characteristics and study the improvement on the determination of the cosmographic parameters: q_0 and H_0. As a interesting result it was shown that, even keeping the current statistical observational uncertainty, an increase in the number of data points increases considerably the figure-of merit for the cosmographic plane (h–q_0), where $h = H_0/100$ (see Fig. 12).

7.3 The Search for New Physics

The SZE/X-ray technique also has been shown to being a powerful tool to investigate fundamental physics. Uzan et al. (2004) argued that this technique is strongly dependent on the validity of the CDDR, $D_L(1+z)^{-2}D_A^{-1} = \eta = 1$, valid for *all* cosmological models based on Riemannian geometry, being dependent neither on Einstein field equations nor on the nature of matter (Etherington 1933 reprinted as Etherington 2007), playing an essential role in modern cosmology. It only requires that source and observer are connected by null geodesics in a Riemannian spacetime and that the number of photons are conserved. However, if $\eta \neq 1$, instead of the real ADD, the measured quantity is $D_A^{data}(z) = D_A(z)\eta^2(z)$. By using different expressions for $\eta(z)$, several authors have tested the CDDR by using ADD of galaxy clusters and D_L from type Ia supernovae observations (Holanda et al. 2010, 2012b; Li et al. 2011b; Nair et al. 2011; Yang et al. 2013; Holanda and Barros 2016). No significant deviation was obtained when an elliptical β model was used in analysis (for results from different cosmological observations see Table I in Holanda et al. (2016).

As was showed by Holanda et al. (2016), this technique depends on the fine structure constant, α. If α is a time-dependent quantity, e.g., $\alpha = \alpha_0\phi(z)$, where ϕ is a function of redshift, the current ADD data do not provide the real ADD to the cluster but instead $D_A^{data}(z) = \eta^2(z)\phi(z)D_A(z)$. Constraints on a possible variation of α for a class of dilaton runaway models was performed considering the sample of $D_A^{data}(z)$ from De Filippis et al. (2005) and estimates of $D_A(z)$ from type Ia supernovae observations. It was found no significant indication of variation of α with the present data.

Finally, it is very important to stress that the SZE/X-ray technique is independent of any calibrator usually adopted in the determinations of the distance scale. The above results, therefore, highlight the cosmological interest in ADD measurements of galaxy clusters.

8 Gamma-Ray Bursts

Gamma-rays bursts are the brightest individual sources in the Universe (see e.g., Hjorth and Bloom 2012; Berger 2014, for recent reviews), therefore it is more than natural to attempt to use them in cosmology. Those intense flashes of the gamma-ray emission lasting from milliseconds to several hours are not easily affected by the medium between an observer and the source, so they can be easily detected at very high redshifts. The record holder is the gamma-ray burst GRB 090429B at $z = 9.4$ (Cucchiara et al. 2011), with photometric redshift determination, or GRB 090429B at $z = 8.2$ (Tanvir et al. 2009), if we limit ourselves to spectroscopically confirmed redshift. Gamma-ray prompted emission is accompanied by afterglow emission at other wavelengths.

However, due to the large burst variety and the critical role of the relativistic boosting of the part of the observed multi-wavelength emission the use of the gamma-ray bursts as distance indicators is rather difficult. The first attempts based on relatively large number of sources was made by Schaefer (2007) who successfully constructed the Hubble diagram in a broad redshift range almost up to $z \sim 7$. The method was based on empirical correlations established for sources at $z < 2$ and then extended to higher redshifts.

Several correlations can be used for that purpose (see e.g., Wang et al. 2015). However, there are two basic problems which are not easy to overcome. First, there is some *circularity problem*, as there are no low redshift gamma-ray bursts and the scaling of the relations has to be done for intermediate redshift sources where cosmological effects already become important. Second, the number of gamma-ray bursts with well measured parameters is still low, and it increases only slowly with time. Samples with high quality data and low scatter in the measured quantities contain less than 50 objects so far (e.g., Dainotti et al. 2016). In addition, the results can be affected by the weak lensing, and the appropriate statistical corrections should be introduced, as discussed by Wang and Dai (2011).

9 Astronomical Distances from Gravitational-Wave Observations

Recent direct detections of gravitational waves from merging binary black hole systems at cosmological distances of $z = 0.1$–0.2, performed by two Advanced LIGO detectors (Abbott et al. 2016d,c, 2017a), first binary black-hole merger observation with the global network of three detectors of Advanced LIGO and the Advanced Virgo (Abbott et al. 2017d), as well as the first Advanced LIGO and Advanced Virgo detection of a nearby (at a distance of 40 Mpc) binary neutron-star merger (Abbott et al. 2017e), followed by a short gamma-ray burst (Abbott et al. 2017c) and broad-band electromagnetic emission observational campaign (Abbott et al. 2017f) create unprecedented opportunities for studying the Universe through a novel, never before explored channel of spacetime fluctuations. Gravitational wave astronomy is often compared to 'listening to' rather than 'looking at' the skies. By design which is motivated by the choice of potential sources, the ground-based gravitational wave detectors, kilometer-size laser interferometer antennas of Advanced LIGO (Aasi et al. 2015) and Advanced Virgo (Acernese et al. 2015), are sensitive in the range of frequencies similar to the audible range of human ears - between 10 Hz and a few kHz. As in the case of an ear, a solitary laser interferometric detector is practically omnidirectional (has a poor angular resolution), and has no imaging capabilities. It registers a coherent signal emitted by a bulk movement of large, rapidly-moving masses. Once emitted, gravitational waves are weakly coupled to the surrounding matter and propagate freely without scattering. This has to be contrasted with the electromagnetic emission which originates at the microscopic level, is strongly coupled to the surroundings and often reprocessed; it carries a reliable information from the last scattering surface only. It seems therefore that gravitational wave detectors are the perfect counterpart to the electromagnetic observatories as they may provide us with information impossible to obtain by other means.

Very shortly after announcing the general theory of relativity in 1915, Albert Einstein realized that a linearized version of his equations resembles the wave equation (Einstein 1916). The solution is interpreted as a short-wavelength, time-varying curvature deformation propagating with the speed of light on an otherwise slowly-varying, large-scale curvature background (a gravitational-wave "ripple" propagating through the four-dimensional spacetime); from the point of view of a metric tensor, it represents a small perturbation of a stationary background metric. Linear approximation corresponds to the waves propagating in the far-field limit. By exploiting the gauge freedom of the theory one may show that the solution has features similar to electromagnetic waves: it is a transverse wave which may be polarized (has two independent polarizations). Over the next 40 years, during which Einstein changed his mind to argue against their genuineness, a controversy persisted over the true nature of gravitational waves. Only in the late 50s and early 60s the works of Felix Pirani (1956), Herman Bondi (1957), Ivor Robertson and Andrzej Trautman (1960) convincingly showed that gravitational waves are indeed physical phenomena that carry and deposit energy.

A realistic wave phenomenon (and not, say, a coordinate artifact) must be capable of transmitting energy from the source to infinity. If the amplitude of an exemplary isotropic field at a radial distance r from the source is $h(r)$, then the flux of energy over a spherical surface at r is $\mathcal{F}(r) \propto h^2(r)$, and the total emitted power (the luminosity) is $\mathcal{L}(r) \propto 4\pi r^2 h^2(r)$. Since the energy has to be conserved, the amplitude $h(r)$ falls like $1/r$, irrespectively of the multipole character of the source (the lowest radiating multipole in the gravity theory is the quadrupole distribution, because for an isolated system a time-changing monopole would correspond to the violation of the mass-energy conservation, and a time-changing dipole would violate the momentum conservation law).

Gravitational waves are related to the changes in the spacetime distance (the proper time interval), therefore they cannot be detected by a local measurement - one has to compare the spacetime positions of remote events (Pirani 1956). The detection principle in the case of the laser interferometric detector is to measure the difference of the relative change in its perpendicular arms' lengths L_x and L_y, $\delta L_x - \delta L_y = \Delta L/L$, by measuring the interference pattern in the output port located at the apex of the device. Due to the quadrupolar nature of a gravitational wave, shortening of one arm corresponds to elongation of the other. This change of lengths is reflected in varying paths that the laser light has to cross before the interference. The dimensionless gravitational-wave amplitude $h = \Delta L/L$ (the "strain") is proportional to the amount of outgoing laser light. The fact that the directly-measurable quantity is the amplitude $h \propto 1/r$, not the energy of the wave as in the electromagnetic antennæ, has a direct consequence for the reach of the observing device: one-order-of-magnitude sensitivity improvement corresponds to one-order-of-magnitude growth of distance reach r, as opposed to the factor of $\sqrt{10}$ in the electromagnetic observations (consequently, the volume of space grows like r^3 in case of gravitational-wave observations, encompassing hundreds of thousands of galaxies for a distance reach of the order of hundreds of Mpc (see Abbott et al. 2016b).

Among promising sources of gravitational radiation are all asymmetric collapses and explosions (supernovæ), rotating deformed stars (gravitational-wave 'pulsars' of continuous and transient nature), and tight binary systems of e.g., neutron stars and black holes. In the following we will focus on the binary systems, since their properties make them the analogues to standard candles of traditional astronomy. The fittingly descriptive term "standard siren" was first used in the work of Holz and Hughes (2005) in the context of gravitational waves from super-massive binary black holes as a target for the planned spaceborne LISA detector (Kea 1996). Dalal et al. (2006) were the first one to point out that short GRBs produced by merging binary neutron-star systems would constitute ideal standard siren. The idea of using well-understood signals to infer the distance and constrain the cosmological parameters, e.g., the Hubble constant, was however proposed much earlier (Schutz 1986, 2002). Recently it was demonstrated in practice for the first time with the observations of a binary neutron-star merger in both gravitational waves and photons (Abbott et al. 2017e,b).

Magnitudes of the gravitational-wave strain h and the luminosity $\mathcal{L}$ may be estimated using dimensional analysis and Newtonian physics. As the waves are generated by the accelerated movement of masses and the mass distribution should be quadrupolar, one may assume that h is proportional to a second time derivative of the quadrupole moment $I_{ij} = \int \rho(\mathbf{x}) x_i x_j d^3x$ for some matter distribution $\rho(\mathbf{x})$. For a binary composed of masses m_1 and m_2, orbiting the center of mass at a separation a with the orbital angular velocity ω, h is proportional to the system's moment of inertia μa^2 and to ω^2, as well as inversely proportional to the distance, $h \propto \mu a^2 \omega^2 / r$, where $\mu = m_1 m_2 / M$ is the reduced mass, and $M = m_1 + m_2$ the total mass. In order to recover the dimensionless h, the characteristic constants of the problem, G and c, are used to obtain

$$h \simeq \frac{G}{c^4}\frac{1}{r}\mu a^2 \omega^2 = \frac{G^{5/3}}{c^4}\frac{1}{r}\mu M^{2/3}\omega^{2/3} \qquad \left(h_{ij} = \frac{2G}{c^4 r}\ddot{I}_{ij} \right), \tag{20}$$

with the use of Kepler's third law ($GM = a^3\omega^2$) in the second equation. The expression in brackets represents the strain tensor h_{ij} in the non-relativistic *quadrupole approximation* (Einstein 1918). Similarly, the luminosity $\mathcal{L}$ (the rate of energy loss in gravitational waves, integrated over a sphere at a distance r) should be proportional to $h^2 r^2$ and some power of ω. From dimensional analysis one has

$$\begin{aligned} &\mathcal{L} = \frac{dE_{GW}}{dt} \propto \frac{G}{c^5} h^2 \omega^2 \propto \frac{G}{c^5}\mu^2 a^4 \omega^6 \\ &\left(\mathcal{L} = \frac{c^3}{16G\pi} \iint \langle \dot{h}_{ij}\dot{h}^{ij} \rangle dS = \frac{G}{5c^5} \langle \dddot{I}_{ij} \dddot{I}^{ij} \rangle \right), \end{aligned} \tag{21}$$

with the proportionality factor of 32/5. Again, the expression in brackets refers to the quadrupole approximation; the angle brackets denote averaging over the orbital period. Waves leave the system at the expense of its orbital energy $E_{orb} = -Gm_1 m_2/(2a)$. Using the time derivative of the Kepler's third law, $\dot{a} = -2a\dot{\omega}/(3\omega)$, one gets the evolution of the orbital frequency driven by the gravitational-wave emission:

$$\frac{dE_{orb}}{dt} \equiv \frac{Gm_1 m_2}{2a^2}\dot{a} = -\frac{dE_{GW}}{dt} \quad \Longrightarrow \quad \dot{\omega} = \frac{96}{5}\frac{\omega^{11/3}}{c^5} G^{5/3} \mathcal{M}^{5/3}. \tag{22}$$

The system changes by increasing its orbital frequency; at the same time the strain amplitude h of emitted waves also grows. This characteristic frequency-amplitude evolution is called the *chirp*, by the similarity to birds' sounds, and the characteristic function of component masses $\mathcal{M} = \left(\mu^3 M^2\right)^{1/5}$ is called the *chirp mass*. Orbital frequency is related in a straightforward manner to the gravitational-wave frequency f_{GW}: from the geometry of the problem it is evident that the frequency of radiation is predominantly at twice the orbital frequency, $f_{GW} = \omega/\pi$. By combining the equations for $\dot{f}_{GW}$ and h, one recovers the distance to the source r. It is a function of the frequency and amplitude parameters, which are *directly* measured by the detector:

$$r = \frac{5}{96\pi^2}\frac{c}{h}\frac{\dot{f}_{GW}}{f_{GW}^3} = 512\frac{1}{h_{21}}\left(\frac{0.01\ \text{s}}{\tau}\right)\left(\frac{100\ \text{Hz}}{f_{GW}}\right)^2 \text{Mpc}, \tag{23}$$

where h_{21} denotes the strain in the units of 10^{-21} and $\tau = f_{GW}/\dot{f}_{GW}$ denotes the rate of change of the gravitational-wave chirp frequency. Note that within the simple Newtonian approximation presented here (at the leading order of the post-Newtonian expansion) the product $h\tau$ is independent of the components' masses (Królak and Schutz 1987). The simplified analysis presented above does not take into account the full post-Newtonian waveform, polarization information, network of detectors analysis etc., but is intended to demonstrate that binary systems are indeed truly extraordinary "standard sirens". Their observations provide absolute, physical distances directly, without the need for a calibration or a 'distance ladder'. Among them, binary black hole systems occupy a special position. In the framework of general relativity, binary black holes waveforms are independent from astrophysical assumptions about the systems' intrinsic parameters and their environment. At cosmological scales the distance r has a true meaning of the *luminosity distance*. However, since the vacuum (black-hole) solutions in general relativity are scale-free, the measurements of their waveforms alone cannot determine the source's redshift. The parameters measured by the detector are related to the rest-frame parameters by the redshift z: $f_{GW} = f_{GW}^{rf}/(1+z)$, $\tau = \tau^{rf}(1+z)$, $\mathcal{M} = \mathcal{M}^{rf}(1+z)$. Independent measurements of the redshift which would facilitate the cosmographic studies of the large-scale Universe requires the collaboration with the electromagnetic observers i.e., the *multi-messenger astronomy*. This may be obtained by assessing the redshift of the galaxy hosting the binary by detecting the electromagnetic counterpart of the event (Bloom et al. 2009; Singer et al. 2016; Abbott et al. 2017f), or by performing statistical study for galaxies' redshifts correlated with the position of the signal's host galaxy (MacLeod and Hogan 2008).The omnidirectional nature of a solitary detector is mitigated by simultaneous data analysis from at least three ground-based detectors in order to perform the triangulation of the source position, hence the crucial need for the LIGO-Virgo collaboration, which will be in the future enhanced by the LIGO-India detector, and the KAGRA detector in Japan (Aso et al. 2013). The above principles were validated by the observations of a binary black-hole merger with three detector network (Abbott et al. 2017d), resulting in improved distance and sky localization.

Compact binaries containing matter are perfect sources to produce the multi-messenger events. Neutron-star binaries are now firmly established as sources of short gamma-ray bursts (idea originally proposed in Paczynski 1986), thanks to simultaneous gravitational and electromagnetic-wave observations from the same event (Abbott et al. 2017f,c and references therein). Although the complete waveform which includes the merger requires the knowledge of the material properties of the components (the presently poorly-known dense-matter equation of state physics), the chirp waveforms are understood well enough to facilitate a firm detection and a distance measurement. Binaries involving neutron stars are the ideal "standard sirens" insofar as they naturally provide both loud gravitational-wave and bright electromagnetic emission. Short gamma-ray bursts occur frequently within the reach of ground-based detectors, at redshifts $z < 0.2$. Nissanke et al. (2013) show that observing a population of the order of 10 of gravitational-wave events related to short gamma-ray bursts would allow to measure the Hubble constant with 5% precision using a network of detectors that includes advanced LIGO and Virgo (30 beamed events could constrain the Hubble constant to better than 1%). In both cases of double black-hole binaries and those involving neutron stars, rapid electromagnetic counterpart observations and precise catalogs of galaxies are needed (Abbott et al. 2016b; Singer et al. 2016).

The main source of error in the distance measurement is the limited sensitivity of the detectors (finite signal-to-noise), which translates to a limited knowledge of the source's direction and orientation (see e.g.,Nissanke et al. 2010 for a short gamma-ray burst related study). This may be improved with the measurements of gravitational-wave polarizations with a network of three or more detectors. Second limiting factor is the detectors' calibration uncertainties. Recent direct detectors of binary black-hole mergers by two Advanced LIGO detectors established the distances with rather large error bars mostly due to these factors (see e.g., Abbott et al. 2016a); the order-of-magnitude improvement in sky localization was evident for the first triple-detection with the global network of Advanced LIGO and Advanced Virgo detectors (Abbott et al. 2017d). Presence of a third detector in the network proved to be crucial for the precise sky localization of a binary neutron-star merger (Abbott et al. 2017f), allowing for rapid electromagnetic follow-up. For redshifts larger than $z = 1$ weak gravitational lensing will contribute to the distortion of the luminosity distance measurements at the order of 10% (Bartelmann and Schneider 2001; Dalal et al. 2006). For a detailed discussion of limiting factors in the case of a network of detectors see Schutz (2011) and references therein.

To conclude, observations of gravitational-waves from cosmological distances with current and planned detectors (e.g., spaceborne LISA, sensitive to low frequencies corresponding to chirping super-massive black hole binaries, Kea 1996, or a third-generation cryogenic underground Einstein Telescope, with an extended low frequency range compared to the Advanced LIGO and Advanced Virgo, Abernathy et al. 2011) promise a wealth of new astrophysical information, as demonstrated by the first multi-messenger gravitational-wave event (Abbott et al. 2017f,c). Future detectors will reach cosmological distances and redshifts of a few, being sensitive to practically all the chirping binaries in their sensitivity band in the Universe and providing precise distance measurements (see e.g., Lang and Hughes 2008). In addition to precisely measuring the Hubble constant, cosmological observations would help determine the distances to galaxies, thus contributing to building the standard 'distance ladder' (calibrating electromagnetic standard candles), establish the distribution of galaxies and voids, characterize the evolution of the dark energy and mass density of the Universe, mass distribution through the gravitational lensing, as well as the chemical evolution effects i.e., establishing the onset of star formation (Królak and Schutz 1987; Sathyaprakash and Schutz 2009). The truly multi-messenger era of astronomy is just beginning.

10 Conclusions

The coming years will bring new large data sets at various wavelengths of the electromagnetic spectrum, accompanied by a new window due to the discovery of gravitational waves. Dedicated projects and generally oriented space and ground-based observatories will allow to make the distance measurements with current well developed methods more precise, and the new emerging methods have a chance to reach maturity. Large data sets will push the statistical errors down significantly, so the possible systematic biases will become more important. Improvements within a given method will be crucial, but the key tests will lie in cross-check of the results based on independent methods. From this point of view, new emerging methods are as important as the further increase in the already impressive accuracy achieved with the use of SN Ia.

Acknowledgements BCZ was supported by the NCN grant 2015/17/B/ST9/03436 and a Fellowship from the Chinese Academy of Sciences, and acknowledges being a part of the network supported by the COST Action TD1403 *Big Data Era in Sky and Earth Observation*. MB was partially supported by the NCN grants UMO-2014/14/M/ST9/00707, UMO-2013/01/ASPERA/ST9/00001 and the COST Action MP1304 "NewCompStar". RFIH was supported by Conselho Nacional de Desenvolvimento Cientifico e Tecnologico (CNPq). SWJ acknowledges support from NASA/Keck JPL RSA 1508337 and Department of Energy award DE-SC0011636. JJ gratefully acknowledges the support of NASA grant GO-14219.003-A. E.L. is supported by a European Union COFUND/Durham Junior Research Fellowship (under EU grant agreement no. 609412). MDO acknowledges partial support by the PRIN-INAF 2014, "Transient Universe: unveiling new types of stellar explosions with PESSTO" (P.I.: A. Pastorello). We thank ISSI-BJ for hospitality and an engaging workshop.

References

J. Aasi, B.P. Abbott, R. Abbott, T. Abbott et al., Advanced LIGO. Class. Quantum Gravity **32**(7), 074001 (2015). 1411.4547. https://doi.org/10.1088/0264-9381/32/7/074001

B.P. Abbott, R. Abbott, T.D. Abbott, M.R. Abernathy, F. Acernese, K. Ackley, C. Adams, T. Adams, P. Addesso, R.X. Adhikari et al., Properties of the binary black hole merger GW150914. Phys. Rev. Lett. **116**(24), 241102 (2016a). 1602.03840. https://doi.org/10.1103/PhysRevLett.116.241102

B.P. Abbott, R. Abbott, T.D. Abbott, M.R. Abernathy, F. Acernese, K. Ackley, C. Adams, T. Adams, P. Addesso, R.X. Adhikari et al., Prospects for observing and localizing gravitational-wave transients with Advanced LIGO and Advanced Virgo. Living Rev. Relativ. **19**, 1 (2016b). 1304.0670. https://doi.org/10.1007/lrr-2016-1

B.P. Abbott, R. Abbott, T.D. Abbott, M.R. Abernathy et al., GW151226: observation of gravitational waves from a 22-solar-mass binary black hole coalescence. Phys. Rev. Lett. **116**(24), 241103 (2016c). 1606.04855. https://doi.org/10.1103/PhysRevLett.116.241103

B.P. Abbott, R. Abbott, T.D. Abbott, M.R. Abernathy et al., Observation of gravitational waves from a binary black hole merger. Phys. Rev. Lett. **116**(6), 061102 (2016d). 1602.03837. https://doi.org/10.1103/PhysRevLett.116.061102

B.P. Abbott, R. Abbott, T.D. Abbott, F. Acernese, K. Ackley, C. Adams, T. Adams, P. Addesso, R.X. Adhikari, V.B. Adya et al., GW170104: observation of a 50-solar-mass binary black hole coalesc ence at redshift 0.2. Phys. Rev. Lett. **118**(22), 221101 (2017a). 1706.01812. https://doi.org/10.1103/PhysRevLett.118.221101

B.P. Abbott, R. Abbott, T.D. Abbott, F. Acernese et al., A gravitational-wave standard siren measurement of the Hubble constant. 1710.05835 (2017b)

B.P. Abbott, R. Abbott, T.D. Abbott, F. Acernese et al., Gravitational waves and gamma-rays from a binary neutron star merger: Gw170817 and grb 170817a. Astrophys. J. Lett. **848**(2), L13 (2017c). http://stacks.iop.org/2041-8205/848/i=2/a=L13

B.P. Abbott, R. Abbott, T.D. Abbott, F. Acernese et al., Gw170814: a three-detector observation of gravitational waves from a binary black hole coalescence. Phys. Rev. Lett. **119**, 141,101 (2017d). https://doi.org/10.1103/PhysRevLett.119.141101

B.P. Abbott, R. Abbott, T.D. Abbott, F. Acernese et al., Gw170817: observation of gravitational waves from a binary neutron star inspiral. Phys. Rev. Lett. **119**, 161,101 (2017e). https://doi.org/10.1103/PhysRevLett.119.161101

B.P. Abbott, R. Abbott, T.D. Abbott, F. Acernese et al., Multi-messenger observations of a binary neutron star merger. Astrophys. J. Lett. **848**(2), L12 (2017f). http://stacks.iop.org/2041-8205/848/i=2/a=L12

M. Abernathy, F. Acernese, A.B.P. Allen et al., Einstein gravitational wave telescope conceptual design study. https://tds.ego-gw.it/ql/?c=7954 (2011)

M.A. Abramowicz, B. Czerny, J.P. Lasota, E. Szuszkiewicz, Slim accretion disks. Astrophys. J. **332**, 646–658 (1988). https://doi.org/10.1086/166683

F. Acernese, M. Agathos, K. Agatsuma, D. Aisa et al., Advanced Virgo: a second-generation interferometric gravitational wave detector. Class. Quantum Gravity **32**(2), 024001 (2015). 1408.3978. https://doi.org/10.1088/0264-9381/32/2/024001

R. Amanullah, A. Goobar, J. Johansson, D.P.K. Banerjee, V. Venkataraman, V. Joshi, N.M. Ashok, Y. Cao, M.M. Kasliwal, S.R. Kulkarni, P.E. Nugent, T. Petrushevska, V. Stanishev, The peculiar extinction law of SN 2014J measured with the Hubble space telescope. Astrophys. J. Lett. **788**, L21 (2014). 1404.2595. https://doi.org/10.1088/2041-8205/788/2/L21

R. Amanullah, J. Johansson, A. Goobar, R. Ferretti, S. Papadogiannakis, T. Petrushevska, P.J. Brown, Y. Cao, C. Contreras, H. Dahle, N. Elias-Rosa, J.P.U. Fynbo, J. Gorosabel, L. Guaita, L. Hangard, D.A. Howell, E.Y. Hsiao, E. Kankare, M. Kasliwal, G. Leloudas, P. Lundqvist, S. Mattila, P. Nugent, M.M. Phillips, A. Sandberg, V. Stanishev, M. Sullivan, F. Taddia, G. Östlin, S. Asadi, R. Herrero-Illana, J.J. Jensen, K. Karhunen, S. Lazarevic, E. Varenius, P. Santos, S.S. Sridhar, S.H.J. Wallström, J. Wiegert, Diversity in extinction laws of type Ia supernovae measured between 0.2 and 2 µm. Mon. Not. R. Astron. Soc. **453**, 3300–3328 (2015). 1504.02101. https://doi.org/10.1093/mnras/stv1505

J.P. Anderson, S. González-Gaitán, M. Hamuy, C.P. Gutiérrez, M.D. Stritzinger, E.F. Olivares, M.M. Phillips, S. Schulze, R. Antezana, L. Bolt, A. Campillay, S. Castellón, C. Contreras, T. de Jaeger, G. Folatelli, F. Förster, W.L. Freedman, L. González, E. Hsiao, W. Krzemiński, K. Krisciunas, J. Maza, P. McCarthy, N.I. Morrell, S.E. Persson, M. Roth, F. Salgado, N.B. Suntzeff, Characterizing the V-band light-curves of hydrogen-rich type II supernovae. Astrophys. J. **786**, 67 (2014). 1403.7091. https://doi.org/10.1088/0004-637X/786/1/67

Y. Andrillat, S. Souffrin, Variations du spectre du noyau de la galaxie de Seyfert NGC 3516. Astrophys. Lett. **1**, 111 (1968)

R. Antonucci, Unified models for active galactic nuclei and quasars. Annu. Rev. Astron. Astrophys. **31**, 473–521 (1993). https://doi.org/10.1146/annurev.aa.31.090193.002353

A.L. Argon, L.J. Greenhill, M.J. Reid, J.M. Moran, E.M.L. Humphreys, Toward a new geometric distance to the active galaxy NGC 4258, I: VLBI monitoring of water maser emission. Astrophys. J. **659**, 1040–1062 (2007). astro-ph/0701396. https://doi.org/10.1086/512718

Y. Aso, Y. Michimura, K. Somiya, M. Ando et al., Interferometer design of the KAGRA gravitational wave detector. Phys. Rev. D **88**(4), 043007 (2013). 1306.6747. https://doi.org/10.1103/PhysRevD.88.043007

Y. Avni, H. Tananbaum, X-ray properties of optically selected QSOs. Astrophys. J. **305**, 83–99 (1986). https://doi.org/10.1086/164230

W. Baade, F. Zwicky, On super-novae. Proc. Natl. Acad. Sci. **20**, 254–259 (1934). https://doi.org/10.1073/pnas.20.5.254

C. Barbarino, M. Dall'Ora, M.T. Botticella, M. Della Valle, L. Zampieri, J.R. Maund, M.L. Pumo, A. Jerkstrand, S. Benetti, N. Elias-Rosa, M. Fraser, A. Gal-Yam, M. Hamuy, C. Inserra, C. Knapic, A.P. LaCluyze, M. Molinaro, P. Ochner, A. Pastorello, G. Pignata, D.E. Reichart, C. Ries, A. Riffeser, B. Schmidt, M. Schmidt, R. Smareglia, S.J. Smartt, K. Smith, J. Sollerman, M. Sullivan, L. Tomasella, M. Turatto, S. Valenti, O. Yaron, D. Young, SN 2012ec: mass of the progenitor from PESSTO follow-up of the photospheric phase. Mon. Not. R. Astron. Soc. **448**, 2312–2331 (2015). 1410.8393. https://doi.org/10.1093/mnras/stv106

E. Baron, P.E. Nugent, D. Branch, P.H. Hauschildt, Type IIP supernovae as cosmological probes: a spectral-fitting expanding atmosphere model distance to SN 1999em. Astrophys. J. Lett. **616**, L91–L94 (2004). astro-ph/0410153. https://doi.org/10.1086/426506

M. Bartelmann, P. Schneider, Weak gravitational lensing. Phys. Rep. **340**, 291–472 (2001). astro-ph/9912508. https://doi.org/10.1016/S0370-1573(00)00082-X

A.J. Barth, V.N. Bennert, G. Canalizo, A.V. Filippenko, E.L. Gates, J.E. Greene, W. Li, M.A. Malkan, A. Pancoast, D.J. Sand, D. Stern, T. Treu, J.H. Woo, R.J. Assef, H.J. Bae, B.J. Brewer, S.B. Cenko, K.I. Clubb, M.C. Cooper, A.M. Diamond-Stanic, K.D. Hiner, S.F. Hönig, E. Hsiao, M.T. Kandrashoff, M.S. Lazarova, A.M. Nierenberg, J. Rex, J.M. Silverman, E.J. Tollerud, J.L. Walsh, The lick AGN monitoring project 2011: spectroscopic campaign and emission-line light curves. Astrophys. J. Suppl. Ser. **217**, 26 (2015). 1503.01146. https://doi.org/10.1088/0067-0049/217/2/26

R. Barvainis, Hot dust and the near-infrared bump in the continuum spectra of quasars and active galactic nuclei. Astrophys. J. **320**, 537–544 (1987). https://doi.org/10.1086/165571

R.L. Beaton, W.L. Freedman, B.F. Madore, G. Bono, E.K. Carlson, G. Clementini, M.J. Durbin, A. Garofalo, D. Hatt, I.S. Jang, J.A. Kollmeier, M.G. Lee, A.J. Monson, J.A. Rich, V. Scowcroft, M. Seibert, L. Sturch, S.C. Yang, The Carnegie-Chicago Hubble program, I: a new approach to the distance ladder using only distance indicators of population II. 1604.01788 (2016)

M.C. Bentz, AGN reverberation mapping. 1505.04805 (2015)

M.C. Bentz, B.M. Peterson, H. Netzer, R.W. Pogge, M. Vestergaard, The radius-luminosity relationship for active galactic nuclei: the effect of host-galaxy starlight on luminosity measurements, II: the full sample of reverberation-mapped AGNs. Astrophys. J. **697**(160–181), doi10.1088/0004–637X/697/1/160 (2009). 0812.2283

M.C. Bentz, K.D. Denney, C.J. Grier, A.J. Barth, B.M. Peterson, M. Vestergaard, V.N. Bennert, G. Canalizo, G. De Rosa, A.V. Filippenko, E.L. Gates, J.E. Greene, W. Li, M.A. Malkan, R.W. Pogge, D. Stern, T. Treu, J.H. Woo, The low-luminosity end of the radius-luminosity relationship for active galactic nuclei. Astrophys. J. **767**, 149 (2013). 1303.1742. https://doi.org/10.1088/0004-637X/767/2/149

M.C. Bentz, M. Batiste, J. Seals, K. Garcia, R. Kuzio de Naray, W. Peters, M.D. Anderson, J. Jones, K. Lester, C. Machuca, A low-mass black hole in the nearby Seyfert galaxy UGC 06728. Astrophys. J. **831**(2), doi10.3847/0004–637X/831/1/2 (2016a). 1608.03893

M.C. Bentz, E.M. Cackett, D.M. Crenshaw, K. Horne, R. Street, B. Ou-Yang, A reverberation-based black hole mass for MCG-06-30-15. Astrophys. J. **830**, 136 (2016b). 1608.01229. https://doi.org/10.3847/0004-637X/830/2/136

E. Berger, Short-duration gamma-ray bursts. Annu. Rev. Astron. Astrophys. **52**, 43–105 (2014). 1311.2603. https://doi.org/10.1146/annurev-astro-081913-035926

M. Betoule, R. Kessler, J. Guy, J. Mosher, D. Hardin, R. Biswas, P. Astier, P. El-Hage, M. Konig, S. Kuhlmann, J. Marriner, R. Pain, N. Regnault, C. Balland, B.A. Bassett, P.J. Brown, H. Campbell, R.G. Carlberg, F. Cellier-Holzem, D. Cinabro, A. Conley, C.B. D'Andrea, D.L. DePoy, M. doi, R.S. Ellis, S. Fabbro, A.V. Filippenko, R.J. Foley, J.A. Frieman, D. Fouchez, L. Galbany, A. Goobar, R.R. Gupta, G.J. Hill, R. Hlozek, C.J. Hogan, I.M. Hook, D.A. Howell, S.W. Jha, L. Le Guillou, G. Leloudas, C. Lidman, J.L. Marshall, A. Möller, A.M. Mourão, J. Neveu, R. Nichol, M.D. Olmstead, N. Palanque-Delabrouille, S. Perlmutter, J.L. Prieto, C.J. Pritchet, M. Richmond, A.G. Riess, V. Ruhlmann-Kleider, M. Sako, K. Schahmaneche, D.P. Schneider, M. Smith, J. Sollerman, M. Sullivan, N.A. Walton, C.J. Wheeler, Improved cosmological constraints from a joint analysis of the SDSS-II and SNLS supernova samples. Astron. Astrophys. **568**, A22 (2014). 1401.4064 https://doi.org/10.1051/0004-6361/201423413

L. Bildsten, K.J. Shen, N.N. Weinberg, G. Nelemans, Faint thermonuclear supernovae from AM Canum Venaticorum binaries. Astrophys. J. Lett. **662**, L95–L98 (2007). astro-ph/0703578. https://doi.org/10.1086/519489

M. Birkinshaw, The Sunyaev-Zel'dovich effect. Phys. Rep. **310**, 97–195 (1999). astro-ph/9808050. https://doi.org/10.1016/S0370-1573(98)00080-5

J.P. Blakeslee, Cosmic distances from surface brightness fluctuations, in *Advancing the Physics of Cosmic Distances*, ed. by R. de Grijs. IAU Symposium, vol. 289 (2013), pp. 304–311. https://doi.org/10.1017/S174392131202159X

J.P. Blakeslee, A. Jordán, S. Mei, P. Côté, L. Ferrarese, L. Infante, E.W. Peng, J.L. Tonry, M.J. West, The ACS Fornax cluster survey, V: measurement and recalibration of surface brightness fluctuations and a precise value of the Fornax-Virgo relative distance. Astrophys. J. **694**, 556–572 (2009). 0901.1138. https://doi.org/10.1088/0004-637X/694/1/556

R.D. Blandford, C.F. McKee, Reverberation mapping of the emission line regions of Seyfert galaxies and quasars. Astrophys. J. **255**, 419–439 (1982). https://doi.org/10.1086/159843

S.I. Blinnikov, A.M. Khokhlov, Development of detonations in degenerate stars. Sov. Astron. Lett. **12**, 131–134 (1986)

S. Blondin, J.L. Prieto, F. Patat, P. Challis, M. Hicken, R.P. Kirshner, T. Matheson, M. Modjaz, A second case of variable Na I D lines in a highly reddened type Ia supernova. Astrophys. J. **693**, 207–215 (2009). 0811.0002. https://doi.org/10.1088/0004-637X/693/1/207

J.S. Bloom, D.E. Holz, S.A. Hughes, K. Menou, A. Adams, S.F. Anderson, A. Becker, G.C. Bower, N. Brandt, B. Cobb, K. Cook, A. Corsi, S. Covino, D. Fox, A. Fruchter, C. Fryer, J. Grindlay, D. Hartmann, Z. Haiman, B. Kocsis, L. Jones, A. Loeb, S. Marka, B. Metzger, E. Nakar, S. Nissanke, D.A. Perley, T. Piran, D. Poznanski, T. Prince, J. Schnittman, A. Soderberg, M. Strauss, P.S. Shawhan, D.H. Shoemaker, J. Sievers, C. Stubbs, G. Tagliaferri, P. Ubertini, P. Wozniak, Astro2010 decadal survey whitepaper: coordinated science in the gravitational and electromagnetic skies. 0902.1527 (2009)

M. Bonamente, M.K. Joy, S.J. LaRoque, J.E. Carlstrom, E.D. Reese, K.S. Dawson, Determination of the cosmic distance scale from Sunyaev-Zel'dovich effect and Chandra X-ray measurements of high-redshift galaxy clusters. Astrophys. J. **647**, 25–54 (2006). astro-ph/0512349. https://doi.org/10.1086/505291

H. Bondi, Plane gravitational waves in general relativity. Nature **179**, 1072–1073 (1957). https://doi.org/10.1038/1791072a0

S. Bose, B. Kumar, Distance determination to eight galaxies using expanding photosphere method. Astrophys. J. **782**, 98 (2014). 1401.5115. https://doi.org/10.1088/0004-637X/782/2/98

M.C.P. Bours, S. Toonen, G. Nelemans, Single degenerate supernova type Ia progenitors. Studying the influence of different mass retention efficiencies. Astron. Astrophys. **552**, A24 (2013). 1302.2629. https://doi.org/10.1051/0004-6361/201220692

J. Braatz, J. Condon, A. Constantin, F. Gao, J. Greene, L. Hao, C. Henkel, V. Impellizzeri, C.Y. Kuo, E. Litzinger, K.Y. Lo, D. Pesce, M. Reid, J. Wagner, W. Zhao, The Megamaser cosmology project: survey completion. IAU Gen. Assem. **22**, 2255730 (2015)

P.J. Brown, P.W.A. Roming, P.A. Milne, The first ten years of Swift supernovae. J. High Energy Astrophys. **7**, 111–116 (2015a). 1504.08017. https://doi.org/10.1016/j.jheap.2015.04.007

P.J. Brown, M.T. Smitka, L. Wang, A. Breeveld, M. de Pasquale, D.H. Hartmann, K. Krisciunas, N.P. Kuin, P.A. Milne, M. Page, Swift ultraviolet observations of supernova 2014J in M82: large extinction from interstellar dust. Astrophys. J. **805**, 74 (2015b). 1408.2381. https://doi.org/10.1088/0004-637X/805/1/74

C.R. Burns, M. Stritzinger, M.M. Phillips, S. Kattner, S.E. Persson, B.F. Madore, W.L. Freedman, L. Boldt, A. Campillay, C. Contreras, G. Folatelli, S. Gonzalez, W. Krzeminski, N. Morrell, F. Salgado, N.B. Suntzeff, The Carnegie supernova project: light-curve fitting with SNooPy. Astron. J. **141**, 19 (2011). 1010.4040. https://doi.org/10.1088/0004-6256/141/1/19

C.R. Burns, M. Stritzinger, M.M. Phillips, E.Y. Hsiao, C. Contreras, S.E. Persson, G. Folatelli, L. Boldt, A. Campillay, S. Castellón, W.L. Freedman, B.F. Madore, N. Morrell, F. Salgado, N.B. Suntzeff, The Carnegie supernova project: intrinsic colors of type Ia supernovae. Astrophys. J. **789**, 32 (2014). 1405.3934. https://doi.org/10.1088/0004-637X/789/1/32

E.M. Cackett, K. Horne, H. Winkler, Testing thermal reprocessing in active galactic nuclei accretion discs. Mon. Not. R. Astron. Soc. **380**, 669–682 (2007). 0706.1464. https://doi.org/10.1111/j.1365-2966.2007.12098.x

E. Cappellaro, M. Riello, G. Altavilla, M.T. Botticella, S. Benetti, A. Clocchiatti, J.I. Danziger, P. Mazzali, A. Pastorello, F. Patat, M. Salvo, M. Turatto, S. Valenti, Death rate of massive stars at redshift about 0.3. Astron. Astrophys. **430**, 83–93 (2005). astro-ph/0407216. https://doi.org/10.1051/0004-6361:20041256

J.A. Cardelli, G.C. Clayton, J.S. Mathis, The relationship between infrared, optical, and ultraviolet extinction. Astrophys. J. **345**, 245–256 (1989). https://doi.org/10.1086/167900

J.E. Carlstrom, G.P. Holder, E.D. Reese, Cosmology with the Sunyaev-Zel'dovich effect. Annu. Rev. Astron. Astrophys. **40**, 643–680 (2002). astro-ph/0208192. https://doi.org/10.1146/annurev.astro.40.060401.093803

A. Cavaliere, R. Fusco-Femiano, The distribution of hot gas in clusters of galaxies. Astron. Astrophys. **70**, 677 (1978)

D. Chelouche, S. Zucker, Quasar cartography: from black hole to broad-line region scales. Astrophys. J. **769**, 124 (2013). 1305.6499. https://doi.org/10.1088/0004-637X/769/2/124

D. Chelouche, S.E. Rafter, G.I. Cotlier, S. Kaspi, A.J. Barth, A tentative size-luminosity relation for the iron emission-line region in quasars. Astrophys. J. Lett. **783**, L34 (2014). 1404.6142. https://doi.org/10.1088/2041-8205/783/2/L34

A.M. Cherepashchuk, V.M. Lyutyi, Rapid variations of Hα intensity in the nuclei of Seyfert galaxies NGC 4151, 3516, 1068. Astrophys. Lett. **13**, 165 (1973)

M. Chevallier, D. Polarski, Accelerating universes with scaling dark matter. Int. J. Mod. Phys. D **10**, 213–223 (2001). gr-qc/0009008. https://doi.org/10.1142/S0218271801000822

L. Chomiuk, SN 2011fe: a laboratory for testing models of type Ia supernovae. Publ. Astron. Soc. Aust. **30**, e046 (2013). 1307.2721. https://doi.org/10.1017/pasa.2013.24

N. Chotard, E. Gangler, G. Aldering, P. Antilogus, C. Aragon, S. Bailey, C. Baltay, S. Bongard, C. Buton, A. Canto, M. Childress, Y. Copin, H.K. Fakhouri, E.Y. Hsiao, M. Kerschhaggl, M. Kowalski, S. Loken, P. Nugent, K. Paech, R. Pain, E. Pecontal, R. Pereira, S. Perlmutter, D. Rabinowitz, K. Runge, R. Scalzo, G. Smadja, C. Tao, R.C. Thomas, B.A. Weaver, C. Wu, Nearby Supernova Factory, The reddening law of type Ia supernovae: separating intrinsic variability from dust using equivalent widths. Astron. Astrophys. **529**, L4 (2011). 1103.5300. https://doi.org/10.1051/0004-6361/201116723

J.S.W. Claeys, O.R. Pols, R.G. Izzard, J. Vink, F.W.M. Verbunt, Theoretical uncertainties of the type Ia supernova rate. Astron. Astrophys. **563**, A83 (2014). 1401.2895. https://doi.org/10.1051/0004-6361/201322714

C. Gea, Gaia data release 1. Testing parallaxes with local Cepheids and RR Lyrae stars. Astron. Astrophys. **605**, A79 (2017). 1705.00688. https://doi.org/10.1051/0004-6361/201629925

A.D. Code, Hubble's study of the stellar contnet of M31. Astrophys. J. **525**, 212 (1999)

S. Collier, K. Horne, I. Wanders, B.M. Peterson, A new direct method for measuring the Hubble constant from reverberating accretion discs in active galaxies. Mon. Not. R. Astron. Soc. **302**, L24–L28 (1999). astro-ph/9811278. https://doi.org/10.1046/j.1365-8711.1999.02250.x

S. Collier, D.M. Crenshaw, B.M. Peterson et al., Multiwavelength monitoring of the narrow-line Seyfert 1 galaxy Arakelian 564, II: ultraviolet continuum and emission-line variability. Astrophys. J. **561**, 146–161 (2001). astro-ph/0107059. https://doi.org/10.1086/323234

S.J. Collier, K. Horne, S. Kaspi et al., Steps toward determination of the size and structure of the broad-line region in active galactic nuclei, XIV: intensive optical spectrophotometric observations of NGC 7469. Astrophys. J. **500**, 162–172 (1998). https://doi.org/10.1086/305720

A. Conley, R.G. Carlberg, J. Guy, D.A. Howell, S. Jha, A.G. Riess, M. Sullivan, Is there evidence for a Hubble bubble? The nature of type Ia supernova colors and dust in external galaxies. Astrophys. J. Lett. **664**, L13–L16 (2007). 0705.0367. https://doi.org/10.1086/520625

A. Conley, M. Sullivan, E.Y. Hsiao, J. Guy, P. Astier, D. Balam, C. Balland, S. Basa, R.G. Carlberg, D. Fouchez, D. Hardin, D.A. Howell, I.M. Hook, R. Pain, K. Perrett, C.J. Pritchet, N. Regnault, SiFTO: an empirical method for fitting SN Ia light curves. Astrophys. J. **681**, 482–498 (2008). 0803.3441. https://doi.org/10.1086/588518

L. Cortese, L.M.R. Fogarty, K. Bekki, J. van de Sande, W. Couch, B. Catinella, M. Colless, D. Obreschkow, D. Taranu, E. Tescari, D. Barat, J. Bland-Hawthorn, J. Bloom, J.J. Bryant, M. Cluver, S.M. Croom, M.J. Dreinkwater, F. d'Eugenio, I.S. Konstantopoulos, A. Lopez-Sanchez, S. Mahajan, N. Scott, C. Tonini, O.I. Wong, J.T. Allen, S. Brough, M. Goodwin, A.W. Green, I.T. Ho, L.S. Kelvin, J.S. Lawrence, N.P.F. Lorente, A.M. Medling, M.S. Owers, S. Richards, R. Sharp, S.M. Sweet, The SAMI galaxy survey: the link between angular momentum and optical morphology. Mon. Not. R. Astron. Soc. **463**, 170–184 (2016). 1608.00291. https://doi.org/10.1093/mnras/stw1891

D.A. Coulter, C.D. Kilpatrick, R.J. Foley, J. Anais, M.M. Phillips, M.R. Drout, B.F. Madore, A.L. Piro, B.J. Shappee, E.Y. Hsiao, Swope photometric observations of SN 2017cbv = DLT17u. Astron. Telegr. **10167**, 167 (2017)

A.P.S. Crotts, Light echoes from supernova 2014J in M82. Astrophys. J. Lett. **804**, L37 (2015). 1409.8671. https://doi.org/10.1088/2041-8205/804/2/L37

A.P.S. Crotts, D. Yourdon, The nature and geometry of the light echo from SN 2006X. Astrophys. J. **689**, 1186–1190 (2008). 0804.2030. https://doi.org/10.1086/592318

A. Cucchiara, A.J. Levan, D.B. Fox, N.R. Tanvir, T.N. Ukwatta, E. Berger, T. Krühler, A. Küpcü Yoldaş, X.F. Wu, K. Toma, J. Greiner, F.E. Olivares, A. Rowlinson, L. Amati, T. Sakamoto, K. Roth, A. Stephens, A. Fritz, J.P.U. Fynbo, J. Hjorth, D. Malesani, P. Jakobsson, K. Wiersema, P.T. O'Brien, A.M. Soderberg, R.J. Foley, A.S. Fruchter, J. Rhoads, R.E. Rutledge, B.P. Schmidt, M.A. Dopita, P. Podsiadlowski, R. Willingale, C. Wolf, S.R. Kulkarni, P. D'Avanzo, A photometric redshift of $z \sim 9.4$ for GRB 090429B. Astrophys. J. **736**, 7 (2011). 1105.4915. https://doi.org/10.1088/0004-637X/736/1/7

B. Czerny, K. Hryniewicz, The origin of the broad line region in active galactic nuclei. Astron. Astrophys. **525**, L8 (2011). 1010.6201. https://doi.org/10.1051/0004-6361/201016025

B. Czerny, K. Hryniewicz, I. Maity, A. Schwarzenberg-Czerny, P.T. Zycki, M. Bilicki, Towards equation of state of dark energy from quasar monitoring: reverberation strategy. Astron. Astrophys. **556**, A97 (2013). 1212.0472. https://doi.org/10.1051/0004-6361/201220832

E. da Cunha, A.M. Hopkins, M. Colless, E.N. Taylor, C. Blake, C. Howlett, C. Magoulas, J.R. Lucey, C. Lagos, K. Kuehn, Y. Gordon, D. Barat, F. Bian, C. Wolf, M.J. Cowley, M. White, I. Achitouv, M. Bilicki, J. Bland-Hawthorn, K. Bolejko, M.J.I. Brown, R. Brown, J. Bryant, S. Croom, T.M. Davis, S.P. Driver, M.D. Filipovic, S.R. Hinton, M. Johnston-Hollitt, D.H. Jones, B. Koribalski, D. Kleiner, J. Lawrence, N. Lorente, J. Mould, M.S. Owers, K. Pimbblet, C.G. Tinney, N.F.H. Tothill, F. Watson, The Taipan galaxy survey: scientific goals and observing strategy. Publ. Astron. Soc. Aust. **34**, e047 (2017). 1706.01246. https://doi.org/10.1017/pasa.2017.41

M.G. Dainotti, S. Postnikov, X. Hernandez, M. Ostrowski, A fundamental plane for long gamma-ray bursts with X-ray plateaus. Astrophys. J. Lett. **825**, L20 (2016). 1604.06840. https://doi.org/10.3847/2041-8205/825/2/L20

N. Dalal, D.E. Holz, S.A. Hughes, B. Jain, Short grb and binary black hole standard sirens as a probe of dark energy. Phys. Rev. D **74**, 063006 (2006). astro-ph/0601275. https://doi.org/10.1103/PhysRevD.74.063006

M. Dall'Ora, M.T. Botticella, M.L. Pumo, L. Zampieri, L. Tomasella, G. Pignata, The type IIP supernova 2012aw in M95: hydrodynamical modeling of the photospheric phase from accurate spectrophotometric monitoring. Astrophys. J. **787**, 139 (2014). 1404.1294. https://doi.org/10.1088/0004-637X/787/2/139

C.B. D'Andrea, M. Sako, B. Dilday, J.A. Frieman, J. Holtzman, R. Kessler, K. Konishi, D.P. Schneider, J. Sollerman, J.C. Wheeler, N. Yasuda, D. Cinabro, S. Jha, R.C. Nichol, H. Lampeitl, M. Smith, D.W. Atlee, B. Bassett, F.J. Castander, A. Goobar, R. Miquel, J. Nordin, L. Östman, J.L. Prieto, R. Quimby, A.G. Riess, M. Stritzinger, Type II-P supernovae from the SDSS-II supernova survey and the standardized candle method. Astrophys. J. **708**, 661–674 (2010). 0910.5597. https://doi.org/10.1088/0004-637X/708/1/661

D. Kea, Lisa: laser interferometer space antenna for gravitational wave measurements. Class. Quantum Gravity **13**, A247–A250 (1996). https://doi.org/10.1088/0264-9381/13/11A/033

T.A. Davis, J. Greene, C.P. Ma, V. Pandya, J.P. Blakeslee, N. McConnell, J. Thomas, The MASSIVE survey, III: molecular gas and a broken Tully-Fisher relation in the most massive early-type galaxies. Mon. Not. R. Astron. Soc. **455**, 214–226 (2016). 1510.00729. https://doi.org/10.1093/mnras/stv2313

E. De Filippis, M. Sereno, M.W. Bautz, G. Longo, Measuring the three-dimensional structure of galaxy clusters, I: application to a sample of 25 clusters. Astrophys. J. **625**, 108–120 (2005). astro-ph/0502153. https://doi.org/10.1086/429401

T. de Jaeger, S. González-Gaitán, J.P. Anderson, L. Galbany, M. Hamuy, M.M. Phillips, M.D. Stritzinger, C.P. Gutiérrez, L. Bolt, C.R. Burns, A. Campillay, S. Castellón, C. Contreras, G. Folatelli, W.L. Freedman, E.Y. Hsiao, K. Krisciunas, W. Krzeminski, H. Kuncarayakti, N. Morrell, E.F. Olivares, S.E. Persson, N. Suntzeff, A Hubble diagram from type II supernovae based solely on photometry: the photometric color method. Astrophys. J. **815**, 121 (2015). 1511.05145. https://doi.org/10.1088/0004-637X/815/2/121

T. de Jaeger, S. González-Gaitán, M. Hamuy, L. Galbany, J.P. Anderson, M.M. Phillips, M.D. Stritzinger, R.G. Carlberg, M. Sullivan, C.P. Gutiérrez, I.M. Hook, D.A. Howell, E.Y. Hsiao, H. Kuncarayakti, V. Ruhlmann-Kleider, G. Folatelli, C. Pritchet, S. Basa, A type II supernova Hubble diagram from the CSP-I, SDSS-II, and SNLS surveys. Astrophys. J. **835**, 166 (2017). 1612.05636. https://doi.org/10.3847/1538-4357/835/2/166

M. della Valle, N. Panagia, Type IA supernovae in late type galaxies—reddening correction, scale height, and absolute maximum magnitude. Astron. J. **104**, 696–703 (1992). https://doi.org/10.1086/116265

S. Dhawan, S.W. Jha, B. Leibundgut, Measuring the Hubble constant with type Ia supernovae as near-infrared standard candles. 1707.00715 (2017)

R. Di Stefano, The progenitors of type Ia supernovae, I: are they supersoft sources? Astrophys. J. **712**, 728–733 (2010). 0912.0757. https://doi.org/10.1088/0004-637X/712/1/728

R. Di Stefano, R. Voss, J.S.W. Claeys, Spin-up/spin-down models for type Ia supernovae. Astrophys. J. Lett. **738**, L1 (2011). 1102.4342. https://doi.org/10.1088/2041-8205/738/1/L1

P.M. Dos Santos, J.R.D. Lepine, Detection of strong H_2O emission from galaxy NGC4945. Nature **278**, 34 (1979). https://doi.org/10.1038/278034a0

A. Dressler, D. Lynden-Bell, D. Burstein, R.L. Davies, S.M. Faber, R. Terlevich, G. Wegner, Spectroscopy and photometry of elliptical galaxies, I: a new distance estimator. Astrophys. J. **313**, 42–58 (1987)

P. Du, C. Hu, K.X. Lu, Y.K. Huang, C. Cheng, J. Qiu, Y.R. Li, Y.W. Zhang, X.L. Fan, J.M. Bai, W.H. Bian, Y.F. Yuan, S. Kaspi, L.C. Ho, H. Netzer, J.M. Wang, SEAMBH Collaboration, Supermassive black holes with high accretion rates in active galactic nuclei, IV: Hβ time lags and implications for super-Eddington accretion. Astrophys. J. **806**, 22 (2015). 1504.01844. https://doi.org/10.1088/0004-637X/806/1/22

P. Du, K.X. Lu, Z.X. Zhang, Y.K. Huang, K. Wang, C. Hu, J. Qiu, Y.R. Li, X.L. Fan, X.E. Fang, J.M. Bai, W.H. Bian, Y.F. Yuan, L.C. Ho, J.M. Wang, SEAMBH Collaboration, Supermassive black holes with high accretion rates in active galactic nuclei, V: a new size-luminosity scaling relation for the broad-line region. Astrophys. J. **825**, 126 (2016a). 1604.06218. https://doi.org/10.3847/0004-637X/825/2/126

P. Du, J.M. Wang, C. Hu, L.C. Ho, Y.R. Li, J.M. Bai, The fundamental plane of the broad-line region in active galactic nuclei. Astrophys. J. Lett. **818**, L14 (2016b). 1601.01391. https://doi.org/10.3847/2041-8205/818/1/L14

R.A. Edelson, T. Alexander, D.M. Crenshaw et al., Multiwavelength observations of short-timescale variability in NGC 4151, IV: analysis of multiwavelength continuum variability. Astrophys. J. **470**, 364 (1996). astro-ph/9605082. https://doi.org/10.1086/177872

R. Edelson, J.M. Gelbord, K. Horne et al., Space telescope and optical reverberation mapping project, II: Swift and HST reverberation mapping of the accretion disk of NGC 5548. Astrophys. J. **806**, 129 (2015). 1501.05951. https://doi.org/10.1088/0004-637X/806/1/129

A. Einstein, Zur allgemeinen Relativitätstheorie, in *Sitzungsberichte der Königlich Preußischen Akademie der Wissenschaften (Berlin)* (1915), pp. 778–786

A. Einstein, Näherungsweise Integration der Feldgleichungen der Gravitation, in *Sitzungsberichte der Königlich Preußischen Akademie der Wissenschaften (Berlin)* (1916), pp. 688–696

A. Einstein, Über Gravitationswellen, in *Sitzungsberichte der Königlich Preußischen Akademie der Wissenschaften (Berlin)* (1918), pp. 154–167

N. Elias-Rosa, S. Benetti, E. Cappellaro, M. Turatto, P.A. Mazzali, F. Patat, W.P.S. Meikle, M. Stehle, A. Pastorello, G. Pignata, R. Kotak, A. Harutyunyan, G. Altavilla, H. Navasardyan, Y. Qiu, M. Salvo, W. Hillebrandt, Anomalous extinction behaviour towards the type Ia SN 2003cg. Mon. Not. R. Astron. Soc. **369**, 1880–1900 (2006). astro-ph/0603316. https://doi.org/10.1111/j.1365-2966.2006.10430.x

N. Elias-Rosa, S. Benetti, M. Turatto, E. Cappellaro, S. Valenti, A.A. Arkharov, J.E. di Beckman, A. Paola, M. Dolci, A.V. Filippenko, R.J. Foley, K. Krisciunas, V.M. Larionov, W. Li, W.P.S. Meikle, A. Pastorello, G. Valentini, W. Hillebrandt, SN 2002cv: a heavily obscured type Ia supernova. Mon. Not. R. Astron. Soc. **384**, 107–122 (2008). 0710.4503. https://doi.org/10.1111/j.1365-2966.2007.12638.x

M. Elitzur, Astronomical masers. Annu. Rev. Astron. Astrophys. **30**, 75–112 (1992). https://doi.org/10.1146/annurev.aa.30.090192.000451

M. Elvis, M. Karovska, Quasar parallax: a method for determining direct geometrical distances to quasars. Astrophys. J. Lett. **581**, L67–L70 (2002). astro-ph/0211385. https://doi.org/10.1086/346015

I.M.H. Etherington, On the definition of distance in general relativity. Philos. Mag. **15** (1933)

I.M.H. Etherington, Republication of: LX. On the definition of distance in general relativity. Gen. Relativ. Gravit. **39**, 1055–1067 (2007). https://doi.org/10.1007/s10714-007-0447-x

S.M. Faber, R.E. Jackson, Velocity dispersions and mass-to-light ratios for elliptical galaxies. Astrophys. J. **204**, 668–683 (1976). https://doi.org/10.1086/154215

M.M. Fausnaugh, C.S. Kochanek, J.R. Gerke, L.M. Macri, A.G. Riess, K.Z. Stanek, The Cepheid distance to the maser-host galaxy NGC 4258: studying systematics with the Large Binocular Telescope. Mon. Not. R. Astron. Soc. **450**, 3597–3619 (2015). 1412.2138. https://doi.org/10.1093/mnras/stv881

M.M. Fausnaugh, K.D. Denney, A.J. Barth et al., Space telescope and optical reverberation mapping project, III: optical continuum emission and broadband time delays in NGC 5548. Astrophys. J. **821**, 56 (2016). 1510.05648. https://doi.org/10.3847/0004-637X/821/1/56

A.V. Filippenko, Optical spectra of supernovae. Annu. Rev. Astron. Astrophys. **35**, 309–355 (1997a). https://doi.org/10.1146/annurev.astro.35.1.309

A.V. Filippenko, Optical spectra of supernovae. Annu. Rev. Astron. Astrophys. **35**, 309–355 (1997b). https://doi.org/10.1146/annurev.astro.35.1.309

G. Fiorentino, I. Musella, M. Marconi, Cepheid theoretical models and observations in HST/WFC3 filters: the effect on the Hubble constant H_0. Mon. Not. R. Astron. Soc. **434**, 2866 (2013). 1306.6276. https://doi.org/10.1093/mnras/stt1193

E.L. Fitzpatrick, D. Massa, An analysis of the shapes of interstellar extinction curves V: The IR-through-UV curve morphology. Astrophys. J. **663**, 320–341 (2007). 0705.0154. https://doi.org/10.1086/518158

G. Folatelli, M.M. Phillips, C.R. Burns, C. Contreras, M. Hamuy, W.L. Freedman, S.E. Persson, M. Stritzinger, N.B. Suntzeff, K. Krisciunas, L. Boldt, S. González, W. Krzeminski, N. Morrell, M. Roth, F. Salgado, B.F. Madore, D. Murphy, P. Wyatt, W. Li, A.V. Filippenko, N. Miller, The Carnegie supernova project: analysis of the first sample of low-redshift type-Ia supernovae. Astron. J. **139**, 120–144 (2010). 0910.3317. https://doi.org/10.1088/0004-6256/139/1/120

M. Foley, R. Foley, D. Scolnic, A. Rest, A.G. Riess, S.W. Jha, R. Kirshner, O. Dosovitz Fox, Y.C. Pan, S. Smartt, Calibration and simulation of the foundation supernova survey, in *American Astronomical Society Meeting Abstracts*, vol. 229 (2017), p. 341.12

R.J. Foley, D. Kasen, Measuring ejecta velocity improves type Ia supernova distances. Astrophys. J. **729**, 55 (2011). 1011.4517. https://doi.org/10.1088/0004-637X/729/1/55

R.J. Foley, A.V. Filippenko, C. Aguilera, A.C. Becker, S. Blondin, P. Challis, A. Clocchiatti, R. Covarrubias, T.M. Davis, P.M. Garnavich, S.W. Jha, R.P. Kirshner, K. Krisciunas, B. Leibundgut, W. Li, T. Matheson, A. Miceli, G. Miknaitis, G. Pignata, A. Rest, A.G. Riess, B.P. Schmidt, R.C. Smith, J. Sollerman, J. Spyromilio, C.W. Stubbs, N.B. Suntzeff, J.L. Tonry, W.M. Wood-Vasey, A. Zenteno, Constraining cosmic evolution of type Ia supernovae. Astrophys. J. **684**, 68 (2008). 0710.2338. https://doi.org/10.1086/589612

R.J. Foley, T. Matheson, S. Blondin, R. Chornock, J.M. Silverman, P. Challis, A. Clocchiatti, A.V. Filippenko, R.P. Kirshner, B. Leibundgut, J. Sollerman, J. Spyromilio, J.L. Tonry, T.M. Davis, P.M. Garnavich, S.W. Jha, K. Krisciunas, W. Li, G. Pignata, A. Rest, A.G. Riess, B.P. Schmidt, R.C. Smith, C.W. Stubbs, B.E. Tucker, W.M. Wood-Vasey, Spectroscopy of high-redshift supernovae from the essence project: the first four years. Astron. J. **137**, 3731–3742 (2009). 0811.4424. https://doi.org/10.1088/0004-6256/137/4/3731

R.J. Foley, N.E. Sanders, R.P. Kirshner, Velocity evolution and the intrinsic color of type Ia supernovae. Astrophys. J. **742**, 89 (2011). 1107.3555. https://doi.org/10.1088/0004-637X/742/2/89

R.J. Foley, O.D. Fox, C. McCully, M.M. Phillips, D.J. Sand, W. Zheng, P. Challis, A.V. Filippenko, G. Folatelli, W. Hillebrandt, E.Y. Hsiao, S.W. Jha, R.P. Kirshner, M. Kromer, G.H. Marion, M. Nelson, R. Pakmor, G. Pignata, F.K. Röpke, I.R. Seitenzahl, J.M. Silverman, M. Skrutskie, M.D. Stritzinger, Extensive HST ultraviolet spectra and multiwavelength observations of SN 2014J in M82 indicate reddening and circumstellar scattering by typical dust. Mon. Not. R. Astron. Soc. **443**, 2887–2906 (2014). 1405.3677. https://doi.org/10.1093/mnras/stu1378

W.L. Freedman, Measuring and modeling the universe (2004)

W.L. Freedman, Cosmology at a crossroads. Nat. Astron. **1**, 0121 (2017). https://doi.org/10.1038/s41550-017-0121

W.L. Freedman, B.F. Madore, The Hubble constant. Annu. Rev. Astron. Astrophys. **48**, 673–710 (2010). 1004.1856. https://doi.org/10.1146/annurev-astro-082708-101829

W.L. Freedman, B.F. Madore, B.K. Gibson, L. Ferrarese, D.D. Kelson, S. Sakai, J.R. Mould, R.C. Kennicutt Jr., H.C. Ford, J.A. Graham, J.P. Huchra, S.M.G. Hughes, G.D. Illingworth, L.M. Macri, P.B. Stetson, Final results from the Hubble space telescope key project to measure the Hubble constant. Astrophys. J. **553**, 47–72 (2001a). astro-ph/0012376. https://doi.org/10.1086/320638

W.L. Freedman, B.F. Madore, B.K. Gibson, L. Ferrarese, D.D. Kelson, S. Sakai, J.R. Mould, R.C Kennicutt Jr., H.C. Ford, J.A. Graham, J.P. Huchra, S.M.G. Hughes, G.D. Illingworth, L.M. Macri, P.B. Stetson,

Final results from the Hubble space telescope key project to measure the Hubble constant. Astrophys. J. **553**, 47–72 (2001b). astro-ph/0012376. https://doi.org/10.1086/320638

W.L. Freedman, C.R. Burns, M.M. Phillips, P. Wyatt, S.E. Persson, B.F. Madore, C. Contreras, G. Folatelli, E.S. Gonzalez, M. Hamuy, E. Hsiao, D.D. Kelson, N. Morrell, D.C. Murphy, M. Roth, M. Stritzinger, L. Sturch, N.B. Suntzeff, P. Astier, C. Balland, B. Bassett, L. Boldt, R.G. Carlberg, A.J. Conley, J.A. Frieman, P.M. Garnavich, J. Guy, D. Hardin, D.A. Howell, R. Kessler, H. Lampeitl, J. Marriner, R. Pain, K. Perrett, N. Regnault, A.G. Riess, M. Sako, D.P. Schneider, M. Sullivan, M. Wood-Vasey, The Carnegie supernova project: first near-infrared Hubble diagram to $z \sim 0.7$. Astrophys. J. **704**, 1036–1058 (2009). 0907.4524. https://doi.org/10.1088/0004-637X/704/2/1036

W.L. Freedman, B.F. Madore, V. Scowcroft, C. Burns, A. Monson, S.E. Persson, M. Seibert, J. Rigby, Carnegie Hubble program: a mid-infrared calibration of the Hubble constant. Astrophys. J. **758**(24) (2012). 1208.3281. https://doi.org/10.1088/0004-637X/758/1/24

J.A. Frieman, M.S. Turner, D. Huterer, Dark energy and the accelerating universe. Annu. Rev. Astron. Astrophys. **46**, 385–432 (2008). 0803.0982. https://doi.org/10.1146/annurev.astro.46.060407.145243

A. Gal-Yam, P.A. Mazzali, I. Manulis, D. Bishop, Supernova discoveries 2010-2011: statistics and trends. Publ. Astron. Soc. Pac. **125**, 749 (2013). 1103.5165. https://doi.org/10.1086/671483

P. Galianni, K. Horne, A test of the failed disc wind scenario for the origin of the broad-line region in active galactic nuclei. Mon. Not. R. Astron. Soc. **435**, 3122–3124 (2013). 1306.2835. https://doi.org/10.1093/mnras/stt1507

E.E.E. Gall, R. Kotak, B. Leibundgut, S. Taubenberger, W. Hillebrandt, M. Kromer, Applying the expanding photosphere and standardized candle methods to type II-Plateau supernovae at cosmologically significant redshifts. The distance to SN 2013eq. Astron. Astrophys. **592**, A129 (2016). 1603.04730. https://doi.org/10.1051/0004-6361/201628333

J. Gao, B.W. Jiang, A. Li, J. Li, X. Wang, Physical dust models for the extinction toward supernova 2014J in M82. Astrophys. J. Lett. **807**, L26 (2015). 1507.00417. https://doi.org/10.1088/2041-8205/807/2/L26

D. García-Senz, C. Badenes, N. Serichol, Is there a hidden hole in type Ia supernova remnants? Astrophys. J. **745**, 75 (2012). 1110.4267. https://doi.org/10.1088/0004-637X/745/1/75

E. Gardner, C. Done, The UV/optical lags in NGC 5548: it's not disc reprocessing and the soft X-ray excess is part of the solution. 1603.09564 (2016)

C.M. Gaskell, L.S. Sparke, Line variations in quasars and Seyfert galaxies. Astrophys. J. **305**, 175–186 (1986). https://doi.org/10.1086/164238

M. Gierliński, M. Nikołajuk, B. Czerny, High-frequency X-ray variability as a mass estimator of stellar and supermassive black holes. Mon. Not. R. Astron. Soc. **383**, 741–749 (2008). 0710.1566. https://doi.org/10.1111/j.1365-2966.2007.12584.x

M. Gilfanov, Á. Bogdán, An upper limit on the contribution of accreting white dwarfs to the type Ia supernova rate. Nature **463**, 924–925 (2010). 1002.3359. https://doi.org/10.1038/nature08685

A. Goobar, Low R_V from circumstellar dust around supernovae. Astrophys. J. Lett. **686**, L103 (2008). 0809.1094. https://doi.org/10.1086/593060

A. Goobar, B. Leibundgut, Supernova cosmology: legacy and future. Annu. Rev. Nucl. Part. Sci. **61**, 251–279 (2011). 1102.1431. https://doi.org/10.1146/annurev-nucl-102010-130434

M.L. Graham, S. Valenti, B.J. Fulton, L.M. Weiss, K.J. Shen, P.L. Kelly, W. Zheng, A.V. Filippenko, G.W. Marcy, D.A. Howell, J. Burt, E.J. Rivera, Time-varying potassium in high-resolution spectra of the type Ia supernova 2014j. Astrophys. J. **801**, 136 (2015). 1412.0653. https://doi.org/10.1088/0004-637X/801/2/136

E.K. Grassberg, V.S. Imshennik, D.K. Nadyozhin, On the theory of the light curves of supernovate. Astrophys. Space Sci. **10**, 28–51 (1971). https://doi.org/10.1007/BF00654604

D. Grupe, S. Komossa, K.M. Leighly, K.L. Page, The simultaneous optical-to-X-ray spectral energy distribution of soft X-ray selected active galactic nuclei observed by Swift. Astrophys. J. Suppl. **187**, 64–106 (2010). 1001.3140. https://doi.org/10.1088/0067-0049/187/1/64

J. Guillochon, J. Parrent, R. Margutti, An open catalog for supernova data. 1605.01054 (2016)

J. Guy, P. Astier, S. Nobili, N. Regnault, R. Pain, SALT: a spectral adaptive light curve template for type Ia supernovae. Astron. Astrophys. **443**, 781–791 (2005). astro-ph/0506583. https://doi.org/10.1051/0004-6361:20053025

J. Guy, P. Astier, S. Baumont, D. Hardin, R. Pain, N. Regnault, S. Basa, R.G. Carlberg, A. Conley, S. Fabbro, D. Fouchez, I.M. Hook, D.A. Howell, K. Perrett, C.J. Pritchet, J. Rich, M. Sullivan, P. Antilogus, E. Aubourg, G. Bazin, J. Bronder, M. Filiol, N. Palanque-Delabrouille, P. Ripoche, V. Ruhlmann-Kleider, SALT2: using distant supernovae to improve the use of type Ia supernovae as distance indicators. Astron. Astrophys. **466**, 11–21 (2007). astro-ph/0701828. https://doi.org/10.1051/0004-6361:20066930

M. Haas, R. Chini, M. Ramolla, F. Pozo Nuñez, C. Westhues, R. Watermann, V. Hoffmeister, M. Murphy, Photometric AGN reverberation mapping—an efficient tool for BLR sizes, black hole masses,

and host-subtracted AGN luminosities. Astron. Astrophys. **535**, A73 (2011). 1109.1848. https://doi.org/10.1051/0004-6361/201117325
I. Hachisu, M. Kato, A theoretical light-curve model for the recurrent nova V394 coronae Australis. Astrophys. J. **540**, 447–451 (2000). astro-ph/0003471. https://doi.org/10.1086/309338
I. Hachisu, M. Kato, K. Nomoto, A new model for progenitor systems of type IA supernovae. Astrophys. J. Lett. **470**, L97 (1996). https://doi.org/10.1086/310303
I. Hachisu, M. Kato, K. Nomoto, Supersoft X-ray phase of single degenerate type Ia supernova progenitors in early-type galaxies. Astrophys. J. Lett. **724**, L212–L216 (2010). 1010.5860. https://doi.org/10.1088/2041-8205/724/2/L212
M. Hamuy, The standard candle method for type II supernovae and the Hubble constant. astro-ph/0309122 (2003)
M. Hamuy, The latest version of the standardized candle method for type II supernovae. Measuring and modeling the universe. astro-ph/0301281 (2004)
M. Hamuy, P.A. Pinto, Type II supernovae as standardized candles. Astrophys. J. Lett. **566**, L63–L65 (2002). astro-ph/0201279. https://doi.org/10.1086/339676
M. Hamuy, M.M. Phillips, J. Maza, N.B. Suntzeff, R.A. Schommer, R. Aviles, A Hubble diagram of distant type IA supernovae. Astron. J. **109**, 1–13 (1995). https://doi.org/10.1086/117251
M. Hamuy, M.M. Phillips, N.B. Suntzeff, R.A. Schommer, J. Maza, R. Aviles, The absolute luminosities of the Calan/Tololo type IA supernovae. Astron. J. **112**, 2391 (1996). astro-ph/9609059. https://doi.org/10.1086/118190
M. Hamuy, G. Folatelli, N.I. Morrell, M.M. Phillips, N.B. Suntzeff, S.E. Persson, M. Roth, S. Gonzalez, W. Krzeminski, C. Contreras, W.L. Freedman, D.C. Murphy, B.F. Madore, P. Wyatt, J. Maza, A.V. Filippenko, W. Li, P.A. Pinto, The Carnegie supernova project: the low-redshift survey. Publ. Astron. Soc. Pac. **118**, 2–20 (2006). astro-ph/0512039. https://doi.org/10.1086/500228
A. Heger, C.L. Fryer, S.E. Woosley, N. Langer, D.H. Hartmann, How massive single stars end their life. Astrophys. J. **591**, 288–300 (2003). astro-ph/0212469. https://doi.org/10.1086/375341
J.R. Herrnstein, J.M. Moran, L.J. Greenhill, A.S. Trotter, The geometry of and mass accretion rate through the maser accretion disk in NGC 4258. Astrophys. J. **629**, 719–738 (2005). astro-ph/0504405. https://doi.org/10.1086/431421
M. Hicken, W.M. Wood-Vasey, S. Blondin, P. Challis, S. Jha, P.L. Kelly, A. Rest, R.P. Kirshner, Improved dark energy constraints from ~ 100 new CfA supernova type Ia light curves. Astrophys. J. **700**, 1097–1140 (2009). 0901.4804. https://doi.org/10.1088/0004-637X/700/2/1097
S. Hilbert, J.R. Gair, L.J. King, Reducing distance errors for standard candles and standard sirens with weak-lensing shear and flexion maps. Mon. Not. R. Astron. Soc. **412**, 1023–1037 (2011). 1007.2468. https://doi.org/10.1111/j.1365-2966.2010.17963.x
J. Hjorth, J.S. Bloom, in *The Gamma-Ray Burst—Supernova Connection* (2012), pp. 169–190
L.C. Ho, Z.Y. Li, A.J. Barth, M.S. Seigar, C.Y. Peng, The Carnegie-Irvine galaxy survey, I: overview and atlas of optical images. Astrophys. J. Suppl. **197**, 21 (2011). 1111.4605. https://doi.org/10.1088/0067-0049/197/2/21
R.F.L. Holanda, K.N.N.O. Barros, Searching for cosmological signatures of the Einstein equivalence principle breaking. Phys. Rev. D **94**(2), 023524 (2016). 1606.07923. https://doi.org/10.1103/PhysRevD.94.023524
R.F.L. Holanda, J.A.S. Lima, M.B. Ribeiro, Testing the distance-duality relation with galaxy clusters and type Ia supernovae. Astrophys. J. Lett. **722**, L233–L237 (2010). 1005.4458. https://doi.org/10.1088/2041-8205/722/2/L233
R.F.L. Holanda, J.V. Cunha, L. Marassi, J.A.S. Lima, Constraining H_0 in general dark energy models from Sunyaev-Zeldovich/X-ray technique and complementary probes. J. Cosmol. Astropart. Phys. **2**, 035 (2012a). 1006.4200. https://doi.org/10.1088/1475-7516/2012/02/035
R.F.L. Holanda, J.A.S. Lima, M.B. Ribeiro, Probing the cosmic distance-duality relation with the Sunyaev-Zel'dovich effect, X-ray observations and supernovae Ia. Astron. Astrophys. **538**, A131 (2012b). 1104.3753. https://doi.org/10.1051/0004-6361/201118343
R.F.L. Holanda, J.S. Alcaniz, J.C. Carvalho, Cosmography with the Sunyaev-Zeldovich effect and X-ray data. J. Cosmol. Astropart. Phys. **6**, 033 (2013). 1303.3307. https://doi.org/10.1088/1475-7516/2013/06/033
R.F.L. Holanda, V.C. Busti, L.R. Colaço, J.S. Alcaniz, S.J. Landau, Galaxy clusters, type Ia supernovae and the fine structure constant. J. Cosmol. Astropart. Phys. **8**, 055 (2016). 1605.02578. https://doi.org/10.1088/1475-7516/2016/08/055
T.W.S. Holoien, J.S. Brown, K.Z. Stanek, C.S. Kochanek, B.J. Shappee, J.L. Prieto, S. Dong, J. Brimacombe, D.W. Bishop, U. Basu, J.F. Beacom, D. Bersier, P. Chen, A.B. Danilet, E. Falco, D. Godoy-Rivera, N. Goss, G. Pojmanski, G.V. Simonian, D.M. Skowron, T.A. Thompson, P.R. Woźniak, C.G. Ávila, G. Bock, J.L.G. Carballo, E. Conseil, C. Contreras, I. Cruz, J.M.F. Andújar, Z. Guo, E.Y. Hsiao, S. Kiyota, R.A. Koff, G. Krannich, B.F. Madore, P. Marples, G. Masi, N. Morrell, L.A.G.

Monard, J.C. Munoz-Mateos, B. Nicholls, J. Nicolas, R.M. Wagner, W.S. Wiethoff, The ASAS-SN bright supernova catalogue, II: 2015. Mon. Not. R. Astron. Soc. **467**, 1098–1111 (2017a). 1610.03061. https://doi.org/10.1093/mnras/stx057

T.W.S. Holoien, J.S. Brown, K.Z. Stanek, C.S. Kochanek, B.J. Shappee, J.L. Prieto, S. Dong, J. Brimacombe, D.W. Bishop, S. Bose, J.F. Beacom, D. Bersier, P. Chen, L. Chomiuk, E. Falco, D. Godoy-Rivera, N. Morrell, G. Pojmanski, J.V. Shields, J. Strader, M.D. Stritzinger, T.A. Thompson, P.R. Woźniak, G. Bock, P. Cacella, E. Conseil, I. Cruz, J.M. Fernandez, S. Kiyota, R.A. Koff, G. Krannich, P. Marples, G. Masi, L.A.G. Monard, B. Nicholls, J. Nicolas, R.S. Post, G. Stone, W.S. Wiethoff, The ASAS-SN bright supernova catalog, III: 2016. 1704.02320 (2017b)

T.W.S. Holoien, P.J. Marshall, R.H. Wechsler, EmpiriciSN: re-sampling observed supernova/host galaxy populations using an XD Gaussian mixture model. Astron. J. **153**, 249 (2017c). 1611.00363. https://doi.org/10.3847/1538-3881/aa68a1

T.W.S. Holoien, K.Z. Stanek, C.S. Kochanek, B.J. Shappee, J.L. Prieto, J. Brimacombe, D. Bersier, D.W. Bishop, S. Dong, J.S. Brown, A.B. Danilet, G.V. Simonian, U. Basu, J.F. Beacom, E. Falco, G. Pojmanski, D.M. Skowron, P.R. Woźniak, C.G. Ávila, E. Conseil, C. Contreras, I. Cruz, J.M. Fernández, R.A. Koff, Z. Guo, G.J. Herczeg, J. Hissong, E.Y. Hsiao, J. Jose, S. Kiyota, F. Long, L.A.G. Monard, B. Nicholls, J. Nicolas, W.S. Wiethoff, The ASAS-SN bright supernova catalogue, I: 2013-2014. Mon. Not. R. Astron. Soc. **464**, 2672–2686 (2017d). 1604.00396. https://doi.org/10.1093/mnras/stw2273

D.E. Holz, S.A. Hughes, Using gravitational-wave standard sirens. Astrophys. J. **629**, 15–22 (2005). astro-ph/0504616. https://doi.org/10.1086/431341

S.F. Hönig, D. Watson, M. Kishimoto, J. Hjorth, A dust-parallax distance of 19 megaparsecs to the supermassive black hole in NGC 4151. Nature **515**, 528–530 (2014). 1411.7032. https://doi.org/10.1038/nature13914

I.M. Hook, Supernovae and cosmology with future European facilities. Philos. Trans. R. Soc. Lond. Ser. A **371**, 20120 (2013). 1211.6586. https://doi.org/10.1098/rsta.2012.0282

A.M. Hopkins, J.F. Beacom, On the normalization of the cosmic star formation history. Astrophys. J. **651**, 142–154 (2006). astro-ph/0601463. https://doi.org/10.1086/506610

G. Hosseinzadeh, D.A. Howell, D. Sand, S. Valenti, L. Tartaglia, C. McCully, K.A. Bostroem, S. Wyatt, I. Arcavi, FLOYDS classification of DLT17u/AT 2017cbv as a very young type Ia supernova. Astron. Telegr. **10164**, 164 (2017)

D.A. Howell, M. Sullivan, K. Perrett, T.J. Bronder, I.M. Hook, P. Astier, E. Aubourg, D. Balam, S. Basa, R.G. Carlberg, S. Fabbro, D. Fouchez, J. Guy, H. Lafoux, J.D. Neill, R. Pain, N. Palanque-Delabrouille, C.J. Pritchet, N. Regnault, J. Rich, R. Taillet, R. Knop, R.G. McMahon, S. Perlmutter, N.A. Walton, Gemini spectroscopy of supernovae from the supernova legacy survey: improving high-redshift supernova selection and classification. Astrophys. J. **6**(34), 1190–1201 (2005). astro-ph/0509195. https://doi.org/10.1086/497119

F. Hoyle, W.A. Fowler, Nucleosynthesis in supernovae. Astrophys. J. **132**, 565 (1960). https://doi.org/10.1086/146963

K. Hryniewicz, B. Czerny, W. Pych, A. Udalski, M. Krupa, A. Świętoń, J. Kaluzny, SALT long-slit spectroscopy of LBQS 2113-4538: variability of the Mg II and Fe II component. Astron. Astrophys. **562**, A34 (2014). 1308.3980. https://doi.org/10.1051/0004-6361/201322487

E.P. Hubble, A spiral nebula as a stellar system, Messier 31. Astrophys. J. **69** (1929). https://doi.org/10.1086/143167

E.M.L. Humphreys, M.J. Reid, J.M. Moran, L.J. Greenhill, A.L. Argon, Toward a new geometric distance to the active galaxy NGC 4258, III: final results and the Hubble constant. Astrophys. J. **775**, 13 (2013). 1307.6031. https://doi.org/10.1088/0004-637X/775/1/13

K. Ichikawa, C. Ricci, Y. Ueda, K. Matsuoka, Y. Toba, T. Kawamuro, B. Trakhtenbrot, M.J. Koss, The complete infrared view of active galactic nuclei from the 70 month Swift/BAT catalog. Astrophys. J. **835**, 74 (2017). 1611.09858. https://doi.org/10.3847/1538-4357/835/1/74

D. Ilić, P. LČ, A.I. Shapovalova, A.N. Burenkov, V.H. Chavushyan, A. Kovačević, Line shape variability in a sample of AGN with broad lines. J. Astrophys. Astron. **36**, 433–445 (2015). 1510.02162. https://doi.org/10.1007/s12036-015-9360-y

N. Itoh, Y. Kohyama, S. Nozawa, Relativistic corrections to the Sunyaev-Zeldovich effect for clusters of galaxies. Astrophys. J. **502**, 7–15 (1998). astro-ph/9712289. https://doi.org/10.1086/305876

N. Ivanova, S. Justham, X. Chen, O. De Marco, C.L. Fryer, E. Gaburov, H. Ge, E. Glebbeek, Z. Han, X.D. Li, G. Lu, T. Marsh, P. Podsiadlowski, A. Potter, N. Soker, R. Taam, T.M. Tauris, E.P.J. van den Heuvel, R.F. Webbink, Common envelope evolution: where we stand and how we can move forward. Astron. Astrophys. Rev. **21**, 59 (2013). 1209.4302. https://doi.org/10.1007/s00159-013-0059-2

D. Jack, M. Mittag, K.P. Schröder, J.H.M.M. Schmitt, A. Hempelmann, J.N. González-Pérez, M.A. Trinidad, J.M. Cabrera Sixto, G. Rauw, Time series of high-resolution spectra of SN 2014J observed with the TIGRE telescope. Mon. Not. R. Astron. Soc. **451**, 4104–4113 (2015). 1506.00938. https://doi.org/10.1093/mnras/stv1238

I.S. Jang, M.G. Lee, The tip of the red giant branch distance to the perfect spiral galaxy M74 hosting three core-collapse supernovae. Astrophys. J. **792**, 52 (2014). 1407.2246. https://doi.org/10.1088/0004-637X/792/1/52

J.B. Jensen, J.P. Blakeslee, Z. Gibson, H.C. Lee, M. Cantiello, G. Raimondo, N. Boyer, H. Cho, Measuring infrared surface brightness fluctuation distances with HST WFC3: calibration and advice. Astrophys. J. **808**, 91 (2015). 1505.00400. https://doi.org/10.1088/0004-637X/808/1/91

S. Jha, R.P. Kirshner, P. Challis, P.M. Garnavich, T. Matheson, A.M. Soderberg, G.J.M. Graves, M. Hicken, J.F. Alves, H.G. Arce, Z. Balog, P. Barmby, E.J. Barton, P. Berlind, A.E. Bragg, C. Briceño, W.R. Brown, J.H. Buckley, N. Caldwell, M.L. Calkins, B.J. Carter, K.D. Concannon, R.H. Donnelly, K.A. Eriksen, D.G. Fabricant, E.E. Falco, F. Fiore, M.R. Garcia, M. Gómez, N.A. Grogin, T. Groner, P.J. Groot, K.E. Haisch Jr., L. Hartmann, C.W. Hergenrother, M.J. Holman, J.P. Huchra, R. Jayawardhana, D. Jerius, S.J. Kannappan, D.W. Kim, J.T. Kleyna, C.S. Kochanek, D.M. Koranyi, M. Krockenberger, C.J. Lada, K.L. Luhman, J.X. Luu, L.M. Macri, J.A. Mader, A. Mahdavi, M. Marengo, B.G. Marsden, B.A. McLeod, B.R. McNamara, S.T. Megeath, D. Moraru, A.E. Mossman, A.A. Muench, J.A. Muñoz, J. Muzerolle, O. Naranjo, K. Nelson-Patel, M.A. Pahre, B.M. Patten, J. Peters, W. Peters, J.C. Raymond, K. Rines, R.E. Schild, G.J. Sobczak, T.B. Spahr, J.R. Stauffer, R.P. Stefanik, A.H. Szentgyorgyi, E.V. Tollestrup, P. Väisänen, A. Vikhlinin, Z. Wang, S.P. Willner, S.J. Wolk, J.M. Zajac, P. Zhao, K.Z. Stanek, UBVRI light curves of 44 type Ia supernovae. Astron. J. **131**, 527–554 (2006). astro-ph/0509234. https://doi.org/10.1086/497989

S. Jha, A.G. Riess, R.P. Kirshner, Improved distances to type Ia supernovae with multicolor light-curve shapes: MLCS2k2. Astrophys. J. **659**, 122–148 (2007). astro-ph/0612666. https://doi.org/10.1086/512054

D.O. Jones, S.A. Rodney, A.G. Riess, B. Mobasher, T. Dahlen, C. McCully, T.F. Frederiksen, S. Casertano, J. Hjorth, C.R. Keeton, A. Koekemoer, L.G. Strolger, T.G. Wiklind, P. Challis, O. Graur, B. Hayden, B. Patel, B.J. Weiner, A.V. Filippenko, P. Garnavich, S.W. Jha, R.P. Kirshner, H.C. Ferguson, N.A. Grogin, D. Kocevski, The discovery of the most distant known type Ia supernova at redshift 1.914. Astrophys. J. **768**, 166 (2013). 1304.0768. https://doi.org/10.1088/0004-637X/768/2/166

D.O. Jones, A.G. Riess, D.M. Scolnic, Reconsidering the effects of local star formation on type Ia supernova cosmology. Astrophys. J. **812**, 31 (2015). 1506.02637. https://doi.org/10.1088/0004-637X/812/1/31

D.W. Just, W.N. Brandt, O. Shemmer, A.T. Steffen, D.P. Schneider, G. Chartas, G.P. Garmire, The X-ray properties of the most luminous quasars from the sloan digital sky survey. Astrophys. J. **665**, 1004–1022 (2007). 0705.3059. https://doi.org/10.1086/519990

S. Justham, Single-degenerate type Ia supernovae without hydrogen contamination. Astrophys. J. Lett. **730**, L34 (2011). 1102.4913. https://doi.org/10.1088/2041-8205/730/2/L34

D. Kasen, S.E. Woosley, On the origin of the type Ia supernova width-luminosity relation. Astrophys. J. **656**, 661–665 (2007). astro-ph/0609540. https://doi.org/10.1086/510375

D. Kasen, S.E. Woosley, Type II supernovae: model light curves and standard candle relationships. Astrophys. J. **703**, 2205–2216 (2009). 0910.1590. https://doi.org/10.1088/0004-637X/703/2/2205

D. Kasen, F.K. Röpke, S.E. Woosley, The diversity of type Ia supernovae from broken symmetries. Nature **460**, 869–872 (2009). 0907.0708. https://doi.org/10.1038/nature08256

S. Kaspi, P.S. Smith, H. Netzer, D. Maoz, B.T. Jannuzi, U. Giveon, Reverberation measurements for 17 quasars and the size-mass-luminosity relations in active galactic nuclei. Astrophys. J. **533**, 631–649 (2000). astro-ph/9911476. https://doi.org/10.1086/308704

S. Kaspi, W.N. Brandt, D. Maoz, H. Netzer, D.P. Schneider, O. Shemmer, Reverberation mapping of high-luminosity quasars: first results. Astrophys. J. **659**, 997–1007 (2007). astro-ph/0612722. https://doi.org/10.1086/512094

K.S. Kawabata, H. Akitaya, M. Yamanaka, R. Itoh, K. Maeda, Y. Moritani, T. Ui, M. Kawabata, K. Mori, D. Nogami, K. Nomoto, N. Suzuki, K. Takaki, M. Tanaka, I. Ueno, S. Chiyonobu, T. Harao, R. Matsui, H. Miyamoto, O. Nagae, A. Nakashima, H. Nakaya, Y. Ohashi, T. Ohsugi, T. Komatsu, K. Sakimoto, M. Sasada, H. Sato, H. Tanaka, T. Urano, T. Yamashita, M. Yoshida, A. Arai, N. Ebisuda, Y. Fukazawa, A. Fukui, O. Hashimoto, S. Honda, H. Izumiura, Y. Kanda, K. Kawaguchi, N. Kawai, D. Kuroda, K. Masumoto, K. Matsumoto, T. Nakaoka, K. Takata, M. Uemura, K. Yanagisawa, Optical and near-infrared polarimetry of highly reddened type Ia supernova 2014J: peculiar properties of dust in M82. Astrophys. J. Lett. **795**, L4 (2014). 1407.0452. https://doi.org/10.1088/2041-8205/795/1/L4

W.E. Kerzendorf, B.P. Schmidt, M. Asplund, K. Nomoto, P. Podsiadlowski, A. Frebel, R.A. Fesen, D. Yong, Subaru high-resolution spectroscopy of star G in the Tycho supernova remnant. Astrophys. J. **701**, 1665–1672 (2009). 0906.0982. https://doi.org/10.1088/0004-637X/701/2/1665

A.L. King, T.M. Davis, K.D. Denney, M. Vestergaard, D. Watson, High-redshift standard candles: predicted cosmological constraints. Mon. Not. R. Astron. Soc. **441**, 3454–3476 (2014). 1311.2356. https://doi.org/10.1093/mnras/stu793

A.L. King, P. Martini, T.M. Davis, K.D. Denney, C.S. Kochanek, B.M. Peterson, A. Skielboe, M. Vestergaard, E. Huff, D. Watson, M. Banerji, R. McMahon, R. Sharp, C. Lidman, Simulations of the OzDES AGN reverberation mapping project. Mon. Not. R. Astron. Soc. **453**, 1701–1726 (2015). 1504.03031. https://doi.org/10.1093/mnras/stv1718

R.P. Kirshner, J. Kwan, Distances to extragalactic supernovae. Astrophys. J. **193**, 27–36 (1974). https://doi.org/10.1086/153123

M. Kishimoto, S.F. Hönig, R. Antonucci, F. Millour, K.R.W. Tristram, G. Weigelt, Mapping the radial structure of AGN tori. Astron. Astrophys. **536**, A78 (2011). 1110.4290. https://doi.org/10.1051/0004-6361/201117367

Y. Kobayashi, Y. Yoshii, B.A. Peterson, S. Miyazaki, T. Aoki, T. Minezaki, K. Kawara, K. Enya, N. Okada, M. Suganuma, B. Greene, M. O'Brien, L.K. Randall, MAGNUM (multicolor active galactic nuclei monitoring) project, in *Advanced Technology Optical/IR Telescopes VI*, ed. by L.M. Stepp. Proc. SPIE, vol. 3352 (1998), pp. 120–128. https://doi.org/10.1117/12.319247

K.T. Korista, M.R. Goad, The variable diffuse continuum emission of broad-line clouds. Astrophys. J. **553**, 695–708 (2001). astro-ph/0101117. https://doi.org/10.1086/320964

S. Koshida, T. Minezaki, Y. Yoshii, Y. Kobayashi, Y. Sakata, S. Sugawara, K. Enya, M. Suganuma, H. Tomita, T. Aoki, B.A. Peterson, Reverberation measurements of the inner radius of the dust torus in 17 Seyfert galaxies. Astrophys. J. **788**, 159 (2014). 1406.2078. https://doi.org/10.1088/0004-637X/788/2/159

S. Koshida, Y. Yoshii, Y. Kobayashi, T. Minezaki, K. Enya, M. Suganuma, H. Tomita, T. Aoki, B.A. Peterson, Calibration of AGN reverberation distance measurements. Astrophys. J. Lett. **842**, L13 (2017). 1705.09757. https://doi.org/10.3847/2041-8213/aa7553

K. Krisciunas, J.L. Prieto, P.M. Garnavich, J.L.G. Riley, A. Rest, C. Stubbs, R. McMillan, Photometry of the type Ia supernovae 1999cc, 1999cl, and 2000cf. Astron. J. **131**, 1639–1647 (2006). astro-ph/0511162. https://doi.org/10.1086/499523

K. Krisciunas, P.M. Garnavich, V. Stanishev, N.B. Suntzeff, J.L. Prieto, J. Espinoza, D. Gonzalez, M.E. Salvo, N. Elias de la Rosa, S.J. Smartt, J.R. Maund, R.P. Kudritzki, The type Ia supernova 2004S, a clone of SN 2001el, and the optimal photometric bands for extinction estimation. Astron. J. **133**, 58–72 (2007). astro-ph/0609268. https://doi.org/10.1086/509126

A. Królak, B.F. Schutz, Coalescing binaries—probe of the universe. Gen. Relativ. Gravit. **19**, 1163–1171 (1987). https://doi.org/10.1007/BF00759095

F. La Franca, S. Bianchi, G. Ponti, E. Branchini, G. Matt, A new cosmological distance measure using active galactic nucleus X-ray variability. Astrophys. J. Lett. **787**, L12 (2014). 1404.2607. https://doi.org/10.1088/2041-8205/787/1/L12

R.N. Lang, S.A. Hughes, Localizing coalescing massive black hole binaries with gravitational waves. Astrophys. J. **677**, 1184–1200 (2008). arXiv:0710.3795. https://doi.org/10.1086/528953

D.C. Leonard, Constraining the type Ia supernova progenitor: the search for hydrogen in nebular spectra. Astrophys. J. **670**, 1275–1282 (2007). 0710.3166. https://doi.org/10.1086/522367

D.C. Leonard, S.M. Kanbur, C.C. Ngeow, N.R. Tanvir, The Cepheid distance to NGC 1637: a direct test of the expanding photosphere method distance to SN 1999em. Astrophys. J. **594**, 247–278 (2003). astro-ph/0305259. https://doi.org/10.1086/376831

K. Lepo, M. van Kerkwijk, A search for rapidly accreting white dwarfs in the Small Magellanic Cloud. Astrophys. J. **771**, 13 (2013). 1301.1368. https://doi.org/10.1088/0004-637X/771/1/13

P. Lesaffre, Z. Han, C.A. Tout, P. Podsiadlowski, R.G. Martin, The C flash and the ignition conditions of type Ia supernovae. Mon. Not. R. Astron. Soc. **368**, 187–195 (2006). astro-ph/0601443. https://doi.org/10.1111/j.1365-2966.2006.10068.x

W. Li, A.V. Filippenko, R.R. Treffers, A.G. Riess, J. Hu, Y. Qiu, A high intrinsic peculiarity rate among type IA supernovae. Astrophys. J. **546**, 734–743 (2001). astro-ph/0006292. https://doi.org/10.1086/318299

W. Li, J.S. Bloom, P. Podsiadlowski, A.A. Miller, S.B. Cenko, S.W. Jha, M. Sullivan, D.A. Howell, P.E. Nugent, N.R. Butler, E.O. Ofek, M.M. Kasliwal, J.W. Richards, A. Stockton, H.Y. Shih, L. Bildsten, M.M. Shara, J. Bibby, A.V. Filippenko, M. Ganeshalingam, J.M. Silverman, S.R. Kulkarni, N.M. Law, D. Poznanski, R.M. Quimby, C. McCully, B. Patel, K. Maguire, K.J. Shen, Exclusion of a luminous red giant as a companion star to the progenitor of supernova SN 2011fe. Nature **480**, 348–350 (2011a). 1109.1593. https://doi.org/10.1038/nature10646

Z. Li, P. Wu, H. Yu, Cosmological-model-independent tests for the distance-duality relation from galaxy clusters and type Ia supernova. Astrophys. J. Lett. **729**, L14 (2011b). 1101.5255. https://doi.org/10.1088/2041-8205/729/1/L14

P. Lira, I. Botti, H. Netzer, S. Kaspi, Reverberation mapping of the most luminous quasars at $z \sim 2$–3, in *Active Galactic Nuclei: What's in a Name?* (2016), p. 10. https://doi.org/10.5281/zenodo.60361

I.I. Litvinova, D.K. Nadezhin, Hydrodynamical models of type II supernovae. Astrophys. Space Sci. **89**, 89–113 (1983). https://doi.org/10.1007/BF01008387

K.Y. Lo, Mega-masers and galaxies. Annu. Rev. Astron. Astrophys. **43**, 625–676 (2005). https://doi.org/10.1146/annurev.astro.41.011802.094927

LSST Science Collaboration, P.A. Abell, J. Allison, S.F. Anderson, J.R. Andrew, J.R.P. Angel, L. Armus, D. Arnett, S.J. Asztalos, T.S. Axelrod et al., LSST science book, version 2.0. 0912.0201 (2009)

E. Lusso, G. Risaliti, The tight relation between X-ray and ultraviolet luminosity of quasars. Astrophys. J. **819**, 154 (2016). 1602.01090. https://doi.org/10.3847/0004-637X/819/2/154

E. Lusso, A. Comastri, C. Vignali, G. Zamorani, M. Brusa, R. Gilli, K. Iwasawa, M. Salvato, F. Civano, M. Elvis, A. Merloni, A. Bongiorno, J.R. Trump, A.M. Koekemoer, E. Schinnerer, E. Le Floc'h, N. Cappelluti, K. Jahnke, M. Sargent, J. Silverman, V. Mainieri, F. Fiore, M. Bolzonella, O. Le Fèvre, B. Garilli, A. Iovino, J.P. Kneib, F. Lamareille, S. Lilly, M. Mignoli, M. Scodeggio, D. Vergani, The X-ray to optical-UV luminosity ratio of X-ray selected type 1 AGN in XMM-COSMOS. Astron. Astrophys. **512**, A34 (2010). 0912.4166. https://doi.org/10.1051/0004-6361/200913298

C.P. Ma, J.E. Greene, N. McConnell, R. Janish, J.P. Blakeslee, J. Thomas, J.D. Murphy, The MASSIVE survey, I: a volume-limited integral-field spectroscopic study of the most massive early-type galaxies within 108 Mpc. Astrophys. J. **795**, 158 (2014). 1407.1054. https://doi.org/10.1088/0004-637X/795/2/158

C.L. MacLeod, C.J. Hogan, Precision of Hubble constant derived using black hole binary absolute distances and statistical redshift information. Phys. Rev. D **77**(4), 043512 (2008). 0712.0618. https://doi.org/10.1103/PhysRevD.77.043512

K. Maeda, T. Nozawa, T. Nagao, K. Motohara, Constraining the amount of circumstellar matter and dust around type Ia supernovae through near-infrared echoes. Mon. Not. R. Astron. Soc. **452**, 3281–3292 (2015). 1411.3778. https://doi.org/10.1093/mnras/stv1498

K. Maeda, A. Tajitsu, K.S. Kawabata, R.J. Foley, S. Honda, Y. Moritani, M. Tanaka, O. Hashimoto, M. Ishigaki, J.D. Simon, M.M. Phillips, M. Yamanaka, D. Nogami, A. Arai, W. Aoki, K. Nomoto, D. Milisavljevic, P.A. Mazzali, A.M. Soderberg, M. Schramm, B. Sato, H. Harakawa, N. Morrell, N. Arimoto, Sodium absorption systems toward SN Ia 2014J originate on interstellar scales. Astrophys. J.816:57 (2016). 1511.05668. https://doi.org/10.3847/0004-637X/816/2/57

K. Maguire, R. Kotak, S.J. Smartt, A. Pastorello, M. Hamuy, F. Bufano, Type II-P supernovae as standardized candles: improvements using near-infrared data. Mon. Not. R. Astron. Soc. **403**, L11–L15 (2010). 0912.3107. https://doi.org/10.1111/j.1745-3933.2009.00804.x

K.S. Mandel, G. Narayan, R.P. Kirshner, Type Ia supernova light curve inference: hierarchical models in the optical and near-infrared. Astrophys. J. **731**, 120 (2011). 1011.5910. https://doi.org/10.1088/0004-637X/731/2/120

K.S. Mandel, D. Scolnic, H. Shariff, R.J. Foley, R.P. Kirshner, The type Ia supernova color-magnitude relation and host galaxy dust: a simple hierarchical Bayesian model. 1609.04470 (2016)

D. Maoz, O. Graur, Star formation, supernovae, iron, and α: consistent cosmic and galactic histories. 1703.04540 (2017)

D. Maoz, F. Mannucci, Type-Ia supernova rates and the progenitor problem: a review. Publ. Astron. Soc. Aust. **29**, 447–465 (2012). 1111.4492. https://doi.org/10.1071/AS11052

D. Maoz, F. Mannucci, T.D. Brandt, The delay-time distribution of type Ia supernovae from Sloan II. Mon. Not. R. Astron. Soc. **426**, 3282–3294 (2012). 1206.0465. https://doi.org/10.1111/j.1365-2966.2012.21871.x

D. Maoz, F. Mannucci, G. Nelemans, Observational clues to the progenitors of type Ia supernovae. Annu. Rev. Astron. Astrophys. **52**, 107–170 (2014). 1312.0628. https://doi.org/10.1146/annurev-astro-082812-141031

P. Massey, E. Waterhouse, K. DeGioia-Eastwood, The progenitor masses of Wolf-Rayet stars and luminous blue variables determined from cluster turnoffs, I: results from 19 OB associations in the Magellanic Clouds. Astron. J. **119**, 2214–2241 (2000). astro-ph/0002233. https://doi.org/10.1086/301345

P. Massey, K. DeGioia-Eastwood, E. Waterhouse, The progenitor masses of Wolf-Rayet stars and luminous blue variables determined from cluster turnoffs, II: results from 12 galactic clusters and OB associations. Astron. J. **121**, 1050–1070 (2001). astro-ph/0010654. https://doi.org/10.1086/318769

S. Mateos, F.J. Carrera, X. Barcons, A. Alonso-Herrero, A. Hernán-Caballero, M. Page, C. Ramos Almeida, A. Caccianiga, T. Miyaji, A. Blain, Survival of the obscuring torus in the most powerful active galactic nuclei. Astrophys. J. Lett. **841**, L18 (2017). 1705.04323. https://doi.org/10.3847/2041-8213/aa7268

T. Matheson, S. Blondin, R.J. Foley, R. Chornock, A.V. Filippenko, B. Leibundgut, R.C. Smith, J. Sollerman, J. Spyromilio, R.P. Kirshner, A. Clocchiatti, C. Aguilera, B. Barris, A.C. Becker, P. Challis, R. Covarrubias, P. Garnavich, M. Hicken, S. Jha, K. Krisciunas, W. Li, A. Miceli, G. Miknaitis, J.L. Prieto, A. Rest, A.G. Riess, M.E. Salvo, B.P. Schmidt, C.W. Stubbs, N.B. Suntzeff, J.L. Tonry, Spectroscopy of high-redshift supernovae from the ESSENCE project: the first 2 years. Astron. J. **129**, 2352–2375 (2005). astro-ph/0411357. https://doi.org/10.1086/429679

T. Matheson, X. Fan, R. Green, A. McConnachie, J. Newman, K. Olsen, P. Szkody, W.M. Wood-Vasey, Spectroscopy in the era of LSST. 1311.2496 (2013)

I.M. McHardy, D.T. Cameron, T. Dwelly, S. Connolly, P. Lira, D. Emmanoulopoulos, J. Gelbord, E. Breedt, P. Arevalo, P. Uttley, Swift monitoring of NGC 5548: X-ray reprocessing and short-term UV/optical variability. Mon. Not. R. Astron. Soc. **444**, 1469–1474 (2014). 1407.6361. https://doi.org/10.1093/mnras/stu1636

S. Mei, J.P. Blakeslee, P. Côté, J.L. Tonry, M.J. West, L. Ferrarese, A. Jordán, E.W. Peng, A. Anthony, D. Merritt, The ACS Virgo Cluster survey, XIII: SBF distance catalog and the three-dimensional structure of the Virgo Cluster. Astrophys. J. **655**, 144–162 (2007). astro-ph/0702510. https://doi.org/10.1086/509598

G. Miknaitis, G. Pignata, A. Rest, W.M. Wood-Vasey, S. Blondin, P. Challis, R.C. Smith, C.W. Stubbs, N.B. Suntzeff, R.J. Foley, T. Matheson, J.L. Tonry, C. Aguilera, J.W. Blackman, A.C. Becker, A. Clocchiatti, R. Covarrubias, T.M. Davis, A.V. Filippenko, A. Garg, P.M. Garnavich, M. Hicken, S. Jha, K. Krisciunas, R.P. Kirshner, B. Leibundgut, W. Li, A. Miceli, G. Narayan, J.L. Prieto, A.G. Riess, M.E. Salvo, B.P. Schmidt, J. Sollerman, J. Spyromilio, A. Zenteno, The ESSENCE supernova survey: survey optimization, observations, and supernova photometry. Astrophys. J. **666**, 674–693 (2007). astro-ph/0701043. https://doi.org/10.1086/519986

R. Minkowski, Spectra of supernovae. Publ. Astron. Soc. Pac. **53**, 224 (1941). https://doi.org/10.1086/125315

R.C. Mitchell, E. Baron, D. Branch, P.H. Hauschildt, P.E. Nugent, P. Lundqvist, S. Blinnikov, C.S.J. Pun, Detailed spectroscopic analysis of SN 1987A: the distance to the Large Magellanic Cloud using the spectral-fitting expanding atmosphere method. Astrophys. J. **574**, 293–305 (2002). astro-ph/0204012. https://doi.org/10.1086/340928

M. Miyoshi, J. Moran, J. Herrnstein, L. Greenhill, N. Nakai, P. Diamond, M. Inoue, Evidence for a black hole from high rotation velocities in a sub-parsec region of NGC4258. Nature **373**, 127–129 (1995). https://doi.org/10.1038/373127a0

J. Modzelewska, B. Czerny, K. Hryniewicz, M. Bilicki, M. Krupa, A. Świętoń, W. Pych, A. Udalski, T.P. Adhikari, F. Petrogalli, SALT long-slit spectroscopy of CTS C30.10: two-component Mg II line. Astron. Astrophys. **570**, A53 (2014). 1408.1520. https://doi.org/10.1051/0004-6361/201424332

A.J. Monson, R.L. Beaton, V. Scowcroft, W.L. Freedman, B.F. Madore, J.A. Rich, M. Seibert, J.A. Kollmeier, G. Clementini, Standard galactic field RR Lyrae, I: optical to mid-infrared phased photometry. Astron. J. **153**, 96 (2017). 1703.01520. https://doi.org/10.3847/1538-3881/153/3/96

R. Nair, S. Jhingan, D. Jain, Observational cosmology and the cosmic distance duality relation. J. Cosmol. Astropart. Phys. **5**, 023 (2011). 1102.1065. https://doi.org/10.1088/1475-7516/2011/05/023

G. Nelemans, S. Toonen, M. Bours, Theoretical delay time distributions, in *Binary Paths to Type Ia Supernovae Explosions*, ed. by R. Di Stefano, M. Orio, M. Moe. IAU Symposium, vol. 281 (2013), pp. 225–231. 1204.2960. https://doi.org/10.1017/S1743921312015098

M.T.B. Nielsen, R. Voss, G. Nelemans, Upper limits on bolometric luminosities of 10 type Ia supernova progenitors from Chandra observations. Mon. Not. R. Astron. Soc. **426**, 2668–2676 (2012). 1109.6605. https://doi.org/10.1111/j.1365-2966.2012.21596.x

M.T.B. Nielsen, C. Dominik, G. Nelemans, R. Voss, Obscuration of supersoft X-ray sources by circumbinary material. A way to hide type Ia supernova progenitors? Astron. Astrophys. **549**, A32 (2013). 1207.6310. https://doi.org/10.1051/0004-6361/201219195

M. Nikolajuk, I.E. Papadakis, B. Czerny, Black hole mass estimation from X-ray variability measurements in active galactic nuclei. Mon. Not. R. Astron. Soc. **350**, L26–L30 (2004). astro-ph/0403326. https://doi.org/10.1111/j.1365-2966.2004.07829.x

S. Nissanke, D.E. Holz, S.A. Hughes, N. Dalal, J.L. Sievers, Exploring short gamma-ray bursts as gravitational-wave standard sirens. Astrophys. J. **725**, 496–514 (2010). 0904.1017. https://doi.org/10.1088/0004-637X/725/1/496

S. Nissanke, D.E. Holz, N. Dalal, S.A. Hughes, J.L. Sievers, C.M. Hirata, Determining the Hubble constant from gravitational wave observations of merging compact binaries. 1307.2638 (2013)

K. Nomoto, Y. Kondo, Conditions for accretion-induced collapse of white dwarfs. Astrophys. J. Lett. **367**, L19–L22 (1991). https://doi.org/10.1086/185922

P. Nugent, M. Sullivan, R. Ellis, A. Gal-Yam, D.C. Leonard, D.A. Howell, P. Astier, R.G. Carlberg, A. Conley, S. Fabbro, D. Fouchez, J.D. Neill, R. Pain, K. Perrett, C.J. Pritchet, N. Regnault, Toward a cosmological Hubble diagram for type II-P supernovae. Astrophys. J. **645**, 841–850 (2006). astro-ph/0603535. https://doi.org/10.1086/504413

J.E. O'Donnell, R_{nu}-dependent optical and near-ultraviolet extinction. Astrophys. J. **422**, 158–163 (1994). https://doi.org/10.1086/173713

V.L. Oknyanskij, On the possibility for measuring the Hubble constant from optical-to-NIR variability time delay in AGNs. Odessa Astron. Publ. **12**, 99 (1999)

E.F. Olivares, M. Hamuy, G. Pignata, J. Maza, M. Bersten, M.M. Phillips, N.B. Suntzeff, A.V. Filippenko, N.I. Morrel, R.P. Kirshner, T. Matheson, The standardized candle method for type II Plateau supernovae. Astrophys. J. **715**, 833–853 (2010). 1004.2534. https://doi.org/10.1088/0004-637X/715/2/833
B. Paczynski, Gamma-ray bursters at cosmological distances. Astrophys. J. Lett. **308**, L43–L46 (1986). https://doi.org/10.1086/184740
R. Pakmor, M. Kromer, F.K. Röpke, S.A. Sim, A.J. Ruiter, W. Hillebrandt, Sub-luminous type Ia supernovae from the mergers of equal-mass white dwarfs with mass $\sim 0.9 M_{solar}$. Nature **463**, 61–64 (2010). 0911.0926. https://doi.org/10.1038/nature08642
R. Pakmor, M. Kromer, S. Taubenberger, V. Springel, Helium-ignited violent mergers as a unified model for normal and rapidly declining type Ia supernovae. Astrophys. J. Lett. **770**, L8 (2013). 1302.2913. https://doi.org/10.1088/2041-8205/770/1/L8
J. Parrent, B. Friesen, M. Parthasarathy, A review of type Ia supernova spectra. Astrophys. Space Sci. **351**, 1–52 (2014). 1402.6337. https://doi.org/10.1007/s10509-014-1830-1
J.T. Parrent, D. Sand, S. Valenti, M.J. Graham, D.A. Howell, Spectroscopic confirmation and classification of PSN J14323388-4413278 in NGC 5643. Astron. (2013). Telegr. **4817**
F. Patat, S. Benetti, S. Justham, P.A. Mazzali, L. Pasquini, E. Cappellaro, M. Della Valle, P. -Podsiadlowski, M. Turatto, A. Gal-Yam, J.D. Simon, Upper limit for circumstellar gas around the type Ia SN 2000cx. Astron. Astrophys. **474**, 931–936 (2007). 0708.3698. https://doi.org/10.1051/0004-6361:20078393
F. Patat, S. Taubenberger, N.L.J. Cox, D. Baade, A. Clocchiatti, P. Höflich, J.R. Maund, E. Reilly, J. Spyromilio, L. Wang, J.C. Wheeler, P. Zelaya, Properties of extragalactic dust inferred from linear polarimetry of type Ia supernovae. Astron. Astrophys. **577**, A53 (2015). 1407.0136. https://doi.org/10.1051/0004-6361/201424507
S. Perlmutter, S. Gabi, G. Goldhaber, A. Goobar, D.E. Groom, I.M. Hook, A.G. Kim, M.Y. Kim, J.C. Lee, R. Pain, C.R. Pennypacker, I.A. Small, R.S. Ellis, R.G. McMahon, B.J. Boyle, P.S. Bunclark, D. Carter, M.J. Irwin, K. Glazebrook, H.J.M. Newberg, A.V. Filippenko, T. Matheson, M. Dopita, W.J. Couch, Measurements of the cosmological parameters Ω and Λ from the first seven supernovae at $z \geq 0.35$. Astrophys. J. **483**, 565–581 (1997). astro-ph/9608192. https://doi.org/10.1086/304265
S. Perlmutter, G. Aldering, M. della Valle, S. Deustua, R.S. Ellis, S. Fabbro, A. Fruchter, G. Goldhaber, D.E. Groom, I.M. Hook, A.G. Kim, M.Y. Kim, R.A. Knop, C. Lidman, R.G. McMahon, P. Nugent, R. Pain, N. Panagia, C.R. Pennypacker, P. Ruiz-Lapuente, B. Schaefer, N. Walton, Discovery of a supernova explosion at half the age of the universe. Nature **391**, 51 (1998). astro-ph/9712212. https://doi.org/10.1038/34124
S. Perlmutter, G. Aldering, G. Goldhaber, R.A. Knop, P. Nugent, P.G. Castro, S. Deustua, S. Fabbro, A. Goobar, D.E. Groom, I.M. Hook, A.G. Kim, M.Y. Kim, J.C. Lee, N.J. Nunes, R. Pain, C.R. Pennypacker, R. Quimby, C. Lidman, R.S. Ellis, M. Irwin, R.G. McMahon, P. Ruiz-Lapuente, N. Walton, B. Schaefer, B.J. Boyle, A.V. Filippenko, T. Matheson, A.S. Fruchter, N. Panagia, H.J.M. Newberg, W.J. Couch, T.S.C. Project, Measurements of Ω and Λ from 42 high-redshift supernovae. Astrophys. J. **517**, 565–586 (1999a). astro-ph/9812133. https://doi.org/10.1086/307221
S. Perlmutter, G. Aldering, G. Goldhaber, R.A. Knop, P. Nugent, P.G. Castro, S. Deustua, S. Fabbro, A. Goobar, D.E. Groom, I.M. Hook, A.G. Kim, M.Y. Kim, J.C. Lee, N.J. Nunes, R. Pain, C.R. Pennypacker, R. Quimby, C. Lidman, R.S. Ellis, M. Irwin, R.G. McMahon, P. Ruiz-Lapuente, N. Walton, B. Schaefer, B.J. Boyle, A.V. Filippenko, T. Matheson, A.S. Fruchter, N. Panagia, H.J.M. Newberg, W.J. Couch, T.S.C. Project, Measurements of Ω and Λ from 42 high-redshift supernovae. Astrophys. J. **517**, 565–586 (1999b). astro-ph/9812133. https://doi.org/10.1086/307221
S. Perlmutter, M.S. Turner, M. White, Constraining dark energy with type Ia supernovae and large-scale structure. Phys. Rev. Lett. **83**, 670–673 (1999c). astro-ph/9901052. https://doi.org/10.1103/PhysRevLett.83.670
B.M. Peterson, Measuring the masses of supermassive black holes. Space Sci. Rev. **183**, 253–275 (2014a). https://doi.org/10.1007/s11214-013-9987-4
B.M. Peterson, Measuring the masses of supermassive black holes. Space Sci. Rev. **183**, 253–275 (2014b). https://doi.org/10.1007/s11214-013-9987-4
B.M. Peterson, A.J. Barth, P. Berlind, R. Bertram, K. Bischoff, N.G. Bochkarev, A.N. Burenkov, F.Z. Cheng, M. Dietrich, A.V. Filippenko, E. Giannuzzo, L.C. Ho, J.P. Huchra, J. Hunley, S. Kaspi, W. Kollatschny, D.C. Leonard, Y.F. Malkov, T. Matheson, M. Mignoli, B. Nelson, P. Papaderos, J. Peters, R.W. Pogge, V.I. Pronik, S.G. Sergeev, E.A. Sergeeva, A.I. Shapovalova, G.M. Stirpe, S. Tokarz, R.M. Wagner, I. Wanders, J.Y. Wei, B.J. Wilkes, H. Wu, S.J. Xue, Z.L. Zou, Steps toward determination of the size and structure of the broad-line region in active galactic nuclei, XV: long-term optical monitoring of NGC 5548. Astrophys. J. **510**, 659–668 (1999). astro-ph/9808236. https://doi.org/10.1086/306604
B.M. Peterson, L. Ferrarese, K.M. Gilbert, S. Kaspi, M.A. Malkan, D. Maoz, D. Merritt, H. Netzer, C.A. Onken, R.W. Pogge, M. Vestergaard, A. Wandel, Central masses and broad-line region sizes of active

galactic nuclei, II: a homogeneous analysis of a large reverberation-mapping database. Astrophys. J. **613**, 682–699 (2004). astro-ph/0407299. https://doi.org/10.1086/423269

M.M. Phillips, The absolute magnitudes of type IA supernovae. Astrophys. J. Lett. **413**, L105–L108 (1993). https://doi.org/10.1086/186970

M.M. Phillips, Near-infrared properties of type Ia supernovae. Publ. Astron. Soc. Aust. **29**, 434–446 (2012). 1111.4463. https://doi.org/10.1071/AS11056

M.M. Phillips, P. Lira, N.B. Suntzeff, R.A. Schommer, M. Hamuy, J. Maza, The reddening-free decline rate versus luminosity relationship for type IA supernovae. Astron. J. **118**, 1766–1776 (1999). https://doi.org/10.1086/301032

M.M. Phillips, J.D. Simon, N. Morrell, C.R. Burns, N.L.J. Cox, R.J. Foley, A.I. Karakas, F. Patat, A. Sternberg, R.E. Williams, A. Gal-Yam, E.Y. Hsiao, D.C. Leonard, S.E. Persson, M. Stritzinger, I.B. Thompson, A. Campillay, C. Contreras, G. Folatelli, W.L. Freedman, M. Hamuy, M. Roth, G.A. Shields, N.B. Suntzeff, L. Chomiuk, I.I. Ivans, B.F. Madore, B.E. Penprase, D. Perley, G. Pignata, G. Preston, A.M. Soderberg, On the source of the dust extinction in type Ia supernovae and the discovery of anomalously strong Na I absorption. Astrophys. J. **779**, 38 (2013). 1311.0147. https://doi.org/10.1088/0004-637X/779/1/38

L. Piersanti, A. Tornambé, L.R. Yungelson, He-accreting white dwarfs: accretion regimes and final outcomes. Mon. Not. R. Astron. Soc. **445**, 3239–3262 (2014). 1409.3589. https://doi.org/10.1093/mnras/stu1885

G. Pietrzyński, D. Graczyk, W. Gieren, I.B. Thompson, B. Pilecki, A. Udalski, I. Soszyński, S. Kozłowski, P. Konorski, K. Suchomska, G. Bono, P.G.P. Moroni, S. Villanova, N. Nardetto, F. Bresolin, R.P. Kudritzki, J. Storm, A. Gallenne, R. Smolec, D. Minniti, M. Kubiak, M.K. Szymański, R. Poleski, Ł. Wyrzykowski, K. Ulaczyk, P. Pietrukowicz, M. Górski, P. Karczmarek, An eclipsing-binary distance to the Large Magellanic Cloud accurate to two per cent. Nature **495**, 76–79 (2013). 1303.2063. https://doi.org/10.1038/nature11878

F.A.E. Pirani, On the physical significance of the Riemann tensor. Acta Phys. Pol. **15**, 389–405 (1956)

Planck Collaboration, P.A.R. Ade, N. Aghanim, M. Arnaud, M. Ashdown, J. Aumont, C. Baccigalupi, A.J. Banday, R.B. Barreiro, J.G. Bartlett et al. Planck 2015 results, XIII: cosmological parameters. Astron. Astrophys. **594**, A13 (2016). 1502.01589. https://doi.org/10.1051/0004-6361/201525830

J. Polshaw, R. Kotak, K.C. Chambers, S.J. Smartt, S. Taubenberger, M. Kromer, E.E.E. Gall, W. Hillebrandt, M. Huber, K.W. Smith, R.J. Wainscoat, A supernova distance to the anchor galaxy NGC 4258. Astron. Astrophys. **580**, L15 (2015). 1509.00507. https://doi.org/10.1051/0004-6361/201526902

J. Polshaw, R. Kotak, L. Dessart, M. Fraser, A. Gal-Yam, C. Inserra, S.A. Sim, S.J. Smartt, J. Sollerman, C. Baltay, D. Rabinowitz, S. Benetti, M.T. Botticella, H. Campbell, T.W. Chen, L. Galbany, R. McKinnon, M. Nicholl, K.W. Smith, M. Sullivan, K. Takáts, S. Valenti, D.R. Young, LSQ13fn: A type II-Plateau supernova with a possibly low metallicity progenitor that breaks the standardised candle relation. Astron. Astrophys. **588**, A1 (2016). 1511.01718. https://doi.org/10.1051/0004-6361/201527682

D. Poznanski, N. Butler, A.V. Filippenko, M. Ganeshalingam, W. Li, J.S. Bloom, R. Chornock, R.J. Foley, P.E. Nugent, J.M. Silverman, S.B. Cenko, E.L. Gates, D.C. Leonard, A.A. Miller, M. Modjaz, F.J.D. Serduke, N. Smith, B.J. Swift, D.S. Wong, Improved standardization of type II-P supernovae: application to an expanded sample. Astrophys. J. **694**, 1067–1079 (2009). 0810.4923. https://doi.org/10.1088/0004-637X/694/2/1067

I.P. Pskovskii, Light curves, color curves, and expansion velocity of type I supernovae as functions of the rate of brightness decline. Sov. Astron. **21**, 675–682 (1977)

M.L. Pumo, L. Zampieri, Radiation-hydrodynamical modeling of core-collapse supernovae: light curves and the evolution of photospheric velocity and temperature. Astrophys. J. **741**, 41 (2011). 1108.0688. https://doi.org/10.1088/0004-637X/741/1/41

E.D. Reese, J.E. Carlstrom, M. Joy, J.J. Mohr, L. Grego, W.L. Holzapfel, Determining the cosmic distance scale from interferometric measurements of the Sunyaev-Zeldovich effect. Astrophys. J. **581**, 53–85 (2002). astro-ph/0205350. https://doi.org/10.1086/344137

A.G. Riess, W.H. Press, R.P. Kirshner, Determining the motion of the local group using type IA supernovae light curve shapes. Astrophys. J. Lett. **445**, L91–L94 (1995). astro-ph/9412017. https://doi.org/10.1086/187897

A.G. Riess, W.H. Press, R.P. Kirshner, A precise distance indicator: type IA supernova multicolor light-curve shapes. Astrophys. J. **473**, 88 (1996). astro-ph/9604143. https://doi.org/10.1086/178129

A.G. Riess, A.V. Filippenko, P. Challis, A. Clocchiatti, A. Diercks, P.M. Garnavich, R.L. Gilliland, C.J. Hogan, S. Jha, R.P. Kirshner, B. Leibundgut, M.M. Phillips, D. Reiss, B.P. Schmidt, R.A. Schommer, R.C. Smith, J. Spyromilio, C. Stubbs, N.B. Suntzeff, J. Tonry, Observational evidence from supernovae for an accelerating universe and a cosmological constant. Astron. J. **116**, 1009–1038 (1998a). astro-ph/9805201. https://doi.org/10.1086/300499

A.G. Riess, A.V. Filippenko, P. Challis, A. Clocchiatti, A. Diercks, P.M. Garnavich, R.L. Gilliland, C.J. Hogan, S. Jha, R.P. Kirshner, B. Leibundgut, M.M. Phillips, D. Reiss, B.P. Schmidt, R.A. Schommer,

R.C. Smith, J. Spyromilio, C. Stubbs, N.B. Suntzeff, J. Tonry, Observational evidence from supernovae for an accelerating universe and a cosmological constant. Astron. J. **116**, 1009–1038 (1998b). astro-ph/9805201. https://doi.org/10.1086/300499

A.G. Riess, R.P. Kirshner, B.P. Schmidt, S. Jha, P. Challis, P.M. Garnavich, A.A. Esin, C. Carpenter, R. Grashius, R.E. Schild, P.L. Berlind, J.P. Huchra, C.F. Prosser, E.E. Falco, P.J. Benson, C. Briceño, W.R. Brown, N. Caldwell, I.P.dell'Antonio, A.V. Filippenko, A.A. Goodman, N.A. Grogin, T. Groner, J.P. Hughes, P.J. Green, R.A. Jansen, J.T. Kleyna, J.X. Luu, L.M. Macri, B.A. McLeod, K.K. McLeod, B.R. McNamara, B. McLean, A.A.E. Milone, J.J. Mohr, D. Moraru, C. Peng, J. Peters, A.H. Prestwich, K.Z. Stanek, A. Szentgyorgyi, P. Zhao, BVRI light curves for 22 type IA supernovae. Astron. J. **117**, 707–724 (1999). astro-ph/9810291. https://doi.org/10.1086/300738

A.G. Riess, P.E. Nugent, R.L. Gilliland, B.P. Schmidt, J. Tonry, M. Dickinson, R.I. Thompson, T. Budavári, S. Casertano, A.S. Evans, A.V. Filippenko, M. Livio, D.B. Sanders, A.E. Shapley, H. Spinrad, C.C. Steidel, D. Stern, J. Surace, S. Veilleux, The farthest known supernova: support for an accelerating universe and a glimpse of the epoch of deceleration. Astrophys. J. **560**, 49–71 (2001). astro-ph/0104455. https://doi.org/10.1086/322348

A.G. Riess, L.G. Strolger, J. Tonry, S. Casertano, H.C. Ferguson, B. Mobasher, P. Challis, A.V. Filippenko, S. Jha, W. Li, R. Chornock, R.P. Kirshner, B. Leibundgut, M. Dickinson, M. Livio, M. Giavalisco, C.C. Steidel, T. Benítez, Z. Tsvetanov, Type Ia supernova discoveries at $z > 1$ from the Hubble Space Telescope: evidence for past deceleration and constraints on dark energy evolution. Astrophys. J. **607**, 665–687 (2004a). astro-ph/0402512. https://doi.org/10.1086/383612

A.G. Riess, L.G. Strolger, J. Tonry, Z. Tsvetanov, S. Casertano, H.C. Ferguson, B. Mobasher, P. Challis, N. Panagia, A.V. Filippenko, W. Li, R. Chornock, R.P. Kirshner, B. Leibundgut, M. Dickinson, A. Koekemoer, N.A. Grogin, M. Giavalisco, Identification of type Ia supernovae at redshift 1.3 and beyond with the advanced camera for surveys on the Hubble Space Telescope. Astrophys. J. Lett. **600**, L163–L166 (2004b). astro-ph/0308185. https://doi.org/10.1086/378311

A.G. Riess, L.G. Strolger, S. Casertano, H.C. Ferguson, B. Mobasher, B. Gold, P.J. Challis, A.V. Filippenko, S. Jha, W. Li, J. Tonry, R. Foley, R.P. Kirshner, M. Dickinson, E. MacDonald, D. Eisenstein, M. Livio, J. Younger, C. Xu, T. Dahlén, D. Stern, New Hubble Space Telescope discoveries of type Ia supernovae at $z \geq 1$: narrowing constraints on the early behavior of dark energy. Astrophys. J. **659**, 98–121 (2007). astro-ph/0611572. https://doi.org/10.1086/510378

A.G. Riess, L. Macri, S. Casertano, M. Sosey, H. Lampeitl, H.C. Ferguson, A.V. Filippenko, S.W. Jha, W. Li, R. Chornock, D. Sarkar, A redetermination of the Hubble constant with the Hubble Space Telescope from a differential distance ladder. Astrophys. J. **699**, 539–563 (2009). 0905.0695. https://doi.org/10.1088/0004-637X/699/1/539

A.G. Riess, L. Macri, S. Casertano, H. Lampeitl, H.C. Ferguson, A.V. Filippenko, S.W. Jha, W. Li, R. Chornock, A 3% solution: determination of the Hubble constant with the Hubble Space Telescope and Wide Field Camera 3. Astrophys. J. **730**, 119 (2011). 1103.2976. https://doi.org/10.1088/0004-637X/730/2/119

A.G. Riess, L.M. Macri, S.L. Hoffmann, D. Scolnic, S. Casertano, A.V. Filippenko, B.E. Tucker, M.J. Reid, D.O. Jones, J.M. Silverman, R. Chornock, P. Challis, W. Yuan, P.J. Brown, R.J. Foley, A 2.4% determination of the local value of the Hubble constant. Astrophys. J. **826**, 56 (2016). 1604.01424. https://doi.org/10.3847/0004-637X/826/1/56

A.G. Riess, S.A. Rodney, D.M. Scolnic, D.L. Shafer, L.G. Strolger, H.C. Ferguson, M. Postman, O. Graur, D. Maoz, S.W. Jha, B. Mobasher, S. Casertano, B. Hayden, A. Molino, J. Hjorth, P.M. Garnavich, D.O. Jones, R.P. Kirshner, A.M. Koekemoer, N.A. Grogin, G. Brammer, S. Hemmati, M. Dickinson, P.M. Challis, S. Wolff, K.I. Clubb, A.V. Filippenko, H. Nayyeri, U. Vivian, D.C. Koo, S.M. Faber, D. Kocevski, L. Bradley, D. Coe, Type Ia supernova distances at $z > 1.5$ from the Hubble Space Telescope multi-cycle treasury programs: the early expansion rate. 1710.00844 (2017)

M. Rigault, Evidence of environmental dependencies of type Ia supernovae from the nearby supernova factory indicated by local Hα. Astron. Astrophys. **560**, A66 (2013). 1309.1182. https://doi.org/10.1051/0004-6361/201322104

M. Rigault, G. Aldering, M. Kowalski, Y. Copin, P. Antilogus, C. Aragon, S. Bailey, C. Baltay, D. Baugh, S. Bongard, K. Boone, C. Buton, J. Chen, N. Chotard, H.K. Fakhouri, U. Feindt, P. Fagrelius, M. Fleury, D. Fouchez, E. Gangler, B. Hayden, A.G. Kim, P.F. Leget, S. Lombardo, J. Nordin, R. Pain, E. Pecontal, R. Pereira, S. Perlmutter, D. Rabinowitz, K. Runge, D. Rubin, C. Saunders, G. Smadja, C. Sofiatti, N. Suzuki, C. Tao, B.A. Weaver, Confirmation of a star formation bias in type Ia supernova distances and its effect on the measurement of the Hubble constant. Astrophys. J. **802**, 20 (2015). 1412.6501. https://doi.org/10.1088/0004-637X/802/1/20

G. Risaliti, E. Lusso, A Hubble diagram for quasars. Astrophys. J. **815**, 33 (2015). 1505.07118. https://doi.org/10.1088/0004-637X/815/1/33

A.M. Ritchey, D.E. Welty, J.A. Dahlstrom, D.G. York, Diffuse atomic and molecular gas in the interstellar medium of m82 toward sn 2014j. Astrophys. J. **799**, 197 (2015). 1407.5723. https://doi.org/10.1088/0004-637X/799/2/197

I. Robinson, A. Trautman, Spherical gravitational waves. Phys. Rev. Lett. **4**, 431–432 (1960). https://doi.org/10.1103/PhysRevLett.4.431

F.K. Röpke, I.R. Seitenzahl, S. Benitez, M. Fink, R. Pakmor, M. Kromer, S.A. Sim, F. Ciaraldi-Schoolmann, W. Hillebrandt, Modeling type Ia supernova explosions. Prog. Part. Nucl. Phys. **66**, 309–318 (2011). https://doi.org/10.1016/j.ppnp.2011.01.026

D. Rubin, R.A. Knop, E. Rykoff, G. Aldering, R. Amanullah, K. Barbary, M.S. Burns, A. Conley, N. Connolly, S. Deustua, V. Fadeyev, H.K. Fakhouri, A.S. Fruchter, R.A. Gibbons, G. Goldhaber, A. Goobar, E.Y. Hsiao, X. Huang, M. Kowalski, C. Lidman, J. Meyers, J. Nordin, S. Perlmutter, C. Saunders, A.L. Spadafora, V. Stanishev, N. Suzuki, L. Wang, Precision measurement of the most distant spectroscopically confirmed supernova Ia with the Hubble Space Telescope. Supernova Cosmology Project T **763**, 35 (2013). Astrophys. J.. 1205.3494. https://doi.org/10.1088/0004-637X/763/1/35

A.J. Ruiter, S.A. Sim, R. Pakmor, M. Kromer, I.R. Seitenzahl, K. Belczynski, M. Fink, M. Herzog, W. Hillebrandt, F.K. Röpke, S. Taubenberger, On the brightness distribution of type Ia supernovae from violent white dwarf mergers. Mon. Not. R. Astron. Soc. **429**, 1425–1436 (2013). 1209.0645. https://doi.org/10.1093/mnras/sts423

P. Ruiz-Lapuente, F. Comeron, J. Méndez, R. Canal, S.J. Smartt, A.V. Filippenko, R.L. Kurucz, R. Chornock, R.J. Foley, V. Stanishev, R. Ibata, The binary progenitor of Tycho Brahe's 1572 supernova. Nature **431**, 1069–1072 (2004). astro-ph/0410673. https://doi.org/10.1038/nature03006

C.L. Sarazin, X-ray emission from clusters of galaxies (1988)

M. Sasdelli, E.E.O. Ishida, W. Hillebrandt, C. Ashall, P.A. Mazzali, S.J. Prentice, Breaking the colour-reddening degeneracy in type Ia supernovae. Mon. Not. R. Astron. Soc. **460**, 373–382 (2016). 1604.03899. https://doi.org/10.1093/mnras/stw900

B. Sathyaprakash, B.F. Schutz, Physics, astrophysics and cosmology with gravitational waves. Living Rev. Relativ. **12**, 2 (2009). https://doi.org/10.1007/lrr-2009-2

B.E. Schaefer, The Hubble diagram to redshift > 6 from 69 gamma-ray bursts. Astrophys. J. **660**, 16–46 (2007). astro-ph/0612285. https://doi.org/10.1086/511742

B.E. Schaefer, A. Pagnotta, An absence of ex-companion stars in the type Ia supernova remnant SNR 0509-67.5. Nature **481**, 164–166 (2012). https://doi.org/10.1038/nature10692

E.F. Schlafly, D.P. Finkbeiner, Measuring reddening with Sloan digital sky survey stellar spectra and recalibrating SFD. Astrophys. J. **737**, 103 (2011). 1012.4804. https://doi.org/10.1088/0004-637X/737/2/103

D.J. Schlegel, D.P. Finkbeiner, M. Davis, Maps of dust infrared emission for use in estimation of reddening and cosmic microwave background radiation foregrounds. Astrophys. J. **500**, 525–553 (1998). astro-ph/9710327. https://doi.org/10.1086/305772

B.P. Schmidt, N.B. Suntzeff, M.M. Phillips, R.A. Schommer, A. Clocchiatti, R.P. Kirshner, P. Garnavich, P. Challis, B. Leibundgut, J. Spyromilio, A.G. Riess, A.V. Filippenko, M. Hamuy, R.C. Smith, C. Hogan, C. Stubbs, A. Diercks, D. Reiss, R. Gilliland, J. Tonry, J. Maza, A. Dressler, J. Walsh, R. Ciardullo, The high-Z supernova search: measuring cosmic deceleration and global curvature of the universe using type IA supernovae. Astrophys. J. **507**, 46–63 (1998a). astro-ph/9805200. https://doi.org/10.1086/306308

B.P. Schmidt, N.B. Suntzeff, M.M. Phillips, R.A. Schommer, A. Clocchiatti, R.P. Kirshner, P. Garnavich, P. Challis, B. Leibundgut, J. Spyromilio, A.G. Riess, A.V. Filippenko, M. Hamuy, R.C. Smith, C. Hogan, C. Stubbs, A. Diercks, D. Reiss, R. Gilliland, J. Tonry, J. Maza, A. Dressler, J. Walsh, R. Ciardullo, The high-Z supernova search: measuring cosmic deceleration and global curvature of the universe using type IA supernovae. Astrophys. J. **507**, 46–63 (1998b). astro-ph/9805200. https://doi.org/10.1086/306308

B.F. Schutz, Determining the Hubble constant from gravitational wave observations. Nature **323**, 310 (1986). https://doi.org/10.1038/323310a0

B.F. Schutz, Lighthouses of gravitational wave astronomy, in *Lighthouses of the Universe: The Most Luminous Celestial Objects and Their Use for Cosmology*, ed. by M. Gilfanov, R. Sunyeav, E. Churazov (2002), p. 207. gr-qc/0111095. https://doi.org/10.1007/10856495_29

B.F. Schutz, Networks of gravitational wave detectors and three figures of merit. Class. Quantum Gravity **28**(12), 125023 (2011). 1102.5421. https://doi.org/10.1088/0264-9381/28/12/125023

D.M. Scolnic, A.G. Riess, R.J. Foley, A. Rest, S.A. Rodney, D.J. Brout, D.O. Jones, Color dispersion and Milky-Way-like reddening among type Ia supernovae. Astrophys. J. **780**, 37 (2014). 1306.4050. https://doi.org/10.1088/0004-637X/780/1/37

M. Sereno, E. De Filippis, G. Longo, M.W. Bautz, Measuring the three-dimensional structure of galaxy clusters, II: are clusters of galaxies oblate or prolate? Astrophys. J. **645**, 170–178 (2006). astro-ph/0602051. https://doi.org/10.1086/503198

S.G. Sergeev, V.T. Doroshenko, Y.V. Golubinskiy, N.I. Merkulova, E.A. Sergeeva, Lag-luminosity relationship for interband lags between variations in B, V, R, and I bands in active galactic nuclei. Astrophys. J. **622**, 129–135 (2005). https://doi.org/10.1086/427820

N.I. Shakura, R.A. Sunyaev, Black holes in binary systems. Observational appearance. Astron. Astrophys. **24**, 337–355 (1973)

H. Shapley, On the existence of external galaxies. Publ. Astron. Soc. Pac. **31**, 261 (1919). https://doi.org/10.1086/122870

B.J. Shappee, J.L. Prieto, D. Grupe, C.S. Kochanek, K.Z. Stanek, G. De Rosa, S. Mathur, Y. Zu, B.M. Peterson, R.W. Pogge, S. Komossa, M. Im, J. Jencson, T.W.S. Holoien, U. Basu, J.F. Beacom, D.M. Szczygieł, J. Brimacombe, S. Adams, A. Campillay, C. Choi, C. Contreras, M. Dietrich, M. Dubberley, M. Elphick, S. Foale, M. Giustini, C. Gonzalez, E. Hawkins, D.A. Howell, E.Y. Hsiao, M. Koss, K.M. Leighly, N. Morrell, D. Mudd, D. Mullins, J.M. Nugent, J. Parrent, M.M. Phillips, G. Pojmanski, W. Rosing, R. Ross, D. Sand, D.M. Terndrup, S. Valenti, Z. Walker, Y. Yoon, The man behind the curtain: X-rays drive the UV through NIR variability in the 2013 active galactic nucleus outburst in NGC 2617. Astrophys. J. **788**, 48 (2014a). 1310.2241. https://doi.org/10.1088/0004-637X/788/1/48

B.J. Shappee, J.L. Prieto, D. Grupe et al., The man behind the curtain: X-rays drive the UV through NIR variability in the 2013 active galactic nucleus outburst in NGC 2617. Astrophys. J. **788**, 48 (2014b). 1310.2241. https://doi.org/10.1088/0004-637X/788/1/48

B.J. Shappee, K.Z. Stanek, C.S. Kochanek, P.M. Garnavich, Whimper of a bang: documenting the final days of the nearby type Ia supernova 2011fe. Astrophys. J. **841**, 48 (2017). 1608.01155. https://doi.org/10.3847/1538-4357/aa6eab

K.J. Shen, L. Bildsten, D. Kasen, E. Quataert, The long-term evolution of double white dwarf mergers. Astrophys. J. **748**, 35 (2012). 1108.4036. https://doi.org/10.1088/0004-637X/748/1/35

Y. Shen, K. Horne, C.J. Grier, B.M. Peterson, K.D. Denney, J.R. Trump, M. Sun, W.N. Brandt, C.S. Kochanek, K.S. Dawson, P.J. Green, J.E. Greene, P.B. Hall, L.C. Ho, L. Jiang, K. Kinemuchi, I.D. McGreer, P. Petitjean, G.T. Richards, D.P. Schneider, M.A. Strauss, C. Tao, W.M. Wood-Vasey, Y. Zu, K. Pan, D. Bizyaev, J. Ge, D. Oravetz, A. Simmons, The Sloan digital sky survey reverberation mapping project: first broad-line Hβ and Mg II lags at z above 0.3 from six-month spectroscopy. Astrophys. J. **818**, 30 (2016). 1510.02802. https://doi.org/10.3847/0004-637X/818/1/30

L.P. Singer, H.Y. Chen, D.E. Holz, W.M. Farr, L.R. Price, V. Raymond, S.B. Cenko, N. Gehrels, J. Cannizzo, M.M. Kasliwal, S. Nissanke, M. Coughlin, B. Farr, A.L. Urban, S. Vitale, J. Veitch, P. Graff, C.P.L. Berry, S. Mohapatra, I. Mandel, Going the distance: mapping host galaxies of LIGO and Virgo sources in three dimensions using local cosmography and targeted follow-up. Astrophys. J. Lett. **829**, L15 (2016). 1603.07333. https://doi.org/10.3847/2041-8205/829/1/L15

S.J. Smartt, Progenitors of core-collapse supernovae. Annu. Rev. Astron. Astrophys. **47**, 63–106 (2009). 0908.0700. https://doi.org/10.1146/annurev-astro-082708-101737

S.J. Smartt, J.J. Eldridge, R.M. Crockett, J.R. Maund, The death of massive stars, I: observational constraints on the progenitors of type II-P supernovae. Mon. Not. R. Astron. Soc. **395**, 1409–1437 (2009). 0809.0403. https://doi.org/10.1111/j.1365-2966.2009.14506.x

D.A. Starkey, K. Horne, C. Villforth, Accretion disc time lag distributions: applying CREAM to simulated AGN light curves. Mon. Not. R. Astron. Soc. **456**, 1960–1973 (2016). 1511.06162. https://doi.org/10.1093/mnras/stv2744

I. Steer, B.F. Madore, J.M. Mazzarella, M. Schmitz, H.G. Corwin, B.H.P. Chan, R. Ebert, G. Helou, K. Baker, X. Chen, C. Frayer, J. Jacobson, T. Lo, P. Ogle, O. Pevunova, S. Terek, Redshift-independent distances in the NASA/IPAC extragalactic database: methodology, content, and use of NED-D. Astron. J. **153**, 37 (2017). 1612.09263. https://doi.org/10.3847/1538-3881/153/1/37

A.T. Steffen, I. Strateva, W.N. Brandt, D.M. Alexander, A.M. Koekemoer, B.D. Lehmer, D.P. Schneider, C. Vignali, The X-ray-to-optical properties of optically selected active galaxies over wide luminosity and redshift ranges. Astron. J. **131**, 2826–2842 (2006). astro-ph/0602407. https://doi.org/10.1086/503627

A. Sternberg, A. Gal-Yam, J.D. Simon, D.C. Leonard, R.M. Quimby, M.M. Phillips, N. Morrell, I.B. Thompson, I. Ivans, J.L. Marshall, A.V. Filippenko, G.W. Marcy, J.S. Bloom, F. Patat, R.J. Foley, D. Yong, B.E. Penprase, D.J. Beeler, C. Allende Prieto, G.S. Stringfellow, Circumstellar material in type Ia supernovae via sodium absorption features. Science **333**, 856 (2011). 1108.3664. https://doi.org/10.1126/science.1203836

M. Stritzinger, C.R. Burns, M.M. Phillips, G. Folatelli, K. Krisciunas, S. Kattner, S.E. Persson, L. Boldt, A. Campillay, C. Contreras, W. Krzeminski, N. Morrell, F. Salgado, W.L. Freedman, M. Hamuy, B.F. Madore, M. Roth, N.B. Suntzeff, The distance to NGC 1316 (Fornax A) from observations of four type Ia supernovae. Astron. J. **140**, 2036–2051 (2010). 1009.4390. https://doi.org/10.1088/0004-6256/140/6/2036

J.W. Sulentic, P. Marziani, D. Dultzin, M. D'Onofrio, A. del Olmo, Fifty years of quasars: physical insights and potential for cosmology. J. Phys. Conf. Ser. **565**, 012018 (2014). 1411.6900. https://doi.org/10.1088/1742-6596/565/1/012018

M. Sullivan, A. Conley, D.A. Howell, J.D. Neill, P. Astier, C. Balland, S. Basa, R.G. Carlberg, D. Fouchez, J. Guy, D. Hardin, I.M. Hook, R. Pain, N. Palanque-Delabrouille, K.M. Perrett, C.J. Pritchet, N. Regnault, J. Rich, V. Ruhlmann-Kleider, S. Baumont, E. Hsiao, T. Kronborg, C. Lidman, S. Perlmutter, E.S. Walker, The dependence of type Ia supernovae luminosities on their host galaxies. Mon. Not. R. Astron. Soc. **406**, 782–802 (2010). 1003.5119. https://doi.org/10.1111/j.1365-2966.2010.16731.x

M. Sullivan, J. Guy, A. Conley, N. Regnault, P. Astier, C. Balland, S. Basa, R.G. Carlberg, D. Fouchez, D. Hardin, I.M. Hook, D.A. Howell, R. Pain, N. Palanque-Delabrouille, K.M. Perrett, C.J. Pritchet, J. Rich, V. Ruhlmann-Kleider, D. Balam, S. Baumont, R.S. Ellis, S. Fabbro, H.K. Fakhouri, N. Fourmanoit, S. González-Gaitán, M.L. Graham, M.J. Hudson, E. Hsiao, T. Kronborg, C. Lidman, A.M. Mourao, J.D. Neill, S. Perlmutter, P. Ripoche, N. Suzuki, E.S. Walker, SNLS3: constraints on dark energy combining the supernova legacy survey three-year data with other probes. Astrophys. J. **737**, 102 (2011). 1104.1444. https://doi.org/10.1088/0004-637X/737/2/102

R.A. Sunyaev, Y.B. Zeldovich, The observations of relic radiation as a test of the nature of X-ray radiation from the clusters of galaxies. Comments Astrophys. Space Phys. **4**, 173 (1972)

N. Suzuki, D. Rubin, C. Lidman, G. Aldering, R. Amanullah, K. Barbary, L.F. Barrientos, J. Botyanszki, M. Brodwin, N. Connolly, K.S. Dawson, A. Dey, M. Doi, M. Donahue, S. Deustua, P. Eisenhardt, E. Ellingson, L. Faccioli, V. Fadeyev, H.K. Fakhouri, A.S. Fruchter, D.G. Gilbank, M.D. Gladders, G. Goldhaber, A.H. Gonzalez, A. Goobar, A. Gude, T. Hattori, H. Hoekstra, E. Hsiao, X. Huang, Y. Ihara, M.J. Jee, D. Johnston, N. Kashikawa, B. Koester, K. Konishi, M. Kowalski, E.V. Linder, L. Lubin, J. Melbourne, J. Meyers, T. Morokuma, F. Munshi, C. Mullis, T. Oda, N. Panagia, S. Perlmutter, M. Postman, T. Pritchard, J. Rhodes, P. Ripoche, P. Rosati, D.J. Schlegel, A. Spadafora, S.A. Stanford, V. Stanishev, D. Stern, M. Strovink, N. Takanashi, K. Tokita, M. Wagner, L. Wang, N. Yasuda, H.K.C. Yee, Supernova Cosmology Project, The Hubble Space Telescope cluster supernova survey, V: improving the dark-energy constraints above $z > 1$ and building an early-type-hosted supernova sample. Astrophys. J. **746**, 85 (2012). . 1105.3470. https://doi.org/10.1088/0004-637X/746/1/85

H. Tananbaum, Y. Avni, G. Branduardi, M. Elvis, G. Fabbiano, E. Feigelson, R. Giacconi, J.P. Henry, J.P. Pye, A. Soltan, G. Zamorani, X-ray studies of quasars with the Einstein observatory. Astrophys. J. Lett. **234**, L9–L13 (1979). https://doi.org/10.1086/183100

N.R. Tanvir, D.B. Fox, A.J. Levan, E. Berger, K. Wiersema, J.P.U. Fynbo, A. Cucchiara, T. Krühler, N. Gehrels, J.S. Bloom, J. Greiner, P.A. Evans, E. Rol, F. Olivares, J. Hjorth, P. Jakobsson, J. Farihi, R. Willingale, R.L.C. Starling, S.B. Cenko, D. Perley, J.R. Maund, J. Duke, R.A.M.J. Wijers, A.J. Adamson, A. Allan, M.N. Bremer, D.N. Burrows, A.J. Castro-Tirado, B. Cavanagh, A. de Ugarte Postigo, M.A. Dopita, T.A. Fatkhullin, A.S. Fruchter, R.J. Foley, J. Gorosabel, J. Kennea, T. Kerr, S. Klose, H.A. Krimm, V.N. Komarova, S.R. Kulkarni, A.S. Moskvitin, C.G. Mundell, T. Naylor, K. Page, B.E. Penprase, M. Perri, P. Podsiadlowski, K. Roth, R.E. Rutledge, T. Sakamoto, P. Schady, B.P. Schmidt, A.M. Soderberg, J. Sollerman, A.W. Stephens, G. Stratta, T.N. Ukwatta, D. Watson, E. Westra, T. Wold, A γ-ray burst at a redshift of $z \sim 8.2$. Nature **461**, 1254–1257 (2009). 0906.1577. https://doi.org/10.1038/nature08459

L. Tartaglia, D. Sand, S. Wyatt, S. Valenti, K.A. Bostroem, D.E. Reichart, V. Haislip, V. Kouprianov, The discovery of DLT17u/AT 2017cbv. Astron. Telegr. **100158** (2017)

F. Thim, J.G. Hoessel, A. Saha, J. Claver, A. Dolphin, G.A. Tammann, Cepheids and long-period variables in NGC 4395. Astron. J. **127**, 2322–2343 (2004). astro-ph/0401558. https://doi.org/10.1086/382244

J. Tonry, D.P. Schneider, A new technique for measuring extragalactic distances. Astron. J. **96**, 807–815 (1988). https://doi.org/10.1086/114847

J.L. Tonry, A. Dressler, J.P. Blakeslee, E.A. Ajhar, A.B. Fletcher, G.A. Luppino, M.R. Metzger, C.B. Moore, The SBF survey of galaxy distances, IV: SBF magnitudes, colors, and distances. Astrophys. J. **546**, 681–693 (2001). astro-ph/0011223. https://doi.org/10.1086/318301

R. Tripp, D. Branch, Determination of the Hubble constant using a two-parameter luminosity correction for type IA supernovae. Astrophys. J. **525**, 209–214 (1999). astro-ph/9904347. https://doi.org/10.1086/307883

R.B. Tully, J.R. Fisher, A new method of determining distances to galaxies. Astron. Astrophys. **54**, 661–673 (1977)

A.V. Tutukov, L.R. Yungelson, Evolutionary scenario for close binary systems of low and moderate masses. Nauchn. Inf. **49**, 3 (1981)

V.P. Utrobin, N.N. Chugai, Progenitor mass of the type IIP supernova 2005cs. Astron. Astrophys. **491**, 507–513 (2008). 0809.3766. https://doi.org/10.1051/0004-6361:200810272

J.P. Uzan, N. Aghanim, Y. Mellier, Distance duality relation from X-ray and Sunyaev-Zel'dovich observations of clusters. Phys. Rev. D **70**(8), 083533 (2004). astro-ph/0405620. https://doi.org/10.1103/PhysRevD.70.083533

F. Vagnetti, S. Turriziani, D. Trevese, M. Antonucci, Variability and the X-ray/UV ratio of active galactic nuclei. Astron. Astrophys. **519**, A17 (2010). 1005.0144. https://doi.org/10.1051/0004-6361/201014320

F. van Leeuwen, M.W. Feast, P.A. Whitelock, C.D. Laney, Cepheid parallaxes and the Hubble constant. Mon. Not. R. Astron. Soc. **379**, 723–737 (2007). 0705.1592. https://doi.org/10.1111/j.1365-2966.2007.11972.x

J.J. Walmswell, J.J. Eldridge, Circumstellar dust as a solution to the red supergiant supernova progenitor problem. Mon. Not. R. Astron. Soc. **419**, 2054–2062 (2012). 1109.4637. https://doi.org/10.1111/j.1365-2966.2011.19860.x

I. Wanders, B.M. Peterson, D. Alloin et al., Steps toward determination of the size and structure of the broad-line region in active galactic nuclei, XI: intensive monitoring of the ultraviolet spectrum of NGC 7469. Astrophys. J. Suppl. **113**, 69–88 (1997). https://doi.org/10.1086/313054

B. Wang, Z. Han, Progenitors of type Ia supernovae. New Astron. Rev. **56**, 122–141 (2012). 1204.1155. https://doi.org/10.1016/j.newar.2012.04.001

B. Wang, X. Meng, X. Chen, Z. Han, The helium star donor channel for the progenitors of type Ia supernovae. Mon. Not. R. Astron. Soc. **395**, 847–854 (2009). 0901.3496. https://doi.org/10.1111/j.1365-2966.2009.14545.x

F.Y. Wang, Z.G. Dai, Weak gravitational lensing effects on cosmological parameters and dark energy from gamma-ray bursts. Astron. Astrophys. **536**, A96 (2011). 1112.4040. https://doi.org/10.1051/0004-6361/201117517

F.Y. Wang, Z.G. Dai, E.W. Liang, Gamma-ray burst cosmology. New Astron. Rev. **67**, 1–17 (2015). 1504.00735. https://doi.org/10.1016/j.newar.2015.03.001

J.M. Wang, P. Du, C. Hu, H. Netzer, J.M. Bai, K.X. Lu, S. Kaspi, J. Qiu, Y.R. Li, F. Wang, SEAMBH Collaboration, Supermassive black holes with high accretion rates in active galactic nuclei, II: the most luminous standard candles in the universe. Astrophys. J. **793**, 108 (2014). 1408.2337. https://doi.org/10.1088/0004-637X/793/2/108

L. Wang, Dust around type Ia supernovae. Astrophys. J. Lett. **635**, L33–L36 (2005). astro-ph/0511003. https://doi.org/10.1086/499053

X. Wang, W. Li, A.V. Filippenko, R.J. Foley, N. Smith, L. Wang, The detection of a light echo from the type Ia supernova 2006X in M100. Astrophys. J. **677**, 1060–1068 (2008a). 0711.2570. https://doi.org/10.1086/529070

X. Wang, W. Li, A.V. Filippenko, K. Krisciunas, N.B. Suntzeff, J. Li, T. Zhang, J. Deng, R.J. Foley, M. Ganeshalingam, T. Li, Optical and near-infrared observations of the highly reddened, rapidly expanding type Ia supernova SN 2006X in M100. Astrophys. J. **675**, 626–643 (2008b). 0708.0140. https://doi.org/10.1086/526413

D. Watson, K.D. Denney, M. Vestergaard, T.M. Davis, A new cosmological distance measure using active galactic nuclei. Astrophys. J. Lett. **740**, L49 (2011). 1109.4632. https://doi.org/10.1088/2041-8205/740/2/L49

T.A. Weaver, S.E. Woosley, Evolution and explosion of massive stars, in *Ninth Texas Symposium on Relativistic Astrophysics*, ed. by J. Ehlers, J.J. Perry, M. Walker. Annals of the New York Academy of Sciences, vol. 336 (1980), pp. 335–357. https://doi.org/10.1111/j.1749-6632.1980.tb15942.x

G. Weigelt, K.H. Hofmann, M. Kishimoto, S. Hönig, D. Schertl, A. Marconi, F. Millour, R. Petrov, D. Fraix-Burnet, F. Malbet, K. Tristram, M. Vannier, VLTI/AMBER observations of the Seyfert nucleus of NGC 3783. Astron. Astrophys. **541**, L9 (2012). 1204.6122. https://doi.org/10.1051/0004-6361/201219213

J.C. Wheeler, D. Pooley, Self-shielding of soft X-rays in type Ia supernova progenitors. Astrophys. J. **762**, 75 (2013). 1208.0858. https://doi.org/10.1088/0004-637X/762/2/75

J. Whelan, I. Iben Jr., Binaries and supernovae of type I. Astrophys. J. **186**, 1007–1014 (1973). https://doi.org/10.1086/152565

O.C. Wilson, Possible applications of supernovae to the study of the nebular red shifts. Astrophys. J. **90**, 634 (1939). https://doi.org/10.1086/144134

W.M. Wood-Vasey, G. Miknaitis, C.W. Stubbs, S. Jha, A.G. Riess, P.M. Garnavich, R.P. Kirshner, C. Aguilera, A.C. Becker, J.W. Blackman, S. Blondin, P. Challis, A. Clocchiatti, A. Conley, R. Covarrubias, T.M. Davis, A.V. Filippenko, R.J. Foley, A. Garg, M. Hicken, K. Krisciunas, B. Leibundgut, W. Li, T. Matheson, A. Miceli, G. Narayan, G. Pignata, J.L. Prieto, A. Rest, M.E. Salvo, B.P. Schmidt, R.C. Smith, J. Sollerman, J. Spyromilio, J.L. Tonry, N.B. Suntzeff, A. Zenteno, Observational constraints on the nature of dark energy: first cosmological results from the ESSENCE supernova survey. Astrophys. J. **666**, 694–715 (2007). astro-ph/0701041. https://doi.org/10.1086/518642

T.E. Woods, M. Gilfanov, He II recombination lines as a test of the nature of SN Ia progenitors in elliptical galaxies. Mon. Not. R. Astron. Soc. **432**, 1640–1650 (2013). 1302.5911. https://doi.org/10.1093/mnras/stt586

S.E. Woosley, D. Kasen, Sub-Chandrasekhar mass models for supernovae. Astrophys. J. **734**, 38 (2011). 1010.5292. https://doi.org/10.1088/0004-637X/734/1/38

S.E. Woosley, T.A. Weaver, Sub-Chandrasekhar mass models for type IA supernovae. Astrophys. J. **423**, 371–379 (1994). https://doi.org/10.1086/173813

X. Yang, H.R. Yu, T.J. Zhang, Constraining smoothness parameter and the DD relation of Dyer-Roeder equation with supernovae. J. Cosmol. Astropart. Phys. **6**, 007 (2013). 1305.6989. https://doi.org/10.1088/1475-7516/2013/06/007

Y. Yang, L. Wang, D. Baade, P.J. Brown, M. Cracraft, P.A. Höflich, J. Maund, F. Patat, W.B. Sparks, J. Spyromilio, H.F. Stevance, X. Wang, J.C. Wheeler, Interstellar-medium mapping in M82 through light echoes around supernova 2014J. Astrophys. J. **834**, 60 (2017). 1610.02458. https://doi.org/10.3847/1538-4357/834/1/60

O. Yaron, D. Prialnik, M.M. Shara, A. Kovetz, An extended grid of nova models, II: the parameter space of nova outbursts. Astrophys. J. **623**, 398–410 (2005). astro-ph/0503143. https://doi.org/10.1086/428435

Y. Yoshii, The MAGNUM project: AGN variability as a new technique for distance determination, in *New Trends in Theoretical and Observational Cosmology*, ed. by K. Sato, T. Shiromizu (2002), p. 235

Y. Yoshii, Y. Kobayashi, T. Minezaki, S. Koshida, B.A. Peterson, A new method for measuring extragalactic distances. Astrophys. J. Lett. **784**, L11 (2014). 1403.1693. https://doi.org/10.1088/2041-8205/784/1/L11

M. Young, M. Elvis, G. Risaliti, The X-ray energy dependence of the relation between optical and X-ray emission in quasars. Astrophys. J. **708**, 1388–1397 (2010). 0911.0474. https://doi.org/10.1088/0004-637X/708/2/1388

G. Zamorani, J.P. Henry, T. Maccacaro, H. Tananbaum, A. Soltan, Y. Avni, J. Liebert, J. Stocke, P.A. Strittmatter, R.J. Weymann, M.G. Smith, J.J. Condon, X-ray studies of quasars with the Einstein Observatory, II. Astrophys. J. **245**, 357–374 (1981). https://doi.org/10.1086/158815

Space Sci Rev (2018) 214:91
https://doi.org/10.1007/s11214-018-0524-3

Cosmological Distance Indicators

**Sherry H. Suyu[1,2,3] · Tzu-Ching Chang[2,4] ·
Frédéric Courbin[5] · Teppei Okumura[2,6]**

Received: 21 January 2018 / Accepted: 2 July 2018 / Published online: 25 July 2018

Abstract We review three distance measurement techniques beyond the local universe: (1) gravitational lens time delays, (2) baryon acoustic oscillation (BAO), and (3) HI intensity mapping. We describe the principles and theory behind each method, the ingredients needed for measuring such distances, the current observational results, and future prospects. Time-delays from strongly lensed quasars currently provide constraints on H_0 with $< 4\%$ uncertainty, and with 1% within reach from ongoing surveys and efforts. Recent exciting discoveries of strongly lensed supernovae hold great promise for time-delay cosmography. BAO features have been detected in redshift surveys up to $z \lesssim 0.8$ with galaxies and $z \sim 2$ with Ly-α forest, providing precise distance measurements and H_0 with $< 2\%$ uncertainty in flat ΛCDM. Future BAO surveys will probe the distance scale with percent-level preci-

Astronomical Distance Determination in the Space Age
Edited by Richard de Grijs and Maurizio Falanga

✉ S.H. Suyu
suyu@mpa-garching.mpg.de

T.-C. Chang
tzu-ching.chang@jpl.nasa.gov

F. Courbin
frederic.courbin@epfl.ch

T. Okumura
tokumura@asiaa.sinica.edu.tw

1 Max-Planck-Institut für Astrophysik, Karl-Schwarzschild-Str. 1, 85748 Garching, Germany

2 Institute of Astronomy and Astrophysics, Academia Sinica, 11F of ASMAB, No. 1, Section 4, Roosevelt Road, Taipei 10617, Taiwan

3 Physik-Department, Technische Universität München, James-Franck-Straße 1, 85748 Garching, Germany

4 Jet Propulsion Laboratory, 4800 Oak Grove Dr, MS 169-237, Pasadena, CA 91109, USA

5 Institute of Physics, Laboratoire d'Astrophysique, Ecole Polytechnique Fédérale de Lausanne (EPFL), Observatoire de Sauverny, 1290 Versoix, Switzerland

6 Kavli Institute for the Physics and Mathematics of the Universe (WPI), UTIAS, The University of Tokyo, Kashiwa, Chiba 277-8583, Japan

sion. HI intensity mapping has great potential to map BAO distances at $z \sim 0.8$ and beyond with precisions of a few percent. The next years ahead will be exciting as various cosmological probes reach 1% uncertainty in determining H_0, to assess the current tension in H_0 measurements that could indicate new physics.

Keywords Cosmology · Gravitational lensing · Baryon acoustic oscillation · Intensity mapping

1 Gravitational Lens Time Delays

1.1 Principles of Gravitational Lens Time Delays

Strong gravitational lensing occurs when a foreground mass distribution is located along the line of sight to a background source such that multiple images of the background source appear around the foreground lens. In cases where the background source intensity varies, such as an active galactic nucleus (AGN) or a supernova (SN), the variability pattern manifests in each of the multiple images and is delayed in time due to the different light paths of the different images. The time delay of image i, relative to the case of no lensing, is

$$t(\boldsymbol{\theta}_i;\boldsymbol{\beta}) = \frac{D_{\Delta t}}{c}\phi(\boldsymbol{\theta}_i;\boldsymbol{\beta}), \tag{1}$$

up to an additive constant,[1] where $\boldsymbol{\theta}_i$ is the position of the lensed image i, $\boldsymbol{\beta}$ is the position of the source, $D_{\Delta t}$ is the so-called "time-delay distance", c is the speed of light, and ϕ is the "Fermat potential" related to the lens mass distribution. The time-delay distance for a lens at redshift z_d and a source at redshift z_s is

$$D_{\Delta t} = (1+z_d)\frac{D_d D_s}{D_{ds}}, \tag{2}$$

where D_d is the angular diameter distance to the lens, D_s is the angular diameter distance to the source, and D_{ds} is the angular diameter distance between the lens and the source. In the ΛCDM cosmology with density parameters Ω_m for matter, Ω_k for spatial curvature, and Ω_Λ for dark energy described by the cosmological constant Λ, the angular diameter distance between two redshifts z_1 and z_2 is

$$D(z_1, z_2) = \frac{1}{1+z_2} f_K\left[\chi(z_1, z_2)\right] \tag{3}$$

where

$$\chi(z_1, z_2) = \frac{c}{H_0}\int_{z_1}^{z_2} dz' \left(\Omega_m(1+z')^3 + \Omega_k(1+z')^2 + \Omega_\Lambda\right)^{-1/2}, \tag{4}$$

and

$$f_K(\chi) = \begin{cases} K^{-1/2}\sin(K^{1/2}\chi) & \text{for } K > 0, \\ \chi & \text{for } K = 0, \\ (-K)^{-1/2}\sinh[(-K)^{1/2}\chi] & \text{for } K < 0, \end{cases} \tag{5}$$

$K = -\Omega_k H_0^2/c^2$ is the spatial curvature, and H_0 is the Hubble constant.

[1] The Fermat potential, being a potential, is defined only up to an additive constant that has no physical consequence. Furthermore, a "mass-sheet transformation" (explained later in Sect. 1.3.2) can also add a term that is independent of $\boldsymbol{\theta}_i$ to the Fermat potential.

By monitoring the variability of the multiple images, we can measure the time delay between the two images i and j:

$$\Delta t_{ij} = t(\boldsymbol{\theta}_i; \boldsymbol{\beta}) - t(\boldsymbol{\theta}_j; \boldsymbol{\beta}) = \frac{D_{\Delta t}}{c} \Delta\phi_{ij}. \tag{6}$$

The Fermat potential ϕ can be determined by modeling the lens mass distribution using observations of the lens system such as the observed lensed image positions, shapes and fluxes. Therefore, with Δt measured and $\Delta\phi$ determined, we can use (6) to infer the value of $D_{\Delta t}$, which is inversely proportional to H_0 ($D_{\Delta t} \propto H_0^{-1}$) through (2) and (3). Being a combination of three angular diameter distances, $D_{\Delta t}$ is mainly sensitive to the Hubble constant H_0 and weakly depends on other cosmological parameters (e.g., Refsdal 1964; Schneider et al. 2006; Suyu et al. 2010).

One can further measure D_{d} from a lens system by measuring the velocity dispersion of the foreground lens, σ_{v}, and combining it with the time delays (Paraficz and Hjorth 2009; Jee et al. 2015). The measurement of D_{d} provides additional constraints on cosmological models (Jee et al. 2016).

In order to measure $D_{\Delta t}$ and D_{d} from a time-delay lens system for cosmography, we need the following

1. spectroscopic redshifts of the lens z_{d} and source z_{s}
2. time delays between the multiple images
3. lens mass model to determine the Fermat potential
4. lens velocity dispersion, which is not only required for D_{d} inference, but also provides additional constraints in breaking lens mass model degeneracies
5. lens environment studies to break lens model degeneracies, such as the mass-sheet degeneracy

In the next sections, we describe the history behind this approach, and detail the advances in recent years in acquiring these ingredients before presenting the latest cosmographic inferences from this approach.

1.2 A Brief History

In his original paper Refsdal (1964) proposed to use gravitationally lensed supernovae to measure the time delays: the light curves associated to each lensed image of a supernova are expected to be seen shifted in time by a value that depends on the potential well of the lensing object and on cosmology. However, due to the shallow limiting magnitude of the telescopes available at the time and due to their restricted field of view, discovering faint and distant supernovae right behind a galaxy or a galaxy cluster was completely out of reach. But Refsdal's idea came right when quasars were discovered (Schmidt 1963; Hazard et al. 1963). These bright, distant and photometrically variable point sources were coming timely, offering a new opportunity to implement the time-delay method: light curves of lensed quasars are constantly displaying new features that can be used to measure the delay.

With the increasing discovery rate of quasars, the first cases of multiply imaged ones also started to grow. The first lensed quasar, Q 0951+567, was found by Walsh et al. (1979), displaying two lensed images. This was followed by the quadruple PG 1115+080 (Young et al. 1981), and a few years later by the discovery of the "Einstein Cross" (Huchra et al. 1985) and of the "cloverleaf" (Magain et al. 1988). The first time-delay measurement became available only in the late 80s with the optical monitoring of Q 0951+567 by Vanderriest et al. (1989) and the radio monitoring of the same object by Lehar et al. (1992). Unfortunately, given the two data sets and methods of analysis to measure the delay, the radio and optical values of the time delay remained in disagreement until new optical data came (e.g. Kundić et al. 1997; Oscoz et al. 1997), allowing to confirm and improve the optical delay of Vanderriest et al. (1989). Further improvement was possible with the "round-the-clock" monitoring of Colley et al. (2003), leading to a time-delay determination to a fraction of a day.

Because of the time and effort it took to solve the "Q0957 controversy", astrophysicists quickly limited their interest in the time-delay method as a cosmological probe. But at least two sets of impressive monitoring data revived the field. The first one is the optical monitoring of the quadruple quasar PG 1115+080 by Schechter et al. (1997), providing time delays to 14% and the second is the radio monitoring, with the VLA, of the quadruple radio source B1608+656 (Fassnacht et al. 1999), reaching similar accuracy. As the uncertainty on the time delay propagates linearly in the error budget on H_0^{-1} this is still not sufficient for precision cosmology to a few percents.

In a large part thanks to the results obtained for PG 1115+080 and B1608+656 several monitoring campaigns were put in place by independent teams in the late 90s and early 2000. Because lensed quasars were more often discovered in the optical and because their variability is faster at these wavelengths due to the smaller source size than in the radio, these new monitoring campaigns took place in the optical. The teams involved used 1 m class telescopes to measure delays to typical accuracies of 10% or slightly better, i.e. a 30% improvement over previous measurements but still too large for cosmological purposes.

Some of the most impressive results were obtained in the years 2000 with the 2.6 m Nordic Optical Telescope (NOT) for FBQ 0951+2635 (Jakobsson et al. 2005), SBS 1520+530 (Burud et al. 2002b), RX J0911+0551 (Hjorth et al. 2002), B1600+434 (Burud et al. 2000), at ESO with the 1.54 m Danish telescope for HE 2149-2745 (Burud et al. 2002a) and at Wise observatory with the 1 m telescope for HE 1104-1805 (Ofek and Maoz 2003). With these new observations and studies it was shown that "mass production" of time delays was possible and not restricted to a few lenses for which the observational situation was particularly favorable. However, the temporal sampling adopted for the observations and the limited signal-to-noise per observing epoch was still limiting the accuracy on the time-delay measurement to 10% in most cases hence limiting H_0^{-1} measurements with individual lenses to this precision.

Fifty years after Refsdal (1964)'s foresight on lensed SN, the first strongly lensed SN was discovered by Kelly et al. (2015) serendipitously in the galaxy cluster MACS J1149.5+2223 with *Hubble Space Telescope* (*HST*) imaging. This core-collapse SN was named "SN Refsdal", and showed 4 multiple images at detection. The predictions (Treu et al. 2016; Grillo et al. 2016; Kawamata et al. 2016; Jauzac et al. 2016) and subsequent detection (Kelly et al. 2016a) of the re-occurrence of the next (time-delayed) image of SN Refsdal provided a true blind test of our understanding of lensing theory and mass modeling. It is reassuring that some teams predicted accurately the re-occurrence (Grillo et al. 2016; Kawamata et al. 2016), and the modeling software GLEE[2] used by Grillo et al. (2016) was also the software employed for cosmography with lensed quasars (e.g., Suyu et al. 2013, 2014; Wong et al. 2017).

In the fall of 2016, the first spatially-resolved multiply-imaged Type Ia SN, iPTF16geu, was discovered by Goobar (2017) in the intermediate Palomar Transient Factory survey. More et al. (2017) independently modeled a single-epoch *HST* image of the system, finding short model-predicted time delays (< 1 day) between the multiple images. Furthermore,

[2]GLEE (Gravitational Lens Efficient Explorer) is a gravitational lens modeling software developed by A. Halkola and S. H. Suyu (Suyu and Halkola 2010; Suyu et al. 2012).

More et al. (2017) found anomalous flux ratios of the SN compared to the smooth model prediction, indicating possible microlensing effects, although Yahalomi et al. (2017) showed that microlensing is unlikely to be the sole cause of the anomalous flux ratios.

Both SN Refsdal and iPTF16geu have been monitored for time-delay measurements (Rodney et al. 2016; Goobar 2017), opening a new window to study cosmology with strongly-lensed SN. Recently Grillo et al. (2018) estimated the time delay of the image SX of SN Refsdal based on the detection presented in Kelly et al. (2016b) (image SX has the longest delay compared to other images of SN Refsdal, so image SX will ultimately provide the most precise time-delay measurement for cosmography from this system), and modeled the mass distribution of the galaxy cluster MACS J1149.5+2223 to infer H_0. This feasibility study shows that H_0 can be measured with $\sim 7\%$ statistical uncertainty, despite the complexity in modeling the cluster lens mass distribution. The full analysis including various systematic uncertainties is forthcoming, after the time delays are measured from the monitoring data. As lensed SNe are only being discovered/observed recently and their utility as a time-delay cosmological probe is just starting, we focus in the rest of the review on the more common lensed quasars as time-delay lenses, but there is a wealth of information to gather with lensed supernovae, both on a cosmological and stellar physics point of view.

1.3 Recent Advances

A 10% error bar on the time delay translates to a similar error on H_0. Improving this further and obtaining H_0 measurements competitive with other techniques, i.e. of 3–4% currently and 1–2% in the near future, requires several ingredients. Time-delay measurements of individual lenses must reach 5% at most and many more systems must be measured. With such precision on individual time delays, and under the assumption that all sources of systematic errors are controlled or negligible, measuring H_0 to 2% is possible with only a handful of lenses and a 1% measurement is not out of reach with of the order of 50 lenses! However, the lens model for the main lensing galaxy must be well constrained and their systematics evaluated and/or mitigated. Third, the contribution of all objects on the line of sight to the overall potential well must be accounted for. Excellent progress has been made on all three fronts in the recent years and there is still room for further (realistic) improvement.

1.3.1 Time Delays

Measuring time delays is hard, but feasible provided telescope time can be guaranteed over long periods of time with stable instrumentation. The main limiting factors are astrophysical, observational or instrumental.

Astrophysical factors include the characteristics of the intrinsic variability of the source quasar and extrinsic variability of its lensed images due to microlensing by stars in the main lensing galaxy. If the source quasar is highly variable intrinsically, both in amplitude and temporal frequency, the time delay is easier to measure. If microlensing variations are strong and/or comparable in frequency to the intrinsic variations then the time-delay value can be degenerate with the properties of the microlensing variations. In some extreme cases microlensing dominates the observed photometric variations to the point the time delay is hardly measurable (e.g. Morgan et al. 2012) even though microlensing itself can be used to infer details properties of the lensed source on micro-arcsec scales, i.e. out of reach of any current and future instrumentation.

The observations needed to measure time delays must be adapted to the intrinsic and extrinsic variations of the selected quasars and of course to the expected time delay for each target. Not surprisingly the shorter the time delay, the finer the temporal sampling is needed. The position of the target on the sky also influences the results: equatorial targets will hardly be visible more than 6 months in a row, but can be followed both from the North and the South, while circumpolar targets can be seen up to 8–9 consecutive months, hence allowing to measure longer time delays and minimizing the effect of the non-visibility gaps between observing seasons.

Finally, instrumental factors strongly impact the results. A key factor with current monitoring campaigns is the availability of telescopes on good sites and with stable instrumentation, i.e. if possible at all with no camera or filter change and with regular temporal sampling. Long gaps in light curves seriously affect the time-delay values in the sense that they make it more difficult to disentangle the microlensing variations from the quasar intrinsic variations. And since angular separation between the lensed images are small, fairly good seeing is required, typically below 1.2 arcsec even though techniques like image deconvolution (e.g. Magain et al. 1998; Cantale et al. 2016) and used, e.g. by the COSMOGRAIL collaboration (see below) help dealing with data sets spanning a broad range of seeing values.

Once long and well-sampled photometric light curves are available, the time delay must be measured. At first glance, this step may be seen as an easy one. However, one has to deal with irregular temporal sampling, gaps in the light curves, variable signal-to-noise and seeing, atmospheric effects (night-to-night calibration) and with microlensing. A number of numerical methods have been devised over the years to carry out the measurement, with different levels of accuracy and precision. They split in different broad categories. Some attempt to cross-correlate the light curves without trying to model/subtract the microlensing variations. Others involve an analytical representation of the intrinsic quasar variations and of the microlensing or involve e.g. Gaussian processes to mimic the microlensing erratic variations. Recent work in this area has been developed in Tewes et al. (2013a), Hojjati et al. (2013), Hojjati and Linder (2014), Rathna Kumar et al. (2015).

These methods (and others, so far unpublished) were tested in an objective way using simulated light curves proposed to the community in the context of the "Time Delay Challenge" (TDC; Dobler et al. 2015). In the TDC, thousands of light curves of different lengths, sampling rate, signal-to-noise, and visibility gaps are proposed to the participants. Once all participants have submitted their time-delay evaluations to the challenge organizers, the time-delay values are revealed and the results are compared objectively using a metrics common to all participating methods. This metrics was known before the start of the challenge. The results of this TDC are summarized in Liao et al. (2015) as well as in individual papers (e.g. Bonvin et al. 2016). A general conclusion of the TDC was that with current lens monitoring data, curve-shifting technique so far in use are sufficient to extract precise and accurate time delays, given the temporal sampling and signal-to-noise of the data.

Following the encouraging results obtained at NOT, ESO and Wise, long-term monitoring campaigns were organized to measure time delays in a systematic way. Two main teams invested effort in this research: the OSU group lead by C.S. Kochanek (OSU, USA) and the COSMOGRAIL (COSmological MOnitoring of GRAvItational Lenses) program led by F. Courbin and G. Meylan at EPFL, Switzerland (e.g. Courbin et al. 2005; Eigenbrod et al. 2005). Both monitoring programs involve 1-m class telescopes with a temporal sampling of 2 to 3 observing epochs per week and a signal-to-noise of typically 100 per quasar image and per epoch. Both projects started in 2004 and are so far the main (but not only) source of time-delay measurements. Early results from the OSU program were obtained in 2006 for HE 0435-1223 (Kochanek et al. 2006) while COSMOGRAIL delivered its first results starting in 2007 for SDSS J1650+4251 (Vuissoz et al. 2007), WFI J2033-4723 (Vuissoz et al. 2008) and HE 0435-1223 (Courbin et al. 2011a). More recent time-delay measurements from COSMOGRAIL were obtained for RX J1131-1231 (Tewes et al. 2013b), HE 0435-1223 (Bonvin et al. 2017) as well as SDSS J1206+4332 and HS 2209+1914 (Eulaers et al. 2013) and SDSS J1001+5027 (Rathna Kumar et al. 2013). An example of COSMOGRAIL light curve is given in Fig. 1. These data are analysed jointly with the H0LiCOW program (see Sect. 1.4).

Other recent studies for specific objects include Giannini et al. (2017) for WFI 2033-4723 and HE 0047-1756, and Shalyapin and Goicoechea (2017) reporting a delay for SDSS J1515+1511. Hainline et al. (2013) measure a tentative time delay for SBS 0909+532, although the curves suffer from strong microlensing. Finally, two (long) time delays have been estimated for two quasars lensed by a galaxy group/cluster: SDSS J1029+2623 (Fohlmeister et al. 2013) and SDSSJ1004+4112 (Fohlmeister et al. 2008). These may not be ideal for cosmological applications though, as a complex lens model for a cluster is harder to constrain than models at galaxy-scale, unless the cluster has additional constraints coming from multiple background sources at different redshifts being strongly lensed.

With the observing cadence of 1 point every 3–4 nights and an SNR of 100 per epoch, the current data can catch quasar variations of the order of 0.1 mag in amplitude, arising on time-scales of months. These time scales are unfortunately of the same order of magnitude as the microlensing variations (see 2nd panel of Fig. 1) making it hard to disentangle between intrinsic and extrinsic variations. For this reason, lensed quasars must be monitored for extended periods of time, typically a decade, to infer any reliable time-delay measurement.

Going beyond current monitoring campaigns like COSMOGRAIL and others is possible, but measuring massively time delays for dozens of lensed quasars requires a new observing strategy to minimize the effect of microlensing and to measure time delays in individual objects in less than 10 years! One solution is to observe at high cadence (1 day^{-1}) and high SNR, of the order of 1000. In this way, very small quasar variations can be caught, on time scales much shorter than microlensing hence allowing the separation of the two signals in frequency. We show with 1000 mock light curves that mimic those of HE 0425-1223 (Fig. 1) that time delays can be measured precisely in only 1 observing season. In doing the simulation, we include realistic microlensing and fast quasar variations with a few mmag amplitude. We then run `PyCS`, the COSMOGRAIL curve-shifting algorithm (Tewes et al. 2013a; Bonvin et al. 2016), to recover the fiducial time delay of 14 days. Figure 2 summarizes our results and provides the length of the monitoring campaign needed to reach a desired relative precision on the time delay, assuming daily observations and an SNR of 1000 per epoch. It appears that a typical 2% precision is achievable in 1 observing season with a 10% failure rate. Doing two seasons allows one to reach the percent precision and a failure rate below 3%. At the time this paper is being written, an intensive lens monitoring program has been started at the 2.2 m MPI telescope at La Silla Observatory, with the above characteristics. Three targets have been observed for 1 semester and time delays have been measured to a few percents for all three! The first of these is presented in Courbin et al. (2017) and features a 1.8% measurement of one of the delays in the newly discovered quadruple lens DES J0408-5354 (Lin et al. 2017; Agnello et al. 2017).

Finally, we note that although quasars have been used so far to implement the time delay method, the original idea of Refsdal was to use lensed supernovae. The first systems have finally be found, as mentioned in Sect. 1.2: SN Refsdal (e.g. Kelly et al. 2015, 2016a; Rodney et al. 2016) and iPTF16geu (e.g. Goobar 2017; More et al. 2017). With the advent of large imaging surveys such as the Zwicky Transient Facility and the Large Synoptic Survey Telescope, prospects to find lensed supernovae are excellent (e.g. Goldstein and Nugent 2017). As supernovae have known light curves, one can measure the time delay by fitting

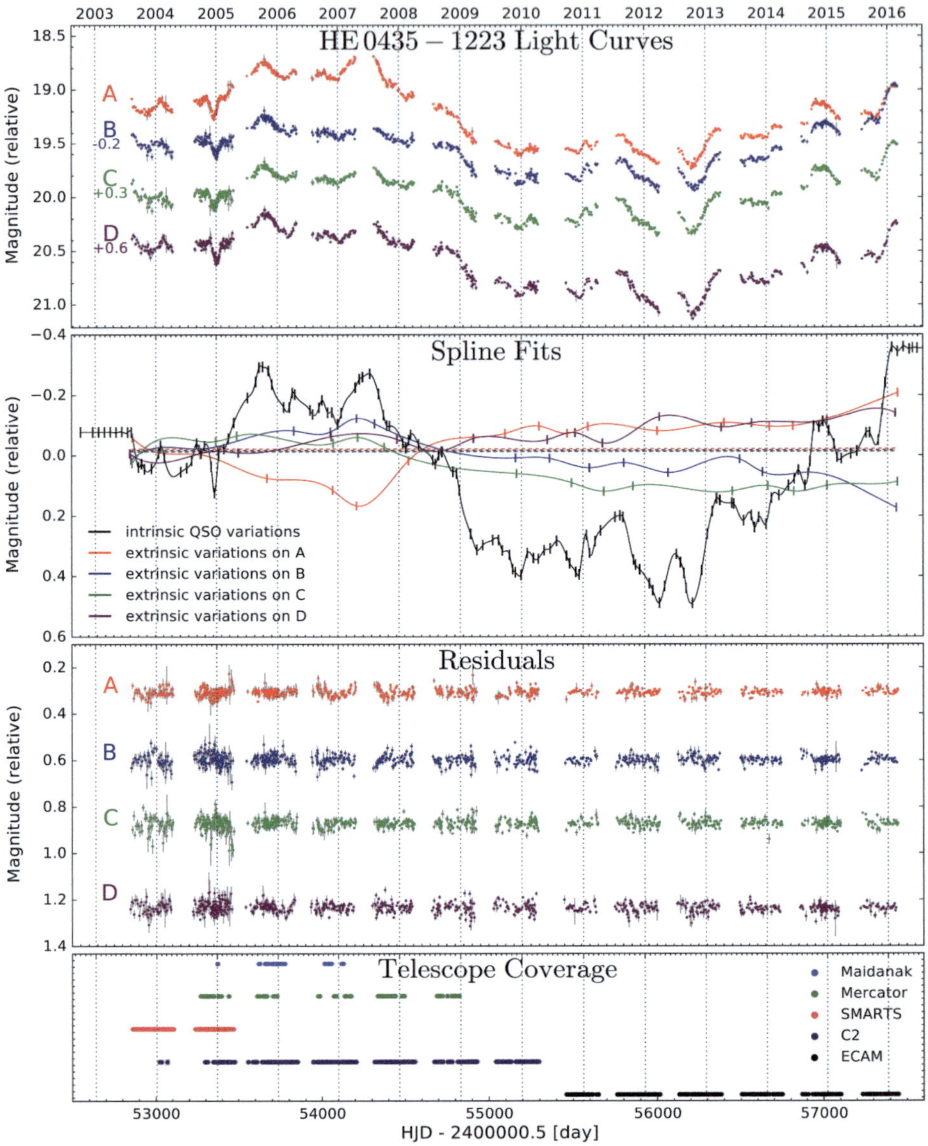

Fig. 1 From top to bottom: example of light curves produced and exploited by the COSMOGRAIL and H0LiCOW programs, here for the quadruply imaged quasar HE 0435-1223. The original light curves are shown on the top. The second panel shows spline fitting to the data including the intrinsic and extrinsic quasar variations. Crucially, long light curves are needed to extract properly the extrinsic variation (microlensing). The residuals to the fit and the journal of the observations with 5 instruments are displayed in the two lower panels (reproduced with permission from Bonvin et al. 2017)

a template to the observed light curves in the lensed images, hence giving much more constraining power than quasars whose photometric variations are close to a random walk. In addition, if the lensed supernova is a Type Ia, then two cosmological probes are available in the same object, hence provide a fantastic cross-check of otherwise completely different

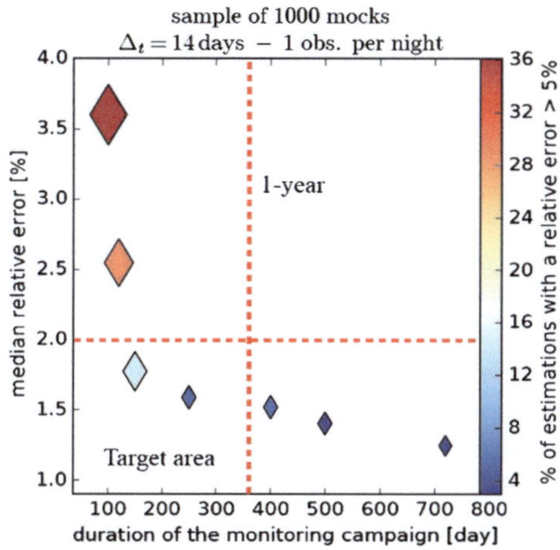

Fig. 2 Expected relative precision on a time delay measurement as a function of the length of the campaign. High-cadence (1 day^{-1}) monitoring is assumed and the fiducial delay in this simulations mimics the longer delay of HE 0435-1223, i.e. 14 days. Clearly, 2% precision can be reached in only 1 observing season. The color code shows the catastrophic failure rate, i.e. the probability of getting a measurement wrong by more than 5%. This probability is about 10% for a 1-season campaign and 3% for a 2-season campaign. (Courtesy: Vivien Bonvin)

methods: standard candles and a geometrical method, provided microlensing effects could be corrected (Dobler and Keeton 2006; Yahalomi et al. 2017; Goldstein et al. 2018).

For all the above reasons, we believe that the future of time-delay cosmography resides in lensed supernovae and in high-cadence monitoring of lensed quasars. However, Tie and Kochanek (2018) recently pointed out that microlensing by stars in the lensing galaxy can introduce a bias in the time-delay measurements. This is due to a combination of differential magnification of different parts of the source and the source geometry itself. The net result is that cosmological time delays can be affected both in a statistical and a systematic way by microlensing. The effect is absolute, with biases on time delays of the order of a day for lensed quasars and tenths of a day for lensed supernovae. Mitigation strategies have been successfully devised (Chen et al. 2018) for lensed quasars and the effect seems less pronounced in lensed supernovae than in lensed quasars (Goldstein et al. 2018; Foxley-Marrable et al. 2018; Bonvin et al. 2018), but clearly this new effect must be accounted for in any future work in the field.

1.3.2 Lens Mass Modeling

To convert the time delays into a measurement of the time-delay distance via (1), one needs to determine the Fermat potential $\phi(\boldsymbol{\theta}_i; \boldsymbol{\beta})$, which depends both on the mass distribution of the main strong-lens galaxy and the mass distribution of other galaxies along the line of sight.

The mass distribution of the main strong-lens galaxy can be modeled using either simply parametrized profiles (e.g., Kormann et al. 1994; Barkana 1998; Golse and Kneib 2002) or grid-based approaches (e.g., Williams and Saha 2000; Blandford et al. 2001; Suyu et al. 2009; Vegetti and Koopmans 2009). The total mass distribution of galaxies appear to be well described by profiles close to isothermal (e.g., Koopmans et al. 2006; Barnabè et al. 2011;

Cappellari et al. 2015), even though neither the baryons nor the dark matter distribution follow isothermal profiles. Even in the complex case of the gravitational lens B1608+656 with two interacting lens galaxies, simply parametrized profiles provide a remarkably good description of the galaxies when compared to the pixelated lens potential reconstruction (Suyu et al. 2009). Therefore, most of the current mass modeling for time-delay cosmography use simply parametrized profiles, either for the total mass distribution (e.g., Koopmans et al. 2003; Fadely et al. 2010; Suyu et al. 2010, 2013; Birrer et al. 2015b) or for separate components of baryons and dark matter (e.g., Courbin et al. 2011b; Schneider and Sluse 2013; Suyu et al. 2014; Wong et al. 2017).

The source (quasar) properties need to be modeled simultaneously with the lens mass distribution to predict the observables. In particular, source position and intensity are needed to predict the positions, fluxes and time delays of the lensed quasar images, whereas the source surface brightness distribution (of the quasar host galaxy) is needed to predict the lensed arcs. These observables (image positions, fluxes and delays of the multiple quasar images, and lensed arcs) are then used to constrain the parameters of the lens mass model and the source. Several softwares are available publicly for modeling lens systems, including GRAVLENS (Keeton 2001), LENSTOOL (Jullo et al. 2007), GLAFIC (Oguri 2010) and LENSVIEW (Wayth and Webster 2006).

Observed quasar image positions, fluxes and delays provide around a dozen of constraints for quads (four-image systems) and even fewer constraints for doubles (two-image systems). Thus lens mass models using only these quasar observables are often not precisely constrained. In particular, the radial profile slope of the lens galaxy is strongly degenerate with $D_{\Delta t}$ (e.g., Wucknitz 2002; Suyu 2012). The time delays depend primarily on the average surface mass density between the multiple images, and thus provide information on the radial profile slope (Kochanek 2002). Nevertheless, even with multiple time delays from quad systems, it is difficult to infer the slope precisely to better than $\sim 10\%$ precision.[3] While mass distribution of massive early-type galaxies, which are the majority of lens galaxies, are close to being isothermal, there is an intrinsic scatter in the slope of about $\sim 15\%$[3] (Koopmans et al. 2006; Barnabè et al. 2011). For precise and accurate $D_{\Delta t}$ measurement of a lens system, it is important to measure directly, at the few percent level, the radial profile slope of the lens galaxy near the lensed images of the quasars. This requires more observations of the lens system, beyond just the multiple point images of lensed quasars.

Over the past decade, multiple methods have been developed to make use of the lensed arcs (lensed host galaxies of the quasars) to constrain the lens mass distribution. The source intensity distribution can be described by simply parametrized profiles, such as Gaussians or Sersic (e.g., Brewer and Lewis 2008; Oguri 2010; Marshall et al. 2007; Oldham et al. 2017), or on a grid of pixels, either regular (e.g., Wallington et al. 1996; Warren and Dye 2003; Koopmans 2005; Suyu et al. 2006) or adaptive (e.g., Dye and Warren 2005; Vegetti and Koopmans 2009; Tagore and Keeton 2014; Nightingale and Dye 2015), or based on basis functions (e.g., Birrer et al. 2015a, Joseph et al. in prep.). These lensed arcs, when imaged with *HST* or ground-based telescopes assisted with adaptive optics, contain thousands of intensity pixels and thus allow the measurement of the radial profile slope of the lens galaxies with a precision of a few percent (e.g., Dye and Warren 2005; Suyu 2012; Chen et al. 2016), that are required for cosmography. In Fig. 3, we show an example of the mass modeling using the full surface brightness distributions of quasar host galaxy.

[3] In terms of impact on $D_{\Delta t}$.

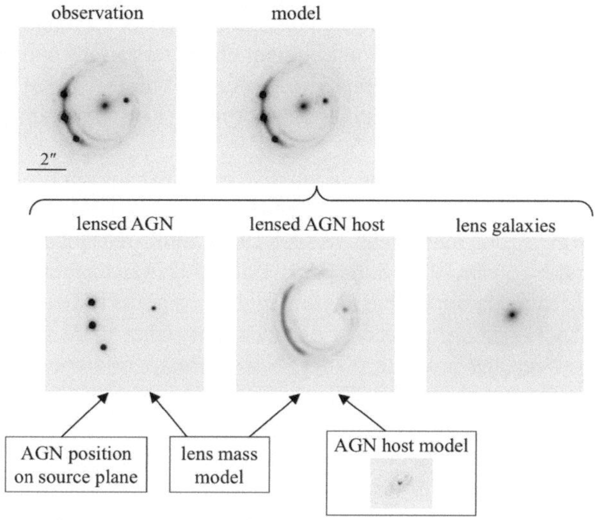

Fig. 3 Illustration of lens mass modeling of the gravitational lens RXJ1131-1231. Top left is the observed *HST* image. Top middle panel is the modeled surface brightness of the lens system, which is composed of three components shown in the second row: lensed AGN images (left), lensed AGN host galaxy (middle), and foreground lens galaxies (right). The bottom row shows that a mass model is required together with the AGN source position and AGN host galaxy surface brightness, to model the lensed AGN and lensed AGN host images. See the text and Suyu et al. (2013, 2014) for more details

Once a model of the surface mass density κ is obtained, lens theory states that the following family of models κ_λ fits equally well to the observed lensing data:

$$\kappa_\lambda = \lambda + (1 - \lambda)\kappa, \tag{7}$$

where λ is a constant. This transformation is analogous to adding a constant mass sheet λ in convergence, and rescaling the mass distribution of the strong lens (to keep the same mass within the Einstein radius); it is therefore called the "mass-sheet degeneracy" (Falco et al. 1985; Schneider and Sluse 2013, 2014; Schneider 2014). Such a transformation corresponds to a rescaling of the background source coordinate by a factor $(1 - \lambda)$, leaving the observed image morphology and brightness invariant. Furthermore, the Fermat potential transforms as

$$\phi_\lambda(\boldsymbol{\theta}; \boldsymbol{\beta}) = (1 - \lambda)\phi(\boldsymbol{\theta}; \boldsymbol{\beta}) + \text{constant that depends only on } \boldsymbol{\beta}. \tag{8}$$

Therefore, for given observed time delays Δt_{ij}, (6) and (8) imply that the time-delay distance $D_{\Delta t}$ would be scaled by $(1 - \lambda)$. The mass-sheet degeneracy has thus a direct impact on cosmography in measuring $D_{\Delta t}$.

While λ so far is simply a constant in this mathematical transformation (7), we can identify it with the physical external convergence, κ_{ext}, due to mass structures along the sight line to the lens system. By gathering additional data sets beyond that of the strong lens system, we can infer κ_{ext} and thus measure $D_{\Delta t}$. Two practical ways to break the mass-sheet degeneracy are (1) studies of the lens environment, to estimate κ_{ext} based on the density of galaxies in the strong-lens line of sight in comparison to random lines of sight (e.g., Momcheva et al. 2006; Fassnacht et al. 2006; Hilbert et al. 2007; Suyu et al. 2010; Greene et al. 2013; Collett et al. 2013; Rusu et al. 2017; Sluse et al. 2017; McCully et al. 2017), and (2) stellar kinematics of the strong lens galaxy, which provides an independent mass measurement within the effective radius to complement the lensing mass enclosed within the Einstein radius (e.g., Grogin and Narayan 1996; Koopmans and Treu 2002; Barnabè et al. 2009; Suyu et al. 2014). The time-delay distance can then be inferred via

$$D_{\Delta t} = \frac{D_{\Delta t}^{\text{model}}}{(1 - \kappa_{\text{ext}})}, \tag{9}$$

where $D_{\Delta t}^{\rm model}$ is the modeled time-delay distance without accounting for the presence of $\kappa_{\rm ext}$. In practice, both lens environment characterisations and stellar kinematics are employed to infer $D_{\Delta t}$ for cosmography (e.g., Suyu et al. 2010, 2013; Birrer et al. 2015b; Wong et al. 2017). The stellar kinematic data further help constrain the strong-lens mass profile (e.g., Suyu et al. 2014).

Lens systems that have massive galaxies close in projection (within $\sim 10''$) to the strong lens, but at a different redshift from the strong lens, will need to be accounted for explicitly in the strong lens model. In such cases, multi-lens plane modeling is needed (e.g., Blandford and Narayan 1986; Schneider et al. 1992; Gavazzi et al. 2008; Wong et al. 2017, Suyu et al., in preparation), but (1) for single-lens plane is then not directly valid. In particular, there is not a single time-delay distance, but rather there are multiple combination of distances between the multiple planes. Nonetheless, for some cases, one could obtain an effective time-delay distance as if it were a single-lens plane system (see, e.g., Wong et al. 2017, for details).

As noted in Sect. 1.1, with stellar kinematic and time-delay data, we can infer the angular diameter distance to the lens, $D_{\rm d}$, in addition to $D_{\Delta t}$ (Paraficz and Hjorth 2009; Jee et al. 2015; Birrer et al. 2015b; Shajib et al. 2017). Measurement of $D_{\rm d}$ is often more sensitive to the dark energy parameters (for typical lens redshifts $\lesssim 1$, see e.g., Fig. 2 of Jee et al. 2016), and can also be used as an inverse distance ladder to infer H_0 (Jee et al., submitted). Currently, the precision in $D_{\rm d}$ is limited by the uncertainty in the single-aperture averaged velocity dispersion measurement and the unknown anisotropy of stellar orbits (Jee et al. 2015). Nonetheless, we anticipate that spatially resolved kinematic data will help to constrain more precisely $D_{\rm d}$.

We have focussed here on the advances in getting $D_{\Delta t}$ and $D_{\rm d}$ from individual lenses with exquisite follow-up data to control the systematic uncertainties. Alternatively, one could analyse a sample of lenses and constrain a global H_0 parameter that is common to all the lenses (e.g., Saha et al. 2006; Oguri 2007; Sereno and Paraficz 2014), assuming that the systematic effects for the lenses average out. For small samples, this assumption might not be valid. Nonetheless, in the future where thousands of lensed quasars are expected (Oguri and Marshall 2010) but most of which will not have exquisite follow-up observations, this large sample of lenses could provide information on the population of lens galaxies as a whole for cosmography (P. J. Marshall & A. Sonnenfeld, priv. comm.). We therefore advocate getting exquisite follow-up observations of a sample of ~ 40 lenses to reach an H_0 measurement with 1% uncertainty (Jee et al. 2016; Shajib et al. 2017), with the other lenses providing information on the profiles of galaxies to use in the mass modeling.

1.4 Distance Measurements and Cosmological Inference

There are so far only a few lensed quasars for which all required data exist to do time-delay cosmography, i.e., with time-delay measurements to a few percent, deep *HST* images showing the lensed image of the host galaxy, deep spectra of the lens to measure the velocity dispersion, and multiband data to map the line of sight contribution to the lensing potential.

Some of the best time-delay measurements available to date include the radio time delay for B1608+656 (Fassnacht et al. 1999, 2002) and the two optical measurements of COSMOGRAIL for RX J1131-1231 (Tewes et al. 2013b), and HE 0435-1223 (Bonvin et al. 2017). These 3 quadruply imaged quasars, for which all the ancillary imaging and spectroscopic data are also available, gave birth to the H0LiCOW program (Suyu et al. 2017), which capitalizes on more than a decade of COSMOGRAIL monitoring as its name reflects:

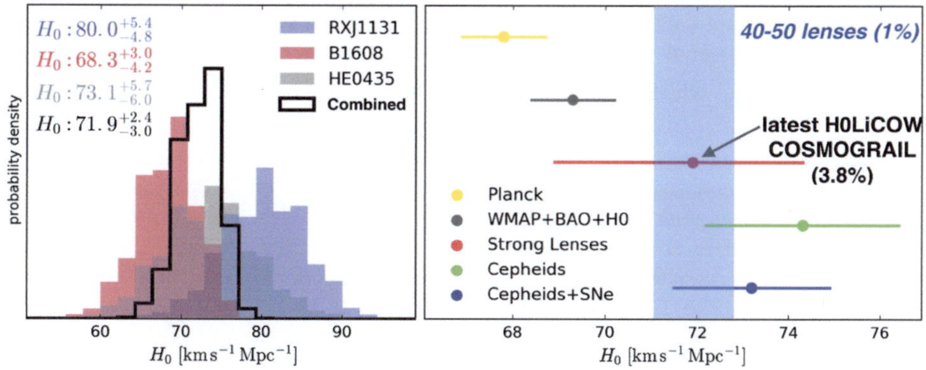

Fig. 4 *Left:* Latest H_0 measurement from quasar time delays from H0LiCOW and COSMOGRAIL for 3 lenses and for their combination in a ΛCDM Universe (reproduced with permission from Bonvin et al. 2017). *Right:* comparison between time delay H_0 measurements and other methods such as CMB shown in yellow (Planck; Ade et al. 2016) and gray (WMAP; Bennett et al. 2013) or local distance estimators such as Cepheids (green; Freedman et al. 2012) and Type Ia supernovae (blue; Riess et al. 2016b). Quasar time delays are so far in agreement with local estimators but higher than Planck. Measurements for 40–50 new time delays will allow one to confirm (or not) the current tension with Planck to more than 5σ

H0 Lenses **i**n **CO**MOGRAIL's **W**ellspring. With the precise time-delay measurements of COSMOGRAIL, H0LiCOW addresses what has been so far limiting the effectiveness of strong lensing in delivering reliable H_0 measurements: the different systematics at work at each step leading to a value for the Hubble constant.

The most recent work by H0LiCOW is summarized in the left panel of Fig. 4 (Bonvin et al. 2017), based on state-of-the-art lens mass modeling and characterisations of mass structures along the line of sight (Sluse et al. 2017; Rusu et al. 2017; Wong et al. 2017; Tihhonova et al. 2017). In the right panel of Fig. 4, we compare the value of H_0 from H0LiCOW with other fully independent cosmological probes such as Type Ia supernovae, Cepheids, and CMB(+BAO) for a ΛCDM cosmology. With the current error bars claimed by each probe there exist a tension between local measurements of H_0 (e.g., Freedman et al. 2012; Efstathiou 2014; Riess et al. 2018) and the value by the Planck team. When completed, H0LiCOW will feature 5 lenses, with an accuracy on H_0 of the order of 3% (Suyu et al. 2017), but reaching close to 1% precision is possible. This will be enabled by working on several fronts simultaneously, by finding more lenses, measuring up to 50 new time delays, and refining the lens modeling tools to mitigate degeneracies between model parameters. Chapter 8 on "Towards a self-consistent astronomical distance scale" provides more details about the (expected) future of quasar time delay cosmography.

2 Baryon Acoustic Oscillations

2.1 BAO as a Standard Ruler

The universe has been expanding, and thus the universe in the earlier stage was much smaller, denser and hotter than today. In such an early universe, electrons interacted with photons via Compton scattering and with protons via Coulomb scattering. Thus, the three components acted as a mixed fluid (Peebles and Yu 1970). They were in the equilibrium state due to the gravity of protons and pressure of photons, and oscillated as sound modes.

These oscillations are called baryon acoustic oscillations (BAO) (see Bassett and Hlozek (2010) and Weinberg et al. (2013) for a comprehensive review). It moved with the speed of sound $c_s = c\sqrt{\frac{1}{3(1+R)}}$, where the ratio of photon density (ρ_r) to baryon density (ρ_b) is defined as $1/R = 4\rho_r/3\rho_b$.

At recombination ($z \sim 1100$), photons decouple from the baryons and start to free stream. We observe the photons as a map of cosmic microwave background (CMB). The left panel of Fig. 5 shows the angular power spectrum of the latest data of the CMB anisotropy probe, Planck satellite (Adam et al. 2016). One can see a clear oscillation feature in the power spectrum, which is characterized by the sound horizon scale at recombination, expressed as (Hu and Sugiyama 1996; Eisenstein and Hu 1998)

$$r_d = \int_{z*}^{\infty} \frac{c_s(z)}{H(z)} dz, \tag{10}$$

where z_* is the redshift at recombination. $H(z)$ is the Hubble parameter,

$$H(z) = H_0\left[\Omega_m(1+z)^3 + \Omega_r(1+z)^4 + \Omega_{DE}(1+z)^{1+w}\right]^{1/2}, \tag{11}$$

where Ω_m (previously introduced in Sect. 1.1), Ω_r and Ω_{DE} are the matter, radiation and dark energy density parameters, respectively, and w is the equation-of-state parameter of dark energy and the simplest candidate for dark energy, the cosmological constant Λ, gives $w = -1$. With standard cosmological models, $r_d \simeq 150$ Mpc. From the Planck observation, it is constrained to $r_d = 144.61 \pm 0.49$ Mpc.

2.2 Probing BAO in Galaxy Distribution

After the recombination, motion of the baryons becomes non-relativistic. The perturbation of baryons then starts to grow at their locations and interact with the perturbation of dark matter. Thus the baryon acoustic feature should be imprinted onto the late-time large-scale structure of the Universe. Theoretically it is predicted to produce the overdensity at the sound horizon scale, ~ 150 Mpc. It is, however, observationally not easy because the observation of BAO signal requires the number of tracers of matter overdensity field to be large enough at the scale to overcome the cosmic variance.

In 2005, detection of the BAO was reported almost simultaneously by two independent groups using the Sloan Digital Sky Survey (SDSS) (Eisenstein et al. 2005) and the 2dF Galaxy Redshift Survey (2dFGRS) (Cole et al. 2005). The right panel of Fig. 5 shows the two-point correlation function obtained from the SDSS galaxy sample by Eisenstein et al. (2005). The 2-point correlation function $\xi(s)$ is defined as an excess of the probability that one can find pairs of galaxies at a given scale s from the case of a random distribution. Thus the scales where $\xi > 0$ and $\xi < 0$ correspond to the statistically overdense and underdense regions respectively. The bump seen around $s \simeq 105h^{-1}$ Mpc ($\simeq 150$ Mpc) is the feature of BAO, and the scale of the bump corresponds to the sound horizon scale at recombination. The inset of the right panel of Fig. 5 zooms into the feature. Unlike observations of the CMB, galaxy redshift surveys are the observation of 3 dimensional space. The peak scale of BAO should be isotropic because the scale corresponds to the sound horizon at recombination. On the other hand, the BAO scale along the line of sight depends on $H^{-1}(z)$ while the BAO scale perpendicular to the line of sight depends on the comoving angular diameter distance $D_M = (1+z)D_A(z)$, where $D_A(z)$ is expressed in flat universe as (3) with $z_1 = 0$, or,

$$D_A(z) = \frac{1}{1+z} \int_0^z \frac{cdz'}{H(z')}. \tag{12}$$

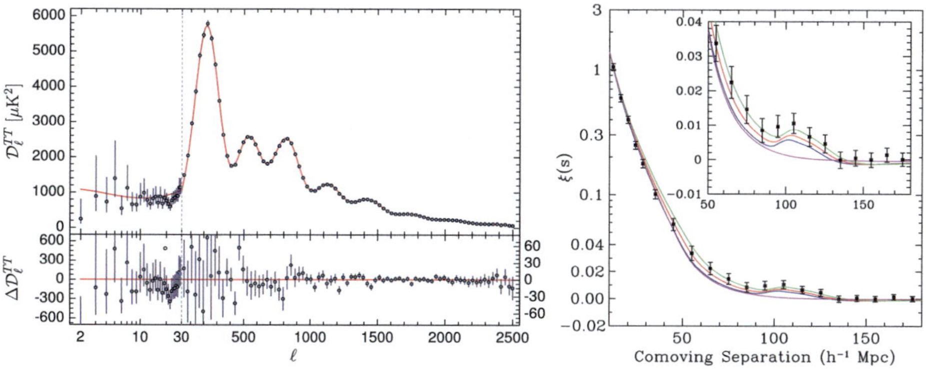

Fig. 5 (*Left*) Angular power spectrum of CMB anisotropies measured from the latest Planck satellite data (© ESA and the Planck Collaboration). The wiggles seen in the spectrum are the feature of BAO, and the oscillation scale corresponds to the sound horizon at recombination. The best-fitting ΛCDM theoretical spectrum is plotted as the solid line in the upper panel. Residuals of the measurement with respect to this model are shown in the lower panel. (*Right*) BAO feature detected in large-scale correlation function of the galaxy distribution of the SDSS (reproduced with permission from Eisenstein et al. 2005). The bump seen at $105h^{-1}$ Mpc ($\simeq 150$ Mpc) corresponds to the sound horizon scale at recombination. The solid lines are the theoretical models with $\Omega_{\rm m}h^2 = 0.12$ (top line), 0.13 (second line) and 0.14 (third line). The bottom line shows a pure CDM model with $\Omega_{\rm m}h^2 = 0.105$, which lacks the acoustic peak. The inset zooms into the BAO peak position

Thus, the scale of BAO probed by a galaxy correlation function has a cosmology dependence of

$$D_{\rm V}(z) = \left[(1+z)D_{\rm A}(z)\right]^{2/3}\left[\frac{cz}{H(z)}\right]^{1/3}. \tag{13}$$

The BAO scale probed by the CMB anisotropy at high redshift and the galaxy distribution at low redshift should be the same. Moreover, the power spectrum with the acoustic features has been precisely determined for the CMB and interpreted using linear cosmological perturbation theory (see the left panel of Fig. 5). Thus the detection of BAO in the galaxy distribution enables us to constrain $D_{\rm V}$, and hence the geometric quantities such as H_0, w, and $\Omega_{\rm DE}$ in (11) (Eisenstein et al. 2005).

2.3 Anisotropy of BAO and Alcock-Paczynski

The method presented above does not use all of the information encoded in BAO. To maximally extract the cosmological information, we need to measure the correlation function as functions of separations of galaxy pairs perpendicular ($s_\perp$) and parallel ($s_\parallel$) to the line of sight, $\xi(s_\perp, s_\parallel)$, where $s = \sqrt{s_\perp^2 + s_\parallel^2}$. In this way, in principle we can constrain $D_{\rm A}(z)$ and $H(z)$ using the transverse and radial BAO measurements, respectively. Given the cosmological dependence of angular and radial distances ((11) and (12)), the shape of the BAO peak is distorted if the wrong cosmology is assumed. This effect was first pointed out by Alcock and Paczynski (1979, AP).

In fact, the AP test using the anisotropy of BAO has additional advantages. Since the BAO scale should correspond to the sound horizon at recombination, it should be a constant. Thus, we can determine the geometric quantities by requiring that the radial BAO scale equals to the angular BAO scale. We no longer need to know the exact value of $r_{\rm d}$ nor need to rely on the CMB experiment (see e.g., Seo and Eisenstein 2003; Matsubara 2004). This is particularly important in the context of measuring H_0 given the tension in H_0 between the local measurements and the inference by the Planck team in flat ΛCDM (see Sect. 1.4).

In galaxy surveys, the distance to each galaxy is measured through redshift, thus it gives the sum of the true distance and the contribution from the peculiar motion of the galaxies and it produces an anisotropy in galaxy distribution along the line of sight, which are called redshift-space distortions (RSD). Since the velocity field of a galaxy is caused by gravity, the anisotropy contains additional cosmological information. This effect is called the Kaiser effect, named after Kaiser (1987) who proposed RSD as a cosmological probe in the linear perturbation theory limit. Since the velocity field is related to the density field through the continuity equation, the anisotropy constrains the quantity f, defined as the logarithmic derivative of the density perturbation, $f = d\ln\delta/d\ln a$. See Okumura et al. (2016) for the constraint on f from RSD as a function of z including the high-z measurement.

By simultaneously measuring the BAO and RSD and combining them with the CMB anisotropy power spectrum, we can obtain a further constraint on additional geometric quantities, such as the time evolution of w, the neutrino mass m_ν, etc. The cosmological results, presented below, correspond to this case.

The correlation function in 2D space including the BAO scales has been measured by Okumura et al. (2008) for the first time using the same galaxy sample as Eisenstein et al. (2005). The right-hand side of the left panel in Fig. 6 shows the measured correlation function, while the left-hand side shows the best fitting model based on linear perturbation theory (Matsubara 2004). The circle shown at the scale $s = (s_\perp^2 + s_\parallel^2)^{1/2} \simeq 105h^{-1}$ Mpc again corresponds to the sound horizon at recombination. The distorted, anisotropic contours shown at the smaller scales are the RSD effect caused by the velocity field.

The right panel of Fig. 6 is the latest measurement of the correlation function using the SDSS-III Baryon Oscillation Spectroscopic Survey (BOSS) DR12 sample (Alam et al. 2016). The anisotropic feature of BAO is more clearly detected due to the improvement of the data, both in the number of galaxies and the survey volume: the data of the DR12 sample used in Alam et al. (2016) comprised 1.2 million galaxies over the volume of 18.7 Gpc^3, whose numbers are respectively 25 times and 9 times larger than those of the DR3 sample used in Okumura et al. (2008).

2.4 Constraints on BAO Distance Scales and H_0

The left panel of Fig. 7 shows the summary of the three distance measures obtained by BAO in various galaxy surveys (Aubourg et al. 2015). The y-axis is the ratio of each distance and r_d, divided by $\sqrt{z}$. The blue points are the measurement of D_V from the angularly-averaged BAO, while the red and green points are respectively the measurements of D_M and D_H obtained from the anisotropy of BAO, where D_H is the radial distance defined as $D_\mathrm{H}(z) = c/H(z)$. The three lines with the same color as the points are the corresponding predictions of the ΛCDM model obtained by Planck (Adam et al. 2016). Nice agreement between BAO measurements from galaxy surveys and Planck cosmology can be seen. However, the agreement with the WMAP cosmology is equivalently good in terms of distance measures (Anderson et al. 2014).

The right panel of Fig. 7 focuses on currently the largest survey, the BOSS survey at $z = 0.5$ (Anderson et al. 2014). Here the joint constraints on $H(z)$ and $D_\mathrm{A}(z)$ are shown. The gray contours are obtained from the 1D BAO analysis (see Sect. 2.2). Because the 1D BAO constrains $D_\mathrm{V} \propto D_\mathrm{A}^{2/3} H^{-1/3}$, there is a perfect degeneracy between D_A and H. On the

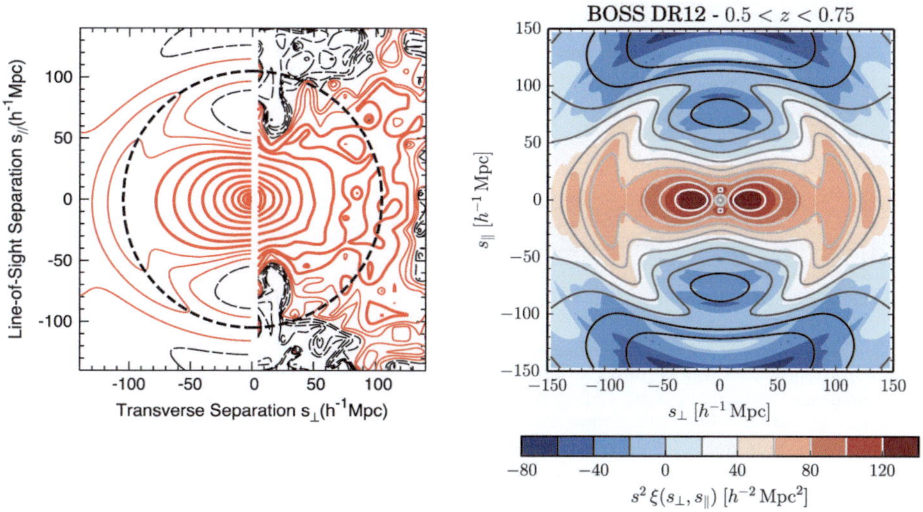

Fig. 6 (*Left*) Contour plots of the correlation function measured from the SDSS galaxy sample as functions of transverse ($s_\perp$) and line-of-sight ($s_\parallel$) separations (*right half*) and the corresponding theoretical prediction (*left half*) (reproduced with permission from Okumura et al. 2008). The dashed thin black lines show $\xi < -0.01$ increasing logarithmically with 0.25 and $-0.01 \le \xi < 0$ linearly with 0.0025. The solid thin lines colored red show $0 \le \xi < 0.01$ increasing linearly with 0.0025 and the solid thick ones colored red are $\xi \ge 0.01$ logarithmically with 0.25. The baryonic feature slightly appears as ridge structures around the scale $s = (s_\perp^2 + s_\parallel^2)^{1/2} \simeq 100h^{-1}$ Mpc (= 150 Mpc), and the dashed circle traces the peaks of the baryon ridges. (*Right*) Similar to the left panel, but the correlation function from currently the largest galaxy sample from BOSS survey (reproduced with permission from Alam et al. 2016). The correlation function is multiplied by the square of the distance, $s^2\xi$, in order to emphasize the BAO feature

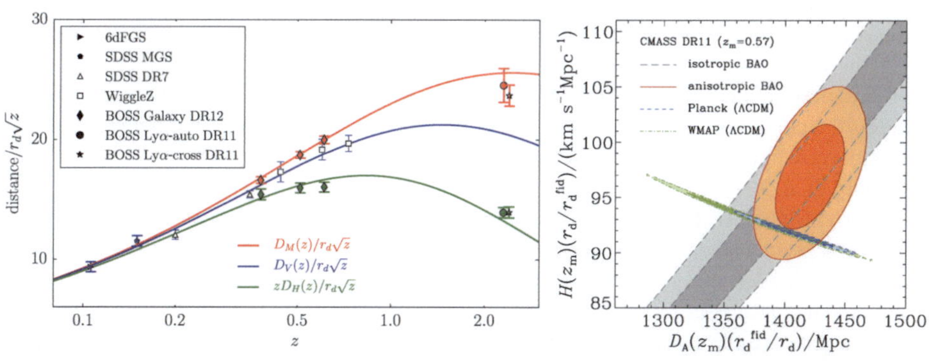

Fig. 7 (*Left*) Three distance measures obtained by BAO in various galaxy surveys (reproduced with permission from Aubourg et al. 2015). The y-axis is the ratio of each distance to $r_{\rm d}$, divided by $\sqrt{z}$. The top-red, middle-blue and bottom-green curves correspond to the distances $D_{\rm M}(z)$, $D_{\rm V}(z)$ and $zD_{\rm H}(z)$, respectively. (*Right*) Joint constraints on the angular diameter distance $D_{\rm A}(z)$ and the Hubble parameter $H(z)$ obtained from the correlation analysis of the BOSS galaxy sample at $z = 0.5$ (reproduced with permission from Anderson et al. 2014). The inner and outer contours correspond to the 68% and 95% confidence levels, respectively. The gray and orange contours are the constraints from 1D and 2D BAO analyses, respectively, while the blue and green contours are from CMB experiments (Planck and WMAP)

other hand, the solid orange contours are from the 2D BAO analysis where the degeneracy is broken to some extent (Sect. 2.3). The obtained constraints on the distance scales are as tight as the flat λCDM constraints from the CMB experiments, Planck (dashed blue) and WMAP

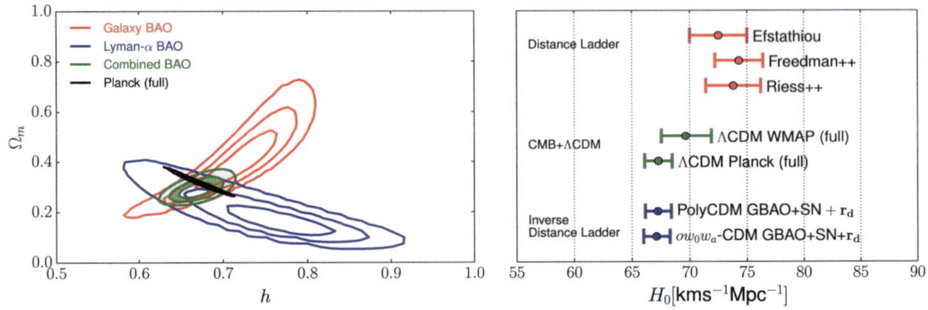

Fig. 8 (*Left*) Joint constraints on the matter density parameter $\Omega_{\rm m}$ and the Hubble constant h in a flat cosmology (reproduced with permission from Aubourg et al. 2015). Each contour shows the 68%, 95% and 99% confidence levels from inward. Galaxy BAO constraints (red) show strong correlations between $\Omega_{\rm m}$ and h, whereas that of Ly-α BAO (blue) show strong anti-correlations. The combination of the two ("Combined BAO" in green) thus breaks the degeneracies, resulting in constraints located at the intersection of the two. Planck CMB constraints (black) show also anti-correlation between $\Omega_{\rm m}$ and h, but are substantially narrower than that of Combined BAO. (*Right*) Comparison of the constraints on H_0 (reproduced with permission from Aubourg et al. 2015) from the distance ladder probes (local measurements, red), the CMB anisotropies (green), and the inverse distance ladder analysis (combination of BAO and supernovae; blue)

(dot-dashed green). Future galaxy surveys will enable us to measure BAO more accurately and determine the cosmic distance scales with higher precision (see Sect. 2.5).

Let us move onto the constraints on cosmological models using BAO observations. The left panel of Fig. 8 shows the joint constraints on the matter density parameter $\Omega_{\rm m}$ and the Hubble constant $h = H_0/(100\ {\rm km\,s^{-1}\,Mpc^{-1}})$ obtained from the measurements of BAO anisotropy (Aubourg et al. 2015). The red and blue contours are the constraints from the BAO measured from the various galaxy samples at $z < 1$ and from the Lyα forest at $z \sim 2$, respectively, as shown in the left panel of Fig. 7. Since these constraints are not very tight, the constrained H_0 from either galaxy BAO or Lyα BAO is consistent with other probes including the local measurements. The combination of these two BAO probes largely tightens the constraint on H_0 and causes a slight, 2σ tension as we will see below. With $\Omega_{\rm m}$ being marginalized over, the Hubble constant is constrained to $h = 0.67 \pm 0.013$ (1σ C. L.) (Aubourg et al. 2015).

The right panel of Fig. 8 summarizes the comparison of H_0 constraints from the BAO measurement and other probes. Augmenting Fig. 4, the top three points are obtained from the distance ladder analysis, showing constraints from the local universe from three independent teams (Riess et al. 2016a; Freedman et al. 2012; Efstathiou 2014). The green, two middle points are the constraints from the two CMB satellite probes, WMAP and Planck. The bottom two points are from the inverse distance ladder analysis, namely the combination of BAO and SN distance measures. As seen from the figure, those from the Planck and the inverse ladder measurements have a mild but non-negligible tension with the distance-ladder measurements, about $\sim 2\sigma$. The discrepancy could be due to just systematics, or a hint of new physics. We will need further observational constraints to resolve these discrepancies.

2.5 Future BAO Surveys

BAO are considered as a probe least affected by systematic biases to measure distance scales, even beyond the local universe ($z > 0$), and thus are a promising tool to reveal the expansion history of the universe and constrain the cosmological model. To improve the precision of

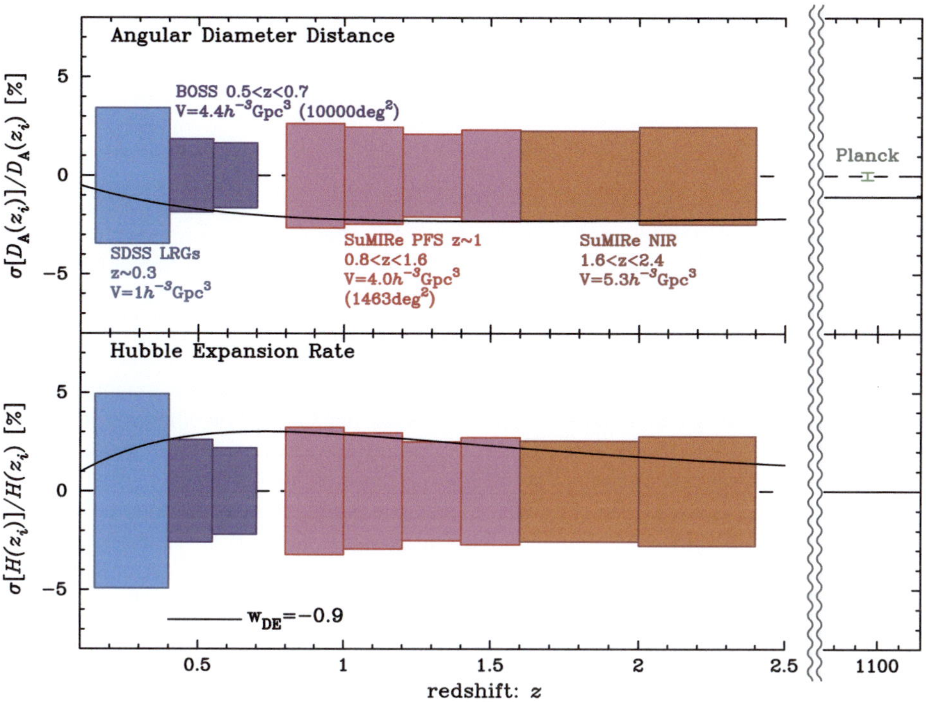

Fig. 9 Fractional errors in the angular diameter distance and the expansion rate through the measurements of BAO (reproduced with permission from Takada et al. 2014). The expected accuracies are compared to the existing SDSS and BOSS surveys at $z < 0.7$. Each panel assumes $w = -1$ as the fiducial model, and when the model is changed to $w = -0.9$, the baseline of the fractional errors is systematically shifted from the dashed line to the solid curve

the distance measurement, a dominant source of error on BAO observations is the cosmic variance. There are larger, ongoing and planned galaxy redshift surveys, such as extended BOSS (eBOSS) (Dawson et al. 2016), Subaru Prime Focus Spectrograph (PFS) (Takada et al. 2014), and Dark Energy Spectroscopic Instrument (DESI) (Aghamousa et al. 2016). With the larger survey volumes, these surveys will measure distance scales using BAO with percent-level precisions. These surveys are also deep and can reveal fainter sources, and hence enable us to extend the distance scales to more distant parts of the universe.

As an example, Fig. 9 presents the accuracies of constraining the angular diameter distance and Hubble expansion rate expected from the analysis of anisotropic BAO (see Sect. 2.3) of the PFS survey at the Subaru Telescope. The PFS will observe the universe at $0.8 < z < 2.4$ by using [OII] emitters as a tracer of the large-scale structure. The survey volume of each redshift slice is on the order of $[h^{-1}\ \mathrm{Gpc}]^3$ and the number density is larger than $10^{-4}[h^{-1}\ \mathrm{Mpc}]^{-3}$, which are comparable to the existing SDSS and BOSS surveys at $z < 0.7$. Hence, one will be able to achieve a few percent constraints on D_{A} and H at high redshifts, almost the same as those obtained from the low-z surveys. Deep galaxy surveys such as the PFS allow for constraining not only the expansion history of the universe but also dark energy (see the solid line of Fig. 9).

Ultimately, we would like to survey the galaxies over the whole sky, which can be achieved by satellite missions, such as Euclid (Amendola et al. 2013) and WFIRST (Spergel et al. 2013). These surveys will measure the cosmic distances with an unprecedentedly high precision.

3 Intensity Mapping

3.1 21cm Intensity Mapping BAO

As we described in Sect. 2, current BAO measurements are enabled by large galaxy spectroscopic surveys, and the resulting constraining power on cosmological parameters generally scales as the effective survey comoving volume. Specifically, the precision of cosmological parameter constraints scales as $\propto 1/\sqrt{N}$, where N is the number of modes, or in the case of 3D map $\propto 1/\sqrt{k_{\rm max}^3 V}$ where V is the comoving volume and $k_{\rm max}$ the maximum useful comoving wavenumber. This scale is often given by the non-linearity scale, $k_{\rm nl}(z=0) \sim 0.2h/{\rm Mpc}$, where the complexity of baryonic astrophysics on galaxy and galaxy-cluster scales limits our ability to extract cosmological parameters. Improving parametric precision will therefore require larger volumes, which requires mapping higher redshift volumes that have the added benefit of increasing $k_{\rm nl}(z)$. The emerging technique of 21 cm Intensity Mapping appears to be a very promising way to reach this goal.

Galaxy redshifts can be measured at radio wavelengths using the 21 cm hyperfine emission of atomic neutral hydrogen (HI). The 21 cm line is unique in cosmology because for $\lambda > 21$ cm it is the dominant astronomical line emission, i.e., for all positive redshifts and all cosmological emission. So to a good approximation the wavelength of a spectral feature can be converted to a Doppler redshift without the uncertainty and ambiguity of having to first identify the atomic transition. The direct determination of redshifts using 21 cm data can be compared to the corresponding optical technique, which requires identifying a suitable subset of target galaxies (photometry), then obtaining an optical spectrum, and finally finding some unique combination of emission and absorption lines that allow an unambiguous determination of the redshift for that galaxy (spectroscopy).

The 21 cm signal has been used to conduct galaxy redshift surveys in the local Universe around $z \sim 0.1$ (Martin et al. 2010; Zwaan et al. 2001) and out to $z \sim 0.4$ (Fernández et al. 2016). Beyond this redshift, current radio telescopes do not have sufficient collecting area or sensitivity to make 21 cm surveys using individual galaxies.

A radically different technique, HI intensity mapping (HIM), has been proposed (Chang et al. 2008; Wyithe et al. 2008). It uses maps of 21 cm emission where individual galaxies are not resolved. Instead, it detects the combined emission from the many galaxies that occupy large (1000 $\rm Mpc^3$) voxels. The technique allows 100 m class telescopes such as the Green Bank Telescope (GBT), which only have angular resolution of several arc-minutes, to rapidly survey enormous comoving volumes at $z \sim 1$ (Abdalla and Rawlings 2005; Peterson et al. 2006; Wyithe et al. 2007; Chang et al. 2008; Wyithe et al. 2008; Tegmark and Zaldarriaga 2009, 2010; Seo et al. 2010). A number of authors (Seo et al. 2010; Xu et al. 2015; Bull et al. 2015) have studied the overall promise of the intensity mapping technique. Chang et al. (2010), Masui et al. (2013) and Switzer et al. (2013) have reported the first detections in cross-correlations and upper limits to the 21 cm auto-power spectrum using the Green Bank Telescope (GBT).

3.1.1 Challenges

One of the major challenges for 21 cm intensity mapping is the mitigation of radio foregrounds, which are predominantly Galatic and extragalactic synchrotron emissions, and are at least $\sim 10^4$ times brighter in intensity than the 21 cm emission. The two can be distinguished because the former have smooth spectra and the latter trace the underlying large-scale structure and have spectral structures. The brightness temperature of the synchrotron foreground typically has a spectral dependence of $\nu^{-2.6}$, or $(1+z)^{2.6}$, and is thus more severe at higher redshifts. Note it has not yet been demonstrated whether the synchrotron radiation is indeed spectrally smooth down to one part of 10^4 or higher and therefore can in principle be suppressed by this factor to reveal the 21 cm fluctuation signals. However, since the BAO wiggles have very specific structures, we can potentially select Fourier modes that are observationally accessible in scale and redshifts (Chang et al. 2008; Seo et al. 2010).

The 21 cm features we are most interested in are the relatively non-smooth BAO 'wiggles'. Unfortunately radio telescopes are diffraction-limited and the beam patterns depend on frequency, which mixes angular and frequency dependence. Since the foregrounds are not smooth in position across the sky it is a nontrivial task to identify and subtract the smooth frequency foregrounds with sufficient accuracy so as to reveal the 21 cm emission. It is easier if we go to very small radial scales, but to get the most cosmological information out of the data we would need to remove the foregrounds over the largest range of scales possible. Shaw et al. (2013, 2015) have demonstrated that this is achievable in principle.

To achieve the foreground subtraction goal and to make accurate 3D maps we need a very accurate model of the beam patterns and characterization of the mapping between the observed and true skies. Liao et al. (2016) have recently demonstrated accurate measurements of the polarized GBT beam to sub-percent level, which is critical for polarized foreground mitigation. Developing and demonstrating the efficacy of methods to model and calibrate large dataset is also necessary to achieve the main objective.

3.1.2 Future Prospects

As discussed in Sect. 2, baryon acoustic oscillations provide a convenient standard ruler in the cosmological large scale structure (LSS) allowing a precise measurement of the distance-redshift relation over cosmic time. This distance redshift relation is measured, whether by BAO, SNe-Ia surveys, or weak lensing, to characterize the dark energy; it is augmented by the growth rate of inhomogeneities as well as redshift-space distortions. All three of these quantities can be measured in HI surveys even though to-date, only optical instruments have detected BAO features in the power spectrum.

Going forward requires the most cost-effective way to map the largest cosmological volume, and this may be radio spectroscopy through intensity mapping. One unique advantage of 21 cm Intensity Mapping is the fact that the 21 cm signal is in principle observable from $z = 0$ up to a redshift of ~ 100, when its spin temperature decouples from the Cosmic Microwave Background radiation. The vast majority of the cosmic volume is only visible during the dark ages via the 21 cm radiation from neutral hydrogen, before the onset of galaxy formation. 21 cm Intensity Mapping thus provides a unique access to measuring LSS during this period (Tegmark and Zaldarriaga 2009, 2010). Besides, 21 cm intensity mapping has a set of observational systematics that should be largely uncorrelated with the systematic effects in optical surveys.

The on-going low-z GBT-HIM survey is a step along the way to a dark ages radio survey. We have made BAO forecasts based on the expected performance of the array. We assume the seven-beam array has a 700–850 MHz frequency coverage with a total system temperature of 33 K. The BAO forecasts are consistent with predictions in Masui et al. (2010) and are in reasonable agreement with those of Bull et al. (2015). We consider three scenarios with different observing depth and sky coverage: 500 or 1000 hours of on-sky GBT observations, covering 100 or 1000 deg^2 of sky areas. The expected errors on the BAO wiggles and the fractional distance constraints are shown in Fig. 10. We anticipate to yield a 3.5% error on the BAO distance at $z \sim 0.8$ with 1000 hours of GBT observing time. The bottom

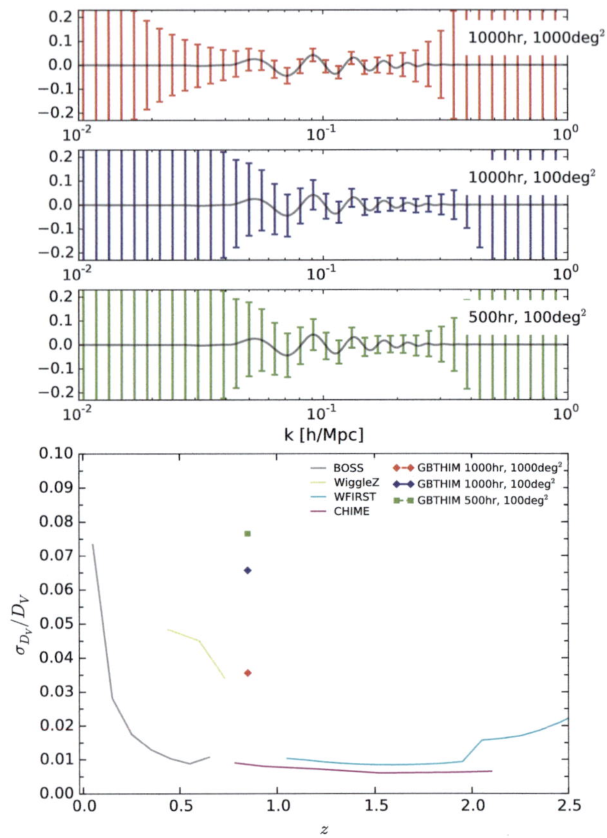

Fig. 10 *Top:* Expected errors on the BAO signature at $z \sim 0.8$ of the GBT-HIM array, assuming different observing depth and sky coverage: (1000 hrs, 1000 deg^2), (1000 hrs, 100 deg^2), and (500 hrs, 100 deg^2). BAO signatures can be detected in all three. *Bottom:* The expected fractional error on the BAO distance scale of the three scenarios. We anticipate a 3.5% error on the BAO distance at $z \sim 0.8$ with 1000 hours of GBT observing time with the array. A detection of BAO will validate HI intensity mapping as a viable tool for large-scale structure and cosmology, and serve as a systematic check and alternative approach to the optical spectroscopic redshift surveys

panel of Fig. 10 also shows recent constraints from WiggleZ (Drinkwater et al. 2010) and the BOSS surveys (Anderson et al. 2014) at lower redshifts, and forecasts for CHIME (Bandura et al. 2014) and WFIRST (Spergel et al. 2015). With the demonstrated results and good understanding of systematic effects at the GBT, and with very different astrophysical and measurement systematics from optical/IR spectroscopic redshift surveys, we anticipate the GBT-HIM array can make a firm detection of the BAO signature at $z \sim 0.8$ with the HI intensity mapping technique, and contribute to the future of large-scale structure surveys and the field of 21-cm cosmology. Other on-going experiments such as CHIME (Bandura et al. 2014) and HIRAX (Newburgh et al. 2016) will reach $z = 0.8$–2.5 and probe even larger cosmological volume.

Acknowledgements Open access funding provided by the Max Planck Society. We thank ISSI-BJ for the kind hospitality and an engaging workshop. We would like to acknowledge Vivien Bonvin for providing some of the figures in this review. T.O. thanks Anze Slosar, Shun Saito and Atsushi Taruya for useful discussions on interpretations of cosmological constraints from the BOSS survey. S.H.S. thanks the Max Planck Society for support through the Max Planck Research Group. F.C. acknowledges support from the Swiss National Science Foundation (SNSF). T.O. acknowledges support from the Ministry of Science and Technology of Taiwan under the grant MOST 106-2119-M-001-031-MY3.

References

F.B. Abdalla, S. Rawlings, Probing dark energy with baryonic oscillations and future radio surveys of neutral hydrogen. Mon. Not. R. Astron. Soc. **360**, 27–40 (2005). https://doi.org/10.1111/j.1365-2966.2005.08650.x. astro-ph/0411342

R. Adam, P.A.R. Ade, N. Aghanim, Y. Akrami, M.I.R. Alves, F. Argüeso, M. Arnaud, F. Arroja, M. Ashdown et al. (Planck Collaboration), Planck 2015 results. I. Overview of products and scientific results. Astron. Astrophys. **594**, A1 (2016). https://doi.org/10.1051/0004-6361/201527101. 1502.01582

P.A.R. Ade, N. Aghanim, M. Arnaud, M. Ashdown, J. Aumont, C. Baccigalupi, A.J. Banday, R.B. Barreiro, J.G. Bartlett et al. (Planck Collaboration), Planck 2015 results. XIII. Cosmological parameters. Astron. Astrophys. **594**, A13 (2016). https://doi.org/10.1051/0004-6361/201525830. 1502.01589

A. Aghamousa, J. Aguilar, S. Ahlen, S. Alam, L.E. Allen, C. Allende Prieto, J. Annis, S. Bailey, C. Balland et al. (DESI Collaboration), The DESI experiment part I: science, targeting, and survey design (2016). ArXiv e-prints 1611.00036

A. Agnello, H. Lin, L. Buckley-Geer, T. Treu, V. Bonvin, F. Courbin, C. Lemon, T. Morishita, A. Amara, M.W. Auger, S. Birrer, J. Chan, T. Collett, A. More, C.D. Fassnacht, J. Frieman, P.J. Marshall, R.G. McMahon, G. Meylan, S.H. Suyu, F. Castander, D. Finley, A. Howell, C. Kochanek, M. Makler, P. Martini, N. Morgan, B. Nord, F. Ostrovski, P. Schechter, D. Tucker, R. Wechsler, T.M.C. Abbott, F.B. Abdalla, S. Allam, A. Benoit-Lévy, E. Bertin, D. Brooks, D.L. Burke, A.C. Rosell, M.C. Kind, J. Carretero, M. Crocce, C.E. Cunha, C.B. D'Andrea, L.N. da Costa S. Desai, J.P. Dietrich, T.F. Eifler, B. Flaugher, P. Fosalba, J. García-Bellido, E. Gaztanaga, M.S. Gill, D.A. Goldstein, D. Gruen, R.A. Gruendl, J. Gschwend, G. Gutierrez, K. Honscheid, D.J. James, K. Kuehn, N. Kuropatkin, T.S. Li, M. Lima, M.A.G. Maia, M. March, J.L. Marshall, P. Melchior, F. Menanteau, R. Miquel, R.L.C. Ogando, A.A. Plazas, A.K. Romer, E. Sanchez, R. Schindler, M. Schubnell, I. Sevilla-Noarbe, M. Smith, R.C. Smith, F. Sobreira, E. Suchyta, M.E.C. Swanson, G. Tarle, D. Thomas, A.R. Walker, Models of the strongly lensed quasar DES J0408-5354. Mon. Not. R. Astron. Soc. **472**, 4038–4050 (2017). https://doi.org/10.1093/mnras/stx2242. 1702.00406

S. Alam, M. Ata, S. Bailey, F. Beutler, D. Bizyaev, J.A. Blazek, A.S. Bolton, J.R. Brownstein, A. Burden, C.H. Chuang, J. Comparat, A.J. Cuesta, K.S. Dawson, D.J. Eisenstein, S. Escoffier, H. Gil-Marín, J.N. Grieb, N. Hand, S. Ho, K. Kinemuchi, D. Kirkby, F. Kitaura, E. Malanushenko, V. Malanushenko, C. Maraston, C.K. McBride, R.C. Nichol, M.D. Olmstead, D. Oravetz, N. Padmanabhan, N. Palanque-Delabrouille, K. Pan, M. Pellejero-Ibanez, W.J. Percival, P. Petitjean, F. Prada, A.M. Price-Whelan, B.A. Reid, S.A. Rodríguez-Torres, N.A. Roe, A.J. Ross, N.P. Ross, G. Rossi, J.A. Rubiño-Martín, A.G. Sánchez, S. Saito, S. Salazar-Albornoz, L. Samushia, S. Satpathy, C.G. Scóccola, D.J. Schlegel, D.P. Schneider, H.J. Seo, A. Simmons, A. Slosar, M.A. Strauss, M.E.C. Swanson, D. Thomas, J.L. Tinker, R. Tojeiro, M. Vargas Magaña, J.A. Vazquez, L. Verde, D.A. Wake, Y. Wang, D.H. Weinberg, M. White, W.M. Wood-Vasey, C. Yèche, I. Zehavi, Z. Zhai, G.B. Zhao, The clustering of galaxies in the completed SDSS-III Baryon Oscillation Spectroscopic Survey: cosmological analysis of the DR12 galaxy sample (2016). ArXiv e-prints 1607.03155

C. Alcock, B. Paczynski, An evolution free test for non-zero cosmological constant. Nature **281**, 358 (1979). https://doi.org/10.1038/281358a0

L. Amendola, S. Appleby, D. Bacon, T. Baker, M. Baldi, N. Bartolo, A. Blanchard, C. Bonvin, S. Borgani, E. Branchini, C. Burrage, S. Camera, C. Carbone, L. Casarini, M. Cropper, C. de Rham, C. Di Porto, A. Ealet, P.G. Ferreira, F. Finelli, J. García-Bellido, T. Giannantonio, L. Guzzo, A. Heavens, L. Heisenberg, C. Heymans, H. Hoekstra, L. Hollenstein, R. Holmes, O. Horst, K. Jahnke, T.D. Kitching, T. Koivisto, M. Kunz, G. La Vacca, M. March, E. Majerotto, K. Markovic, D. Marsh, F. Marulli, R. Massey, Y. Mellier, D.F. Mota, N.J. Nunes, W. Percival, V. Pettorino, C. Porciani, C. Quercellini, J. Read, M. Rinaldi, D. Sapone, R. Scaramella, C. Skordis, F. Simpson, A. Taylor, S. Thomas, R. Trotta, L. Verde, F. Vernizzi, A. Vollmer, Y. Wang, J. Weller, T. Zlosnik, Cosmology and fundamental physics with the Euclid satellite. Living Rev. Relativ. **16**, 6 (2013). https://doi.org/10.12942/lrr-2013-6. 1206.1225

L. Anderson, É. Aubourg, S. Bailey, F. Beutler, V. Bhardwaj, M. Blanton, A.S. Bolton, J. Brinkmann, J.R. Brownstein, A. Burden, C.H. Chuang, A.J. Cuesta, K.S. Dawson, D.J. Eisenstein, S. Escoffier, J.E. Gunn, H. Guo, S. Ho, K. Honscheid, C. Howlett, D. Kirkby, R.H. Lupton, M. Manera, C. Maraston, C.K. McBride, O. Mena, F. Montesano, R.C. Nichol, S.E. Nuza, M.D. Olmstead, N. Padmanabhan, N. Palanque-Delabrouille, J. Parejko, W.J. Percival, P. Petitjean, F. Prada, A.M. Price-Whelan, B. Reid, N.A. Roe, A.J. Ross, N.P. Ross, C.G. Sabiu, S. Saito, L. Samushia, A.G. Sánchez, D.J. Schlegel, D.P. Schneider, C.G. Scoccola, H.J. Seo, R.A. Skibba, M.A. Strauss, M.E.C. Swanson, D. Thomas, J.L. Tinker, R. Tojeiro, M.V. Magaña, L. Verde, D.A. Wake, B.A. Weaver, D.H. Weinberg, M. White, X. Xu, C. Yèche, I. Zehavi, G.B. Zhao, The clustering of galaxies in the SDSS-III Baryon Oscillation Spectroscopic Survey: baryon acoustic oscillations in the Data Releases 10 and 11 Galaxy samples. Mon. Not. R. Astron. Soc. **441**, 24–62 (2014). https://doi.org/10.1093/mnras/stu523. 1312.4877

É. Aubourg, S. Bailey, J.E. Bautista, F. Beutler, V. Bhardwaj, D. Bizyaev, M. Blanton, M. Blomqvist, A.S. Bolton, J. Bovy, H. Brewington, J. Brinkmann, J.R. Brownstein, A. Burden, N.G. Busca, W. Carithers, C.H. Chuang, J. Comparat, R.A.C. Croft, A.J. Cuesta, K.S. Dawson, T. Delubac, D.J. Eisenstein, A. Font-Ribera, J. Ge, J.M. Le Goff, S.G.A. Gontcho, J.R. Gott, J.E. Gunn, H. Guo, J. Guy, J.C. Hamilton, S. Ho, K. Honscheid, C. Howlett, D. Kirkby, F.S. Kitaura, J.P. Kneib, K.G. Lee, D. Long, R.H. Lupton, M.V. Magaña, V. Malanushenko, E. Malanushenko, M. Manera, C. Maraston, D. Margala, C.K. McBride, J. Miralda-Escudé, A.D. Myers, R.C. Nichol, P. Noterdaeme, S.E. Nuza, M.D. Olmstead, D. Oravetz, I. Pâris, N. Padmanabhan, N. Palanque-Delabrouille, K. Pan, M. Pellejero-Ibanez, W.J. Percival, P. Petitjean, M.M. Pieri, F. Prada, B. Reid, J. Rich, N.A. Roe, A.J. Ross, N.P. Ross, G. Rossi, J.A. Rubiño-Martín, A.G. Sánchez, L. Samushia, R.T.G. Santos, C.G. Scóccola, D.J. Schlegel, D.P. Schneider, H.J. Seo, E. Sheldon, A. Simmons, R.A. Skibba, A. Slosar, M.A. Strauss, D. Thomas, J.L. Tinker, R. Tojeiro, J.A. Vazquez, M. Viel, D.A. Wake, B.A. Weaver, D.H. Weinberg, W.M. Wood-Vasey, C. Yèche, I. Zehavi, G.B. Zhao (BOSS Collaboration), Cosmological implications of baryon acoustic oscillation measurements. Phys. Rev. D **92**(12), 123516 (2015). https://doi.org/10.1103/PhysRevD.92.123516. 1411.1074

K. Bandura, G.E. Addison, M. Amiri, J.R. Bond, D. Campbell-Wilson, L. Connor, J.F. Cliche, G. Davis, M. Deng, N. Denman, M. Dobbs, M. Fandino, K. Gibbs, A. Gilbert, M. Halpern, D. Hanna, A.D. Hincks, G. Hinshaw, C. Höfer, P. Klages, T.L. Landecker, K. Masui, J. Mena Parra, L.B. Newburgh, P. Ul, J.B. Peterson, A. Recnik, J.R. Shaw, K. Sigurdson, M. Sitwell, G. Smecher, R. Smegal, K. Vanderlinde, D. Wiebe, Canadian Hydrogen Intensity Mapping Experiment (CHIME) pathfinder, in *Ground-Based and Airborne Telescopes V*. Proc. SPIE, vol. 9145 (2014), p. 914522. https://doi.org/10.1117/12.2054950. 1406.2288

R. Barkana, Fast calculation of a family of elliptical mass gravitational lens models. Astrophys. J. **502**, 531 (1998). https://doi.org/10.1086/305950. astro-ph/9802002

M. Barnabè, O. Czoske, L.V.E. Koopmans, T. Treu, A.S. Bolton, R. Gavazzi, Two-dimensional kinematics of SLACS lenses—II. Combined lensing and dynamics analysis of early-type galaxies at $z = 0.08$–0.33. Mon. Not. R. Astron. Soc. **399**, 21–36 (2009). https://doi.org/10.1111/j.1365-2966.2009.14941.x. 0904.3861

M. Barnabè, O. Czoske, L.V.E. Koopmans, T. Treu, A.S. Bolton, Two-dimensional kinematics of SLACS lenses—III. Mass structure and dynamics of early-type lens galaxies beyond $z \sim 0.1$. Mon. Not. R. Astron. Soc. **415**, 2215–2232 (2011). https://doi.org/10.1111/j.1365-2966.2011.18842.x. 1102.2261

B. Bassett, R. Hlozek, Baryon acoustic oscillations (2010), p. 246

C.L. Bennett, D. Larson, J.L. Weiland, N. Jarosik, G. Hinshaw, N. Odegard, K.M. Smith, R.S. Hill, B. Gold, M. Halpern, E. Komatsu, M.R. Nolta, L. Page, D.N. Spergel, E. Wollack, J. Dunkley, A. Kogut, M. Limon, S.S. Meyer, G.S. Tucker, E.L. Wright, Nine-year Wilkinson Microwave Anisotropy Probe (WMAP) observations: final maps and results. Astrophys. J. Suppl. Ser. **208**, 20 (2013). https://doi.org/10.1088/0067-0049/208/2/20. 1212.5225

S. Birrer, A. Amara, A. Refregier, Gravitational lens modeling with basis sets. Astrophys. J. **813**, 102 (2015a). https://doi.org/10.1088/0004-637X/813/2/102. 1504.07629

S. Birrer, A. Amara, A. Refregier, The mass-sheet degeneracy and time-delay cosmography: analysis of the strong lens RXJ1131-1231 (2015b). ArXiv e-prints 1511.03662

R. Blandford, R. Narayan, Fermat's principle, caustics, and the classification of gravitational lens images. Astrophys. J. **310**, 568–582 (1986). https://doi.org/10.1086/164709

R. Blandford, G. Surpi, T. Kundić, Modeling galaxy lenses, in *ASP Conf. Ser. 237: Gravitational Lensing: Recent Progress and Future Goals*, ed. by T.G. Brainerd, C.S. Kochanek. (Astron. Soc. Pac., San Francisco, 2001), p. 65

V. Bonvin, M. Tewes, F. Courbin, T. Kuntzer, D. Sluse, G. Meylan, COSMOGRAIL: the COSmological MOnitoring of GRAvItational Lenses. XV. Assessing the achievability and precision of time-delay measurements. Astron. Astrophys. **585**, A88 (2016). https://doi.org/10.1051/0004-6361/201526704. 1506.07524

V. Bonvin, F. Courbin, S.H. Suyu, P.J. Marshall, C.E. Rusu, D. Sluse, M. Tewes, K.C. Wong, T. Collett, C.D. Fassnacht, T. Treu, M.W. Auger, S. Hilbert, L.V.E. Koopmans, G. Meylan, N. Rumbaugh, A. Sonnenfeld, C. Spiniello, H0LiCOW—V. New COSMOGRAIL time delays of HE 0435-1223: H_0 to 3.8 per cent precision from strong lensing in a flat ΛCDM model. Mon. Not. R. Astron. Soc. **465**, 4914–4930 (2017). https://doi.org/10.1093/mnras/stw3006. 1607.01790

V. Bonvin, O. Tihhonova, M. Millon, J.H.H. Chan, E. Savary, S. Huber, F. Courbin, Impact of the 3D source geometry on time-delay measurements of lensed type-Ia Supernovae (2018). ArXiv e-prints 1805.04525

B.J. Brewer, G.F. Lewis, Unlensing HST observations of the Einstein ring 1RXS J1131-1231: a Bayesian analysis. Mon. Not. R. Astron. Soc. **390**, 39–48 (2008). https://doi.org/10.1111/j.1365-2966.2008.13715.x. 0807.2145

P. Bull, P.G. Ferreira, P. Patel, M.G. Santos, Late-time cosmology with 21 cm intensity mapping experiments. Astrophys. J. **803**(1), 21 (2015). https://doi.org/10.1088/0004-637X/803/1/21. 1405.1452

I. Burud, J. Hjorth, A.O. Jaunsen, M.I. Andersen, H. Korhonen, J.W. Clasen, J. Pelt, F.P. Pijpers, P. Magain, R. Østensen, An optical time delay estimate for the double gravitational lens system B1600+434. Astrophys. J. **544**, 117–122 (2000). https://doi.org/10.1086/317213. astro-ph/0007136

I. Burud, F. Courbin, P. Magain, C. Lidman, D. Hutsemékers, J.P. Kneib, J. Hjorth, J. Brewer, E. Pompei, L. Germany, J. Pritchard, A.O. Jaunsen, G. Letawe, G. Meylan, An optical time-delay for the lensed BAL quasar HE 2149-2745. Astron. Astrophys. **383**, 71–81 (2002a). https://doi.org/10.1051/0004-6361:20011731. astro-ph/0112225

I. Burud, J. Hjorth, F. Courbin, J.G. Cohen, P. Magain, A.O. Jaunsen, A.A. Kaas, C. Faure, G. Letawe, Time delay and lens redshift for the doubly imaged BAL quasar SBS 1520+530. Astron. Astrophys. **391**, 481–486 (2002b). https://doi.org/10.1051/0004-6361:20020856. astro-ph/0206084

N. Cantale, F. Courbin, M. Tewes, P. Jablonka, G. Meylan, Firedec: a two-channel finite-resolution image deconvolution algorithm. Astron. Astrophys. **589**, A81 (2016). https://doi.org/10.1051/0004-6361/201424003. 1602.02167

M. Cappellari, A.J. Romanowsky, J.P. Brodie, D.A. Forbes, J. Strader, C. Foster, S.S. Kartha, N. Pastorello, V. Pota, L.R. Spitler, C. Usher, J.A. Arnold, Small scatter and nearly isothermal mass profiles to four half-light radii from two-dimensional stellar dynamics of early-type galaxies. Astrophys. J. Lett. **804**, L21 (2015). https://doi.org/10.1088/2041-8205/804/1/L21. 1504.00075

T.C. Chang, U.L. Pen, J.B. Peterson, P. McDonald, Baryon acoustic oscillation intensity mapping of dark energy. Phys. Rev. Lett. **091**, 303 (2008). https://doi.org/10.1103/PhysRevLett.100.091303. 0709.3672

T.C. Chang, U.L. Pen, K. Bandura, J.B. Peterson, An intensity map of hydrogen 21-cm emission at redshift $z \approx 0.8$. Nature **466**, 463–465 (2010). https://doi.org/10.1038/nature09187

G.C.F. Chen, S.H. Suyu, K.C. Wong, C.D. Fassnacht, T. Chiueh, A. Halkola, I.S. Hu, M.W. Auger, L.V.E. Koopmans, D.J. Lagattuta, J.P. McKean, S. Vegetti, SHARP—III. First use of adaptive-optics imaging to constrain cosmology with gravitational lens time delays. Mon. Not. R. Astron. Soc. **462**, 3457–3475 (2016). https://doi.org/10.1093/mnras/stw991. 1601.01321

G.C.F. Chen, C.D. Fassnacht, J.H.H. Chan, V. Bonvin, K. Rojas, M. Millon, F. Courbin, S.H. Suyu, K.C. Wong, D. Sluse, T. Treu, A.J. Shajib, J.W. Hsueh, D.J. Lagattuta, J.P. McKean, Constraining the microlensing effect on time delays with new time-delay prediction model in H_0 measurements (2018). ArXiv e-prints 1804.09390

S. Cole, W.J. Percival, J.A. Peacock, P. Norberg, C.M. Baugh, C.S. Frenk, I. Baldry, J. Bland-Hawthorn, T. Bridges, R. Cannon, M. Colless, C. Collins, W. Couch, N.J.G. Cross, G. Dalton, V.R. Eke, R. De Propris, S.P. Driver, G. Efstathiou, R.S. Ellis, K. Glazebrook, C. Jackson, A. Jenkins, O. Lahav, I. Lewis, S. Lumsden, S. Maddox, D. Madgwick, B.A. Peterson, W. Sutherland, K. Taylor, The 2dF Galaxy Redshift Survey: power-spectrum analysis of the final data set and cosmological implications. Mon. Not. R. Astron. Soc. **362**, 505–534 (2005). https://doi.org/10.1111/j.1365-2966.2005.09318.x. astro-ph/0501174

T.E. Collett, P.J. Marshall, M.W. Auger, S. Hilbert, S.H. Suyu, Z. Greene, T. Treu, C.D. Fassnacht, L.V.E. Koopmans, M. Bradač, R.D. Blandford, Reconstructing the lensing mass in the universe from photometric catalogue data. Mon. Not. R. Astron. Soc. **432**, 679–692 (2013). https://doi.org/10.1093/mnras/stt504. 1303.6564

W.N. Colley, R.E. Schild, C. Abajas, D. Alcalde, Z. Aslan, I. Bikmaev, V. Chavushyan, L. Chinarro, J.P. Cournoyer, R. Crowe, V. Dudinov, A.K.D. Evans, Y.B. Jeon, L.J. Goicoechea, O. Golbasi, I. Khamitov, K. Kjernsmo, H.J. Lee, J. Lee, K.W. Lee, M.G. Lee, O. Lopez-Cruz, E. Mediavilla, A.F.J. Moffat, R. Mujica, A. Ullan, J. Muñoz, A. Oscoz, M.G. Park, N. Purves, O. Saanum, N. Sakhibullin, M. Serra-Ricart, I. Sinelnikov, R. Stabell, A. Stockton, J. Teuber, R. Thompson, H.S. Woo, A. Zheleznyak, Around-the-clock observations of the Q0957+561A,B gravitationally lensed quasar. II. Results for the second observing season. Astrophys. J. **587**, 71–79 (2003). https://doi.org/10.1086/368076. astro-ph/0210400

F. Courbin, A. Eigenbrod, C. Vuissoz, G. Meylan, P. Magain, COSMOGRAIL: the COSmological MOnitoring of GRAvItational Lenses, in *Gravitational Lensing Impact on Cosmology*, ed. by Y. Mellier, G. Meylan. IAU Symposium, vol. 225 (2005), pp. 297–303. https://doi.org/10.1017/S1743921305002097

F. Courbin, V. Chantry, Y. Revaz, D. Sluse, C. Faure, M. Tewes, E. Eulaers, M. Koleva, I. Asfandiyarov, S. Dye, P. Magain, H. van Winckel, J. Coles, P. Saha, M. Ibrahimov, G. Meylan, COSMOGRAIL: the COSmological MOnitoring of GRAvItational Lenses. IX. Time delays, lens dynamics and baryonic fraction in HE 0435-1223. Astron. Astrophys. **536**, A53 (2011a). https://doi.org/10.1051/0004-6361/201015709. 1009.1473

F. Courbin, V. Chantry, Y. Revaz, D. Sluse, C. Faure, M. Tewes, E. Eulaers, M. Koleva et al., COSMOGRAIL: the COSmological MOnitoring of GRAvItational Lenses. IX. Time delays, lens dynamics and baryonic fraction in HE 0435-1223. Astron. Astrophys. **536**, A53 (2011b). https://doi.org/10.1051/0004-6361/201015709. 1009.1473

F. Courbin, V. Bonvin, E. Buckley-Geer, C.D. Fassnacht, J. Frieman, H. Lin, P.J. Marshall, S.H. Suyu, T. Treu, T. Anguita, V. Motta, G. Meylan, E. Paic, M. Tewes, A. Agnello, D.C.Y. Chao, M. Chijani, D. Gilman, K. Rojas, P. Williams, A. Hempel, S. Kim, R. Lachaume, M. Rabus, T.M.C. Abbott, S. Allam, J. Annis, M. Banerji, K. Bechtol, A. Benoit-Lévy, D. Brooks, D.L. Burke, A. Carnero Rosell, M. Carrasco Kind, J. Carretero, C.B.D'Andrea, L.N. da Costa, C. Davis, D.L. DePoy, S. Desai, B. Flaugher, P. Fosalba, J. Garcia-Bellido, E. Gaztanaga, D.A. Goldstein, D. Gruen, R.A. Gruendl, J. Gschwend, G. Gutierrez, K. Honscheid, D.J. James, K. Kuehn, S. Kuhlmann, N. Kuropatkin, O. Lahav, M. Lima, M.A.G. Maia, M. March, J.L. Marshall, R.G. McMahon, F. Menanteau, R. Miquel, B. Nord, A.A. Plazas, E. Sanchez, V. Scarpine, R. Schindler, M. Schubnell, I. Sevilla-Noarbe, M. Smith, M. Soares-Santos, F. Sobreira, E. Suchyta, G. Tarle, D.L. Tucker, A.R. Walker, W. Wester, COSMOGRAIL XVI: Time delays for the quadruply imaged quasar DES J0408-5354 with high-cadence photometric monitoring (2017). ArXiv e-prints 1706.09424

K.S. Dawson, J.P. Kneib, W.J. Percival, S. Alam, F.D. Albareti, S.F. Anderson, E. Armengaud, É Aubourg, S. Bailey, J.E. Bautista, A.A. Berlind, M.A. Bershady, F. Beutler, D. Bizyaev, M.R. Blanton, M. Blomqvist, A.S. Bolton, J. Bovy, W.N. Brandt, J. Brinkmann, J.R. Brownstein, E. Burtin, N.G. Busca, Z. Cai, C.H. Chuang, N. Clerc, J. Comparat, F. Cope, R.A.C. Croft, I. Cruz-Gonzalez, L.N. da Costa, M.C. Cousinou, J. Darling, A. de la Macorra, S. de la Torre, T. Delubac, H. du Mas des Bourboux, T. Dwelly, A. Ealet, D.J. Eisenstein, M. Eracleous, S. Escoffier, X. Fan, A. Finoguenov, A. Font-Ribera, P. Frinchaboy, P. Gaulme, A. Georgakakis, P. Green, H. Guo, J. Guy, S. Ho, D. Holder, J. Huehnerhoff, T. Hutchinson, Y. Jing, E. Jullo, V. Kamble, K. Kinemuchi, D. Kirkby, F.S. Kitaura, M.A. Klaene, R.R. Laher, D. Lang, P. Laurent, J.M. Le Goff, C. Li, Y. Liang, M. Lima, Q. Lin, W. Lin, Y.T. Lin, D.C. Long, B. Lundgren, N. MacDonald, M.A. Geimba Maia, E. Malanushenko, V. Malanushenko, V. Mariappan, C.K. McBride, I.D. McGreer, B. Ménard, A. Merloni, A. Meza, A.D. Montero-Dorta, D. Muna, A.D. Myers, K. Nandra, T. Naugle, J.A. Newman, P. Noterdaeme, P. Nugent, R. Ogando, M.D. Olmstead, A. Oravetz, D.J. Oravetz, N. Padmanabhan, N. Palanque-Delabrouille, K. Pan, J.K. Parejko, I. Pâris, J.A. Peacock, P. Petitjean, M.M. Pieri, A. Pisani, F. Prada, A. Prakash, A. Raichoor, B. Reid, J. Rich, J. Ridl, S. Rodriguez-Torres, A. Carnero Rosell, A.J. Ross, G. Rossi, J. Ruan, M. Salvato, C. Sayres, D.P. Schneider, D.J. Schlegel, U. Seljak, H.J. Seo, B. Sesar, S. Shandera, Y. Shu, A. Slosar, F. Sobreira, A. Streblyanska, N. Suzuki, D. Taylor, C. Tao, J.L. Tinker, R. Tojeiro, M. Vargas-Magaña, Y. Wang, B.A. Weaver, D.H. Weinberg, M. White, W.M. Wood-Vasey, C. Yeche, Z. Zhai, C. Zhao, G.B. Zhao, Z. Zheng, G. Ben Zhu, H. Zou, The SDSS-IV extended baryon oscillation spectroscopic survey: overview and early data. Astron. J. **151**, 44 (2016). https://doi.org/10.3847/0004-6256/151/2/44. 1508.04473

G. Dobler, C.R. Keeton, Microlensing of lensed supernovae. Astrophys. J. **653**, 1391–1399 (2006). https://doi.org/10.1086/508769. astro-ph/0608391

G. Dobler, C.D. Fassnacht, T. Treu, P. Marshall, K. Liao, A. Hojjati, E. Linder, N. Rumbaugh, Strong lens time delay challenge. I. Experimental design. Astrophys. J. **799**, 168 (2015). https://doi.org/10.1088/0004-637X/799/2/168

M.J. Drinkwater, R.J. Jurek, C. Blake, D. Woods, K.A. Pimbblet, K. Glazebrook, R. Sharp, M.B. Pracy, S. Brough, M. Colless, W.J. Couch, S.M. Croom, T.M. Davis, D. Forbes, K. Forster, D.G. Gilbank, M. Gladders, B. Jelliffe, N. Jones, I.H. Li, B. Madore, D.C. Martin, G.B. Poole, T. Small, E. Wisnioski, T. Wyder, H.K.C. Yee, The WiggleZ dark energy survey: survey design and first data release. Mon. Not. R. Astron. Soc. **401**, 1429–1452 (2010). https://doi.org/10.1111/j.1365-2966.2009.15754.x. 0911.4246

S. Dye, S.J. Warren, Decomposition of the visible and dark matter in the Einstein ring 0047-2808 by semi-linear inversion. Astrophys. J. **623**, 31–41 (2005). https://doi.org/10.1086/428340

G. Efstathiou, H_0 revisited. Mon. Not. R. Astron. Soc. **440**, 1138–1152 (2014). https://doi.org/10.1093/mnras/stu278. 1311.3461

A. Eigenbrod, F. Courbin, C. Vuissoz, G. Meylan, P. Saha, S. Dye, COSMOGRAIL: the COSmological MOnitoring of GRAvItational Lenses. I. How to sample the light curves of gravitationally lensed quasars to measure accurate time delays. Astron. Astrophys. **436**, 25–35 (2005). https://doi.org/10.1051/0004-6361:20042422. astro-ph/0503019

D.J. Eisenstein, W. Hu, Baryonic features in the matter transfer function. Astrophys. J. **496**, 605 (1998). https://doi.org/10.1086/305424. astro-ph/9709112

D.J. Eisenstein, I. Zehavi, D.W. Hogg, R. Scoccimarro, M.R. Blanton, R.C. Nichol, R. Scranton, H. Seo, M. Tegmark, Z. Zheng, S.F. Anderson, J. Annis, N. Bahcall, J. Brinkmann, S. Burles, F.J. Castander, A. Connolly, I. Csabai, M. doi, M. Fukugita, J.A. Frieman, K. Glazebrook, J.E. Gunn, J.S. Hendry, G. Hennessy, Z. Ivezić, S. Kent, G.R. Knapp, H. Lin, Y. Loh, R.H. Lupton, B. Margon, T.A. McKay, A. Meiksin, J.A. Munn, A. Pope, M.W. Richmond, D. Schlegel, D.P. Schneider, K. Shimasaku, C. Stoughton, M.A. Strauss, M. SubbaRao, A.S. Szalay, I. Szapudi, D.L. Tucker, B. Yanny, D.G. York, Detection of the baryon acoustic peak in the large-scale correlation function of SDSS luminous red galaxies. Astrophys. J. **633**, 560–574 (2005). https://doi.org/10.1086/466512. astro-ph/0501171

E. Eulaers, M. Tewes, P. Magain, F. Courbin, I. Asfandiyarov, S. Ehgamberdiev, S. Rathna Kumar, C.S. Stalin, T.P. Prabhu, G. Meylan, H. Van Winckel, COSMOGRAIL: the COSmological MOnitoring of GRAvItational Lenses. XII. Time delays of the doubly lensed quasars SDSS J1206+4332 and HS 2209+1914. Astron. Astrophys. **553**, A121 (2013). https://doi.org/10.1051/0004-6361/201321140. 1304.4474

R. Fadely, C.R. Keeton, R. Nakajima, G.M. Bernstein, Improved constraints on the gravitational lens Q0957+561. II. Strong lensing. Astrophys. J. **711**, 246–267 (2010). https://doi.org/10.1088/0004-637X/711/1/246. 0909.1807

E.E. Falco, M.V. Gorenstein, I.I. Shapiro, On model-dependent bounds on H(0) from gravitational images application of Q0957 + 561A,B. Astrophys. J. Lett. **289**, L1–L4 (1985). https://doi.org/10.1086/184422

C.D. Fassnacht, T.J. Pearson, A.C.S. Readhead, I.W.A. Browne, L.V.E. Koopmans, S.T. Myers, P.N. Wilkinson, A determination of H_0 with the CLASS gravitational lens B1608+656. I. Time delay measurements with the VLA. Astrophys. J. **527**, 498–512 (1999). https://doi.org/10.1086/308118. astro-ph/9907257

C.D. Fassnacht, E. Xanthopoulos, L.V.E. Koopmans, D. Rusin, Astrophys. J. **581**, 823 (2002)

C.D. Fassnacht, R.R. Gal, L.M. Lubin, J.P. McKean, G.K. Squires, A.C.S. Readhead, Mass along the line of sight to the gravitational lens B1608+656: galaxy groups and implications for H_0. Astrophys. J. **642**, 30–38 (2006). https://doi.org/10.1086/500927. astro-ph/0510728

X. Fernández, H.B. Gim, J.H. van Gorkom, M.S. Yun, E. Momjian, A. Popping, L. Chomiuk, K.M. Hess, L. Hunt, K. Kreckel, D. Lucero, N. Maddox, T. Oosterloo, D.J. Pisano, M.A.W. Verheijen, C.A. Hales, A. Chung, R. Dodson, K. Golap, J. Gross, P. Henning, J. Hibbard, Y.L. Jaffé, J. Donovan Meyer, M. Meyer, M. Sanchez-Barrantes, D. Schiminovich, A. Wicenec, E. Wilcots, M. Bershady, N. Scoville, J. Strader, E. Tremou, R. Salinas, R. Chávez, Highest redshift image of neutral hydrogen in emission: a CHILES detection of a starbursting galaxy at $z = 0.376$. Astrophys. J. Lett. **824**, L1 (2016). https://doi.org/10.3847/2041-8205/824/1/L1. 1606.00013

J. Fohlmeister, C.S. Kochanek, E.E. Falco, C.W. Morgan, J. Wambsganss, The rewards of patience: an 822 day time delay in the gravitational lens SDSS J1004+4112. Astrophys. J. **676**, 761–766 (2008). https://doi.org/10.1086/528789. 0710.1634

J. Fohlmeister, C.S. Kochanek, E.E. Falco, J. Wambsganss, M. Oguri, X. Dai, A two-year time delay for the lensed quasar SDSS J1029+2623. Astrophys. J. **764**, 186 (2013). https://doi.org/10.1088/0004-637X/764/2/186. 1207.5776

M. Foxley-Marrable, T.E. Collett, G. Vernardos, D.A. Goldstein, D. Bacon, The impact of microlensing on the standardisation of strongly lensed type Ia supernovae (2018). ArXiv e-prints 1802.07738

W.L. Freedman, B.F. Madore, V. Scowcroft, C. Burns, A. Monson, S.E. Persson, M. Seibert, J. Rigby, Carnegie Hubble program: a mid-infrared calibration of the Hubble constant. Astrophys. J. **758**, 24 (2012). https://doi.org/10.1088/0004-637X/758/1/24. 1208.3281

R. Gavazzi, T. Treu, L.V.E. Koopmans, A.S. Bolton, L.A. Moustakas, S. Burles, P.J. Marshall, The Sloan lens ACS survey. VI. Discovery and analysis of a double Einstein ring. Astrophys. J. **677**, 1046–1059 (2008). https://doi.org/10.1086/529541. 0801.1555

E. Giannini, R.W. Schmidt, J. Wambsganss, K. Alsubai, J.M. Andersen, T. Anguita, V. Bozza, D.M. Bramich, P. Browne, S. Calchi Novati, Y. Damerdji, C. Diehl, P. Dodds, M. Dominik, A. Elyiv, X. Fang, R. Figuera Jaimes, F. Finet, T. Gerner, S. Gu, S. Hardis, K. Harpsøe, T.C. Hinse, A. Hornstrup, M. Hundertmark, J. Jessen-Hansen, U.G. Jørgensen, D. Juncher, N. Kains, E. Kerins, H. Korhonen, C. Liebig, M.N. Lund, M.S. Lundkvist, G. Maier, L. Mancini, G. Masi, M. Mathiasen, M. Penny, S. Proft, M. Rabus, S. Rahvar, D. Ricci, G. Scarpetta, K. Sahu, S. Schäfer, F. Schönebeck, J. Skottfelt, C. Snodgrass, J. Southworth, J. Surdej, J. Tregloan-Reed, C. Vilela, O. Wertz, F. Zimmer, MiNDSTEp differential photometry of the gravitationally lensed quasars WFI 2033-4723 and HE 0047-1756: microlensing and a new time delay. Astron. Astrophys. **597**, A49 (2017). https://doi.org/10.1051/0004-6361/201527422. 1610.03732

D.A. Goldstein, P.E. Nugent, How to find gravitationally lensed type Ia supernovae. Astrophys. J. Lett. **834**, L5 (2017). https://doi.org/10.3847/2041-8213/834/1/L5. 1611.09459

D.A. Goldstein, P.E. Nugent, D.N. Kasen, T.E. Collett, Precise time delays from strongly gravitationally lensed type Ia supernovae with chromatically microlensed images. Astrophys. J. **855**, 22–30 (2018). https://doi.org/10.3847/1538-4357/aaa975. 1708.00003

G. Golse, J.P. Kneib, Pseudo elliptical lensing mass model: application to the NFW mass distribution. Astron. Astrophys. **390**, 821–827 (2002). https://doi.org/10.1051/0004-6361:20020639. astro-ph/0112138

A. Goobar, iPTF16geu: a multiply imaged, gravitationally lensed type Ia supernova. Science **356**, 291–295 (2017). https://doi.org/10.1126/science.aal2729. 1611.00014

Z.S. Greene, S.H. Suyu, T. Treu, S. Hilbert, M.W. Auger, T.E. Collett, P.J. Marshall, C.D. Fassnacht, R.D. Blandford, M. Bradač, L.V.E. Koopmans, Improving the precision of time-delay cosmography with observations of galaxies along the line of sight. Astrophys. J. **768**, 39 (2013). https://doi.org/10.1088/0004-637X/768/1/39. 1303.3588

C. Grillo, W. Karman, S.H. Suyu, P. Rosati, I. Balestra, A. Mercurio, M. Lombardi, T. Treu, G.B. Caminha, A. Halkola, S.A. Rodney, R. Gavazzi, K.I. Caputi, The story of supernova refsdal told by muse. Astrophys. J. **822**, 78 (2016). https://doi.org/10.3847/0004-637X/822/2/78. 1511.04093

C. Grillo, P. Rosati, S.H. Suyu, I. Balestra, G.B. Caminha, A. Halkola, P.L. Kelly, M. Lombardi, A. Mercurio, S.A. Rodney, T. Treu, Measuring the value of the Hubble constant "à la Refsdal". Astrophys. J. (2018, in press). https://doi.org/10.3847/1538-4357/aac2c9

N.A. Grogin, R. Narayan, A new model of the gravitational lens 0957+561 and a limit on the Hubble constant. Astrophys. J. **464**, 92 (1996). https://doi.org/10.1086/177302

L.J. Hainline, C.W. Morgan, C.L. MacLeod, Z.D. Landaal, C.S. Kochanek, H.C. Harris, T. Tilleman, L.J. Goicoechea, V.N. Shalyapin, E.E. Falco, Time delay and accretion disk size measurements in the lensed quasar SBS 0909+532 from multiwavelength microlensing analysis. Astrophys. J. **774**, 69 (2013). https://doi.org/10.1088/0004-637X/774/1/69. 1307.3254

C. Hazard, M.B. Mackey, A.J. Shimmins, Investigation of the radio source 3C 273 by the method of lunar occultations. Nature **197**, 1037–1039 (1963). https://doi.org/10.1038/1971037a0

S. Hilbert, S.D.M. White, J. Hartlap, P. Schneider, Strong lensing optical depths in a ΛCDM universe. Mon. Not. R. Astron. Soc. **382**, 121–132 (2007). https://doi.org/10.1111/j.1365-2966.2007.12391.x. astro-ph/0703803

J. Hjorth, I. Burud, A.O. Jaunsen, P.L. Schechter, J.P. Kneib, M.I. Andersen, H. Korhonen, J.W. Clasen, A.A. Kaas, R. Østensen, J. Pelt, F.P. Pijpers, The time delay of the quadruple quasar RX J0911.4+0551. Astrophys. J. Lett. **572**, L11–L14 (2002). https://doi.org/10.1086/341603. astro-ph/0205124

A. Hojjati, E.V. Linder, Next generation strong lensing time delay estimation with Gaussian processes. Phys. Rev. D **90**(12), 123501 (2014). https://doi.org/10.1103/PhysRevD.90.123501. 1408.5143

A. Hojjati, A.G. Kim, E.V. Linder, Robust strong lensing time delay estimation. Phys. Rev. D **87**(12), 123512 (2013). https://doi.org/10.1103/PhysRevD.87.123512. 1304.0309

W. Hu, N. Sugiyama, Small-scale cosmological perturbations: an analytic approach. Astrophys. J. **471**, 542 (1996). https://doi.org/10.1086/177989. astro-ph/9510117

J. Huchra, M. Gorenstein, S. Kent, I. Shapiro, G. Smith, E. Horine, R. Perley, 2237 + 0305: a new and unusual gravitational lens. Astron. J. **90**, 691–696 (1985). https://doi.org/10.1086/113777

P. Jakobsson, J. Hjorth, I. Burud, G. Letawe, C. Lidman, F. Courbin, An optical time delay for the double gravitational lens system FBQ 0951+2635. Astron. Astrophys. **431**, 103–109 (2005). https://doi.org/10.1051/0004-6361:20041432. astro-ph/0409444

M. Jauzac, J. Richard, M. Limousin, K. Knowles, G. Mahler, G.P. Smith, J.P. Kneib, E. Jullo, P. Natarajan, H. Ebeling, H. Atek, B. Clément, D. Eckert, E. Egami, R. Massey, M. Rexroth, Hubble frontier fields: predictions for the return of SN refsdal with the MUSE and GMOS spectrographs. Mon. Not. R. Astron. Soc. **457**, 2029–2042 (2016). https://doi.org/10.1093/mnras/stw069. 1509.08914

I. Jee, E. Komatsu, S.H. Suyu, Measuring angular diameter distances of strong gravitational lenses. J. Cosmol. Astropart. Phys. **11**, 033 (2015). https://doi.org/10.1088/1475-7516/2015/11/033. 1410.7770

I. Jee, E. Komatsu, S.H. Suyu, D. Huterer, Time-delay cosmography: increased leverage with angular diameter distances. J. Cosmol. Astropart. Phys. (2016, in press). https://doi.org/10.1088/1475-7516/2016/04/031

E. Jullo, J. Kneib, M. Limousin, Á Elíasdóttir, P.J. Marshall, T. Verdugo, A Bayesian approach to strong lensing modelling of galaxy clusters. New J. Phys. **9**, 447 (2007). https://doi.org/10.1088/1367-2630/9/12/447. 0706.0048

N. Kaiser, Clustering in real space and in redshift space. Mon. Not. R. Astron. Soc. **227**, 1–21 (1987)

R. Kawamata, M. Oguri, M. Ishigaki, K. Shimasaku, M. Ouchi, Precise strong lensing mass modeling of four Hubble frontier field clusters and a sample of magnified high-redshift galaxies. Astrophys. J. **819**, 114 (2016). https://doi.org/10.3847/0004-637X/819/2/114. 1510.06400

C.R. Keeton, Computational methods for gravitational lensing (2001). ArXiv e-prints astro-ph/0102340

P.L. Kelly, S.A. Rodney, T. Treu, R.J. Foley, G. Brammer, K.B. Schmidt, A. Zitrin, A. Sonnenfeld, L.G. Strolger, O. Graur, A.V. Filippenko, S.W. Jha, A.G. Riess, M. Bradac, B.J. Weiner, D. Scolnic, M.A. von Malkan, A. der Linden, M. Trenti, J. Hjorth, R. Gavazzi, A. Fontana, J.C. Merten, C. McCully, T. Jones, M. Postman, A. Dressler, B. Patel, S.B. Cenko, M.L. Graham, B.E. Tucker, Multiple images of a highly magnified supernova formed by an early-type cluster galaxy lens. Science **347**, 1123–1126 (2015). https://doi.org/10.1126/science.aaa3350. 1411.6009

P.L. Kelly, S.A. Rodney, T. Treu, L.G. Strolger, R.J. Foley, S.W. Jha, J. Selsing, G. Brammer, M. Bradač, S.B. Cenko, O. Graur, A.V. Filippenko, J. Hjorth, C. McCully, A. Molino, M. Nonino, A.G. Riess, K.B. Schmidt, B. von Tucker, A. der Linden, B.J. Weiner, A. Zitrin, Deja vu all over again: the reappearance of supernova refsdal. Astrophys. J. Lett. **819**, L8 (2016a). https://doi.org/10.3847/2041-8205/819/1/L8. 1512.04654

P.L. Kelly, S.A. Rodney, T. Treu, L.G. Strolger, R.J. Foley, S.W. Jha, J. Selsing, G. Brammer, M. Bradač, S.B. Cenko, O. Graur, A.V. Filippenko, J. Hjorth, C. McCully, A. Molino, M. Nonino, A.G. Riess, K.B.

Schmidt, B. von Tucker, A. der Linden, B.J. Weiner, A. Zitrin, Deja vu all over again: the reappearance of supernova refsdal. Astrophys. J. Lett. **819**, L8 (2016b). https://doi.org/10.3847/2041-8205/819/1/L8. 1512.04654

C.S. Kochanek, What do gravitational lens time delays measure? Astrophys. J. **578**, 25–32 (2002). https://doi.org/10.1086/342476. astro-ph/0205319

C.S. Kochanek, N.D. Morgan, E.E. Falco, B.A. McLeod, J.N. Winn, J. Dembicky, B. Ketzeback, The time delays of gravitational lens HE 0435-1223: an early-type galaxy with a rising rotation curve. Astrophys. J. **640**, 47–61 (2006). https://doi.org/10.1086/499766. astro-ph/0508070

L.V.E. Koopmans, Gravitational imaging of cold dark matter substructures. Mon. Not. R. Astron. Soc. **363**, 1136–1144 (2005). https://doi.org/10.1111/j.1365-2966.2005.09523.x

L.V.E. Koopmans, T. Treu, The stellar velocity dispersion of the lens galaxy in MG 2016+112 at $z = 1.004$. Astrophys. J. Lett. **568**, L5–L8 (2002). https://doi.org/10.1086/340143. astro-ph/0201017

L.V.E. Koopmans, T. Treu, C.D. Fassnacht, R.D. Blandford, G. Surpi, The Hubble constant from the gravitational lens B1608+656. Astrophys. J. **599**, 70–85 (2003). https://doi.org/10.1086/379226. astro-ph/0306216

L.V.E. Koopmans, T. Treu, A.S. Bolton, S. Burles, L.A. Moustakas, The Sloan lens ACS survey. III. The structure and formation of early-type galaxies and their evolution since $z \sim 1$. Astrophys. J. **649**, 599–615 (2006). https://doi.org/10.1086/505696. astro-ph/0601628

R. Kormann, P. Schneider, M. Bartelmann, Isothermal elliptical gravitational lens models. Astron. Astrophys. **284**, 285–299 (1994)

T. Kundić, E.L. Turner, W.N. Colley, J.R. Gott III, J.E. Rhoads, Y. Wang, L.E. Bergeron, K.A. Gloria, D.C. Long, S. Malhotra, J. Wambsganss, A robust determination of the time delay in 0957+561A, B and a measurement of the global value of Hubble's constant. Astrophys. J. **482**, 75–82 (1997). https://doi.org/10.1086/304147. astro-ph/9610162

J. Lehar, J.N. Hewitt, B.F. Burke, D.H. Roberts, The radio time delay in the double quasar 0957 + 561. Astrophys. J. **384**, 453–466 (1992). https://doi.org/10.1086/170887

K. Liao, T. Treu, P. Marshall, C.D. Fassnacht, N. Rumbaugh, G. Dobler, A. Aghamousa, V. Bonvin, F. Courbin, A. Hojjati, N. Jackson, V. Kashyap, S. Rathna Kumar, E. Linder, K. Mandel, X.L. Meng, G. Meylan, L.A. Moustakas, T.P. Prabhu, A. Romero-Wolf, A. Shafieloo, A. Siemiginowska, C.S. Stalin, H. Tak, M. Tewes, D. van Dyk, Strong lens time delay challenge. II. Results of TDC1. Astrophys. J. **800**, 11 (2015). https://doi.org/10.1088/0004-637X/800/1/11. 1409.1254

Y.W. Liao, T.C. Chang, C.Y. Kuo, K.W. Masui, N. Oppermann, U.L. Pen, J.B. Peterson, Accurate polarization calibration at 800 MHz with the green bank telescope. Astrophys. J. **833**, 289 (2016). https://doi.org/10.3847/1538-4357/833/2/289. 1610.04365

H. Lin, E. Buckley-Geer, A. Agnello, F. Ostrovski, R.G. McMahon, B. Nord, N. Kuropatkin, D.L. Tucker, T. Treu, J.H.H. Chan, S.H. Suyu, H.T. Diehl, T. Collett, M.S.S. Gill, A. More, A. Amara, M.W. Auger, F. Courbin, C.D. Fassnacht, J. Frieman, P.J. Marshall, G. Meylan, C.E. Rusu, T.M.C. Abbott, F.B. Abdalla, S. Allam, M. Banerji, K. Bechtol, A. Benoit-Lévy, E. Bertin, D. Brooks, D.L. Burke, A. Carnero Rosell, M. Carrasco Kind, J. Carretero, F.J. Castander, M. Crocce, C.B. D'Andrea, L.N. da Costa, S. Desai, J.P. Dietrich, T.F. Eifler, D.A. Finley, B. Flaugher, P. Fosalba, J. García-Bellido, E. Gaztanaga, D.W. Gerdes, D.A. Goldstein, D. Gruen, R.A. Gruendl, J. Gschwend, G. Gutierrez, K. Honscheid, D.J. James, K. Kuehn, O. Lahav, T.S. Li, M. Lima, M.A.G. Maia, M. March, J.L. Marshall, P. Martini, P. Melchior, F. Menanteau, R. Miquel, R.L.C. Ogando, A.A. Plazas, A.K. Romer, E. Sanchez, R. Schindler, M. Schubnell, I. Sevilla-Noarbe, M. Smith, R.C. Smith, F. Sobreira, E. Suchyta, M.E.C. Swanson, G. Tarle, D. Thomas, A.R. Walker (DES Collaboration), Discovery of the lensed quasar system DES J0408-5354. Astrophys. J. Lett. **838**, L15 (2017). https://doi.org/10.3847/2041-8213/aa624e. 1702.00072

P. Magain, J. Surdej, J.P. Swings, U. Borgeest, R. Kayser, Discovery of a quadruply lensed quasar—the 'clover leaf' H1413 + 117. Nature **334**, 325–327 (1988). https://doi.org/10.1038/334325a0

P. Magain, F. Courbin, S. Sohy, Deconvolution with correct sampling. Astrophys. J. **494**, 472–477 (1998). https://doi.org/10.1086/305187. astro-ph/9704059

P.J. Marshall, T. Treu, J. Melbourne, R. Gavazzi, K. Bundy, S.M. Ammons, A.S. Bolton, S. Burles, J.E. Larkin, D. Le Mignant, D.C. Koo, L.V.E. Koopmans, C.E. Max, L.A. Moustakas, E. Steinbring, S.A. Wright, Superresolving distant galaxies with gravitational telescopes: Keck laser guide star adaptive optics and Hubble space telescope imaging of the lens system SDSS J0737+3216. Astrophys. J. **671**, 1196–1211 (2007). https://doi.org/10.1086/523091. 0710.0637

A.M. Martin, E. Papastergis, R. Giovanelli, M.P. Haynes, C.M. Springob, S. Stierwalt, The arecibo legacy fast ALFA survey. X. The H I mass function and Ω_H I from the 40% ALFALFA survey. Astrophys. J. **723**, 1359–1374 (2010). https://doi.org/10.1088/0004-637X/723/2/1359. 1008.5107

K.W. Masui, P. McDonald, U.L. Pen, Near-term measurements with 21 cm intensity mapping: neutral hydrogen fraction and BAO at $z < 2$. Phys. Rev. D **81**(10), 103527 (2010). https://doi.org/10.1103/PhysRevD.81.103527. 1001.4811

K.W. Masui, E.R. Switzer, N. Banavar, K. Bandura, C. Blake, L.M. Calin, T.C. Chang, X. Chen, Y.C. Li, Y.W. Liao, A. Natarajan, U.L. Pen, J.B. Peterson, J.R. Shaw, T.C. Voytek, Measurement of 21 cm brightness fluctuations at $z \sim 0.8$ in cross-correlation. Astrophys. J. Lett. **763**, L20 (2013). https://doi.org/10.1088/2041-8205/763/1/L20. 1208.0331

T. Matsubara, Correlation function in deep redshift space as a cosmological probe. Astrophys. J. **615**, 573–585 (2004). https://doi.org/10.1086/424561. astro-ph/0408349

C. McCully, C.R. Keeton, K.C. Wong, A.I. Zabludoff, Quantifying environmental and line-of-sight effects in models of strong gravitational lens systems. Astrophys. J. **836**, 141 (2017). https://doi.org/10.3847/1538-4357/836/1/141. 1601.05417

I. Momcheva, K. Williams, C. Keeton, A. Zabludoff, A spectroscopic study of the environments of gravitational lens galaxies. Astrophys. J. **641**, 169–189 (2006). https://doi.org/10.1086/500382. astro-ph/0511594

A. More, S.H. Suyu, M. Oguri, S. More, C.H. Lee, Interpreting the strongly lensed supernova iPTF16geu: time delay predictions, microlensing, and lensing rates. Astrophys. J. Lett. **835**, L25 (2017). https://doi.org/10.3847/2041-8213/835/2/L25. 1611.04866

C.W. Morgan, L.J. Hainline, B. Chen, M. Tewes, C.S. Kochanek, X. Dai, S. Kozlowski, J.A. Blackburne, A.M. Mosquera, G. Chartas, F. Courbin, G. Meylan, Further evidence that quasar X-ray emitting regions are compact: X-ray and optical microlensing in the lensed quasar Q J0158-4325. Astrophys. J. **756**, 52 (2012). https://doi.org/10.1088/0004-637X/756/1/52. 1205.4727

L.B. Newburgh, K. Bandura, M.A. Bucher, T.C. Chang, H.C. Chiang, J.F. Cliche, R. Davé, M. Dobbs, C. Clarkson, K.M. Ganga, T. Gogo, A. Gumba, N. Gupta, M. Hilton, B. Johnstone, A. Karastergiou, M. Kunz, D. Lokhorst, R. Maartens, S. Macpherson, M. Mdlalose, K. Moodley, L. Ngwenya, J.M. Parra, J. Peterson, O. Recnik, B. Saliwanchik, M.G. Santos, J.L. Sievers, O. Smirnov, P. Stronkhorst, R. Taylor, K. Vanderlinde, G. Van Vuuren, A. Weltman, A. Witzemann, HIRAX: a probe of dark energy and radio transients, in *Ground-Based and Airborne Telescopes VI*. Proc. SPIE, vol. 9906 (2016), p. 99065X. https://doi.org/10.1117/12.2234286. 1607.02059

J.W. Nightingale, S. Dye, Adaptive semi-linear inversion of strong gravitational lens imaging. Mon. Not. R. Astron. Soc. **452**, 2940–2959 (2015). https://doi.org/10.1093/mnras/stv1455. 1412.7436

E.O. Ofek, D. Maoz, Time-delay measurement of the lensed quasar HE 1104-1805. Astrophys. J. **594**, 101–106 (2003). https://doi.org/10.1086/376903. astro-ph/0305200

M. Oguri, Gravitational lens time delays: a statistical assessment of lens model dependences and implications for the global Hubble constant. Astrophys. J. **660**, 1–15 (2007). https://doi.org/10.1086/513093. astro-ph/0609694

M. Oguri, The mass distribution of SDSS J1004+4112 revisited. Publ. Astron. Soc. Jpn. **62**, 1017–1024 (2010). https://doi.org/10.1093/pasj/62.4.1017. 1005.3103

M. Oguri, P.J. Marshall, Gravitationally lensed quasars and supernovae in future wide-field optical imaging surveys. Mon. Not. R. Astron. Soc. **405**, 2579–2593 (2010). https://doi.org/10.1111/j.1365-2966.2010.16639.x. 1001.2037

T. Okumura, T. Matsubara, D.J. Eisenstein, I. Kayo, C. Hikage, A.S. Szalay, D.P. Schneider, Large-scale anisotropic correlation function of SDSS luminous red galaxies. Astrophys. J. **676**, 889–898 (2008). https://doi.org/10.1086/528951. 0711.3640

T. Okumura, C. Hikage, T. Totani, M. Tonegawa, H. Okada, K. Glazebrook, C. Blake, P.G. Ferreira, S. More, A. Taruya, S. Tsujikawa, M. Akiyama, G. Dalton, T. Goto, T. Ishikawa, F. Iwamuro, T. Matsubara, T. Nishimichi, K. Ohta, I. Shimizu, R. Takahashi, N. Takato, N. Tamura, K. Yabe, N. Yoshida, The Subaru FMOS galaxy redshift survey (FastSound). IV. New constraint on gravity theory from redshift space distortions at z 1.4. Publ. Astron. Soc. Jpn. **68**, 38 (2016). https://doi.org/10.1093/pasj/psw029. 1511.08083

L. Oldham, M.W. Auger, C.D. Fassnacht, T. Treu, B.J. Brewer, L.V.E. Koopmans, D. Lagattuta, P. Marshall, J. McKean, S. Vegetti, Red nuggets grow inside-out: evidence from gravitational lensing. Mon. Not. R. Astron. Soc. **465**, 3185–3202 (2017). https://doi.org/10.1093/mnras/stw2832. 1611.00008

A. Oscoz, E. Mediavilla, L.J. Goicoechea, M. Serra-Ricart, J. Buitrago, Time delay of QSO 0957+561 and cosmological implications. Astrophys. J. Lett. **479**, L89–L92 (1997). https://doi.org/10.1086/310599

D. Paraficz, J. Hjorth, Gravitational lenses as cosmic rulers: Ω_m, Ω_Λ from time delays and velocity dispersions. Astron. Astrophys. **507**, L49–L52 (2009). https://doi.org/10.1051/0004-6361/200913307. 0910.5823

P.J.E. Peebles, J.T. Yu, Primeval adiabatic perturbation in an expanding universe. Astrophys. J. **162**, 815 (1970). https://doi.org/10.1086/150713

J.B. Peterson, K. Bandura, U.L. Pen, The Hubble sphere hydrogen survey (2006). ArXiv e-prints astro-ph/0606104

S. Rathna Kumar, M. Tewes, C.S. Stalin, F. Courbin, I. Asfandiyarov, G. Meylan, E. Eulaers, T.P. Prabhu, P. Magain, H. Van Winckel, S. Ehgamberdiev, COSMOGRAIL: the COSmological MOnitoring of GRAv-

itational lenses. XIV. Time delay of the doubly lensed quasar SDSS J1001+5027. Astron. Astrophys. **557**, A44 (2013). https://doi.org/10.1051/0004-6361/201322116. 1306.5105
S. Rathna Kumar, C.S. Stalin, T.P. Prabhu, H_0 from ten well-measured time delay lenses. Astron. Astrophys. **580**, A38 (2015). https://doi.org/10.1051/0004-6361/201423977. 1404.2920
S. Refsdal, On the possibility of determining Hubble's parameter and the masses of galaxies from the gravitational lens effect. Mon. Not. R. Astron. Soc. **128**, 307 (1964)
A.G. Riess, L.M. Macri, S.L. Hoffmann, D. Scolnic, S. Casertano, A.V. Filippenko, B.E. Tucker, M.J. Reid, D.O. Jones, J.M. Silverman, R. Chornock, P. Challis, W. Yuan, P.J. Brown, R.J. Foley, A 2.4% determination of the local value of the Hubble constant. Astrophys. J. **826**, 56 (2016a). https://doi.org/10.3847/0004-637X/826/1/56. 1604.01424
A.G. Riess, L.M. Macri, S.L. Hoffmann, D. Scolnic, S. Casertano, A.V. Filippenko, B.E. Tucker, M.J. Reid, D.O. Jones, J.M. Silverman, R. Chornock, P. Challis, W. Yuan, R.J. Foley, A 2.4% determination of the local value of the Hubble constant (2016b). ArXiv e-prints 1604.01424
A.G. Riess, S. Casertano, W. Yuan, L. Macri, J. Anderson, J.W. Mackenty, J.B. Bowers, K.I. Clubb, A.V. Filippenko, D.O. Jones, B.E. Tucker, New parallaxes of galactic Cepheids from spatially scanning the Hubble space telescope: implications for the Hubble constant (2018). ArXiv e-prints 1801.01120
S.A. Rodney, L.G. Strolger, P.L. Kelly, M. Bradač, G. Brammer, A.V. Filippenko, R.J. Foley, O. Graur, J. Hjorth, S.W. Jha, C. McCully, A. Molino, A.G. Riess, K.B. Schmidt, J. Selsing, K. Sharon, T. Treu, B.J. Weiner, A. Zitrin, SN refsdal: photometry and time delay measurements of the first Einstein cross supernova. Astrophys. J. **820**, 50 (2016). https://doi.org/10.3847/0004-637X/820/1/50. 1512.05734
C.E. Rusu, C.D. Fassnacht, D. Sluse, S. Hilbert, K.C. Wong, K.H. Huang, S.H. Suyu, T.E. Collett, P.J. Marshall, T. Treu, L.V.E. Koopmans, H0LiCOW—III. Quantifying the effect of mass along the line of sight to the gravitational lens HE 0435-1223 through weighted galaxy counts. Mon. Not. R. Astron. Soc. **467**, 4220–4242 (2017). https://doi.org/10.1093/mnras/stx285. 1607.01047
P. Saha, J. Coles, A.V. Macciò, L.L.R. Williams, The Hubble time inferred from 10 time delay lenses. Astrophys. J. Lett. **650**, L17–L20 (2006). https://doi.org/10.1086/507583. astro-ph/0607240
P.L. Schechter, C.D. Bailyn, R. Barr, R. Barvainis, C.M. Becker, G.M. Bernstein, J.P. Blakeslee, S.J. Bus, A. Dressler, E.E. Falco, R.A. Fesen, P. Fischer, K. Gebhardt, D. Harmer, J.N. Hewitt, J. Hjorth, T. Hurt, A.O. Jaunsen, M. Mateo, D. Mehlert, D.O. Richstone, L.S. Sparke, J.R. Thorstensen, J.L. Tonry, G. Wegner, D.W. Willmarth, G. Worthey, The quadruple gravitational lens PG 1115+080: time delays and models. Astrophys. J. Lett. **475**, L85–L88 (1997). https://doi.org/10.1086/310478. astro-ph/9611051
M. Schmidt, 3C 273: a star-like object with large red-shift. Nature **197**, 1040 (1963). https://doi.org/10.1038/1971040a0
P. Schneider, Generalized multi-plane gravitational lensing: time delays, recursive lens equation, and the mass-sheet transformation (2014). ArXiv e-prints 1409.0015
P. Schneider, D. Sluse, Mass-sheet degeneracy, power-law models and external convergence: impact on the determination of the Hubble constant from gravitational lensing. Astron. Astrophys. **559**, A37 (2013). https://doi.org/10.1051/0004-6361/201321882. 1306.0901
P. Schneider, D. Sluse, Source-position transformation: an approximate invariance in strong gravitational lensing. Astron. Astrophys. **564**, A103 (2014). https://doi.org/10.1051/0004-6361/201322106. 1306.4675
P. Schneider, J. Ehlers, E.E. Falco, *Gravitational Lenses* vol. XIV, (Springer, Berlin, 1992), 560 pp., 112 figs
P. Schneider, C.S. Kochanek, J. Wambsganss, *Gravitational Lensing: Strong, Weak and Micro* (Springer, Berlin, 2006). https://doi.org/10.1007/978-3-540-30310-7
H.J. Seo, D.J. Eisenstein, Probing dark energy with baryonic acoustic oscillations from future large galaxy redshift surveys. Astrophys. J. **598**, 720–740 (2003). https://doi.org/10.1086/379122. astro-ph/0307460
H.J. Seo, S. Dodelson, J. Marriner, D. Mcginnis, A. Stebbins, C. Stoughton, A. Vallinotto, A ground-based 21 cm baryon acoustic oscillation survey. Astrophys. J. **721**, 164–173 (2010). https://doi.org/10.1088/0004-637X/721/1/164. 0910.5007
M. Sereno, D. Paraficz, Hubble constant and dark energy inferred from free-form determined time delay distances. Mon. Not. R. Astron. Soc. **437**, 600–605 (2014). https://doi.org/10.1093/mnras/stt1938. 1310.2251
A.J. Shajib, T. Treu, A. Agnello, Improving time-delay cosmography with spatially resolved kinematics (2017). ArXiv e-prints 1709.01517
V.N. Shalyapin, L.J. Goicoechea, Doubly imaged quasar SDSS J1515+1511: time delay and lensing galaxies. Astrophys. J. **836**, 14 (2017). https://doi.org/10.3847/1538-4357/836/1/14. 1701.04272
J.R. Shaw, K. Sigurdson, U.L. Pen, A. Stebbins, M. Sitwell, All-sky interferometry with spherical harmonic transit telescopes. Astrophys. J. (2013). https://doi.org/10.1088/0004-637X/781/2/57. 1302.0327
J.R. Shaw, K. Sigurdson, M. Sitwell, A. Stebbins, U.L. Pen, Coaxing cosmic 21 cm fluctuations from the polarized sky using m-mode analysis. Phys. Rev. D **083**, 514 (2015). https://doi.org/10.1103/PhysRevD.91.083514. 1401.2095

D. Sluse, A. Sonnenfeld, N. Rumbaugh, C.E. Rusu, C.D. Fassnacht, T. Treu, S.H. Suyu, K.C. Wong, M.W. Auger, V. Bonvin, T. Collett, F. Courbin, S. Hilbert, L.V.E. Koopmans, P.J. Marshall, G. Meylan, C. Spiniello, M. Tewes, H0LiCOW—II. Spectroscopic survey and galaxy-group identification of the strong gravitational lens system HE 0435-1223. Mon. Not. R. Astron. Soc. **470**, 4838–4857 (2017). https://doi.org/10.1093/mnras/stx1484. 1607.00382

D. Spergel, N. Gehrels, J. Breckinridge, M. Donahue, A. Dressler, B.S. Gaudi, T. Greene, O. Guyon, C. Hirata, J. Kalirai, N.J. Kasdin, W. Moos, S. Perlmutter, M. Postman, B. Rauscher, J. Rhodes, Y. Wang, D. Weinberg, J. Centrella, W. Traub, C. Baltay, J. Colbert, D. Bennett, A. Kiessling, B. Macintosh, J. Merten, M. Mortonson, M. Penny, E. Rozo, D. Savransky, Y. Stapelfeldt, K. Zu, C. Baker, E. Cheng, D. Content, J. Dooley, M. Foote, R. Goullioud, K. Grady, C. Jackson, J. Kruk, M. Levine, M. Melton, C. Peddie, J. Ruffa, S. Shaklan, Wide-Field InfraRed Survey Telescope-Astrophysics Focused Telescope Assets WFIRST-AFTA final report (2013). ArXiv e-prints 1305.5422

D. Spergel, N. Gehrels, C. Baltay, D. Bennett, J. Breckinridge, M. Donahue, A. Dressler, B.S. Gaudi, T. Greene, O. Guyon, C. Hirata, J. Kalirai, N.J. Kasdin, B. Macintosh, W. Moos, S. Perlmutter, M. Postman, B. Rauscher, J. Rhodes, Y. Wang, D. Weinberg, D. Benford, M. Hudson, W.S. Jeong, Y. Mellier, W. Traub, T. Yamada, P. Capak, J. Colbert, D. Masters, M. Penny, D. Savransky, D. Stern, N. Zimmerman, R. Barry, L. Bartusek, K. Carpenter, E. Cheng, D. Content, F. Dekens, R. Demers, K. Grady, C. Jackson, G. Kuan, J. Kruk, M. Melton, B. Nemati, B. Parvin, I. Poberezhskiy, C. Peddie, J. Ruffa, J.K. Wallace, A. Whipple, E. Wollack, F. Zhao, Wide-Field InfrarRed Survey Telescope-Astrophysics Focused Telescope Assets WFIRST-AFTA 2015 report (2015). ArXiv e-prints 1503.03757

S.H. Suyu, Cosmography from two-image lens systems: overcoming the lens profile slope degeneracy. (2012). ArXiv e-prints 1202.0287

S.H. Suyu, A. Halkola, The halos of satellite galaxies: the companion of the massive elliptical lens SL2S J08544-0121. Astron. Astrophys. **524**, A94 (2010). https://doi.org/10.1051/0004-6361/201015481. 1007.4815

S.H. Suyu, P.J. Marshall, M.P. Hobson, R.D. Blandford, A Bayesian analysis of regularized source inversions in gravitational lensing. Mon. Not. R. Astron. Soc. **371**, 983–998 (2006). https://doi.org/10.1111/j.1365-2966.2006.10733.x. astro-ph/0601493

S.H. Suyu, P.J. Marshall, R.D. Blandford, C.D. Fassnacht, L.V.E. Koopmans, J.P. McKean, T. Treu, Dissecting the gravitational lens B1608+656. I. Lens potential reconstruction. Astrophys. J. **691**, 277–298 (2009). https://doi.org/10.1088/0004-637X/691/1/277. 0804.2827

S.H. Suyu, P.J. Marshall, M.W. Auger, S. Hilbert, R.D. Blandford, L.V.E. Koopmans, C.D. Fassnacht, T. Treu, Dissecting the gravitational lens B1608+656. II. Precision measurements of the Hubble constant, spatial curvature, and the dark energy equation of state. Astrophys. J. **711**, 201–221 (2010). https://doi.org/10.1088/0004-637X/711/1/201. 0910.2773

S.H. Suyu, S.W. Hensel, J.P. McKean, C.D. Fassnacht, T. Treu, A. Halkola, M. Norbury, N. Jackson, P. Schneider, D. Thompson, M.W. Auger, L.V.E. Koopmans, K. Matthews, Disentangling baryons and dark matter in the spiral gravitational lens B1933+503. Astrophys. J. **750**, 10 (2012). https://doi.org/10.1088/0004-637X/750/1/10. 1110.2536

S.H. Suyu, M.W. Auger, S. Hilbert, P.J. Marshall, M. Tewes, T. Treu, C.D. Fassnacht, L.V.E. Koopmans, D. Sluse, R.D. Blandford, F. Courbin, G. Meylan, Two accurate time-delay distances from strong lensing: implications for cosmology. Astrophys. J. **766**, 70 (2013). https://doi.org/10.1088/0004-637X/766/2/70. 1208.6010

S.H. Suyu, T. Treu, S. Hilbert, A. Sonnenfeld, M.W. Auger, R.D. Blandford, T. Collett, F. Courbin, C.D. Fassnacht, L.V.E. Koopmans, P.J. Marshall, G. Meylan, C. Spiniello, M. Tewes, Cosmology from gravitational lens time delays and Planck data. Astrophys. J. Lett. **788**, L35 (2014). https://doi.org/10.1088/2041-8205/788/2/L35. 1306.4732

S.H. Suyu, V. Bonvin, F. Courbin, C.D. Fassnacht, C.E. Rusu, D. Sluse, T. Treu, K.C. Wong, M.W. Auger, X. Ding, S. Hilbert, P.J. Marshall, N. Rumbaugh, A. Sonnenfeld, M. Tewes, O. Tihhonova, A. Agnello, R.D. Blandford, G.C.F. Chen, T. Collett, L.V.E. Koopmans, K. Liao, G. Meylan, C. Spiniello, H0LiCOW—I. H_0 lenses in COSMOGRAIL's wellspring: program overview. Mon. Not. R. Astron. Soc. **468**, 2590–2604 (2017). https://doi.org/10.1093/mnras/stx483. 1607.00017

E.R. Switzer, K.W. Masui, K. Bandura, L.M. Calin, T.C. Chang, X.L. Chen, Y.C. Li, Y.W. Liao, A. Natarajan, U.L. Pen, J.B. Peterson, J.R. Shaw, T.C. Voytek, Determination of $z \sim 0.8$ neutral hydrogen fluctuations using the 21 cm intensity mapping autocorrelation. Mon. Not. R. Astron. Soc. **434**, L46–L50 (2013). https://doi.org/10.1093/mnrasl/slt074. 1304.3712

A.S. Tagore, C.R. Keeton, Statistical and systematic uncertainties in pixel-based source reconstruction algorithms for gravitational lensing. Mon. Not. R. Astron. Soc. **445**, 694–710 (2014). https://doi.org/10.1093/mnras/stu1671. 1408.6297

M. Takada, R.S. Ellis, M. Chiba, J.E. Greene, H. Aihara, N. Arimoto, K. Bundy, J. Cohen, O. Doré, G. Graves, J.E. Gunn, T. Heckman, C.M. Hirata, P. Ho, J.P. Kneib, O.L. Fèvre, L. Lin, S. More, H. Murayama,

T. Nagao, M. Ouchi, M. Seiffert, J.D. Silverman, L. Sodré, D.N. Spergel, M.A. Strauss, H. Sugai, Y. Suto, H. Takami, R. Wyse, Extragalactic science, cosmology, and galactic archaeology with the Subaru prime focus spectrograph. Publ. Astron. Soc. Jpn. **66**, R1 (2014). https://doi.org/10.1093/pasj/pst019. 1206.0737

M. Tegmark, M. Zaldarriaga, Fast Fourier transform telescope. Phys. Rev. D **79**(8), 083530 (2009). https://doi.org/10.1103/PhysRevD.79.083530. 0805.4414

M. Tegmark, M. Zaldarriaga, Omniscopes: large area telescope arrays with only NlogN computational cost. Phys. Rev. D **82**(10), 103501 (2010). https://doi.org/10.1103/PhysRevD.82.103501. 0909.0001

M. Tewes, F. Courbin, G. Meylan, COSMOGRAIL: the COSmological MOnitoring of GRAvItational lenses. XI. Techniques for time delay measurement in presence of microlensing. Astron. Astrophys. **553**, A120 (2013a). https://doi.org/10.1051/0004-6361/201220123. 1208.5598

M. Tewes, F. Courbin, G. Meylan, C.S. Kochanek, E. Eulaers, N. Cantale, A.M. Mosquera, P. Magain, H. Van Winckel, D. Sluse, G. Cataldi, D. Vörös, S. Dye, COSMOGRAIL: the COSmological MOnitoring of GRAvItational lenses. XIII. Time delays and 9-yr optical monitoring of the lensed quasar RX J1131-1231. Astron. Astrophys. **556**, A22 (2013b). https://doi.org/10.1051/0004-6361/201220352. 1208.6009

S.S. Tie, C.S. Kochanek, Microlensing makes lensed quasar time delays significantly time variable. Mon. Not. R. Astron. Soc. **473**, 80–90 (2018). https://doi.org/10.1093/mnras/stx2348. 1707.01908

O. Tihhonova, F. Courbin, D. Harvey, S. Hilbert, C.E. Rusu, C.D. Fassnacht, V. Bonvin, P.J. Marshall, G. Meylan, D. Sluse, S.H. Suyu, T. Treu, K.C. Wong, H0LiCOW VIII. A weak lensing measurement of the external convergence in the field of the lensed quasar HE 0435-1223 (2017). ArXiv e-prints 1711.08804

T. Treu, G. Brammer, J.M. Diego, C. Grillo, P.L. Kelly, M. Oguri, S.A. Rodney, P. Rosati, K. Sharon, A. Zitrin, I. Balestra, M. Bradač, T. Broadhurst, G.B. Caminha, A. Halkola, A. Hoag, M. Ishigaki, T.L. Johnson, W. Karman, R. Kawamata, A. Mercurio, K.B. Schmidt, L.G. Strolger, S.H. Suyu, A.V. Filippenko, R.J. Foley, S.W. Jha, B. Patel, "Refsdal" meets popper: comparing predictions of the re-appearance of the multiply imaged supernova behind MACSJ1149.5+2223. Astrophys. J. **817**, 60 (2016). https://doi.org/10.3847/0004-637X/817/1/60. 1510.05750

C. Vanderriest, J. Schneider, G. Herpe, M. Chevreton, M. Moles, G. Wlerick, The value of the time delay Delta t(A, B) for the 'double' quasar 0957+561 from optical photometric monitoring. Astron. Astrophys. **215**, 1–13 (1989)

S. Vegetti, L.V.E. Koopmans, Bayesian strong gravitational-lens modelling on adaptive grids: objective detection of mass substructure in galaxies. Mon. Not. R. Astron. Soc. **392**, 945–963 (2009). https://doi.org/10.1111/j.1365-2966.2008.14005.x. 0805.0201

C. Vuissoz, F. Courbin, D. Sluse, G. Meylan, M. Ibrahimov, I. Asfandiyarov, E. Stoops, A. Eigenbrod, L. Le Guillou, H. van Winckel, P. Magain, COSMOGRAIL: the COSmological MOnitoring of GRAv-Itational Lenses. V. The time delay in SDSS J1650+4251. Astron. Astrophys. **464**, 845–851 (2007). https://doi.org/10.1051/0004-6361:20065823. astro-ph/0606317

C. Vuissoz, F. Courbin, D. Sluse, G. Meylan, V. Chantry, E. Eulaers, C. Morgan, M.E. Eyler, C.S. Kochanek, J. Coles, P. Saha, P. Magain, E.E. Falco, COSMOGRAIL: the COSmological MOnitoring of GRAvItational Lenses. VII. Time delays and the Hubble constant from WFI J2033-4723. Astron. Astrophys. **488**, 481–490 (2008). https://doi.org/10.1051/0004-6361:200809866. 0803.4015

S. Wallington, C.S. Kochanek, R. Narayan, LensMEM: a gravitational lens inversion algorithm using the maximum entropy method. Astrophys. J. **465**, 64 (1996). https://doi.org/10.1086/177401

D. Walsh, R.F. Carswell, R.J. Weymann, 0957 + 561 A, B—twin quasistellar objects or gravitational lens. Nature **279**, 381–384 (1979). https://doi.org/10.1038/279381a0

S.J. Warren, S. Dye, Semilinear gravitational lens inversion. Astrophys. J. **590**, 673–682 (2003). https://doi.org/10.1086/375132

R.B. Wayth, R.L. Webster, LENSVIEW: software for modelling resolved gravitational lens images. Mon. Not. R. Astron. Soc. **372**, 1187–1207 (2006). https://doi.org/10.1111/j.1365-2966.2006.10922.x. astro-ph/0609542

D.H. Weinberg, M.J. Mortonson, D.J. Eisenstein, C. Hirata, A.G. Riess, E. Rozo, Observational probes of cosmic acceleration. Phys. Rep. **530**, 87–255 (2013). https://doi.org/10.1016/j.physrep.2013.05.001. 1201.2434

L.L.R. Williams, P. Saha, Pixelated lenses and H_0 from time-delay quasars. Astron. J. **119**, 439–450 (2000). https://doi.org/10.1086/301234

K.C. Wong, S.H. Suyu, M.W. Auger, V. Bonvin, F. Courbin, C.D. Fassnacht, A. Halkola, C.E. Rusu, D. Sluse, A. Sonnenfeld, T. Treu, T.E. Collett, S. Hilbert, L.V.E. Koopmans, P.J. Marshall, N. Rumbaugh, H0LiCOW—IV. Lens mass model of HE 0435-1223 and blind measurement of its time-delay distance for cosmology. Mon. Not. R. Astron. Soc. **465**, 4895–4913 (2017). https://doi.org/10.1093/mnras/stw3077. 1607.01403

O. Wucknitz, Degeneracies and scaling relations in general power-law models for gravitational lenses. Mon. Not. R. Astron. Soc. **332**, 951–961 (2002). https://doi.org/10.1046/j.1365-8711.2002.05426.x. astro-ph/0202376

S. Wyithe, A. Loeb, P. Geil, Baryonic acoustic oscillations in 21 cm emission: a probe of dark energy out to high redshifts (2007). ArXiv e-prints 0709.2955

S. Wyithe, A. Loeb, P. Geil, Baryonic acoustic oscillations in 21 cm emission: a probe of dark energy out to high redshifts. Mon. Not. R. Astron. Soc. **383**, 1195–1209 (2008). 0709.2955

Y. Xu, X. Wang, X. Chen, Forecasts on the dark energy and primordial non-gaussianity observations with the Tianlai cylinder array. Astrophys. J. **798**(1), 40 (2015). https://doi.org/10.1088/0004-637X/798/1/40. 1410.7794

D.A. Yahalomi, P.L. Schechter, J. Wambsganss, A quadruply lensed SN Ia: gaining a time-delay...losing a standard candle (2017). ArXiv e-prints 1711.07919

P. Young, R.S. Deverill, J.E. Gunn, J.A. Westphal, J. Kristian, The triple quasar Q1115+080A, B, C—a quintuple gravitational lens image. Astrophys. J. **244**, 723–735 (1981). https://doi.org/10.1086/158750

M.A. Zwaan, P.G. van Dokkum, M.A.W. Verheijen, Hydrogen 21-centimeter emission from a galaxy at cosmological distance. Science **293**, 1800–1803 (2001). https://doi.org/10.1126/science.1063034. astro-ph/0109108

Space Sci Rev (2017) 212:1743–1785
DOI 10.1007/s11214-017-0395-z

Toward an Internally Consistent Astronomical Distance Scale

Richard de Grijs[1,2] · Frédéric Courbin[3] · Clara E. Martínez-Vázquez[4,5] · Matteo Monelli[4,5] · Masamune Oguri[6] · Sherry H. Suyu[7,8,9]

Received: 7 April 2017 / Accepted: 26 June 2017 / Published online: 22 August 2017

Abstract Accurate astronomical distance determination is crucial for all fields in astrophysics, from Galactic to cosmological scales. Despite, or perhaps because of, significant efforts to determine accurate distances, using a wide range of methods, tracers, and techniques, an internally consistent astronomical distance framework has not yet been established. We review current efforts to homogenize the Local Group's distance framework, with particular emphasis on the potential of RR Lyrae stars as distance indicators, and attempt to extend this in an internally consistent manner to cosmological distances. Calibration based on Type Ia supernovae and distance determinations based on gravitational lensing repre-

Astronomical Distance Determination in the Space Age
Edited by Richard de Grijs and Maurizio Falanga

✉ R. de Grijs
grijs@pku.edu.cn

F. Courbin
frederic.courbin@epfl.ch

C.E. Martínez-Vázquez
clara.marvaz@gmail.com

M. Monelli
monelli@iac.es

M. Oguri
masamune.oguri@ipmu.jp

S.H. Suyu
suyu@mpa-garching.mpg.de

1 Kavli Institute for Astronomy & Astrophysics and Department of Astronomy, Peking University, Yi He Yuan Lu 5, Hai Dian District, Beijing 100871, China

2 International Space Science Institute—Beijing, 1 Nanertiao, Zhongguancun, Hai Dian District, Beijing 100190, China

3 Institute of Physics, Laboratoire d'Astrophysique, Observatoire de Sauverny, Ecole Polytechnique Fédérale de Lausanne (EPFL), 1290 Versoix, Switzerland

4 Instituto de Astrofísica de Canarias, Calle Vía Láctea sn, 38205 La Laguna, Spain

5 Dpto. Astrofísica, Universidad de La Laguna (ULL), 38206 La Laguna, Tenerife, Spain

sent particularly promising approaches. We provide a positive outlook to improvements to the status quo expected from future surveys, missions, and facilities. Astronomical distance determination has clearly reached maturity and near-consistency.

Keywords Gravitational lensing: strong · Stars: distances · Stars: variables: RR Lyrae · Galaxy: center · Galaxy: fundamental parameters · Galaxies: distances and redshifts · Local Group · Magellanic Clouds · Distance scale

1 Introduction

Distance determination from the nearest stars to the edge of the observable Universe crucially depends on accurate calibration of every successive rung of the so-called astronomical 'distance ladder' (for a modern version of the distance ladder, see de Grijs 2013). Local benchmark objects often used to optimally constrain the uncertainties inherent to using a stepwise calibration approach include the distances to the Galactic Center and a number of well-studied galaxies in the Local Group.

R_0, the distance from the solar circle to the Galactic Center is particularly instrumental to achieve a basic calibration for a wide range of methods used for distance determination, across an extensive range of distance scales. Among other key physical parameters, the distances, masses, and luminosities of Galactic objects, as well as the Milky Way's total mass and luminosity, depend directly on knowing R_0 accurately. For instance, most luminosity and a large number of mass estimators scale as R_0^2, whereas masses based on total densities or orbital modeling scale as R_0^3.

This intrinsic dependence therefore often requires adoption of an integrated mass and/or a rotation model for the Milky Way, which in turn would also require us to accurately know the Sun's circular velocity, Θ_0. As R_0 estimates are refined, so are the estimated distances, masses, and luminosities of numerous Galactic and extragalactic objects, as well as our best estimates of the rate of Galactic rotation and the size of the Milky Way. In addition, a highly accurate direct Galactic Center distance determination is paramount for a reliable recalibration of the zero points of numerous commonly used distance calibrators, including of Cepheids, RR Lyrae (see below), and Mira variable stars.

Beyond Galactic distance tracers, distance measurements to the Large Magellanic Cloud (LMC) have played an important role in constraining the value of the Hubble constant, H_0. The '*Hubble Space Telescope (HST)* Key Project (HSTKP) on the Extragalactic Distance Scale' (Freedman et al. 2001) resulted in an H_0 estimate of $H_0 = 72 \pm 3$ (statistical) $\pm$ 7 (systematic) $\mathrm{km\,s^{-1}\,Mpc^{-1}}$. The latter, systematic uncertainty was thought to be dominated by the remaining systematic uncertainties in the assumed distance to the LMC (Freedman et al. 2001; Schaefer 2008; Pietrzyński et al. 2013), which contributed of order $\pm$3–4 $\mathrm{km\,s^{-1}\,Mpc^{-1}}$ to the tally.

On a number of occasions, the accuracy of LMC distance determinations has been called into question by claims of 'publication bias' (e.g., Schaefer 2008, 2013; Rubele et al. 2012;

[6] Department of Physics, Graduate School of Science, University of Tokyo, 7-3-1 Hongo, Bunkyo-ku, Tokyo 113-0033, Japan

[7] Max-Planck-Institut für Astrophysik, Karl-Schwarzschild-Str. 1, 85748 Garching, Germany

[8] Institute of Astronomy and Astrophysics, Academia Sinica, P.O. Box 23-141, Taipei 10617, Taiwan

[9] Physik-Department, Technische Universität München, James-Franck-Straße 1, 85748 Garching, Germany

Table 1 Internally consistent distance moduli (as a function of increasing distance) to selected Local Group galaxies, comprising a robust local framework (adapted from de Grijs and Bono 2015)

Galaxy	$(m-M)_0^{\rm best}$ (mag)	Tracer(s)	$(m-M)_0^{\rm TRGB}$ (mag)
LMC	18.49 ± 0.09	Cepheids, RR Lyrae, CMD	18.54–18.69
SMC	18.96 ± 0.02	EBs, Cepheids, RR Lyrae, TRGB, RC	19.00 ± 0.04
NGC 185	24.00 ± 0.12	TRGB, RR Lyrae	24.03 ± 0.33
NGC 147	24.11 ± 0.11	TRGB, RR Lyrae	24.16 ± 0.22
IC 1613	24.34 ± 0.05	Cepheids, RR Lyrae, TRGB	24.29 ± 0.12
IC 10	24.36 ± 0.45	TRGB	24.36 ± 0.45
M32	24.43 ± 0.07	SBF, TRGB, RR Lyrae	24.32 ± 0.20
M31	24.45 ± 0.10	Cepheids, RR Lyrae, TRGB	24.47 ± 0.01
NGC 205	24.56 ± 0.15	TRGB, RR Lyrae	24.45 ± 0.20
M33	24.67 ± 0.07	Cepheids, RR Lyrae, TRGB	24.70 ± 0.11
NGC 4258	29.29 ± 0.08	H_2O masers	29.24–29.44

Notes: CMD: color–magnitude diagram; EBs: eclipsing binaries; RC: red clump; SBF: surface-brightness fluctuations; TRGB: tip of the red giant branch

Walker 2012). Therefore, in de Grijs et al. (2014) we re-analyzed the full body of LMC distance measurements published between 1990 and 2013, embarking on the most extensive data-mining effort of this type done to date. Perhaps somewhat surprisingly, we concluded that strong publication bias is unlikely to have been the main driver of the clustering of many published LMC distance moduli. However, we found that many of the published values were based on highly non-independent tracer samples and analysis methods. In turn, this interdependence appears to have led to significant correlations among the body of LMC distances we considered in de Grijs et al. (2014). Our final, recommended true distance modulus for the LMC is $(m-M)_0 = 18.49 \pm 0.09$ mag (de Grijs et al. 2014, see also Crandall and Ratra 2015). In an effort to provide a firm mean distance estimate to the Small Magellanic Cloud (SMC), and thus place it within the internally consistent Local Group distance framework (that is, a distance framework where all combinations of distances lead to the same underlying scaling), de Grijs and Bono (2015) similarly performed extensive analysis of the published literature to compile the largest database available to date containing SMC distance estimates. For the first time, we provided estimates of the mean SMC distance based on large numbers of distance tracers, without imposing any a priori preferences. We derived a true distance modulus of $(m-M)_0^{\rm SMC} = 18.96 \pm 0.02$ mag, corresponding to a distance of $61.9^{+0.6}_{-0.5}$ kpc.

Indeed, the nearest galaxies in the Local Group contain numerous objects that can be used to determine robust distances, that is, distances with well-defined uncertainties that are mutually consistent among the different tracers. In de Grijs et al. (2014) and de Grijs and Bono (2014, 2015, 2016), we aimed at establishing an internally consistent local distance framework by reference to the subset of distances to the Galactic Center, the LMC and SMC, M31, M32, and M33, as well as a number of well-known dwarf galaxies. We aimed at reaching consensus on the best, most homogeneous, and internally most consistent set of Local Group distance moduli.

At the present time, we are in a good position to make recommendations for the use of robust distance measurements to a set of key Local Group galaxies: see Table 1 and Fig. 1.

In this review, we will first consider the distances to the Galactic Center as well as to a select number of Local Group galaxies resulting from large-scale data mining of the literature

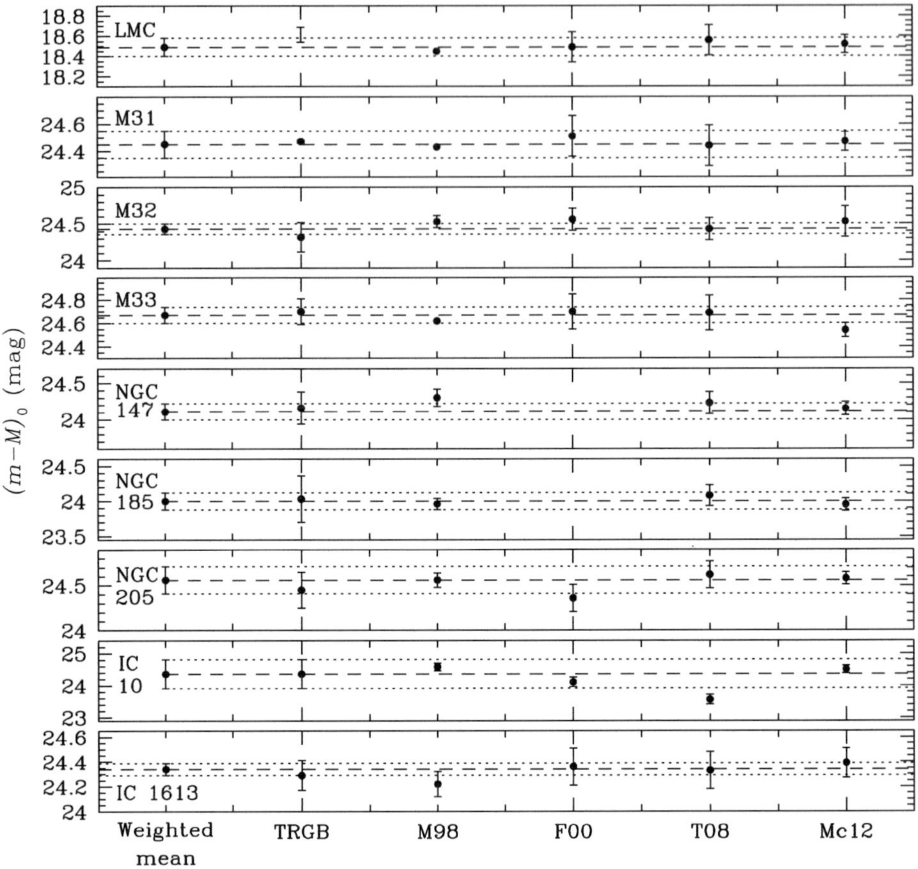

Fig. 1 Comparison of our set of benchmark distances to the sample of Local Group galaxies (indicated in the individual panels) discussed in this article with those from a number of recent distance compilations. 'Weighted mean' and TRGB: de Grijs and Bono (2015), except where indicated in Table 1; M98: Mateo (1998); F00: Ferrarese et al. (2000); T08: Tammann et al. (2008); Mc12: McConnachie (2012). From de Grijs and Bono (2015)

(Sects. 1–3). Our aim is to establish an internally consistent distance framework in the Local Group using multiple, independent methods. We will then proceed to review the distances to Local Group galaxies based specifically on space-based data of variable stars, with emphasis on the uniformity in distance measurement across the nearby galaxies afforded by RR Lyrae stars (Sects. 4–6). Local Group RR Lyrae represent a powerful tool, because they are found in old stellar populations that are present in galaxies of any Hubble type and luminosity class (with the possible exception of the ultra-faint dwarf galaxies). We will discuss a large number of case studies aimed at their internally consistent calibration as a single, homogenized distance indicator well beyond the Local Group. At greater distances, we focus on efforts to obtain improved distances and particularly on a number of alternate techniques to constraining the Hubble constant (Sect. 7), with particular emphasis on gravitational lensing as a promising tool (Sect. 8). The resulting distances are not cross-calibrations of other distance indicators; instead, they encompass a derived quantity—the Hubble law—from which distances are computed.

We emphasize that an internally consistent distance scale exists naturally, for a given cosmology, in the Hubble flow owing to the relationship between distance and redshift. On these cosmological scales, the key to consistency is therefore not distance as such, but re-enforcing the cosmological parameters from which these distances are derived. At the present time, these cosmological parameters are predominantly derived on the basis of observations of Type Ia supernovae (SNe Ia) and the cosmic microwave background. Our discussion in Sects. 7 and 8 focuses on alternative means of measuring the Hubble constant, not necessarily on the objects' specific distances. Therefore, gravitational lensing and galaxy cluster-based techniques are not direct cross-checks on the SNe Ia-derived distances, but on the Hubble constant measured from SNe Ia. The latter distance framework is currently calibrated on the basis of Cepheid distances, which are directly tied into the internally consistent distance scales discussed in Sects. 1–3.

2 Distance to the Galactic Center

In recent years various authors have speculated that some degree of publication bias may have affected Galactic Center distance determinations (e.g., Reid 1989, 1993; Nikiforov 2004; Foster and Cooper 2010; Malkin 2013a,b; Francis and Anderson 2014). Aiming at deriving a statistically well-justified Galactic Center distance based on a large variety of tracers and reducing any occurrence of publication bias, de Grijs and Bono (2016) therefore undertook an extensive data-mining effort of the literature.

They separated the compilation of R_0 determinations into direct and indirect distance measurements. The former include distances such as those based on orbital modeling of the 'S stars' orbiting Sgr A*, the closest visual counterpart of the Milky Way's central supermassive black hole, as well as those relying on statistical parallaxes of Galactic Center tracers. Careful assessment of the body of published R_0 estimates based on these methods resulted in a Galactic Center distance recommendation of $R_0 = 8.3 \pm 0.2$ (stat.) $\pm\, 0.4$ (syst.) kpc.

A much larger body of Galactic Center distance determinations is based on indirect methods, either relying on determinations of the centroids defined by the three-dimensional distributions of a range of different tracer populations (e.g., globular clusters, Cepheid, RR Lyrae, or Mira variables, or red clump stars) or measurements based on kinematic observations of objects at the solar circle. The latter approaches are affected by significantly larger uncertainties than the former; the central, mean Galactic Center distances based on the kinematic methods are systematically smaller than those based on centroid determinations. Most centroid-based distances are in good agreement with those resulting from the direct methods. de Grijs and Bono (2016) did not find any conclusive evidence of the presence of publication bias in the post-1990 R_0 measurements.

3 Internally Consistent Distances to the Magellanic Clouds

As the nearest, irregular galaxies to our Milky Way, the Magellanic Clouds represent the first, important rung of the extragalactic distance ladder. They cover sizeable areas on the sky, thus providing access to large samples of tracer objects. For instance, the VISTA Survey of the Magellanic Clouds (Cioni et al. 2011), which covers the LMC, the SMC, the Magellanic Bridge, and a few fields in the Magellanic Stream, is in currently the process of compiling a 184 deg^2 mosaicked map of the entire system.

Indeed, the LMC is the nearest extragalactic environment that hosts statistically significant samples of the tracer populations for distance determination, such as Cepheid and RR Lyrae variable stars, eclipsing binaries (EBs), and red-giant-branch (RGB) stars, as well as supernova (SN) 1987A. These could thus potentially link the local (i.e., solar-neighborhood and Galactic) tracers to their counterparts in more distant and more poorly resolved galaxies. At a distance of approximately 50 kpc, the LMC represents the only well-studied environment linking Galactic distance tracers to those in other large spiral and elliptical galaxies at greater distances. Its three-dimensional, thin disk-like morphology renders it an ideal target galaxy where off-center distances can easily be projected onto the galaxy's center through simple projection.

Lingering systematic uncertainties remain in the distance to the LMC. This has obvious ramifications in the context of using the LMC distance as a calibrator, specifically to reduce the uncertainties in the Hubble constant (cf. Freedman et al. 2001; Schaefer 2008; Pietrzyński et al. 2013). Although this has also led to persistent claims of 'publication bias' affecting published distances to the galaxy (cf. Schaefer 2008, 2013; Rubele et al. 2012; Walker 2012), de Grijs et al. (2014) made a case of highly correlated results rather than publication bias. Nevertheless, the LMC distance moduli in common use today are likely still affected by poorly understood systematic uncertainties.

Fortunately, the LMC is sufficiently close that some geometric distance tracers are readily available, including an increasing sample of EB systems. Highly promising, Pietrzyński et al. (2013) determined the direct distances to eight late-type EB systems in the LMC, resulting in an average distance modulus of $(m - M)_0 = 18.493$ stat. $\pm\, 0.008 \pm 0.047$ (syst.) mag. This value is accurate to 2.2% and confirms that the statistically justified distance modulus of $(m - M)_0 = 18.49 \pm 0.09$ mag is indeed reasonable. Pietrzyński et al. (2013) discovered giant stars in EB systems, which uniquely allowed them to use the very well calibrated surface-brightness–$(V - K)$ color relation for such stars to determine their angular sizes, so that their error estimates are more robust and reproducible, and less affected by lingering systematic effects than competing results based on early-type EBs. Some of the most recent analyses, e.g., Inno et al. (2016)'s study of the three-dimensional distribution of classical Cepheids in the LMC's disk, have derived true distance moduli that are in excellent agreement with both the statistical average and the geometric distance determination of Pietrzyński et al. (2013).

In this context it is important to note that Riess et al. (2016) recently used Cepheid distances to independently obtain an internally consistent distance framework including the Milky Way, the LMC, M31, and the maser host galaxy NGC 4258. This underscores the importance of the LMC as the lowest rung of the extragalactic distance ladder, and this also links the local distance framework naturally to our discussion of derivations of the Hubble constant in Sect. 7 and beyond, particularly those based on SNe Ia observations.

de Grijs and Bono (2015) performed a similar analysis for the SMC as they did for the LMC. Obtaining a clear-cut distance to the SMC is not as straightforward as for the LMC, however. First, given the SMC's bar-like main body with hints of spiral arms and a very extended 'Wing' to the East, defining the position of the galaxy's center is troublesome. Rubele et al. (2015) recently derived distances to different areas across the galaxy. They report distances projected onto both the SMC's kinematic and stellar density centers, $(m - M)_0^{\rm kin} = 18.97 \pm 0.01$ mag and $(m - M)_0^{\rm stars} = 18.91 \pm 0.02$ mag, respectively. This implies that the choice of SMC center could introduce systematic uncertainties of order 0.05–0.1 mag in the resulting distance modulus.

Second, the SMC is very extended along the line of sight. Its depth could range from 6–12 kpc (Crowl et al. 2001) up to 20 kpc (Groenewegen 2000, for recent discussions, see e.g., Kapakos and Hatzidimitriou 2012; Subramanian and Subramaniam 2012; Cignoni et al. 2013; Kalirai et al. 2013; Nidever et al. 2013; Scowcroft et al. 2016), although the Cepheid population associated with the main body implies a shallower depth of 1.76 ± 0.6 kpc (Subramanian and Subramaniam 2015). Scowcroft et al. (2016) recently also used Cepheids to study the SMC's line-of-sight depth, but based on observations in nine photometric passbands simultaneously, including four near- to mid-infrared *Spitzer Space Telescope* filters. They confirmed that the Cepheid distribution is inconsistent with a significant depth, although they pointed out that the southwestern side of the galaxy is up to 20 kpc more distant than the SMC's northeast.

Third, the SMC's measured inclination appears to depend on the stellar tracer (and thus the age of the stellar population) used (Caldwell and Coulson 1986; Laney and Stobie 1986; Groenewegen 2000; Haschke et al. 2012; Subramanian and Subramaniam 2012; Rubele et al. 2015). Careful geometric corrections of individual objects back to the galaxy's center will reduce the scatter in the calibration relations, but this is not always possible. A warped disk (the presence of which is still debated) will introduce additional systematic uncertainties, of order 0.1 mag in the distance modulus (de Grijs and Bono 2015).

de Grijs (2011, his Chap. 6.1.1) provides a detailed discussion of the systematic uncertainties associated with the effects of extinction in the context of distance determinations. These include uncertainties related to our knowledge of (i) the most likely extinction law, (ii) the intrinsic photometric properties of one's calibration objects, and (iii) the dust geometry. The choice of extinction law is particularly important when comparing similar types of objects drawn from Galactic and Magellanic Cloud samples, since 'the' Galactic extinction law (which may vary along different lines of sight) differs systematically from that in the Magellanic Clouds (for recent studies, see e.g., Dobashi et al. 2009; Bot et al. 2010, and references therein). Nevertheless, the differences are generally $\lesssim 0.05$ mag at near-infrared and longer wavelengths and shortward of $\lambda = 0.8$ μm.

4 RR Lyrae Variables in the Local Group

RR Lyrae stars are low-mass (~ 0.6–$0.8 M_{\odot}$) core helium-burning stars in their horizontal-branch (HB) evolutionary phase. They are radially pulsating variable stars with periods ranging from 0.2 to 1.0 days and V-band amplitudes between 0.2 and $\lesssim 2$ mag. Their mean absolute V-band magnitude is $M_V \sim +0.6$ mag, they are of moderate brightness ($\approx 40 L_{\odot}$), and their effective mean temperatures range from 6000 to 7250 K (Catelan et al. 2004). RR Lyrae stars are found in stellar systems hosting old old stellar populations ($t > 10$ Gyr; Walker 1989; Smith 1995; Catelan and Smith 2015); they can currently be observed out to distances of ~ 2 Mpc (Da Costa et al. 2010; Yang et al. 2014). Additionally, RR Lyrae stars are primary distance indicators, since they obey well-defined optical/near-infrared period–luminosity relations (PLRs; see, e.g., Bono et al. 2001, 2003; Clementini et al. 2003; Catelan et al. 2004; Marconi et al. 2015, see for more details also R. Beaton et al., in prep.).

The Local Group is home two 'normal' spiral galaxies, the Milky Way and the Andromeda galaxy, M31, shepherded by a number of dwarf galaxies. Thanks to large photometric surveys, the number of known satellites of both systems has been increasing rapidly in the last few years (e.g. Willman et al. 2006; Martin et al. 2015, see also http://www.astro.uvic.ca/~alan/Nearby_Dwarf_Database.html). From an observational point of view, the properties of Local Group galaxies span the most diverse ranges (Mateo 1998; McConnachie 2012). In particular, observational approaches to characterize the population of RR Lyrae have to be adapted to the different systems. Faint nearby galaxies (often called 'ultra-faint dwarfs,' with $-8 \lesssim M_V \lesssim -1.5$ mag) are characterized by extremely low surface brightnesses. They typically represent concentrations of a few hundred stars in a relatively small projected volume covering a few square arcminutes on the sky. These objects typically have low mean metallicities ([Fe/H] ~ -2 dex), exhibit blue HB morphology, and host very few RR Lyrae stars (e.g. Vivas et al. 2016).

On the other hand, more massive 'classical' dwarf satellites ($-13 \lesssim M_V \lesssim -9$ mag) are very rich in RR Lyrae stars (Kinemuchi et al. 2008; Stetson et al. 2014; Coppola et al. 2015; Martínez-Vázquez et al. 2015), but they cover large areas on the sky, up to many square degrees, and require wide-field facilities on medium-sized, ground-based telescopes for follow-up studies. At present, a complete census and analysis of the population of variable stars, and of RR Lyrae stars in particular, is not yet available for any bright Local Group galaxy (with the possible exception of Leo I and the Carina dwarf spheroidal galaxy, Stetson et al. 2014; Coppola et al. 2015). A case in point in this context is illustrated by the monitoring of two of the most massive Milky Way satellites, the Magellanic Clouds, performed by the Optical Gravitational Lensing Experiment, which collected ~ 45000 RR Lyrae variables in the Magellanic System in their latest release (Soszyński et al. 2016).

Another crucial point to bear in mind is that Local Group galaxies represent an enormous variety of intrinsic properties, e.g., in terms of their sizes, morphologies, gas content, chemical evolution, and stellar populations. As a consequence, they host a variety of families of variable stars, whose properties reflect the evolutionary history of the host system. As an example, RR Lyrae stars carry direct information of the chemical properties of the environment in which they formed during the first stages of a galaxy's lifecycle. An intrinsic spread in metallicity in the population of RR Lyrae stars in a dwarf galaxy implies chemical enrichment which not only occurred at early times, but which was sufficiently rapid to be imprinted in the stellar population that we observe today as RR Lyrae stars (Bernard et al. 2008; Martínez-Vázquez et al. 2015). Moreover, the number of RR Lyrae stars in a galaxy depends on the level of star formation at early times. Leo A ($M_V = -12.1$ mag), which has very little early star-formation activity (Cole et al. 2007), hosts a handful of RR Lyrae stars (Bernard et al. 2013), while Sculptor ($M_V = -11.1$ mag), is a predominantly old system (de Boer et al. 2012) with a population of RR Lyrae stars that outnumbers that of Leo A by two orders of magnitude, despite the latter galaxy's lower mass (Martínez-Vázquez et al. 2016).

RR Lyrae stars are primary distance indicators, which can be detected beyond the limits of the Local Group (Da Costa et al. 2010; Yang et al. 2014; McQuinn et al. 2015). Different methods can be applied to derive distances based on the properties of RR Lyrae stars, including (i) the luminosity–metallicity relation (LMR) in the V band, (ii) PLRs and period–Wesenheit relations (PWRs), and (iii) the First Overtone Blue Edge (FOBE) method. Nevertheless, compiling a homogeneous set of distances to nearby galaxies using the properties of their variable stellar populations is not an easy task. Below, we summarize the main aspects which must be taken into account.

1. *Photometric calibration:* As regards data pertaining to nearby satellite galaxies, large numbers of data sets are available in the literature, collected over decades by different observers with different telescopes, equipped with different filters and cameras. This poses a fundamental problem to derive homogeneous photometry in a well-established system (Stetson 2000). Photometric measurements based on HST observations are typically provided in the natural VEGAMAG photometric system, and they may be recalibrated to the Johnson–Cousins system a posteriori. However, different conversion relations have been proposed (Bedin et al. 2004; Sirianni et al. 2005; Bernard et al. 2009), many of which are affected by significant color dependences.

2. *Data quality and derivation of the pulsation parameters:* Deriving good pulsation properties for variable stars requires adequate sampling of their light curves. Especially for existing *HST* data, this is not always the case: rarely more than 25 phase points are available. The parameters derived from sparsely populated light curves may depend on the approach used for the analysis, the fitting algorithm, and/or the template adopted. Moreover, long-term and continuous monitoring of variable stars has revealed complex behavior (including period doubling, the Blazhko effect, period changes, and non-radial modes, among others) which is impossible to detect and characterize on the basis of sparsely sampled observations.
3. *The effects of metallicity:* It is worth recalling that most relations used to derive distances to RR Lyrae stars require one to assume the relevant metallicity. The only notable exceptions (Marconi et al. 2015; Martínez-Vázquez et al. 2015) are the PWR in $(V, B-V)$ and $(V, B-I)$. Marconi et al. (2015, their Tables 7 and 8) performed an elaborate analysis of optical, optical–near-infrared, and near-infrared PWRs. They found that the coefficient of the metallicity term was less than ~ 0.05 dex for the $(V, B-V)$ and $(V, B-I)$ PWRs, indicating that the V-band metallicity effects are counteracted by the convolution of the reddening vector and the $(B-V)$ or $(B-I)$ colors. Since high-resolution spectroscopy for RR Lyrae stars in external galaxies is generally not available, other indicators or methods must be established to estimate the chemical composition of RR Lyrae stars. Typically, the metallicity distribution of bright red-giant-branch (RGB) stars is available from spectroscopic studies, most commonly low-resolution calcium triplet spectra. Since different authors may use different metallicity scales or calibrations, deriving homogeneous distances is hence not a trivial task.
Moreover, RGB stars cover an age range of $\gtrsim$1.5 Gyr, and the mix of stellar populations of different ages and metallicities is highly degenerate in color–magnitude space. The metallicity distribution derived from RGB stars is a good approximation of a galaxy's global population, but it does not properly describe the RR Lyrae stars or the old populations (> 10 Gyr). As a consequence, assuming a mean metallicity for the RR Lyrae stars that is too high would introduce a systematic error in distance-modulus estimates, at the level of ~ 0.2 mag. Therefore, assumptions have to be made regarding the RR Lyrae metallicity distribution, or alternatively this distribution must be determined based on other methods (Martínez-Vázquez et al. 2016).
4. *Calibration:* Despite its extensive use in the literature and its simple form, the LMR relation cannot yet boast a solid calibration, neither of its zero point nor of its slope. A number of different calibrations exist (Chaboyer 1999; Bono et al. 2003; Clementini et al. 2003; Carretta et al. 2009), and using one rather than another can introduce a difference on the order of 0.2 mag in the distance determination.
5. *Reddening:* Correction for interstellar extinction is a crucial step required to derive the absolute magnitudes of one's target stars—and in turn their distance moduli. Systems with significant internal reddening may be not adequately corrected by employing any of the widely used reddening maps (Schlegel et al. 1998; Schlafly and Finkbeiner 2011). Moreover, differential reddening introduces significant scatter in the mean magnitudes and colors of stars, which may strongly affect the analysis (e.g., see Sarajedini et al. 2006). To (at least partially) overcome this problem, one can use a Wesenheit pseudo-magnitude (van den Bergh 1968), which is reddening-free by construction (under assumption of a reddening law), or move to near-infrared wavelengths, where the effect of extinction is strongly reduced.

RR Lyrae-based distance estimates in the literature are, therefore, not homogeneous for a large number of reasons: use of different algorithms for the photometry or of different recipes for the photometric calibration, the existence of different calibrations for the same method of distance determination, metallicities which may be defined on a different scale and/or which may not be appropriate for the sample of RR Lyrae stars, and the assumed reddening, among others. Revision of existing data in an effort to provide a homogeneous distance framework is not a trivial task either. We will next revise the data and the distance estimates available for galaxies in the M31 system, for isolated galaxies, and even to some extent for galaxies outside the Local Group. We will subsequently proceed, using already available data, to provide a homogeneous RR Lyrae-based distance framework. In particular, we will rely on data for six M31 satellite galaxies, which we use as test cases to probe the reliability of different methods, and we will also provide new distance estimates to 16 dwarf galaxies based on *HST* data.

5 RR Lyrae in the Local Group and Beyond: Status

The search for RR Lyrae stars in the M31 system has long been limited because of (i) their (relatively) faint apparent magnitudes ($V \sim 25$ mag) and (ii) crowding effects. The first successful attempt to identify RR Lyrae stars in the M31 halo was achieved by Pritchet and van den Bergh (1987), using Canada–France–Hawaii Telescope data. A few years later, Saha and Hoessel (1990) and Saha et al. (1990) detected candidate RR Lyrae stars in the M31 dwarf elliptical (dE) satellites NGC 185 and NGC 147, respectively. Nevertheless, it was only thanks to the *HST* and its spatial resolution that it was possible to reach well below the HB and achieve the first solid determination of the properties of RR Lyrae stars in the M31 field and its satellites (e.g., Clementini et al. 2001).

5.1 M31 Dwarf Spheroidals

Based on *HST*/Wide Field and Planetary Camera-2 (WFPC2) data, in a series of papers devoted to the properties of the stellar populations in M31 dwarfs, discoveries of RR Lyrae stars were reported in And I (Da Costa et al. 1996), And II (Da Costa et al. 2000), and And III (Da Costa et al. 2002). The populations of variable stars detected in these three galaxies were later analyzed in detail by Pritzl et al. (2004, And II) and Pritzl et al. (2005a, And I and And III). Moreover, And VI was studied by Pritzl et al. (2002a) on the basis of data of comparable quality. The distances and distance moduli were derived for the four galaxies using the LMR relation of Lee (1990), assuming a mean metallicity derived from the mean color of the RGB.

Interestingly, following these pioneering studies, the number of known satellites of M31 has dramatically increased in the last 10 years, mostly thanks to the PAndAS survey (McConnachie et al. 2009). However, few surveys specifically include studies of variable stars. *HST*/WFCP2 data were used by Yang and Sarajedini (2012) to study two low-mass M31 satellites, And XI and And XIII, where they detected 17 and 9 bona fide RR Lyrae candidates, respectively. Their distance moduli were derived using Chaboyer (1999)'s calibration of the LMR relation. The RR Lyrae metallicity was derived using the Alcock et al. (2000) relation, and the mean of the distribution was used to derive the distance modulus.

Other recently discovered satellites have been studied using wide-field, ground-based Large Binocular Telescope data. The results have been presented in a series of papers which discuss the RR Lyrae stellar populations in And XIX (Cusano et al. 2013), And XXI (Cusano et al. 2015), and And XXV (Cusano et al. 2015). In these studies, the authors adopted a consistent method to derive the distances to the three galaxies, based on the LMR relation calibrated by Clementini et al. (2003). In these three cases, the adopted metallicity was based on low-resolution RGB spectroscopy (Collins et al. 2013), which is reasonable for the RR Lyrae stars in these galaxies as well, given the low mass of the three galaxies.

5.2 RR Lyrae Stars in the M31 Field

The availability of the *HST*/Advanced Camera for Surveys (ACS) resulted in a new and significant step forward. New *HST*/ACS data targeted two of the brightest M31 satellites, M33 (Sarajedini et al. 2006) and M32 (Sarajedini et al. 2012). The metallicity of individual RR Lyrae stars was derived using a relation between [Fe/H] and the period (Sarajedini et al. 2006) and a relation between [Fe/H], the period, and the amplitude (Alcock et al. 2000). Once the mean metallicity of the sample was known, the distance modulus was derived assuming the LMR as calibrated by Chaboyer (1999). A similar approach was used by Fiorentino et al. (2010) and Fiorentino et al. (2012a) for M32, who used the relation of Alcock et al. (2000) to derive the metallicity distribution of their M32 sample RR Lyrae stars. However, Fiorentino et al. (2010) used two different approaches to derive the distance modulus to the galaxy. In addition to the LMR relation (Cacciari and Clementini 2003), they also applied the FOBE method. Given the sizeable sample or RRc-type stars (first-overtone pulsators), they derived an independent distance modulus, in agreement with the other LMR method.

The most comprehensive analysis of RR Lyrae stars in M31 has been published by Jeffery et al. (2011), who analyzed different ACS fields in the galaxy's disk and inner halo, as well as a region overlapping with the Giant Stellar Stream (GSS) that is being accreted onto M31 (Ibata et al. 2001). Jeffery et al. (2011) performed a detailed analysis of the metallicity determination based on different methods, as well as their effect on the derived distance modulus. In particular, they compared the period–metallicity relation (Sarajedini et al. 2006), the period–amplitude–metallicity relation (Alcock et al. 2000), Bono et al. (2007)'s method based on theoretical period–amplitude–metallicity relations, and the period–metallicity relation for RRc-type stars (Sandage 1993). A comparison of the metallicities derived photometrically using the first three methods with respect of spectroscopic measurements reveals good agreement between the Alcock et al. (2000) and Bono et al. (2007) methods, although the latter exhibits a significantly smaller dispersion.

Overall, very little is known about the global properties of the RR Lyrae stars in M31. The *HST* data available only cover a tiny fraction of its body, and they are all concentrated in the inner halo (Ferguson and Mackey 2016). Moreover, since many satellite objects (such as NGC 205, the GSS, or And I) are embedded in, or projected along the line of sight, it is often difficult to unequivocally distinguish variables associated with either M31 or its satellite galaxies.

5.3 Isolated Dwarfs

At present, few Local Group galaxies are really isolated and located outside the virial radii of either the Milky Way or M31 (McConnachie 2012). Nevertheless, a handful (DDO 210, VV 124, Sagittarius dwarf irregular galaxy) are truly isolated systems in the sense that they are currently on their first approach of the innermost region of the Local Group, and they never experienced any strong interactions with either the Milky Way or M31. For some other galaxies, the situation is not as clear, because they are currently located outside the tidal radius of either major spiral galaxy, but we cannot exclude the possibility of past interactions (Fraternali et al. 2009). Detailed color–magnitude analysis reaching the old main-sequence turn-off of relatively and truly isolated Local Group dwarf galaxies requires use of the *HST*. The first systematic investigation of a representative sample of such galaxies has been performed by the LCID (Local Cosmology from Isolated Dwarfs) team (Gallart et al. 2015). Recently, more data have been acquired for DDO 210 (Cole et al. 2014), WLM, and Pegasus dwarf irregular (dIrr) galaxy (as yet unpublished). Studies of variable stars have been published by Bernard et al. (2009, Cetus and Tucana), Bernard et al. (2010, IC 1613), Bernard et al. (2013, Leo A), Clementini et al. (2012, Leo T), Ordoñez et al. (2014, Phoenix), and Ordoñez and Sarajedini (2016, DDO 210).

5.3.1 Purely Old Systems: Cetus and Tucana

Cetus and Tucana represent remarkable examples of isolated dwarf spheroidal (dSph) galaxies, since they are the only two gas-poor, purely old systems that do not follow the so-called density–morphology relation. In fact, the general tendency is that dSph systems are more clustered around large spirals, while dIrr galaxies are usually found in more isolated environments. However, Cetus and Tucana are the only two known purely old dwarf galaxies located at many hundreds of kpc from both the Milky Way and M31 (Monelli et al. 2010b). Nevertheless, their recessional velocities suggest that they may have been involved in an interaction a few Gyr ago (Sales et al. 2007; D'Onghia et al. 2009). Bernard et al. (2009) analyzed the detailed properties of RR Lyrae stars in both systems. Distance moduli were homogeneously derived by means of the LMR relation based on the Clementini et al. (2003) calibration and the FOBE method.

Despite many similarities among the general properties of their stellar populations, their star-formation histories (SFHs) were different during the oldest epochs. In particular, although the initial epoch of star formation was the same, the peak of the star-formation activity occurred at an earlier epoch in Tucana than in Cetus (Monelli et al. 2010a). This apparently subtle difference clearly emerges from the properties of their RR Lyrae stars. On the one hand, Monelli et al. (2012) presented a sample of more than 600 RR Lyrae stars in Cetus using wide-field Very Large Telescope (VLT)/VIMOS data and concluded that the properties of the RR Lyrae stars in Cetus are homogeneous over the entire galaxy body. The mean period, magnitude, and amplitude of the RR Lyrae stars do not significantly change as a function of radius, and overall, RR Lyrae stars seem to originate from a stellar population with similar chemical properties.

On the other hand, Tucana represents a much more complex case. Bernard et al. (2008) discussed the complex properties of a few hundred RR Lyrae stars, showing that the samples of bright and faint RR Lyrae stars have different pulsation properties and spatial distributions: faint stars are more centrally concentrated and their periods are shorter, at fixed amplitude. These results can be explained naturally if RR Lyrae stars exhibit an intrinsic spread in their metallicity distribution: faint RR Lyrae stars belong to a slightly more metal-rich population, consistent with them being preferentially formed in the innermost region of the galaxy, as commonly found in most nearby dSph galaxies.

These two examples strongly suggest that the detailed properties of RR Lyrae stars must be well-known in order to safely use them as distance indicators. If the population of RR Lyrae stars presents an intrinsic spread in metallicity (e.g., IC 1613, Bernard et al. 2010; Sculptor, Martínez-Vázquez et al. 2015; NGC 185, Monelli et al. submitted), it is risky to blindly adopt a mean luminosity and a mean metallicity for the entire sample. Bernard et al. (2009) derived the distance modulus of Tucana using the bright and the faint samples of RR Lyrae stars separately, assuming an appropriate metallicity constraint resulting from their SFH calculation, and they obtained consistent distance moduli, within the uncertainties.

5.3.2 Galaxies with Extended SFHs

1. *IC 1613 and Leo A:* Whenever the SFH of a galaxy covers the full cosmic time, it is possible to use both old RR Lyrae stars and young classical Cepheids (CCs) to simultaneously obtain distance estimates. Examples of such analyses have been presented for IC 1613 (Bernard et al. 2010) and Leo A (Bernard et al. 2013). These authors used consistent assumptions and five different relations to derive their distance moduli, including period–luminosity and period–luminosity–color relations for CCs, combined with the LMR and FOBE methods for the RR Lyrae stars. Interestingly, while good agreement was found for IC 1613, the distance derived based on CCs was somewhat longer than that derived from RR Lyrae stars.
   In addition, a number of issues should be taken into account in assessing similar analyses. The periods of Cepheids can span from a few to hundreds of days. Therefore, it is very difficult, from an observational point of view, to collect a data set which allows for a good characterization of the light curves over the full period range, especially using *HST*. Nevertheless, coupling extensive ground-based data sets with dedicated and optimized space-borne observations has provided an excellent basis for such investigations. Scowcroft et al. (2013) provided a multiwavelength view of Cepheids in IC 1613, combining optical results from the OGLE survey (Udalski et al. 2001) with multi-epoch observations in JHK_s, obtained with the Magellan telescope, and at longer wavelengths using the *Spitzer Space Telescope*. The well-characterized optical light curves allowed them to detect and constrain the infrared properties of a Cepheid sample in IC 1613 and derive PLRs. The inclusion of deep *HST*/WFC3 near-infrared data allowed Hatt et al. (2017) to extend the comparison to RR Lyrae stars. These latter authors obtained very good agreement between Population I and Population II distance indicators. The same approach was used by Rich et al. (2014) to study the Cepheid population of NGC 6822. Combining optical OGLE detections with ground-based near-infrared and *Spitzer* mid-infrared data, Rich et al. (2014) derived six PLRs, as well as self-consistent distance and reddening estimates.
2. *DDO 210 and Phoenix:* DDO 210 is a low-mass, isolated galaxy with properties similar to those of Leo A (Cole et al. 2014). Its variable-star content has been studied by Ordoñez and Sarajedini (2016), who derived a distance modulus using the period–luminosity–metallicity relation in the I band of Catelan et al. (2004), once they had derived individual metallicities using the Alcock et al. (2000) relation.
   Gallart et al. (2004) presented a large number of candidate RR Lyrae stars in Phoenix, but the data were not good enough for a detailed analysis. Such an analysis was, however, performed using *HST*/WFPC2 images by Ordoñez et al. (2014), who used RR Lyrae stars to study the early chemical evolution of Phoenix. Unfortunately, neither paper used RR Lyrae stars to derive distances.
   Cepheids have been identified in both galaxies. However, the identification of CCs, especially in the short-period tail of the distribution, can be complicated by the presence of anomalous Cepheids (ACs). ACs and CCs both cross the instability strip during the central helium-burning phase of their evolution. The key difference with CCs is that ACs ignite helium in a degenerate core, while CCs are sufficiently massive that this occurs under non-degenerate conditions. Therefore, ACs have masses $\lesssim 2.3 M_\odot$ and low metallicities ([Fe/H] $\lesssim -1.5$ dex; Fiorentino et al. 2006), since in this mass range, stars with higher metallicity evolve on the red side of the instability strip. Phoenix was the first system where CCs and ACs were found to coexist (Gallart et al. 2004). The coexistence of the two types of Cepheids has since been confirmed in other low-metallicity systems, including IC 1613 (Bernard et al. 2010), Leo A (Bernard et al. 2013), Leo I (Fiorentino et al.

2012b), and DDO 210 (Ordoñez and Sarajedini 2016). Also, both Magellanic Clouds contain a small sample of ACs (Soszyński et al. 2008, 2016).

5.3.3 Leo T

Leo T is an intriguing low-mass, star-forming, relatively isolated dwarf galaxy. Using *HST*/WFPC2 data, Clementini et al. (2012) detected one RR Lyrae star and 12 ACs. This is because of the combination of its small total baryonic mass and its SFH, which is characterized by a dominant episode at intermediate ages and a low rate at early epochs. Nevertheless, the properties of the only RR Lyrae star discovered allowed an estimate of the distance modulus, based on the Clementini et al. (2003) LMR, adopting a proxy metallicity, spectroscopically derived for the galaxy's RGB stars.

5.4 RR Lyrae Beyond the Local Group

The *HST* has allowed the discovery of RR Lyrae stars in a handful of dwarf galaxies outside the Local Group. Da Costa et al. (2010) first reported detections of RR Lyrae in two dwarf galaxies in the Sculptor group. Variables were later analyzed in detail by Yang et al. (2014). Individual metallicities were derived using the Alcock et al. (2000) method, and the mean metallicity of the distribution was used with the Chaboyer (1999) LMR relation. The derived distances, close to 2 Mpc, are in agreement, within the uncertainties, with other estimates based on the TRGB.

The only other galaxy external to the Local Group where RR Lyrae have been discovered is Leo P (McQuinn et al. 2015). This is a very interesting low-mass, gas-rich galaxy with an extended SFH, similar to Leo T. A distance modulus based on the 10 RR Lyrae discovered was obtained using the Carretta et al. (2000) LMR relation. The metallicity pertaining to the RR Lyrae stars was estimated by comparing the distribution of stars in the color–magnitude diagram with theoretical isochrones.

6 Toward Homogeneous RR Lyrae-Based Distance Determinations

6.1 The ISLAndS Project

ISLAndS is an ongoing project targeting six M31 satellites with as main aim to compare the SFHs in the Milky Way and the M31 systems (Weisz et al. 2014; Monelli et al. 2016; Skillman et al. 2017). Full analysis of the variable stars will be presented in a series of forthcoming papers (Martínez-Vázquez et al., in prep). Key results regarding distance estimates are summarized here. The sample of galaxies includes And I, And II, And III, And XV, And XVI, and And XXVIII, thus spanning a wide range in mass, luminosity, and distance from M31. Next, we will use the ISLAndS galaxies as test cases to compare the performance of four different methods to derive their distance moduli. Three of the methods are based on the properties of the RR Lyrae stars, including (i) the reddening-free PWR (Marconi et al. 2015), (ii) the LMR (Bono et al. 2003; Clementini et al. 2003), and (iii) the FOBE relation (Caputo et al. 2000); these are supplemented by (iv) the TRGB method.

As a first step, the metallicity values were homogenized to Carretta et al. (2009) scale. For those galaxies with metallicity estimates based on theoretical spectra, we applied a correction to take into account the updated solar iron abundance, rescaling to $\log \epsilon_{\rm Fe} = 7.54$ dex.

The second step is meant to ensure homogeneous assumptions for the metallicity. The metallicity estimates available in the literature were derived using calcium-triplet spectroscopy of bright RGB stars. The mean metallicity of the RGB stars in And III (Kalirai et al. 2010), And XV (Letarte et al. 2009), And XVI (Collins et al. 2015), and And XXVIII (Slater et al. 2015), is close to [Fe/H] ~ -1.8 dex (or less). In agreement with the limited age and metallicity spreads characteristic of the dominant population, derived from the SFHs (Skillman et al. 2017), we assume that the spectroscopic mean metallicity is representative of the old population as well.

However, And I and And II exhibit higher mean metallicities and larger metallicity spreads (Ho et al. 2012, 2015). Nevertheless, the small number of high-amplitude, short-period (fundamental-mode) RRab-type stars in these galaxies (Fiorentino et al. 2015) suggests that, even if the tail of the RR Lyrae metallicity distribution reaches such relatively high values, the bulk of the RR Lyrae stars must have a lower metallicity ([Fe/H] < -1.5 dex; Fiorentino et al. 2015). Therefore, we adopted [Fe/H] $= -1.8$ dex as representative of the mean metallicity of the RR Lyrae stars in And I and And II.

Once the metallicity had been fixed, we estimated the distance moduli to the ISLAndS galaxies as outlined below. The results are summarized in Fig. 2 and Table 2.

1. *Period–Wesenheit relations:* Since the data were originally calibrated according to the VEGAMAG system, individual phase points were recalibrated to the Johnson system using the transformation provided by Bernard et al. (2009) and optimized for the RR Lyrae color range. Since the F475W filter is characterized by a wide passband, similar to the g band, measurements were calibrated in both the B and the V filters, while the F814W band was transformed to the Johnson–Cousins I filter. This allowed us to use two PWRs from Marconi et al. (2015), the $(I, B-I)$ and $(I, V-I)$ relations, but not the metallicity-independent $(V, B-V)$ PWR, because the B and V measurements are not independent. For the adopted relations, a metallicity dependence is present, but it is weak: a change of 0.3 dex translates into a change in the distance modulus of order 0.03 mag.
   We applied both relations to the samples of RRab- and RRc-type variables, as well as to the total sample (RRab + fundamentalized RRc: $\log P_{\rm fund} = \log P_{\rm RRc} + 0.127$; Bono et al. 2001). The comparison shows very good agreement among the different determinations, on average within ± 0.04 mag. Since the number of stars in the full sample is the largest, we selected as final adopted distance modulus the mean value of the distance moduli calculated using both PWRs and that for the full sample. Table 2 presents for the six galaxies (Col. 1) the range of distance moduli available in the literature (Col. 2), and our final estimate based on the PWR of the full sample of RR Lyrae stars (Col. 3).
2. *Luminosity–metallicity relation:* We adopted two different relations, i.e., those proposed by Clementini et al. (2003) and Bono et al. (2003). The zero-point of the former was modified such that the distance modulus to the LMC is in agreement with the distance modulus obtained by Pietrzyński et al. (2013), assuming [Fe/H] $= -1.5$ dex for the RR Lyrae stars (Gratton et al. 2004). The resulting equation is therefore

$$M_{V,\rm C03} = 0.892 \pm 0.052 + (0.214 \pm 0.047)\ [\rm Fe/H]. \quad (1)$$

   The second calibration adopted was a double-linear relation with a break in the slope at [Fe/H] $= -1.6$ dex. We adopted the lower metallicity, appropriate for the six ISLAndS galaxies ([Fe/H] $= -1.8$ or less).

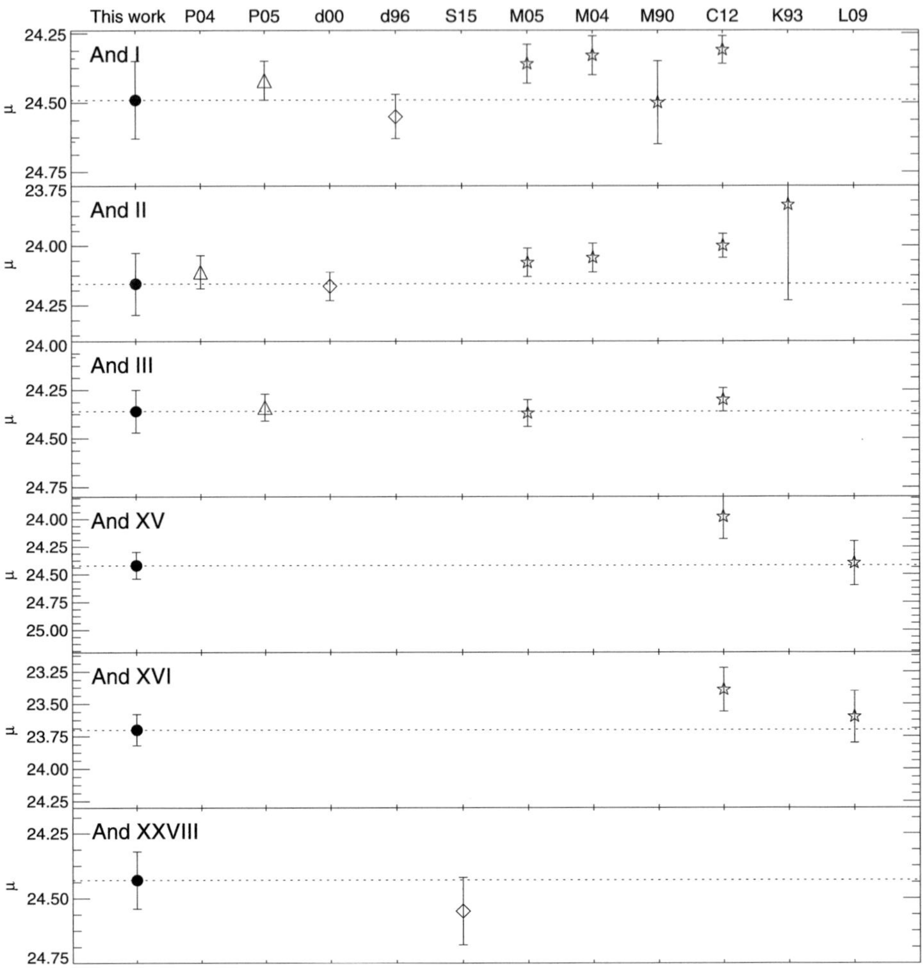

Fig. 2 Summary of our distance modulus estimates (*solid symbols*) and comparison with literature values (*open symbols*). Estimates based on the TRGB are provided only for the most massive galaxies, for which the TRGB could be estimated reliably. The *solid circles* and *dotted lines* show the measurements based on the PWR relations. Other symbols show values taken from the literature, for comparison. In particular, we report values based on RR Lyrae stars (*asterisks*: Pritzl et al. 2004), the HB luminosity (*open squares*: Da Costa et al. 1996, 2000; Slater et al. 2015), and the TRGB (*open diamonds*: Mould and Kristian 1990; Koenig et al. 1993; McConnachie et al. 2004, 2005; Conn et al. 2012)

3. *The FOBE method:* This method is based on the predicted period–luminosity–metallicity relation for pulsators located along the FOBE of the instability strip (see Caputo et al. 2000):

$$M_{V,\mathrm{FOBE}} = -0.685 - 2.255 \log(P_{\mathrm{FOBE}}) - 1.259 \log(M/M_{\odot}) + 0.058 \log(Z). \quad (2)$$

This is a well-defined technique for stellar systems with a large number of RRc-type stars, in particular if the blue side of the first-overtone instability strip is well-populated. For this reason, we applied it only to five of our six galaxies, since the small number of RRc stars in And XVI may yield inconsistent results (Monelli et al. 2016). To derive

Table 2 True distance moduli for the six M31 satellite dSphs observed as part of the ISLAndS project

Galaxy	True distance moduli	
	Literature value (mag)	PWR (mag)
And I	24.31–24.55	24.51 ± 0.08
And II	23.87–24.17	24.17 ± 0.07
And III	24.30–24.37	24.38 ± 0.06
And XV	23.98–24.4	24.43 ± 0.07
And XVI	23.39–23.6	23.71 ± 0.07
And XXVIII	24.55 ± 0.13	24.45 ± 0.05

the stellar masses, we adopted the BaSTI evolutionary models (Pietrinferni et al. 2004). For the selected metallicity, for a star with an effective temperature typical of RRc stars ($\log T_{\rm eff} \approx 3.86$ [K]), we derived values of $M \sim 0.7 M_{\odot}$. Distance moduli obtained for each galaxy using this method are included in Fig. 2.

4. *Tip of the RGB:* The TRGB is a well-studied, work-horse standard candle by virtue of its weak dependence on both age (Salaris and Weiss 2002) and, particularly in the I band, metallicity (at least for relatively metal-poor systems; Da Costa and Armandroff 1990; Lee et al. 1993). The TRGB is frequently used to obtain distance estimates to galaxies of all morphological types (e.g., Rizzi et al. 2007; Bellazzini et al. 2011; Wu et al. 2014). However, determining the cut-off in the luminosity function at the bright end of the RGB is not straightforward in low-mass systems because of the inherently small number of bright RGB stars (Madore and Freedman 1995; Bellazzini et al. 2001, 2008). A total of more than about 100 stars in the top magnitude range of the RGB is considered a safe threshold for reliable estimates. In our galaxies, this condition is met only by And I ($N > 200$), And II ($N > 150$), and nearly in And III ($N \sim 90$). The small numbers of such stars in the other three galaxies prevent us from deriving reliable measurements of the apparent magnitudes of their TRGBs. We adopted the calibrations of Rizzi et al. (2007), Bellazzini et al. (2011), and Cassisi and Salaris (2013) to compile Fig. 2.

Figure 2 summarizes the distance modulus determinations derived in this paper. In particular, the solid circles and the dotted line show the adopted final distance moduli based on averaging the values from the two PWRs. Filled diamonds show the results from the other methods. This figure shows that the agreement among the different methods employed here is remarkably good, since most of the derived distance moduli agree within 1σ. Taking as reference the PWR distance, some general trends can be discerned. The distance derived using the LMR based on the Bono et al. (2003) calibration is marginally longer with respect to that resulting from the Clementini et al. (2003) calibration. The FOBE distance is longer than the PWR distance in three cases (And III, And XV, and And XVI) and shorter for And II. Nevertheless, this method is most sensitive to the sampling of the instability strip, and in particular the lack of RR Lyrae close to the blue edge of the instability strip introduces a bias toward greater distances.

Asterisks and open symbols in Fig. 2 represent distance modulus estimates available in the literature, derived using different techniques, including those based on RR Lyrae stars (asterisks: Pritzl et al. 2004), the HB luminosity (open squares: Da Costa et al. 1996, 2000; Slater et al. 2015), and the TRGB (open diamonds: Mould and Kristian 1990; Koenig et al. 1993; McConnachie et al. 2004, 2005; Conn et al. 2012). Figure 2 shows generally good agreement with our estimates, within the uncertainties. Note that the TRGB tends to yield

Table 3 True distance moduli for 10 Local Group dwarf galaxies observed with the HST

Galaxy	Distance moduli		
	Literature (mag)	PWR (mag)	LMR (mag)
Cetus	24.46 ± 0.12^{a}	$24.47 \pm 0.02 \pm 0.04$	24.43 ± 0.08
DDO 210	25.07 ± 0.12^{b}	$25.07 \pm 0.03 \pm 0.05$	–
IC 1613	24.44 ± 0.09^{c}	$24.47 \pm 0.02 \pm 0.04$	23.41 ± 0.08
Leo A	24.57 ± 0.13^{d}	$24.46 \pm 0.07 \pm 0.10$	24.41 ± 0.08
Tucana	24.74 ± 0.12^{e}	$24.78 \pm 0.01 \pm 0.05$	24.72 ± 0.08
And XI	24.33 ± 0.05^{f}	$24.47 \pm 0.05 \pm 0.16$	24.55 ± 0.08
And XIII	24.62 ± 0.05^{g}	$24.52 \pm 0.07 \pm 0.14$	24.72 ± 0.08
ESO410-G005	26.33 ± 0.07^{h}	$26.36 \pm 0.01 \pm 0.23$	26.36 ± 0.08
ESO294-G010	26.40 ± 0.07^{i}	$26.45 \pm 0.01 \pm 0.20$	26.42 ± 0.08
Phoenix	23.09 ± 0.10^{j}	$23.14 \pm 0.02 \pm 0.15$	23.17 ± 0.08

[a]Bernard et al. (2009), [b]Ordoñez and Sarajedini (2015), [c]Bernard et al. (2013), [d]Bernard et al. (2013), [e]Bernard et al. (2009), [f]Yang and Sarajedini (2012), [g]Yang and Sarajedini (2012), [h]Yang et al. (2014), [i]Yang et al. (2014), [j]Hidalgo et al. (2009)

shorter distances than the RR Lyrae and the HB luminosity-based methods, although they are still compatible within 1.5σ. A number of discrepant cases (And XV and And XVI; see Conn et al. 2012) can be ascribed to the sparsely populated bright part of the RGB in these galaxies.

6.2 Revising the Distances to Local Group Galaxies

Using existing data to construct a homogeneous distance framework for a large number of Local Group galaxies is still far from having been achieved. Ideally, one would want to derive homogeneous pulsation properties and metallicities, and then uniformly use RR Lyrae stars to derive distances. However, the highly inhomogeneous data set available (in terms of data quality, filter passbands used, and completeness) renders this a complicated task. Moreover, different authors often use different methods of data analysis and, critical in this case, different approaches and assumptions as regards the metallicity.

In an incomplete attempt to resolve this situation, based on a similar approach as that adopted for the ISLAndS sample, we re-analyzed the available data for a sample of Local Group galaxies for which high-quality *HST* data are available. This includes galaxies characterized by different levels of isolation and extends to systems in the Sculptor group. We selected those galaxies for which photometry in the Johnson V and I or in the B and I bands is available in the literature.[1] We assumed a metallicity of $[\mathrm{Fe/H}] = -1.8$ dex as representative of the RR Lyrae stars in all galaxies.

New distances were derived using the $(I, V - I)$ PWR, with the exception of DDO 210, for which only B and I catalogs are available. A second estimate was obtained using the LMR, based on the Clementini et al. (2003) calibration and updated as described in the previous section. Table 3 summarizes the results. Column 2 presents the original literature values;

[1]We emphasize that we did not attempt to use the original photometry in the VEGAMAG system and transform it, since the original data are often not provided. Therefore, the photometric conversion to the Johnson system may have been performed using different methods by different authors. Although we do not expect this to introduce significant systematics in the distance determination, this issue must be taken into account for future investigations.

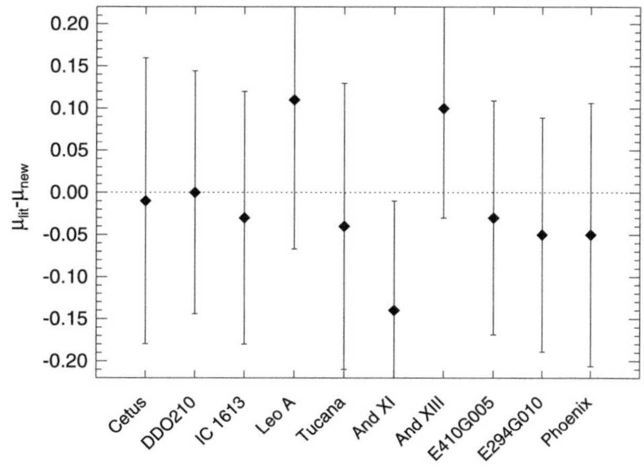

Fig. 3 Differences between distance-modulus estimates obtained based on RR Lyrae stars and literature values for a number of galaxies. These updated values represent homogenized metallicities, employing the Marconi et al. (2015) PWR

we included only estimates based on RR Lyrae stars, with the exception of Phoenix (Hidalgo et al. 2009; Ordoñez et al. 2014). Columns 3 and 4 present the same distance estimates recalculated using the PWR and the updated LMR (Eq. (1)). Both methods show excellent agreement in most cases. Figure 3 shows the differences between the literature values and our updated values based on the PWR. Note that, overall, the literature values are based on slightly higher mean metallicities than adopted for our recalibration ([Fe/H] $= -1.5$ dex for Leo A, values between [Fe/H] $= -1.6$ and -1.8 dex for the other galaxies). This can account for up to a few hundredths of a magnitude in distance differences. Nevertheless, the change is small, at the level of at most $\sim 1\sigma$, where the error bars reflect realistic estimates of systematic and random uncertainties associated with the method.

7 Internally Consistent Distance Determination Beyond the Nearest Galaxies

Internally consistent determinations of the Hubble constant well beyond the Local Group, based on cross-correlation of a variety of independent techniques, are crucially important for establishing the three-dimensional morphology of the Universe on the largest scales. By association, such measurements provide indirect tests of the consistency of the cosmological framework implied by SNe Ia measurements. Distance measurements at these distances are particularly important in the context of reducing the effects of galaxies' peculiar velocities as well as regarding the inhomogeneity of the matter distribution, which is the root cause of the difference between the locally measured and the true, global values of the Hubble constant (e.g., Turner et al. 1992).

SNe Ia are bright standard candles and, hence, in principle they enable distance measurements out to cosmological scales ($z \sim 1$). However, to use SNe Ia as absolute distance indicators, their absolute magnitudes must be calibrated using carefully selected primary distance indicators such as Cepheid variables (e.g., Riess et al. 2016). To check for the possible effects of unknown systematics, those distances should be compared with a set of independent distance measurements that do not rely on mutually dependent calibrations of the distance ladder. Megamasers can be used to directly measure distances to galaxies out to ~ 50 Mpc, which is sufficiently distant to reduce the effects of peculiar velocities (Reid et al. 2009). Here, we review several other independent techniques that may potentially serve as important tests of the absolute distance scale beyond the nearest galaxies.

7.1 Clusters of Galaxies

Combined with X-ray data, observations of the Sunyaev–Zel'dovich effect (SZE) affecting massive clusters of galaxies can be used to infer the Hubble constant (Silk and White 1978). This is because the X-ray surface brightness depends on $S_X \propto \int n_e^2 \Lambda \mathrm{d}\ell$, where Λ is the X-ray cooling function determined from the gas temperature T_e, and the SZE depends on $y \propto \int n_e T_e \mathrm{d}\ell$. Assuming spherically shaped galaxy clusters, line-of-sight distances can be expressed as $\mathrm{d}\ell \sim D_\mathrm{A} \mathrm{d}\theta$, where D_A is the angular diameter distance and θ the cluster's angular size. By eliminating the electron density, n_e, the angular diameter distance can be expressed in terms of X-ray and SZ observables as $D_\mathrm{A} \propto y^2/S_X$, assuming spherical symmetry. The effects of non-spherical cluster morphologies can be reduced by averaging H_0 measurements for many clusters.

Reese et al. (2002) determined distances to 18 clusters at $z = 0.14$–0.78 to constrain the Hubble constant to $H_0 = 60^{+4}_{-3}$(statistical)$^{+13}_{-18}$(systematic) km s^{-1} Mpc^{-1}. Bonamente et al. (2006) obtained an updated result, $H_0 = 76.9^{+3.9}_{-3.4}$(stat.)$^{+10.0}_{-8.0}$(syst.) km s^{-1} Mpc^{-1} using 38 clusters at $z = 0.14 - 0.89$. These analyses suggest that the dominant uncertainty affecting this method is already determined by various systematic errors, such as inhomogeneities in the intracluster medium (e.g., Kawahara et al. 2008). This may imply that a better understanding of the physical state of the intracluster medium (e.g., Aharonian et al. 2016) is crucial to obtain more accurate distance measurements based on this approach.

7.2 Relative Ages of Old Galaxies

Jimenez and Loeb (2002) proposed to use relative ages of passively evolving galaxies at a range of redshifts to directly constrain the Hubble parameter, $H(z)$. Spectra of passively evolving galaxies are dominated by main-sequence stars of $M \sim 1 M_\odot$, for which the relevant stellar evolution processes are well-understood. In addition, relative ages can be determined more easily and with higher accuracy than absolute ages, because several systematic effects are factored out. Once the relative ages at different redshifts have been determined, one can derive the Hubble parameter by employing the relation $H(z) = -(\mathrm{d}z/\mathrm{d}t)/(1+z)$. Jimenez et al. (2003) applied this technique to Sloan Digital Sky Survey (SDSS) galaxies at $z < 0.17$ to obtain $H_0 = 69 \pm 12$ km s^{-1} Mpc^{-1}, which is consistent with other H_0 measurements. Further theoretical and observational studies aimed at improving our understanding of galaxy spectra have the potential to reduce the remaining uncertainties in the current-best values of H_0.

7.3 Gravitational Waves as Standard Sirens

Gravitational waves from inspiraling compact binaries are well placed to become useful as absolute distance indicators (Schutz 1986; Holz and Hughes 2005). They are sometimes referred to as 'standard sirens.' This is because masses of inspiraling and merging objects, such as neutron stars and black holes, can be inferred from the shape of the waveform (e.g., from the frequency and its time evolution), which also determines the absolute strain amplitude (a dimensionless parameter that depends on the strength of the tidal gravitational field between both binary components). We can then directly measure the luminosity distance to a gravitational-wave source, because the observed strain is inversely proportional to the luminosity distance.

We can measure luminosity distances using gravitational-wave standard sirens, but not the associated redshifts. Therefore, independent observations of the objects' electromagnetic counterparts are usually required to locate the host galaxy of a gravitational-wave event and measure its redshift to constrain the Hubble constant. For instance, Dalal et al. (2006) discussed the possibility of simultaneous observations of short gamma-ray bursts and gravitational waves from merging neutron-star binaries. They argued that several tens of such observations can constrain the Hubble constant to the $\sim 2\%$ level.

A major advantage of this method is its 'clean' physics. Assuming that general relativity is valid, the waveforms of such merging events can be computed easily from first principles. On the other hand, obtaining observations of gravitational waves has been a significant challenge for decades. Recently, the first observation of gravitational waves, referred to as GW150914, from a pair of black holes has been reported (Abbott et al. 2016). This has now finally opened up the possibility of using gravitational waves as a tool to benchmark the cosmic distance scale. Whereas it is not yet clear whether binary black hole mergers produce any detectable electromagnetic counterparts for redshift measurements, it may also be possible to use the spatial cross-correlation of gravitational-wave sources and galaxies with known redshifts to constrain the Hubble constant without any observations of electromagnetic counterparts (Oguri 2016). For this cross-correlation technique to become viable, we need observations of large numbers of gravitational waves over the entire sky, as well as decent localizations of gravitational-wave sources, with an accuracy of $\sim 1^\circ$. This will become possible with the next-generation of gravitational-wave detectors.

8 Gravitational Lensing as a Promising Tool

Time delays of strongly lensed quasars provide a one-step measurement of a cosmological distance, namely the time-delay distance to the lens system (e.g., Refsdal 1964; Suyu et al. 2010). With additional information from the stellar velocity dispersion of the lens, we can further constrain the associated angular diameter distance (Paraficz and Hjorth 2009; Jee et al. 2015). In a companion review, we have described the recent cosmological constraints based on this approach, particularly those based on the COSMOGRAIL (Courbin et al. 2005, 2011; Eigenbrod et al. 2005; Tewes et al. 2013) and H0LiCOW (Sluse et al. 2016; Bonvin et al. 2017; Rusu et al. 2017; Suyu et al. 2017; Wong et al. 2017) projects. Here, we describe the future prospects for this approach, focusing in particular on (i) expanding the time-delay lens sample and (ii) follow-up observations, in light of upcoming missions and surveys.

8.1 Where Are the Time-Delay Lenses?

Quadruply lensed quasars ('quads') with ancillary data can each provide a time-delay distance measurement of ~ 5–8%. However, currently only a handful of quads are known. For example, there are only four quads in the H0LiCOW sample of lenses for cosmography (Suyu et al. 2017). There are several times more doubly lensed quasars ('doubles'), but accurate and precise[2] distance measurements from these systems would be more challenging compared to the equivalent determinations based on quads, since doubles have fewer observational constraints compared with quads. For example, quad systems can yield four image positions and flux measurements of quasars as constraints, whereas double systems have half of that number as constraints. In addition, quads can have three time-delay measurements between the multiple images, which is substantially better for cosmographic studies

[2] Accuracy is a measure of how close the values are to the target or standard value; precision reflects their spread around the mean.

than the single time-delay measurement available in doubles. Unless we can find more time-delay lenses, especially quads, the statistical power of this approach will be limited.

Fortunately, discoveries of hundreds if not thousands of time-delay lenses are expected in current and future surveys and missions, with $\sim 13\%$ of them being quads (Oguri and Marshall 2010; Liao et al. 2015). Below, we describe three ongoing imaging surveys that will likely expand the current sample of quads by an order of magnitude:

1. *Dark Energy Survey (DES):* The DES (Abbott et al. 2016) uses the 4 meter Mayall telescope at Cerro Tololo observatory in Chile. Its large-format CCD array of 74 detectors covers, in one single snapshot, an area of 2.2 degrees on a side (Flaugher et al. 2015). After five years of operation the full survey will cover ~ 5000 deg^2 in the g, r, i, z, and Y bands down to $r = 24.3$ mag (10σ). DES has already discovered lensed quasars. Two were reported in Agnello et al. (2015b) in the context of the SRIDES program (Agnello and Treu 2015). In addition, a doubly imaged quasar (Ostrovski et al. 2017) was found in the data taken during the first year of the survey. Many more are found at the time of writing this review, as the data obtained during the second year are analyzed. Given the area and depth of the survey, Oguri and Marshall (2010) predicted that a total of ~ 1000 new lensed quasars should be found upon completion of the survey.
2. *Hyper Suprime-Cam (HSC) Survey:* The HSC survey is a Subaru Strategic Program, using the newly installed HSC (Miyazaki et al. 2012) on the Subaru 8.2 meter telescope, which is capable of imaging a large area of the sky in a single pointing.[3] The survey started in 2014 and is divided into three layers characterized by different areas and depths, i.e., wide, deep, and ultra-deep. In particular, we expect to find most of the new lenses in the wide survey, which will cover ~ 1400 deg^2, mostly in equatorial regions, to $i \sim 26$ mag in *grizy* broad bands with excellent seeing ($\sim 0.6''$ in the i band). The first-year HSC survey data covering over 100 deg^2 in the five broad bands have recently been released to the public (Aihara et al. 2017). The expected number of lensed quasars in the HSC survey is ~ 600 (Oguri and Marshall 2010).
3. *Kilo-Degree Survey (KiDS):* The KiDS program is running at the European Southern Observatory's Paranal observatory in Chile (Kuijken et al. 2015). It uses the 2.6 meter VLT Survey Telescope (VST) in the *ugri* optical bands and covers 10-degree wide strips, including an equatorial one between $10^h20^m < \mathrm{RA} < 15^h50^m$ and a southern strip going through the Galactic South Pole, $12^h00^m < \mathrm{RA} < 03^h30^m$. The total area covered is ~ 1500 deg^2, i.e., similar to the HSC survey. The observing strategy is designed to ensure the best seeing in the r band, which has a median seeing of $\sim 0.7''$. The depth in this band is $r = 24.9$ mag (AB, 5σ). Overall, KiDS should be as efficient as HSC in finding bright, strongly lensed quasars.

Various new algorithms have been developed for finding galaxy-scale lenses, many of which have been applied to these imaging surveys. Here we focus on automated algorithms that are aimed at finding lensed quasars, although there are also methods to find lensed galaxies without quasars (e.g., Bolton et al. 2006; Gavazzi et al. 2014; Joseph et al. 2014; Napolitano et al. 2016; Paraficz et al. 2016; Shu et al. 2016; Sonnenfeld et al. 2017) or lens systems in general through visual inspection by citizen scientists (e.g., Marshall et al. 2016; More et al. 2016b). While most of the first lensed quasars were found at radio wavelengths through the Cosmic Lens All-Sky Survey (Myers et al. 2003), the SDSS has yielded dozens of lensed quasars through the SDSS Quasar Lens Search (e.g., Oguri et al. 2006; Kayo et al. 2010; Inada et al. 2012; Oguri et al. 2012) and, more recently, another 13 two-image lensed

[3] http://hsc.mtk.nao.ac.jp/ssp/.

quasars (More et al. 2016a). The advantage of using SDSS is the availability of a spectroscopic quasar sample from which one could search for lensed objects. Consequently, lenses obtained from such a search will guaranteed be lensed quasars, rather than lensed galaxies (without quasars), which are more numerous than lensed quasars but not time-varying. Current imaging surveys such as HSC, DES, and KiDS do not include a spectroscopic component as part of the surveys. Therefore, recent searches necessarily rely on color information by selecting objects with colors and morphologies compatible with lensed active galactic nuclei (e.g., Jackson et al. 2012; Agnello et al. 2015a,b; Ostrovski et al. 2017; Schechter et al. 2017). In addition, as advocated by Marshall et al. (2009), for an object to qualify as a lens, it must have a physical lens mass model. This is the basis of the model-based search algorithms such as CHITAH (Chan et al. 2015) and LensTractor (Marshall et al., in prep.). Moreover, Kochanek et al. (2006) have proposed to use quasar variability to find lens systems by looking for clustered (possibly blended) variable sources corresponding to the multiple quasar images. Our expectation is that these surveys will each contain hundreds to about a thousand lensed quasars (Oguri and Marshall 2010), and efforts are underway to find these.

Two surveys planned for execution the 2020s should yield yet another order-of-magnitude increase in the number of lenses compared with current surveys:

1. *Euclid:* The first space-based, wide-field cosmological survey will be the ESA/NASA *Euclid* 1.2 meter telescope (Laureijs et al. 2011). Once placed at the L2 Sun–Earth Lagrange point in 2020, this satellite will be equipped with a 0.5×0.5 deg^2 optical imager, which uses a single broad filter covering the $R + I + Z$ bands, as well as a near-infrared imager covering the same field of view to perform imaging in the Y, J, and H bands. In total, *Euclid* will cover ~ 15000 deg^2 of extragalactic sky down to a 10σ optical AB magnitude of 24.5 mag and down to a 5σ AB magnitude of 24.0 mag in each of the three near-infrared bands. An additional deep survey will include 40 deg^2 of sky two magnitudes deeper than the wide survey. *Euclid* will image a total of 12 billion sources (3σ) and it will obtain near-infrared slitless spectra of 35 million of these. With its excellent image quality and point-spread-function (PSF) stability, *Euclid* is optimized for weak-lensing tomography and it will also implementation of other cosmological probes such as baryon acoustic oscillations, the integrated Sachs–Wolfe effect, galaxy cluster counts, and redshift space distortions. With its sharp PSF ($0.18''$ FWHM), *Euclid* will also be a superb machine to discover gravitationally lensed quasars with an image quality that, on its own, already allows one to constrain lens models.
2. *Large Synoptic Survey Telescope (LSST):*[4] The LSST is an ambitious new survey telescope that is under construction in Chile. The telescope has a primary mirror diameter of 8.4 meters, with an effective aperture of 6.7 meters (because of the tertiary mirror area in the middle of the primary–tertiary mirror and some obscuration). It will rapidly survey the entire visible southern sky of ~ 20000 deg^2 twice each week, for 10 years. Each patch of sky it will be visited 1000 times during the survey, in six filters, *ugrizy*. The survey will reach 24.5 mag depth in a single visit (30 seconds) and 27.5 mag co-added depth. The four primary science objectives of the LSST that drive the survey design are (i) constraining dark energy and dark matter, (ii) taking an inventory of the solar system, (iii) exploring the transient optical sky, and (iv) mapping the Milky Way. LSST will be revolutionary in the area of gravitational lensing. Thousands of lensed quasars will be found by LSST (Oguri and Marshall 2010), and at least hundreds of them will have well-defined time delays measurable directly from the survey data alone (Liao et al. 2015).

[4]https://www.lsst.org/.

8.2 Observational Follow-up Requirements and Analysis

8.2.1 Confirmation

Both high angular-resolution images and spectroscopy would help reveal whether the lens candidates, found in ground-based imaging surveys, are gravitational lenses or other astrophysical systems.

With high-resolution images, one can check whether the multiple image features are consistent with lensing morphology. While *Euclid* images are of high-resolution by default, space-based or adaptive-optics ground-based images could be used to confirm candidates from ground-based imaging surveys. Nonetheless, some systems observed with high-resolution imaging could still be difficult to decipher, for example, a doubly-imaged quasar and a star–quasar–galaxy chance alignment could look similar in high-resolution images, in which case spectroscopy is needed for confirmation.

Identical spectra of multiple components/features of an object would confirm the object to be a strongly lensed system, with the multiple components originating from the same source. Spectroscopic confirmation could further yield spectroscopic redshift measurements of the lens and/or the source, which are crucial pieces of information to convert the angular quantities which we measure to physical quantities.

8.2.2 Follow-up Observations

With ground-based and upcoming space-based wide-field surveys, not only does finding lensed quasars become a fairly easy task, but also some of the follow-up observations required to confirm (or not) the candidates come for free. KiDS, DES, LSST, and all deep, multi-band ground-based surveys have at least one of their bands optimized for weak-lensing measurements. As a consequence, measuring the colors of the individual quasar images can be done directly from the survey data. In many cases, detecting the lensing galaxy or galaxies among the quasar images is possible from the weak-lensing quality images. However, additional ingredients are necessary to use the confirmed lensed quasars for cosmological purposes.

1. *The lens redshift and velocity dispersion:* In the era of modern wide-field surveys, the survey data on their own offer the required material to deblend the lensed images of the quasar and fit a crude lens model. More difficult are measurements of the lens redshifts, since they are often hidden in the glare of the quasar images, so that this requires long exposure times on 10 meter class telescopes. In COSMOGRAIL, Eigenbrod et al. (2006, 2007) measured 15 lens redshifts based on 2 hour exposure times and slit spectroscopy with the VLT. To separate the spectra of the quasar images from that of the lens galaxy, the data are spatially deconvolved (Courbin et al. 2000) using the spatial information contained in the spectra of PSF stars, obtained in multi-slit mode. The same technique can be used to measure the velocity dispersion of the lensing galaxy. This requires, however, longer exposure times and an instrumental setup with higher spectral resolution. With the advent of integral-field spectrographs, in combination with adaptive optics, or mounted on the *James Webb Space Telescope* (*JWST*), it will become possible to map the full velocity fields and the velocity-dispersion fields of lens galaxies, hence providing a new wealth of observational constraints useful in breaking the degeneracies inherent to lens models. Figure 4 illustrates what is possible already with the near-infrared two-dimensional SINFONI spectrograph mounted on the VLT.

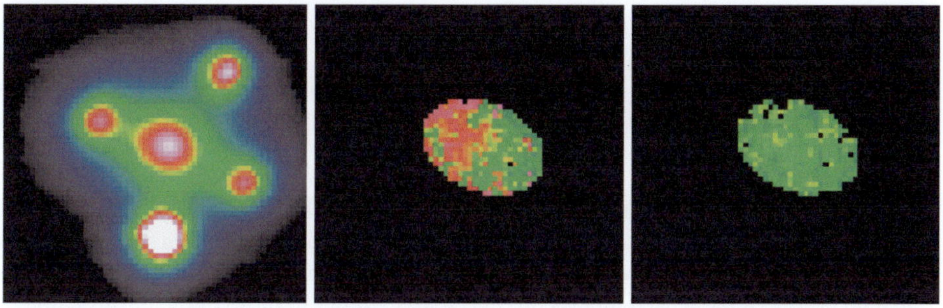

Fig. 4 (*Left*) Reconstructed image of the Einstein Cross, Q2237+080, seen with the near-infrared SINFONI integral-field spectrograph mounted on the VLT. The field of view is 3″ on a side. North is up and East to the left. No adaptive optics was used to obtain this 6 hour integration with 0.5″ seeing. (*Middle*) Radial-velocity field showing a small rotation signal. The maximum velocity in this 1″ field is 60 $\mathrm{km\,s^{-1}}$. (*Right*) Line-of-sight velocity-dispersion field, which is relatively flat with a mean value of $\sigma_r = 160 \pm 20\ \mathrm{km\,s^{-1}}$ (from Eigenbrod 2009, PhD Thesis)

2. *High-resolution imaging of the lensed quasar host:* Quasar time delays depend on the slope of the lens mass at the position of the quasar images, which is the quantity we need to constrain with observations. One route to such a mass-slope measurement is use of the Einstein ring formed by the quasar host galaxy. If the latter is radially extended, the leverage on the mass slope of the lens is sufficiently good to fit simple lens models such as power-laws or more physically justified models involving stellar and dark matter components (Suyu et al. 2014, their Fig. 2). Fortunately, such rings are visible in most lensed quasars with time delays, provided that deep and high-resolution images can be taken, as in the H0LiCOW project (e.g. Suyu et al. 2017). Notably, the cosmological constraints derived using *HST* imaging can be further improved with Keck adaptive optics (Chen et al. 2016). *JWST* and the new generation of ground-based 30–40 meter-class telescopes have the potential to revolutionize the field thanks to the levels of detail achievable in the images of Einstein rings (Meng et al. 2015).
3. *Line-of-sight effect:* While the multiple images of lensed quasars are formed by individual galaxies, the details of the image configuration and the precise slope of the total potential well at the image positions depend on the contributions from all massive objects up to the quasar redshift. Lens galaxies are often found in groups, i.e., they may share the dark matter halo of the group they belong to, in addition to their own halo. The contribution of the group to the total deflection field must be evaluated. If it is massive enough, deriving a mass map is possible from a weak-lensing analysis (e.g. Kneib et al. 2000). However, one must also account for the contribution of all other masses along the line of sight. Currently, this is done by comparing galaxy counts measured in the field of view of strongly lensed quasars (e.g. Fassnacht et al. 2011) with galaxy counts in cosmological simulations (Hilbert et al. 2007, 2009). Lines of sight are chosen in the simulations with the same *relative* galaxy density as in the observations and used to derive a probability distribution function for the mass density toward a given lensed quasar (Suyu et al. 2010; Greene et al. 2013; Suyu et al. 2013; Rusu et al. 2016). Other, more sophisticated methods (Collett et al. 2013; McCully et al. 2017) have the potential to reduce the uncertainty along the line of sight to that dominated by statistical errors only. All the above methods require multiband imaging to measure photometric redshifts and stellar masses. Ideally, multi-object spectroscopy is desirable to infer in detail the mass

contributions from all galaxies in the immediate vicinity of the lens ($< 15''$), as done recently for HE 0435–1223 (Sluse et al. 2016).

4. *Time-delay measurements:* Large-scale sky surveys and discoveries of thousands of new lensed quasars potentially allow a drastic reduction in the random uncertainties affecting distance estimates using time delays. However, current time-delay measurements, e.g., based on COSMOGRAIL, need years of photometric monitoring. The reason for this is that the time-delay precision is limited by microlensing by stars in the lensing galaxy, which act as secondary lenses in each quasar image. As stars move along the line of sight to the quasar images, they affect the quasar images' apparent brightnesses. This results in a flickering of the lensed images on time-scales similar to those of the intrinsic quasar variations and with similar (or sometimes larger) amplitudes. Our ability to measure time delays therefore depends on our effectiveness in disentangling microlensing (different in each quasar image) from intrinsic quasar variations (identical in each quasar image). Obviously, it would be much easier to separate the microlensing variations from the intrinsic quasar variations if they were acting on different temporal time-scales. This is, in fact, possible! Quasars show variations on very short time-scales, much shorter than those typical of microlensing. The *Kepler* light curves of active galactic nuclei reveal that essentially all quasars exhibit variations on time-scales of a week or two (Mushotzky et al. 2011). Since microlensing is present on scales of a few years, the microlensing versus quasar variations can be separated in frequency, provided that the low-amplitude and fast intrinsic quasar variations can be detected and measured. Typical slow quasar variations over time-scales of months have amplitudes on the order of 0.5 mag, whereas faster, week-long variations usually only span a few mmag (r.m.s.). For this reason, the future of lens monitoring may reside in high signal-to-noise and high-cadence monitoring, e.g., with daily observations of lenses with a 2–4 meter-class telescope. Such observations allow one to catch the quasar variations, almost insensitive to long time-scale microlensing. Time-delay measurements become possible in 1 or 2 years per object instead of 10–15 years, as is currently the case for smaller telescopes and a coarser temporal sampling of one or two epochs per week. However, given the signal-to-noise ratio needed (1000 per quasar image and per epoch), it will be difficult to monitor more than 20 lenses per night at any given telescope. In addition, measuring time delays longer than 90 days is not possible with this high-cadence mode over 1 or 2 years, since the time delay becomes comparable to the object's visibility period. Long-term monitoring with large telescopes capable of following hundreds of lenses per night, like the LSST, is therefore the other complementary route to enable large numbers of precise and accurate time-delay measurements.

8.2.3 Analysis and Lens Modeling

Modeling the lens-mass distribution and accounting for the mass structure along the line of sight to the quasar source are crucial to convert time delays into distance measurements. Deep imaging which reveals the Einstein ring is important to precisely constrain the lens-mass distribution near the location of multiple images, where it matters for cosmography. While this has been done mostly in the optical/infrared with thousands of intensity pixels (e.g., Kochanek et al. 2001; Koopmans et al. 2003; Dye and Warren 2005; Suyu et al. 2009, 2013; Fadely et al. 2010; Birrer et al. 2016; Wong et al. 2017), interferometric radio data have also been employed (e.g., Kochanek and Narayan 1992; Ellithorpe et al. 1996; Wucknitz et al. 2004). The recent completion of the Atacama Large Millimeter/submillimeter Array (ALMA), which provides high-sensitivity and high angular-resolution imaging, opens up a new avenue for modeling the Einstein rings (e.g., Hezaveh et al. 2013; Bussmann et al. 2015; Rybak et al. 2015; Hezaveh et al. 2016).

To model the observed images, simultaneous determination of the source surface brightness and the lens-mass distributions is needed. Numerous modeling approaches have been developed over the past decade (e.g., Warren and Dye 2003; Suyu et al. 2006, 2009; Barnabè and Koopmans 2007; Jullo et al. 2007; Jullo and Kneib 2009; Vegetti and Koopmans 2009; Suyu and Halkola 2010; Oguri 2010; Collett and Auger 2014; Tagore and Keeton 2014; Birrer et al. 2015; Nightingale and Dye 2015; Rybak et al. 2015; Hezaveh et al. 2016; Tagore and Jackson 2016), some of which have been used recently to measure distances to lensed quasars (e.g., Suyu et al. 2010, 2013, 2014; Birrer et al. 2016; Wong et al. 2017). In order to avoid confirmation bias, the cosmographic analysis of the H0LiCOW collaboration (Suyu et al. 2017) is blinded (Suyu et al. 2013; Bonvin et al. 2017; Wong et al. 2017). For cosmographic inference, each lens takes an expert modeler weeks of analysis owing to the high computational load caused by the large numbers of data points (image pixels) and parameters (dozens) involved. While parallel computing can reduce the computational time, frequent manual input from the expert modeler is often required. Automating the full modeling procedure is challenging given the complexity of lens modeling, but it is worth pursuing in the era of hundreds/thousands of lenses.

To break lens-modeling degeneracies caused by the mass-sheet degeneracy and source-position transformation (Falco et al. 1985; Gorenstein et al. 1988; Schneider and Sluse 2013, 2014; Xu et al. 2016), additional information such as the stellar kinematics of the lens galaxy is needed. However, most of the lens modeling software suites tend to make use of only the lensing information and incorporate the kinematic information separately. Barnabè and Koopmans (2007) and Barnabè et al. (2011) provide self-consistent modeling of three-dimensional axisymmetric lens-mass distributions, although the cosmological implications of such a modeling strategy are yet to be explored. We expect that two-dimensional kinematic maps of lens galaxies would provide complementary information to break lens-parameter degeneracies, especially for two-image systems. Future lens modeling would benefit from including such kinematic information. As a demonstration, Suyu et al. (2014) found a $\sim 4\%$ discrepancy in the time-delay distance measurements between different lens mass models of a strong lens system owing to lensing degeneracies. Incorporating an aperture-averaged velocity dispersion measurement partially breaks the lensing degeneracy and reduces the discrepancy to $\sim 1\%$. In addition, kinematic data are the key ingredient for measuring the angular diameter distance to the lens (Jee et al. 2015). This distance measurement substantially improves the cosmographic constraints from time-delay distances alone, making time-delay lenses an even more competitive cosmographic probe (Jee et al. 2016).

In terms of quantifying line-of-sight structures, current approaches for cosmography are based on relative galaxy counts around a strong lens compared with those of simulations. With sufficient photometric and spectroscopic information of the structures along the line of sight, one could (in principle) attempt a full light-cone reconstruction of the mass. Steps into this direction have been carried out by, e.g., Collett et al. (2013), McCully et al. (2014), and McCully et al. (2017). This has the potential of achieving more precise and accurate inferences about the external convergence.

9 Future Outlook

The detailed narrative provided in this review should have impressed upon the reader that efforts to establish an internally consistent distance framework from the nearest galaxies to the highest redshifts are indeed coming together on all scales. Analysis, theoretical, and observational developments are indeed very promising, with significant improvements in both accuracy and precision anticipated in the next few years.

We are at the dawn of a new era for variability studies. Current and future surveys are about to yield an unprecedented amount of information on the variable stars populating the Local Group. The European Space Agency's *Gaia* mission will result in the discovery of thousands of new RR Lyrae stars in the Milky Way halo, and it will nail down the thorny systematics affecting the calibration of the luminosity–metallicity relation.

Going to near-infrared and longer wavelengths may also enable us to reduce the uncertainties in the distances to Local Group galaxies. At present, 2–3% distance accuracy is already achievable for distance estimates to the LMC, which may soon be improved to $\sim 1\%$. For instance, the Carnegie Hubble Program, using data from the warm *Spitzer* mission, derived $(m - M)_0^{\rm LMC} = 18.477 \pm 0.034$ mag (Freedman et al. 2012). Ripepi et al. (2012) used near-infrared *VISTA* observations to derive $(m - M)_0^{\rm LMC} = 18.46 \pm 0.03$ mag, while Inno et al. (2013) found $(m - M)_0^{\rm LMC} = 18.45 \pm 0.02$ (stat.) ± 0.10 (syst.) mag based on optical/near-infrared PLR analysis of a large sample of fundamental-mode LMC Cepheids. These distances are within the mutual uncertainties of the direct, geometric distance determination based on eclipsing binaries obtained by Pietrzyński et al. (2013). The most recent SMC distance determination, which is based on mid-infrared analysis of fundamental-mode Cepheids, yields $(m - M)_0^{\rm mid-IR} = 18.96 \pm 0.01$ (stat.) ± 0.03 (syst.) mag (Scowcroft et al. 2016, see also Graczyk et al. 2014). Extant and upcoming data will soon allow a systematic characterization of variable stars, thus reducing error bars and fixing the first rungs of the distance ladder.

Water maser measurements, first applied to obtain a geometric distance to NGC 4258 (Herrnstein et al. 1999, see also Table 1), have since been extended to other nearby systems. Preliminary efforts to determine a water-maser-based distance to M33 have thus far resulted in $D_{\rm M33} = 750 \pm 140 \pm 50$ kpc—$(m - M)_0^{\rm M33} = 24.38^{+0.49}_{-0.64}$ mag (total uncertainty)—where the first uncertainty in the linear distance determination is related to uncertainties in the HI rotation model adopted and the second uncertainty comes from proper-motion measurements (cf. de Grijs 2013).

The LSST will produce a flood of data that will allow us to discover a close-to-complete sample of RR Lyrae stars out to hundreds of kpc, mapping the Milky Way's halo and its substructures. Moreover, the next generation of extremely large telescopes will push the discovery of RR Lyrae stars to few Mpc, and of classical Cepheids out to ~ 100 Mpc.

Very Long Baseline Interferometry will enable geometric distance determination out to more than 100 Mpc, including to NGC 4258, M33, UGC 3789, and NGC 6264. Combined with *a priori* information on a galaxy's rotation curve and its inclination with respect to our line of sight, we can construct a slightly warped 'tilted-ring' model of the galaxy's dynamical structure. This, in turn, allows correlation of the angular proper motion measurements with the rotational velocity information obtained in linear units and, thus, provides an independent distance measurement.

Simultaneously, the Megamaser Cosmology Project (e.g., Reid et al. 2009, 2013; Braatz et al. 2010) aims at using extragalactic maser sources to directly measure H_0 in the Hubble flow, which is clearly very challenging at distances beyond 100 Mpc. The project team's preliminary results look promising, however: using NGC 6264 ($D = 144 \pm 19$ Mpc) as a benchmark, they find $H_0 = 68 \pm 9$ km s^{-1} Mpc^{-1} (Kuo et al. 2013), which is indeed very close to the current best determinations of H_0 based on a variety of independent measures. This looks like a promising way forward to eventually construct an internally consistent distance ladder with well-defined uncertainties out to the Hubble flow (see also Thévenin et al. 2017).

With the newly discovered lenses expected from upcoming surveys, the anticipated observational follow-up, and analysis developments, we expect that measuring distances to a sample of ~ 50 lenses should be feasible within the next few years. Jee et al. (2016) showed that with a sample of 55 bright lenses from current and future imaging surveys, the resulting cosmological constraints will be highly complementary and competitive to other cosmological probes, including the cosmic microwave background, baryon acoustic oscillations, and SNe.

Acknowledgements This research was partially supported by the National Natural Science Foundation of China (NSFC; grants U1631102, 11373010, and 11633005 to R.d.G.), the Spanish Ministry of Economy and Competitiveness (MINECO; grant AYA2014-56795-P to M.M. and C.E.M.-V.), the Max Planck Society through the Max Planck Research Group (S.H.S.), and the Swiss National Science Foundation (SNSF; F.C.). We thank ISSI-BJ for hospitality and an engaging workshop. We also acknowledge both referees for their constructive reviews.

References

B.P. Abbott et al. (LIGO Scientific Collaboration Virgo Collaboration), Observation of gravitational waves from a binary black hole merger. Phys. Rev. Lett. **116**, 061102 (2016). doi:10.1103/PhysRevLett.116.061102

A. Agnello, T. Treu, STRIDES: galaxy evolution over cosmic time from new samples of gravitationally lensed quasars, in *Int'l Astron. Union General Assembly*, vol. 29 (2015), p. 57367

A. Agnello, B.C. Kelly, T. Treu, P.J. Marshall, Data mining for gravitationally lensed quasars. Mon. Not. R. Astron. Soc. **448**, 1446–1462 (2015a). doi:10.1093/mnras/stv037

A. Agnello, T. Treu, F. Ostrovski, P.L. Schechter, E.J. Buckley-Geer, H. Lin, M.W. Auger, F. Courbin, C.D. Fassnacht, J. Frieman, N. Kuropatkin, P.J. Marshall, R.G. McMahon, G. Meylan, A. More, S.H. Suyu, C.E. Rusu, D. Finley, T. Abbott, F.B. Abdalla, S. Allam, J. Annis, M. Banerji, A. Benoit-Lévy, E. Bertin, D. Brooks, D.L. Burke, A.C. Rosell, M.C. Kind, J. Carretero, C.E. Cunha, C.B. D'Andrea, L.N. da Costa, S. Desai, H.T. Diehl, J.P. Dietrich, P. Doel, T.F. Eifler, J. Estrada, A.F. Neto, B. Flaugher, P. Fosalba, D.W. Gerdes, D. Gruen, G. Gutierrez, K. Honscheid, D.J. James, K. Kuehn, O. Lahav, M. Lima, M.A.G. Maia, M. March, J.L. Marshall, P. Martini, P. Melchior, C.J. Miller, R. Miquel, R.C. Nichol, R. Ogando, A.A. Plazas, K. Reil, A.K. Romer, A. Roodman, M. Sako, E. Sanchez, B. Santiago, V. Scarpine, M. Schubnell, I. Sevilla-Noarbe, R.C. Smith, M. Soares-Santos, F. Sobreira, E. Suchyta, M.E.C. Swanson, G. Tarle, J. Thaler, D. Tucker, A.R. Walker, R.H. Wechsler, Y. Zhang, Discovery of two gravitationally lensed quasars in the dark energy survey. Mon. Not. R. Astron. Soc. **454**, 1260–1265 (2015b). doi:10.1093/mnras/stv2171

H. Aihara, R. Armstrong, S. Bickerton et al., First data release of the Hyper Suprime-Cam Subaru Strategic Program. Publ. Astron. Soc. Jpn. (2017, submitted). arXiv:1702.08449

C. Alcock, R.A. Allsman, D.R. Alves, T.S. Axelrod, A. Basu, A.C. Becker, D.P. Bennett, K.H. Cook, A.J. Drake, K.C. Freeman, M. Geha, K. Griest, L. King, M.J. Lehner, S.L. Marshall, D. Minniti, C.A. Nelson, B.A. Peterson, P. Popowski, M.R. Pratt, P.J. Quinn, C.W. Stubbs, W. Sutherland, A.B. Tomaney, T. Vandehei, D.L. Welch, The MACHO project 9 million star color-magnitude diagram of the Large Magellanic Cloud. Astron. J. **119**, 2194–2213 (2000). doi:10.1086/301326

M. Barnabè, L.V.E. Koopmans, A unifying framework for self-consistent gravitational lensing and stellar dynamics analyses of early-type galaxies. Astrophys. J. **666**, 726–746 (2007). doi:10.1086/520495

M. Barnabè, O. Czoske, L.V.E. Koopmans, T. Treu, A.S. Bolton, Two-dimensional kinematics of SLACS lenses—III. Mass structure and dynamics of early-type lens galaxies beyond $z \sim 0.1$. Mon. Not. R. Astron. Soc. **415**, 2215–2232 (2011). doi:10.1111/j.1365-2966.2011.18842.x

L.R. Bedin, G. Piotto, J. Anderson, S. Cassisi, I.R. King, Y. Momany, G. Carraro, ω Centauri: the population puzzle goes deeper. Astrophys. J. Lett. **605**, L125–L128 (2004). doi:10.1086/420847

M. Bellazzini, F.R. Ferraro, E. Pancino, Multiple stellar populations in the sextans dwarf spheroidal galaxy? Mon. Not. R. Astron. Soc. **327**, L15–L20 (2001). doi:10.1046/j.1365-8711.2001.04889.x

M. Bellazzini, R.A. Ibata, S.C. Chapman, A.D. Mackey, L. Monaco, M.J. Irwin, N.F. Martin, G.F. Lewis, E. Dalessandro, The nucleus of the Sagittarius Dsph galaxy and M54: a window on the process of galaxy nucleation. Astron. J. **136**, 1147–1170 (2008). doi:10.1088/0004-6256/136/3/1147

M. Bellazzini, G. Beccari, T.A. Oosterloo, S. Galleti, A. Sollima, M. Correnti, V. Testa, L. Mayer, M. Cignoni, F. Fraternali, S. Gallozzi, An optical and Hi study of the dwarf local group galaxy VV124

= UGC4879. A gas-poor dwarf with a stellar disk? Astron. Astrophys. **527**, A58 (2011). doi:10.1051/0004-6361/201016159

E.J. Bernard, C. Gallart, M. Monelli, A. Aparicio, S. Cassisi, E.D. Skillman, P.B. Stetson, A.A. Cole, I. Drozdovsky, S.L. Hidalgo, M. Mateo, E. Tolstoy, The ACS LCID project: RR Lyrae stars as tracers of old population gradients in the isolated dwarf spheroidal galaxy Tucana. Astrophys. J. Lett. **678**, L21–L24 (2008). doi:10.1086/588285

E.J. Bernard, M. Monelli, C. Gallart, I. Drozdovsky, P.B. Stetson, A. Aparicio, S. Cassisi, L. Mayer, A.A. Cole, S.L. Hidalgo, E.D. Skillman, E. Tolstoy, The ACS LCID project. I. Short-period variables in the isolated dwarf spheroidal galaxies Cetus and Tucana. Astrophys. J. **699**, 1742–1764 (2009). doi:10.1088/0004-637X/699/2/1742

E.J. Bernard, M. Monelli, C. Gallart, A. Aparicio, S. Cassisi, I. Drozdovsky, S.L. Hidalgo, E.D. Skillman, P.B. Stetson, The ACS LCID project. II. Faint variable stars in the isolated dwarf irregular galaxy IC 1613. Astrophys. J. **712**, 1259–1276 (2010). doi:10.1088/0004-637X/712/2/1259

E.J. Bernard, M. Monelli, C. Gallart, G. Fiorentino, S. Cassisi, A. Aparicio, A.A. Cole, I. Drozdovsky, S.L. Hidalgo, E.D. Skillman, P.B. Stetson, E. Tolstoy, The ACS LCID project—VIII. The short-period Cepheids of Leo A. Mon. Not. R. Astron. Soc. **432**, 3047–3061 (2013). doi:10.1093/mnras/stt655

S. Birrer, A. Amara, A. Refregier, Gravitational lens modeling with basis sets. Astrophys. J. **813**, 102 (2015). doi:10.1088/0004-637X/813/2/102

S. Birrer, A. Amara, A. Refregier, The mass-sheet degeneracy and time-delay cosmography: analysis of the strong lens RXJ1131-1231. J. Cosmol. Astropart. Phys. **8**, 020 (2016). doi:10.1088/1475-7516/2016/08/020

J.S. Bolton, M.G. Haehnelt, M. Viel, R.F. Carswell, Spatial fluctuations in the spectral shape of the ultraviolet background at $2 < z < 3$ and the reionization of helium. Mon. Not. R. Astron. Soc. **366**, 1378–1390 (2006). doi:10.1111/j.1365-2966.2006.09927.x

M. Bonamente, M.K. Joy, S.J. LaRoque, J.E. Carlstrom, E.D. Reese, K.S. Dawson, Determination of the cosmic distance scale from Sunyaev–Zel'dovich effect and Chandra X-ray measurements of high-redshift galaxy clusters. Astrophys. J. **647**, 25–54 (2006). doi:10.1086/505291

G. Bono, F. Caputo, V. Castellani, M. Marconi, J. Storm, Theoretical insights into the RR Lyrae K-band period-luminosity relation. Mon. Not. R. Astron. Soc. **326**, 1183–1190 (2001). doi:10.1046/j.1365-8711.2001.04655.x

G. Bono, F. Caputo, V. Castellani, M. Marconi, J. Storm, S. Degl'Innocenti, A pulsational approach to near-infrared and visual magnitudes of RR Lyr stars. Mon. Not. R. Astron. Soc. **344**, 1097–1106 (2003). doi:10.1046/j.1365-8711.2003.06878.x

G. Bono, F. Caputo, M. Di Criscienzo, RR Lyrae stars in galactic globular clusters. VI. The period-amplitude relation. Astron. Astrophys. **476**, 779–790 (2007). doi:10.1051/0004-6361:20078206

V. Bonvin, F. Courbin, S.H. Suyu, P.J. Marshall, C.E. Rusu, D. Sluse, M. Tewes, K.C. Wong, T. Collett, C.D. Fassnacht, T. Treu, M.W. Auger, S. Hilbert, L.V.E. Koopmans, G. Meylan, N. Rumbaugh, A. Sonnenfeld, C. Spiniello, H0LiCOW V. New COSMOGRAIL time delays of HE0435-1223: H_0 to 3.8% precision from strong lensing in a flat ΛCDM model. Mon. Not. R. Astron. Soc. **465**, 4914–4930 (2017). doi:10.1093/mnras/stw3006

C. Bot, N. Ysard, D. Paradis, J.P. Bernard, G. Lagache, F.P. Israel, W.F. Wall, Submillimeter to centimeter excess emission from the Magellanic clouds. II. On the nature of the excess. Astron. Astrophys. **523**, A20 (2010). doi:10.1051/0004-6361/201014986

J.A. Braatz, M.J. Reid, E.M.L. Humphreys, C. Henkel, J.J. Condon, K.Y. Lo, The Megamaser Cosmology Project. II. The angular-diameter distance to UGC 3789. Astrophys. J. **718**, 657–665 (2010). doi:10.1088/0004-637X/718/2/657

R.S. Bussmann, D. Riechers, A. Fialkov, J. Scudder, C.C. Hayward, W.I. Cowley, J. Bock, J. Calanog, S.C. Chapman, A. Cooray, F. De Bernardis, D. Farrah, H. Fu, R. Gavazzi, R. Hopwood, R.J. Ivison, M. Jarvis, C. Lacey, A. Loeb, S.J. Oliver, I. Pérez-Fournon, D. Rigopoulou, I.G. Roseboom, D. Scott, A.J. Smith, L. Vieira, J.D. Wang, J. Wardlow, HerMES: ALMA imaging of Herschel-selected dusty star-forming galaxies. Astrophys. J. **812**, 43 (2015). doi:10.1088/0004-637X/812/1/43

C. Cacciari, G. Clementini, Globular cluster distances from RR Lyrae stars, in *Stellar Candles for the Extragalactic Distance Scale*, ed. by D. Alloin, W. Gieren. Lect. Notes Phys., vol. 635 (Springer, Berlin, 2003), pp. 105–122. doi:10.1007/978-3-540-39882-0_6

J.A.R. Caldwell, I.M. Coulson, The geometry and distance of the Magellanic clouds from Cepheid variables. Mon. Not. R. Astron. Soc. **218**, 223–246 (1986). doi:10.1093/mnras/218.2.223

F. Caputo, V. Castellani, M. Marconi, V. Ripepi, Pulsational M_V versus [Fe/H] relation(s) for globular cluster RR Lyrae variables. Mon. Not. R. Astron. Soc. **316**, 819–826 (2000). doi:10.1046/j.1365-8711.2000.03591.x

E. Carretta, R.G. Gratton, G. Clementini, F. Fusi Pecci, Distances, ages, and epoch of formation of globular clusters. Astrophys. J. **533**, 215–235 (2000). doi:10.1086/308629

E. Carretta, A. Bragaglia, R. Gratton, V. D'Orazi, S. Lucatello, Intrinsic iron spread and a new metallicity scale for globular clusters. Astron. Astrophys. **508**, 695–706 (2009). doi:10.1051/0004-6361/200913003

S. Cassisi, M. Salaris, *Old Stellar Populations: How to Study the Fossil Record of Galaxy Formation* (Wiley, New York, 2013)

M. Catelan, B.J. Pritzl, H.A. Smith, The RR Lyrae period–luminosity relation. I. Theoretical calibration. Astrophys. J. Suppl. Ser. **154**, 633–649 (2004). doi:10.1086/422916

M. Catelan, H.A. Smith, *Pulsating Stars* (Wiley, New York, 2015)

B. Chaboyer, Globular cluster distance determinations, in *Post-Hipparcos Cosmic Candles*. Astrophys. Space Sci. Libr., vol. 237 (1999), pp. 111–124. doi:10.1007/978-94-011-4734-7_7

J.H.H. Chan, S.H. Suyu, T. Chiueh, A. More, P.J. Marshall, J. Coupon, M. Oguri, P. Price, Chitah: strong-gravitational-lens hunter in imaging surveys. Astrophys. J. **807**, 138 (2015). doi:10.1088/0004-637X/807/2/138

G.C.F. Chen, S.H. Suyu, K.C. Wong, C.D. Fassnacht, T. Chiueh, A. Halkola, I. Hu, M.W. Auger, L.V.E. Koopmans, D.J. Lagattuta, J.P. McKean, S. Vegetti, SHARP—III: first use of adaptive optics imaging to constrain cosmology with gravitational lens time delays. Mon. Not. R. Astron. Soc. **462**, 3457–3475 (2016). doi:10.1093/mnras/stw991

M. Cignoni, A.A. Cole, M. Tosi, J.S. Gallagher, E. Sabbi, J. Anderson, E.K. Grebel, A. Nota, Mean age gradient and asymmetry in the star formation history of the Small Magellanic Cloud. Astrophys. J. **775**, 83 (2013). doi:10.1088/0004-637X/775/2/83

M.-R.L. Cioni, G. Clementini, L. Girardi, R. Guandalini, M. Gullieuszik, B. Miszalski, M.-I. Moretti, V. Ripepi, S. Rubele, G. Bagheri, K. Bekki, N. Cross, W.J.G. de Blok, R. de Grijs, J.P. Emerson, C.J. Evans, B. Gibson, E. Gonzales-Solares, M.A.T. Groenewegen, M. Irwin, V.D. Ivanov, J. Lewis, M. Marconi, J.-B. Marquette, C. Mastropietro, B. Moore, R. Napiwotzki, T. Naylor, J.M. Oliveira, M. Read, E. Sutorius, J.T. van Loon, M.I. Wilkinson, P.R. Wood, Astron. Astrophys. **527**, A116 (2011). doi:10.1051/0004-6361/201016137

G. Clementini, L. Federici, C. Corsi, C. Cacciari, M. Bellazzini, H.A. Smith, RR Lyrae variables in the globular clusters of M31: a first detection of likely candidates. Astrophys. J. Lett. **559**, L109–L112 (2001). doi:10.1086/323973

G. Clementini, R. Gratton, A. Bragaglia, E. Carretta, L. Di Fabrizio, M. Maio, Distance to the Large Magellanic Cloud: the RR Lyrae stars. Astron. J. **125**, 1309–1329 (2003). doi:10.1086/367773

G. Clementini, M. Cignoni, R. Contreras Ramos, L. Federici, V. Ripepi, M. Marconi, M. Tosi, I. Musella, Variability and star formation in Leo T, the lowest luminosity star-forming galaxy known today. Astrophys. J. **756**, 108 (2012). doi:10.1088/0004-637X/756/2/108

A.A. Cole, E.D. Skillman, E. Tolstoy, J.S. Gallagher iii, A. Aparicio, A.E. Dolphin, C. Gallart, S.L. Hidalgo, A. Saha, P.B. Stetson, D.R. Weisz, A. Leo, A late-blooming survivor of the epoch of reionization in the local group. Astrophys. J. Lett. **659**, L17–L20 (2007). doi:10.1086/516711

A.A. Cole, D.R. Weisz, A.E. Dolphin, E.D. Skillman, A.W. McConnachie, A.M. Brooks, R. Leaman, Delayed star formation in isolated dwarf galaxies: Hubble space telescope star formation history of the Aquarius dwarf irregular. Astrophys. J. **795**, 54 (2014). doi:10.1088/0004-637X/795/1/54

T.E. Collett, M.W. Auger, Cosmological constraints from the double source plane lens SDSSJ0946+1006. Mon. Not. R. Astron. Soc. **443**, 969–976 (2014). doi:10.1093/mnras/stu1190

T.E. Collett, P.J. Marshall, M.W. Auger, S. Hilbert, S.H. Suyu, Z. Greene, T. Treu, C.D. Fassnacht, L.V.E. Koopmans, M. Bradač, R.D. Blandford, Reconstructing the lensing mass in the Universe from photometric catalogue data. Mon. Not. R. Astron. Soc. **432**, 679–692 (2013). doi:10.1093/mnras/stt504

M.L.M. Collins, S.C. Chapman, R.M. Rich, R.A. Ibata, N.F. Martin, M.J. Irwin, N.F. Bate, G.F. Lewis, J. Peñarrubia, N. Arimoto, C.M. Casey, A.M.N. Ferguson, A. Koch, A.W. McConnachie, N. Tanvir, A kinematic study of the Andromeda dwarf spheroidal system. Astrophys. J. **768**, 172 (2013). doi:10.1088/0004-637X/768/2/172

M.L.M. Collins, N.F. Martin, R.M. Rich, R.A. Ibata, S.C. Chapman, A.W. McConnachie, A.M. Ferguson, M.J. Irwin, G.F. Lewis, Comparing the observable properties of dwarf galaxies on and off the Andromeda plane. Astrophys. J. Lett. **799**, L13 (2015). doi:10.1088/2041-8205/799/1/L13

A.R. Conn, R.A. Ibata, G.F. Lewis, Q.A. Parker, D.B. Zucker, N.F. Martin, A.W. McConnachie, M.J. Irwin, N. Tanvir, M.A. Fardal, A.M.N. Ferguson, S.C. Chapman, D. Valls-Gabaud, A Bayesian approach to locating the red giant branch tip magnitude. II. Distances to the satellites of M31. Astrophys. J. **758**, 11 (2012). doi:10.1088/0004-637X/758/1/11

G. Coppola, M. Marconi, P.B. Stetson, G. Bono, V.F. Braga, V. Ripepi, M. Dall'Ora, I. Musella, R. Buonanno, M. Fabrizio, I. Ferraro, G. Fiorentino, G. Iannicola, M. Monelli, M. Nonino, F. Thévenin, A.R. Walker, The Carina project IX: on hydrogen and helium burning variables. Astrophys. J. **814**, 71 (2015). doi:10.1088/0004-637X/814/1/71

F. Courbin, P. Magain, M. Kirkove, S. Sohy, A method for spatial deconvolution of spectra. Astrophys. J. **529**, 1136–1144 (2000). doi:10.1086/308291

F. Courbin, A. Eigenbrod, C. Vuissoz, G. Meylan, P. Magain, COSMOGRAIL: the COSmological MOnitoring of GRAvItational Lenses, in *Gravitational Lensing Impact on Cosmology*, ed. by Y. Mellier, G. Meylan. Int'l Astron. Union Symp., vol. 225 (2005), pp. 297–303. doi:10.1017/S1743921305002097

F. Courbin, V. Chantry, Y. Revaz, D. Sluse, C. Faure, M. Tewes, E. Eulaers, M. Koleva, I. Asfandiyarov, S. Dye, P. Magain, H. van Winckel, J. Coles, P. Saha, M. Ibrahimov, G. Meylan, COSMOGRAIL: the COSmological MOnitoring of GRAvItational Lenses. IX. Time delays, lens dynamics and baryonic fraction in HE 0435-1223. Astron. Astrophys. **536**, A53 (2011). doi:10.1051/0004-6361/201015709

S. Crandall, B. Ratra, Non-Gaussian error distributions of LMC distance moduli measurements. Astrophys. J. **815**, 87 (2015). doi:10.1088/0004-637X/815/2/87

H.H. Crowl, A. Sarajedini, A.E. Piatti, D. Geisler, E. Bica, J.J. Clariá, J.F.C. Santos Jr., The line-of-sight depth of populous clusters in the Small Magellanic Cloud. Astron. J. **122**, 220–231 (2001). doi:10.1086/321128

F. Cusano, G. Clementini, A. Garofalo, M. Cignoni, L. Federici, M. Marconi, I. Musella, V. Ripepi, K. Boutsia, M. Fumana, S. Gallozzi, V. Testa, Dwarf spheroidal satellites of M31. I. Variable stars and stellar populations in Andromeda XIX. Astrophys. J. **779**, 7 (2013). doi:10.1088/0004-637X/779/1/7

F. Cusano, A. Garofalo, G. Clementini, M. Cignoni, L. Federici, M. Marconi, I. Musella, V. Ripepi, R. Speziali, E. Sani, R. Merighi, Variable stars and stellar populations in Andromeda XXI. II. Another merged galaxy satellite of M31? Astrophys. J. **806**, 200 (2015). doi:10.1088/0004-637X/806/2/200

G.S. Da Costa, T.E. Armandroff, Standard globular cluster giant branches in the $(M_I, V - I_0)$ plane. Astron. J. **100**, 162–181 (1990). doi:10.1086/115500

G.S. Da Costa, T.E. Armandroff, N. Caldwell, P. Seitzer, The dwarf spheroidal companions to M31: WFPC2 observations of Andromeda I. Astron. J. **112**, 2576–2595 (1996). doi:10.1086/118204

G.S. Da Costa, T.E. Armandroff, N. Caldwell, P. Seitzer, The dwarf spheroidal companions to M31: WFPC2 observations of Andromeda II. Astron. J. **119**, 705–726 (2000). doi:10.1086/301223

G.S. Da Costa, T.E. Armandroff, N. Caldwell, The dwarf spheroidal companions to M31: WFPC2 observations of Andromeda III. Astron. J. **124**, 332–348 (2002). doi:10.1086/340965

G.S. Da Costa, M. Rejkuba, H. Jerjen, E.K. Grebel, Ancient stars beyond the local group: RR Lyrae variables and blue horizontal branch stars in sculptor group dwarf galaxies. Astrophys. J. Lett. **708**, L121–L125 (2010). doi:10.1088/2041-8205/708/2/L121

N. Dalal, D.E. Holz, S.A. Hughes, B. Jain, Short GRB and binary black hole standard sirens as a probe of dark energy. Phys. Rev. D **74**, 063006 (2006). doi:10.1103/PhysRevD.74.063006

T. Abbott, F.B. Abdalla, J. Aleksić, S. Allam, A. Amara, D. Bacon, E. Balbinot, M. Banerji, K. Bechtol, A. Benoit-Lévy, G.M. Bernstein, E. Bertin, J. Blazek, C. Bonnett, S. Bridle, D. Brooks, R.J. Brunner, E. Buckley-Geer, D.L. Burke, G.B. Caminha, D. Capozzi, J. Carlsen, A. Carnero-Rosell, M. Carollo, M. Carrasco-Kind, J. Carretero, F.J. Castander, L. Clerkin, T. Collett, C. Conselice, M. Crocce, C.E. Cunha, C.B. D'Andrea, L.N. da Costa, T.M. Davis, S. Desai, H.T. Diehl, J.P. Dietrich, S. Dodelson, P. Doel, A. Drlica-Wagner, J. Estrada, J. Etherington, A.E. Evrard, J. Fabbri, D.A. Finley, B. Flaugher, R.J. Foley, P. Fosalba, J. Frieman, J. García-Bellido, E. Gaztanaga, D.W. Gerdes, T. Giannantonio, D.A. Goldstein, D. Gruen, R.A. Gruendl, P. Guarnieri, G. Gutierrez, W. Hartley, K. Honscheid, B. Jain, D.J. James, T. Jeltema, S. Jouvel, R. Kessler, A. King, D. Kirk, R. Kron, K. Kuehn, N. Kuropatkin, O. Lahav, T.S. Li, M. Lima, H. Lin, M.A.G. Maia, M. Makler, M. Manera, C. Maraston, J.L. Marshall, P. Martini, R.G. McMahon, P. Melchior, A. Merson, C.J. Miller, R. Miquel, J.J. Mohr, X. Morice-Atkinson, K. Naidoo, E. Neilsen, R.C. Nichol, B. Nord, R. Ogando, F. Ostrovski, A. Palmese, A. Papadopoulos, H.V. Peiris, J. Peoples, W.J. Percival, A.A. Plazas, S.L. Reed, A. Refregier, A.K. Romer, A. Roodman, A. Ross, E. Rozo, E.S. Rykoff, I. Sadeh, M. Sako, C. Sánchez, E. Sanchez, B. Santiago, V. Scarpine, M. Schubnell, I. Sevilla-Noarbe, E. Sheldon, M. Smith, R.C. Smith, M. Soares-Santos, F. Sobreira, M. Soumagnac, E. Suchyta, M. Sullivan, M. Swanson, G. Tarle, J. Thaler, D. Thomas, R.C. Thomas, D. Tucker, J.D. Vieira, V. Vikram, A.R. Walker, R.H. Wechsler, J. Weller, W. Wester, L. Whiteway, H. Wilcox, B. Yanny, Y. Zhang, J. Zuntz (Dark Energy Survey Collaboration), The dark energy survey: more than dark energy—an overview. Mon. Not. R. Astron. Soc. **460**, 1270–1299 (2016). doi:10.1093/mnras/stw641

T.J.L. de Boer, E. Tolstoy, V. Hill, A. Saha, K. Olsen, E. Starkenburg, B. Lemasle, M.J. Irwin, G. Battaglia, The star formation and chemical evolution history of the sculptor dwarf spheroidal galaxy. Astron. Astrophys. **539**, A103 (2012). doi:10.1051/0004-6361/201118378

R. de Grijs, *An Introduction to Distance Measurement in Astronomy* (Wiley, New York, 2011)

R. de Grijs, Advancing the physics of cosmic distances: conference summary, in *Advancing the Physics of Cosmic Distances*. Int'l Astron. Union Symp., vol. 289 (2013), pp. 351–360. doi:10.1017/S1743921312021709

R. de Grijs, J.E. Wicker, G. Bono, Clustering of local group distances: publication bias or correlated measurements? I. The Large Magellanic Cloud. Astron. J. **147**, 122 (2014). doi:10.1088/0004-6256/147/5/122

R. de Grijs, G. Bono, Clustering of local group distances: publication bias or correlated measurements? II. M31 and beyond. Astron. J. **148**, 117 (2014). doi:10.1088/0004-6256/148/1/17

R. de Grijs, G. Bono, Clustering of local group distances: publication bias or correlated measurements? III. The Small Magellanic Cloud. Astron. J. **149**, 179 (2015). doi:10.1088/0004-6256/149/6/179

R. de Grijs, G. Bono, Clustering of local group distances: publication bias or correlated measurements? IV. The galactic center. Astrophys. J. Suppl. Ser. **227**, 5 (2016). doi:10.3847/0067-0049/227/1/5

K. Dobashi, J.-P. Bernard, A. Kawamura, F. Egusa, A. Hughes, D. Paradis, C. Bot, W.T. Reach, Extinction map of the Small Magellanic Cloud based on the SIRIUS and 6X 2MASS point source catalogs. Astron. J. **137**, 5099–5109 (2009). doi:10.1088/0004-6256/137/6/5099

E. D'Onghia, G. Besla, T.J. Cox, L. Hernquist, Resonant stripping as the origin of dwarf spheroidal galaxies. Nature **460**, 605–607 (2009). doi:10.1038/nature08215

S. Dye, S.J. Warren, Decomposition of the visible and dark matter in the Einstein ring 0047-2808 by semi-linear inversion. Astrophys. J. **623**, 31–41 (2005). doi:10.1086/428340

A. Eigenbrod, F. Courbin, C. Vuissoz, G. Meylan, P. Saha, S. Dye, COSMOGRAIL: the COSmological MOnitoring of GRAvItational Lenses. I. How to sample the light curves of gravitationally lensed quasars to measure accurate time delays. Astron. Astrophys. **436**, 25–35 (2005). doi:10.1051/0004-6361:20042422

A. Eigenbrod, F. Courbin, G. Meylan, C. Vuissoz, P. Magain, COSMOGRAIL: the COSmological MOnitoring of GRAvItational Lenses. III. Redshift of the lensing galaxy in eight gravitationally lensed quasars. Astron. Astrophys. **451**, 759–766 (2006). doi:10.1051/0004-6361:20054454

A. Eigenbrod, F. Courbin, G. Meylan, COSMOGRAIL: the COSmological MOnitoring of GRAvItational Lenses. VI. Redshift of the lensing galaxy in seven gravitationally lensed quasars. Astron. Astrophys. **465**, 51–56 (2007). doi:10.1051/0004-6361:20066939

J.D. Ellithorpe, C.S. Kochanek, J.N. Hewitt, Visibility LensClean and the reliability of deconvolved radio images. Astrophys. J. **464**, 556–567 (1996). doi:10.1086/177346

R. Fadely, C.R. Keeton, R. Nakajima, G.M. Bernstein, Improved constraints on the gravitational lens Q0957+561. II. Strong lensing. Astrophys. J. **711**, 246–267 (2010). doi:10.1088/0004-637X/711/1/246

E.E. Falco, M.V. Gorenstein, I.I. Shapiro, On model-dependent bounds on H_0 from gravitational images application of Q0957+561A,B. Astrophys. J. Lett. **289**, L1–L4 (1985). doi:10.1086/184422

C.D. Fassnacht, L.V.E. Koopmans, K.C. Wong, Galaxy number counts and implications for strong lensing. Mon. Not. R. Astron. Soc. **410**, 2167–2179 (2011). doi:10.1111/j.1365-2966.2010.17591.x

A.M.N. Ferguson, A.D. Mackey, Substructure and tidal streams in the Andromeda galaxy and its satellites, in *Astrophys. Space Sci. Libr.*, vol. 420, ed. by H.J. Newberg, J.L. Carlin (2016), pp. 191–217. doi:10.1007/978-3-319-19336-6_8

L. Ferrarese, J.R. Mould, R.C. Kennicutt Jr., J. Huchra, H.C. Ford, W.L. Freedman, P.B. Stetson, B.F. Madore, S. Sakai, B.K. Gibson, J.A. Graham, S.M. Hughes, G.D. Illingworth, D.D. Kelson, L. Macri, K. Sebo, N.A. Silbermann, The Hubble space telescope key project on the extragalactic distance scale. XXVI. The calibration of population II secondary distance indicators and the value of the Hubble constant. Astrophys. J. **529**, 745–767 (2000). doi:10.1086/308309

G. Fiorentino, M. Limongi, F. Caputo, M. Marconi, Synthetic properties of bright metal-poor variables. I. 'Anomalous' Cepheids. Astron. Astrophys. **460**, 155–166 (2006). doi:10.1051/0004-6361:20065349

G. Fiorentino, A. Monachesi, S.C. Trager, T.R. Lauer, A. Saha, K.J. Mighell, W. Freedman, A. Dressler, C. Grillmair, E. Tolstoy, RR Lyrae variables in M32 and the disk of M31. Astrophys. J. **708**, 817–833 (2010). doi:10.1088/0004-637X/708/1/817

G. Fiorentino, R. Contreras Ramos, E. Tolstoy, G. Clementini, A. Saha, The ancient stellar population of M32: RR Lyrae variable stars confirmed. Astron. Astrophys. **539**, A138 (2012a). doi:10.1051/0004-6361/201117434

G. Fiorentino, P.B. Stetson, M. Monelli, G. Bono, E.J. Bernard, A. Pietrinferni, On the central helium-burning variable stars of the LeoI dwarf spheroidal galaxy. Astrophys. J. Lett. **759**, L12 (2012b). doi:10.1088/2041-8205/759/1/L12

G. Fiorentino, G. Bono, M. Monelli, P.B. Stetson, E. Tolstoy, C. Gallart, M. Salaris, C.E. Martínez-Vásquez, E.J. Bernard, Weak galactic halo-dwarf spheroidal connection from RR Lyrae stars. Astrophys. J. Lett. **798**, L12 (2015). doi:10.1088/2041-8205/798/1/L12

B. Flaugher, H.T. Diehl, K. Honscheid, T.M.C. Abbott, O. Alvarez, R. Angstadt, J.T. Annis, M. Antonik, O. Ballester, L. Beaufore, G.M. Bernstein, R.A. Bernstein, B. Bigelow, M. Bonati, D. Boprie, D. Brooks, E.J. Buckley-Geer, J. Campa, L. Cardiel-Sas, F.J. Castander, J. Castilla, H. Cease, J.M. Cela-Ruiz, S. Chappa, E. Chi, C. Cooper, L.N. da Costa, E. Dede, G. Derylo, D.L. DePoy, J. de Vicente, P. Doel, A. Drlica-Wagner, J. Eiting, A.E. Elliott, J. Emes, J. Estrada, A. Fausti Neto, D.A. Finley, R. Flores, J. Frieman, D. Gerdes, M.D. Gladders, B. Gregory, G.R. Gutierrez, J. Hao, S.E. Holland, S. Holm, D. Huffman, C. Jackson, D.J. James, M. Jonas, A. Karcher, I. Karliner, S. Kent, R. Kessler, M. Kozlovsky, R.G. Kron, D. Kubik, K. Kuehn, S. Kuhlmann, K. Kuk, O. Lahav, A. Lathrop, J. Lee, M.E. Levi, P. Lewis, T.S. Li,

I. Mandrichenko, J.L. Marshall, G. Martinez, K.W. Merritt, R. Miquel, F. Muñoz, E.H. Neilsen, R.C. Nichol, B. Nord, R. Ogando, J. Olsen, N. Palaio, K. Patton, J. Peoples, A.A. Plazas, J. Rauch, K. Reil, J.P. Rheault, N.A. Roe, H. Rogers, A. Roodman, E. Sanchez, V. Scarpine, R.H. Schindler, R. Schmidt, R. Schmitt, M. Schubnell, K. Schultz, P. Schurter, L. Scott, S. Serrano, T.M. Shaw, R.C. Smith, M. Soares-Santos, A. Stefanik, W. Stuermer, E. Suchyta, A. Sypniewski, G. Tarle, J. Thaler, R. Tighe, C. Tran, D. Tucker, A.R. Walker, G. Wang, M. Watson, C. Weaverdyck, W. Wester, R. Woods, B. Yanny (DES Collaboration), The dark energy camera. Astron. J. **150**, 150 (2015). doi:10.1088/0004-6256/150/5/150

T. Foster, B. Cooper, Structure and dynamics of the Milky Way: the evolving picture, in *The Dynamic Interstellar Medium: A Celebration of the Canadian Galactic Plane Survey*, ed. by R. Kothes, T.L. Landecker, A.G. Willis. Astron. Soc. Pac. Conf. Proc., vol. 438 (2010), pp. 16–30

C. Francis, E. Anderson, Two estimates of the distance to the galactic centre. Mon. Not. R. Astron. Soc. **441**, 1105–1114 (2014). doi:10.1093/mnras/stu631

F. Fraternali, E. Tolstoy, M.J. Irwin, A.A. Cole, Life at the periphery of the local group: the kinematics of the Tucana dwarf galaxy. Astron. Astrophys. **499**, 121–128 (2009). doi:10.1051/0004-6361/200810830

W.L. Freedman, B.F. Madore, B.K. Gibson, L. Ferrarese, D.D. Kelson, S. Sakai, J.R. Mould, R.C. Kennicutt Jr., H.C. Ford, J.A. Graham, J.P. Huchra, S.M.G. Hughes, G.D. Illingworth, L.M. Macri, P.B. Stetson, Final results from the Hubble space telescope key project to measure the Hubble constant. Astrophys. J. **553**, 47–72 (2001). doi:10.1086/320638

W.L. Freedman, B.F. Madore, V. Scowcroft, C. Burns, A. Monson, S.E. Persson, M. Seibert, J. Rigby, Carnegie Hubble program: a mid-infrared calibration of the Hubble constant. Astrophys. J. **758**, 24 (2012). doi:10.1088/0004-637X/758/1/24

C. Gallart, A. Aparicio, W.L. Freedman, B.F. Madore, D. Martínez-Delgado, P.B. Stetson, The variable-star population in Phoenix: coexistence of anomalous and short-period classical Cepheids and detection of RR Lyrae variables. Astron. J. **127**, 1486–1501 (2004). doi:10.1086/381955

C. Gallart, M. Monelli, L. Mayer, A. Aparicio, G. Battaglia, E.J. Bernard, S. Cassisi, A.A. Cole, A.E. Dolphin, I. Drozdovsky, S.L. Hidalgo, J.F. Navarro, S. Salvadori, E.D. Skillman, P.B. Stetson, D.R. Weisz, The ACS LCID project: on the origin of dwarf galaxy types—a manifestation of the halo assembly bias? Astrophys. J. Lett. **811**, L18 (2015). doi:10.1088/2041-8205/811/2/L18

R. Gavazzi, P.J. Marshall, T. Treu, A. Sonnenfeld, RINGFINDER: automated detection of galaxy-scale gravitational lenses in ground-based multi-filter imaging data. Astrophys. J. **785**, 144 (2014). doi:10.1088/0004-637X/785/2/144

M.V. Gorenstein, I.I. Shapiro, E.E. Falco, Degeneracies in parameter estimates for models of gravitational lens systems. Astrophys. J. **327**, 693–711 (1988). doi:10.1086/166226

D. Graczyk, G. Pietrzyński, I.B. Thompson, W. Gieren, B. Pilecki, P. Konorski, A. Udalski, I. Soszyński, S. Villanova, M. Górski, K. Suchomska, P. Karczmarek, R.-P. Kudritzki, F. Bresolin, A. Gallenne, The Araucaria Project. The distance to the Small Magellanic Cloud from late-type eclipsing binaries. Astrophys. J. **780**, 59 (2014). doi:10.1088/0004-637X/780/1/59

R.G. Gratton, A. Bragaglia, G. Clementini, E. Carretta, L. Di Fabrizio, M. Maio, E. Taribello, Metal abundances of RR Lyrae stars in the bar of the Large Magellanic Cloud. Astron. Astrophys. **421**, 937–952 (2004). doi:10.1051/0004-6361:20035840

Z.S. Greene, S.H. Suyu, T. Treu, S. Hilbert, M.W. Auger, T.E. Collett, P.J. Marshall, C.D. Fassnacht, R.D. Blandford, M. Bradač, L.V.E. Koopmans, Improving the precision of time-delay cosmography with observations of galaxies along the line of sight. Astrophys. J. **768**, 39 (2013). doi:10.1088/0004-637X/768/1/39

M.A.T. Groenewegen, LMC and SMC Cepheids: combining OGLE with DENIS and 2MASS infrared data. Astron. Astrophys. **363**, 901–916 (2000)

R. Haschke, E.K. Grebel, S. Duffau, Three-dimensional maps of the Magellanic clouds using RR Lyrae stars and Cepheids. II. The Small Magellanic Cloud. Astron. J. **144**, 107 (2012). doi:10.1088/0004-6256/144/4/107

D. Hatt, R.L. Beaton, W.L. Freedman, B.F. Madore, I.S. Jang, T.J. Hoyt, M.G. Lee, A.J. Monson, J.A. Rich, V. Scowcroft, M. Seibert, The Carnegie–Chicago Hubble Program. II. The Distance to IC 1613: The Tip of the Red Giant Branch and RR Lyrae Period–Luminosity Relations. Astrophys. J. (2017, submitted). arXiv:1703.06468

J.R. Herrnstein, J.M. Moran, L.J. Greenhill, P.J. Diamond, M. Inoue, N. Nakai, M. Miyoshi, C. Henkel, A. Riess, A geometric distance to the galaxy NGC4258 from orbital motions in a nuclear gas disk. Nature **400**, 539–541 (1999). doi:10.1038/22972

Y.D. Hezaveh, D.P. Marrone, C.D. Fassnacht, J.S. Spilker, J.D. Vieira, J.E. Aguirre, K.A. Aird, M. Aravena, M.L.N. Ashby, M. Bayliss, B.A. Benson, L.E. Bleem, M. Bothwell, M. Brodwin, J.E. Carlstrom, C.L. Chang, S.C. Chapman, T.M. Crawford, A.T. Crites, C. De Breuck, T. de Haan, M.A. Dobbs, E.B. Fomalont, E.M. George, M.D. Gladders, A.H. Gonzalez, T.R. Greve, N.W. Halverson, F.W. High, G.P. Holder, W.L. Holzapfel, S. Hoover, J.D. Hrubes, K. Husband, T.R. Hunter, R. Keisler, A.T. Lee, E.M.

Leitch, M. Lueker, D. Luong-Van, M. Malkan, V. McIntyre, J.J. McMahon, J. Mehl, K.M. Menten, S.S. Meyer, L.M. Mocanu, E.J. Murphy, T. Natoli, S. Padin, T. Plagge, C.L. Reichardt, A. Rest, J. Ruel, J.E. Ruhl, K. Sharon, K.K. Schaffer, L. Shaw, E. Shirokoff, B. Stalder, Z. Staniszewski, A.A. Stark, K. Story, K. Vanderlinde, A. Weiß, N. Welikala, R. Williamson, ALMA observations of SPT-discovered, strongly lensed, dusty, star-forming galaxies. Astrophys. J. **767**, 132 (2013). doi:10.1088/0004-637X/767/2/132

Y.D. Hezaveh, N. Dalal, D.P. Marrone, Y.Y. Mao, W. Morningstar, D. Wen, R.D. Blandford, J.E. Carlstrom, C.D. Fassnacht, G.P. Holder, A. Kemball, P.J. Marshall, N. Murray, L. Perreault Levasseur, J.D. Vieira, R.H. Wechsler, Detection of lensing substructure using ALMA observations of the dusty galaxy SDP.81. Astrophys. J. **823**, 37 (2016). doi:10.3847/0004-637X/823/1/37

S.L. Hidalgo, A. Aparicio, D. Martínez-Delgado, C. Gallart, On the extended structure of the Phoenix dwarf galaxy. Astrophys. J. **705**, 704–716 (2009). doi:10.1088/0004-637X/705/1/704

S. Hilbert, S.D.M. White, J. Hartlap, P. Schneider, Strong lensing optical depths in a ΛCDM universe. Mon. Not. R. Astron. Soc. **382**, 121–132 (2007). doi:10.1111/j.1365-2966.2007.12391.x

S. Hilbert, J. Hartlap, S.D.M. White, P. Schneider, Ray-tracing through the millennium simulation: born corrections and lens-lens coupling in cosmic shear and galaxy-galaxy lensing. Astron. Astrophys. **499**, 31–43 (2009). doi:10.1051/0004-6361/200811054

F. Aharonian et al. (Hitomi Collaboration), The quiescent intracluster medium in the core of the Perseus cluster. Nature **535**, 117–121 (2016). doi:10.1038/nature18627

N. Ho, M. Geha, R.R. Munoz, P. Guhathakurta, J. Kalirai, K.M. Gilbert, E. Tollerud, J. Bullock, R.L. Beaton, S.R. Majewski, Stellar kinematics of the Andromeda II dwarf spheroidal galaxy. Astrophys. J. **758**, 124 (2012). doi:10.1088/0004-637X/758/2/124

N. Ho, M. Geha, E.J. Tollerud, R. Zinn, P. Guhathakurta, L.C. Vargas, Metallicity evolution of the six most luminous M31 dwarf satellites. Astrophys. J. **798**, 77 (2015). doi:10.1088/0004-637X/798/2/77

D.E. Holz, S.A. Hughes, Using gravitational-wave standard sirens. Astrophys. J. **629**, 15–22 (2005). doi:10.1086/431341

R. Ibata, M. Irwin, G. Lewis, A.M.N. Ferguson, N. Tanvir, A giant stream of metal-rich stars in the halo of the galaxy M31. Nature **412**, 49–52 (2001)

N. Inada, M. Oguri, M.S. Shin, I. Kayo, M.A. Strauss, T. Morokuma, C.E. Rusu, M. Fukugita, C.S. Kochanek, G.T. Richards, D.P. Schneider, D.G. York, N.A. Bahcall, J.A. Frieman, P.B. Hall, R.L. White, The sloan digital sky survey quasar lens search. V. Final catalog from the seventh data release. Astron. J. **143**, 119 (2012). doi:10.1088/0004-6256/143/5/119

L. Inno, N. Matsunaga, G. Bono, F. Caputo, R. Buonanno, K. Genovali, C.D. Laney, M. Marconi, A.M. Piersimoni, F. Primas, M. Romaniello, On the distance of the Magellanic clouds using Cepheid NIR and optical–NIR period–Wesenheit relations. Astrophys. J. **764**, 84 (2013). doi:10.1088/0004-637X/764/1/84

L. Inno, G. Bono, N. Matsunaga, G. Fiorentino, M. Marconi, B. Lemasle, R. da Silva, I. Soszyński, A. Udalski, M. Romaniello, H.-W. Rix, The panchromatic view of the Magellanic clouds from classical Cepheids. I. Distance, reddening, and geometry of the Large Magellanic Cloud disk. Astrophys. J. **832**, 176 (2016). doi:10.3847/0004-637X/832/2/176

N. Jackson, H. Rampadarath, E.O. Ofek, M. Oguri, M.S. Shin, New lensed quasars from the MUSCLES survey. Mon. Not. R. Astron. Soc. **419**, 2014–2024 (2012). doi:10.1111/j.1365-2966.2011.19857.x

I. Jee, E. Komatsu, S.H. Suyu, Measuring angular diameter distances of strong gravitational lenses. J. Cosmol. Astropart. Phys. **2015**(11), 033 (2015). doi:10.1088/1475-7516/2015/11/033

I. Jee, E. Komatsu, S.H. Suyu, D. Huterer, Time-delay cosmography: increased leverage with angular diameter distances. J. Cosmol. Astropart. Phys. **2016**(04), 031 (2016). doi:10.1088/1475-7516/2016/04/031

E.J. Jeffery, E. Smith, T.M. Brown, A.V. Sweigart, J.S. Kalirai, H.C. Ferguson, P. Guhathakurta, A. Renzini, R.M. Rich, HST/ACS observations of RR Lyrae stars in six ultra-deep fields of M31. Astron. J. **141**, 171 (2011). doi:10.1088/0004-6256/141/5/171

R. Jimenez, A. Loeb, Constraining cosmological parameters based on relative galaxy ages. Astrophys. J. **573**, 37–42 (2002). doi:10.1086/340549

R. Jimenez, L. Verde, T. Treu, D. Stern, Constraints on the equation of state of dark energy and the Hubble constant from stellar ages and the cosmic microwave background. Astrophys. J. **593**, 622–629 (2003). doi:10.1086/376595

R. Joseph, F. Courbin, R.B. Metcalf, C. Giocoli, P. Hartley, N. Jackson, F. Bellagamba, J.P. Kneib, L. Koopmans, G. Lemson, M. Meneghetti, G. Meylan, M. Petkova, S. Pires, A PCA-based automated finder for galaxy-scale strong lenses. Astron. Astrophys. **566**, A63 (2014). doi:10.1051/0004-6361/201423365

E. Jullo, J.P. Kneib, Multiscale cluster lens mass mapping—I. Strong lensing modelling. Mon. Not. R. Astron. Soc. **395**, 1319–1332 (2009). doi:10.1111/j.1365-2966.2009.14654.x

E. Jullo, J. Kneib, M. Limousin, Á. Elíasdóttir, P.J. Marshall, T. Verdugo, A Bayesian approach to strong lensing modelling of galaxy clusters. New J. Phys. **9**, 447 (2007). doi:10.1088/1367-2630/9/12/447

J.S. Kalirai, R.L. Beaton, M.C. Geha, K.M. Gilbert, P. Guhathakurta, E.N. Kirby, S.R. Majewski, J.C. Ostheimer, R.J. Patterson, J. Wolf, The SPLASH survey: internal kinematics, chemical abundances, and masses of the Andromeda I, II, III, VII, X, and XIV dwarf spheroidal galaxies. Astrophys. J. **711**, 671–692 (2010). doi:10.1088/0004-637X/711/2/671

J.S. Kalirai, J. Anderson, A. Dotter, H.B. Richer, G.G. Fahlman, B.M.S. Hansen, J. Hurley, I.N. Reid, R.M. Rich, M.M. Shara, Ultra-deep Hubble space telescope imaging of the Small Magellanic Cloud: the initial mass function of stars with $M \lesssim 1 M_{\odot}$. Astrophys. J. **763**, 110 (2013). doi:10.1088/0004-637X/763/2/110

E. Kapakos, D. Hatzidimitriou, RR Lyrae variables in the Small Magellanic Cloud—II. The extended area: chemical and structural analysis. Mon. Not. R. Astron. Soc. **426**, 2063–2077 (2012). doi:10.1111/j.1365-2966.2012.21834.x

H. Kawahara, T. Kitayama, S. Sasaki, Y. Suto, Systematic errors in the Hubble constant measurement from the Sunyaev-Zel'dovich effect. Astrophys. J. **674**, 11–21 (2008). doi:10.1086/524132

I. Kayo, N. Inada, M. Oguri, T. Morokuma, P.B. Hall, C.S. Kochanek, D.P. Schneider, Eight new quasar lenses from the sloan digital sky survey quasar lens search. Astron. J. **139**, 1614–1621 (2010). doi:10.1088/0004-6256/139/4/1614

K. Kinemuchi, H.C. Harris, H.A. Smith, N.A. Silbermann, L.A. Snyder, A.P. La Cluyzé, C.L. Clark, The variable stars of the draco dwarf spheroidal galaxy: revisited. Astron. J. **136**, 1921–1939 (2008). doi:10.1088/0004-6256/136/5/1921

J.P. Kneib, J.G. Cohen, J. Hjorth, RX J0911+05: a massive cluster lens at $z = 0.769$. Astrophys. J. Lett. **544**, L35–L39 (2000). doi:10.1086/317285

C.S. Kochanek, R. Narayan, LensClean: an algorithm for inverting extended, gravitationally lensed images with application to the radio ring lens PKS 1830-211. Astrophys. J. **401**, 461–473 (1992). doi:10.1086/172078

C.S. Kochanek, C.R. Keeton, B.A. McLeod, The importance of Einstein rings. Astrophys. J. **547**, 50–59 (2001). doi:10.1086/318350

C.S. Kochanek, B. Mochejska, N.D. Morgan, K.Z. Stanek, A simple method to find all lensed quasars. Astrophys. J. Lett. **637**, L73–L76 (2006). doi:10.1086/500559

C.H.B. Koenig, J.M. Nemec, J.R. Mould, G.G. Fahlman, Color-magnitude diagram for the Andromeda II dwarf galaxy. Astron. J. **106**, 1819–1825 (1993). doi:10.1086/116767

L.V.E. Koopmans, T. Treu, C.D. Fassnacht, R.D. Blandford, G. Surpi, The Hubble constant from the gravitational lens B1608+656. Astrophys. J. **599**, 70–85 (2003). doi:10.1086/379226

K. Kuijken, C. Heymans, H. Hildebrandt, R. Nakajima, T. Erben, J.T.A. de Jong, M. Viola, A. Choi, H. Hoekstra, L. Miller, E. van Uitert, A. Amon, C. Blake, M. Brouwer, A. Buddendiek, I.F. Conti, M. Eriksen, A. Grado, J. Harnois-Déraps, E. Helmich, R. Herbonnet, N. Irisarri, T. Kitching, D. Klaes, F. La Barbera, N. Napolitano, M. Radovich, P. Schneider, C. Sifón, G. Sikkema, P. Simon, A. Tudorica, E. Valentijn, G. Verdoes Kleijn, L. van Waerbeke, Gravitational lensing analysis of the kilo-degree survey. Mon. Not. R. Astron. Soc. **454**, 3500–3532 (2015). doi:10.1093/mnras/stv2140

C.Y. Kuo, J.A. Braatz, M.J. Reid, K.Y. Lo, J.J. Condon, C.M.V. Impellizzeri, C. Henkel, The megamaser cosmology project. V. An angular-diameter distance to NGC 6264 at 140 Mpc. Astrophys. J. **767**, 155 (2013). doi:10.1088/0004-637X/767/2/155

C.D. Laney, R.S. Stobie, Infrared photometry of Magellanic cloud Cepheids—intrinsic properties of Cepheids and the spatial structure of clouds. Mon. Not. R. Astron. Soc. **222**, 449–472 (1986). doi:10.1093/mnras/222.3.449

R. Laureijs, J. Amiaux, S. Arduini, J. Augueres, J. Brinchmann, R. Cole, M. Cropper, C. Dabin, L. Duvet, A. Ealet et al. Euclid Definition Study Report (2011). arXiv:1110.3193

M.G. Lee, W.L. Freedman, B.F. Madore, The tip of the red giant branch as a distance indicator for resolved galaxies. Astrophys. J. **417**, 553–559 (1993). doi:10.1086/173334

Y.W. Lee, On the Sandage period shift effect among field RR Lyrae stars. Astrophys. J. **363**, 159–167 (1990). doi:10.1086/169326

B. Letarte, S.C. Chapman, M. Collins, R.A. Ibata, M.J. Irwin, A.M.N. Ferguson, G.F. Lewis, N. Martin, A. McConnachie, N. Tanvir, A Keck/DEIMOS spectroscopic survey of the faint M31 satellites And XV and And XVI. Mon. Not. R. Astron. Soc. **400**, 1472–1478 (2009). doi:10.1111/j.1365-2966.2009.15546.x

K. Liao, T. Treu, P. Marshall, C.D. Fassnacht, N. Rumbaugh, G. Dobler, A. Aghamousa, V. Bonvin, F. Courbin, A. Hojjati, N. Jackson, V. Kashyap, S. Rathna Kumar, E. Linder, K. Mandel, X.L. Meng, G. Meylan, L.A. Moustakas, T.P. Prabhu, A. Romero-Wolf, A. Shafieloo, A. Siemiginowska, C.S. Stalin, H. Tak, M. Tewes, D. van Dyk, Strong lens time delay challenge. II. Results of TDC1. Astrophys. J. **800**, 11 (2015). doi:10.1088/0004-637X/800/1/11

B.F. Madore, W.L. Freedman, The tip of the red giant branch as a distance indicator for resolved galaxies. II: Computer simulations. Astron. J. **109**, 1645–1652 (1995). doi:10.1086/117391

Z. Malkin, Statistical analysis of the determinations of the Sun's Galactocentric distance, in *Advancing the Physics of Cosmic Distances*, ed. by R. de Grijs. Int'l Astron. Union Symp., vol. 289 (2013a), pp. 406–409. doi:10.1017/S1743921312021825

Z.M. Malkin, Analysis of determinations of the distance between the sun and the galactic center. Astron. Rep. **47**, 128–133 (2013b). doi:10.1134/S1063772913020078

M. Marconi, G. Coppola, G. Bono, V. Braga, A. Pietrinferni, R. Buonanno, M. Castellani, I. Musella, V. Ripepi, R.F. Stellingwerf, On a new theoretical framework for RR Lyrae stars. I. The metallicity dependence. Astrophys. J. **808**, 50 (2015). doi:10.1088/0004-637X/808/1/50

P.J. Marshall, D.W. Hogg, L.A. Moustakas, C.D. Fassnacht, M. Bradač, T. Schrabback, R.D. Blandford, Automated detection of galaxy-scale gravitational lenses in high-resolution imaging data. Astrophys. J. **694**, 924–942 (2009). doi:10.1088/0004-637X/694/2/924

P.J. Marshall, A. Verma, A. More, C.P. Davis, S. More, A. Kapadia, M. Parrish, C. Snyder, J. Wilcox, E. Baeten, C. Macmillan, C. Cornen, M. Baumer, E. Simpson, C.J. Lintott, D. Miller, E. Paget, R. Simpson, A.M. Smith, R. Küng, P. Saha, T.E. Collett, SPACE WARPS—I. Crowdsourcing the discovery of gravitational lenses. Mon. Not. R. Astron. Soc. **455**, 1171–1190 (2016). doi:10.1093/mnras/stv2009

N.F. Martin, D.L. Nidever, G. Besla, K. Olsen, A.R. Walker, A.K. Vivas, R.A. Gruendl, C.C. Kaleida, R.R. Muñoz, R.D. Blum, A. Saha, B.C. Conn, E.F. Bell, Y.H. Chu, M.-R.L. Cioni, T.J.L. de Boer, C. Gallart, S. Jin, A. Kunder, S.R. Majewski, D. Martínez-Delgado, A. Monachesi, M. Monelli, L. Monteagudo, N.E.D. Noël, E.W. Olszewski, G.S. Stringfellow, R.P. van der Marel, D. Zaritsky, Hydra II: a faint and compact Milky Way dwarf galaxy found in the survey of the Magellanic stellar history. Astrophys. J. **804**, L5 (2015). doi:10.1088/2041-8205/804/1/L5

C.E. Martínez-Vázquez, M. Monelli, G. Bono, P.B. Stetson, I. Ferraro, E.J. Bernard, C. Gallart, G. Fiorentino, G. Iannicola, A. Udalski, Variable stars in local group galaxies—I. Tracing the early chemical enrichment and radial gradients in the Sculptor dSph with RR Lyrae stars. Mon. Not. R. Astron. Soc. **454**, 1509–1516 (2015). doi:10.1093/mnras/stv2014

C.E. Martínez-Vázquez, P.B. Stetson, M. Monelli, E.J. Bernard, G. Fiorentino, C. Gallart, G. Bono, S. Cassisi, M. Dall'Ora, I. Ferraro, G. Iannicola, A.R. Walker, Variable stars in local group galaxies—II. Sculptor dSph. Mon. Not. R. Astron. Soc. **462**, 4349–4370 (2016). doi:10.1093/mnras/stw1895

M.L. Mateo, Dwarf galaxies of the local group. Annu. Rev. Astron. Astrophys. **36**, 435–506 (1998). doi:10.1146/annurev.astro.36.1.435

A.W. McConnachie, The observed properties of dwarf galaxies in and around the local group. Astron. J. **144**, 4 (2012). doi:10.1088/0004-6256/144/1/4

A.W. McConnachie, M.J. Irwin, A.M.N. Ferguson, R.A. Ibata, G.F. Lewis, N. Tanvir, Determining the location of the tip of the red giant branch in old stellar populations: M33, Andromeda I and II. Mon. Not. R. Astron. Soc. **350**, 243–252 (2004). doi:10.1111/j.1365-2966.2004.07637.x

A.W. McConnachie, M.J. Irwin, A.M.N. Ferguson, R.A. Ibata, G.F. Lewis, N. Tanvir, Distances and metallicities for 17 local group galaxies. Mon. Not. R. Astron. Soc. **356**, 979–997 (2005). doi:10.1111/j.1365-2966.2004.08514.x

A.W. McConnachie, M.J. Irwin, R.A. Ibata, J. Dubinski, L.M. Widrow, N.F. Martin, P. Côté, A.L. Dotter, J.F. Navarro, A.M.N. Ferguson, T.H. Puzia, G.F. Lewis, A. Babul, P. Barmby, O. Bienaymé, S.C. Chapman, R. Cockcroft, M.L.M. Collins, M.A. Fardal, W.E. Harris, A. Huxor, A.D. Mackey, J. Peñarrubia, R.M. Rich, H.B. Richer, A. Siebert, N. Tanvir, D. Valls-Gabaud, K.A. Venn, The remnants of galaxy formation from a panoramic survey of the region around M31. Nature **461**, 66–69 (2009). doi:10.1038/nature08327

A.W. McConnachie, The observed properties of dwarf galaxies in and around the local group. Astron. J. **144**, 4 (2012). doi:10.1088/0004-6256/144/1/4

C. McCully, C.R. Keeton, K.C. Wong, A.I. Zabludoff, A new hybrid framework to efficiently model lines of sight to gravitational lenses. Mon. Not. R. Astron. Soc. **443**, 3631–3642 (2014). doi:10.1093/mnras/stu1316

C. McCully, C.R. Keeton, K.C. Wong, A.I. Zabludoff, Quantifying environmental and line-of-sight effects in models of strong gravitational lens systems. Astrophys. J. **836**, 141 (2017). doi:10.3847/1538-4357/836/1/141

K.B.W. McQuinn, E.D. Skillman, A. Dolphin, J.M. Cannon, J.J. Salzer, K.L. Rhode, E.A.K. Adams, D. Berg, R. Giovanelli, L. Girardi, M.P. Haynes, P. Leo, An unquenched very low-mass galaxy. Astrophys. J. **812**, 158 (2015). doi:10.1088/0004-637X/812/2/158

X.L. Meng, T. Treu, A. Agnello, M.W. Auger, K. Liao, P.J. Marshall, Precision cosmology with time delay lenses: high resolution imaging requirements. J. Cosmol. Astropart. Phys. **2015**(09), 059 (2015). doi:10.1088/1475-7516/2015/09/059

S. Miyazaki, Y. Komiyama, H. Nakaya, Y. Kamata, Y. Doi, T. Hamana, H. Karoji, H. Furusawa, S. Kawanomoto, T. Morokuma, Y. Ishizuka, K. Nariai, Y. Tanaka, F. Uraguchi, Y. Utsumi, Y. Obuchi, Y. Okura, M. Oguri, T. Takata, D. Tomono, T. Kurakami, K. Namikawa, T. Usuda, H. Yamanoi, T.

Terai, H. Uekiyo, Y. Yamada, M. Koike, H. Aihara, Y. Fujimori, S. Mineo, H. Miyatake, N. Yasuda, J. Nishizawa, T. Saito, M. Tanaka, T. Uchida, N. Katayama, S.Y. Wang, H.Y. Chen, R. Lupton, C. Loomis, S. Bickerton, P. Price, J. Gunn, H. Suzuki, Y. Miyazaki, M. Muramatsu, K. Yamamoto, M. Endo, Y. Ezaki, N. Itoh, Y. Miwa, H. Yokota, T. Matsuda, R. Ebinuma, K. Takeshi, Hyper Suprime-Cam, in *Ground-Based and Airborne Instrumentation for Astronomy IV*. SPIE Conf. Ser., vol. 8446 (2012), p. 84460Z. doi:10.1117/12.926844

M. Monelli, C. Gallart, S.L. Hidalgo, A. Aparicio, E.D. Skillman, A.A. Cole, D.R. Weisz, L. Mayer, E.J. Bernard, S. Cassisi, A.E. Dolphin, I. Drozdovsky, P.B. Stetson, The ACS LCID project. VI. The star formation history of the Tucana dSph and the relative ages of the isolated dSph galaxies. Astrophys. J. **722**, 1864–1878 (2010a). doi:10.1088/0004-637X/722/2/1864

M. Monelli, S.L. Hidalgo, P.B. Stetson, A. Aparicio, C. Gallart, A.E. Dolphin, A.A. Cole, D.R. Weisz, E.D. Skillman, E.J. Bernard, L. Mayer, J.F. Navarro, S. Cassisi, I. Drozdovsky, E. Tolstoy, The ACS LCID Project. III. The star formation history of the Cetus dSph galaxy: a post-reionization fossil. Astrophys. J. **720**, 1225–1245 (2010b). doi:10.1088/0004-637X/720/2/1225

M. Monelli, E.J. Bernard, C. Gallart, G. Fiorentino, I. Drozdovsky, A. Aparicio, G. Bono, S. Cassisi, E.D. Skillman, P.B. Stetson, Variable stars in the Cetus dwarf spheroidal galaxy: population gradients and connections with the star formation history. Mon. Not. R. Astron. Soc. **422**, 89–105 (2012). doi:10.1111/j.1365-2966.2012.20539.x

M. Monelli, C.E. Martínez-Vázquez, E.J. Bernard, C. Gallart, E.D. Skillman, D.R. Weisz, A.E. Dolphin, S.L. Hidalgo, A.A. Cole, N.F. Martin, A. Aparicio, S. Cassisi, M. Boylan-Kolchin, L. Mayer, A. McConnachie, K.B.W. McQuinn, J.F. Navarro, The ISLANDS project. I. Andromeda XVI, an extremely low mass galaxy not quenched by reionization. Astrophys. J. **819**, 147 (2016). doi:10.3847/0004-637X/819/2/147

A. More, M. Oguri, I. Kayo, J. Zinn, M.A. Strauss, B.X. Santiago, A.M. Mosquera, N. Inada, C.S. Kochanek, C.E. Rusu, J.R. Brownstein, L.N. da Costa, J.P. Kneib, M.A.G. Maia, R.M. Quimby, D.P. Schneider, A. Streblyanska, D.G. York, The SDSS-III BOSS quasar lens survey: discovery of 13 gravitationally lensed quasars. Mon. Not. R. Astron. Soc. **456**, 1595–1606 (2016a). doi:10.1093/mnras/stv2813

A. More, A. Verma, P.J. Marshall, S. More, E. Baeten, J. Wilcox, C. Macmillan, C. Cornen, A. Kapadia, M. Parrish, C. Snyder, C.P. Davis, R. Gavazzi, C.J. Lintott, R. Simpson, D. Miller, A.M. Smith, E. Paget, P. Saha, R. Küng, T.E. Collett, SPACE WARPS—II. New gravitational lens candidates from the CFHTLS discovered through citizen science. Mon. Not. R. Astron. Soc. **455**, 1191–1210 (2016b). doi:10.1093/mnras/stv1965

J. Mould, J. Kristian, The dwarf spheroidal galaxy Andromeda I. Astrophys. J. **354**, 438–445 (1990). doi:10.1086/168706

R.F. Mushotzky, R. Edelson, W. Baumgartner, P. Gandhi, Kepler observations of rapid optical variability in active galactic nuclei. Astrophys. J. Lett. **743**, L12 (2011). doi:10.1088/2041-8205/743/1/L12

S.T. Myers, N.J. Jackson, I.W.A. Browne, A.G. de Bruyn, T.J. Pearson, A.C.S. Readhead, P.N. Wilkinson, A.D. Biggs, R.D. Blandford, C.D. Fassnacht, L.V.E. Koopmans, D.R. Marlow, J.P. McKean, M.A. Norbury, P.M. Phillips, D. Rusin, M.C. Shepherd, C.M. Sykes, The cosmic lens all-sky survey—I. Source selection and observations. Mon. Not. R. Astron. Soc. **341**, 1–12 (2003). doi:10.1046/j.1365-8711.2003.06256.x

N.R. Napolitano, G. Covone, N. Roy, C. Tortora, F. La Barbera, M. Radovich, F. Getman, M. Capaccioli, A. Colonna, M. Paolillo, G.A. Verdoes Kleijn, L.V.E. Koopmans (KiDS collaboration), Strong lens search in the ESO public survey KiDS, in *The Universe of Digital Sky Surveys*. Astrophys. Space Sci. Proc., vol. 42 (2016), pp. 129–134. doi:10.1007/978-3-319-19330-4_20

D.L. Nidever, A. Monachesi, E.F. Bell, S.R. Majewski, R.R. Muñoz, R.L. Beaton, A tidally stripped stellar component of the Magellanic bridge. Astrophys. J. **779**, 145 (2013). doi:10.1088/0004-637X/779/2/145

J.W. Nightingale, S. Dye, Adaptive semi-linear inversion of strong gravitational lens imaging. Mon. Not. R. Astron. Soc. **452**, 2940–2959 (2015). doi:10.1093/mnras/stv1455

I. Nikiforov, The distance to the center of the galaxy: the current state-of-the-art in measuring R_0, in *Order and Chaos in Stellar and Planetary Systems*, ed. by G.G. Byrd, K.V. Kholshevnikov, A.A. Myllri, I.I. Nikiforov, V.V. Orlov. Astron. Soc. Pac. Conf. Proc., vol. 316 (2004), pp. 199–208

M. Oguri, The mass distribution of SDSS J1004+4112 revisited. Publ. Astron. Soc. Jpn. **62**, 1017–1024 (2010). doi:10.1093/pasj/62.4.1017

M. Oguri, Measuring the distance-redshift relation with the cross-correlation of gravitational wave standard sirens and galaxies. Phys. Rev. D **93**, 083511 (2016). doi:10.1103/PhysRevD.93.083511

M. Oguri, P.J. Marshall, Gravitationally lensed quasars and supernovae in future wide-field optical imaging surveys. Mon. Not. R. Astron. Soc. **405**, 2579–2593 (2010). doi:10.1111/j.1365-2966.2010.16639.x

M. Oguri, N. Inada, B. Pindor, M.A. Strauss, G.T. Richards, J.F. Hennawi, E.L. Turner, R.H. Lupton, D.P. Schneider, M. Fukugita, J. Brinkmann, The sloan digital sky survey quasar lens search. I. Candidate selection algorithm. Astron. J. **132**, 999–1013 (2006). doi:10.1086/506019

M. Oguri, N. Inada, M.A. Strauss, C.S. Kochanek, I. Kayo, M.S. Shin, T. Morokuma, G.T. Richards, C.E. Rusu, J.A. Frieman, M. Fukugita, D.P. Schneider, D.G. York, N.A. Bahcall, R.L. White, The sloan digital sky survey quasar lens search. VI. Constraints on dark energy and the evolution of massive galaxies. Astron. J. **143**, 120 (2012). doi:10.1088/0004-6256/143/5/120

A.J. Ordoñez, A. Sarajedini, Population effects on the metallicity distribution function derived from the red giant branch. Astron. J. **149**, 201 (2015). doi:10.1088/0004-6256/149/6/201

A.J. Ordoñez, A. Sarajedini, The pulsating variable star population in DDO210. Mon. Not. R. Astron. Soc. **455**, 2163–2177 (2016). doi:10.1093/mnras/stv2494

A.J. Ordoñez, S.C. Yang, A. Sarajedini, The RR Lyrae variable population in the Phoenix dwarf galaxy. Astrophys. J. **786**, 147 (2014). doi:10.1088/0004-637X/786/2/147

F. Ostrovski, R.G. McMahon, A.J. Connolly, C.A. Lemon, M.W. Auger, M. Banerji, J.M. Hung, S.E. Koposov, C.E. Lidman, S.L. Reed, S. Allam, A. Benoit-Lévy, E. Bertin, D. Brooks, E. Buckley-Geer, A. Carnero Rosell, M. Carrasco Kind, J. Carretero, C.E. Cunha, L.N. da Costa, S. Desai, H.T. Diehl, J.P. Dietrich, A.E. Evrard, D.A. Finley, B. Flaugher, P. Fosalba, J. Frieman, D.W. Gerdes, D.A. Goldstein, D. Gruen, R.A. Gruendl, G. Gutierrez, K. Honscheid, D.J. James, K. Kuehn, N. Kuropatkin, M. Lima, H. Lin, M.A.G. Maia, J.L. Marshall, P. Martini, P. Melchior, R. Miquel, R. Ogando, A. Plazas Malagón, K. Reil, K. Romer, E. Sanchez, B. Santiago, V. Scarpine, I. Sevilla-Noarbe, M. Soares-Santos, F. Sobreira, E. Suchyta, G. Tarle, D. Thomas, D.L. Tucker, A.R. Walker, VDES J2325-5229 a $z = 2.7$ gravitationally lensed quasar discovered using morphology independent supervised machine learning. Mon. Not. R. Astron. Soc. **465**, 4325–4334 (2017). doi:10.1093/mnras/stw2958

D. Paraficz, J. Hjorth, Gravitational lenses as cosmic rulers: Ω_m, Ω_Λ from time delays and velocity dispersions. Astron. Astrophys. **507**, L49–L52 (2009). doi:10.1051/0004-6361/200913307

D. Paraficz, F. Courbin, A. Tramacere, R. Joseph, R.B. Metcalf, J.P. Kneib, P. Dubath, D. Droz, F. Filleul, D. Ringeisen, C. Schäfer, The PCA lens-finder: application to CFHTLS. Astron. Astrophys. **592**, A75 (2016). doi:10.1051/0004-6361/201527971

A. Pietrinferni, S. Cassisi, M. Salaris, F. Castelli, A large stellar evolution database for population synthesis studies. I. Scaled solar models and isochrones. Astrophys. J. **612**, 168–190 (2004). doi:10.1086/422498

G. Pietrzyński, D. Graczyk, W. Gieren, I.B. Thompson, B. Pilecki, A. Udalski, I. Soszyński, S. Kozłowski, P. Konorski, K. Suchomska, G. Bono, P.G.P. Moroni, S. Villanova, N. Nardetto, F. Bresolin, R.P. Kudritzki, J. Storm, A. Gallenne, R. Smolec, D. Minniti, M. Kubiak, M.K. Szymański, R. Poleski, Ł. Wyrzykowski, K. Ulaczyk, P. Pietrukowicz, M. Górski, P. Karczmarek, An eclipsing-binary distance to the Large Magellanic Cloud accurate to two per cent. Nature **495**, 76–79 (2013). doi:10.1038/nature11878

C.J. Pritchet, S. van den Bergh, Observations of RR Lyrae stars in the halo of M31. Astrophys. J. **316**, 517–529 (1987). doi:10.1086/165223

B.J. Pritzl, T.E. Armandroff, G.H. Jacoby, G.S. Da Costa, The dwarf spheroidal companions to M31: variable stars in Andromeda VI. Astron. J. **124**, 1464–1485 (2002a). doi:10.1086/341823

B.J. Pritzl, H.A. Smith, M. Catelan, A.V. Sweigart, Variable stars in the unusual, metal-rich globular cluster NGC 6388. Astron. J. **124**, 949–976 (2002b). doi:10.1086/341381

B.J. Pritzl, T.E. Armandroff, G.H. Jacoby, G.S. Da Costa, The dwarf spheroidal companions to M31: variable stars in Andromeda II. Astron. J. **127**, 318–333 (2004). doi:10.1086/380613

B.J. Pritzl, T.E. Armandroff, G.H. Jacoby, G.S. Da Costa, The dwarf spheroidal companions to M31: variable stars in Andromeda I and Andromeda III. Astron. J. **129**, 2232–2256 (2005a). doi:10.1086/428372

B.J. Pritzl, K.A. Venn, M. Irwin, A comparison of elemental abundance ratios in globular clusters, field stars, and dwarf spheroidal galaxies. Astron. J. **130**, 2140–2165 (2005b). doi:10.1086/432911

E.D. Reese, J.E. Carlstrom, M. Joy, J.J. Mohr, L. Grego, W.L. Holzapfel, Determining the cosmic distance scale from interferometric measurements of the Sunyaev–Zel'dovich effect. Astrophys. J. **581**, 53–85 (2002). doi:10.1086/344137

S. Refsdal, On the possibility of determining Hubble's parameter and the masses of galaxies from the gravitational lens effect. Mon. Not. R. Astron. Soc. **128**, 307–310 (1964). doi:10.1093/mnras/128.4.307

M.J. Reid, The distance to the galactic center: R_0, in *The Center of the Galaxy*, ed. by M. Morris. Int'l Astron. Union Symp., vol. 136 (1989), pp. 37–46

M.J. Reid, The distance to the center of the galaxy. Annu. Rev. Astron. Astrophys. **31**, 345–372 (1993). doi:10.1146/annurev.aa.31.090193.002021

M.J. Reid, J.A. Braatz, J.J. Condon, L.J. Greenhill, C. Henkel, K.Y. Lo, The megamaser cosmology project. I. Very long baseline interferometric observations of UGC 3789. Astrophys. J. **695**, 287–291 (2009). doi:10.1088/0004-637X/695/1/287

M.J. Reid, J.A. Braatz, J.J. Condon, K.Y. Lo, C.Y. Kuo, C.M.V. Impellizzeri, C. Henkel, The megamaser cosmology project. IV. A direct measurement of the Hubble constant from UGC 3789. Astrophys. J. **767**, 154 (2013). doi:10.1088/0004-637X/767/2/154

J.A. Rich, S.E. Persson, W.L. Freedman, B.F. Madore, A.J. Monson, V. Scowcroft, M. Seibert, A new Cepheid distance measurement and method for NGC 6822. Astrophys. J. **794**, 107 (2014). doi:10.1088/0004-637X/794/2/107

A.G. Riess, L.M. Macri, S.L. Hoffmann, D. Scolnic, S. Casertano, A.V. Filippenko, B.E. Tucker, M.J. Reid, D.O. Jones, J.M. Silverman, R. Chornock, P. Challis, W. Yuan, P.J. Brown, R.J. Foley, A 2.4% determination of the local value of the Hubble constant. Astrophys. J. **826**, 56 (2016). doi:10.3847/0004-637X/826/1/56

V. Ripepi, M.I. Moretti, M. Marconi, G. Clementini, M.-R.L. Cioni, J.B. Marquette, L. Girardi, S. Rubele, M.A.T. Groenewegen, R. de Grijs, B.K. Gibson, J.M. Oliveira, J.T. van Loon, J.P. Emerson, The VMC survey—V. First results for classical Cepheids. Mon. Not. R. Astron. Soc. **424**, 1807–1816 (2012). doi:10.1111/j.1365-2966.2012.21274.x

L. Rizzi, E.V. Held, I. Saviane, R.B. Tully, M. Gullieuszik, The distance to the Fornax dwarf spheroidal galaxy. Mon. Not. R. Astron. Soc. **380**, 1255–1260 (2007). doi:10.1111/j.1365-2966.2007.12196.x

S. Rubele, L. Kerber, L. Girardi, M.-R. Cioni, P. Marigo, S. Zaggia, K. Bekki, R. de Grijs, J. Emerson, M.A.T. Groenewegen, M. Gullieuszik, V. Ivanov, B. Miszalski, J.M. Oliveira, B. Tatton, J.T. van Loon, The VMC survey. IV. The LMC star formation history and disk geometry from four VMC tiles. Astron. Astrophys. **537**, A106 (2012). doi:10.1051/0004-6361/201117863

S. Rubele, L. Girardi, L. Kerber, M.-R.L. Cioni, A.E. Piatti, S. Zaggia, K. Bekki, A. Bressan, G. Clementini, R. de Grijs, J.P. Emerson, M.A.T. Groenewegen, V.D. Ivanov, M. Marconi, P. Marigo, M.-I. Moretti, V. Ripepi, S. Subramanian, B.L. Tatton, J.T. van Loon, The VMC survey—XIV. First results on the look-back time star formation rate tomography of the Small Magellanic Cloud. Mon. Not. R. Astron. Soc. **449**, 639–661 (2015). doi:10.1093/mnras/stv141

C.E. Rusu, C.D. Fassnacht, D. Sluse, S. Hilbert, K.C. Wong, K.H. Huang, S.H. Suyu, T.E. Collett, P.J. Marshall, T. Treu, L.V.E. Koopmans, H0LiCOW III. Quantifying the effect of mass along the line of sight to the gravitational lens HE 0435-1223 through weighted galaxy counts. Mon. Not. R. Astron. Soc. **467**, 4220–4242 (2017). doi:10.1093/mnras/stx285

C.E. Rusu, M. Oguri, Y. Minowa, M. Iye, N. Inada, S. Oya, I. Kayo, Y. Hayano, M. Hattori, Y. Saito, M. Ito, T.S. Pyo, H. Terada, H. Takami, M. Watanabe, Subaru telescope adaptive optics observations of gravitationally lensed quasars in the sloan digital sky survey. Mon. Not. R. Astron. Soc. **458**, 2–55 (2016). doi:10.1093/mnras/stw092

M. Rybak, J.P. McKean, S. Vegetti, P. Andreani, S.D.M. White, ALMA imaging of SDP.81—I. A pixelated reconstruction of the far-infrared continuum emission. Mon. Not. R. Astron. Soc. **451**, L40–L44 (2015). doi:10.1093/mnrasl/slv058

A. Saha, J.G. Hoessel, RR Lyrae stars in local group galaxies. I—NGC 185. Astron. J. **99**, 97–148 (1990). doi:10.1086/115316

A. Saha, J.G. Hoessel, A.E. Mossman, RR Lyrae stars in local group galaxies. II—NGC 147. Astron. J. **100**, 108–126 (1990). doi:10.1086/115495

M. Salaris, A. Weiss, Homogeneous age dating of 55 galactic globular clusters. Clues to the galaxy formation mechanisms. Astron. Astrophys. **388**, 492–503 (2002). doi:10.1051/0004-6361:20020554

L.V. Sales, J.F. Navarro, M.G. Abadi, M. Steinmetz, Cosmic ménage à trois: the origin of satellite galaxies on extreme orbits. Mon. Not. R. Astron. Soc. **379**, 1475–1483 (2007). doi:10.1111/j.1365-2966.2007.12026.x

A. Sandage, The Oosterhoff period-metallicity relation for RR Lyrae stars at the blue fundamental edge of the instability strip. Astron. J. **106**, 687–702 (1993). doi:10.1086/116675

A. Sarajedini, M.K. Barker, D. Geisler, P. Harding, R. Schommer, RR Lyrae variables in M33. I. Evidence for a field halo population. Astron. J. **132**, 1361–1371 (2006). doi:10.1086/506152

A. Sarajedini, S.C. Yang, A. Monachesi, T.R. Lauer, S.C. Trager, An ancient metal-poor population in M32, and halo satellite accretion in M31, identified by RR Lyrae stars. Mon. Not. R. Astron. Soc. **425**, 1459–1472 (2012). doi:10.1111/j.1365-2966.2012.21609.x

B.E. Schaefer, A problem with the clustering of recent measures of the distance to the Large Magellanic Cloud. Astron. J. **135**, 112–119 (2008). doi:10.1088/0004-6256/135/1/112

B.E. Schaefer, An accurate distance to the nearest galaxy. Nature **495**, 51–52 (2013). doi:10.1038/495051a

P.L. Schechter, N.D. Morgan, B. Chehade, N. Metcalfe, T. Shanks, M. McDonald, First lensed quasar system(s) from the VST-ATLAS survey: one quad and three nearly identical pairs. Astron. J. (2017, in press). arXiv:1607.07476. doi:10.3847/1538-3881/aa6899

E.F. Schlafly, D.P. Finkbeiner, Measuring reddening with sloan digital sky survey stellar spectra and recalibrating SFD. Astrophys. J. **737**, 103 (2011). doi:10.1088/0004-637X/737/2/103

D.J. Schlegel, D.P. Finkbeiner, M. Davis, Maps of dust infrared emission for use in estimation of reddening and cosmic microwave background radiation foregrounds. Astrophys. J. **500**, 525–553 (1998). doi:10.1086/305772

P. Schneider, D. Sluse, Mass-sheet degeneracy, power-law models and external convergence: impact on the determination of the Hubble constant from gravitational lensing. Astron. Astrophys. **559**, A37 (2013). doi:10.1051/0004-6361/201321882

P. Schneider, D. Sluse, Source-position transformation: an approximate invariance in strong gravitational lensing. Astron. Astrophys. **564**, A103 (2014). doi:10.1051/0004-6361/201322106

B.F. Schutz, Determining the Hubble constant from gravitational wave observations. Nature **323**, 310–311 (1986). doi:10.1038/323310a0

V. Scowcroft, W.L. Freedman, B.F. Madore, A.J. Monson, S.E. Persson, M. Seibert, J.R. Rigby, J. Melbourne, The Carnegie Hubble program: the infrared leavitt law in IC 1613. Astrophys. J. **773**, 106 (2013). doi:10.1088/0004-637X/773/2/106

V. Scowcroft, W.L. Freedman, B.F. Madore, A. Monson, S.E. Persson, J. Rich, M. Seibert, J.R. Rigby, The Carnegie Hubble program: the distance and structure of the SMC as revealed by mid-infrared observations of Cepheids. Astrophys. J. **816**, 49 (2016). doi:10.3847/0004-637X/816/2/49

Y. Shu, A.S. Bolton, C.S. Kochanek, M. Oguri, I. Pérez-Fournon, Z. Zheng, S. Mao, A.D. Montero-Dorta, J.R. Brownstein, R. Marques-Chaves, B. Ménard, The BOSS emission-line lens survey. III. Strong lensing of Lyα emitters by individual galaxies. Astrophys. J. **824**, 86 (2016). doi:10.3847/0004-637X/824/2/86

J. Silk, S.D.M. White, The determination of Q_0 using X-ray and microwave observations of galaxy clusters. Astrophys. J. Lett. **226**, L103–L106 (1978). doi:10.1086/182841

M. Sirianni, M.J. Jee, N. Benítez, J.P. Blakeslee, A.R. Martel, G. Meurer, M. Clampin, G. De Marchi, H.C. Ford, R. Gilliland, G.F. Hartig, G.D. Illingworth, J. Mack, W.J. McCann, The photometric performance and calibration of the Hubble space telescope advanced camera for surveys. Publ. Astron. Soc. Pac. **117**, 1049–1112 (2005). doi:10.1086/444553

E.D. Skillman, M. Monelli, D.R. Weisz, S.L. Hidalgo, A. Aparicio, E.J. Bernard, M. Boylan-Kolchin, S. Cassisi, A.A. Cole, A.E. Dolphin, H.C. Ferguson, C. Gallart, M.J. Irwin, N.F. Martin, C.E. Martínez-Vázquez, L. Mayer, A.W. McConnachie, K.B.W. McQuinn, J.F. Navarro, P.B. Stetson, The ISLAndS project II: the lifetime star formation histories of six Andromeda dSphs. Astrophys. J. **837**, 102 (2017). doi:10.3847/1538-4357/aa60c5

C.T. Slater, E.F. Bell, N.F. Martin, E.J. Tollerud, N. Ho, A deep study of the dwarf satellites Andromeda XXVIII and Andromeda XXIX. Astrophys. J. **806**, 230 (2015). doi:10.1088/0004-637X/806/2/230

D. Sluse, A. Sonnenfeld, N. Rumbaugh, C.E. Rusu, C.D. Fassnacht, T. Treu, S.H. Suyu, K.C. Wong, M.W. Auger, V. Bonvin, T. Collett, F. Courbin, S. Hilbert, L.V.E. Koopmans, P.J. Marshall, G. Meylan, C. Spiniello, M. Tewes, H0LiCOW II. Spectroscopic survey and galaxy-group identification of the strong gravitational lens system HE0435-1223. Mon. Not. R. Astron. Soc. (2016, submitted) arXiv:1607.00382

H.A. Smith, *RR Lyrae Stars*. Cambridge Astrophys. Ser., vol. 27 (Cambridge University Press, Cambridge, 1995)

A. Sonnenfeld, J.H.H. Chan, Y. Shu, A. More, M. Oguri, S.H. Suyu, K.C. Wong, C.-H. Lee, J. Coupon, A. Yonehara, A.S. Bolton, A.T. Jaelani, M. Tanaka, S. Miyazaki, Y. Komiyama, Survey of Gravitationally-lensed Objects in HSC Imaging (SuGOHI). I. Automatic search for galaxy-scale strong lenses. Publ. Astron. Soc. Jpn. (2017, submitted). arXiv:1704.01585

I. Soszyński, A. Udalski, M.K. Szymański, M. Kubiak, G. Pietrzyński, Ł. Wyrzykowski, O. Szewczyk, K. Ulaczyk, R. Poleski, The optical gravitational lensing experiment. The OGLE-III catalog of variable stars. II. Type II Cepheids and anomalous Cepheids in the Large Magellanic Cloud. Acta Astron. **58**, 293–312 (2008)

I. Soszyński, A. Udalski, M.K. Szymański, Ł. Wyrzykowski, K. Ulaczyk, R. Poleski, P. Pietrukowicz, S. Kozłowski, D.M. Skowron, J. Skowron, P. Mróz, M. Pawlak, The OGLE collection of variable stars. Over 45 000 RR Lyrae stars in the Magellanic system. Acta Astron. **66**, 131–147 (2016)

P.B. Stetson, Homogeneous photometry for star clusters and resolved galaxies. II. Photometric standard stars. Publ. Astron. Soc. Pac. **112**, 925–931 (2000). doi:10.1086/316595

P.B. Stetson, G. Fiorentino, G. Bono, E.J. Bernard, M. Monelli, G. Iannicola, C. Gallart, I. Ferraro, Homogeneous photometry. VI: variable stars in the Leo I dwarf spheroidal galaxy. Publ. Astron. Soc. Pac. **126**, 616–641 (2014). doi:10.1086/677352

S. Subramanian, A. Subramaniam, The three-dimensional structure of the Small Magellanic Cloud. Astrophys. J. **744**, 128 (2012). doi:10.1088/0004-637X/744/2/128

S. Subramanian, A. Subramaniam, Disk of the Small Magellanic Cloud as traced by Cepheids. Astron. Astrophys. **573**, A135 (2015). doi:10.1051/0004-6361/201424248

S.H. Suyu, A. Halkola, The halos of satellite galaxies: the companion of the massive elliptical lens SL2S J08544-0121. Astron. Astrophys. **524**, A94 (2010). doi:10.1051/0004-6361/201015481

S.H. Suyu, P.J. Marshall, M.P. Hobson, R.D. Blandford, A Bayesian analysis of regularized source inversions in gravitational lensing. Mon. Not. R. Astron. Soc. **371**, 983–998 (2006). doi:10.1111/j.1365-2966.2006.10733.x

S.H. Suyu, P.J. Marshall, R.D. Blandford, C.D. Fassnacht, L.V.E. Koopmans, J.P. McKean, T. Treu, Dissecting the gravitational lens B1608+656. I. Lens potential reconstruction. Astrophys. J. **691**, 277–298 (2009). doi:10.1088/0004-637X/691/1/277
S.H. Suyu, P.J. Marshall, M.W. Auger, S. Hilbert, R.D. Blandford, L.V.E. Koopmans, C.D. Fassnacht, T. Treu, Dissecting the gravitational lens B1608+656. II. Precision measurements of the Hubble constant, spatial curvature, and the dark energy equation of state. Astrophys. J. **711**, 201–221 (2010). doi:10.1088/0004-637X/711/1/201
S.H. Suyu, M.W. Auger, S. Hilbert, P.J. Marshall, M. Tewes, T. Treu, C.D. Fassnacht, L.V.E. Koopmans, D. Sluse, R.D. Blandford, F. Courbin, G. Meylan, Two accurate time-delay distances from strong lensing: implications for cosmology. Astrophys. J. **766**, 70 (2013). doi:10.1088/0004-637X/766/2/70
S.H. Suyu, T. Treu, S. Hilbert, A. Sonnenfeld, M.W. Auger, R.D. Blandford, T. Collett, F. Courbin, C.D. Fassnacht, L.V.E. Koopmans, P.J. Marshall, G. Meylan, C. Spiniello, M. Tewes, Cosmology from gravitational lens time delays and Planck data. Astrophys. J. Lett. **788**, L35 (2014). doi:10.1088/2041-8205/788/2/L35
S.H. Suyu, V. Bonvin, F. Courbin, C.D. Fassnacht, C.E. Rusu, D. Sluse, T. Treu, K.C. Wong, M.W. Auger, X. Ding, S. Hilbert, P.J. Marshall, N. Rumbaugh, A. Sonnenfeld, M. Tewes, O. Tihhonova, A. Agnello, R.D. Blandford, G.C.F. Chen, T. Collett, L.V.E. Koopmans, K. Liao, G. Meylan, C. Spiniello, H0LiCOW I. H_0 lenses in COSMOGRAIL's Wellspring: program overview. Mon. Not. R. Astron. Soc. (2017, in press). arXiv:1607.00017. doi:10.1093/mnras/stx483
A.S. Tagore, N. Jackson, On the use of shapelets in modelling resolved, gravitationally lensed images. Mon. Not. R. Astron. Soc. **457**, 3066–3075 (2016). doi:10.1093/mnras/stw057
A.S. Tagore, C.R. Keeton, Statistical and systematic uncertainties in pixel-based source reconstruction algorithms for gravitational lensing. Mon. Not. R. Astron. Soc. **445**, 694–710 (2014). doi:10.1093/mnras/stu1671
G.A. Tammann, A. Sandage, B. Reindl, The expansion field: the value of H_0. Astron. Astrophys. Rev. **15**, 289–331 (2008). doi:10.1007/s00159-008-0012-y
M. Tewes, F. Courbin, G. Meylan, C.S. Kochanek, E. Eulaers, N. Cantale, A.M. Mosquera, P. Magain, H. Van Winckel, D. Sluse, G. Cataldi, D. Vörös, S. Dye, COSMOGRAIL: the COSmological MOnitoring of GRAvItational lenses. XIII. Time delays and 9-yr optical monitoring of the lensed quasar RX J1131-1231. Astron. Astrophys. **556**, A22 (2013). doi:10.1051/0004-6361/201220352
F. Thévenin, M. Falanga, C.Y. Kuo, G. Pietrzyński, M. Yamaguchi, Modern geometric methods of distance determination. Space Sci. Rev. (2017, submitted)
E.L. Turner, R. Cen, J.P. Ostriker, The relation of local measures of Hubble's constant to its global value. Astron. J. **103**, 1427–1437 (1992). doi:10.1086/116156
A. Udalski, Ł. Wyrzykowski, G. Pietrzyński, O. Szewczyk, M. Szymański, M. Kubiak, I. Soszyński, K. Żebrún, The optical gravitational lensing experiment. Cepheids in the galaxy IC 1613: no dependence of the period–luminosity relation on metallicity. Acta Astron. **51**, 221–245 (2001)
S. van den Bergh, The galaxies of the local group (to be concluded). J. R. Astron. Soc. Can. **62**, 145–180 (1968)
S. Vegetti, L.V.E. Koopmans, Bayesian strong gravitational-lens modelling on adaptive grids: objective detection of mass substructure in galaxies. Mon. Not. R. Astron. Soc. **392**, 945–963 (2009). doi:10.1111/j.1365-2966.2008.14005.x
A.K. Vivas, R. Zinn, J. Farmer, S. Duffau, Y. Ping, Disentangling the Virgo overdensity with RR Lyrae stars. Astrophys. J. **831**, 165 (2016). doi:10.3847/0004-637X/831/2/165
A.R. Walker, CCD photometry of the RR Lyrae variables in the LMC cluster NGC 2257 and the adjacent field. Astron. J. **98**, 2086–2123 (1989). doi:10.1086/115282
A.R. Walker, The Large Magellanic Cloud and the distance scale. Astrophys. Space Sci. **341**, 43–49 (2012). doi:10.1007/s10509-011-0961-x
S.J. Warren, S. Dye, Semilinear gravitational lens inversion. Astrophys. J. **590**, 673–682 (2003). doi:10.1086/375132
D.R. Weisz, E.D. Skillman, S.L. Hidalgo, M. Monelli, A.E. Dolphin, A. McConnachie, E.J. Bernard, C. Gallart, A. Aparicio, M. Boylan-Kolchin, S. Cassisi, A.A. Cole, H.C. Ferguson, M. Irwin, N.F. Martin, L. Mayer, K.B.W. McQuinn, J.F. Navarro, P.B. Stetson, Comparing M31 and Milky Way satellites: the extended star formation histories of Andromeda II and Andromeda XVI. Astrophys. J. **789**, 24 (2014). doi:10.1088/0004-637X/789/1/24
B. Willman, M. Masjedi, D.W. Hogg, J.J. Dalcanton, D. Martinez-Delgado, M. Blanton, A.A. West, A. Dotter, B. Chaboyer, Willman 1—A Galactic Satellite at 40 kpc With Multiple Stellar Tails (2006). arXiv:astro-ph/0603486
K.C. Wong, S.H. Suyu, M.W. Auger, V. Bonvin, F. Courbin, C.D. Fassnacht, A. Halkola, C.E. Rusu, D. Sluse, A. Sonnenfeld, T. Treu, T.E. Collett, S. Hilbert, L.V.E. Koopmans, P.J. Marshall, N. Rumbaugh,

H0LiCOW IV. Lens mass model of HE 0435-1223 and blind measurement of its time-delay distance for cosmology. Mon. Not. R. Astron. Soc. **465**, 4895–4913 (2017). doi:10.1093/mnras/stw3077

P.F. Wu, R.B. Tully, L. Rizzi, A.E. Dolphin, B.A. Jacobs, I.D. Karachentsev, Infrared tip of the red giant branch and distances to the Maffei/IC 342 group. Astron. J. **148**, 7 (2014). doi:10.1088/0004-6256/148/1/7

O. Wucknitz, A.D. Biggs, I.W.A. Browne, Models for the lens and source of B0218+357: a LENSCLEAN approach to determine H_0. Mon. Not. R. Astron. Soc. **349**, 14–30 (2004). doi:10.1111/j.1365-2966.2004.07514.x

D. Xu, D. Sluse, P. Schneider, V. Springel, M. Vogelsberger, D. Nelson, L. Hernquist, Lens galaxies in the Illustris simulation: power-law models and the bias of the Hubble constant from time delays. Mon. Not. R. Astron. Soc. **456**, 739–755 (2016). doi:10.1093/mnras/stv2708

S.C. Yang, A. Sarajedini, HST/WFPC2 imaging of the dwarf satellites And XI and And XIII: horizontal branch morphology and RR Lyraes. Mon. Not. R. Astron. Soc. **419**, 1362–1375 (2012). doi:10.1111/j.1365-2966.2011.19792.x

S.C. Yang, R. Wagner-Kaiser, A. Sarajedini, S.C. Kim, J. Kyeong, The early chemical enrichment histories of two sculptor group dwarf galaxies as revealed by RR Lyrae variables. Astrophys. J. **784**, 76 (2014). doi:10.1088/0004-637X/784/1/76

MIX
Papier aus verantwortungsvollen Quellen
Paper from responsible sources
FSC® C105338

If you have any concerns about our products,
you can contact us on
ProductSafety@springernature.com

In case Publisher is established outside the EU,
the EU authorized representative is:
Springer Nature Customer Service Center GmbH
Europaplatz 3, 69115 Heidelberg, Germany

Printed by Libri Plureos GmbH
in Hamburg, Germany